W9-ASY-381

Introductory
Plant Biology

Introductory
Plant Biology

fourth edition

Kingsley R. Stern
California State University, Chico

wcb
Wm. C. Brown Publishers
Dubuque, Iowa

Book Team

Editor *Kevin Kane*
Developmental Editor *Mary J. Porter*
Production Editor *Sherry Padden*
Designers *Mark Elliot Christianson/Tara L. Bazata*
Photo Research Editor *Michelle Oberhoffer*
Visuals Processor *Reneé Pins*
Marketing Manager *Matt Shaughnessy*

wcb

Chairman of the Board *Wm. C. Brown*
President and Chief Executive Officer *Mark C. Falb*

wcb

Wm. C. Brown Publishers, College Division

President *G. Franklin Lewis*
Vice President, Editor-in-Chief *George Wm. Bergquist*
Vice President, Director of Production *Beverly Kolz*
National Sales Manager *Bob McLaughlin*
Director of Marketing *Thomas E. Doran*
Marketing Information Systems Manager *Craig S. Marty*
Executive Editor *Edward G. Jaffe*
Production Editorial Manager *Julie A. Kennedy*
Manager of Design *Marilyn A. Phelps*
Photo Research Manager *Faye M. Schilling*

Cover Image

The cover features one of several species of native North American blueberries (*Vaccinium* spp.), which are scattered from coast to coast, primarily in northern and mountainous regions. Native Americans prized them for food, either fresh or dried, and drank blueberry leaf tea to treat dysentery and diarrhea. Today many tons of wild blueberries are still harvested annually and eaten by Americans and Canadians, as well as by visitors from all over the world. In addition, horticultural selections of hybrids between the two hardiest species, *V. corymbosum* and *V. pennsylvanicum,* are grown commercially. Significant plantings have been established in New Jersey, North Carolina, Michigan, and western Washington.

All photographs not credited belong to Kingsley R. Stern ©

Cover photo by © Michel Bourque/Valan Photos

Cover and interior design by Kay D. Fulton

Library of Congress Catalog Card Number: 87–70728

ISBN 0–697–05128–5 (perfect) 0–697–05295–8 (cloth)

Printed in the United States of America by Wm. C. Brown Publishers
2460 Kerper Boulevard, Dubuque, IA 52001

10 9 8 7 6 5 4

To the memory of
Franklin Charles Lane 1928–1971
Botanist, Teacher, Friend

Contents

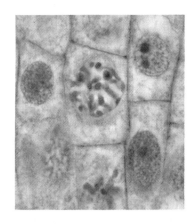

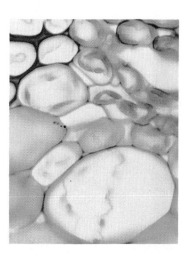

Appendices

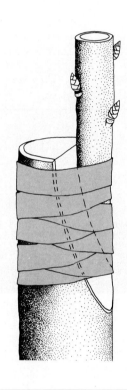

Preface

This book is designed as an introductory text in botany. It assumes little knowledge of the sciences on the part of the student. It includes sufficient information for some shorter introductory botany courses open to both majors and nonmajors, but it is arranged so that certain sections—for example, "Early History and Development of Plant Study," "Soils," "Division Psilophyta"—can be omitted without disrupting the overall continuity of the course.

Botany instructors vary greatly in their opinions as to the depth of coverage needed for the topics of photosynthesis and respiration in a text of this type. Some feel that nonmajors, in particular, should have a brief introduction only, whereas others consider a more detailed discussion essential. In this text photosynthesis and respiration are discussed at three levels. Some may find one or two levels sufficient, and others may wish their students to become familiar with the processes at all three levels.

Despite eye-catching chapter titles and headings, many texts for majors and nonmajors relegate the current interests of a significant number of students to comparative obscurity. This text emphasizes current interests without giving short shrift to botanical principles. Present interests of students include subjects such as organic gardening, Native American and pioneer uses of plants, pollution and recycling, houseplants, backyard vegetable gardens, natural dye plants, poisonous and hallucinogenic plants, and the nutritional values of edible plants. The rather perfunctory coverage or absence of such topics in many botany texts has occurred partly because botanists previously have tended to believe that the topics are more appropriately covered in anthropology and horticulture courses. I have found, however, that both majors and nonmajors in botany

who may be initially disinterested in the subject matter of a required course frequently become engrossed if the material is repeatedly related to such topics. Accordingly, a considerable amount of ecological and ethnobotanical material has been included with traditional botany throughout the book—without, however, resorting to excessive use of technical terms.

ORGANIZATION OF THE TEXT

A relatively conventional sequence of botanical subjects is included. Chapters 1 and 2 cover introductory and background information; chapters 3 through 11 deal with structure and function; chapters 12 and 13 introduce meiosis, genetics, and evolution. Chapter 14 presents a five-kingdom system of classification; chapters 15 through 21 stress, in phylogenetic sequence, the diversity of organisms traditionally regarded as plants, and chapter 22 deals with ethnobotanical aspects and information of general interest pertaining to 16 major families of flowering plants. Chapter 23 is an overview of the vast topic of ecology, although ecological topics and applied botany are included in most of the preceding chapters as well. Some of these subjects are broached in anecdotes that introduce the chapters, whereas others are mentioned in the human and ecological relevance sections (with which most of the chapters in the latter half of the book conclude).

AIDS TO THE READER

Each chapter begins with an **overview, some learning goals** and an **outline**. These features were designed to provide an organized, structured presentation of concepts and information. Within the chapter, new terms are defined as they are introduced, and those used more than once are boldfaced and included in a pronouncing glossary. The use of scientific names throughout the body of the text has been held to a minimum, but a complete list of the scientific names of all organisms mentioned is given in appendix 1. Each chapter concludes with a **summary, review questions, discussion questions,** and an **additional reading** list. These aids provide a review of the material presented within the chapter. The **additional reading** list directs readers to a combination of articles and books to further their understanding and to enable them to pursue interesting topics.

The end of the book features five appendices and a glossary. As indicated above, **appendix 1** contains a list of scientific names of all the organisms mentioned in the text. **Appendix 2** deals with biological controls and companion planting; **appendix 3** lists wild edible plants, poisonous plants, hallucinogenic plants, spices, and natural dye plants; **appendix 4** discusses vegetative propagation and pruning, gives horticultural information on houseplants and on the cultivation and nutritional value of vegetables; and **appendix 5** gives some metric equivalents. At the very end of the text is a **glossary** of terms. For each term, the glossary indicates pronunciation, a complete definition, and a page reference to where a more complete discussion can be found.

NEW TO THIS EDITION

The discussion of DNA has been moved to chapter 2, which itself has been revised and slightly expanded. The classification of tissues in chapter 4 has been simplified. Chapters 5 and 6 on stems and roots have been reversed. Chapter 8 is new; it incorporates part of the former chapter 20 on flowering plants with the former chapter 21 on fruits and seeds. Chapters 9–11 on water, metabolism, and growth in plants have been revised and updated, with some new material being added. The material on evolution in chapter 13 has been revised and expanded. The five-kingdom classification and key in chapter 14 have been revised to reflect recent research on the subject, and the new classification is used throughout the kingdom surveys. The discussion of genetic engineering in chapter 15 has been revised and expanded. Chapters 15–23 have all been updated and revised to varying degrees. The appendices have been slightly revised, with the order of topics in appendix 4 having been changed. Many photos and line drawings have been revised or replaced. More than 50% of all illustrations are now in full color.

SUPPLEMENTARY MATERIALS

Instructor's Manual

The Instructor's Manual available with *Introductory Plant Biology* offers a variety of course schedules while providing overviews, goals, suggested answers, film sources, and examination questions for each text chapter.

Laboratory Manual

The Laboratory Manual that accompanies the *Introductory Plant Biology* text has been revised throughout. It is written for the student entering the study of botany for the first time. The exercises utilize plants to introduce biological principles and the scientific method. The exercises are written to allow for maximum flexibility in sequencing.

Transparencies

Seventy-six acetate two- and four-color transparencies are available with the *Introductory Plant Biology* text. The transparencies are taken from the text and represent the important figures that merit extra visual review and discussion.

TestPak

wcb TestPak, a free, computerized testing service, simplifies testing while offering you flexibility. There are two convenient TestPak options available:
  —Use your Apple® *II*e, IIc, Macintosh, or IBM PC to pick and choose your test questions, edit them, and add your own questions. We can send you program and test item diskettes for this purpose.
  —If you do not have a microcomputer, you can still pick and choose your questions via our call-in/mail-in service. Within two working days of your request, we'll put a test master, a student answer sheet, and an answer key in the mail to you. Call-in hours are 8:30–5:00 CST, Monday through Friday.

QuizPak

wcb QuizPak, a student self-testing program on the microcomputer, is available free of charge to adopters. QuizPak lets students choose a chapter to review, then the computer quizzes them with questions selected for that chapter. QuizPak may be used at a number of work stations simultaneously, runs on the Apple® *II*e, IIc, or Macintosh, and on the IBM PC, and requires only one disk drive.

GradePak

wcb GradePak is a computerized service available free to qualifying adopters of *Introductory Plant Biology*. GradePak makes calculating and reporting your students' grades an easy task! A single disk holds data for classes of up to 500 students with room for 60 scores per student. GradePak calculates class averages on all scores and overall class averages. Plus, it gives you a grade profile report showing what each student requires to earn an A, B, C, or D for the term. GradePak allows numeric input as well as letter grades. It is available for the Apple® II+, *IIe,* and IIc, and for the IBM PC.

ACKNOWLEDGMENTS

The contributions of many individuals to the development of this book are gratefully acknowledged. Critical reviewers, who provided many valuable suggestions for improving and updating the text, include:

Lydia C. Arciszewski, *Bergen Community College*
Rolf W. Benseler, *California State University—Hayward*
Linda R. Berg, *University of Maryland*
Maynard C. Bowers, *Northern Michigan University*
Hayle Buchanan, *Weber State College*
Laura Chunosoff, *Pace University*
Rebecca McBride DiLiddo, *Suffolk University*
Michael Gardiner, *University of Puget Sound*
Linda E. Graham, *University of Wisconsin—Madison*
J. M. Herr, Jr., *University of South Carolina*
Lawrence A. Kapustka, *Miami University*
Robert C. Lommasson, *University of Nebraska*
James E. Marler, *Louisiana State University*
Robert S. Mellor, *University of Arizona*
Harvey A. Miller, *University of Central Florida*
Roger del Moral, *University of Washington*

Lloyd Ohl, *University of Wisconsin—Eau Claire*
P. C. Pendse, *California Polytechnic State University—San Luis Obispo*
Robert R. Robbins, *University of Wisconsin*
James L. Seago, *State University of New York—Oswego*
B. Dwain Vance, *North Texas State University*
H. B. Ward, *University of Mississippi*
Henry Webert, *Nicholls State University*
Terry F. Werner, *Harris-Stowe State College*
Louis Westmoreland, *San Jacinto College—Central*
Lorraine Wiley, *California State University—Fresno*

MARKET RESEARCH RESPONDENTS

We would like to thank the following adopters of the third edition for their help in preparing the current one. Each contributed greatly to our understanding of the relative strengths and weaknesses of the third edition by responding to a user's survey of the text.

Isabella A. Abbott, *University of Hawaii*
Douglas S. Allen, *Louisiana State University—Alexandria*
Delia Anderson, *William Carey College*
Edward F. Anderson, *Whitman College*
Philip J. Arnholt, *Concordia College*
William R. Ball, *Southeastern Community College*
Dr. M. James Barrier, *Baptist College at Charleston*
R. Glenn Bellah, *Bethany College*
Thomas J. Belzer, *Pasadena City College*
Rolf W. Benseler, *California State University—Hayward*
Roland A. Bergthold, *Sierra College*
Dr. Harry Bischoff, *Texas Lutheran College*
Jim Blassingame, *South Plains College*
Sr. Helen T. Bleistine, *Chestnut Hill College*
Dr. Cecile Boehmer, *Alice Lloyd College*
Dr. Gary A. Borger, *University of Wisconsin Center—Marathon County*
Robert P. Borgman, *Buena Vista College*
Cynthia A. Bottrell, *Scott Community College*
Maynard C. Bowers, *Northern Michigan University*

Eugene G. Bozniak, *Weber State College*

William J. Bramlage, *University of Massachusetts*

Roger H. Brown, *Berkshire Community College*

Robert Buckman, *Dakota State College*

Frances M. Cardillo, *Manhattan College*

Laura Chunosoff, *Pace University*

Dr. Richard Churchill, *SMVTI*

W. Wade Cooper, *Shelton State Community College*

Raymond R. Crawford, Jr., *San Jacinto College*

Opal H. Dakin, *Hinds Junior College*

Robert Davidson, *SUNY, Agricultural and Technical College at Delhi*

Gary Davis, *Spokane Community College*

Johnnie Driessner, *Concordia College*

Rev. Damian DuQuesnay, *St. Leo College*

Robert I. Ediger, *California State University—Chico*

Jerry L. Faulkner, *Tennessee Temple University*

Linda R. Finke, *Xavier University*

John L. Finneran, *Northern Essex Community College*

Katharine Floyd, *Wesleyan College*

Kenneth G. Foote, *University of Wisconsin—Eau Claire*

Gerald Fuhrmann, *Concordia College*

Dr. Eugene C. Gasiorkiewicz, *University of Wisconsin—Parkside*

Dennis J. George, *Johnson County Community College*

Cathy Haas, *Hartnell Community College*

L. W. Hagener, *Northern Montana College*

David Hartsell, *Francis Marion College*

Richard L. Hauke, *University of Rhode Island*

Glen D. Hegstad, *Northwestern College*

Richard H. Hevly, *Northern Arizona University*

L. Michael Hill, *Bridgewater College*

David Hodgson, *Adirondack Community College*

Barbara Joe Hoshisake, *Los Angeles City College*

Lyle T. Hubbard, Jr., *University of Alaska—Juneau*

Ray H. Hughes, *Lee College*

Richard W. Ikenberry, *Kearney State College*

John D. Jackson, *North Hennepin Community College*

Daryl H. Johnson, *Mississippi County Community College*

Wendel J. Johnson, *University of Wisconsin Center—Marinette*

William H. Kinch, *Santa Monica College*

William A. Kinnison, *Central Arizona College*

Duane Klarich, *Northern Montana College*

Penelope M. Koines, *University of Maryland*

Martin LaBar, *Central Wesleyan College*

Timothy I. Ladd, *Saint Mary's University*

James Lampky, *Central Michigan University*

Alex Lasseigne, *Nicholls State University*

David D. Lowrie, *Clarendon College*

Donnis C. Lyon, *Jones Jr. College*

Larry Joe McCumber, *Francis Marion College*

J. Philip McLaren, *Eastern Nazarene College*

Robert C. McReynolds, *San Jacinto College*

L. Maynard Moe, California State College—Bakersfield

Howard C. Monroe, *College of San Mateo*

Roger del Moral, *University of Washington*

Sharon J. Morton, *Miami University*

Gerald W. Naylor, *Carson-Newman College*

Norton H. Nickerson, *Tufts University*

Paul Nighswonger, *Northwestern Oklahoma State University*

Thomas E. Oldfield, *Ferris State College*

Dr. Damon R. Olszowy, *SUNY at Farmingdale*

Carl M. Oney, *Onondaga Community College*

James D. Perry, *University of North Carolina at Asheville—(UNAC)*

R. Gordon Perry, *Fairleigh Dickinson University*

Robert A. F. Pool, *Spokane Community College*

Martha J. Powell, *Miami University*

Daniel J. Prochaska, *Miami University—Oxford, OH*

Ronald A. Pursell, *Pennsylvania State University*

James A. Raines, *North Harris County College*

Rod Rakowicz, *South Puget Sound Community College*

Serafin Ramon, *Panhandle State University*

Joe Russo, *Kings River Community College*

David J. Schimpf, *University of Minnesota—Duluth*

Walter F. Scott, *Arkansas State University—Beebe*

Frank Seabury, *The Citadel*

Jack B. Secor, *Eastern New Mexico University*

Harry E. Sheally, Jr., *University of South Carolina—Aiken*

Gary Shield, *Kirkwood Community College*

Alden E. Smith, *State University College at Buffalo*

Charles E. Smith, Jr., *Ball State University*

Kenneth J. Smith, *Montealm Community College*

Milton R. Sommerfeld, *Arizona State University*

Marschall C. Stevens, *Citrus College*

Robert J. Swanson, *North Hennepin Community College*

Phillip H. Theis, *Butler County Community College*

R. Dale Thomas, *Northeast Louisiana University*

Dr. John D. Tiftickjian, *Delta State University*

Robert L. Tumey, *Catawba Valley Technical College*

Maura Geens Tyrrell, *Stonehill College*

Jane Glasgow Vance, *Northeast Alabama State Junior College*

John F. Vanderploeg, *Ferris State College*

Donald F. Van Dyke, *Riverside Community College*

Lawrence F. Virkaitis, *SUNY Morrisville Agricultural & Technical College*

Ray Watson, *Mississippi State University*

Henry Webert, *Nicholls State University*

Dr. Melvin J. Wentland, *St. John Fisher College*

Fred Winkler, *L. B. Wallace Junior College*

James A. Winsor, *Pennsylvania State University*

Bernard L. Woodhouse, *Savannah State College*

Donald E. Wujek, *Oakland Community College*

Dianne Yang, *Fresno City College & College of the Sequoias*

Dr. Charles Yarish, *University of Connecticut at Stamford*

Todd C. Yetter, *Miami University*

Additional persons who read parts of the manuscript and made many helpful criticisms and suggestions include Richard S. Demaree, Jr., Robert B. McNairn, the late George E. Corson, Jr., William L. Stephens, Donald T. Kowalski, and Robert A. Schlising. Others whose encouragement and contributions are deeply appreciated include W. T. Stearn, E. Gibbes Patton, Donald B. Joley, Paul C. Silva, Donald E. Brink, Jr., Suzanne Costanza, the faculty and staff of the Department of Biological Sciences, California State University, Chico (especially Timothy B. Devine); my many inspiring students, the Lyon Arboretum of the University of Hawaii, the editorial, production, and design staffs of the Wm. C. Brown Publishers, and most of all my wife Janet, and my children, Kevin and Sharon, who sacrificed a great deal to bring about each edition of the book. Special thanks are due the artists, Denise Robertson Devine, Elly Simmons, and Susan Juanarena.

Kingsley R. Stern
Chico, California

Introductory
Plant Biology

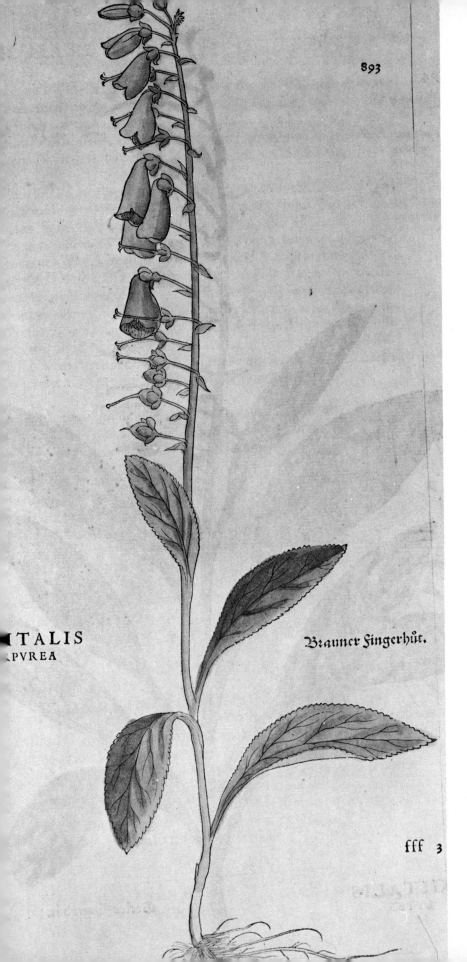

893

‹TALIS
‹PVREA

Brauner Fingerhůt.

fff 3

Overview

*T*his chapter introduces you to botany: what it is, how it developed, how it relates to our everyday lives, and what its potential is for the future. The discussion includes a brief introduction to the beginning of plant study in ancient civilizations, an examination of the scientific method, and a look at the development of botany after the invention of the microscope. It concludes with a brief survey of the major disciplines within the field of botany.

1 *The Development of Plant Study*

INTRODUCTION

Plants may not often make the front page of the newspaper, but hardly a day goes by without their at least being mentioned by the time one reaches the sports section. Some recent headlines I have noted, for example, include "Herbs Blamed in Accidental Poisonings," "Ugly Desert Jojoba May Keep Bay Area Transit Buses Rolling," "Hot Future Seen for Algae," "UA Brewing Fuels of Tomorrow from Common Plants of Today," "State Owes Bushels to 'U' Plant Breeders," and "Plants and Flowers Make You Sexy."

Actually, all of us have been totally dependent on green organisms such as plants and algae since before we were born. An extravagant statement? Consider that all green organisms are unique in being capable of transforming the sun's energy into forms usable by animal life and that such energy is vital to the very existence of animal life. Consider further that virtually all the oxygen in the air we breathe is produced by such green organisms and that they alone can remove the carbon dioxide waste we give off into the atmosphere. If some major disease were to kill off all or a large portion of the vegetation on earth (including that found in the oceans), all the animals on land, in the sea, or in the air would soon starve, but even if some alternative source of energy were available, they would suffocate within 11 years—the amount of time it is estimated for all the earth's oxygen to be completely depleted if it were not replenished.

Apart from these aspects, green organisms and plant products are so interwoven with the fabric of human society that we largely take them for granted. We are aware, of course, that corn, potatoes, and other vegetables are plants; but all foods, including meat, fish, poultry, eggs, cheese, and milk, to mention but a few, owe their existence to plants. Condiments such as spices, mustard, and pepper and luxuries such as perfumes are produced by plants, as are dyes, adhesives, digestible surgical gut, food stabilizers, beverages, and emulsifiers. Our houses are constructed with lumber from trees, which also furnish the cellulose for paper, cardboard, and synthetic fibers. Our clothing, camping equipment, bedding, draperies, and other textile goods are made from fibers of many different plant families. Coal is fossilized plant material, and oil may have come from microscopic green organisms or animals that were plant consumers either directly or indirectly. All medicines and drugs at one time came from plants, fungi, or bacteria, and many important ones, including some of the antibiotics, still do. Microscopic organisms are responsible for recycling both plant and animal wastes and for aiding in the building of healthy soils. Others are responsible for human diseases and allergies.

Whether or not gluts or shortages of oil and other fossil fuels are politically or economically manipulated, their finite nature is indisputable and these fuels eventually will disappear. Accordingly, increased attention is being paid to alternative energy sources. Methane gas, which can be used as a substitute for natural gas, has been produced from animal manures and decomposed plants in numerous villages in India and elsewhere for many years, and after several years of trial on a small scale in the United States, plans are now under way for production of methane on a larger scale from human sewage. Potatoes, grains, and other sources of carbohydrates are currently used in the manufacture of alcohols, some of which are being blended with gasoline ("gasohol"), and such uses may increase in the future.

What of plants and the future? It is estimated that by the year 2000, the population of the earth will double, approaching 7 billion persons, every one of whom will need food, clothing, and shelter in order to survive. To ensure survival, a majority of us may eventually need to learn not only how to cultivate food plants but also how to use plants in rendering polluted water and land productive again and how to use them in renewing urban areas; in addition, everyone may need to learn how to reverse the trend of plants shrinking in number, a trend caused by the staggering increase in the number of earth inhabitants.

Some have suggested that the odds are against humanity saving itself from itself and have even indicated that it might become necessary to emigrate to another planet. If so, microscopic algae could play a vital role in space exploration. For a number of years, experiments with portable "oxygen generators" have been in progress. Tanks of water teeming with tiny green algae are taken aboard a spacecraft and installed so that they will be exposed to light for at least part of the time. The algae not only produce oxygen that the spacecraft inhabitants can breathe, but they also utilize the waste carbon dioxide end product of respiration. In addition, as they multiply, excess algae can be fed to a special kind of shrimp, which in turn multiply and become food for the space travelers. Other wastes are recycled by different microscopic organisms. When this self-supporting arrangement, called a *closed system* is perfected, the range of spacecraft should greatly increase because heavy oxygen tanks will not be necessary, and the amount of food reserves needed will be reduced.

As in the past, primitive peoples use plants not only for food, shelter, clothing, and medicine but also in hunting and fishing. Today small teams of botanists, anthropologists, and medical doctors are interviewing medicine men and herbal healers in remote tropical regions and taking notes on various uses of plants by primitive peoples. These scientists are doing so in order to preserve those with potential for modern civilization before disruption of their habitats causes them to become extinct.

Genetic engineering (see chapter 15), which involves the introduction of desirable genes from one organism into another, holds enormous potential for improvements in the quality and yield of crop plants and in the suppression or elimination of human defects and diseases. Recent genetic engineering developments include bacteria that can prevent frost damage on plants up to 8° F below freezing; other bacteria that can break up crude oil spills and break down refinery, sewage, food, and paper pulping wastes; and still others that can remove grease and fat clogs in drains and sewers. Plant breeders—who have already given us a wide variety of hybrid plants with greater vigor, disease resistance, and yield than the parent plants—are engaged in trying to produce plants that will inhibit weeds, or grow in areas that are inhospitable to crop plants presently available; still others that naturally repel insects are being investigated and improved. Plant breeders at Cornell University, for example, are working on developing a potato plant with sticky hairs that will trap insects, while other scientists are seeking to produce plants with proportionately higher protein and vitamin content. Plant breeders also are developing varieties of plants that can thrive in relatively dry or salty soil, and others they call "green glue" that can bind and stabilize soils and even reclaim land that has become desert. The next decade or two will undoubtedly view plants with various properties that were completely unforeseen by our grandparents.

EARLY HISTORY AND DEVELOPMENT OF PLANT STUDY

On a number of occasions I have received visits from anxious mothers or from pediatricians wanting to know if parts of backyard plants that young children have consumed are poisonous. I have also been consulted by various law enforcement officials, landscape architects, pollen collectors, students interested in wildflowers, vegetarians, organic and traditional gardeners, and a variety of other professional and amateur persons with a wide range of interest in plant life. Some have wanted plants identified; others have wanted suggestions for treating diseased plants. Still others have wanted to know about grafting techniques, the effect of "gray water" (water that has been used for bathing or washing dishes, etc.) on plants, the suitability of plants for specific locations, the preservation of plants, the edibility of wild plants, and a host of other plant-related subjects.

Knowledge about plant life throughout the world has now become so vast that it is impossible for any person to be an authority on more than a tiny fraction of it. But our libraries contain thousands of books dealing with virtually every facet of botanical investigation, and research journals publish the latest discoveries from around the world on an almost daily basis. Why and how has all this knowledge accumulated? To answer this question we need to take a brief look at the early history and development of plant study.

Plants and Primitive Peoples

Fossil evidence indicates that between 10,000 and 35,000 years ago humans migrated to the Americas and Australia from Africa, India, and Indonesia. The migrations to the Americas evidently occurred between Siberia and Alaska across the Bering Strait, which was a land bridge during parts of the Pleistocene era. These early humans were primarily hunters. Indeed, the extinction of many large land mammals in North America coincides with the appearance and activities of humans some 13,000 to 10,000 B.C.

If you are a hunter or a fisherman, however, you are well aware that success in hunting or fishing can vary considerably, depending on various environmental and other circumstances. By 8000 B.C. our ancestors had begun to develop more reliable sources of food through primitive forms of agriculture.

Compelling archaeological evidence obtained from the walls of tombs, mummy wrappings, hieroglyphics, cave paintings, and carvings indicates the cultivation of grains, legumes and certain fruits (e.g., figs, olives, pomegranates, dates) was well-established in the Near East by 6500 B.C. The Near Eastern center and other major centers of origin of cultivated plants are discussed in chapter 22.

By 4000 B.C. the date had become one of the most important crops to the Assyrians and Egyptians. The dates were eaten fresh or dried, and the sap of date palms was fermented for wine. Although they knew nothing of the details, the Assyrians were aware of sexuality in the date palm and pollinated the female trees by hand. By the seventh century B.C., they had produced a systematically arranged list of medicinal plants, which suggests that the physicians and pharmacists of the day had a noteworthy knowledge of plants and their uses.

The Chinese have been cultivating medicinal and other useful plants for at least 4,500 years. Records from that far back in history are fragmentary, and it is often impossible to separate fact from legend; some authorities agree, however, that the founder of Chinese agriculture was an emperor by the name of Shen Nung, who was born in 2737 B.C. Shen Nung is said to have invented the plow and is also believed to have established an annual seed-sowing ceremony, during which seeds of soybeans, wheat, rice, millet, and sorghum were planted by royalty. He appears to have been an authority on poisons and antidotes. He also wrote a book on drugs and medicines that was incorporated into the *Pun-tsao,* a Chinese *pharmacopoeia* (an officially recognized book describing drugs and medicines) of 40 volumes, published during the seventeenth century. During the Han dynasties, which lasted from about 200 B.C. until the birth of Christ, gardens became very extensive in China and many ornamental plants were cultivated. Plants such as primroses, poppies, and chrysanthemums were brought to the Western world from China over 2,000 years ago.

The Egyptians cultivated primitive forms of wheat and barley from about 5000 to 3400 B.C., although some authorities place the cultivation of these two cereals as far back as 10,000 to 15,000 B.C. More modern forms of cereals, such as six-rowed barley, may have been under cultivation by 2000 B.C. By the fifth century B.C., the Egyptians apparently were brewing *booza,* a beer, from barley.

The Emergence of Botany as a Science

The study of plants, called **botany** from the French word *botanique* (botanical) and three Greek words *botanikos* (botanical), *botanē* (plant or herb), and *boskein* (to feed), appears to have had its origins with Stone Age peoples who sought to modify their surroundings and feed themselves. Initially, the primary interest in plants was practical, centering around how plants might provide food, fibers, fuel, and medicine. Eventually, however, an intellectual interest arose; individuals became curious about how plants reproduced and how they were put together. This inquisitiveness led to plant study becoming a **science,** which broadly defined is simply "a search for knowledge of the natural world."

A science may be distinguished from other fields of intellectual endeavor by several features. It involves the observation, recording, organization, and classification of facts, and more importantly, it involves what is done with the facts. Scientific procedure involves experimentation, observation, and the verifying or discarding of information, chiefly through inductive reasoning from known samples. There is no universal agreement on the precise details of the process. A few decades ago, scientific procedure was considered to involve a routine series of steps. This series of steps came to be known as the *scientific method* and there are still instances where such a structured approach works well. In general, however, the scientific method now describes the procedures of assuming and testing *hypotheses.* A **hypothesis** is simply a tentative, unproven explanation for something that has been observed. It may not be the correct explanation—testing will determine whether it is correct or incorrect. The nature of the testing will vary according to the circumstances and materials. For example, we may *observe* that apples are red fruits that taste sweet. We may then make the *hypothesis* that all red fruits taste sweet. We may *test* the hypothesis by tasting red fruits, and as a result of our testing (since red crab apples are bitter), we may *modify the hypothesis* to state that only some red fruits are sweet.

When a hypothesis is tested, *data* (bits of information) are accumulated and may lead to the formulation of a useful generalization called a *principle.* Several related principles may lend themselves to grouping into a *theory,* which is not simply a guess. A theory is a group of generalizations (principles) that help us understand something. We reject or modify theories only when new principles increase our understanding of a phenomenon. To be accepted by scientists the results of any experiments designed to test the hypothesis must be repeatable and capable of being duplicated by others.

However one defines science, it is clear that plant science existed in ancient Greece. As in even older cultures, the study of plants started when the early Greeks developed a practical interest in food and drug plants, but they slowly became curious as well about the structure and function of plants. As the physicians and the pharmacists of the era gathered and used medicinal plants, they studied the variations and forms and came to recognize apparent relationships among them.

One of these Greek herbal physicians had a son in 384 B.C., who became one of the most renowned philosophers of all time—Aristotle the Stagirite. Although Aristotle is perhaps better known for his philosophical works, he was an accomplished mathematician and also acquired extensive knowledge in nearly all aspects of natural history. In fact, he combined philosophical and scientific interests as few other philosophers have done. At the age of 17, Aristotle went to Athens, where he met and became a pupil of Plato. He left Athens after Plato's death in 347 B.C. and studied marine animals at a coastal area for several years, eventually returning to Athens to found the first botanical garden of which there is any record.

When Aristotle died, he willed the botanical garden and its associated library to his pupil and assistant, Theophrastus of Eresus (figure 1.1). Theophrastus was an extraordinary man, who not only acquired all the knowledge Aristotle had accumulated on plants but added prodigiously to it from his own observations. It is said that he had 2,000 disciples and wrote 200 treatises. The most important of the latter to have survived are two books entitled *History of Plants* and *Causes of Plants*. So great were his contributions to botany as a science that the famous eighteenth-century Swedish botanist Linnaeus gave him the title "Father of Botany"; few, if any, dispute his right to the honor.

FIGURE 1.1   Theophrastus. (Courtesy National Library of Medicine)

Herbals Appear

During the second century A.D., two books that had a significant influence on botanical studies appeared. Pliny's *Historia Naturalis* contained lists of food or medicinal plants; the other, Dioscorides' *Materia Medica,* was the first book to contain illustrations of plants, all laboriously copied by hand. Many of the common names used by Dioscorides are still used today. European scholars who followed Dioscorides continued, by hand, to copy these books, which became known as **herbals,** and held them in such high esteem that it was considered heresy to question anything in their contents; consequently few new ideas were added during the Dark and Middle Ages that lasted from 400 to 1400 A.D.

With the advent of the printing press in the middle of the fifteenth century, the number of herbals mushroomed, and the period from about 1500 to 1700 A.D. became known as the Age of Herbals. These botanical works were primarily the products of German botanists, although some Italian and English botanists made their own contributions between 1470 and 1670. The *herbalists,* as they were called, were mostly concerned with medicinal plants, which they studied in the botanical gardens that had become numerous and extensive in Europe by this

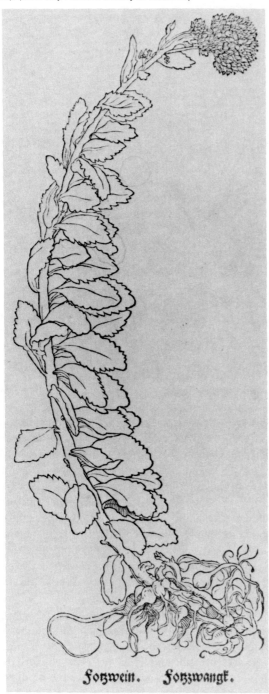

Foßwein.   Foßzwangł.

time. They produced elaborate and intriguing illustrations for the herbals, occasionally accompanied by outlandish stories and descriptions. Some of the stories became legends and developed into the *Doctrine of Signatures*. According to this doctrine, if a part of a plant, such as the root, had the shape of a part of the human body, it would be useful in treating a disease of the human part it most closely resembled. Thus, the meat of a walnut, which somewhat resembles a miniature brain, was used in treating brain diseases, and hepatica leaves, which have lobes reminiscent of those of the liver, were used to treat ailments of that organ. One of the more famous herbalists was Otto Brunfels, who published a three-volume herbal in 1530 (figure 1.2). His work had excellent illustrations and is considered to be a link between ancient and modern botany.

THE FIRST MICROSCOPES

The microscope had and continues to have a profound effect not only on plant studies but on the biological sciences and related fields as a whole. In 1590, Zacharias and Francis Janssen, Dutch brothers who were spectacle makers, drew on the experience of their father, Hans, who was famous for his optical work. They discovered how to combine two convex lenses in the interior of a tube, and produced the first instrument for magnifying minute objects. Because of this, Zacharias Janssen, in particular, is often referred to as the inventor of the compound microscope, although it was Faber of Bamberg, a physician serving Pope Urban VII, who first applied the term *microscope* to the instrument during the first half of the seventeenth century.

A Dutch draper by the name of Anton van Leeuwenhoek (1632–1723), who ground lenses and made microscopes in his spare time, is best known for his development of primitive microscopes. Leeuwenhoek was the first to describe bacteria, sperms, and other tiny cells he observed with his microscopes, some of which could magnify as much as 200 times. In his will, he left 26 of his 400 handmade microscopes to the Royal Society of London.

DIVERSIFICATION OF PLANT STUDY

Before the invention of the microscope, plant study had been dominated by investigations based primarily on the external features of plants. The magnification of the early microscopes was not very great by present standards, but these instruments nevertheless led to the discovery of *cells* (discussed in chapter 3) and opened up whole new areas of study.

Plant anatomy, which is concerned chiefly with the internal structure of plants, was established through the efforts of several scientific pioneers. Early plant anatomists of note included Marcello Malpighi (1628–1694), an Italian who discovered various tissues in stems and roots, and Nehemiah Grew (1628–1711), an Englishman who described the structure of wood more precisely than any of his predecessors.

Plant physiology, which is concerned with plant function, was established by J. B. van Helmont (1577–1644), a Flemish physician and chemist, who was the first to demonstrate that plants do not have the same nutritional needs as animals. In a classic experiment, van Helmont planted a willow branch weighing 5 pounds in an earthenware tub filled with 200 pounds of dry soil. He covered the soil to prevent dust settling on it from the air, and after five years he reweighed the willow and the soil. He found that the soil weighed only 2 ounces less than it had at the beginning of the experiment, but that the willow had gained 164 pounds. He concluded that the tree had added to its bulk and size from the water it had absorbed. We know now that most of the weight came as a result of photosynthetic activity (see chapter 10), but van Helmont deserves credit for landmark experimentation along these lines.

The seventeenth century saw a marked increase in botanical explorations to various parts of the globe. In the fifteenth century when Columbus visited Cuba, he found local Indian tribes cultivating corn (maize). This important food plant had apparently been in use by the pre-Incas of Peru some 5,000 years earlier. By the time explorers ventured into the Americas in the 1600s, they found that maize culture had spread from Argentina in the south to the St. Lawrence River area in the north; American Indians had also domesticated the white potato, and in Mexico Indians were cultivating flowers and medicinal plants.

The explorers took large numbers of plants back to Europe with them, and it soon became clear to those working with the plants that some sort of formalized system was necessary just to keep the collections straight. Several *plant taxonomists* (botanists who specialize in the identifying, naming, and classifying of plants) proposed ways of accom-

FIGURE 1.3 Linnaeus. (Courtesy National Library of Medicine)

plishing this, but we owe our present system of naming and classifying plants to the Swedish botanist Carolus Linnaeus (1707–1778) (figure 1.3). **Plant taxonomy** (also called *plant systematics*), which is the oldest branch of plant study, began in antiquity, but Linnaeus did more for the field than any other person in history. Thousands of plant names in use today are those originally recorded by this remarkable man in his book *Species Plantarum,* published in 1753. An expanded account of Linnaeus and his system of classification is given in chapter 14.

The nineteenth century saw the development of **plant geography,** the study of how and why plants are distributed where they are, and of the allied field of **plant ecology,** which is the study of the interaction of plants with one another and with their environment. Public awareness of the field of ecology as a whole increased considerably after 1962, following publication of a best-selling book entitled *Silent Spring* by Rachel Carson. In this book, based on more than four years of literature research, the author called attention to the fact that more than

FIGURE 1.4 Sir Joseph D. Hooker. (Courtesy National Library of Medicine)

500 new toxic chemicals annually are put to use as pesticides in the United States alone, and detailed how these chemicals and other pollutants were having an insidious impact on all facets of human life and environment.

Among the most noteworthy of the early plant geographers were two natives of Berlin, Germany, Carl Willdenow (1765–1812) and Alexander von Humboldt (1769–1859), who published books on the relationship of seed dispersal to plant distribution and on the associations of various plants with one another in tropical and temperate climates. These studies were brought to a climax by Sir Joseph D. Hooker (1817–1911) (figure 1.4), who eventually became director of the Royal Botanic Gardens in Kew, England. Hooker traveled widely studying plant life in both the northern and southern hemispheres, and he also published floras (accounts of the plants of a specific region) of India and Antarctica. Charles Darwin, whose books (particularly his *Origin of Species*) revolutionized some basic biological concepts of the adaptation of organisms to their environment, said of Hooker's *Flora Antarctica*, "It is by far the grandest and most interesting essay on subjects of nature I have ever read."

The study of the form and structure of plants, **plant morphology,** also developed during the nineteenth century, and by the beginning of the twentieth century much of the basic information incorporated in the plant sciences today had been discovered and elucidated. The number of scientists engaged in investigating plants had also increased conspicuously. **Genetics,** the science of heredity, had been founded by the Austrian monk Gregor Mendel (1822–1884) through his classic experiments with peas, and **cytology,** the science of cell structure and function (now called **cell biology**), had received great impetus from the discovery of how cells multiply and function in sexual reproduction. The midtwentieth-century development of *electron microscopes* (see chapter 3) further spurred cell research and led to a vast new insight into cells and new forms of cell research that continues to the present. Many other significant twentieth-century discoveries are discussed at appropriate junctures throughout the chapters to follow.

PLANT SCIENCES AND THE FUTURE

There is still a vast amount of information to be discovered, and new discoveries continue to be made daily. As far back as 1938, for example, an investigator by the name of Wellensiek noted that 11,000 papers on botanical subjects had been published in that year alone; the number in recent years is undoubtedly many times greater. Further, it appears probable that at least one-third of all the organisms traditionally regarded as plants (particularly algae and fungi) have yet to be named, let alone thoroughly investigated.

Wild plants and animals are becoming extinct at a rapidly accelerating rate as their natural habitats are destroyed through development and pollution; in fact, there is strong evidence that many undescribed organisms are becoming extinct before we have learned anything about them. Efforts must be upgraded to educate the general public on the necessity of preserving natural habitats so that the numerous tangible and aesthetic benefits of doing so will be available to succeeding generations. Also, both basic and applied research in botany must be given additional support if the earth's burgeoning human population is to continue to be fed, clothed, and housed.

SUMMARY

We are totally dependent on green plants because they alone can convert the sun's energy into forms that are usable by, and vital to the very existence of, animal life. Plants and plant products are so interwoven with the fabric of human society that we largely take them for granted. We know that vegetables are plants, but animals, animal products, many luxuries and condiments, and other useful substances such as fibers, lumber, coal, medicines, and drugs either depend on plants or are produced by them.

In a future of proliferating world populations, all persons, to ensure human survival, may need to acquire some knowledge of plants and how to use them. If it should become necessary to emigrate to another planet, plants will undoubtedly play a vital role in space exploration as portable oxygen generators. Today, teams of scientists are interviewing medicine men and herbal healers in the tropics to try to discover hitherto little-known plants used by primitive peoples before the plants become extinct.

Our libraries contain thousands of books dealing with botanical subjects. All this knowledge has accumulated over the span of human existence. Humans migrated to the Americas between 10,000 and 35,000 years ago; they originally were hunters, but gradually they learned how to cultivate crops. The Assyrians and Egyptians were cultivating fruits and cereals by 4000 B.C., and the Chinese practiced primitive agriculture at least 4,500 years ago. Botany, the study of plants, apparently began with Stone Age peoples' practical uses of plants. Eventually botany became a science, as intellectual curiosity about plants arose.

A science involves observation, recording, organization, and classification of facts. The verifying or discarding of facts is done chiefly from known samples through inductive reasoning. The scientific method involves specifically following a routine series of steps and generally assuming and testing hypotheses (tentative explanations for something that has been observed). Botanists believe that plant science existed in ancient Greece. A third-century B.C. Greek, Theophrastus, especially contributed much to the field of botany.

During the second century A.D., Pliny and Dioscorides produced books on food or medicinal plants; these books were copied by hand and became known as herbals. During the fifteenth century the number of herbals increased greatly with the advent of the printing press. The herbals were illustrated and sometimes contained bizarre stories about plants; these stories resulted in the development of the Doctrine of Signatures. The doctrine held that plant parts resembling human body parts were useful in treating ailments of those particular human organs.

The compound microscope, which was first produced by Zacharias and Francis Janssen in 1590 and developed by Leeuwenhoek in the 1600s, had a profound effect on studies in the biological sciences and led to the discovery of cells. Plant anatomy and plant physiology developed during the seventeenth century. Van Helmont was the first to demonstrate that plants have nutritional needs different from those of animals. He did so by careful experimentation and observation of a willow grown in a tub. Also during the seventeenth century, Europeans engaged in botanical exploration on other continents and took plants back to Europe. During the eighteenth century the Swedish botanist Linnaeus produced the elements of our present system of naming and classifying plants. During the nineteenth century plant ecology, plant geography and plant morphology developed, and by the beginning of the twentieth century genetics and cell biology had received their great impetus from the discovery of how cells multiply and function in sexual reproduction. Much remains yet to be discovered and investigated.

REVIEW QUESTIONS

1. Briefly indicate contributions to plant science made by the following: Shen Nung, the ancient Egyptians, Theophrastus, Grew, Linnaeus.
2. What is meant by the scientific method?
3. Distinguish among herb, herbal, and herbalist.
4. What was the significance of van Helmont's experiment with the willow tree?
5. What is the oldest branch of botany, and why did it precede other branches?
6. What is the thrust of each of the other branches of botany?
7. What is the Doctrine of Signatures?

DISCUSSION QUESTIONS

1. Since humans survived on wild plants for thousands of years, might it be desirable to return to that practice?
2. On the basis of what you have read, would you single out any one individual as having contributed the most to the development of plant study? Why?
3. How would you guess that Stone Age peoples discovered medicinal uses for plants?
4. Many of the early botanists were also doctors. Why do you suppose this is no longer so?
5. Consider the following hypothesis: "The majority of mushrooms that grow in grassy areas are not poisonous." How could you go about testing this hypothesis scientifically?

ADDITIONAL READING

Anderson, F. J. 1977. *An illustrated history of the herbals.* New York: Cambridge University Press.

Baumel, H. B. 1978. *Biology: its historical development.* New York: Philosophical Library.

Carson, R. 1962. *Silent spring.* Boston: Houghton Mifflin Co.

Ewan, J., ed. 1969. *A short history of botany in the United States.* Forestburgh, NY: Lubrecht & Cramer.

Gardner, E. J. 1972. *History of biology,* 3d ed. Minneapolis: Burgess Publishing Co.

Gerard, J. 1974 reprint of 1597 ed. *The herball or general historie of plants,* 2 vols. Norwood, NJ: Walter J. Hohnson.

Greene, E. L. 1983. *Landmarks of botanical history,* 2 vols. Egerton, F. N., ed. Palo Alto, CA: Stanford University Press.

Grew, N. 1965 reprint of 1682 ed. *The anatomy of plants with an idea of a philosophical history of plants and several other lectures.* New York: Johnson Reprint Corp.

Harvey-Gibson, R. J. 1981. *Outlines of the history of botany.* Cohen, I. B., ed. Salem, NH: Ayer Company Publishers.

Lindroth, S. et al. 1983. *Linnaeus: the man and his work.* Frangsmyr, T., ed. Tr. by Srigley, M., Vowles, B. Berkeley, CA: University of California Press.

Morton, A. G. 1981. *History of botanical science: an account of the development of botany from the ancient time to the present.* Orlando, FL: Academic Press.

Nordenskiold, E. 1935. *The history of biology.* St. Clair Shores, MI: Scholarly Press.

Swift, L. H. 1974 reprint of 1970 ed. *Botanical bibliographies: a guide to bibliographic materials applicable to botany.* Forestburgh, NY: Lubrecht & Cramer.

Overview

*T*his chapter begins with a discussion of the attributes of living organisms. These include growth, reproduction, response to stimuli, metabolism, movement, complexity of organization, and adaptation to the environment. Then it examines the chemical and physical bases of life. A brief look at the elements and their atoms is followed by a discussion of compounds, molecules, bonds, ions, valence, mixtures, acids, bases, and salts. Forms of energy and the chemical components of protoplasm are examined next. The chapter concludes with an introduction to carbohydrates, lipids, proteins and protein synthesis, and nucleic acids.

2   *The Nature of Life*

Some Learning Goals

1. Learn the attributes of living organisms.

2. Define matter; describe its basic state.

3. Distinguish compounds from mixtures and describe acids, bases, and salts.

4. Know the various forms of energy.

5. Learn the elements found in protoplasm.

6. Understand the nature of carbohydrates, lipids, and proteins.

7. Discuss protein synthesis and the nature of DNA.

Outline

INTRODUCTION

Have you ever dropped a pellet of dry ice (frozen carbon dioxide) into a pan of water and watched what happens? As the solid pellet is rapidly converted to a gas due to its contact with the warmer water, it darts randomly about the surface looking like a highly energetic bug waterskiing. Does all that motion make the dry ice alive? Hardly; yet one of the attributes of living things is movement. But if living things move, what about plants? Is a tree not alive because it does not crawl down the sidewalk? Again the answer is no, but these questions do serve to point out some of the difficulties encountered in defining *life.* In fact, some contend that there is no such thing as life—only living organisms—and that life is a concept based on the collective attributes of living organisms.

The activities of living organisms emanate from *protoplasm,* the physical basis, or the "stuff of life," contained in tiny structural units called *cells,* which are unique to living things. Both protoplasm and cells are discussed in chapter 3. Other attributes, or features, of living things are discussed in the following sections.

ATTRIBUTES OF LIVING ORGANISMS

Growth

The complex phenomenon of **growth** has been described simply as an increase in mass, which is usually also correlated with an increase in volume. Growth, which results primarily from the production of new protoplasm, includes variations in *form*— some the result of inheritance, some the result of environmental response. As an example of environmental response, consider what might happen if you were to plant two apple seeds of the same variety in poor soil and subsequently give them unequal treatment. If you were to give one just barely enough water to allow it to germinate and grow, while you not only gave the other an ample water supply but also worked fertilizers and conditioners into the soil around it, you might expect the second one to grow much larger and be much more productive than the first. In other words, although your two apple trees grew from the same variety of apple seed, they would differ in form, following patterns of growth dictated by the protoplasm and the environment. Various aspects of growth are discussed in chapter 11.

Reproduction

Dinosaurs were abundant 160 million years ago, but none exists today. Numerous mammals, birds, reptiles, plants, and other organisms are now on lists of endangered or threatened species, and many species are doomed to extinction within the next decade or two. All these once-living or living things have one feature in common: it became impossible or it has become difficult for them to reproduce. **Reproduction** is such an obvious feature of living organisms that we take it for granted—until it is lost.

When reproduction occurs, the offspring are always similar to the parents: guppies never have puppies—just more guppies, and a petunia seed, when planted, will not develop into a polyanthus. In addition, offspring of one kind tend to resemble their parents more than they do other individuals of the same kind. The laws governing these aspects of inheritance are discussed in chapter 12.

Response to Stimuli

If you stick a pin into a pillow, you certainly do not expect any reaction from the pillow, but if you were to stick the same pin into a friend, you know the reaction would be instantaneous (assuming the friend was conscious) because response to stimuli is a characteristic of all living things. You might argue, however, that when you stuck a pin into your geranium nothing happened, even though you were fairly certain the plant was alive. What you might not have been aware of is that the geranium did indeed respond, but in a manner very different from that of a human. Plant responses to stimuli are generally much slower than those of animals and usually are of a different nature. If the geranium's food-conducting tissue was pierced, it probably responded by producing a plugging substance called **callose** in the affected cells. Some studies have shown that callose may form within as little as five seconds after wounding. In addition, an unorganized tissue called **callus,** which forms much more slowly, may be produced at the site of the wound. Responses of plants to injury and to other stimuli such as light, temperature, and gravity are discussed in chapters 9 through 11.

Metabolism

Metabolism has been defined as the "sum total of all the biochemical reactions taking place within an organism." All living organisms undergo various metabolic activities that include the production of new protoplasm, the repair of damage, and normal maintenance. The most important activities include **respiration,** an energy-releasing process that takes place in all living cells; **photosynthesis,** an energy-harnessing process in green cells that is, in turn, associated with energy storage; **digestion,** the conversion of large or insoluble food molecules to smaller soluble ones; and **assimilation,** the conversion of raw materials into protoplasm and other cell substances. These topics are discussed in chapters 9 through 11.

Movement

As observed at the beginning of this chapter, plants generally do not move from one place to another (although their reproductive cells may do so). This does not mean, however, that plants do not exhibit movement, a universal characteristic of living things. The leaves of sensitive plants (*Mimosa pudica*) fold within a few seconds after being disturbed or subjected to sudden environmental changes, and the tiny underwater traps of bladderworts snap shut in less than one-hundredth of a second. But most plant movements are slow and imperceptible, when compared with those of animals, and are primarily related to growth phenomena. They become obvious only when demonstrated experimentally or when shown by time-lapse photography. The latter often reveals many types and directions of motion, particularly in young organisms. Movement is not confined to the organism as a whole but occurs down to the cellular level. Cyclosis, a streaming motion of protoplasm, occurs constantly in living cells. The streaming tends to resemble a river flowing clockwise or counterclockwise within the boundaries of each cell, but movement may actually be in various directions.

Complexity of Organization

Living organisms are composed of large numbers of **molecules** (the smallest unit of an element or compound retaining its own identity). The molecules are not simply mixed, like the ingredients of a cake or the concrete in a sidewalk, but are organized into cells and tissues. Even the most complex nonliving object has only a tiny fraction of the types of molecules of the simplest living organism, and in the living organism the arrangements of these molecules are highly structured and complex. Bacteria, for example, are considered to have the simplest cells known, yet each cell contains a minimum of 600 different kinds of protein in addition to hundreds of other substances, and each component has a specific place or structure within the cell. When larger living objects such as flowering plants are examined, the complexity of organization is overwhelming, and the number of molecule types can run into the millions.

Adaptation to the Environment

Assume that you skip a flat stone across a body of water and it lands on the opposite shore. The stone is not affected by the change from air to water to land during its quick journey; it does not respond to its environment. Living organisms, however, do respond to the air, light, water, and soil of their environment, as will be explained in later chapters. In addition, they are genetically adapted to their environment in many subtle ways. Some weeds (e.g., dandelions) can thrive in a wide variety of soils and climates, whereas many species now threatened with extinction have adaptations to their environment that are so specific they cannot tolerate even relatively minor changes.

CHEMICAL AND PHYSICAL BASES OF LIFE

The Elements: Units of Matter

The basic "stuff of the universe" is called *matter.* On earth, matter occurs in three states—*solid, liquid,* and *gas.* In simple terms, matter's characteristics are as follows:

1. It occupies space.
2. It has mass (with which we commonly associate weight).
3. It is composed of **elements,** of which there are at present 107 known (103 officially recognized)—17 have been produced artificially and 90 occur naturally on our planet. Only a few of the natural elements (e.g., nitrogen, oxygen, gold, silver, copper) occur in pure form; the others are found combined together chemically in various ways.

The smallest stable subdivision of an element that can exist is called an **atom.** Atoms are so minute that individual atoms were not indirectly rendered visible to us by the most powerful electron microscopes until the mid-1980s. We do know from experimental evidence, however, that each atom has a tiny **nucleus** consisting of particles called **protons,** which have positive electrical charges, and other particles called **neutrons,** which have no electrical charges. If the nucleus, which contains nearly all of the atom's mass, were enlarged so that it was as big as a bowling ball, the atom, which is mostly space, would be larger than an aircraft carrier. Each kind of atom has a specific number of protons in its nucleus, ranging from 1 in the lightest element, hydrogen, to 92 in the heaviest element, uranium. Each element has an *atomic number,* which is based on the number of protons present in a single atom. The *atomic mass* of an element is determined by the number of protons and neutrons present in a single atom (see table 2.1).

An atom's protons and neutrons equal each other in mass, but both are about 1,840 times heavier than associated whirling negative electric charges called **electrons,** whose masses are so infinitesimal that they are generally disregarded. The electrons whirl around the nucleus in a region called an **orbital,** which has an imaginary axis and is somewhat cloudlike, but is without a precise boundary, so that we cannot be certain of an electron's position within the orbital at any time. This has led to an orbital being defined as a *volume of space in which a given electron occurs 90% of the time.* Each orbital is limited to 2 electrons, with the orbital closest to the nucleus being more or less spherical. This spherical orbital lies so close to the nucleus that it is usually not shown on diagrams of atoms; the one to several additional orbitals, which are shaped mostly like the tips of cotton swabs, generally occupy much more space. The electrons of each orbital tend to repel those of other orbitals, so that the axes of all the orbitals of an atom are oriented as far apart from each other as possible, although the outer parts of the orbitals actually overlap more than shown in diagrams of them (figure 2.1).

TABLE 2.1
Atomic Numbers and Masses of Some
Elements Found in Plants

Element	Atomic Number	Usual Atomic Mass
Hydrogen (H)	1	1
Boron (B)	5	11
Carbon (C)	6	12
Nitrogen (N)	7	14
Oxygen (O)	8	16
Magnesium (Mg)	12	24
Phosphorus (P)	15	31
Sulphur (S)	16	32
Chlorine (Cl)	17	35
Potassium (K)	19	39
Calcium (Ca)	20	40
Iron (Fe)	26	56

FIGURE 2.1 Models of orbitals. *A.* The 2 electrons closest to the atom's nucleus occupy a single spherical orbital. *B.* Additional orbitals are dumbbell-shaped, with axes that are perpendicular to one another. The atom's nucleus is at the intersection of the axes.

A.

B.

Electrons usually equal the protons in number, so that the positive electric charges of the protons balance the negative charges of the electrons, making the atom electrically neutral. The number of neutrons in the atoms of an element can vary slightly so that the element may occur in forms having different weights, but with all forms behaving alike chemically. Such variations of an element are called **isotopes.** The element oxygen (figure 2.2), for example, has 7 known isotopes. The nucleus of 1 of these isotopes contains 8 protons and 8 neutrons; the nucleus of another isotope holds 8 protons and 10 neutrons, and the nucleus of a third isotope consists of 8 protons and 9 neutrons. If the number of neutrons in an isotope of a particular element varies too greatly from the average number of neutrons for its atoms, the isotope may be unstable and split into smaller parts, with the release of a great deal of energy. Such an isotope is said to be *radioactive.*

Molecules: Combinations of Elements

The atoms of most elements have the property of binding to other atoms of the same or different elements and forming new combinations; in fact, most elements do not exist independently as single atoms. When two or more elements are united in a definite ratio by chemical bonds, the substance is called a **compound;** the smallest independently existing particle of a compound or element consisting of 2 or more atoms bound together is called a **molecule.** The molecules of the gases oxygen and hydrogen, for example, exist in nature as combinations of 2 atoms of oxygen or 2 atoms of hydrogen, respectively. Water molecules consist of 2 atoms of hydrogen and 1 atom of oxygen (figure 2.3).

Bonds and Ions

Bonds that hold atoms together in molecules form in different ways. In nature many molecules lose or gain electrons and become positively or negatively charged particles called **ions.** When two oppositely charged ions are in contact they are said to be bound by an *ionic bond.* Ions are shown with their charges as superscripts. For example, table salt (sodium chloride) is formed by ionic bonding between an ion of sodium (Na^+) and an ion of chlorine (Cl^-). The sodium becomes a positively charged ion when it loses 1 of its electrons, which is gained by an atom of chlorine. This extra electron makes the chlorine ion negatively charged, and the sodium ion and chlorine ion become bonded together by the force of the opposite charge. Most biologically important molecules carry positive and/or negative charges and therefore exist as ions in living matter.

In *covalent bonds* pairs of electrons hold together 2 or more atomic nuclei and travel between them, keeping them at a stable distance from each other. Covalent bonds are the principal force holding together atoms that make up the biologically important molecules to be mentioned later in this chapter. In another type of bond, negatively charged oxygen and/or nitrogen atoms of 1 molecule may form an attracting force between positively but weakly charged hydrogen atoms of other molecules,

FIGURE 2.3   Models of oxygen, water, and hydrogen molecules.

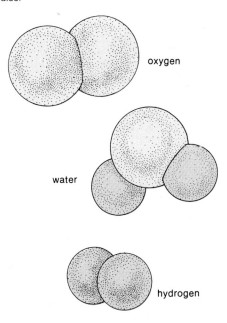

oxygen

water

hydrogen

FIGURE 2.2   Isotopes of oxygen portrayed two-dimensionally.

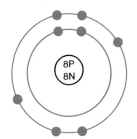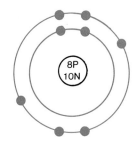

thus forming a weak bond. Since such bonds are always associated with hydrogen, they are called *hydrogen bonds*. Hydrogen bonds, despite their having only about 5% of the strength of covalent bonds, are abundant in many biologically significant molecules and therefore very important in nature. Other types of bonds besides ionic, covalent, and hydrogen bonds need not be considered here.

The combining capacity of an atom or ion is called *valence*. Atoms of the element calcium, for example, have a valence of 2, whereas those of the element chlorine have a valence of 1. In order for the atoms of these two elements to combine, there must be a balance between electrons lost or gained (i.e., the valences must balance); it thus takes *2* chlorine atoms to combine with *1* calcium atom. The compound formed by the union of calcium and chlorine is called *calcium chloride*. It is customary to use standard abbreviations taken from the Latin names of the elements when giving chemical formulas or equations. Thus calcium chloride is $CaCl_2$, indicating that 1 atom of calcium (Ca^{++}) required 2 atoms of chlorine (Cl^-) to form a calcium chloride molecule.

A **mixture** differs from a compound in not having all of its molecules or atoms united in definite ratios. For example, granite is composed of several different materials that vary in proportion to one another throughout the rock; likewise, a cake consists of ingredients that can vary in proportion to one another. Accordingly, granite, cakes, and a myriad of other substances with variable proportions of molecules are mixtures.

Acids, Bases, and Salts

Water molecules are electrically neutral, but they are asymmetrical in shape (see figure 2.3). This results in the molecules having very slight positive charges at one end and very slight negative charges at the other end. Such molecules are said to be *polar*. When the negatively charged end of 1 water molecule comes close to the positively charged end of another water molecule, weak hydrogen bonds hold the molecules together; water molecules also, however, sometimes separate into H^+ and OH^- (hydroxyl) ions, with the number of H^+ ions precisely equaling the number of OH^- ions in pure water. *Acids,* which taste sour like cranberry or lemon juice, are defined as substances that release hydrogen (H^+) ions when dissolved in water, with the result that there are proportionally more hydrogen than hydroxyl ions

present. Conversely, *bases* (also referred to as alkaline compounds), which usually feel slippery or soapy, are defined as compounds that release negatively charged hydroxyl (OH^-) ions when dissolved in water. Bases may also be defined as compounds that accept H^+ ions. The concentration of H^+ ions present is used to define degrees of acidity or alkalinity on a specific scale, called the **pH** scale. The scale ranges from 1 to 14, with each unit representing a tenfold change in H^+ concentration. Pure water has a pH of 7—the point on the scale where the number of H^+ and OH^- ions is exactly the same or the neutral point.[1] The lower a number is below 7, the higher the degree of acidity, and conversely, the higher a number is above 7, the higher the degree of alkalinity. Vinegar, for example, has a pH of 3, tomato juice has a pH of 4.3, and egg white has a pH of 8.

When an acid and a base are mixed, the H^+ ions of the acid bond with the OH^- ions of the base, forming water (H_2O). The remaining ions bond together, forming a *salt*. For example, if hydrochloric acid (HCl) is mixed with sodium hydroxide (NaOH), which is a base, water (H_2O) and sodium chloride (NaCl)—a salt—are formed. The reaction is represented by symbols in an equation that shows what occurs:

$$HCl + NaOH \longrightarrow H_2O + NaCl$$

Energy

The activities of all living things, from molecules to cells to whole organisms, require **energy** to take place; the source of that energy on earth is the sun. Energy, which can be defined as the "ability to produce a change in motion or matter," or simply as the "ability to do work," takes a number of forms. *The first law of thermodynamics* states that energy is constant—it cannot be increased or diminished—but it can be converted from one form to another. Among its forms are *chemical, electrical, heat,* and *light* energy. *The second law of thermodynamics* states that when energy is converted from one form to another in a given system in which no energy enters or leaves, it flows in one direction and the amount of

---

1. Note that although distilled water is theoretically "pure," its pH is always less than 7 because carbon dioxide from the air with which it is in contact dissolves in it, forming carbonic acid (H_2CO_3); the actual pH of distilled water is usually approximately 5.7.

useful energy remaining after the conversion will always be less than before the conversion. For example, heat will always flow from a hot iron to cold clothing but never from the cold clothing to the hot iron. Such energy-yielding reactions are vital to the normal functions of cells and provide the energy needed for other cell reactions that require energy. Both types of reactions are discussed in chapter 10. Forms of energy include *kinetic* (motion) and *potential* energy. Potential energy is defined as the "capacity to do work owing to the position or state of a particle." For example, a ball resting at the top of a hill possesses potential energy that is converted to kinetic energy if the ball rolls down the hill. Some chemical reactions release energy and others require an input of energy. Some of the numerous energy exchanges occurring in living cells are discussed in later chapters.

Chemical Components of Protoplasm

The living substance of all cells is called **protoplasm.** Protoplasm is organized into numerous bodies of various sizes, most of which are discussed in chapter 3. About 96% of protoplasm is composed of the elements carbon, hydrogen, oxygen, and nitrogen; 3% consists of phosphorus and potassium. The remaining 1% includes calcium, iron, magnesium, sodium, chlorine, copper, manganese, cobalt, zinc, and minute quantities of other elements. When a plant first absorbs these elements from the soil or atmosphere, or when it utilizes breakdown products within the cell, the elements are in the form of simple molecules or ions. These simple forms may be converted to very large, complex molecules through the metabolism of the cells. The large molecules invariably have chains of carbon atoms within them and are said to be **organic,** as opposed to **inorganic** molecules, which usually contain no carbon atoms. The name organic was given to most of the chemicals of living things when it was believed that only living organisms could produce molecules containing carbon. Today, many organic compounds can be produced artificially in the laboratory, and scientists sometimes hesitate to classify as either organic or inorganic some of the 4 million carbon-containing compounds thus far identified. Most scientists, nevertheless, agree that inorganic compounds usually do not contain carbon. Four of the most important classes of organic compounds found in protoplasm are carbohydrates, lipids, proteins, and nucleic acids.

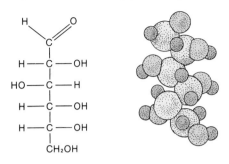

FIGURE 2.4 Structural formula and model of a glucose molecule. The molecule can exist either as a straight chain (as shown) or in the form of a ring (not shown).

Carbohydrates

Carbohydrates, which are the most abundant organic compounds in nature, have a number of roles in living organisms. Two of the most important are as energy sources and structural components. They contain carbon, hydrogen, and oxygen in, or close to, a ratio of 1 C:2 H:1 O. Examples of carbohydrates include monosaccharides (simple sugars), disaccharides, and polysaccharides. A monosaccharide such as *glucose* has a skeleton chain of 6 carbon atoms, with hydrogen and oxygen atoms attached, giving it the formula $C_6H_{12}O_6$ (figure 2.4). Disaccharides consist of 2 simple sugars joined together. *Sucrose,* or common table sugar, is a disaccharide made up of the 2 simple sugars glucose and fructose; sucrose is the form in which sugar is usually transported by plants. Polysaccharides sometimes consist of hundreds of simple sugars attached to one another. Examples of polysaccharides are starches, which consist of 300 to 1,000 glucose units attached to one another, and the structural substance *cellulose,* which consists of about 2,000 slightly different glucose units attached to one another. When glucose molecules are attached to one another, forming a starch molecule, each of the glucose molecules gives up 1 molecule of water. The *n* in the formula for starch—$(C_6H_{10}O_5)$ *n*—represents many units. In order for a starch molecule to become available as an energy source in cells, it has to be broken up into its individual simple sugar units through the restoration of the water molecule for each unit. This addition of water molecules is called **hydrolysis;** it takes place during the process of **digestion.**

FIGURE 2.5   Partial structural formula and partial model of a fat molecule. The two fatty acids not depicted are similar to fatty acid A. H = hydrogen, C = carbon, O = oxygen.

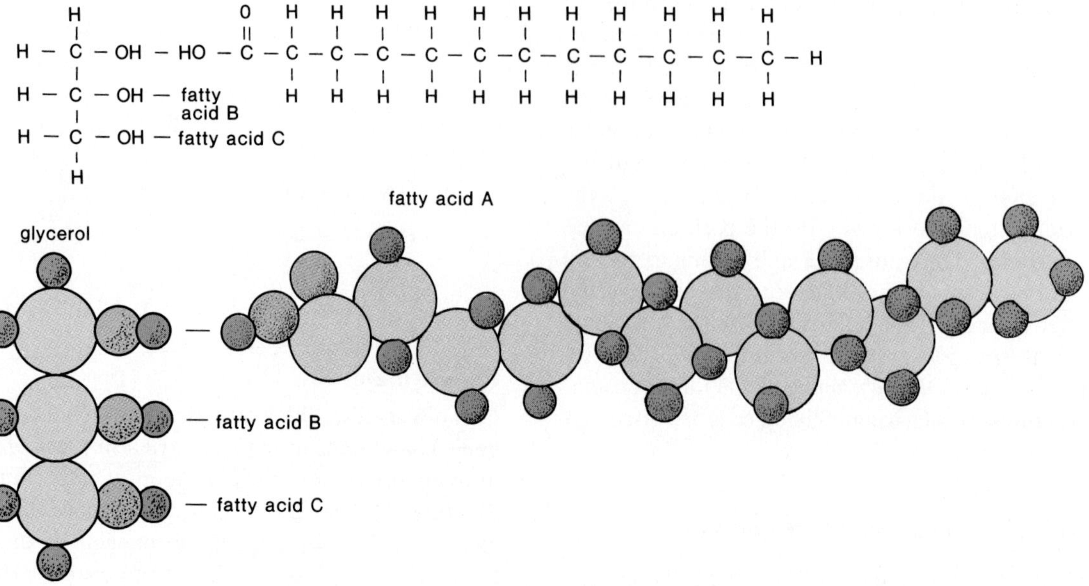

glycerol

fatty acid A

— fatty acid B

— fatty acid C

Lipids

Lipids are fatty or oily substances that are insoluble in water. They store more energy than carbohydrates, and play an important role in the energy reserves and structural components of cells. Like carbohydrates, their molecules contain carbon, hydrogen, and oxygen, but there is proportionately much less oxygen present. Examples of lipids include **fats** (figure 2.5) and **oils,** whose molecules are manufactured from sugars and are composed of a unit of glycerol (an alcohol) with 3 fatty acids attached. The fatty acids have carbon atoms to which hydrogen atoms can become attached. If hydrogen atoms are attached to every available attachment point of these fatty acid carbon atoms, the fat is said to be *saturated;* if, however, there are very few places for the hydrogen atoms to attach themselves, as is the case with some highly advertised vegetable oils, the fat is said to be *polyunsaturated.* Like polysaccharides and proteins (discussed in the next section), lipids are broken down by hydrolysis. Other examples of lipids are waxes, which have the fatty acids attached to an alcohol molecule other than glycerol, and *phospholipids,* which are similar in structure to fats but have at least a phosphate group replacing 1 of the 3 fatty acids. Phospholipids are important components of cell membranes.

FIGURE 2.6   Structural formula and model of the amino acid glycine.

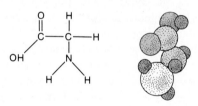

Proteins

A single cell may contain several thousand different kinds of **proteins.** These are important molecules that are usually very large and are composed of subunits called **amino acids** (figure 2.6). Proteins consist of carbon, hydrogen, oxygen, and nitrogen atoms, and sometimes also sulphur atoms. Many structural components of protoplasm are proteins.

There are about 20 different kinds of amino acids, and from 50 to 50,000 or more of them are present in various combinations in each protein molecule. Each amino acid has two special groups of atoms: an amino group, $-NH_2$, and a carboxyl group (the acid portion) $-COOH$, both of which are attached to the same carbon atom. The $-NH_2$

FIGURE 2.7　A model of cytochrome *c*, a relatively small protein. (From Richard E. Dickerson and Irving Geis. 1982. *Proteins: Structure, Function and Evolution*, 2d ed. Benjamin Cummings Publishing Co., Menlo Park, Calif. Illustration copyright Irving Geis).

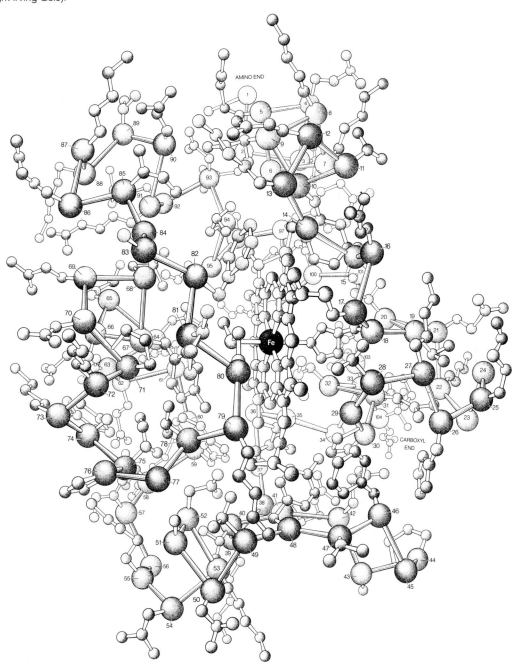

and the −COOH groups of a protein's amino acid building blocks bond together to form chains. The bonds are called **peptide bonds,** and the chains they produce are called *polypeptide chains.*

The polypeptide chains of a protein frequently coil in a *helix;* the helix itself, which resembles a spiral staircase, may fold into a globular form, as illustrated by the protein cytochrome *c* shown in figure 2.7.

Since proteins vary in the total number of amino acids they contain, in the relative number of different kinds of amino acids present, and in the sequence of the amino acids in their polypeptide chains, the number of different kinds of protein possible staggers the imagination. Among living things it is no wonder that every single species of the millions in existence possesses at least one unique protein.

Enzymes

Enzymes, which are produced in all living cells, are large complex proteins that function as organic *catalysts* (a catalyst is a substance that speeds up a chemical reaction without being used up in the reaction). Each of the roughly 2,000 known enzymes catalyzes, even at very low concentrations, a specific reaction in the cells in which it is produced. They are essential to the normal metabolic activities of cells, and, because they are unchanged in the reactions they accelerate, they are often used repeatedly. The names of enzymes normally end in *-ase*. One of the most common is maltase, which catalyzes the hydrolysis of maltose to glucose.

Nucleic Acids

Nucleic acids are complex substances originally thought to be limited to the nuclei of cells, but now known also to be associated with other cell parts. They are vital to the normal internal communication and functioning of all living cells; the two examples—deoxyribonucleic acid (DNA) and ribonucleic acid (RNA)—will be discussed in some detail.

Deoxyribonucleic acid (DNA) molecules of *chromosomes* (see chapter 3) consist of double helical (spiral) coils of repeating subunits called *nucleotides,* each composed of a base, a sugar, and a phosphate. The molecules are very large and contain, in units known as *genes,* the coded information that precisely determines the nature and proportions of the myriad substances found in cells and also the ultimate form and structure of the organism itself. If this coded information were written out, it would fill over 1,000 books of 300 pages each—at least for the more complex organisms. DNA molecules can replicate (duplicate themselves) in precise fashion. When a cell divides, the hereditary information contained in the DNA of the new cells is an

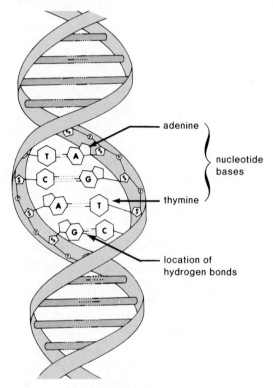

FIGURE 2.8   Structure of a DNA molecule. In this enlargement of a small portion of a DNA molecule, the rungs of the twisted ladder formed by the two entwined spiraling strands consist of nitrogen-containing bases supported by alternating units of sugar (S) and phosphate (P) molecules. The purines adenine or guanine (A or G) occur opposite the pyrimidines thymine or cytosine (T or C). The purines and the pyrimidines opposite each other are held together by hydrogen bonds linking the nitrogenous bases of the paired molecules. (After drawing from Volpe, E. Peter. 1981. *Understanding Evolution,* 4th ed. Wm. C. Brown Publishers, Dubuque, Iowa)

exact copy of the original and can be passed on from generation to generation without change, except in the event of a *mutation* (discussed in chapter 13).

Structure of DNA

The thousands of atoms in a DNA molecule are found in chains of building blocks called *nucleotides.* Each nucleotide consists of three parts: (1) a nitrogenous base, (2) a 5-carbon sugar, deoxyribose, and (3) a phosphate group. Both the nitrogenous base and the phosphate group are bonded to the sugar (figure 2.8). Four kinds of nucleotides occur in DNA. Each kind has a unique nitrogenous base, but all have the same types of phosphate groups and sugar. Two of the nucleotide bases—*adenine* and *guanine*—are called *purines;* they have a molecular structure that resembles two linked rings. The other two nucleotide bases—*cytosine* and *thymine*—are called *pyrimidines;* they have a molecular structure consisting of a single ring.

Each nucleotide in a DNA molecule is bonded to the next one in such a way that the nucleotides form a chain, with the sugar of one attached to the phosphate group of the next; the nitrogenous bases are oriented as side groups on the chain. Each species has its own unique DNA, with the nucleotides occurring in a different sequence for each kind of DNA molecule. The possibilities for variety in the sequences seem virtually unlimited (figure 2.9).

In 1951 two cell biologists, James D. Watson and Francis Crick, initiated a series of brilliant investigations that eventually led to the construction of a model of a DNA molecule. When they began their work they had certain information on which to start building their conclusions. They knew, for example, that Linus Pauling had postulated that the structure of DNA might be similar to the structure of protein, and that he had shown part of the structure of some proteins to be helical and maintained by hydrogen bonds between the amino acids. They also knew, from studies by Erwin Chargraff and others at Columbia University the previous year, that there was a ratio of 1:1 between nucleotides with adenine and nucleotides with thymine in DNA molecules, and that there was also a ratio of 1:1 between nucleotides with guanine and those with cytosine. Furthermore, they knew, from studies the same year by Rosalind Franklin and Maurice H. F. Wilkins, in which specialized X-ray equipment was utilized, that a DNA molecule was composed of regularly repeating units that appeared to be helically arranged.

When they had pieced together all the facts, Watson and Crick concluded that a DNA molecule consisted of a huge, entwined, double helix exactly 20 angstrom units wide, the strands of which appeared to be wrapped around an invisible pole—one strand spiraling in one direction and the other strand spiraling at the same angle in the opposite direction. They also concluded that the nucleotides were linked

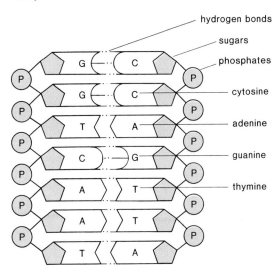

FIGURE 2.9   How nucleotides are paired in a tiny portion of a strand of DNA. The variations in sequences of pairs are virtually unlimited.

in ladderlike fashion between the two strands. They knew, however, that if purines were linked across the strands of the helix, the size of the combined molecules would exceed the 20 angstrom width of the DNA molecule, and that 2 linked pyrimidines would not be wide enough to reach all the way across. If a purine and a pyrimidine were linked to each other, however, the width would fit the model perfectly. Accordingly, they concluded that the ladder rungs had to consist exclusively of purine–pyrimidine pairs. This meant that adenine–cytosine and guanine–thymine pairings could not occur, but adenine–thymine (or thymine–adenine) and guanine–cytosine (or cytosine–guanine) could occur. This nicely explained the 1:1 ratios shown by Chargraff and his colleagues, and such a model was proposed in 1953. Watson, Crick, and Wilkins shared a Nobel Prize for their work, Franklin having died before she, too, could be a part of the honor. Subsequent research has continued to support their conclusions to the extent that their model of a DNA molecule is now accepted as an authentic representation.

FIGURE 2.10   Replication (duplication) of DNA. A molecule replicates by "unzipping" along the hydrogen bonds, which link the pairs of nucleotides in the middle, and then each half serves as a template for nucleotides available in the cell. As the new bases fit into the appropriate places and are linked together, a double helix is re-formed for each original half.

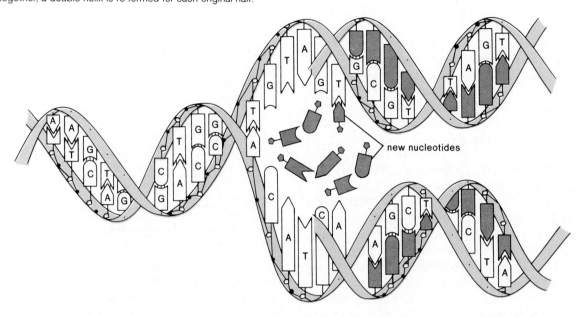

new nucleotides

Replication (Duplication) of DNA

If DNA is indeed the determiner of inheritable characteristics, it must be able to duplicate itself precisely in order to pass along information from generation to generation. Since all DNA molecules have only 4 different nucleotides, the differences between the DNA of one organism and that of another must lie in the sequence of the four possible types of ladder rungs and their total number, as well as in the ratios of adenine and thymine to guanine and cytosine. If these 4 nucleotides are synthesized in living cells, what tells the cell exactly how to put them together to form the appropriate DNA molecules? Watson and Crick observed that if the two strands of DNA are "unzipped" by an enzyme breaking the hydrogen bonds between the nucleotides down the middle of the ladder, each separated chain provides all the information needed for putting together a new ladder (figure 2.10). In other words, since guanine can pair only with cytosine and vice versa, a guanine nucleotide synthesized by the cell can lock on, or bond, only to a cytosine, and an adenine can bond only to a thymine. Thus, if a cell "unzips" the two strands in its DNA molecules, separate nucleotides can line up next to each of the single chains in precise sequence and be bonded together to form a new chain, with the separate chains functioning as molds, or templates, for self-duplication. This theory of Watson and Crick, also proposed in 1953, has since been supported by evidence obtained by the use of nucleotides containing isotopes of carbon and nitrogen atoms.

RNA and Synthesis of Proteins

Once the chemical nature of DNA and the way in which it duplicates itself became understood, attention was focused on the manner in which DNA determines a cell's or an organism's function and structure. It was known that proteins are large complex molecules composed of 20 different kinds of amino acids linked together. If each of the 4 nucleotides in DNA coded only 1 amino acid, there could not be more than 4 amino acids. If all possible combinations of nucleotide pairs (16) were used to code amino acids, the number would still not be enough to code the 20 known amino acids. If a minimum of 3 nucleotides were involved for each amino acid, however, up to 64 combinations would be possible—more than enough to code 20 different amino acids. Such an arrangement was postulated and later confirmed; each trio of nucleotides along a DNA strand does indeed code a specific amino acid. It was also known that most proteins are synthesized in the cytoplasm, whereas DNA is largely confined to the nucleus. Obviously, then, the DNA could not function as a direct template for protein synthesis. Some intermediary agent or agents had to be involved.

In the early 1940s, it was suspected that ribonucleic acid (RNA) was involved in protein synthesis because it had been found that cells producing large amounts of protein contained large amounts of RNA and vice versa. RNA is similar to DNA, but it differs in that its sugar component has an additional oxygen atom and it has a different pyrimidine, *uracil,* instead of thymine. RNA usually occurs as a single strand and only very rarely as a completely double strand.

Specific Types of RNA and Their Functions

Messenger RNA (mRNA) Messenger RNA is a large molecule composed of several hundred to several thousand nucleotides. As with a new strand of DNA, a strand of mRNA forms along a portion of a DNA molecule like a row of jigsaw pieces fitting together with another row, with uracil taking the place of thymine. A sequence of 3 nucleotide bases, known as a *codon,* codes 1 amino acid. Once the mRNA molecule has been formed within the nucleus, enzymes split off and modify portions of the mRNA molecule. The enzymes are then carried out into the cytoplasm (probably through a nuclear pore). In the cytoplasm each mRNA unit becomes attached by one end to a ribosome, which is itself composed of another form of RNA (rRNA) and protein. Usually several ribosomes become associated with the mRNA molecule and move along in sequence,

"reading" the coded information and participating in the bonding together of amino acids available in the cytoplasm. One mRNA molecule apparently is "read" by two or three such groups of ribosomes and then destroyed. It has been estimated that the average life of an mRNA molecule in the much-studied, digestive-tract bacterium *Escherichia coli* is about 2 minutes.

Transfer RNA (tRNA) A molecule of tRNA consists of roughly 80 nucleotides bonded together in a single strand that doubles back on itself. There is at least one form of tRNA for each of the 20 amino acids. Each specific form of tRNA has an anticodon loop, where the strand doubles back on itself. This anticodon is a sequence of 3 nucleotides that will "recognize" a sequence of 3 specific amino acid nucleotides on a strand of mRNA. An amino acid is bound on the open, or acceptor end, of the tRNA strand with the aid of an enzyme that evidently identifies with both a specific tRNA strand and a specific amino acid (figure 2.11).

Once the tRNA molecule has "captured" an amino acid, the tRNA becomes bound to a *ribosome* (see chapter 3) at the same point to which an mRNA molecule has become attached. Then, as the ribosome moves along the mRNA strand, the tRNA releases its amino acid and becomes detached, with another tRNA amino acid complex taking its place. In this way, the amino acids are lined up in sequence according to the mRNA code, and each is bonded to the amino acid of the next tRNA molecule until a complete protein molecule has been assembled. The tRNA thus relays amino acids in response to a specific "blueprint" coded by the DNA and carried by the mRNA. The amino acids of the various proteins of the cell are attached to each other precisely according to the information delivered from the nucleus, with specific enzymes ensuring that the right amino acids become designated proteins.

After a tRNA strand becomes detached and releases its amino acid, it is then available for reuse in "capturing" another amino acid molecule. Apparently tRNA molecules can go through this cycle repeatedly, although enzymes eventually may break them down.

FIGURE 2.11   How a protein is synthesized. *1.* A strand of DNA in the nucleus serves as a template for the synthesis of a strand of mRNA. *2.* The mRNA moves out into the cytoplasm where it becomes associated with ribosomes. *3.* The tRNA strands each pick up a specific amino acid from the supply available in the cytoplasm. *4.* The tRNA briefly couples with the mRNA at a precise spot where its anticodon complements a codon on the mRNA. *5.* A ribosome moves along the mRNA, adding amino acids to the polypeptide chain of the protein being synthesized. *6.* The tRNA can now repeat the process of picking up an amino acid from the supply available. (Note: The molecules and the structures are not drawn to scale with respect to one another or the cell.)

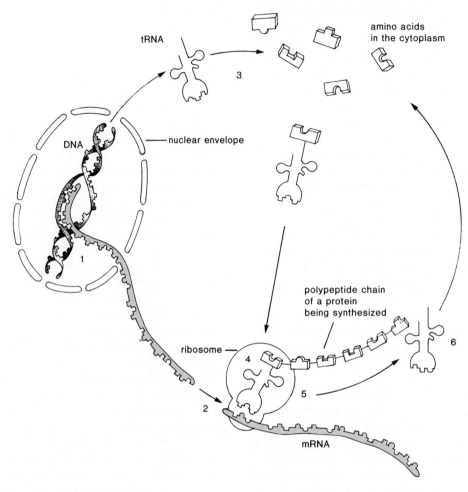

Ribosomal RNA (rRNA)   As indicated in chapter 3, *ribosomes* are composed of two more-or-less spherical subunits, each having its own specific kinds of proteins and RNA. The larger of the two subunits contains a "slot" into which a tRNA molecule can fit. How one tRNA molecule replaces another in this slot has not yet been determined. Ribosomal RNA is formed along one strand of the DNA helix in the same manner as mRNA and tRNA molecules. The role of rRNA in protein synthesis is not certain, but it is presumed that the ribosome aligns, in precise sequence, mRNA, tRNA, and amino acids for the protein being synthesized. Ribosomal RNA, like transfer RNA, has a long life.

SUMMARY

Activities of living organisms stem from protoplasm, the physical basis, or the "stuff of life." Growth is among the attributes of living organisms. Growth has been described as an increase in volume; it results primarily from the production of new protoplasm. Variations in form may be inherited or result from response to the environment. Reproduction involves offspring that are always similar in form to their parents; if reproduction ceases, the organisms become extinct. Plants generally respond to stimuli more slowly and in a different fashion from animals. All living organisms exhibit metabolic activities, including production of new protoplasm, respiration, photosynthesis, digestion, and assimilation; they also all exhibit movement to varying degrees. Cyclosis is the streaming motion of protoplasm within living cells. Living organisms have a much more complex structure than nonliving objects and are adapted to their individual environments.

The basic "stuff of the universe" is called matter; it occurs in three states: solid, liquid, and gas. It is composed of elements, the smallest stable subdivision of which is an atom. Atoms contain, in a tiny nucleus, positively charged protons, and uncharged neutrons, surrounded by much larger orbitals or regions of whirling, negatively charged electrons. Isotopes are forms of elements that have slight variations in the number of neutrons in their atoms. Atoms of elements can bond to other atoms, and those of most elements do not exist independently; compounds are substances composed of 2 or more elements combined in a definite ratio by chemical bonds; molecules are the smallest independently existing particles. If a molecule loses or gains electrons it becomes an ion, which may form an ionic bond with another ion. In a covalent bond pairs of electrons link 2 or more atomic nuclei; nitrogen and/or oxygen atoms of 1 molecule may form weak hydrogen bonds with hydrogen atoms of other molecules. The combining capacities of atoms or ions are called valence. The atoms of mixtures are not all chemically united in definite ratios. Acids release positively charged hydrogen ions when dissolved in water. Bases release negatively charged hydroxyl ions when dissolved in water. The pH scale is used to measure degrees of acidity or alkalinity. Salts and water are formed when acids and bases are mixed.

Energy can be defined as "ability to produce a change in motion or matter" or simply as "ability to do work." Its forms include chemical, electrical, heat, light, kinetic, and potential.

Protoplasm is composed primarily of carbon, hydrogen, oxygen, and nitrogen, with a little phosphorus and potassium, plus trace amounts of other elements. A plant may convert the simple molecules or ions it recycles or absorbs from the soil to very large, complex molecules. Organic molecules are usually large and have chains of carbon atoms within them. Carbohydrates contain carbon, hydrogen, and oxygen in a ratio of 1 C:2 H:1 O. Carbohydrates occur as monosaccharides (simple sugars, e.g., glucose); disaccharides, consisting of 2 simple sugars joined together (e.g., sucrose); and polysaccharides, consisting of many simple sugars joined together (e.g., starch, cellulose). Simple sugars each give up a molecule of water when they are attached to one another to form starch. Hydrolysis is the process of restoring a water molecule to each simple sugar when starch is broken down during digestion. Lipids (e.g., fats and oils) consist of a unit of glycerol with 3 fatty acids attached. They are insoluble in water and contain carbon, hydrogen, and oxygen, with proportionately much less oxygen than found in carbohydrates. Saturated fats are those that have hydrogen atoms attached to every available attachment point of their carbon atoms; if there are very few places for hydrogen atoms to attach themselves, the fat is said to be polyunsaturated. Proteins are usually large molecules that are composed of subunits called amino acids. Each amino acid has two special groups of atoms: an amino group ($-NH_2$) and a carboxyl group ($-COOH$); these groups bond amino acids together to form polypeptide chains; the bonds are called peptide bonds. Enzymes are large protein molecules that function as organic catalysts.

There are 2 nucleic acids (DNA and RNA) associated primarily with cell nuclei. Chromosomes contain helical coils of DNA, which contain coded information determining the nature and proportions of substances in cells and the ultimate form and structure of the organism. DNA is composed of building blocks called nucleotides, which occur in four forms. Watson, Crick, and Wilkins were awarded a Nobel Prize in 1953 for developing a model of DNA that has since come to be accepted as authentic. DNA replicates (duplicates itself) by "unzipping" down its nucleotide "ladder rungs" and having new nucleotides replace the old ones until two new chains of DNA have been formed. New proteins are formed precisely according to coded information contained in the DNA of the nucleus by means of RNA, which occurs in three forms. Messenger RNA forms along part of a DNA molecule and is carried out of the nucleus into the cytoplasm, where it becomes attached to a ribosome. Ribosomes "read" the coded information and participate in assembling amino acids together to form a protein, with the aid of transfer RNA molecules. Ribosomes themselves are largely composed of a third form of RNA called ribosomal RNA.

REVIEW QUESTIONS

1. What distinguishes a living organism from a nonliving object such as a rock or a tin can?
2. What is meant by the term *organic?*
3. How are acids, bases, and salts distinguished from one another?
4. Distinguish among carbohydrates, lipids, and proteins.
5. What is energy, and what forms does it take?
6. Describe the structure and duplication of DNA.
7. What is the nature and function of each form of RNA?

DISCUSSION QUESTIONS

1. Can part of an organism be alive while another part is dead? Explain.
2. What is the difference between inherited form and form resulting from response to the environment?
3. Does DNA really control the synthesis of proteins? Explain.

ADDITIONAL READING

Adler, I. 1977. *How life began,* rev. ed. New York: Harper & Row.

Ambrose, E. J. 1982. *The nature and origin of the biological world.* New York: Halsted Press.

Baker, J. J. W., and G. E. Allen. 1965. *Matter, energy and life.* Reading, MA: Addison-Wesley Publishing Co.

Lehninger, A. L. 1982. *Principles of biochemistry.* New York: Worth Publishers.

Margulis, L. 1982. *Early life.* Portola Valley, CA: Jones & Bartlett.

Oparin, A. I. 1953. *Origin of life,* 2d ed. Mineola, NY: Dover Publications.

Raven, P. H., R. F. Evert, and S. E. Eichhorn. 1986. *Biology of plants,* 4th ed. New York: Worth Publishers.

Ray, P. M., T. A. Steeves, and S. A. Fultz. 1983. *Botany.* Philadelphia: Saunders College Publishing.

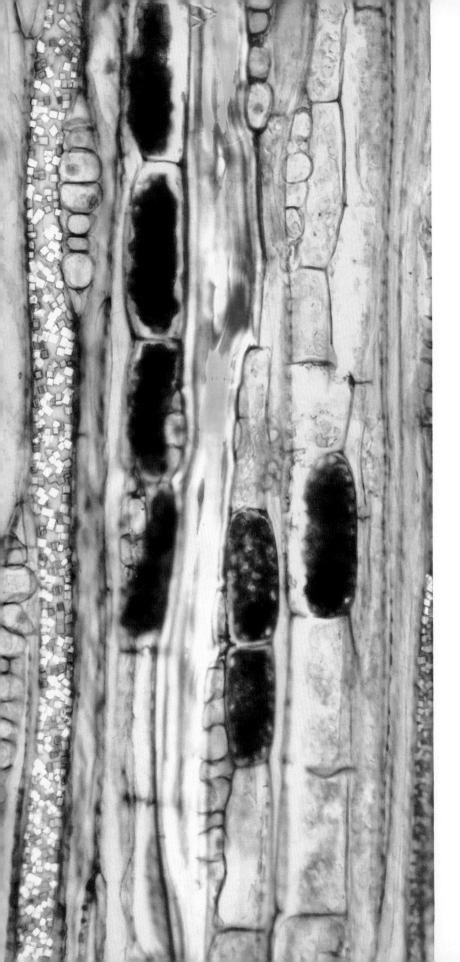

Photo by G. S. Ellmore

Overview

*T*his chapter gives a brief review of the history of the discovery of cells and the development of the cell theory. After observations on cell size and structure, it discusses each of a cell's particulates, beginning with the cell wall. Included are the plasma membrane, endoplasmic reticulum, ribosomes, mitochondria, Golgi bodies, plastids, microtubules, microfilaments, the nucleus, microbodies, vacuoles, and vacuolar membranes. Distinctions between prokaryotic and eukaryotic cells are then given. The chapter next discusses mitosis and cytokinesis, and a simple review of intercellular communication. The chapter concludes with the differences between plant and animal cells.

3   Cells

Some Learning Goals

1. Learn five historical figures associated with the development of the cell theory and name their contributions.

2. Know the following cell structures and organelles, and indicate the function of each: plasma membrane, mitochondria, endoplasmic reticulum, ribosomes, Golgi bodies, plastids, and vacuoles.

3. Describe the components of a nucleus and understand the function of each component.

4. Understand the cell cycle and the events that take place in each phase of mitosis.

5. Contrast plant cells with animal cells.

Outline

INTRODUCTION

All living organisms are composed of cells, be they aardvarks or apple trees, zebras or zamias—or even microscopic microbes. All living organisms, including each of us, also began life as a single cell; this cell may have divided repeatedly until it developed into an organism consisting of billions of cells. During the first few hours of any organism's development the cells all resemble each other, but changes soon occur, not only in the appearance of the cells but also in their function. The modifications of some, for example, permit them to serve as conducting cells, while others come to function in secretion or support. Some cells live and function for many years; others mature and degenerate in just a few days. Even as you read this, millions of new cells are being produced in your body. Some cells add to your total body mass (if you have not yet stopped growing), but most replace the millions of older cells that are destroyed every second you remain alive. The variety and form of cells seems almost infinite, but certain features are shared by most of them. A discussion of these features forms the body of this chapter.

CELLS

History

The discovery of cells is associated with the development of the microscope in the seventeenth century (see chapter 1). In 1665, the English physicist Robert Hooke, using a primitive microscope (figure 3.1), examined thin slices of cork he had cut with a sharp penknife. Hooke compared the boxlike compartments he saw to the surface of a honeycomb, and is credited with applying the term *cell* to those compartments. He also estimated that a cubic inch of cork would contain approximately 1,259 million such cells. What Hooke saw in the cork were really only the walls of dead cells, but he also observed "juices" in living cells of elderberry plants and thought he had found something similar to the veins and arteries of animals.

FIGURE 3.1   Robert Hooke's microscope as illustrated in one of his works. (Courtesy National Library of Medicine)

During the next 50 years Anton van Leeuwenhoek, along with Hooke's compatriot Nehemiah Grew, reported at frequent intervals on the existence of cells in a variety of plant tissues and on the form and structure of single-celled organisms that were referred to as "animalcules."

After this period, little more was reported on cells until the early 1800s. This lack of progress was due in large part to the imperfections of the primitive microscopes and also to the crude methods of tissue preparation used. But both microscopes and tissue preparations slowly improved, and by 1809 the famous French biologist Jean Baptiste de Lamarck had seen a wide enough variety of cells and tissues to conclude that "no body can have life if its constituent parts are not cellular tissue or are not formed by cellular tissue." In 1824, another Frenchman, René J. H. Dutrochet, reinforced Lamarck's conclusions that all animal and plant tissues are composed of cells of various kinds. Neither of them, however, realized that each cell could, in most cases, reproduce itself and exist independently. In 1831, the English botanist Robert Brown discovered that all cells contain a relatively large body that he called the *nucleus;* shortly thereafter, the German botanist Matthias Schleiden observed a smaller body within the nucleus that he called the *nucleolus*. Schleiden and a German zoologist, Theodor Schwann, were not the first to grasp the significance of cells, but they explained them with greater clarity and perception than others before them had done. They are generally credited with developing the *cell theory,* beginning with their publications of 1838–1839. In essence, this theory holds that all living organisms are composed of cells and that cells form a unifying structural basis of organization.

In 1858 an important augmentation of the cell theory appeared in a classic textbook by another German scientist, Rudolf Virchow. He argued cogently that every cell comes from a preexisting cell ("*omnis cellula e cellula*") and that there is no spontaneous generation of cells. Virchow's publication stirred up a great controversy, because prior to this time there was a widespread belief among scientists and nonscientists alike that animals could originate spontaneously from dust. Many who had microscopes were thoroughly convinced they could see "animalcules" appearing in decomposing substances. The controversy became so heated that in 1860, the Paris Academy of Sciences offered a prize to anyone who could, through experiments, shed light on the matter. Just two years later the brilliant French scientist Louis Pasteur was awarded the prize. Pasteur, using swan-necked flasks, demonstrated conclusively that boiled media remained sterile indefinitely if microorganisms from the air were excluded from the media (figure 3.2).

In 1871 Pasteur proved that natural alcoholic fermentation always involves the activity of yeast cells. In 1897 the German scientist Eduard Buchner accidentally discovered that the yeast cells did not need to be alive for fermentation to occur. He found that extracts from the yeast cells would convert sugar to alcohol. This discovery was a major surprise to the biologists of the time and quickly led to the identification and description of **enzymes** (see chapter 2), the organic catalysts (substances that aid chemical reactions without themselves being changed) found in all living cells; it also led to the belief that cells were little more than miniature packets of enzymes. During the first half of the twentieth century, however, great advances were made in the refinement of microscopes and in tissue preparation techniques.

FIGURE 3.2   Pasteur's experiment. Nutrient broths were sterilized by boiling after they had been added to both straight-necked and swan-necked flasks. The broths were then left in their openmouthed containers for several weeks. Microorganisms such as bacteria entered the flasks from the atmosphere, but moisture in the bends of those flasks with curved necks trapped the organisms, causing the broth to remain sterile, while the broth in those flasks with straight necks became contaminated.

organisms trapped here

Numerous structures and bodies, in addition to the nucleus, were observed in cells, and the relationship between structure and function came to be realized and understood on a much broader scale than previously had been possible.

Modern Microscopes

Without microscopes very little would be known about cells, and, as indicated earlier, our present vast knowledge of cells and all aspects of biological investigations associated with them is directly related to the development of these instruments.

Light microscopes increase magnification as light passes through a series of transparent lenses, currently made of various types of glass or calcium fluoride crystals. The curvatures of the lense materials and their composition are designed to minimize distortion of image shapes and colors. Light microscopes are of two different basic types: *compound microscopes,* which require the material being examined to be sliced thinly enough for light to pass through and *dissecting microscopes,* which permit the viewing of opaque objects. The best compound microscopes in use today can produce useful magnifications of up to 1,500 times under ideal conditions; most dissecting microscopes magnify up to 30 times. Such microscopes will continue to be useful, particularly for observing living cells, into the foreseeable future. The production and development of modern electron microscopes since the 1950s, however, has resulted in observation of much greater detail than is possible with light microscopes. Instead of light, electron microscopes use a beam of electrons produced when electricity of high voltage is passed through a wire. This electron beam is directed through a vacuum in a large tube or column. When the beam passes through a specimen an image is formed on a plate; magnification is controlled by powerful magnets located on the column. Like light microscopes, electron microscopes are made in two different forms. *Transmission electron microscopes* (figure 3.3) permit magnfications of 200,000 or more times, but the material to be viewed must be very thinly sliced and introduced into the column's vacuum, so that living objects cannot be observed.

Scanning electron microscopes usually do not achieve such high magnifications (3,000 to 10,000 times is the usual range), but opaque objects can be observed as a scanner renders the object visible on a cathode tube like a television screen. The techniques for such observation have become so refined that even the preserved material can appear exceptionally lifelike. Significant new discoveries by cell biologists using both electron and light microscopes in their research have now become frequent events.

CELL SIZE AND STRUCTURE

Most plant cells, and the vast majority of animal cells, are so tiny they are invisible to the unaided eye. Cells of higher plants generally vary in length between 10 and 100 micrometers.[1] Since there are roughly 25,000 micrometers to the inch, it would take about 500 average-sized cells to extend across 2.54 centimeters (1 inch) of space; 30 of them could

---

1. See appendix 5 for metric conversion tables.

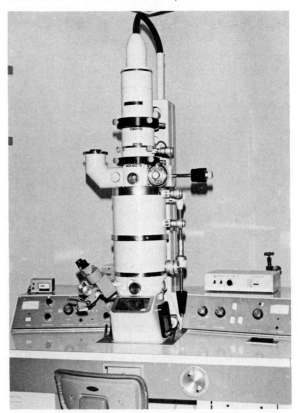

FIGURE 3.3   An electron microscope.

easily be placed across the head of a pin. Some bacterial cells are less than 0.5 micrometer in diameter, whereas cells of the green alga *mermaid's wineglass* (*Acetabularia*) are mostly between 2 and 5 centimeters in length, and fiber cells of some nettles are about 20 centimeters long.

Because cells are so minute, the numbers occurring in full-grown organisms are astronomical. For example, it has been calculated that a single mature leaf of a pear tree contains 50 million cells and that the total number of cells in the roots, stem, branches, leaves, and fruit of a full-grown pear tree exceeds 15 trillion. Can you imagine how many cells there are in one of the 3,000-year-old redwood trees of California, which reach heights of 90 meters (300 feet) and measure up to 4.5 meters (15 feet) in diameter near the base?

FIGURE 3.4   A cross section through leaf mesophyll cells of a garden bean seedling seen with the aid of an electron microscope. (Courtesy Jean Whatley)

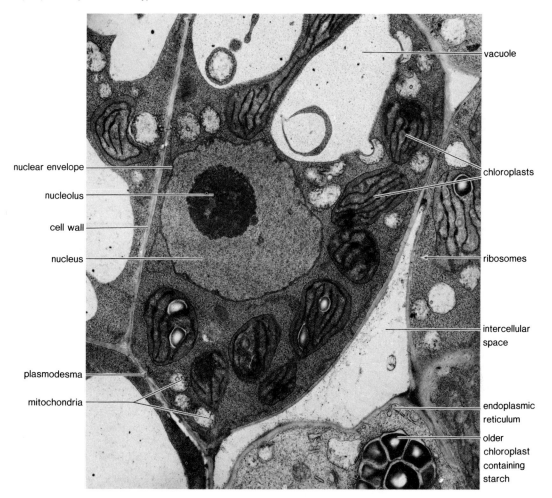

nuclear envelope

nucleolus

cell wall

nucleus

plasmodesma

mitochondria

vacuole

chloroplasts

ribosomes

intercellular space

endoplasmic reticulum

older chloroplast containing starch

Some cells are boxlike with six walls, but others assume a wide variety of shapes, depending on their location and function. The most abundant cells in the younger parts of plants and fruits may be more or less spherical when they are first formed, but they are packed together in such a way that they commonly have 14 sides by the time they are mature. These are discussed in the next chapter.

As indicated at the beginning of chapter 2, the living part of the cell within the wall is called *protoplasm.* Two main components of protoplasm are readily discernible: the *nucleus,* which controls the cell's activities, and the **cytoplasm,** which is a soup-like fluid containing water, dissolved substances, and many, small **organelles** (membrane-bound bodies of various sizes and shapes). The organelles are the sites of many different activities that take place within the cell. A brief examination of each of the various cell components follows (figures 3.4 and 3.5 show these various components.)

FIGURE 3.5 *A.* A leaf cell diagrammed with the aid of a
light microscope. *B.* The same cell greatly enlarged to show
submicroscopic features.

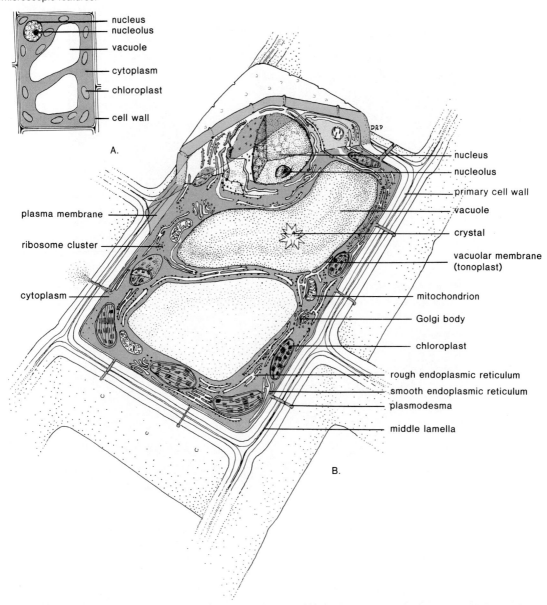

The Cell Wall

A popular song of three or more decades ago had
several verses listing food items the author purport-
edly disliked, with each verse ending, "But I like bananas, because they have no bones!" Indeed, bananas
and all parts of plants differ from animals in that
they have no bones or similar internal skeletal structures. Yet large trees support branches and leaves
weighing many tons. They are able to do this because most plant cells have semirigid or rigid walls
that perform the functions of bones; that is, they

provide strength and support for the plants (and also
protect the delicate cell contents within). When millions of these cells function together as a tissue, their
collective strength is enormous. The largest trees
alive today, the redwoods, exceed the mass, or
volume, of the largest land animals, the elephants,
by more than a hundred times. The wood of one tree
could support the combined weight of a thousand
elephants.

When new **cell walls** are first formed, a **middle
lamella,** consisting primarily of **pectin** (the complex
organic material that gives stiffness to fruit jellies),
appears. This middle lamella is normally shared by

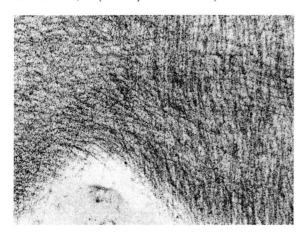

two adjacent cells and is so thin that it may not be visible with an ordinary light microscope unless it is specially stained. A fine network of *cellulose*—a substance composed of simple glucose molecules attached end to end (figure 3.6); it is used in the production of paper and other products—is laid down on either side of the middle lamella. Sometimes the cellulose is deposited in two stages, forming a *primary cell wall* and then a *secondary cell wall* inside the primary wall. When this happens, the secondary cell wall is usually the more extensive of the two structures. Most of the cell wall consists of cellulose and, depending on the type of cell involved, other

substances such as sugars and **lignin** (a complex organic substance that adds mechanical strength to cell walls) may be impregnated into the wall. An externally applied pain reliever, dimethyl sulfoxide (DMSO), is manufactured from lignin. The thickness of the wall can also vary, occupying more than 95% of the volume of the cell in some instances and less than 5% in others. Cells that function in food storage or manufacture usually have thin walls, while those primarily involved in support usually have walls of moderate to extensive thickness.

Protoplasm

The Plasma Membrane

The outer boundary of the living part of the cell, the **plasma membrane,** is roughly eight-millionths of a millimeter thick. To get an idea of how incredibly thin that is, consider that it would take 12,500 such membranes neatly stacked in a pile to achieve the thickness of an ordinary piece of writing paper. Yet this delicate structure is of vital importance in regulating what substances enter and leave the cell, and in the production and assembly of cellulose for cell walls. Evidence obtained since the early 1970s indicates that this and other cell membranes are mosaics composed of lipids, with proteins interspersed throughout (figure 3.7); covalent bonds link carbohydrates to both the lipids and the proteins on the outer surfaces of the membranes. When plasma membranes are stained and magnified by an electron microscope as much as a million times, they appear as two dark parallel lines with lighter material

FIGURE 3.7 A model of a small portion of a plasma membrane, showing its fluid mosaic nature. The proteins shown here as blocks (they are actually coiled chains of polypeptides), occur either on the surfaces or are embedded.

Some of the embedded proteins extend all the way through the continuous part of the membrane, which is a double layer of lipids. The pairs of vertical lines extending inward from each lipid represent long chain fatty acids.

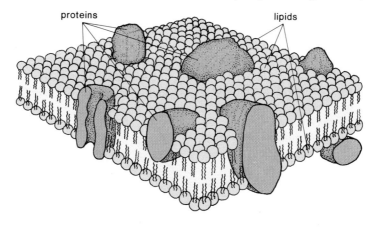

proteins   lipids

FIGURE 3.8   A small portion of the endoplasmic reticulum
and ribosomes in a young leaf cell of corn (*Zea mays*).
(Courtesy Jean Whatley)

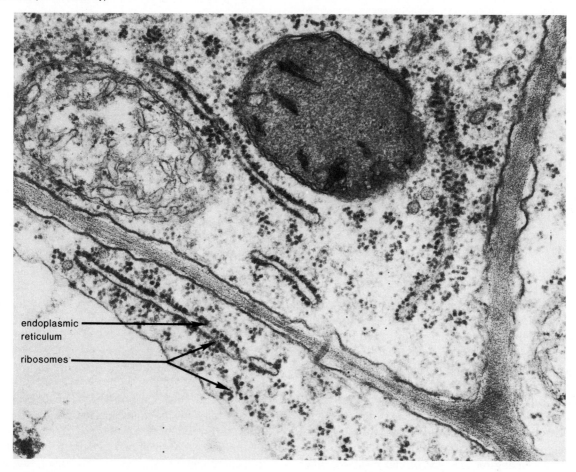

between them; such "dark-light-dark" membranes
have become known as *unit membranes*. Some proteins extend across the entire width of the membrane, whereas others are embedded or apparently
are loosely bound to the outer surface. The remainder of the cell contents usually push the plasma
membrane up against the cell wall because of pressures developed by osmosis (see chapter 9), but the
membrane is quite flexible and often forms folds,
which may in turn become little hollow spheres or
vesicles that float off into the cell. In fact, experiments have shown that by adding detergents to a
unit membrane it can be broken up and dispersed,
yet it can reform (albeit imperfectly) when the detergents are removed. The membrane may even
shrink away from the wall temporarily, but if it ever
ruptures the cell soon dies.

The Endoplasmic Reticulum

The **endoplasmic reticulum** is a complex system of
unit membrane channels that occurs throughout the
cytoplasm, with the amount and form varying considerably from cell to cell. It appears, in section, as
a series of parallel membranes that resemble long,
narrow bags, sacs, or tubes. The nucleus, which directs the various activities of the cell, is connected
to the endoplasmic reticulum, and many of the most
important activities occur either on the surface of
the endoplasmic reticulum or between its compartments. *Ribosomes* (see the section that follows) may
line the outer surfaces of the endoplasmic reticulum. Such endoplasmic reticulum is said to be
"rough," and is primarily associated with the synthesis, secretion, or storage of proteins (figure 3.8).

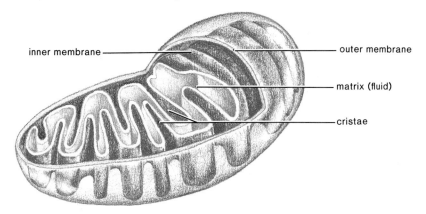

inner membrane

outer membrane

matrix (fluid)

cristae

This contrasts with "smooth" endoplasmic reticulum, which has no ribosomes lining the surface, and which is associated with lipid secretion. Both types of endoplasmic reticulum can occur in the same cell. Many enzymes involved in the process of respiration are synthesized on the endoplasmic reticulum and transported through it to the mitochondria. The endoplasmic reticulum also appears to be the primary site of membrane synthesis within the cell.

Ribosomes

As indicated, **ribosomes** may line the endoplasmic reticulum (see figure 3.8), but they may also occur free in the cytoplasm, nucleus, chloroplasts, or other organelles. They are tiny, averaging about 20 nanometers in diameter in most plant cells. The unattached ribosomes often occur in clusters of 5 to 100, particularly when they are involved in performing their function of linking amino acids together to form the large, complex protein molecules that are a basic part of all living organisms. Ribosomes are roughly ellipsoidal in shape although recent evidence suggests the surface topography is varied and complex. Each ribosome is composed of two subunits, which in turn are made up of RNA and proteins. About 55 kinds of protein are found in each ribosome of prokaryotic cells and a slightly higher number in those of eukaryotic cells (see the discussion of various types of RNA at the end of chapter 2). Because ribosomes are not bound by membranes, they technically are not true organelles.

Mitochondria

Mitochondria are often referred to as the "powerhouses" of the cell, for it is within them that energy is released from organic molecules by the process of respiration (the role of mitochondria in respiration is further discussed in chapter 10). This energy is needed to keep the individual cells and the plant functioning as a whole. Mitochondria are numerous and tiny, typically measuring from 1 to 3 or more micrometers in length and having a width of roughly one-half micrometer; they thus are barely visible with light microscopes. In living cells they are in constant motion, and tend to accumulate in groups where energy is needed. They often divide in two or fuse together, and assume various shapes such as those of gherkins or paddles. A sectioned mitochondrion resembles a scooped-out watermelon with inward extensions of the rind forming mostly incomplete partitions at right angles to the surface (figure 3.9). The appearance of incomplete partitions results from the fact that each mitochondrion is bound by two unit membranes, with the inner membrane forming numerous platelike folds called *cristae* that greatly increase the surface area available to the enzymes contained in a matrix fluid; this fluid also contains DNA, RNA, ribosomes, proteins, and dissolved substances.

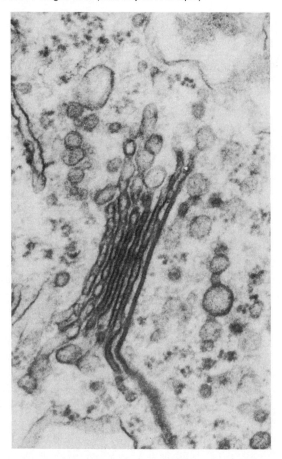

The Golgi Apparatus (Dictyosomes)

The *Golgi apparatus,* is a term applied collectively to all of the **Golgi bodies** or **dictyosomes** (figure 3.10) of a cell, the latter term more often being applied to plant cells. They appear as groups of flat, roundish sacs, frequently bound by branching tubules. A number of them are usually scattered throughout the cytoplasm of a living cell. The sacs are often organized into stacks of 5 to 8, but up to 30 or more are not uncommon in simpler organisms. Carbohydrates such as cellulose apparently collect in small vesicles (blisterlike bodies) that are pinched off from the margins of the dictyosome or Golgi body sacs. These vesicles migrate to the plasma membrane and fuse with it. There they discharge their contents to the outside, where they become a part of the cell wall. The enzymes needed for the "packaging" process are produced or contained within the dictyosomes or Golgi bodies themselves. Thus, one might describe them as collecting and packaging centers.

Plastids

Several kinds of **plastids** are generally found in living cells, with the **chloroplasts** (figure 3.11) of green organisms usually being the most conspicuous kind of plastid visible. They occur in a variety of shapes and sizes, from the beautiful corkscrewlike ribbons found in cells of the green alga *Spirogyra* (see figure 16.11), to the star-shaped chloroplasts of other green algae such as *Zygnema* (see figure 16.7D). The chloroplasts of higher plants, however, tend to be shaped somewhat like two Frisbees glued together along their edges, and when they are sliced in median section they resemble the outline of a football. Although a number of the algae and a few other plants have only 1 or 2 chloroplasts per cell, the number of chloroplasts is usually much greater in a green cell of higher plants, 75 to 125 being quite common. The chloroplasts may be from 2 to 10 micrometers in diameter, and each is bound by an envelope consisting of two delicate unit membranes; within is a colorless, enzyme-containing matrix, the **stroma.**

Grana (singular: **granum**), which are stacks of coin-shaped double membranes called **thylakoids** (figure 3.11B), are suspended in the stroma. The thylakoids contain green **chlorophyll** and other pigments. These "coin stacks" of grana are vital to life as we know it on our planet today, for it is within the thylakoids that the first steps of the all-important process of *photosynthesis* (see chapter 10) occurs. In photosynthesis, green plants convert carbon dioxide (from the air) and water to simple food substances, harnessing energy from the sun in the process. Thus, the existence of humans and all other animal life depends on the activities of the chloroplasts.

In each chloroplast there are usually about 40 to 60 grana linked together by arms, and each granum may contain from 2 or 3 to more than 100 stacked thylakoids (see figure 3.11). There are usually 4 or 5 *starch grains* in the stroma, in addition to oil droplets and enzymes. Like the matrix fluid in mitochondria, the stroma contains strands of DNA and ribosomes, and chloroplasts are also involved in the synthesis of amino acids and fatty acids.

Another type of plastid found in some cells of more complex plants is the **chromoplast.** Although chromoplasts are similar to chloroplasts in size, they vary considerably in shape, often being somewhat angular. They can develop from chloroplasts through internal changes, including the disappearance of

FIGURE 3.11 *A.* A chloroplast of a leaf cell of a garden bean. (Courtesy Jean Whatley). *B.* Interlocking thylakoids. (After Weier, T. E., T. Bisalputra, and A. Harrison. 1966. "Subunits in chloroplast membranes of *Scenedesmus quadricola." Journal of Ultrastructure Research* 15:38–56. Copyright 1966 by Academic Press, Inc.)

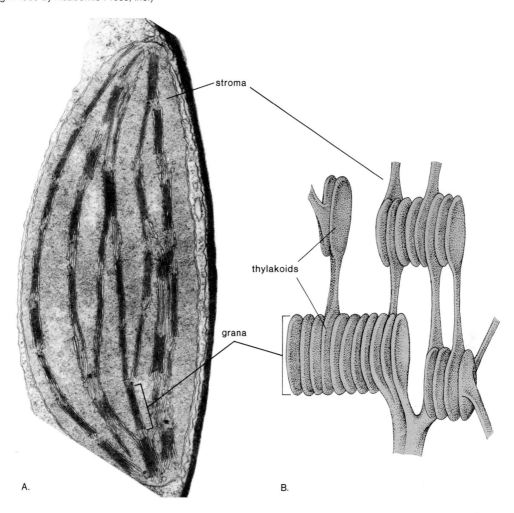

stroma

thylakoids

grana

A.

B.

chlorophyll. Chromoplasts are yellow, orange, or red in color due to the presence of carotenoid pigments, which they synthesize and accumulate. They are most abundant in the orange or red parts of plants, such as ripe tomatoes, red peppers, or flower petals.

Leucoplasts are a third type of plastid common to cells of higher plants. They are essentially colorless and include *amyloplasts,* which are known to synthesize starches, and *elaioplasts,* which synthesize oils. If exposed to light, some leucoplasts will develop into chloroplasts.

Plastids of all types develop from *proplastids,* which are small, pale green or colorless organelles having roughly the size and form of mitochondria. They are simpler in internal structure than plastids and have fewer thylakoids, the thylakoids not being arranged in grana stacks. Proplastids frequently divide and become distributed throughout the cell; after a cell itself divides, each daughter cell has a proportionate share. Plastids may also arise through the division of existing mature plastids.

Microtubules and Microfilaments

Microtubules are unbranched, thin, hollow, tubelike structures composed of protein; they are of varying lengths and tend to be between 15 and 25 nanometers in diameter. They are most commonly found just inside the plasma membrane, where they apparently control the addition of cellulose to the cell wall. Other functions of microtubules include the steering of vesicles containing cell wall components synthesized by dictyosomes (Golgi bodies) to the cell wall, and aiding movement of the tiny whiplike *flagella* and *cilia* possessed by some cells (see section on *plant movements* in chapter 11). Microtubules are also found in the special fibers that form the *spindles* and *cell plates* of dividing cells discussed later in this chapter. *Microfilaments* occur in nearly all cells as long protein filaments with an average diameter of 6 nanometers. They are often in bundles, and appear to play a role in *cytoplasmic streaming* (discussed on p. 46). Microtubules and microfilaments together form a flexible framework within the cytoplasm.

The Nucleus

The **nucleus** is frequently the most conspicuous organelle in a living cell, although it may be obscured by chloroplasts in green cells when they are observed with the aid of a light microscope. In cells without chloroplasts, the nucleus may appear as a grayish lump, somewhat spherical or ellipsoidal in shape and often lying against the plasma membrane to one side of the cell or in a corner. Some nuclei are irregular in form and, like most other organelles, they can vary greatly in size; they are, however, generally from 2 to 15 micrometers or larger in diameter. Certain fungi and algae have numerous nuclei within a single extensively branched cell, but more complex plants usually have a single nucleus located within the cytoplasm of each living cell.

The nucleus is the control center of the cell. In some ways it functions like a combination of a DNA-programmed computer and a dispatcher. The rest of the cell receives coded messages or "blueprints" from the nucleus and assembles items called for from the raw materials available to it. These raw materials are either absorbed by the plant from the soil or recycled from other areas. The nucleus not only directs the myriad activities of the complex cell

FIGURE 3.12   The nuclear envelope of a barley seed embryo cell nucleus showing the nuclear pores. X ca. 3,000. (Courtesy K. A. Platt-Aloia)

"factory" but also stores hereditary information, which is passed from cell to cell as new cells are formed.

Each nucleus is bound by two unit membranes, which constitute the **nuclear envelope.** Structurally complex pores occupy up to one-third of the total surface area of the nuclear envelope (figure 3.12). These pores apparently permit only certain kinds of molecules to pass between the nucleus and the cytoplasm. The outer membrane of the nuclear envelope is connected to parts of the endoplasmic reticulum.

Within the nuclear envelope is a granular-appearing fluid called the *nucleoplasm*. The nucleoplasm is packed with short fibers that are about 10 nanometers in diameter, and several different larger bodies are suspended within it. The most noticeable of these larger bodies are **nucleoli** (singular: **nucleolus**), which superficially resemble miniature nuclei; there may be from one to several nucleoli in each nucleus. Nucleoli are composed primarily of protein and are not bound by membranes. They are involved with the synthesis and export of a form of RNA (see chapter 2) to other parts of the cell.

Other important nuclear structures, which are not apparent with **light microscopy** unless the cell is stained or is in the process of dividing, include thin strands of **chromatin.** When a nucleus divides, the chromatin strands become shorter and thicker, and in their condensed condition, they are called **chromosomes.** Chromatin is composed of protein and DNA (see chapter 2). Each species of plant or animal has its own fixed number and composition of chromosomes in each of its cells. The cells involved in sexual reproduction have half the number found in other cells of the same organism. The number of chromosomes present in a nucleus normally bears no relation to the size and complexity of the organism. Each body cell of a radish, for example, has 18 chromosomes in its nucleus, while a cell of one species of goldenweed has 4, and a cell of a tropical adder's tongue fern has over 1,000.

Other Organelles

Various small bodies distributed throughout the cytoplasm tend to give it a granular appearance. Examples of such components include two types of small spherical organelles called *microbodies,* which are bound by a single membrane and found in the cytoplasm. One contains enzymes involved in a phase of photosynthesis and photorespiration (see chapter 10); the other contains enzymes that aid in the conversion of fats to carbohydrates. At one time oil droplets, which are common in cytoplasm, were believed to be bound by a membrane, and were called *spherosomes;* recent evidence suggests no membrane is present, and they therefore do not constitute true organelles. One organelle, called a *lysosome,* stores digestive enzymes, but is apparently confined to animal cells. The digestive activities of lysosomes are similar to those of the *vacuoles* of plant cells (discussed next).

Vacuoles

In a mature living plant cell, as much as 90% or more of the volume may be taken up by one or two large central **vacuoles** that are bound by **vacuolar membranes** (tonoplasts). The vacuolar membranes, which constitute the inner boundaries of the living part of the cell, are similar in structure and function to plasma membranes.

The vacuole was evidently so-called because of a belief that it was just an empty space; hence its name has the same Latin root as the word *vacuum* (from *vacuus*—meaning "empty"). Vacuoles, however, are filled with a watery fluid called **cell sap,** which is usually slightly to significantly acidic, and plays a role in maintaining pressures within the cell (see the discussion of *osmosis* in chapter 9). Cell sap contains dissolved substances such as salts, sugars, organic acids, and small quantities of soluble proteins. It also frequently contains water-soluble pigments called **anthocyanins.** Anthocyanins are responsible for many of the red, blue, or purple colors of flowers and some reddish leaves. In some instances, anthocyanins accumulate to a greater extent in response to cold temperatures in the fall. They should not be confused, however, with the red and orange carotenoid pigments confined to the chromoplasts. Carotenoid pigments are not soluble in water at all, but the yellow carotenes also play a role in fall leaf coloration (see chapter 7).

Sometimes large crystals of waste products form within the cell sap after certain ions have become concentrated there. Vacuoles in newly formed cells are usually tiny and numerous. They increase in size and unite as the cell matures. In addition to accumulating the various substances and ions mentioned above, vacuoles are apparently also involved in the recycling of certain materials within the cell, and even aid in the breakdown and digestion of organelles such as mitochondria and plastids.

Eukaryotic versus Prokaryotic Cells

With the few exceptions just mentioned, nearly all higher plant and animal cells exhibit the various features discussed so far. There are some very primitive organisms, however, whose cells lack a number of these features (e.g., membrane-bound nuclei and organelles). Such cells, called **prokaryotic,** to distinguish them from the typical **eukaryotic** cells just discussed, may have been the origin of plastids, mitochondria, and other organelles. This and other aspects of prokaryotic cells are covered in chapter 15.

Cyclosis

Cyclosis, or **cytoplasmic streaming,** occurs in living cells. When a living cell is examined with the light microscope, the organelles may appear to be moving as a current within the protoplasm carries them around within the walls. This streaming probably facilitates exchanges of materials within the cell and plays a role in the movement of substances from cell to cell. The precise nature and origin of cyclosis is still not known, but there is evidence that bundles of microfilaments may be responsible for it; other evidence suggests that it may be related to the transport of cellular substances by microtubules.

CELLULAR REPRODUCTION

The Cell Cycle

When cells divide they go through an orderly series of events known as the **cell cycle** (figure 3.13). This cycle is usually divided into *interphase* and *mitosis,* mitosis itself being subdivided into four phases. The length of the cell cycle varies with the kind of organism involved, the type of cell within an organism, and with temperature and other environmental factors. In most instances, however, interphase may occupy up to 90% or more of the time it takes to complete the cycle.

Interphase

Living cells that are not dividing are said to be in *interphase,* and it is such cells that have been discussed up to this point.

For many years cells were considered to be "resting" when they were not actually dividing, but we know now that three consecutive periods of intense activity take place during interphase. Immediately after a nucleus has divided, a period

FIGURE 3.13   A diagram of a cell cycle.

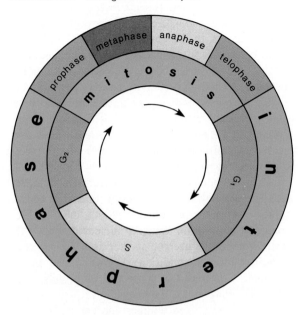

designated as G_1 begins. During this G_1 period, substances that either stimulate or inhibit the next period, designated S, are produced, and various organelles divide or grow. During the S period the unique process of DNA replication (duplication) takes place. Details of this process and of DNA structure are discussed in chapter 2. The final period of interphase, designated as G_2, includes the formation of substances and structures involved directly in mitosis.

Mitosis

All organisms begin life as a single cell. Almost immediately, however, this initial cell may begin to divide, producing new cells; these, in turn, divide and produce more cells. The process of dividing, called **mitosis** (see figure 3.14), involves the precisely equal allocation to two daughter cells of DNA and certain other substances duplicated during interphase; it goes on in at least parts of the organism until death. Strictly speaking, mitosis refers to the division of the nucleus alone, but the division of the remainder of the cell, called **cytokinesis,** usually accompanies or follows mitosis; both processes will be considered together here.

FIGURE 3.14   The phases of mitosis as seen in onion root-tip cells. (Photomicrographs by G. S. Ellmore)

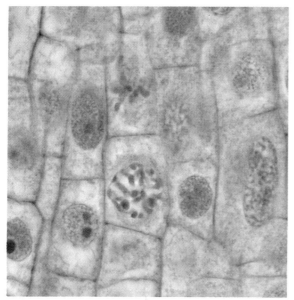

A. Cell (center) in prophase

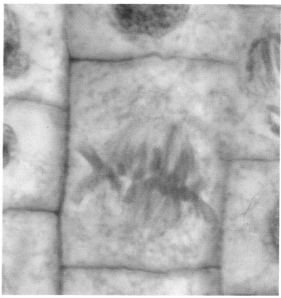

B. Cell (center) in metaphase

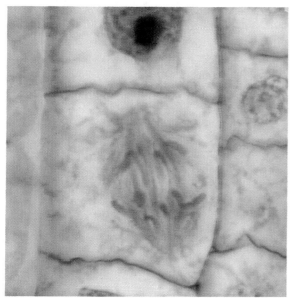

C. Cell (center) in anaphase

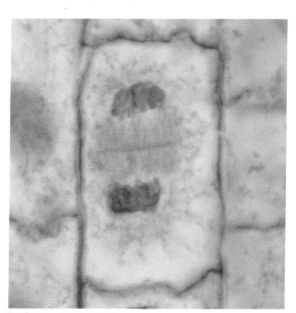

D. Cell (center) in telophase

In higher plants such as ferns, conifers, and flowering plants, mitosis occurs in specific regions, or tissues, called **meristems** (see figure 4.1). Meristems are found in the root and stem tips and also in a thin, perforated and branching cylinder of tissue called the **vascular cambium** (often referred to simply as the **cambium**), located in the interior of stems a short distance from the surface. In some herbaceous and most woody plants, a second meristem similar in form to the cambium lies between the cambium and the outer bark. This second meristem is called the **cork cambium.** These specific tissues are discussed in chapters 4, 5, and 6.

When mitosis takes place, it makes no difference how many chromosomes are in the nucleus. The daughter cells that result from the process have exactly the same number as the parent cell. It is a continuous process, which may take as little as 5 minutes or as long as several hours from start to finish. Typically, however, it takes from 30 minutes to 2 or 3 hours. The process, which is initiated with the appearance of a ringlike *preprophase band* of microtubules just beneath the plasma membrane, is usually divided into four arbitrary phases, primarily for convenience. Descriptions of the phases follow.

Prophase

The main features of *prophase* (figure 3.14A) are (1) the chromosomes become shorter and thicker, and their double nature becomes apparent; and (2) the nucleolus and the nuclear envelope disappear.

Prophase takes up about as much time as the remaining three phases combined. The beginning of this phase is marked by the appearance of the chromosomes as faint threads in the nucleus. These chromosomes gradually coil or fold into thicker and shorter structures, and soon two threads, or *chromatids,* can be distinguished for each chromosome. The chromatids are themselves independently coiled; these coils appear to tighten and condense until the chromosomes have become relatively short, thick, and rodlike, with constricted areas called **centromeres** holding each pair of chromatids together (see figure 3.15); a dense granule called a **kinetochore,** to which spindle fibers become attached, is located

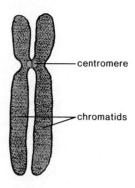

FIGURE 3.15   Parts of a chromosome.

near each centromere. When examined with the aid of a light microscope, the centromeres appear to be single structures, but they actually have become double by the G_2 stage of interphase and simply function as a single unit at this point. They may be located almost anywhere on a chromosome, but tend to be toward the middle. Sometimes other constrictions may appear on individual chromosomes, usually toward one end, giving them the appearance of having extra knobs. These knobs are referred to as *satellites;* the constrictions at the base of the satellites have no known function, but the satellites themselves are useful in helping to distinguish certain chromosomes from others in a nucleus.

As prophase progresses, the nucleolus gradually becomes less distinct and eventually disappears. The end of prophase is marked by the disappearance of the nuclear envelope.

In certain simpler organisms such as fungi and algae, and in virtually all animal cells, the cytoplasm just outside the nucleolus contains dual pairs of tiny keg-shaped organelles called *centrioles.* The centrioles are surrounded by microtubules, which radiate out from them and arrange cytoplasmic particles in the vicinity into starlike rays, collectively called an *aster.* At the beginning of prophase, the aster divides into two parts; one part remains at its original location, while the other part migrates around the nuclear envelope to the opposite side. Centrioles and asters have not been detected in the cells of most of the more complex members of the Plant Kingdom.

Metaphase

The main features of *metaphase* (figure 3.14B) are (1) a **spindle,** made up of *spindle fibers,* forms in the area previously occupied by the nucleus; and (2) the chromosomes become aligned in the center of the cell.

After the nuclear envelope breaks down, spindle fibers can be seen in the area previously occupied by the nucleus. These are oriented so that they extend in arcs between the two *poles* (invisible points toward each end of the cell), forming a structure that looks like an old-fashioned spinning top made of fine threads. The spindle fibers consist of groups of microtubules and are referred to collectively as the *spindle.* The chromosomes migrate to the center of the spindle and become aligned so that their centromeres are in a plane roughly in the center of the cell. This invisible line or plate is called the *equator.* During their movement to the equator each chromosome becomes attached at its centromere to an individual spindle fiber. Most of the spindle fibers remain independently extended between the poles.

Anaphase

The main feature of *anaphase* (figure 3.14C) is that the two chromatids of each chromosome separate and move to opposite poles. At the end of metaphase the two chromatids of each chromosome are still joined together at their centromeres. Anaphase begins with these centromeres splitting or separating lengthwise, and continues with the chromatids of each chromosome separating in unison and moving toward the poles. The chromatids, which after separation at their centromeres are now called *daughter chromosomes,* give the appearance of being pulled toward the poles by the contraction or gradual shortening of the spindle fibers attached to their centromeres. If the centromere is in the center of the chromatid, it leads the way, with the rest of the chromatid assuming a V shape as it appears to drag in the cytoplasm. All of the chromatids separate and move at the same time. Although experiments have shown that a chromatid will not migrate to a pole if the fiber attached to its centromere is severed, other experiments have shown that the chromatids will separate from one another but not move to the poles, even if no spindle is present. The forces behind both the separation of the chromatids and their movement to the poles have been the subject of much speculation, but these forces have not yet been positively identified.

Telophase

The main features of *telophase* (figure 3.14D) are (1) each group of daughter chromosomes becomes surrounded by a new nuclear envelope, (2) the daughter chromosomes become longer and thinner and are finally indistinguishable, (3) new nucleoli appear, (4) many of the spindle fibers disappear, and (5) a cell plate forms.

The transition from anaphase to telophase is not distinct, but telophase is definitely in progress when elements of new nuclear envelopes appear around each group of daughter chromosomes at the poles. These elements gradually form intact envelopes as the daughter chromosomes return to the diffuse, indistinct threads seen at the onset of prophase. The new nucleoli appear on specific regions of certain chromosomes.

During telophase the spindle gradually disappears and a set of shorter fibers (fibrils), composed of microtubules develops in the region of the equator between the daughter nuclei. This set of fibrils, which appears somewhat keg-shaped, is called a **phragmoplast.** Small vesicles, which have the appearance of droplets when viewed with a light microscope, appear along the plane of the equator on the phragmoplast fibrils. These vesicles, which are derived from dictyosomes or Golgi bodies, evidently contain pectin. The vesicles increase in number from the center outward and merge, forming a disc-shaped **cell plate,** which continues to grow outward until it reaches the walls of the cell. *Plasmodesmata* (see figures 3.4 and 3.5), which are minute strands of cytoplasm that extend between cells through the walls, are apparently formed as portions of the endoplasmic reticulum are "trapped" between fusing vesicles of the cell plate.

The mature cell plate becomes the middle lamella between the daughter cells. New plasma membranes develop on either side of the cell plate as it forms, and the protoplasm of each daughter cell deposits new cell wall materials between the middle lamella and the plasma membranes. These new walls are relatively flexible and remain so until the cells increase to their mature size. At that time additional cellulose and other substances may be added, forming a secondary cell wall interior to the primary wall.

COMMUNICATION BETWEEN CELLS

Although cells are independent units capable of carrying on complex activities, it is essential that all living cells be coordinated in the activities of the organism as a whole through some system of communication. In plants, living cells are in contact with one another via fine strands of cytoplasm called *plasmodesmata* (see figures 3.4 and 3.5). These extend between adjacent cells through the walls. The minute hole through which the plasmodesma extends is often located within a pair of doughnut-shaped "blisters" or roughly circular depressions (one in each wall) where the adjacent walls are very thin or nonexistent; these areas, where little more than the middle lamella separates abutting cells, are called **pits** (see figure 4.10). The translocation of sugars, amino acids, and other substances between cells apparently takes place by means of the plasmodesmata.

DIFFERENCES BETWEEN THE CELLS OF HIGHER PLANTS AND ANIMALS

All animals have either internal or external support for their tissues from a skeleton of some kind. Animal cells do not have cell walls; instead, the plasma membrane, called the *cell membrane* by most zoologists (animal scientists), forms the outer boundary of animal cells. Higher plant cells have walls that are thickened and rigid to varying degrees, with a framework of cellulose fibrils. Higher plant cells also have minute strands of cytoplasm called **plasmodesmata** (see the previous paragraph) connecting the protoplasts with each other through microscopic holes in the walls; animal cells lack plasmodesmata since they have no walls. When higher plant cells divide, a *cell plate* is formed (see telophase of mitosis), whereas no cell plates form in animal cells, the cells instead dividing by pinching in two.

Other differences, which are minor, pertain to the presence or absence of certain organelles. *Centrioles,* for example, which are tiny paired organelles found just outside the nucleus, occur in all animal cells but are generally absent from higher plant cells. Centrioles are discussed in the section dealing with *mitosis*. Plastids, which are common in plant cells, are not found in animal cells. Vacuoles, which are often large in plant cells, are either small or absent in animal cells.

SUMMARY

All living organisms are composed of cells. Cells are modified according to the functions they perform; some cells live for a few days whereas others live for many years.

The discovery of cells is associated with the development of the microscope. Robert Hooke was the first to use the word cells for boxlike compartments he saw in cork with the aid of a primitive microscope in 1665. Leeuwenhoek and Grew reported frequently during the next 50 years on the existence of cells in a variety of tissues. In 1809 Lamarck concluded that all living tissue is composed of cells, and in 1824 Dutrochet reinforced Lamarck's conclusions. In 1833 Brown discovered that all cells contain a nucleus, and shortly thereafter Schleiden observed a nucleolus within a nucleus. Schleiden and Schwann are generally credited with developing the cell theory in 1838–1839. The theory holds not only that all living organisms are composed of cells but that cells also form a unifying structural basis of organization. In 1858 Virchow argued cogently that every cell comes from a preexisting cell and that there is no spontaneous generation of cells from dust, as was the belief of the time. Pasteur confirmed Virchow's contentions in 1862, with experiments involving microorganisms in swan-necked flasks. Pasteur later proved that fermentation involves activity of yeast cells, and Buchner, in 1897, discovered that yeast cells do not need to be alive for fermentation to occur. This led to the discovery of organic catalysts called enzymes.

Light microscopes utilize light and glass or calcium fluoride lenses to achieve magnification, which is generally limited to 1,500 times for compound microscopes that require viewed material to be sliced thinly enough for the light to pass through. Dissecting microscopes can be used to view opaque objects, and usually magnify up to 30 times. Electron microscopes utilize magnets and a beam of electrons within a vacuum to achieve magnification; transmission electron microscopes magnify up to 200,000 or more times; scanning electron microscopes, which can be used with opaque objects, usually magnify up to 10,000 times.

Cells are minute; most vary in diameter between 10 and 100 micrometers. They number into the billions in larger organisms such as trees. Plant cells are bound by walls. The living part of the cell within the walls, protoplasm, has two main components: the nucleus, which functions as a control center, and cytoplasm, a fluid in which various membrane-bound organelles are suspended.

A pectic middle lamella is sandwiched between the primary cell walls of adjacent cells. The primary wall and also the secondary cell wall, often added inside the primary wall, are composed mostly of cellulose, but secondary cell walls may also contain lignin and other substances. A flexible plasma membrane, which is double (a unit membrane) and often forms folds, constitutes the outer boundary of the protoplasm; it regulates the substances that enter and leave the cell. The endoplasmic reticulum is a system of membrane channels associated with the storing and transporting of protein and other cell products. Granular particles called ribosomes may line the outer surfaces of the endoplasmic reticulum; they function in protein synthesis. Mitochondria are tiny, numerous organelles that are bound by a double membrane with inner platelike folds called cristae; they are associated with respiration. Dictyosomes or Golgi bodies, which constitute the Golgi apparatus, are organelles that appear as groups of sacs and function as collecting and packaging centers for the cell. Plastids are larger organelles that are green (chloroplasts), orange or red (chromoplasts), or colorless (leucoplasts). Chloroplasts contain enzymes, in a matrix called the stroma, and grana, which are stacks of coin-shaped membranes (thylakoids) containing green chlorophyll pigments; photosynthesis occurs in the thylakoids. Plastids develop from proplastids, which divide frequently, and also arise from the division of mature plastids. Microtubules are tiny tubelike structures that are commonly found just inside the plasma membrane. They apparently control the addition of cellulose to the cell wall. Microfilaments are associated with cytoplasmic streaming. The nucleus is bound by a nuclear envelope, consisting of two unit membranes that are perforated by numerous pores. Within the nucleus is a fluid called nucleoplasm, one or more spherical nucleoli, and thin strands of chromatin, which condense to become chromosomes when nuclei divide. Each species of organism has a specific number of chromosomes in each cell.

As much as 90% of the volume of a mature cell may be taken up with a vacuole or vacuoles that are bound by a vacuolar membrane and contain a watery fluid called cell sap. Cell sap contains dissolved substances and frequently also water-soluble pigments called anthocyanins, which are red or blue. Prokaryotic cells, which lack some of the features of the eukaryotic cells (the subject of this chapter), are discussed in chapter 15. Movement of cytoplasm within a cell in a streaming motion is called cyclosis.

In the cell cycle, cells that are not dividing are said to be in interphase, which is subdivided into three periods of intense activity that precede mitosis or division of the nucleus. Mitosis is usually accompanied by division of the rest of the cell and takes place in plant regions known as meristems. Mitosis is arbitrarily divided into four phases: (1) prophase, in which the chromosomes become apparent and the nuclear envelope breaks down; (2) metaphase, in which the chromosomes become aligned at the equator of the cell, and a system of spindle fibers, some of which are attached to the chromosomes at their centromeres, is fully developed; (3) anaphase, in which the chromosomes each separate lengthwise (the half chromosomes are called chromatids), with each group of chromatids migrating to opposite poles of the cell; and (4) telophase, in which each group of chromatids becomes a new nucleus, and a wall dividing the daughter nuclei forms, creating two daughter cells. The development of the dividing wall is initiated by the appearance of a set of short fibrils constituting the phragmoplast. Droplets, or vesicles, of pectin merge, forming a cell plate that grows to become the middle lamella of the new cell wall.

Living cells are in contact with one another via fine strands of cytoplasm called plasmodesmata, which often extend through minute holes in wall depressions called pits.

Animal cells differ from those of higher plants in not having a wall, plastids, or large vacuoles. Also, they have keg-shaped centrioles in pairs just outside the nucleus, and pinch in two instead of forming a cell plate when they divide.

REVIEW QUESTIONS

1. How is cellular structure beneficial to plants and animals?
2. Of what importance to the cell is the wall?
3. What is the difference between protoplasm and cytoplasm?
4. How can you distinguish between cytoplasm and vacuoles?
5. Of what are chloroplasts composed? What is the function of each component?
6. What is the function of a cell nucleus? How does it perform its function?
7. What are plasmodesmata, and what is their importance to living plants?
8. What are pits? Where are they located?
9. Are prophase and telophase of mitosis exactly the reverse of one another?
10. What are the differences and similarities between plant and animal cells?

DISCUSSION QUESTIONS

1. Would you consider any one type of cell more useful than another? Why?
2. After you have completed your introductory plant science course, do you believe you would be able to determine the function of each of a cell's organelles in a laboratory? Explain.

ADDITIONAL READING

Alberts, B. et al. 1983. *Molecular biology of the cell.* New York: Garland Publishing, Inc.

Baker, N. R., and J. Barber, eds. 1984. *Chloroplast biogenesis.* New York: Elsevier Science Publishing Co.

Bryant, J. A., and D. Francis, eds. 1985. *The cell division cycle in plants.* New Rochelle, NY: Cambridge University Press.

Burgess, J. 1985. *An introduction to plant cell development.* New Rochelle, NY: Cambridge University Press.

Darnell, J. E. et al. 1986. *Molecular cell biology.* New York: W. H. Freeman & Co.

DeRobertis, E. D., and E. M. DeRobertis, Jr. 1980. *Cell and molecular biology,* 7th ed. Philadelphia: Saunders College Publishing.

Dyson, R. D. 1985. *Cell biology: a molecular approach,* 3d ed. Newton, MA: Allyn and Bacon.

Gunning, B. E., and A. W. Robards, eds. 1976. *Intercellular communication in plants: studies on plasmodesmata.* New York: Springer-Verlag.

Hall, J. L. 1979. *Electron microscopy and cytochemistry of plant cells.* New York: Elsevier Science Publishing Co.

John, P. C., ed. 1981. *The cell cycle.* Society for experimental biology seminar series No. 10. New Rochelle, NY: Cambridge University Press.

Karp, G. 1984. *Cell biology,* 2d ed. Hightstown, NJ: McGraw-Hill Book Co.

Loewenstein, W. R. May 1970 "Intercellular communication." *Scientific American.* (Offprint 1178).

Robinson, D. G. 1985. *Plant membranes.* New York: John Wiley & Sons, Inc.

Sheeler, P., and D. E. Bianchi. 1983. *Cell biology: structure, biochemistry and function.* New York: John Wiley & Sons.

Wolfe, S. L. 1981. *Biology of the cell,* 2d ed. Belmont, CA: Wadsworth Publishing Co.

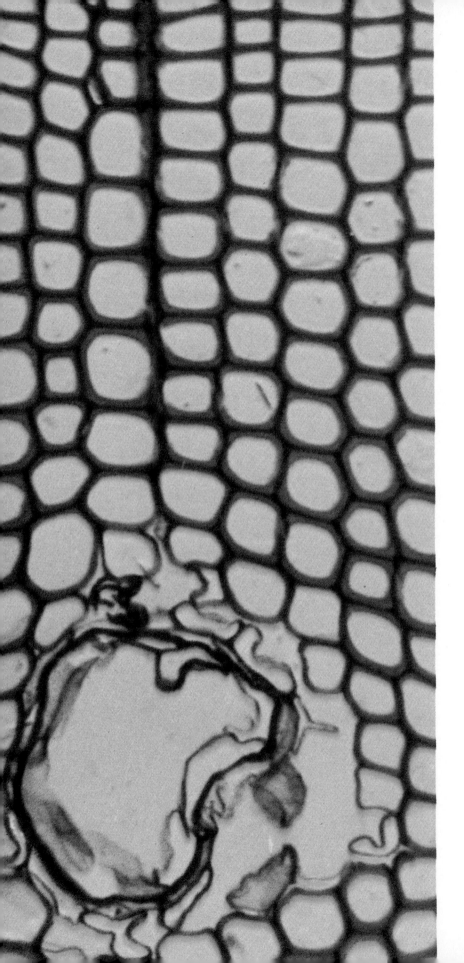

Overview

A discussion of meristems
(apical meristems, vascular
cambium, cork cambium,
intercalary meristems); permanent
tissues (parenchyma, collenchyma,
sclerenchyma, secretory tissues,
epidermis); and complex tissues
(xylem, phloem, periderm) forms the
body of this chapter.

4 Tissues

Some Learning Goals

1. Know the meristems present in plants, and where they are found.

2. Learn the conducting tissues of plants and the function of each cell component.

3. Learn tissues of plants that are neither meristematic nor function in conduction at maturity.

INTRODUCTION

There are many interesting modifications of higher plants discussed in the three chapters that follow this one but, regardless of the outer form, most plants have three major groups of *organs*—**roots, stems,** and **leaves.** Each of these organs is composed of **tissues,** which are defined as "groups of cells performing a common function." Any plant organ may be composed of a number of different tissues, each of which is classified according to structure, origin, or function. In addition, three basic tissue patterns occur in stems and roots (see *woody dicots, herbaceous dicots,* and *monocots,* discussed in chapters 5 and 6). The following are major kinds of tissues found in higher plants. The specific types of cells associated with each tissue, as well as illustrations of them, are included in the discussions that follow the classification.

I. Meristematic Tissues (figure 4.1)
   1. *Apical meristems*
   2. *Vascular cambium*
   3. *Cork cambium*
   4. *Intercalary meristems*
II. Permanent Tissues
   1. *Parenchyma*
   2. *Collenchyma*
   3. *Sclerenchyma*
   4. *Secretory tissues*
   5. *Epidermis*
   6. *Xylem*
   7. *Phloem*
   8. *Periderm*

I. MERISTEMATIC TISSUES

Meristems or *meristematic tissues* are tissues in which cells actively divide. As new cells are produced, they typically are small, six-sided boxlike structures, each with a proportionately large nucleus in the center and tiny vacuoles, or no vacuoles

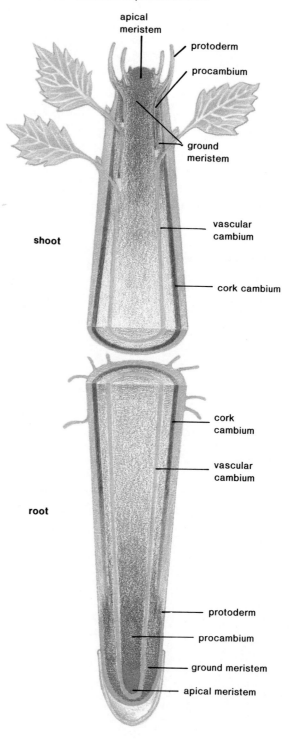

FIGURE 4.1   Locations of plant meristems.

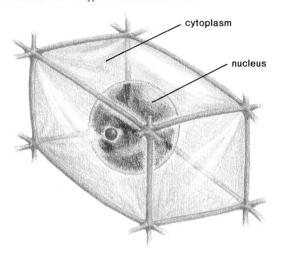

FIGURE 4.2   A typical meristematic cell.

cytoplasm

nucleus

at all (figure 4.2). As they mature, however, they assume many different shapes and sizes, each related to the cell's ultimate function.

Apical Meristems

Apical meristems are found at, or near, the tips of roots and shoots, which increase in length as the apical meristems produce new cells. Three *primary meristems,* as well as embryo leaves and buds, develop from each apical meristem; they are called **protoderm, ground meristem,** and **procambium.** The tissues they produce are called *primary tissues.* Note their locations in figure 4.1; they are discussed in chapters 5 and 6.

Vascular Cambium

The **vascular cambium** forms a thin, often branching cylinder that, except at the tips, runs the length of the stems and roots of most perennial plants and many herbaceous annuals. It is primarily responsible for the production of tissues that increase the girth of a plant. The individual, self-perpetuating cells of the vascular cambium are referred to as *initials;* both the cambium and its initials are discussed in chapters 5 and 6.

Cork Cambium

The **cork cambium,** like the vascular cambium, forms a thin cylinder that runs the length of stems and roots of woody plants. It is located to the exterior of the vascular cambium, however, so that in cross section the vascular cambium appears as a tube of smaller diameter lying within the tube of the cork cambium. The cork cambium is primarily responsible for producing the outer bark of woody plants; it is discussed in chapters 5 and 6. The tissues laid down by the vascular cambium and the cork cambium are produced *after* the primary tissues have matured, and are called *secondary tissues.*

Intercalary Meristems

Grasses and related plants have neither a vascular cambium nor a cork cambium, but they do have apical meristems, and other meristematic tissues called *intercalary meristems,* in the vicinity of **nodes** (leaf attachment areas; see chapter 7), which occur at intervals along stems. Intercalary meristems, like apical meristems, produce increases in the length of stems.

II. PERMANENT TISSUES

Permanent tissues are composed of cells that, after being produced by meristems, have assumed various shapes and sizes related to their functions as they developed and matured. Some permanent tissues consist of only one kind of cell, whereas others may have two to several kinds of cells. Simpler basic types of permanent tissues are discussed first, followed by those that are more complex.

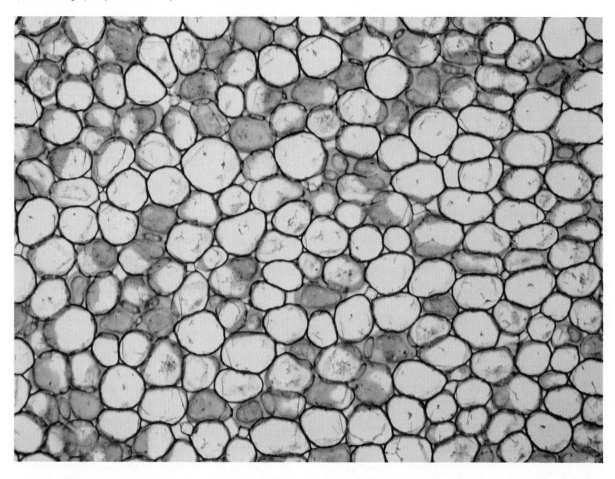

Parenchyma

Parenchyma cells (figure 4.3) are the most abundant of the cell types, being found in almost all major parts of higher plants. They are more or less spherical in shape when they are first produced, but when all the spherical cells of parenchyma tissue push up against one another, their thin, pliable walls are flattened at the points of contact. As a result parenchyma cells assume various shapes and sizes, with the majority having 14 sides. They tend to have large vacuoles and may contain starch grains, oils, tannins (tanning or dyeing substances), crystals, and various other secretions. Spaces commonly occur between parenchyma cells; in fact, in water lilies and other aquatic plants, the intercellular spaces are quite extensive and form a network throughout the entire plant. This type of parenchyma tissue—with extensive connected air spaces—is referred to as *aerenchyma,* while parenchyma cells containing numerous chloroplasts (as found in leaves) form **chlorenchyma** tissue. The chief function of chlorenchyma tissue is photosynthesis, whereas parenchyma tissues without chloroplasts function primarily in food or water storage. Some parenchyma cells develop irregular extensions of the inner wall that greatly increase the surface area. Such cells, called *transfer*

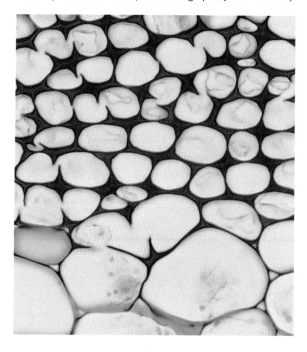

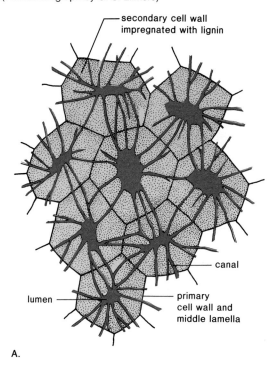

A.

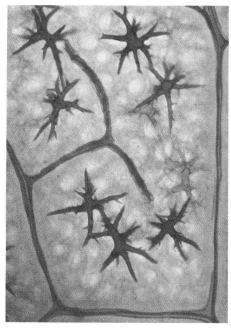

B.

cells, apparently play a role in transferring dissolved substances between adjacent cells. Many parenchyma cells live a long time; in some cacti, for example, they may live to be over 100 years old.

Collenchyma

Collenchyma cells (figure 4.4), like parenchyma cells, have living protoplasm and may remain alive a long time. They are distinguished from parenchyma cells primarily by the thicker walls, which are also usually uneven in thickness. Collenchyma cells often occur just beneath the epidermis; typically they are longer than they are wide and their walls are pliable as well as strong. They provide flexible support for both growing organs and mature organs such as leaves and floral parts.

Sclerenchyma

Sclerenchyma tissue consists of cells that have thick, tough walls, normally impregnated with **lignin.** Most sclerenchyma cells are dead at maturity and function in support. Two types of sclerenchyma occur: **sclereids** and **fibers.** Sclereids (see figure 4.5 for illustrations of two types of sclereids) may be randomly distributed in other tissues. For example, the

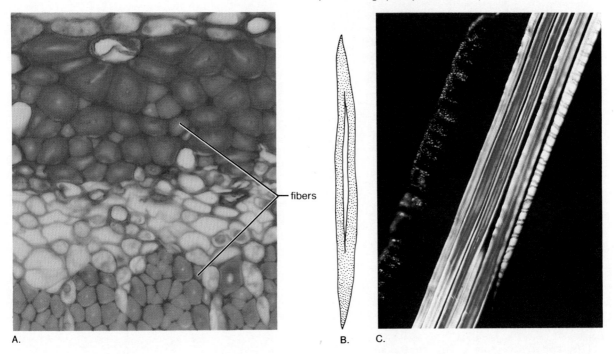

FIGURE 4.6   Fibers. *A.* A cross section of a portion of stem tissue from a linden tree. Note the thickness of the walls of the darker fibers. *B.* A single fiber in longitudinal section. *C.* A longitudinal section through fibers in a *Welwitschia* leaf. (Photomicrographs by G. S. Ellmore)

fibers

A.

B.   C.

slightly gritty texture of pears is due to the presence of groups of sclereids, or *stone cells,* as they are sometimes called. They tend to be about as long as they are wide, and sometimes occur in specific zones (e.g., the margins of camellia leaves) rather than being scattered within other tissues. *Fibers* (figure 4.6) may be found in association with a number of different tissues in roots, stems, leaves, and fruits. They are usually much longer than they are broad and have a proportionately tiny cavity, or *lumen,* in the center of the cell. At the present time, fibers from more than 40 different families of plants are in commercial use in the manufacture of textile goods, ropes, string, canvas, and similar products. Archaeological evidence indicates that humans have been using fibers for at least 10,000 years.

Secretory Tissues

Secretory tissues or individual **secretory cells** release substances that have been produced within the protoplasm and moved to the outside of the cells in which they originate. Often the substances consist of waste products that are of no further use to the plant, but some substances such as **hormones** (see chapter 11) are vital to normal plant functions. Secretory tissues or cells can occur in a wide variety of places in a plant. Among the most common secretory tissues are those that secrete nectar in flowers; oils in citrus, mint, and many other leaves; mucilage in the glandular hairs of sundews and other insect-trapping plants; latex in members of several plant families such as the Spurge Family; and resins in coniferous plants such as pine trees. Latex and resins are usually secreted by cells lining tubelike ducts that form networks throughout certain plant species (see figure 5.10). Some plant secretions such as pine resin, rubber, mint oil, and opium have considerable commercial value.

Epidermis

The outermost layer of cells of all young plant organs is called the **epidermis.** Since it is in direct contact with the environment it is subject to modification by environmental factors and often includes several different kinds of cells. The epidermis is usually one cell thick, but a few plants produce aerial roots called

FIGURE 4.7 A small portion of a kaffir lily leaf, showing the thick cuticle secreted by the upper epidermis. (Photomicrograph by G. S. Ellmore)

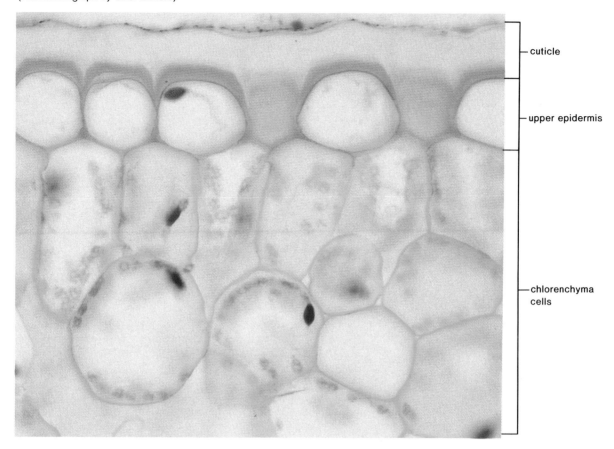

cuticle

upper epidermis

chlorenchyma cells

velamen roots (e.g., orchids), in which the epidermis may be several cells thick, with the outer cells functioning something like a sponge. Such a multiple-layered epidermis also occurs in the leaves of some tropical figs and members of the Pepper Family. Most epidermal cells secrete a fatty substance called **cutin** within and on the surface of the outer walls. Cutin forms a protective layer called the **cuticle** (figure 4.7). The thickness of the cuticle (or, more importantly, wax secreted by the epidermis on top of the cuticle) to a large extent determines how much water is lost through the cell walls by evaporation. The cuticle is also exceptionally resistant to bacteria and other disease organisms and has been recovered from fossil plants millions of years old. The waxes deposited on the cuticle in a number of plants (see figure 7.5) apparently reach the surface through microscopic channels in the cell walls. The susceptibility of a plant to herbicides may depend on the

thickness of these wax layers. Some wax deposits are extensive enough to have commercial value. Carnauba wax, for example, is deposited on the leaves of the wax palm. It and other waxes are harvested for use in polishes and phonograph records.

In leaves, the epidermal cell walls at right angles to the surface often assume bizarre shapes, with convolutions that, under the microscope, give them the appearance of pieces of a jigsaw puzzle. Epidermal cells of roots produce tubular extensions called *root hairs* (see figure 6.4) a short distance behind the growing tips; these greatly increase the absorptive area of the surface. Hairs of a different nature occur on the epidermis of above-ground parts of plants. These form outgrowths consisting of one

FIGURE 4.8   *A.* Tack-shaped glands and epidermal hairs of various sizes on the surface of flower bracts of a western tarweed. (Scanning electron micrograph Courtesy Robert L. Carr).

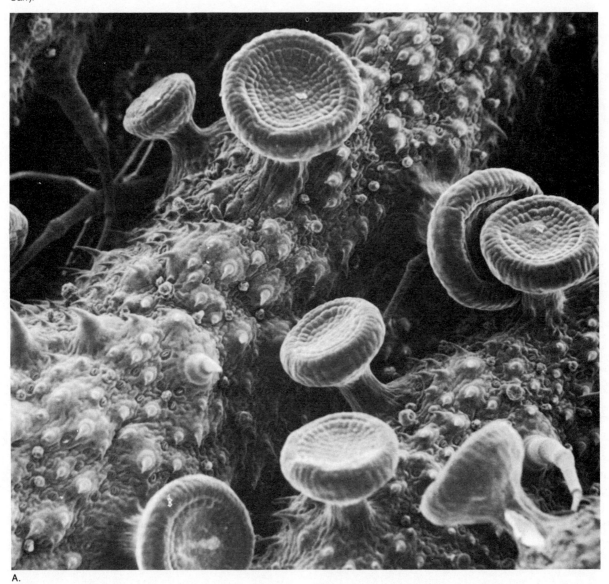

A.

to several cells (figure 4.8B). Leaves also have numerous small pores, the **stomata,** bordered by pairs of specialized epidermal cells called **guard cells** (see figures 7.6 and 9.7). Guard cells differ in shape from other epidermal cells; they also differ in that chloroplasts are present within them. Stomata and guard cells are discussed in chapters 7 and 9. Some epidermal cells may be modified as **glands** that secrete protective or other substances (figure 4.8A).

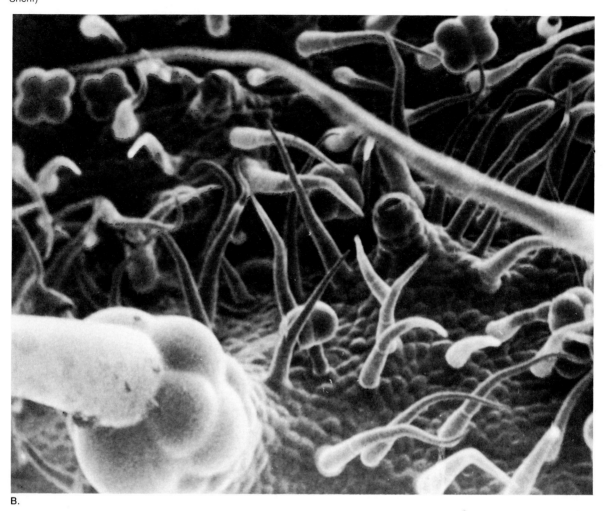

B.

Most of the tissues thus far discussed consist mainly of one kind of cell, but a few important tissues are always composed of more than one kind of cell, and are sometimes referred to in a general sense as *complex tissues*. The two most important complex tissues in plants, *xylem* and *phloem,* function primarily in the transport of water, ions, and soluble food substances throughout the plant. Some complex tissues are produced by apical meristems, but the bulk of such tissues in woody plants are produced by the vascular cambium and are often referred to as *vascular tissues. Periderm,* which comprises the outer bark of woody plants, mostly consists of *cork* cells but is included in this discussion because it also contains pockets of parenchyma cells.

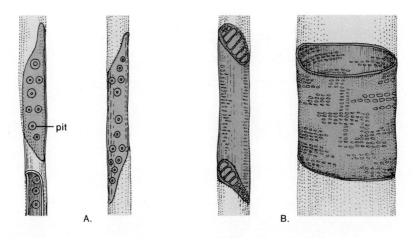

A.   B.

FIGURE 4.10   Pits are depressions or cavities in cell walls
where the secondary wall does not form. There may be from
one or two to several thousand in a cell. They always occur in
pairs, with one on each side of the middle lamella. Some,
called *bordered pits* (left), bulge out from the wall and
resemble doughnuts in surface view, while others, called
simple pits (right), do not bulge.

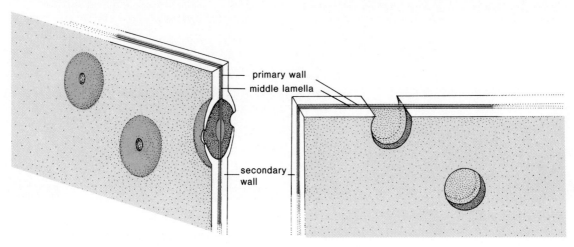

Xylem

Xylem, which is an important component of the
"plumbing" system of a plant, conducts water and
dissolved substances throughout all organs; it consists of a combination of parenchyma cells, fibers,
vessels, tracheids, and *ray cells* (figure 4.9). **Vessels**
are long tubes made up of individual cells called
vessel elements that are open at each end, with bar-
like strips of wall material extending across the open
areas in some instances. The cells are joined end to
end, forming the tubes. **Tracheids,** which, like vessel

elements, are dead at maturity, are tapered at each
end and have no openings similar to those of vessels.
The ends overlap with those of other tracheids, and
wherever two tracheids are in contact with one another, *pits* (figure 4.10) are usually present. As indicated in chapter 3, pit pairs permit water to pass
from cell to cell. In certain kinds of plants, such as
cone-bearing trees, the xylem is composed almost
entirely of tracheids. The walls of many tracheids
have spiral thickenings on them that are easily seen
with the light microscope. Most conduction is up and
down, but some lateral (sideways) conduction also
occurs. The lateral conduction takes place in the

FIGURE 4.11   Phloem. A longitudinal section through a
portion of stem phloem of tobacco. (After a drawing from
Cronshaw, J. and Esau, K. *Journal of Cell Biology,* 38:298.
July–September, 1968)

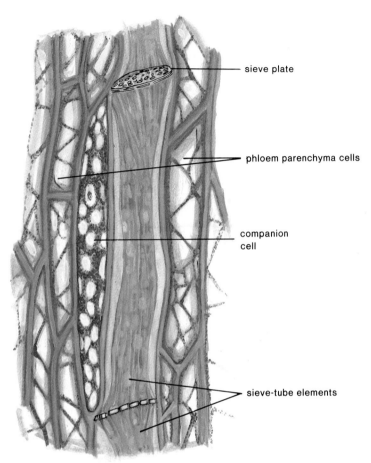

rays; ray cells, which also function in food storage, are actually long-lived parenchyma cells that are produced in horizontal rows by special *ray initials* of the vascular cambium. In woody plants the rays radiate out from the center of stems and roots like the spokes of a wheel (see figures 5.8 and 5.14).

Phloem

Phloem tissue (figure 4.11), which functions primarily in the conduction throughout the plant of dissolved food substances, is mostly composed of **sieve-tube elements** and **companion cells.** Phloem is derived from common parent cells of the cambium that also produce xylem cells; it also usually includes fibers, parenchyma, and ray cells. Sieve-tube elements, like vessel elements, are laid end to end, forming **sieve tubes.** Unlike vessel elements, however, the end walls have no large openings; instead,

the walls are full of small pores through which the cytoplasm extends from cell to cell. The porous regions of sieve elements are called **sieve plates.** Sieve-tube elements have no nuclei at maturity, despite the fact that their cytoplasm is very active in the conduction of food materials in solution throughout the plant. Apparently the adjacent companion cells form a very close relationship with the sieve tubes next to them, and function in some manner that brings about the conduction of the food.

Sieve cells, which are found in ferns and cone-bearing trees, are similar to sieve-tube elements but tend to overlap at their ends rather than form continuous tubes. Like sieve-tube elements, they have no nuclei at maturity, and they also have no adjacent companion cells. They do have equivalent adjacent parenchyma cells, called *albuminous cells,* which apparently function in the same manner as companion cells.

FIGURE 4.12   Periderm and a lenticel. A cross section
through a small portion of elderberry periderm, showing a
large lenticel. (Photomicrograph by G. S. Ellmore)

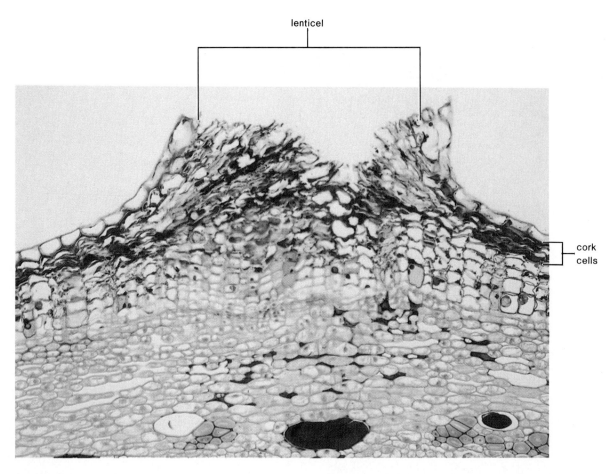

lenticel

cork
cells

Periderm

In woody plants, the epidermis is sloughed off and
replaced by a **periderm** after the cork cambium begins producing new tissues that increase the girth of
the stem or root. The periderm constitutes the outer
bark and is primarily composed of somewhat rectangular and boxlike **cork** cells, which are dead at
maturity (figure 4.12). While the protoplasm of cork
cells is still functioning, it secretes a fatty substance,
suberin, into the walls. This makes cork cells water-
proof and helps them protect the tissues beneath the
bark. Some cork tissues, such as those produced by
the cork oak, are harvested commercially and are
used for bottle corks and in the manufacture of linoleum and gaskets.

Some parts of a cork cambium form pockets of
loosely arranged parenchyma cells that are not impregnated with suberin. These pockets of tissue protrude through the surface of the periderm; they are
called **lenticels** (figure 4.12) and function in gas exchange between the air and the interior of the stem.
The fissures in the bark of trees have lenticels at their
bases.

The various tissues discussed are shown as they
occur in a woody stem in figure 5.6.

SUMMARY

A group of cells performing a common function is called a tissue. Apical meristems are found in the vicinity of the tips of stems and roots; the vascular cambium and the cork cambium occur as lengthwise cylinders within stems and roots; intercalary meristems occur in the vicinity of nodes of grasses and related plants.

Permanent tissues are produced by meristems and each consists of one to several kinds of cells. They include parenchyma, collenchyma, sclerenchyma, secretory tissues, epidermis, xylem, phloem, and periderm.

Parenchyma cells are thin-walled, while collenchyma cells have unevenly thickened walls that provide flexible support for various plant organs. Two types of sclerenchyma—fibers (which are longer and tapering) and sclereids (which are short in length)—occur; both types have thick walls and are usually dead at maturity. Secretory tissues occur in various places in plants; they secrete substances such as nectar, oils, mucilage, latex, and resins.

Epidermis is usually one cell thick, with fatty cutin (forming the cuticle) within and on the surface of the outer walls; epidermis may include guard cells that border pores called stomata; root hairs, which are tubular extensions of single cells; other hairs that consist of one to several cells; and glands that secrete protective substances.

Complex tissues have more than one kind of cell. The principal types are xylem, phloem, and periderm. Xylem conducts water and dissolved substances throughout the plant. It consists of a combination of parenchyma, fibers, vessels (tubular channels composed of cells called vessel elements, which are attached end to end), tracheids (cells with tapering end walls that overlap), and ray cells (involved in lateral conduction). Phloem conducts dissolved food materials throughout the plant. It is composed of sieve tubes (made up of cells called sieve-tube elements), companion cells (which apparently regulate adjacent sieve-tube elements), parenchyma, ray cells, and fibers. Sieve cells, which have overlapping end walls, and adjacent albuminous cells take the place of sieve-tube elements and companion cells in ferns and cone-bearing trees.

Periderm, which consists of cork cells and loosely arranged groups of cells comprising lenticels involved in gas exchange, constitutes the outer bark of woody plants.

REVIEW QUESTIONS

1. What is the function of meristems? Where are they located?
2. How are parenchyma, collenchyma, and sclerenchyma distinguished from one another?
3. Distinguish between epidermis and periderm.
4. What types of substances do secretory cells secrete?
5. What are the functions of xylem and phloem? What cells are involved in their normal activities?

DISCUSSION QUESTIONS

1. Most plant meristems are located at the tips of shoots and roots and in cylindrical layers within stems and roots. What could happen if they were present in leaves?
2. The cambium produces xylem toward the center of a tree and phloem toward the outside. Do you think it would make any difference if the positions of the xylem and phloem were reversed? Why?

ADDITIONAL READING

Cutler, E. F., and K. L. Alvin. 1982. *The plant cuticle.* Orlando, FL: Academic Press.

Cutter, E. G. 1978. *Plant anatomy, Part I: Cells and tissues,* 2d ed. Reading, MA: Addison-Wesley Publishing Co., Inc.

Esau, K. 1977. *Anatomy of seed plants,* 2d ed. New York: John Wiley & Sons, Inc.

Fahn, A. 1982. *Plant anatomy,* 3d ed. Elmsford, NY: Pergamon Press.

Lloyd, C. W., ed. 1983. *The plant cytoskeleton in growth and development.* Orlando, FL: Academic Press.

Metcalfe, C. R., ed. 1969–1982. *Anatomy of the monocotyledons,* 7 vols. Fair Lawn, NY: Oxford University Press.

Metcalfe, C. R., and L. Chalk, eds. 1950 (vol. 1, 2d ed., 1980). *Anatomy of the dicotyledons,* 2 vols. Fair Lawn, NY: Oxford University Press.

Roland, J. C., and F. Roland. 1981. *Atlas of flowering plant structure.* Translated from the French by D. Baker. New York: Longman, Inc.

Overview

*A*fter a brief introduction, this chapter discusses in general the origin and development of stems. Items such as the apical meristem and the tissues derived from it, leaf gaps, cambia, secondary tissues, and lenticels are included. This general discussion is followed by notes on the distinctions between herbaceous and woody dicot stems, and monocot stems. This section covers annual rings, rays, heartwood and sapwood, resin canals, bark, laticifers, and vascular bundles.

Next comes a survey of specialized stems (rhizomes, stolons, tubers, bulbs, corms, cladophylls, and others). The chapter concludes with a discussion of the economic importance of wood and stems.

5   *Stems*

Some Learning Goals

1. Know the tissues that develop from shoot apices and the meristems from which each tissue is derived. Distinguish between primary tissues and secondary tissues.

2. Learn and give the function of each of the following: vascular cambium, cork cambium, stomata, and lenticels.

3. Contrast the stems of herbaceous and woody dicots with the stems of monocots.

4. Understand the composition of wood and its annual rings, sapwood, heartwood, and bark. Explain how a log is sawed for commercial use.

5. Distinguish among rhizomes, stolons, tubers, bulbs, corms, cladophylls, and tendrils.

6. Learn at least ten human uses of wood and stems in general.

Outline

INTRODUCTION

Whether you order a dozen long-stemmed roses for someone special, use a toothpick or chopsticks, build a wood-framed house, sit in a wooden chair and read a newspaper, or graft one variety of fruit tree onto another, you may or may not be conscious of the fact that these and literally hundreds of other activities either directly or indirectly involve plant stems, and that these particular plant organs have been an integral part of human life since cave men first used wooden clubs to kill for food.

One of the activities just mentioned—*grafting*—usually involves artificially uniting stems, or parts of stems, of different but related varieties of plants. The careful matching of certain tissues is critical to its success, as is seen in the discussion of grafting in appendix 4. To understand how and why grafts may or may not be successful, however, and to identify which parts of stems are useful for food, clothing, construction lumber, furniture, paper, etc., we first need to examine the structure of stems and learn the basic functions of the various tissues.

Unlike animals, some plants can grow indefinitely, with the meristems at their tips increasing their length and other meristems increasing their girth for hundreds or even thousands of years. In stems, the cells produced by the meristems usually become the familiar, erect, aerial *shoot system* with branches and leaves. In certain plants, such as ferns or perennial grasses, however, this shoot system may develop horizontally beneath or at the surface of the ground; in other plants the stem may be so short and inconspicuous as to appear nonexistent. In a number of plants, modifications of the stem permit specialized functions such as climbing or the storage of food or water.

EXTERNAL FORM OF A WOODY TWIG

A woody twig consists basically of an axis with leaves attached (figure 5.1). The leaves may be attached to the twig in a spiral around the stem (they are then said to be *alternate,* or *alternately arranged*), or they may occur in pairs (*opposite,* or *oppositely arranged*), or sometimes they occur in groups of three

---

FIGURE 5.1   A woody twig. *A.* In its winter condition. *B.* The same twig as it appeared the summer before.

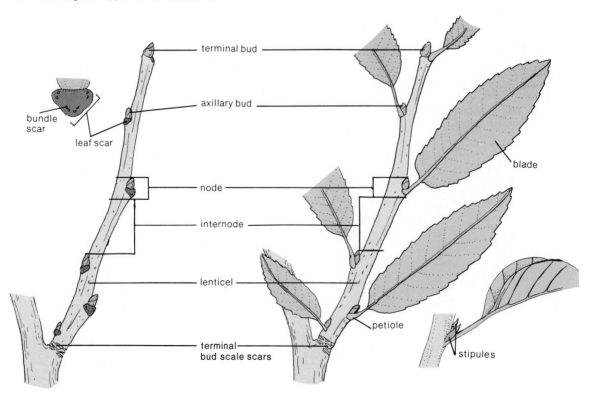

or more called *whorls* (*whorled arrangement*). The area, or region (*not* structure), of a stem where a leaf or leaves are attached is called a **node,** and a stem region between nodes is called an **internode.** A leaf usually has a flattened blade and a stalk, called the **petiole,** by which it is attached to the twig.

A bud occurs in the angle between a petiole and the stem. This angle is called an **axil,** and the bud located in the axil is an *axillary bud.* Axillary buds may become branches or they may contain tissues that will develop into the next season's flowers. Most, but not all, buds are protected by one or several *bud scales,* which fall off when bud tissue growth begins. In addition to axillary buds, a *terminal bud* is often present at the tip of each twig. The terminal bud usually resembles an axillary bud, but it is frequently a little larger. It usually produces tissues that extend the length of the twig during the growing season rather than forming a separate branch. The bud scales of a terminal bud leave tiny scars around the twig when they fall off in the spring. A twig's age can be determined by counting the number of groups of *bud scale scars* on it.

Sometimes scars with a different origin also occur on a twig. These scars originate from a leaf that has paired appendages called **stipules** at the base of the petiole. The stipules may remain throughout the life of the leaf, but in some plants they fall off as the buds expand in the spring, leaving tiny *stipule scars,* which may resemble a fine line encircling the twig or may be very inconspicuous small scars on either side of the petiole base.

Deciduous trees and shrubs (those that lose their leaves annually in the fall) have characteristic **leaf scars,** with dormant axillary buds above them after the leaves fall. Tiny **bundle scars,** which mark the location of the water-conducting and food-conducting tissues, can be seen within the leaf scars. There are frequently three bundle scars present, but the number can vary from one to many. The shape and size of the leaf scars, and the arrangement and numbers of the bundle scars, are characteristic of each species; it is often possible to determine the identity of a woody plant in its winter condition by means of these structures.

ORIGIN AND DEVELOPMENT OF STEMS

As indicated in chapter 4, there is an *apical meristem* at the tip of each stem, and it is this meristem that produces the tissues resulting in the stem's increase in length. Before the onset of the growing season, the apical meristem is dormant. It is protected by bud scales of the bud in which it is located and also to a certain extent by leaf **primordia** (singular: **primordium**), the tiny embryo leaves that will develop into mature leaves after the bud scales drop off and growth begins. The apical meristem in the embryo stem of a seed is also dormant until the seed begins to germinate.

When a bud begins to expand, or a seed germinates, the cells of the apical meristem undergo mitosis, and soon three primary meristems develop from it (see figure 4.1). The outermost of these primary meristems, the **protoderm,** gives rise to the *epidermis.* As noted in chapter 4, the epidermis is typically one cell thick and usually becomes coated with a thin, fatty protective layer, the *cuticle.* A cylinder of strands constituting the **procambium** appears to the interior of the protoderm. (The procambium produces water-conducting *primary xylem cells* and food-conducting *primary phloem cells*). The remainder of the meristematic tissue, called **ground meristem,** produces two tissues composed mostly of parenchyma cells. The parenchyma tissue in the center of the stem is the **pith.** The cells of this area are very large and may break down shortly after they are formed, leaving a cylindrical hollow area; even if they do not break down early, they may eventually be crushed as new tissues produced by other meristems add to the girth of the stem, particularly in woody plants. The other tissue produced by the ground meristem is the **cortex.** The cortex may become more extensive than the pith, but in woody plants it, too, eventually will be crushed and replaced by new tissues produced from within. The parenchyma of both the pith and the cortex function in storing food or sometimes, if chloroplasts are present, in manufacturing it.

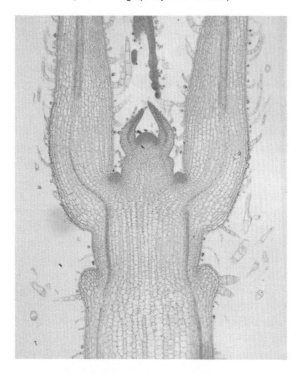

FIGURE 5.2 A longitudinal section through the tip of a *Coleus* stem. (Photomicrograph by G. S. Ellmore)

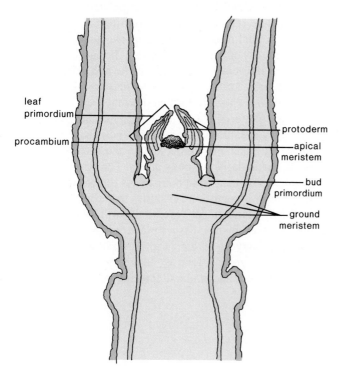

leaf primordium

procambium

protoderm

apical meristem

bud primordium

ground meristem

All five of the tissues produced by this apical meristem complex (epidermis, primary xylem, primary phloem, pith, and cortex) arise while the stem is increasing in length and are called *primary tissues.* As these primary tissues are produced, the leaf primordia and the *bud primordia* (embryo buds in the axils of the leaf primordia) develop into mature leaves and buds (figure 5.2). As each leaf and each bud develops, a strand of xylem and phloem, called a *trace,* branches off from the cylinder of xylem and phloem extending up and down the stem and enters the leaf or the bud. As the traces branch from the main cylinder of xylem and phloem, each trace leaves a little thumbnail-shaped gap in the cylinder of tissue. These gaps are called **leaf gaps** and *bud gaps* (figure 5.3).

A narrow band of cells between the primary xylem and the primary phloem may retain its meristematic nature and become the *vascular cambium,* often referred to simply as the *cambium.* These cambial cells continue to divide indefinitely, with the divisions taking place mostly in a plane parallel to the surface of the plant. The *secondary tissues* produced by the vascular cambium thus add to the girth of the stem instead of to its length (figure 5.4).

If a nail is driven into the side of a tree and observed periodically over a number of years, differences between the activities of the apical meristem complex and those of the vascular cambium will readily become apparent. The nail may eventually become embedded as the stem increases in girth, but it will always remain at the same height above the ground, as the cells that increase the length of a stem are produced only at the tips.

Cells produced by the vascular cambium become *tracheids, vessel elements, fibers,* or other components of *secondary xylem* (*inside* of the meristem toward the center), or become sieve-tube elements, companion cells, or other components of *secondary phloem* (*outside* of the meristem toward the surface). The functions of these secondary tissues are the same as those of their primary counterparts—secondary xylem conducts *water* and *soluble nutrients,* whereas secondary phloem conducts food manufactured by the plant through photosynthesis, in soluble form throughout the plant.

In many plants, especially woody species, a second cambium arises within the cortex or, in some instances, develops from the epidermis or phloem. This is called the **cork cambium,** or **phellogen.** The cork cambium produces boxlike **cork cells,** which become impregnated with **suberin,** a fatty substance that makes the cells impervious to moisture. The cork cells, which are produced annually in cylindrical

FIGURE 5.3 A portion of a young stem showing leaf gaps and bud gaps in the cylinder of vascular tissue. (After Esau, K. 1965. *Plant Anatomy,* 2d ed. John Wiley & Sons, Inc., New York. Copyright 1965 John Wiley & Sons, Inc. Redrawn by permission of John Wiley & Sons, Inc.)

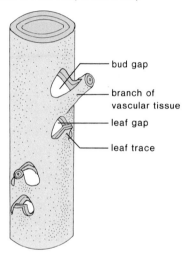

- bud gap

- branch of vascular tissue

- leaf gap

- leaf trace

layers, die shortly after they are formed. The cork cambium may also produce parenchymalike **phelloderm** cells to the inside. Cork tissue makes up the outer bark of woody plants; it functions in reducing water loss and in protecting the stem against mechanical injury. The tissue cuts off water and food supplies to the epidermis, which soon dies and is sloughed off. In fact, if the cork were to be formed as a solid cylinder covering the entire stem, vital gas exchange with the interior of the stem would not be possible. In young stems, such gas exchange takes place through the *stomata* located in the epidermis (see figures 7.6 and 9.7). As woody stems age, **lenticels** (see figure 4.12) develop beneath the stomata. As cork is produced, the parenchyma cells of the lenticels remain, so that exchange of gases (e.g., oxygen, carbon dioxide) can continue through spaces between the cells. As indicated in chapter 4, lenticels occur in the fissures of the bark of older trees and often appear as small bumps on younger bark. In birch and cherry trees, the lenticels form conspicuous horizontal lines.

TISSUE PATTERNS OF STEMS

Steles

Primary xylem, primary phloem, and the pith, if present, make up a central cylinder called the **stele** in at least younger stems and roots. The simplest form of stele, called a *protostele,* consists of a solid

FIGURE 5.4 An illustration of how a cell of the vascular cambium produces new secondary xylem cells to the inside and new secondary phloem cells to the outside. Note, in cross section, that the cambium gradually becomes shifted away from the center as new cells are produced.

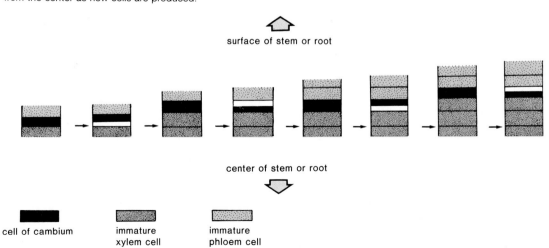

surface of stem or root

center of stem or root

cell of cambium   immature xylem cell   immature phloem cell

FIGURE 5.5 A cross section of an alfalfa (*Medicago*) stem, showing the arrangement of tissues typical of herbaceous dicot stems. (Photomicrograph by G. S. Ellmore)

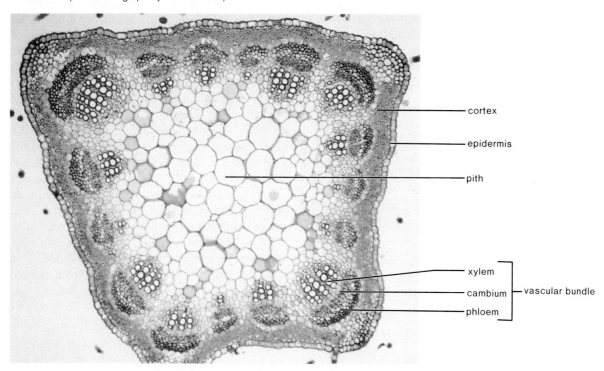

cortex

epidermis

pith

xylem

cambium — vascular bundle

phloem

core of conducting tissues in which the phloem usually surrounds the xylem. Protosteles were common in primitive seed plants that are now extinct, and are also found in relatives of ferns, such as whisk ferns and club mosses (see chapter 19).

Siphonosteles, which are tubular with pith in the center, occur in most ferns.

Most present-day coniferous and flowering plants have *eusteles* in which the primary xylem and primary phloem are in *vascular bundles,* as discussed below.

Flowering plants develop from seeds that have either one or two "seed leaves" called **cotyledons** attached to the embryo stem (see chapters 11 and 21), whereas the seeds of cone-bearing trees such as pines have several (usually eight) cotyledons. The cotyledons may function in storing food needed by the young seedling until its first true leaves can produce food themselves. Flowering plants that develop from seeds having two cotyledons are called **dicotyledons** (often abbreviated to *dicot*), whereas those developing from seeds with a single cotyledon are called

monocotyledons (abbreviated to *monocot*). Dicots and monocots differ from one another in several other respects; differences in stem structure are noted in the following sections, and a summary of these and other differences is given in table 8.1

Herbaceous Dicotyledonous Stems

In general, plants that complete their life cycles within one year (**annuals**) have green herbaceous (nonwoody) stems. Their tissues are largely primary, although cambia (plural of cambium) may develop some secondary tissues. Herbaceous dicot stems (figure 5.5) have discrete patches of xylem and phloem called **vascular bundles,** which occur in a ring that separates the cortex from the pith, although in a few plants (e.g., foxgloves), the xylem and the phloem are produced as continuous rings. As previously noted, the procambium produces only primary xylem and phloem, but later a vascular cambium arises between these two primary tissues and adds secondary xylem and phloem to the vascular bundles. In some plants the cambium extends

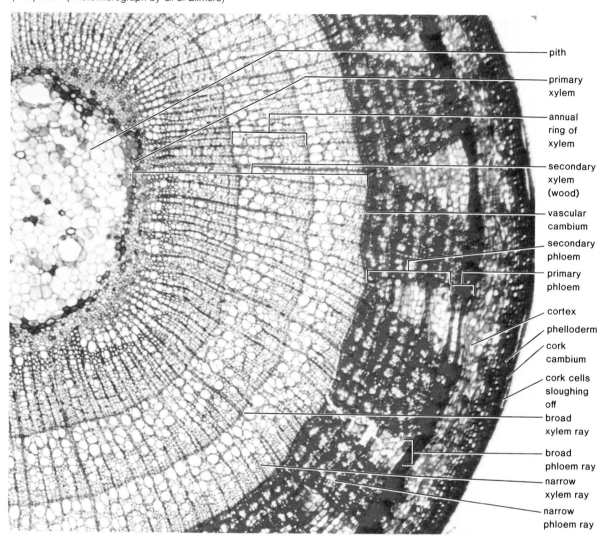

pith

primary xylem

annual ring of xylem

secondary xylem (wood)

vascular cambium

secondary phloem

primary phloem

cortex

phelloderm

cork cambium

cork cells sloughing off

broad xylem ray

broad phloem ray

narrow xylem ray

narrow phloem ray

between the vascular bundles, appearing as a narrow ring producing the conducting tissues within the bundles and the parenchyma cells between them. In other plants the cambium is confined to the bundles, each of which has its own small band of this meristematic tissue between the xylem and phloem.

Woody Dicotyledonous Stems

The arrangement of primary tissues in woody dicot stems (and also in those of cone-bearing trees) is very similar to that found in herbaceous dicot stems

during the early stages of growth. As soon as the vascular cambium and the cork cambium start functioning, however, obvious differences begin to appear, the most conspicuous of which involve the secondary xylem, or *wood,* as it is best known (figure 5.6). Some tropical trees (e.g., ebony), in which both the vascular cambium and the cork cambium are active all year, produce an ungrained, uniform wood. The wood of most trees, however, exhibits seasonal growth. In trees of temperate climates, virtually all growth takes place during the spring and summer and then ceases for the year. When the vascular cambium of a typical broadleaf tree first becomes active in the spring, it usually produces relatively large vessel elements of secondary xylem; such xylem is referred to as *spring wood*. As the season progresses, the vascular cambium may produce vessel elements whose diameters become progressively smaller in each succeeding series of cells produced, or there may be fewer vessel elements in proportion to tracheids produced until tracheids (and sometimes fibers) predominate. This xylem, which is produced after the spring wood, and which has smaller or fewer vessel elements and larger numbers of tracheids, is referred to as *summer wood*. Over a period of years, the result of this type of switch between the early spring and the summer growth is a series of alternating concentric rings of light and dark cells. One year's growth of xylem is called an **annual ring.** Note that an annual ring normally may contain many layers of xylem cells and it is all the layers produced in one growing season that constitute an annual ring—not just dark layers.

The vascular cambium produces more secondary xylem than it does phloem. In addition, xylem cells have stronger, more rigid walls than those of phloem cells and thus are less subject to collapse under pressure. As a result, the bulk of a tree trunk consists of annual rings of wood. The annual rings not only indicate the age of the tree (since normally only one is produced each year), but they can also tell something of the climate and other conditions occurring during the tree's lifetime (figure 5.7). For example, if the rainfall during a particular year is higher than normal, the annual ring for that year will be wider than usual. Sometimes caterpillars or locusts will strip the leaves of a tree shortly after they have appeared. This usually results in two annual rings being very close together, since very little growth can occur under such conditions. If a fire not resulting in the death of the tree occurs, it may be possible to determine the year of its occurrence, since the burn scar may appear next to a given ring. The most recent season's growth is directly adjacent to the vascular cambium, and one need only count the rings back from the cambium to determine the actual year of the fire. It is not necessary to cut down a tree to determine its age. Botanists and foresters

FIGURE 5.7 Climatic history illustrated by a cross section of a 62-year-old tree. (Courtesy St. Regis Paper Company)

This tree is 62 years old. It's been through fire and drought, plague and plenty. And all of this is recorded in its rings.

Each spring and summer a tree adds new layers of wood to its trunk. The wood formed in spring grows fast, and is lighter because it consists of large cells. In summer, growth is slower; the wood has smaller cells and is darker. So when the tree is cut, the layers appear as alternating rings of light and dark wood.

Count the dark rings, and you know the tree's age. Study the rings, and you can learn much more. Many things affect the way the tree grows, and thus alter the shape, thickness, color and evenness of the rings.

For St. Regis these rings have a special significance. They record the steady accumulation of those fibers we use to create noteworthy printing papers, kraft paper and boards, fine papers, packaging products, building materials, and products for consumers.

Essentially, then, the life of the forest is St. Regis' life. That is why we—together with the other members of the forest products industry—are vitally concerned with maintaining the beauty and utility of America's forests for the generations to come.

1904
The tree—a loblolly pine—is born.

1909
The tree grows rapidly, with no disturbance. There is abundant rainfall and sunshine in spring and summer. The rings are relatively broad, and are evenly spaced.

Two

1914
When the tree was 6 years old, something pushed against it, making it lean. The rings are now wider on the lower side, as the tree builds "reaction wood" to help support it.

1924
The tree is growing straight again. But its neighbors are growing too, and their crowns and root systems take much of the water and sunshine the tree needs.

1927
The surrounding trees are harvested. The larger trees are removed and there is once again ample nourishment and sunlight. The tree can now grow rapidly again.

1930
A fire sweeps through the forest. Fortunately, the tree is only scarred, and year by year more and more of the scar is covered over by newly formed wood.

1942
These narrow rings may have been caused by a prolonged dry spell. One or two dry summers would not have dried the ground enough to slow the tree's growth this much.

1957
Another series of narrow rings may have been caused by an insect like the larva of the sawfly. It eats the leaves and leafbuds of many kinds of coniferous trees.

FIGURE 5.8   A three-dimensional, magnified view of a block of hardwood.

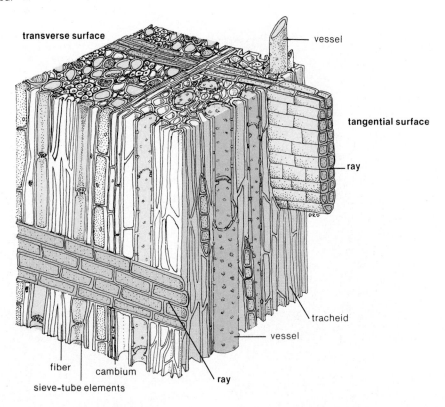

some of the phloem rays, when observed in cross section, flare out from a width of 2 or 3 cells near the cambium to many cells wide in the part next to the cortex (see figure 5.6). In longitudinal section, rays may be from 2 or 3 cells to 50 or more cells deep, but the majority of rays in both xylem and phloem are 1 or 2 cells wide. Rays consist of parenchyma cells that may remain alive for ten years or more. Their function is the lateral conduction of nutrients and water, with some cells also functioning in food storage. Ray cells can be observed in cross section if a woody stem is cut or split lengthwise along a ray (figure 5.8). Another view of rays (in tangential section) is obtained when the stem is cut at a tangent (i.e., cut lengthwise and off center).

As a tree ages, the protoplasts of some of the parenchyma cells that surround the vessels and tracheids grow through the pits in the walls of these conducting cells and balloon out into the cavities. The growth continues until much of the cavity of the vessel or tracheid has been filled. Such protrusions, called **tyloses** (singular: **tylosis**), prevent further

employ an *increment borer* for this purpose. This device removes a plug of wood from the tree at right angles to the trunk. The annual rings can then be counted in the plug; the small hole left in the tree can be treated with a disinfectant to prevent disease and covered up without harm to the tree.

A count of annual rings has produced some red faces on at least one occasion. The Hooker Oak, located in the community of Chico, California, and named in honor of a famous British botanist who once examined it, was a huge, much-visited tree until its demise in 1977. Beneath the tree was a plaque indicating the tree to be over 1,000 years old. A count of rings after its death, however, revealed it to be less than 300 years old!

When a tree trunk is examined in transverse, or cross, section, lighter streaks or lines called **rays** can be seen radiating out from the center across the annual rings (see figure 5.7). That part of a ray within the xylem is called a *xylem ray;* its extension through the phloem is called a *phloem ray.* In basswood trees

FIGURE 5.9 An Australian baobab tree estimated to be
over 2,000 years old. Its hollow trunk was once used for a
town jail. (Courtesy The Age, Melbourne)

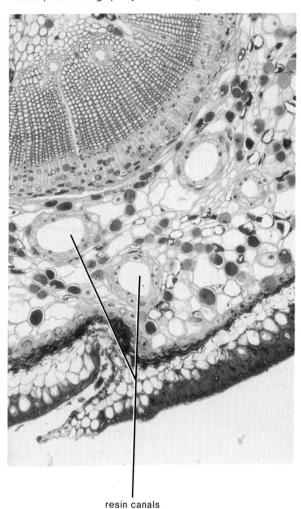

resin canals

conduction of water and dissolved substances. When
this occurs, resins, gums, and tannins begin to accumulate, along with pigments that darken the color
of the wood. This older, darker wood at the center
is called **heartwood,** while the lighter, still functioning xylem closest to the cambium is called **sapwood.** Except for giving strength and support, the
heartwood is not of much use to the tree since it can
no longer conduct materials. Thus, a tree may live
and function perfectly well after the heartwood has
rotted away and left the interior hollow (figure 5.9).
It is even possible to remove part of the sapwood and
other tissues and apparently not affect the tree very
much, as has been done with giant trees such as the
coastal redwoods of California, where holes big
enough to drive a car through have been cut out
without killing the trees.

Sapwood forms at approximately the same rate
as heartwood develops, so there is always sufficient
"plumbing" for the vital conducting functions. The
relative widths of the two types of wood, however,
vary considerably from species to species. For example, in the golden chain tree (a native of Europe

and a member of the legume family), the sapwood
is usually only one or two rings wide, whereas in several North American trees (e.g., maple, ash, and
beech), the sapwood may be many rings wide.

Cone-bearing trees (see chapter 20) such as pine
trees have xylem that consists primarily of tracheids; no fibers or vessel elements are produced.
Since it has no fibers, the wood tends to be softer
than that of trees with fibers. In many cone-bearing
trees, **resin canals** are scattered not only through the
xylem but throughout other tissues as well. These
canals are tubelike and may or may not be branched;
they are lined with specialized cells that secrete *resin*
(see chapter 20) into their cavities (figure 5.10). Al-

though resin canals are commonly associated with cone-bearing trees, they are not confined to them. Tropical flowering plants such as olibanum and myrrh trees, for example, have resin ducts in the bark that produce the soft resins frankincense and myrrh of biblical note.

While the vascular cambium is producing secondary xylem to the inside, it is also producing secondary phloem to the outside. The term **bark** is usually applied to all the tissues outside the cambium, including the phloem. Some scientists distinguish between the *inner bark,* consisting of primary and secondary phloem, and the *outer bark* (periderm), consisting of cork tissue and cork cambium. Despite the presence of fibers, the thin-walled conducting cells of the phloem are not usually able to withstand for many seasons the pressure of thousands of new cells added to their interior, and the older layers become crushed and functionless. The parenchyma cells of the cortex to the outside of the phloem also function only briefly because they too become crushed or sloughed off. Before they disappear, however, the cork cambium begins its production of cork, and since new cambia arise inwardly in the phloem, the older bark may consist of alternating layers of crushed phloem and cork. The younger layers of phloem, nearest to the cambium, transport sugars and other substances in solution from their points of origin in the leaves to various parts of the plant, where they are either stored or used in the process of *respiration* (see chapter 10). This sugar content of the phloem was recognized by North American Indian tribes. Some stripped the young phloem and cambium from Douglas fir trees and used the dried strips as food for winter and in emergencies.

Specialized cells or ducts called **laticifers** are found in about 20 families of herbaceous and woody flowering plants. These cells are most common in the phloem but occur throughout all parts of the plants. The laticifers, which resemble vessels, form extensive branched networks of latex-secreting cells originating from rows of meristematic cells. Unlike vessels, however, the cells remain living and may have many nuclei. *Latex* is a thick fluid that is white, yellow, orange, or red in color and consists of gums, proteins, sugars, oils, salts, alkaloidal drugs, enzymes, and other substances. Its function in the plant is not clear, although some believe it aids in closing wounds. Some forms of latex have considerable commercial value (see the discussion under *Spurge Family* in chapter 22). Of these, rubber is the most important. Amazon Indians utilized rubber for making balls and containers hundreds of years before Pará rubber trees were cultivated for their latex. The chicle tree produces a latex used in the making of chewing gum. Several poppies, notably the opium poppy, produce a latex containing important medicinal drugs such as morphine and other drug complexes such as heroin. Other well-known latex producers include milkweeds, dogbanes, and dandelions.

Monocotyledonous Stems

Most monocots (e.g., grasses, lilies) are herbaceous plants that do not attain great size. The stems have neither a vascular cambium nor a cork cambium, and thus produce no secondary vascular tissues or cork. As in herbaceous dicots, the surfaces of the stems are covered by an epidermis, but the xylem and phloem tissues produced by the procambium appear in cross section as discrete vascular bundles, scattered throughout the stem instead of being arranged in a ring (figure 5.11). Each bundle, regardless of its specific location, is oriented so that its xylem is closest to the center of the stem and its phloem is closest to the surface. In a typical monocot such as corn, a bundle's xylem usually contains two large vessels with several small vessels between them (figure 5.11). The first-formed xylem cells usually stretch and collapse under the stresses of early growth and leave an irregularly shaped air space toward the base of the bundle; the remnants of a vessel are often present in this air space. The phloem consists entirely of sieve tubes and companion cells, and the entire bundle is surrounded by a sheath of sclerenchyma cells. The parenchyma tissue between the vascular bundles is not separated into cortex and pith in monocots; it is usually referred to as *ground tissue,* or *fundamental tissue,* although its function and appearance are the same as those of the parenchyma cells in cortex and pith.

The bundles of a corn stem are more numerous just beneath the surface than they are toward the center. Also, a band of sclerenchyma cells, usually 2 or 3 cells thick, occurs immediately beneath the epidermis, and parenchyma cells in the area develop thicker walls as the stem matures. The concentration of bundles, combined with the additional band

FIGURE 5.11   *A.* A portion of a cross section of a monocot
(*Zea mays*) stem. *B.* A single vascular bundle (enlarged).
(Photomicrographs by G. S. Ellmore)

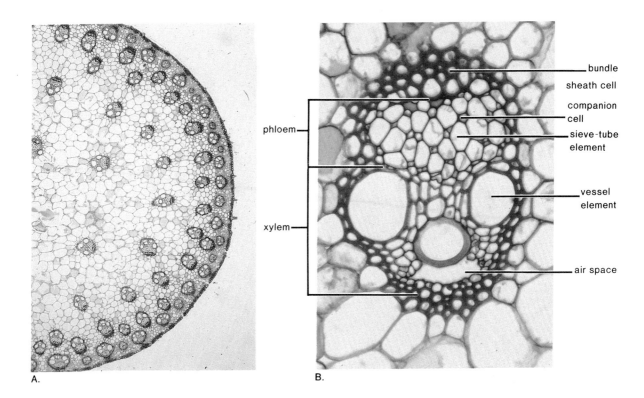

A.   B.

of sclerenchyma cells beneath the epidermis and the
thicker-walled parenchyma cells, all contribute to
giving the stem the capacity to withstand stresses
resulting from summer storms and the weight of the
leaves and the ears of corn as they mature.

In wheat, rice, barley, oats, rye, and other
grasses, there is an intercalary meristem (see chapter
4) at the base of each internode; like the apical meristem, it contributes to elongation in growth. The
stems of such plants elongate rapidly during the
growing season, but because there is no vascular
cambium producing tissues that would add to the
girth of the stems, growth is columnar, with little
difference in diameter between the top and the
bottom.

Palm trees differ from most monocots in that
they attain considerable size, but they do so primarily as a result of their parenchyma cells con-
tinuing to divide and enlarge without a true cambium
developing. Several popular houseplants (e.g., ti

plants, *Dracaena, Sansevieria*) are monocots in
which a secondary meristem develops as a cylinder
that extends throughout the stem. Unlike the vascular cambium of dicots and conifers, this secondary meristem produces only parenchyma cells to
the outside and secondary vascular bundles to the
inside.

Several commercially important cordage fibers
(e.g., broom corn, Mauritius and Manila hemps,
sisal) come from the stems and leaves of monocots,
but the individual cells are not separated from one
another by *retting* (a process that utilizes the rotting
power of microorganisms thriving under moist conditions to break down thin-walled cells) as they are
when fibers from dicots are obtained. Instead, during
commercial preparation, entire vascular bundles are
scraped free of the surrounding parenchyma cells by
hand; the individual bundles then serve as unit "fibers." If such fibers are treated with chemicals or

FIGURE 5.12   Types of specialized stems.

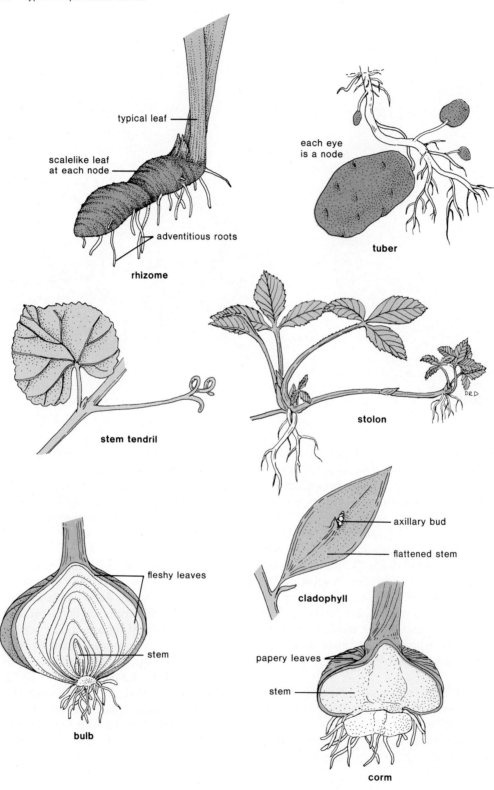

typical leaf

scalelike leaf
at each node

adventitious roots

rhizome

each eye
is a node

tuber

stem tendril

stolon

axillary bud

flattened stem

cladophyll

fleshy leaves

stem

bulb

papery leaves

stem

corm

bleached, the cementing middle lamella between the cells breaks down. Monocot fibers are not as strong or as durable as most dicot fibers.

SPECIALIZED STEMS

While an erect shoot system is characteristic of most higher plants, many species have stems that perform specialized functions; these stems are modified accordingly (figure 5.12). Although the overall appearance of specialized stems may differ markedly from that of the stems discussed so far, they all have *nodes, internodes,* and *axillary buds;* these features distinguish them from roots and leaves, which do not have them. The leaves at the nodes of these specialized stems are often small and scalelike. They are seldom green, but full-size functioning leaves are also produced. Descriptions of some of the specialized stems follow.

Rhizomes

Rhizomes (figure 5.12) are horizontal stems that grow below ground, often near the surface of the soil. Superficially they resemble roots, but a close examination will reveal scalelike leaves and axillary buds at each node, at least during some stage of development, with short to long internodes in between. **Adventitious** roots are produced all along the rhizome, mainly on the lower surface. The word *adventitious* refers to structures arising at unusual places, such as roots growing from stems, or leaves or buds appearing at places other than leaf axils and tips of stems. A rhizome may be a relatively thick, fleshy food-storage organ, as in irises, or it may be quite slender, as in many perennial grasses or some ferns.

Stolons

Stolons, more or less horizontal stems that are also called *runners,* differ from rhizomes in growing above ground, generally along the surface; they also have long internodes (figure 5.12). In strawberries, stolons are usually produced after the first flowering of the season has occurred. Several may radiate out from the parent plant and grow along the surface of the ground, attaining lengths of up to 1 meter (3 feet) or more within a few weeks. Adventitious buds appear at alternate nodes along the stolons and develop into new strawberry plants, which can be separated and grown independently. In some houseplants, such as the saxifrages, stolons may grow out

and hang over the edge of the pot, producing new plants at intervals. In Irish potatoes, stolons or rhizomes grow underground and produce potatoes at their tips.

Tubers

As food accumulates at the tips of underground stolons, such as those produced by Irish or white potato plants, several internodes swell, becoming **tubers** (figure 5.12). After the tuber is mature, the stolon dies, isolating it. The "eyes" of the potato are actually nodes formed in a spiral around the modified stem. Each eye consists of an axillary bud in the axil of a scalelike leaf, although the latter is visible only in very young tubers, the small ridge being a leaf scar.

Bulbs

Bulbs (figure 5.12) are actually large buds with a small stem at the lower end surrounded by numerous fleshy leaves. Adventitious roots grow from the bottom of the stem, but the fleshy leaves comprise the bulk of the bulb tissue, which functions in food storage. In onions, the fleshy leaves usually are surrounded by the scalelike leaf bases of long, green, aboveground leaves. Other plants producing bulbs include lilies, hyacinths, and tulips.

Corms

Corms, which superficially resemble bulbs, differ from them in being composed almost entirely of stem tissue except for the few papery scalelike leaves sparsely covering the outside (figure 5.12). Adventitious roots are produced at the base, and corms, like bulbs, function in food storage. Well-known plants producing corms include crocuses and gladioli.

Cladophylls

In the butcher's-broom plant, stems are flattened and very leaflike in appearance. Such stems are called **cladophylls** (figure 5.12). In the center of each butcher's-broom cladophyll is a node bearing very small scalelike leaves with axillary buds. The feathery appearance of asparagus is due to numerous small cladophylls, which also occur in greenbriers and certain orchids.

Other Specialized Stems

In many cacti and in some of the spurges, the stems are stout and fleshy. Such stems are modified for water and food storage. Other stems may be modified in the form of *spines,* as in the honey locust whose branched spines may exceed 3 decimeters (1 foot) in length but not all spines are modified stems. For example, the black locust has a pair of spines at the base of each petiole of most leaves (these are parts of the leaf called *stipules,* mentioned in the discussion of twigs and discussed further in chapter 7). The prickles of raspberries, and the thorns of roses, both of which originate from the epidermis, are technically not true spines. Tiger lilies produce small aerial bulblets in the axils of their leaves.

Climbing plants have stems modified in various ways that adapt them for their growth habit. Some stems, called *ramblers,* simply rest on the tops of other plants, but many produce **tendrils.** These are specialized stems in the grape and Boston ivy but are modified leaves or leaf parts in plants like peas and cucumbers. In Boston ivy, the tendrils have adhesive disks. In English ivy, the stems climb with the aid of adventitious roots that arise along the sides of the stem and become embedded in the bark or other support material over which the plant is growing.

WOOD AND ITS USES

The use of wood by humans dates back into antiquity, and present uses are so numerous that it would be impossible to list in a work of this type more than the most important ones. Before looking at the economic importance of wood, a brief discussion of its properties is in order.

In a living tree, up to 50% of the weight of the wood comes from the water content. Before the wood can be used, the moisture content is reduced to 10% or less through *seasoning,* either by air-drying the wood in ventilated piles or stacks or by drying it in special ovens known as kilns. The seasoning has to be done gradually and under carefully controlled conditions or the timber may warp and split along the rays, making it unfit for most uses. The dry part of wood is composed of 60% to 75% *cellulose* and about 15% to 25% *lignin,* an organic substance that is deposited in the walls of xylem cells and makes the walls tough and hard. Other substances present in smaller amounts include resins, gums, oils, dyes, tannins, and starch. The proportions and amounts of these and other substances determine how various woods will be used (figure 5.13).

FIGURE 5.13   Uses of some common North American woods. (Courtesy St. Regis Paper Company)

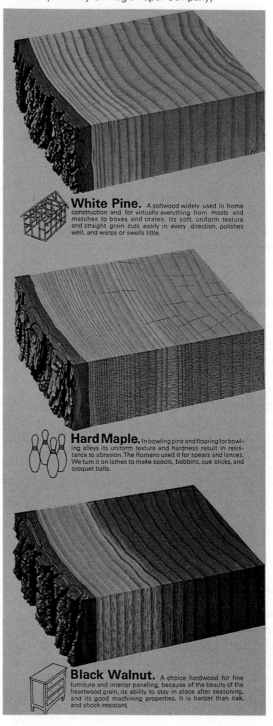

White Pine. A softwood widely used in home construction and for virtually everything from masts and matches to boxes and crates. Its soft, uniform texture and straight grain cuts easily in every direction, polishes well, and warps or swells little.

Hard Maple. In bowling pins and flooring for bowling alleys its uniform texture and hardness result in resistance to abrasion. The Romans used it for spears and lances. We turn it on lathes to make spools, bobbins, cue sticks, and croquet balls.

Black Walnut. A choice hardwood for fine furniture and interior paneling, because of the beauty of the heartwood grain, its ability to stay in place after seasoning, and its good machining properties. It is harder than oak, and shock-resistant.

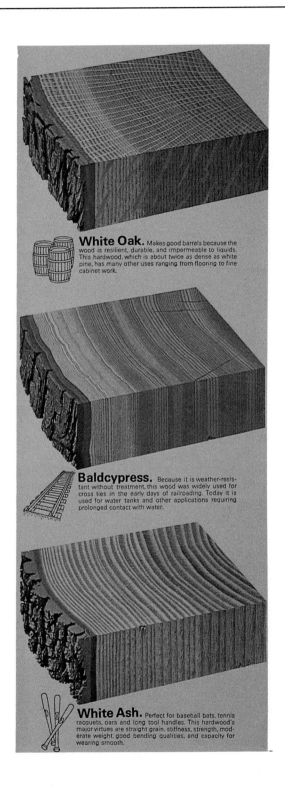

White Oak. Makes good barrels because the wood is resilient, durable, and impermeable to liquids. This hardwood, which is about twice as dense as white pine, has many other uses ranging from flooring to fine cabinet work.

Baldcypress. Because it is weather-resistant without treatment, this wood was widely used for cross ties in the early days of railroading. Today it is used for water tanks and other applications requiring prolonged contact with water.

White Ash. Perfect for baseball bats, tennis racquets, oars and long tool handles. This hardwood's major virtues are straight grain, stiffness, strength, moderate weight, good bending qualities, and capacity for wearing smooth.

Red Spruce. A favorite for violin sounding boards because of its high resonant qualities. A softwood, it is easy to work, and is light in relation to its strength and stiffness. These qualities also make it eminently suitable for ladder rails, canoe paddles, and oars.

Hemlock. This relatively soft, light, straight-grained, resin-free wood, with its uniformly long fibers, is becoming one of the most important species for paper pulp. It is also used for structural lumber and plywood, and for boxes, barrels, and concrete forms.

Hickory. A hardwood unsurpassed for the handles of impact tools like axes and hammers, and for skis, because of its hardness, strength, toughness, and resiliency. In horse-and-buggy days it was widely used for wheel spokes and rims, singletrees, and buggy shafts.

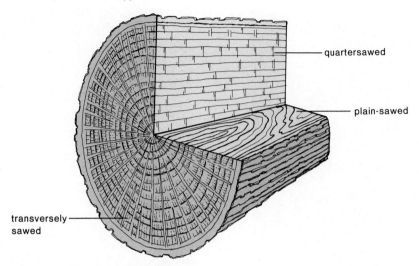

FIGURE 5.14   How the surfaces of plain-sawed, quartersawed, and transversely sawed wood appear.

quartersawed

plain-sawed

transversely sawed

Density

Among the most important physical properties of wood is its *density*. Technically, the density is the weight per unit volume. The weight is compared with that of an equal volume of water and is stated as a fraction of 1.0. Because of the considerable air space within the cells, most woods have a *specific gravity,* as the comparative density is called, of less than 1.0. The range of specific gravities of known woods varies from 0.04 to 1.40, the lightest commercially used wood being *balsa* with a specific gravity of about 0.12. Woods with specific gravities of less than 0.50 are considered light; those with specific gravities of above 0.70 are considered heavy. Among the heaviest woods are the South American *ironwood* and *lignum vitae,* with specific gravities of over 1.25. Lignum vitae, obtained from West Indian trees, is extremely hard wood and is used instead of metal in the manufacture of main bearings for drive shafts of submarines because it is self-lubricating and less noisy.

Durability

A wood's ability to withstand decay organisms and insects is referred to as its *durability*. Moisture is needed for the enzymatic breakdown of cellulose and other wood substances by decay organisms, but the seasoning process usually reduces the moisture to a level below that necessary for the fungi and other decay organisms to survive. Other natural constituents of wood that repel decay organisms include tannins and oils. Wood with a tannin content of 15% or more may survive on a forest floor for many years after the tree has been felled by diseases of the phloem or other causes. Among the most durable of American woods are cedar, catalpa, black locust, red mulberry, and Osage orange. The least durable woods include cottonwood, willow, fir, and basswood.

Types of Sawing

Logs are usually cut longitudinally in one of two ways: along the radius or at right angles to the rays (figure 5.14). Radially cut or **quartersawed** boards show the annual rings in side view; they appear as longitudinal streaks and are the most conspicuous feature of the wood. Only a few perfect quartersawed boards can be obtained from a log, making them quite expensive. Boards cut at right angles to the rays (tangentially cut boards) are more common. In these, the annual rings appear as irregular bands of light and dark alternating streaks or patches, with the ends of the rays visible as narrower and less conspicuous vertical streaks. Lumber cut tangentially is referred to as being *plain-sawed,* or *slab cut* (figure 5.14). Slabs are the boards with rounded sides at the outside of the log; they are usually made into chips for pulping.

Knots

Knots are the bases of lost branches that have become covered, over a period of time, by new annual rings of wood produced by the cambium of the trunk. They are found in greater concentration in the older parts of the log toward the center, because in the forest the lowermost branches of a tree (produced while the trunk was small in girth) often die from lack of sufficient light. When a branch dies and falls off, the cambium at its base also dies, but the cambium of the trunk remains alive and increases the girth of the tree, slowly enveloping the dead tissue of the branch base until it may be completely buried and not visible from the surface. Knots usually weaken the boards in which they occur.

Wood Products

In the United States, about half of the wood produced is used as lumber, primarily for construction; the sawdust and other waste formed in processing the boards is converted to particle board and pulp. A considerable amount of lumber goes into the making of furniture, which may be constructed of solid wood or covered with a *veneer*. A veneer is a very thin sheet of desirable wood that is glued to cheaper lumber; it is carefully cut so as to produce the best possible view of the grain (see figure 5.15 on page 88).

The next most extensive use of wood is for *pulp,* which among other things is converted by various processes to paper, synthetic fibers, plastics, and linoleum. In recent years, it has been added as a filler to commercial ice cream and bread. Some hardwoods are treated chemically or heated under controlled conditions to yield a number of chemicals such as wood alcohol and acetic acid, but other sources of these products are now usually considered more economical than they were in the past. Charcoal, excelsior, cooperage (kegs, casks, and barrels), railroad ties, boxes and crates, musical instruments, bowling pins, tool handles, pilings, cellophane, photographic film, and Christmas trees are but a few of the additional wood products worth billions of dollars annually on the world market (see figure 5.13).

In developing countries, approximately half of the timber cut is used for fuel; in the United States, a little less than 10% is currently used for that purpose. In colonial times, wood was the almost exclusive source of heating energy. In the 1980s, Brazil's major cities were still using scrub timber from the surrounding forests to energize their utilities, but rapidly depleting supplies have pointed to the need for alternate sources of energy. Many types of coal are wood that has been compressed for millions of years until nearly pure carbon remains. The formation of coal and other fossils is discussed in chapter 19. Although the world's supply of coal is still plentiful, the rate at which this fossil fuel is being consumed makes it obvious that resources will eventually be exhausted unless our energy demands find renewable or less destructive alternatives.

Some of the vast array of secondary products from stems, including dyes, medicines, spices, and foods are discussed in later chapters and in the appendices.

SUMMARY

To understand the how and why of grafting, the propagation of plants from stems, and which stem parts are useful, an examination of stem structure and function is needed.

The shoot system of plants, with branches and leaves, is usually erect, but some stems may be horizontal or have modifications that permit climbing or storage of food or water. Woody twigs have leaves arranged alternately, oppositely, or in a whorl. Nodes are stem regions where leaves are attached; internodes occur between nodes. Most leaves have petioles and blades. Axillary buds occur in leaf axils. Most buds are protected by bud scales. Terminal buds occur at twig tips. When terminal bud scales fall off they leave groups of bud scale scars that can be used in determining the age of the twig. Stipules are paired appendages present at the base of some leaves; when they fall off, they leave small scars on the twig. When whole leaves fall they cause leaf scars on the twig, with tiny bundle scars within the leaf-scar surfaces.

Each stem has an apical meristem at its tip that produces tissues resulting in increase in length. Leaf primordia develop into mature leaves when growth begins. Three primary meristems develop from an apical meristem: the protoderm gives rise to the epidermis, which becomes coated with a cuticle; the procambium produces primary xylem and primary phloem, and the ground meristem produces pith and cortex. As each leaf and each bud develops from a primordium, a trace of xylem and phloem branches off from the main cylinder, leaving a leaf gap or a

FIGURE 5.15   How a log is used. (Courtesy St. Regis
Paper Company)

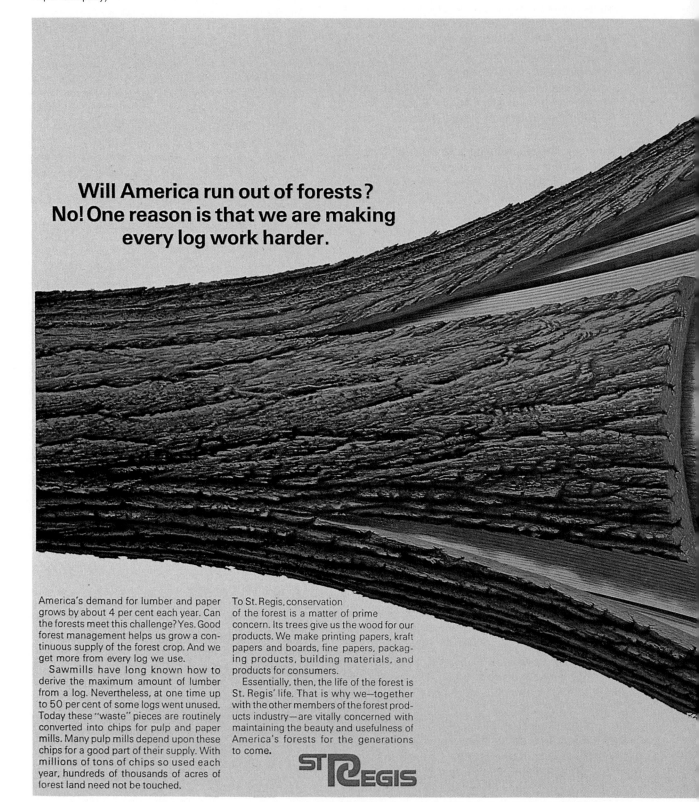

**Will America run out of forests?
No! One reason is that we are making
every log work harder.**

America's demand for lumber and paper grows by about 4 per cent each year. Can the forests meet this challenge? Yes. Good forest management helps us grow a continuous supply of the forest crop. And we get more from every log we use.

Sawmills have long known how to derive the maximum amount of lumber from a log. Nevertheless, at one time up to 50 per cent of some logs went unused. Today these "waste" pieces are routinely converted into chips for pulp and paper mills. Many pulp mills depend upon these chips for a good part of their supply. With millions of tons of chips so used each year, hundreds of thousands of acres of forest land need not be touched.

To St. Regis, conservation of the forest is a matter of prime concern. Its trees give us the wood for our products. We make printing papers, kraft papers and boards, fine papers, packaging products, building materials, and products for consumers.

Essentially, then, the life of the forest is St. Regis' life. That is why we—together with the other members of the forest products industry—are vitally concerned with maintaining the beauty and usefulness of America's forests for the generations to come.

ST REGIS

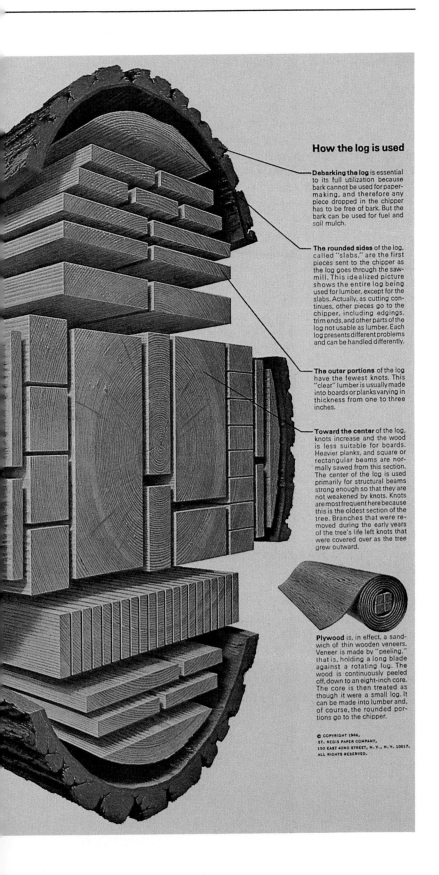

How the log is used

Debarking the log is essential to its full utilization because bark cannot be used for paper-making, and therefore any piece dropped in the chipper has to be free of bark. But the bark can be used for fuel and soil mulch.

The rounded sides of the log, called "slabs," are the first pieces sent to the chipper as the log goes through the sawmill. This idealized picture shows the entire log being used for lumber, except for the slabs. Actually, as cutting continues, other pieces go to the chipper, including edgings, trim ends, and other parts of the log not usable as lumber. Each log presents different problems and can be handled differently.

The outer portions of the log have the fewest knots. This "clear" lumber is usually made into boards or planks varying in thickness from one to three inches.

Toward the center of the log, knots increase and the wood is less suitable for boards. Heavier planks, and square or rectangular beams are normally sawed from this section. The center of the log is used primarily for structural beams strong enough so that they are not weakened by knots. Knots are most frequent here because this is the oldest section of the tree. Branches that were removed during the early years of the tree's life left knots that were covered over as the tree grew outward.

Plywood is, in effect, a sandwich of thin wooden veneers. Veneer is made by "peeling," that is, holding a long blade against a rotating log. The wood is continuously peeled off, down to an eight-inch core. The core is then treated as though it were a small log. It can be made into lumber and, of course, the rounded portions go to the chipper.

bud gap. A vascular cambium, producing secondary tissues, may arise between primary xylem and phloem. Secondary xylem cells include tracheids, vessel elements, and fibers. Secondary phloem cells include sieve-tube elements and companion cells.

In many plants, a cork cambium producing cork and phelloderm cells develops near the surface of the stem. Cork cells, which constitute the outer bark, have suberin in their walls; suberin is impervious to moisture and, therefore, the outer bark aids in protection. Lenticels in the bark permit gas exchange.

Primary vascular tissues and the pith, if present, constitute the stele. Protosteles have a solid core of xylem, usually surrounded by phloem; siphonosteles are tubular, with pith in the center; eusteles have the vascular tissues in discrete bundles.

Dicotyledons (dicots) are plants whose seeds have two seed leaves (cotyledons), whereas monocotyledons (monocots) have seeds with one seed leaf. Herbaceous dicots have vascular bundles arranged in a ring in the stem. Woody dicots have most of their tissues arranged in concentric layers. The most conspicuous tissue is wood (secondary xylem). Spring wood usually has relatively large vessel elements, whereas summer wood has smaller vessels and/or a predominance of tracheids. An annual ring is one year's growth of xylem. A tree's age and other aspects of its history can be determined from annual rings. Rays, which function in lateral conduction, radiate from the center of the trunk. Older wood toward the center (heartwood) ceases to function when its cells become plugged with tyloses. Younger, functioning wood (sapwood) is closer to the surface. A tree's functions are not particularly affected by the rotting of its heartwood. The wood of cone-bearing trees consists primarily of tracheids, and resin canals are often present.

In woody plants older tissues composed of thin-walled cells become crushed and functionless, and some are sloughed off. Laticifers are latex-secreting cells or ducts found in various flowering plants. The latex of some plants has considerable commercial value.

Monocot stems have neither a vascular cambium nor a cork cambium; their vascular bundles are scattered. The parenchyma tissue is not divided into pith and cortex; it is called ground, or fundamental tissue, instead. Each vascular bundle is surrounded by a sheath of sclerenchyma cells. The stems of monocots such as corn can withstand stresses because of the presence of numerous bundles and a band of sclerenchyma cells and thicker-walled parenchyma cells just beneath the surface.

Palm trees are monocots that become large because their parenchyma cells continue to divide. Other monocots develop a secondary meristem that produces parenchyma cells and secondary vascular bundles. Grasses have intercalary meristems at the base of each internode that contribute to rapid increases in length. Several commercially important cordage fibers are obtained from monocots.

Specialized stems include rhizomes, stolons, tubers, bulbs, corms, cladophylls, and tendrils. Such stems may have adventitious roots.

The dry part of wood consists primarily of cellulose and lignin. Resins, gums, oils, dyes, tannins, and starch are also present. Properties of wood that play a role in its use include density, specific gravity, and durability. Logs are usually cut longitudinally along the radius (quartersawed) or at right angles to the rays (tangentially, plain-sawed, or slab cut). Knots are bases of branches that have become covered over by new wood; they usually weaken the boards in which they occur.

About half the timber produced in the United States is used as lumber. Sawdust and waste is converted to particle board and pulp for paper, synthetics, and linoleum. Other timber is used for cooperage, charcoal, railroad ties, boxes, tool handles, etc. Developing countries use a greater proportion of their timber for fuel.

REVIEW QUESTIONS

1. What is the function of bud scales?
2. How can you tell the age of a twig?
3. Distinguish among procambium, vascular cambium, and cork cambium.
4. How can you tell, when you look at a cross section of a young stem, whether it is a dicot or a monocot?
5. What are laticifers?
6. An Irish or white potato is a stem, but a sweet potato is a root. How can you tell?
7. Distinguish among corms, bulbs, and tubers.
8. If you were examining the top of a wooden desk, how could you tell if the wood had been radially or tangentially cut (quartersawed or plain-sawed)?
9. What differences are there between heartwood and sapwood?
10. What is meant by the specific gravity of wood?

DISCUSSION QUESTIONS

1. If the cambium of a tropical tree were active all year long, how would its wood differ from that of a typical temperate climate tree?
2. It was mentioned that a nail driven into the side of a tree will remain at exactly the same distance from the ground for the life of the tree. Why?
3. Do climbing plants have any advantages over erect plants? Any disadvantages?
4. If two leaves are removed from a plant and one is coated with petroleum jelly while the other is not, the uncoated leaf will shrivel considerably sooner than the coated one. Would it be helpful to coat the stems of young trees with petroleum jelly? Explain.
5. Suggest some reasons for heartwood being preferred to sapwood for making furniture.

ADDITIONAL READING

Barefoot, A. C., and F. W. Hankins. 1982. *Identification of modern tertiary woods.* Fair Lawn, NY: Oxford University Press.

Core, H. A., W. A. Cote, and A. C. Day. 1979. *Wood structure and identification,* 2d ed. Syracuse, NY: Syracuse University Press.

Cutter, E. G. 1978. *Plant anatomy: experiment and interpretation. Part I: Cells and tissues,* 2d ed. Reading, MA: Addison-Wesley Publishing Co.

Cutter, E. G. 1971. *Plant anatomy: experiment and interpretation. Part II: Organs.* Reading, MA: Addison-Wesley Publishing Co.

Esau, K. 1977. *Anatomy of seed plants,* 2d ed. New York: John Wiley & Sons.

Fahn, A. 1982. *Plant anatomy,* 3d ed. Elmsford, NY: Pergamon Press.

Fritts, H. C. 1976. *Tree rings and climate.* New York: Academic Press.

Metcalfe, C. R., and L. Chalk, eds. 1950. (vol. 1, 2d ed., 1980). *Anatomy of the dicotyledons,* 2 vols. Fair Lawn, NY: Oxford University Press.

Meylan, B. A., and B. G. Butterfield. 1972. *Three-dimensional structure of wood.* Syracuse, NY: Syracuse University Press.

Ray, P. M. 1972. *The living plant,* 2d ed. New York: Holt, Rinehart and Winston.

Schery, R. W. 1972. *Plants for man,* 2d ed. Englewood Cliffs, NJ: Prentice-Hall.

Steeves, T. A., and I. M. Sussex. 1972. *Patterns in plant development.* Englewood Cliffs, NJ: Prentice-Hall.

Zimmerman, M. H., and C. L. Brown. 1975. *Trees: structure and function.* New York: Springer-Verlag.

Overview

*T*his chapter discusses roots, beginning with the functions and development of roots from a seed. It covers the function and structure of the root cap, region of cell division, region of elongation, and region of maturation (with its tissues). The endodermis and pericycle are also discussed.

Specialized roots (food-storage roots, water-storage roots, propagative roots, pneumatophores, aerial roots, contractile roots, buttress roots, parasitic roots, mycorrhizae, root nodules) are given brief treatment. This is followed by some observations on the economic importance of roots.

6   Roots

Some Learning Goals

1. Know the primary functions and forms of roots.

2. Learn the root regions, including the root cap, region of cell division, region of elongation, and region of maturation (including root hairs and all tissues), and know the function of each.

3. Discuss the specific functions of the endodermis and the pericycle.

4. Understand the differences among the various types of specialized roots.

5. Know at least ten practical uses of roots.

Outline

INTRODUCTION

You have probably seen pictures, at least, of the destruction caused by a tornado as it cut a swath through a village or a city, but have you ever seen what a twister can do to a forest? Large trees may be snapped off above the ground or knocked down, and branches may be stripped bare of leaves. Depending on the type of forest and the composition of the soil, however, you will probably see relatively few trees completely torn up by the roots and blown elsewhere. And in the tropics it is indeed rare to find healthy palm trees uprooted after a hurricane.

Roots anchor trees firmly in the soil, usually by forming an extensive branching network that constitutes about one-third of the total dry weight of the plant. The roots of most plants do not usually extend down into the earth beyond a depth of 3 to 5 meters (10 to 16 feet); those of many herbaceous species are confined to the upper 6 to 9 decimeters (2 to 3 feet). The roots of a few plants such as alfalfa, however, frequently extend more than 6 meters (20 feet) into the earth. When the Suez Canal was being built, workers encountered roots of tamarisk at depths of nearly 30 meters (100 feet), and mesquite roots have been seen 53.4 meters (175 feet)

deep in a pit mine in the southwestern United States. Some plants, such as cacti, form very shallow root systems but still effectively achieve *anchorage* by means of a densely branching mass of roots radiating out in all directions as far as 15 meters (50 feet) from the stem. In addition to anchorage, roots function extensively in the *absorption of water and minerals in solution,* with the bulk of such "feeder" roots being confined to the upper meter (3.3 feet) of soil. The roots of some plants have specialized functions such as food or water storage. These functions are discussed in the section on specialized roots.

Although some aquatic plants (e.g., duckweeds) normally produce roots in water, and others (e.g., many orchids) produce aerial roots, the great majority of vascular plants develop their root systems in soils. Soils vary considerably in composition, texture, and other characteristics. These features of soils are discussed in the second half of chapter 9.

When a seed germinates, a part of the **embryo** within it, the **radicle,** grows out and develops into the first root. This may develop further either into a *taproot,* from which branch roots arise, or it may give way to the formation of numerous *adventitious roots,* becoming a *fibrous* root system with numerous roots of similar diameter (figure 6.1). Many

FIGURE 6.1   Root systems. *A.* A fibrous root system of a grass. *B.* A taproot system of a dandelion.

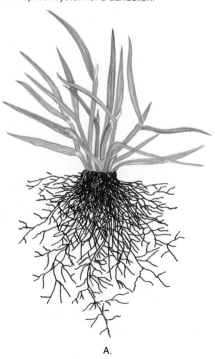

A.

B.

mature plants exhibit a combination of taproot and fibrous root systems. The number of roots produced by a single plant may be prodigious. For example, a single mature ryegrass plant may have as many as 15 million individual roots and branch roots, with a combined length of 644 kilometers (400 miles) and a total surface area larger than a volleyball court, all contained within 57 cubic decimeters (2 cubic feet) of soil. Significant additional surface area is provided by *root hairs* discussed in the Region of Maturation section.

Most *dicot* plants (plants with two "seed leaves") have taproot systems with one to several predominant roots from which secondary roots develop, whereas *monocots* (plants with one "seed leaf") have fibrous roots. Other types of roots, such as adventitious roots (see chapter 5), may develop in both dicots and monocots. In English and other ivies, lateral adventitious roots that aid in climbing appear along the aerial stems, and in certain plants with modified stems (e.g., rhizomes, corms, and bulbs; see figure 5.12), adventitious roots are the only kind produced.

ROOT STRUCTURE

Botanists have traditionally recognized four regions or zones in developing young roots. Three of the regions are not sharply defined at their boundaries. The cells of each region gradually develop the form of those of the next region; the extent of each region varies considerably, depending on the species involved. These regions are called (1) the *root cap,* (2) the *region of cell division,* (3) the *region of elongation,* and (4) the *region of maturation* (figure 6.2).

The Root Cap

The **root cap** is composed of a thimble-shaped mass of parenchyma cells covering the tip of each root. It is quite large and obvious in some plants, while in others it is nearly invisible. One of its functions is to *protect* from damage the delicate tissues behind it as the young root tip pushes through often angular and abrasive soil particles. The root cap has no equivalent in stems. Its outer cells secrete mucilage in the walls; these cells break down and are constantly being sloughed off (and replaced from the

FIGURE 6.2   A longitudinal section through a dicot root tip. *A.* Regions of the root. *B.* Locations of the primary meristems of the root.

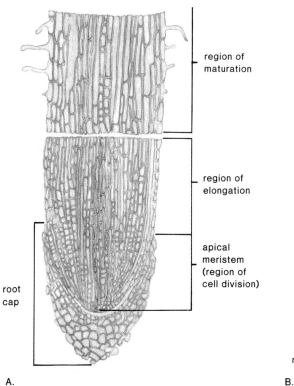

region of maturation

region of elongation

apical meristem (region of cell division)

root cap

A.

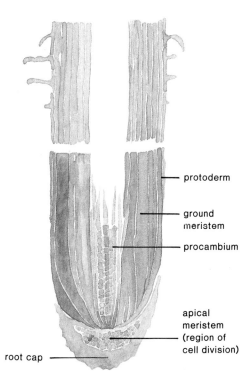

protoderm

ground meristem

procambium

apical meristem (region of cell division)

root cap

B.

FIGURE 6.3   A cross section of a portion of a root of greenbrier (*Smilax*—a monocot). (Photomicrograph by G. S. Ellmore)

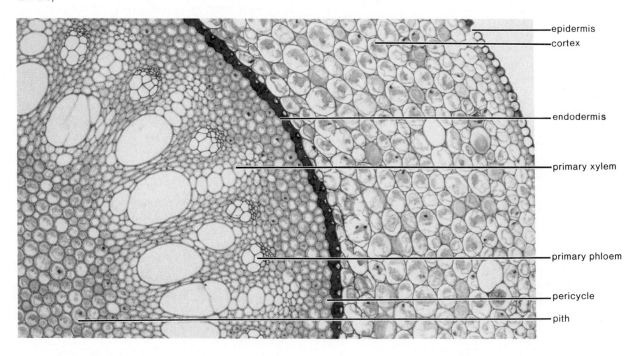

— epidermis
— cortex
— endodermis
— primary xylem
— primary phloem
— pericycle
— pith

inside), forming a slimy lubricant that facilitates movement through the soil. This mucilaginous lubricant also provides a medium favorable to the growth of beneficial bacteria.

The root cap, whose cells have an average life of less than a week, can be slipped off or cut off, and when this is done, a new root cap is produced. Until the root cap has been renewed, however, the root seems to grow randomly instead of downward, thus suggesting that the root cap has another function, namely, the *perception of gravity*. It is known that *amyloplasts* (plastids containing starch grains) collect on the sides of root cap cells facing the direction of gravitational force. When a root that has been growing vertically is artificially tipped horizontally, the amyloplasts tumble or float down to the "bottom" of the cells in which they occur and within 30 minutes to a few hours the root begins growing downward again. The exact nature of this gravitational response is not known, but there is some evidence that calcium ions known to be present in the amyloplasts influence the distribution of growth hormones in the cells.

The Region of Cell Division

The root cap is produced by cells in the *region of cell division*, which is in the center of the root tip and surrounded by the root cap. It is composed of an **apical meristem** similar to that of stem tips. Most of the cell divisions take place at the edges of an inverted cup-shaped zone a short distance behind the actual base of the meristem just adjacent to the root cap. Here the cells divide every 12 to 36 hours, whereas at the base of the meristem, they may divide only once in every 200 to 500 hours. Also, the divisions are often rhythmic, reaching a peak once or twice each day, usually toward noon and midnight, with relatively quiescent intermediate periods. Cells in this region are mostly cuboidal in shape, with relatively large centrally located nuclei and few, if any, small vacuoles. The apical meristem soon subdivides into three meristematic areas, in both roots and stems: (1) the **protoderm** gives rise to an outer layer of cells, the *epidermis;* (2) the **ground meristem,** to the inside of the protoderm, produces parenchyma cells of the *cortex;* (3) the **procambium,** which appears as a solid cylinder in the center of the root, produces *primary xylem* and *phloem* (figure 6.3). *Pith* tissue, which is seen in most stems, is absent in most dicot roots but is found in those of many monocots such as grasses.

The Region of Elongation

The *region of elongation,* which merges with the apical meristem, usually extends about 1 centimeter (0.4 inch) or less from the tip of the root. Here the cells become several times their original length and also somewhat wider, while the tiny vacuoles merge and grow until one or two large vacuoles, occupying up to 90% or more of the volume of each cell, have been formed. No further increase in cell size takes place above this region, so only the root cap and apical meristem are actually pushing through the soil. Except for the tips and for those roots where there is gradual increase in girth through the addition of *secondary tissues,* the often very extensive remainder of each root remains stationary for the life of the plant.

The Region of Maturation

Most of the cells mature into the various distinctive cell types of the primary tissues in this region, which is sometimes called the *region of differentiation,* or *root-hair zone.* It is given this last name because of the numerous, hairlike, delicate protuberances that develop from many of the epidermal cells. These **root hairs,** as they are called, are not separate cells; in fact, the nucleus of the epidermal cell to which each is attached often moves out into the protuberance. They are so numerous that they appear as fine down to the naked eye, typically numbering more than 38,000 per square centimeter (250,000 per square inch) of surface area in roots of plants such as corn, and they seldom exceed 1 centimeter (0.4 inch) in length. A single ryegrass plant occupying less than 0.6 cubic meter (2 cubic feet) of soil has been found to have more than 14 billion root hairs, with a total surface area almost the size of a football field. This obviously greatly increases the absorptive surface of the root, as the root hairs adhere tightly to soil particles (figure 6.4) with the aid of microscopic fibers that they produce.

When a seedling or plant is moved, many of the delicate root hairs are torn off or die within seconds if exposed to the sun, thus greatly reducing the

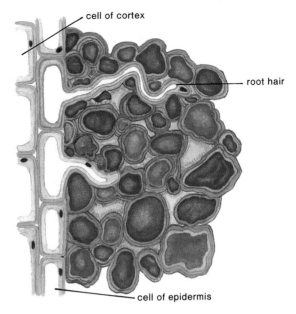

FIGURE 6.4   Root hairs in contact with soil particles.

cell of cortex

root hair

cell of epidermis

plant's capacity to absorb water and minerals in solution. This is why plants should be watered, shaded, and pruned after transplanting until new root hairs have formed. In any growing root, the extent of the root-hair zone remains fairly constant, with new root hairs being formed toward the root cap and older root hairs dying back in the more mature regions. The life of the average root hair is usually not more than a few days, although a few live for a maximum of perhaps three weeks.

If the region of maturation is examined in cross section (figure 6.5), it will be seen that, with the exception of the pith in dicots, most of the tissues of young stems are also present in roots. This is hardly surprising, since roots and stems are connected to each other. There are some differences, however. The cuticle, which may be relatively thick on the epidermal cells of stems, is very thin on the root hairs and epidermal cells of roots. Presumably any significant amount of fatty substance present in or on the walls of water absorbing cells would interfere with their function. The **cortex,** a tissue composed of parenchyma cells adjacent to the epidermis, functions primarily in food storage. It may be many cells

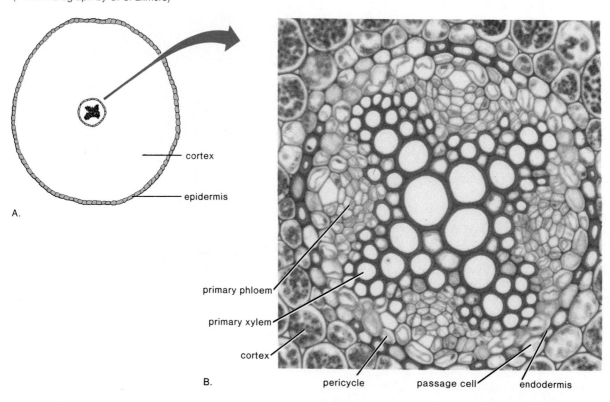

FIGURE 6.5   A cross section through the region of
maturation of a root of a buttercup (*Ranunculus*—a dicot).
(Photomicrograph by G. S. Ellmore)

cortex

epidermis

A.

primary phloem

primary xylem

cortex

B.

pericycle   passage cell   endodermis

thick and is similar to that of stems except for the presence of an **endodermis** at its inner boundary. The endodermis does not generally occur in stems, whereas it is nearly universally present in roots. It consists of a single layer of cells, most of which have thickened walls. Young endodermal cells also have a coating of fatty suberin, either covering all the inner wall faces or, more commonly, forming bands around the radial and transverse walls. These suberin bands are called **Casparian strips.** The protoplasts of the endodermal cells are fused to the Casparian strips, creating a barrier to the passage of water through the otherwise permeable cell walls. This barrier forces water to pass through the protoplasm of the endodermal cells and thus assists the root in regulating the movement of water and dissolved substances entering or leaving the central vascular cylinder. In some roots, the epidermis, the cortex, and the endodermis are sloughed off as girth increases, but in those roots where the endodermis is retained, all the inner walls of the cells eventually become coated with suberin. Thin-walled **passage**

cells are scattered singly or in small groups throughout the thicker-walled endodermal cells found in the younger portions of the root.

On the inside of the endodermis and immediately adjacent to it is a cylinder of parenchyma cells called the **pericycle.** This is usually one cell wide, although in some plants it may extend for several cells. It is a very important tissue, for the cells retain their capacity to divide even after they have matured, and it is within the pericycle that the *branch roots* and part of the *vascular cambium* of dicots arise (figure 6.6).

The *primary xylem,* with its water-conducting cells such as tracheids, forms a solid core in the center of most dicot and conifer roots, whereas in many monocot and some dicot roots it surrounds pith parenchyma cells. Root xylem, however, unlike stem xylem, first forms with short spirelike ridges, or arms, pointing toward the pericycle, at least in dicot roots. There are often four of these arms, but there can be from two to several. *Primary phloem* forms in discrete patches between the xylem arms, and branch roots usually arise in the pericycle opposite the xylem

FIGURE 6.6 A cross section through a willow (*Salix*—a dicot) root showing the origin of a branch root. (Photomicrograph by G. S. Ellmore)

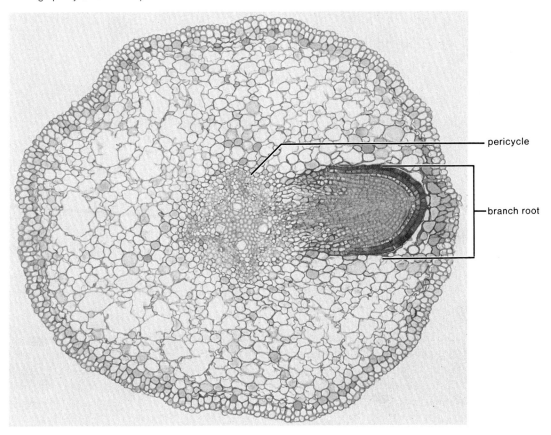

pericycle

branch root

arms. The primary phloem, containing food-conducting cells, also forms in patches in monocot roots, but in these roots the xylem arms may not be as strongly pronounced.

In woody dicots, some herbaceous dicots, and conifers, parts of the pericycle and parenchyma cells between the xylem arms and the phloem patches become a vascular cambium that starts producing secondary phloem to the outside and secondary xylem to the inside. Soon the cambium follows the outline of the primary xylem, and eventually, instead of appearing as patches and arms, the secondary conducting tissues appear as concentric cylinders. As is the case in stems, the primary tissues are crushed out of existence or sloughed off and lost. In woody plants, a cork cambium normally arises in the pericycle and gives rise to *cork* tissue (*periderm*) similar to that of stems. In fact, the cross section of a woody root several years old, with its annual rings of xylem along with its sapwood and heartwood, may be distinguished from that of a stem of the same species

only by the relatively larger proportion of parenchyma cells present in the root. Although there are exceptions, monocot roots, like monocot stems, generally have no secondary meristems and therefore no secondary growth.

Natural grafting may occur between roots of different trees of the same species, particularly in the tropics. When the roots come in contact with one another, they unite through secondary growth, but the details of the uniting process are not yet known. One unfortunate aspect of this grafting is that if one tree becomes diseased, the disease can be transmitted through these grafts to all the other trees connected to it.

In roots, there are no nodes or internodes, and branch or lateral roots arise internally in the pericycle, pushing their way out to the surface through the endodermis, cortex, and epidermis (figure 6.6). This process is in contrast with that in stems, where

FIGURE 6.7   A sweet potato plant. Note the food-storage roots.

branches develop from axillary buds. Some specialized roots, which at first glance may appear similar to specialized stems (see chapter 5) such as rhizomes, corms, and bulbs, can be distinguished by the absence of nodes and internodes.

SPECIALIZED ROOTS

As indicated earlier, most plants produce either a fibrous root system, a taproot system, or, more commonly, combinations of the two types. Some plants, however, have roots with modifications that permit specific functions in addition to the absorption of water and minerals in solution.

Food-Storage Roots

Most roots and stems store some food, but in certain plants the roots are enlarged and store large quantities of starch and other carbohydrates (figure 6.7). In sweet potatoes and yams, for example, extra cambial cells develop in parts of the xylem of branch roots and produce numerous parenchyma cells that cause the organs to swell and provide storage areas for considerable quantities of carbohydrates. Similar food-storage roots are found in the deadly poisonous water hemlocks, in dandelions, and in salsify.

Some food-storage organs, such as those of carrots, beets, turnips, and radishes, are actually a combination of root and stem. In carrots, for example, approximately 2 centimeters (0.8 inch) at the top is derived from stem tissue, yet a casual inspection will not reveal any differences of origin.

Water-Storage Roots

Some members of the Pumpkin Family, particularly those that grow in arid regions or in those areas where there may be no precipitation for several months of the year, produce huge water-storage roots. In certain manroots, for example, roots weighing 30 kilograms (66 pounds) or more are frequently produced (figure 6.8), and a major root of one calabazilla plant was found to weigh 72.12 kilograms (159 pounds). The water in the roots is apparently used by the plants when the supply in the soil is inadequate.

Propagative Roots

Many plants produce *adventitious buds* along the roots that grow near the surface of the ground. The buds develop into aerial stems called *suckers*, which have additional rootlets at their bases; these rooted suckers can be separated from the original root and

FIGURE 6.9   Pneumatophores (foreground) of tropical
mangroves rising above the sand at low tide. The
pneumatophores are spongy outgrowths from the roots
beneath the surface; they facilitate the exchange of oxygen
and carbon dioxide for the roots, which grow in areas where
little oxygen is otherwise available to them. (Courtesy Lani
Stemmerman)

Pneumatophores

Water, even after air has been bubbled through it,
contains less than one-thirtieth of the amount of free
oxygen found in the air. Thus, plants growing with
their roots in water may encounter less than the
amount of oxygen necessary for normal respiration.
Some swamp plants, such as the black mangrove and
the yellow water weed, develop special "spongy"
roots, called *pneumatophores,* which rise above the
surface and facilitate gas exchange between the atmosphere and the subsurface roots to which they are
connected (figure 6.9). Some older texts attributed
a similar function to the woody "knees" of the bald
cypress that occurs in southern swamps (see figure
20.18), but there is no conclusive evidence for this
theory.

Aerial Roots

The various kinds of aerial roots produced by plants
include the *velamen* roots of orchids, *prop* roots of
corn and banyan trees (figure 6.10), *adventitious*
roots of ivies, and *photosynthetic* roots of certain orchids. Velamen roots have an epidermis several cells
thick. It has long been assumed that this thick epidermis aids in the absorption of rainwater, but some
now suspect it may function more in preventing loss
of moisture from the root. Corn prop roots, produced toward the base of the stems, support the
plants in a high wind. A number of tropical plants,
including the screw pines and various mangroves,
produce sizeable prop roots extending for several feet
above the surface of the ground or water. Debris collects between them and helps to create additional
soil. Many of the tropical figs or banyan trees produce roots that grow down from the branches until
they contact the soil. Once they are established, they
continue secondary growth and look just like additional trunks. Some of these banyan trees attain great
age and dimension. In India and southeast Asia,
several such trees display almost 1,000 root-trunks
and have circumferences approaching 450 meters
(1,476 feet). The oldest is estimated to be about
2,000 years of age. The vanilla orchid, from which
we obtain vanilla flavoring, produces chlorophyll in

grown individually. A number of fruit trees, such as
cherries and pears, frequently produce such roots.
The adventitious roots of horseradish and rice-paper
plants can become a nuisance in gardens, the latter
often producing propagative roots within a radius of
10 meters (33 feet) or more from the parent plant.
Canada thistles and some other weeds have a remarkable facility to reproduce in this fashion as well
as by means of seeds. This ability made it difficult
to control them in the past, but some **biological controls** now being investigated (see appendix 2) may
be an answer to the problem in the future.

its aerial roots and thus can manufacture food with
them through the process of photosynthesis. The adventitious roots of English ivy, Boston ivy, and Virginia creeper appear along the stem and aid the
plants in climbing.

Contractile Roots

A number of herbaceous dicots and monocots have
contractile roots that pull the plant deeper into the
soil. Many lily bulbs are pulled a little deeper into
the soil each year as additional sets of contractile
roots are developed (figure 6.11). The bulbs continue to be pulled down until an area of relatively
stable temperatures is reached. Further, plants like
dandelions always seem to have the leaves coming
out of the ground as the top of the stem is pulled
down a small amount each year when the root con-
tracts. The contractile part of the root may lose as

much as two-thirds of its length within a few weeks.
Mechanisms of contraction include the thickening
and constriction of parenchyma cells, causing the
xylem elements to spiral somewhat like a corkscrew.

Buttress Roots

Some tropical trees produce huge buttresslike roots
toward the base of the trunk, giving them great stability (figure 6.12). Except for their angular appearance, these roots look like a part of the trunk.

Parasitic Roots

A number of plants, such as the dodders, broomrapes, and pinedrops, have no chlorophyll (necessary for photosynthesis) and have become dependent
on plants with chlorophyll for their nutrition. They
parasitize their host plants via somewhat rootlike
projections called **haustoria** (singular: **haustorium**)

FIGURE 6.11   How a lily bulb, over three seasons, is drawn deeper into the soil through the action of contractile roots. *A.* A seed germinates. *B.* Contractile roots pull the newly formed bulb down several millimeters during the first season. *C.* The bulb is pulled down further the second season. *D.* The bulb is pulled down even further the third season. The bulb will continue to be pulled down in succeeding seasons until it reaches an area of relatively stable soil temperatures.

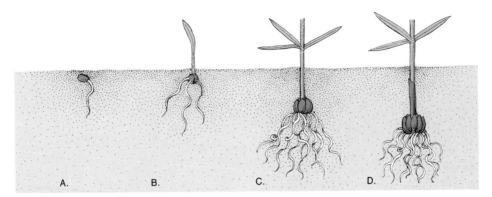

FIGURE 6.12   Buttress roots of a tropical fig tree.

FIGURE 6.13   Pale, twining stems of a dodder plant that is parasitizing grasses.

(figure 6.13), which develop along the stem in contact with the host. The haustoria penetrate the outer tissues and establish connections with the water-conducting and food-conducting tissues. Some green plants, such as Indian warrior and the mistletoes, also form haustoria, but these apparently function primarily in obtaining water and dissolved minerals from the host plants, since the partially parasitic plants are capable of manufacturing at least some of their own food through photosynthesis.

Mycorrhizae

Members of a sizeable majority of flowering plant families have various fungi associated with their roots. The association is *mutualistic* (mutualism is a form of *symbiosis;* see page 276); that is, both the fungus and the root benefit from it but are also dependent upon the association for normal development. The fungus often forms a mantle of millions of threadlike strands around the root and seldom penetrates deeper than the cortex. If fungal threads should happen to penetrate any deeper, they are apparently digested and broken down by the host plant. The fungus evidently breaks down complex molecules in the humus of the soil; the smaller molecules are absorbed by the root, while the root furnishes the fungus with sugars and amino acids. These "fungus-roots," or **mycorrhizae** (figure 6.14), are essential to the normal growth and development of some plants such as forest trees and orchids. Orchid seeds simply will not germinate until a mycorrhizal fungus invades their cells. Many plants seem to have considerable difficulty in absorbing essential elements such as phosphorus, even when the elements are abundant in the soil, if mycorrhizal fungi have been killed by fumigation or are otherwise absent. Other plants do not seem to need the mycorrhizae unless essential elements in the soil are present in amounts barely sufficient for healthy growth. Plants with mycorrhizae develop few root hairs as compared with those growing without an associated fungus. Mycorrhizae have proved to be particularly susceptible to acid rain (see chapter 23); this may portend major problems for our coniferous forests in the future if the problem of acid rain is not solved.

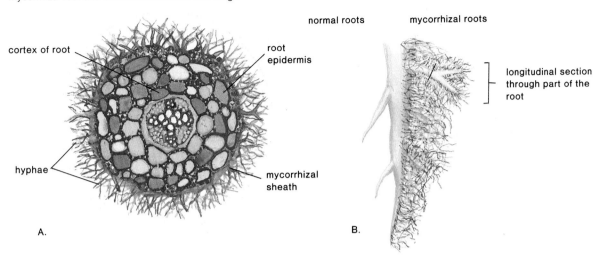

FIGURE 6.14 Mycorrhizae. *A*. A mycorrhizal root in cross section. *B*. Comparison, in longitudinal section, between a mycorrhizal root and one not associated with fungi.

cortex of root

root epidermis

hyphae

mycorrhizal sheath

normal roots   mycorrhizal roots

longitudinal section through part of the root

A.

B.

Root Nodules

When legume plants (e.g., peas and beans), alders, and a few other plants are uprooted, numerous small swellings can be seen along the roots. These *root nodules* contain bacteria that supplement the plant's nitrogen supply. This subject is detailed in chapter 23 in the discussion of the nitrogen cycle. Root nodules should not be confused with *root knots,* which are also swellings and are seen in the roots of tomatoes and many other plants. Root knots develop in response to the invasion of tissue by small, parasitic roundworms (nematodes). Unlike bacterial nodules, root knots are not beneficial to the plant and the activities of the parasites within them can eventually lead to the premature death of the plant.

HUMAN RELEVANCE OF ROOTS

Roots have been important sources of food for humans since prehistoric times, and some, such as the carrot, have been in cultivation in Europe for at least 2,000 years. A number of cultivated root crops involve **biennials** (i.e., plants that complete their life cycles from seed to flowering and back to seed in two seasons). Such plants store food in a swollen taproot during the first year of growth and then utilize the stored food to produce flowers in the second season.

Among the best-known biennial root crops are sugar beets, beets, turnips, rutabagas, parsnips, radishes, horseradishes, and carrots. Other important root crops include sweet potatoes, yams, and cassava. Cassava (figure 6.15), from which tapioca is made, forms a major part of the basic diet for millions of inhabitants of the tropics. It yields more starch per hectare (about 45 metric tons, the equivalent of 20 tons per acre) than any other cultivated crop, with a minimum of human labor. A number of minor root crops are cultivated in South America and other parts of the world. Among such plants are relatives of wild mustards, nasturtiums, and sorrel.

A number of spices, such as sassafras, sarsaparilla, licorice, and angelica, are obtained from roots. Sweet potatoes are used in the production of alcohol in Japan. Several important red to brownish dyes are obtained from roots of members of the Madder Family, to which coffee plants belong. Drugs obtained from roots include aconite, ipecac, ginseng, gentian, and reserpine, a tranquilizer. A valuable insecticide, rotenone, is obtained from the barbasco plant, which has been cultivated for centuries as a fish poison by primitive South American

tribes. When thrown into a dammed stream, the roots containing rotenone cause the fish to float but in no way poison them for human consumption. Other uses of roots are discussed in chapter 22.

SUMMARY

Roots function primarily in anchorage and in the absorption of water and minerals in solution. The radicle of a germinating seed develops into the first root, from which develops either a taproot with branch roots, or a fibrous root system in which the roots are of similar diameter. Many plants have combinations of both systems, with some also developing adventitious roots as they mature. In developing young roots, four zones or regions are traditionally recognized: (1) A thimble-shaped root cap at the tip, which protects the root as it pushes through the soil and also aids in the perception of gravity. (2) A region of cell division that is surrounded by the root cap. It is composed of an apical meristem that subdivides into a protoderm, which produces the epidermis; a ground meristem that produces the cortex; and a procambium in the center that produces primary xylem and primary phloem. (3) A region of elongation in which the cells produced by the apical meristem become considerably longer and slightly wider. (4) A region of maturation in which the cells mature into the distinctive cell types of primary tissues.

Some of the epidermal cells in the region of maturation develop protuberances called root hairs; the root hairs greatly increase the absorptive surface of the root. The tissues that mature in the region of maturation are similar to those of stem tips, except that pith is absent in most dicot roots, and of a different origin (the procambium) in monocot roots. The cortex has an endodermis at its inner boundary. This consists of a cylinder, one cell wide, of cells that have suberin bands called Casparian strips around the radial and tranverse walls, and an occasional thin-walled passage cell interspersed throughout the cells with suberin. Immediately adjacent to the endodermis toward the center of the root are parenchyma cells constituting the pericycle. Branch roots and the vascular cambium arise in the pericycle. The primary xylem initially frequently forms a solid core in the center of the root;

this core usually develops with two to several (usually four) arms, at least in dicot roots. At first, primary phloem is produced in discrete patches between the arms of the primary xylem, but as secondary tissues are added, the tissues eventually appear as concentric cylinders. In woody plants, a cork cambium usually arises in the pericycle and produces cork tissues similar to those of stems.

Roots may graft together naturally. There are no nodes or internodes in roots. Roots may be specialized as food-storage or water-storage roots; as pneumatophores, which facilitate respiration in swamps; as aerial roots, which function in an aerial habitat (velamen roots, prop roots, photosynthetic roots, adventitious roots that aid in climbing); as contractile roots, which pull a plant deeper into the soil; as buttress roots, which add stability; as haustoria, which have a parasitic function; and as mycorrhizal roots, which involve fungal associations. Some plants (e.g., legumes, alders) have nitrogen-fixing bacteria in small swellings (nodules) on their roots.

Root crops include sugar beets, beets, turnips, rutabagas, parsnips, radishes, carrots, sweet potatoes, yams, and cassava. Spices such as sassafras, sarsaparilla, licorice, and angelica are obtained from roots. Other uses of roots include the production of alcohol from sweet potatoes, and the extraction of dyes, drugs, insecticides, and poisons.

REVIEW QUESTIONS

1. Distinguish between a tiny root and a root hair. What is the function of a root hair?
2. What is the difference between parasitic roots and mycorrhizae?
3. If you were shown cross sections of a young root and a young stem from the same dicot plant, how could you tell them apart?
4. What is the function of the root cap and from which meristem does it originate?
5. How do endodermal cells differ from other types of cells?
6. Where do branch roots originate?
7. Indicate some spices and drugs obtained from roots.

DISCUSSION QUESTIONS

1. Japanese gardeners regularly trim away parts of the root system to assist in dwarfing a plant. A plant's food is obtained through photosynthesis in the leaves; can you suggest why trimming the roots can cause dwarfing?
2. It was suggested that roots perceive gravity through the root cap. Would it really matter if roots grew randomly in the soil instead of responding to gravity?
3. From the viewpoint of the plant, can you suggest a practical reason for branch roots originating internally instead of at the surface?
4. When you eat a yam or a sweet potato, what kinds of compounds and cells are you consuming?

ADDITIONAL READING

Cronquist, A. 1981. *Basic botany,* 2d ed. New York: Harper and Row, Publishers.

Cutter, E. G. 1971. *Plant anatomy: experiment and interpretation. Part II: Organs.* Reading, MA: Addison-Wesley Publishing Co.

Epstein, E. 1973. "Roots." *Scientific American* 228: 48–58.

Esau, K. 1977. *Anatomy of seed plants,* 2d ed. New York: John Wiley & Sons.

Fahn, A. 1982. *Plant anatomy,* 3d ed. Elmsford, NY: Pergamon Press.

Miller, R. H. 1974. *Root anatomy and morphology: a guide to the literature.* Hamden, CT: Shoe String Press.

Northcote, D. H. 1980. *Differentiation in higher plants.* Carolina Biology Readers Series. Burlington, NC: Carolina Biological Supply Co.

Russell, R. S. 1977. *Plant root systems: their function and interaction with the soil.* New York: McGraw-Hill.

Torrey, J. G., and D. Clarkson, eds. 1975. *The development and function of roots.* Orlando FL: Academic Press.

Overview

*T*his chapter introduces leaves by comparing them with solar panels and by discussing their general functions, morphology, and dimensions. This is followed by descriptive information on basic leaf types and specific forms and arrangements. The chapter next discusses the internal structure of leaves, including epidermis and cuticle, stomata, glands, mesophyll, and veins.

Specialized leaves, including tendrils, spines, flower pot leaves, window leaves, reproductive leaves, floral leaves, and insectivorous leaves, are then examined. The chapter concludes with an explanation of autumnal color changes and abscission and some observations on the human and ecological relevance of leaves.

7   Leaves

Some Learning Goals

1. Learn the external forms and parts of leaves. Know the functions of a typical leaf, and the specific tissues and cells that contribute to those functions.

2. Understand the differences among pinnate, palmate, and dichotomous venation, and also the differences between simple and compound leaves.

3. Contrast tendrils, spines, storage leaves, flower pot leaves, window leaves, reproductive leaves, floral leaves, and different types of insect-trapping leaves.

4. Explain why deciduous leaves turn various colors in the fall and how such leaves are shed.

5. Know at least fifteen uses of leaves by humans.

Outline

INTRODUCTION

During the "energy crisis" of the 1970s most persons in industrialized countries became acutely aware that fossil fuels, our largest source of energy, are limited and costly, and that despite temporary gluts, the supplies eventually will be exhausted. One means of counteracting this problem, the use of solar energy, has been increasingly employed by the construction industry, particularly in the southwestern United States. A number of new houses are being built with flat panels and windows inclined at angles to maximize the amount of energy captured from the sun's rays. Some buildings have mechanical devices that slowly move the solar panels so that they will follow the sun in its daily course across the sky. Variations of such devices are undoubtedly destined to become commonplace in the future.

Plants had their own solar panels, which captured the sun's energy, many aeons before it became obvious to modern civilization that fossil fuel supplies eventually would be exhausted. The remarkably constructed plant organs called *leaves* are far more efficient than their comparatively clumsy counterparts on buildings. Their flattened surfaces, which are completely covered with a transparent protective layer of cells, the *epidermis,* admit light to all parts of the interior. Many leaves daily become twisted on their stalks, or *petioles,* so that their upper surfaces are inclined at right angles to the sun's rays throughout daylight hours.

Green leaves capture the light energy available to them by means of the most important process for life on earth, at least life as we know it today. This process, called *photosynthesis* (discussed in chapter 10), involves the trapping of energy in sugar molecules that are constructed from ordinary water and from carbon dioxide present in the atmosphere. All the energy needs of living organisms ultimately depend on photosynthesis, from the first day of their existence to the last.

The lower surfaces of leaves (and sometimes the upper surfaces as well) are dotted with tiny pores called *stomata,* which not only permit air circulation but are also involved with cooling the interior of the leaf, as water evaporates from the moist cell surfaces and passes out of them into the air in vapor form.

Leaves also perform other functions. For example, all living cells *respire* (respiration is also discussed in chapter 10), and in the process of this and other metabolic activities, waste products are produced. These wastes are deposited outside the plant when the leaves are shed, mostly in the fall, after being sealed off at the bases of their petioles (see discussion of leaf colors and *abscission* later in this chapter). The following season the discarded leaves are replaced with new ones. Leaves are also involved in the movement of the water absorbed by the roots and transported throughout the plant. Most of the water reaching the leaves evaporates into the atmosphere by a process known as **transpiration** (discussed in chapter 9). In some plants, *root pressure* (see page 163) forces water out of *hydathodes,* which are special openings at the tips of leaf veins, usually at night when transpiration is not occurring (see page 166). The water thus expelled may contain ions secreted by root cells. Additional functions of leaves are discussed throughout this chapter.

LEAF ARRANGEMENTS AND TYPES

Many of the roughly 275,000 different kinds of plants that produce leaves can be distinguished from one another by their leaves alone. In addition, many plants such as lilies, ferns, and pine trees produce different forms of leaves, including tiny papery *scales,* on the same plant. The variety of shapes, sizes, and textures of leaves seems to be almost infinite. Some of the smaller duckweeds have leaves less than 1 millimeter (0.04 inch) wide. The leaves of the Seychelles Island palm attain spans of 6 meters (20 feet), and the floating leaves of a giant water lily, which reach 2 meters (6.5 feet) in diameter (figure 7.1), can support, without sinking, weights

FIGURE 7.1 Floating leaves of a giant water lily, which
sometimes attain a diameter of 2 meters (6.5 feet), the larger
leaves being capable of supporting the weight of a child
without sinking.

of more than 45 kilograms (100 pounds) distributed over their surface. In addition to flattened, variously shaped leaves there are others that are tubular, feathery, cup-shaped, spinelike, or needlelike; in fact, leaves may assume virtually any form (figure 7.2). They may be smooth or hairy, slippery or sticky, waxy or glossy, pleasantly fragrant or foul-smelling, edible or poisonous. They also may be of almost every color of the rainbow, and of exquisite beauty, especially when viewed with a microscope.

As indicated in chapter 5, all leaves originate as *primordia* in the buds, regardless of their ultimate size or form. At maturity most leaves consist of a stalk, called the **petiole,** and a flattened **blade,** or *lamina*. The blade has a network of *veins* (*vascular bundles*), with the larger veins being parallel to one another in monocots (figure 7.3) and diverging from one another in various ways in dicots (see figure 7.7). A pair of appendages, called **stipules** (see chapter 5), are sometimes present at the base of the petiole. Leaves of flowering plants are associated with *leaf gaps* (see figure 5.3 and associated text) and may be *simple* or *compound*.

The arrangement of leaves may be in a *spiral* or *alternate* pattern on the stem, or two may be attached at the same node, providing an *opposite* arrangement. When three or more leaves occur at a node, they are said to be *whorled* (see chapter 5). The leaf itself may be a **simple leaf,** with an undivided blade, or it may be a **compound leaf,** with the blade divided into *leaflets* in various ways (see figure

FIGURE 7.2   Types of leaves and leaf arrangements.
A. Palmately compound leaf of a buckeye. *B.* Pinnately
compound leaf of a black walnut. *C.* Alternate, simple but
lobed leaves of a tulip tree. *D.* Opposite, simple leaves of a
dogwood. *E.* Palmately veined leaf of a maple. *F.* Globe-
shaped succulent leaves of string-of-pearls. *G.* Pinnately
veined, lobed leaf of an oak. *H.* A grass leaf. *I.* Whorled leaves
of a bedstraw. *J.* Linear leaves of a yew. *K.* Fan-shaped leaf
of a *Ginkgo* tree, showing dichotomous venation.

FIGURE 7.3   A portion of a monocot leaf showing the parallel veins.

7.2). **Pinnately compound** leaves have the leaflets in pairs along a central stalklike **rachis,** whereas **palmately compound** leaves have all the leaflets attached at the same point at the top of the petiole. Sometimes the leaflets of a pinnately compound leaf may be subdivided into still smaller leaflets, forming a *bipinnately compound* leaf. The arrangement of veins in a leaf or leaflet blade (*venation*) may also be either pinnate or palmate. In **pinnately veined** leaves there is a main vein, called the **midrib,** with secondary veins branching from it, but in **palmately veined** leaves several veins fan out from the base of the blade. In a few leaves (e.g., *Ginkgo*) no midrib or other large veins are present; instead the veins fork evenly and progressively from the base of the blade to the opposite margin. This is called *dichotomous venation* (see figure 7.2K).

INTERNAL STRUCTURE OF LEAVES

If a typical leaf is cut transversely and examined with the aid of a microscope, three regions stand out: *epidermis, mesophyll,* and *veins* (referred to as *vascular bundles* in our discussion of roots and stems) (figure 7.4). The **epidermis** is a single layer of cells

FIGURE 7.4   A stereoscopic view of a portion of a typical leaf.

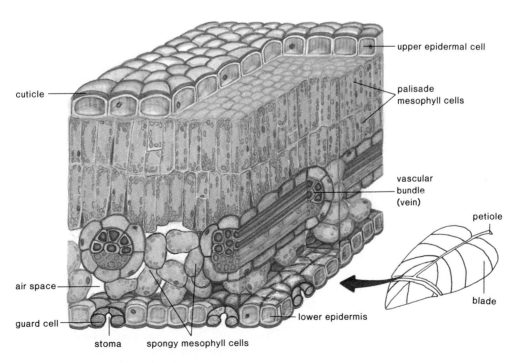

cuticle

upper epidermal cell

palisade mesophyll cells

vascular bundle (vein)

petiole

blade

air space

guard cell

stoma   spongy mesophyll cells

lower epidermis

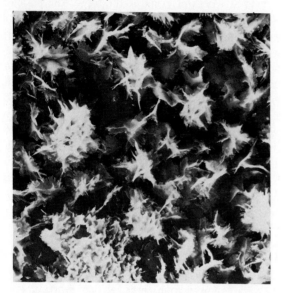

covering the entire surface of the leaf. The epidermis on the lower surface of the blade can sometimes be distinguished from the upper epidermis by the presence of *stomata,* which are discussed in the paragraph that follows. When seen from the top, the wavy, undulating epidermal cells frequently resemble pieces of a jigsaw puzzle fitted together. Except for *guard cells,* the upper epidermal cells contain no chloroplasts, their function being primarily protection of the delicate tissues to the interior. A thin coating of waxy **cutin** (the **cuticle**) is normally present, although it may not be visible with ordinary light microscopes without being specially stained. In addition to the cuticle, many plants produce other waxy substances on their surfaces (figure 7.5). In studies of the effects on plants of smog and auto exhaust fumes, it was found that these waxes may be produced in abnormal fashion on beet leaves within as little as 24 hours after exposure to the pollutants. Beet leaves also respond to aphid damage by producing wax around each tiny puncture. Occasionally the upper epidermal cells may contain crystals of waste materials. Different types of **glands** may also be present in the epidermis. Glands occur in the form of depressions, protuberances, or appendages either directly on the leaf surface or on the ends of hairs (see figure 4.8). Glands often secrete sticky substances.

Stomata

The lower epidermis of most plants generally resembles the upper epidermis, but it is perforated by numerous tiny pores, the **stomata** (figure 7.6). They also occur in both leaf surfaces of some plants (e.g., alfalfa, corn), exclusively on the upper epidermis of other plants like water lilies where the lower epidermis is in contact with water, and are absent altogether from the submerged leaves of aquatic plants. Stomata are very numerous, ranging from about 1,000 to more than 1.2 million per square centimeter (6,300 to 8 million per square inch) of surface. An average-sized sunflower leaf has about 2 million of these pores throughout its lower epidermis. Each stoma is bordered by two sausage- or dumbbell-shaped cells that usually are smaller than most of the neighboring epidermal cells. These are called **guard cells,** and unlike other cells of either epidermis, they contain chloroplasts. The photosynthesis that takes place in the guard cells aids in the functioning of the cells. The guard cell walls are distinctly thickened but quite flexible on the side adjacent to the pore. As the guard cells expand or contract with changes in the amount of water within the cells, their unique construction causes the stomata to open or close. When the guard cells are expanded, the stomata are open; when the water content of the guard cells decreases, the cells contract and the stomata close (for more detailed discussions of this stomatal mechanism see *Regulation of Transpiration* in chapter 9 and *Turgor Movements* in chapter 11). Stomata function in permitting gas exchange between the interior of the leaf and the atmosphere and in regulating the evaporation of most of the water entering the plant at the roots (see chapter 9). The stomata of water lilies and other plants with floating leaves are confined to the upper epidermis.

Mesophyll and Veins

Most photosynthesis takes place in the **mesophyll** between the two epidermal layers. When two regions of the mesophyll are distinguishable, as is often the case, the uppermost consists of tightly packed, somewhat barrel-shaped parenchyma cells, commonly in two rows. This region is called the **palisade mesophyll** and may contain more than 80% of the leaf's chloroplasts. The lower region, consisting of loosely arranged parenchyma cells with abundant air spaces between them, is called the **spongy mesophyll.** Its cells also contain numerous chloroplasts.

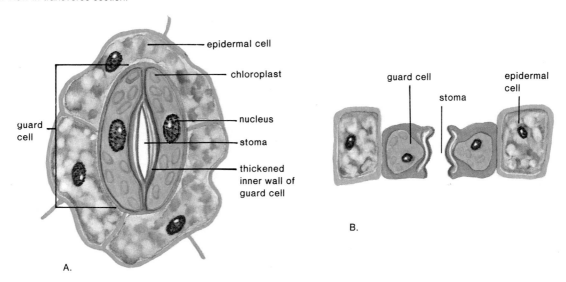

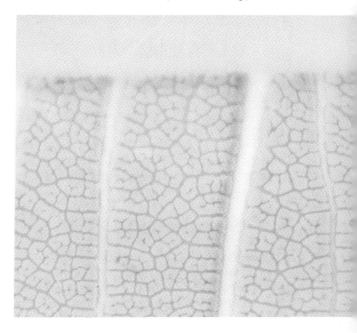

Parenchyma tissue containing numerous chloroplasts is also referred to as *chlorenchyma* tissue. Besides being present in leaves, chlorenchyma tissue is
found in the outer parts of the cortex in the stems
of herbaceous plants. Inside the leaf, the surfaces of
mesophyll cells in contact with the air are moist. If
moisture decreases below a certain level, the stomata close, thus significantly reducing further
drying.

FIGURE 7.7   Net venation of a portion of a frangipanni leaf.

Veins (*vascular bundles*) of various sizes are
scattered throughout the mesophyll (figure 7.7).
They consist of xylem and phloem tissues surrounded by a jacket of fibers called the **bundle sheath.**
The veins give the leaf its "skeleton." The carbohydrates produced in the mesophyll cells are transported in solution throughout the plant by the
phloem. Water, sometimes located more than 100
meters (330 feet) away in the ground below, is
brought up to the leaf by the xylem, which, like the
phloem, is part of a vast network of "plumbing"
throughout the plant. Since veins run in all directions in the network, particularly in dicots, it is
common, when examining a cross section of a leaf
under the microscope, to see veins cut transversely,
lengthwise, and at a tangent, all in the same section.

Monocot leaves, in addition to having *parallel
veins,* usually do not have the mesophyll differentiated into palisade and spongy layers. Some monocot leaves (e.g., those of grasses) have large, thin-
walled *bulliform cells* on either side of the main

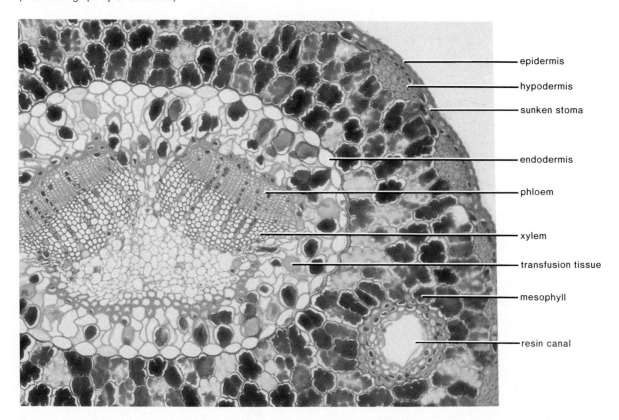

- epidermis
- hypodermis
- sunken stoma
- endodermis
- phloem
- xylem
- transfusion tissue
- mesophyll
- resin canal

central vein (midrib) toward the upper surface. The bulliform cells partly collapse under dry conditions, causing the leaf blade to fold or roll, thus reducing transpiration (see figure 11.12).

SPECIALIZED LEAVES

If the leaves of all plants could function normally under any environmental conditions, various leaf modifications would provide no special benefits to a plant. But the form and structure of plants of tropical rain forests do not permit them to thrive in a desert, and cacti soon die if planted in a creek because their structure, form, and life cycles are attuned to specific combinations of environmental factors such as temperature, humidity, light, water, and soil conditions. In addition, the modifications of leaves occupying any single ecological niche may be very diverse, resulting in such a rich variety of leaf forms and specializations throughout the Plant Kingdom that only a few may be mentioned here.

Shade Leaves

Some leaves with inconspicuous modifications may occur along with unmodified leaves on the same plant. For example, leaves in the shade tend to be thinner and have fewer hairs than leaves on the same tree that are exposed to direct light. They also tend to be larger and to have fewer well-defined mesophyll layers and fewer chloroplasts than their counterparts in the sun.

Leaves of Arid Regions

In different climatic zones or habitats, however, the modifications are generally more pronounced. Thus, plants growing in arid regions may have thick and leathery leaves, and fewer stomata or stomata sunken below the surface in special depressions. They also may have succulent water-retaining leaves or no leaves at all (with the stems taking over the function of photosynthesis), or they may have dense, hairy coverings. Pine trees, whose water supply may be

FIGURE 7.9   Tendrils at the tip of a garden pea plant leaf.

severely restricted in the winter when the soil is frozen, have some leaf modifications similar to those of desert plants. The modifications include sunken stomata, a thick cuticle, and a layer of thick-walled cells (the **hypodermis**) beneath the epidermis (figure 7.8). The leaves of the compass plant face east and west, with the blades at right angles to the ground, so that when the sun is overhead it strikes only the thin edge of the leaf, minimizing moisture loss. In plants that grow in water, the submerged leaves usually have considerably less xylem than phloem. Large spaces are found in the mesophyll, which is not differentiated into palisade and spongy layers. Other modifications are described in the following sections.

Tendrils

There are many plants whose leaves are partly or completely modified as **tendrils.** These leaves, when curled tightly around more rigid objects, help the plant in climbing or in supporting weak stems. In garden peas, the leaves are compound (figure 7.9), and the upper leaflets are reduced to whiplike strands that, like all tendrils, are very sensitive to contact. In fact, if a healthy tendril is lightly stroked, there is a sudden rapid growth of cells on the opposite side, and it starts curling in the direction of the contact within a minute or two. If the contact is very brief

the tendril undergoes a reverse movement and straightens out again. If, however, the tendril encounters a suitable solid support (e.g., a twig), the stimulation is continuous, and the tendril coils tightly around the support as it grows.

In the yellow vetchling, the whole leaf is modified as a tendril, and photosynthesis is carried on by the leaflike stipules at the base. In the potato vine and the garden nasturtium, the petioles serve as tendrils, while in some greenbriers, stipules are modified as tendrils. In *Clematis,* the rachises of some of the compound leaves serve very effectively as tendrils. Members of the Pumpkin Family, which include squashes, melons, and cucumbers, produce tendrils that may be up to 3 decimeters (1 foot) in length.

As the tendrils develop, they become coiled like a spring. When contact with a support is made, the tip not only curls around it, but the direction of the coil reverses (see figure 11.7); sclerenchyma cells then develop in the vicinity of contact. This makes a very strong but flexible attachment that protects the plant from damage during high winds. It should be noted that the tendrils of many other plants (e.g., grapes), are not modified leaves but have developed from stems.

FIGURE 7.10 Spines. *A.* Stipules of a mesquite leaf modified as spines. *B.* Leaves of a common barberry modified as spines.

A.

B.

Spines

Many desert plants have their leaves modified as **spines.** This reduction in leaf surface correspondingly reduces water loss from the plants, and the spines also apparently protect the plants from browsing animals. Photosynthesis in such desert plants, which would otherwise take place in leaves, takes place instead in the green stems. These spines, along with those of barberries (figure 7.10B), are modifications of the whole leaf, in which most of the normal leaf tissue is replaced with sclerenchyma. In a number of woody plants (e.g., mesquite, black locust), the stipules at the bases of the leaves are modified as short, paired spines (figure 7.10A). As is the case with tendrils, many spines are modified stems rather than modified leaves. The thorns of roses and prickles of raspberries originate at the surface of the stems.

Storage Leaves

As previously mentioned, desert plants may have *succulent leaves* (i.e., leaves that are modified so that they retain water). The modifications that permit water storage involve large, thin-walled parenchyma cells without chloroplasts to the interior of chlorenchyma tissue just beneath the epidermis. These non-photosynthetic cells contain large vacuoles that can store proportionately substantial quantities of water. If removed from the plant and set aside, the leaves will often retain much of the water for several weeks. Many plants with succulent leaves carry on a special form of photosynthesis discussed in chapter 10. Other fleshy leaves, such as those that comprise the bulk of onion and lily bulbs, store large amounts of carbohydrates, which are utilized by the plant in the subsequent growing season.

FIGURE 7.11 Flower pot leaves of *Dischidia*. One leaf is sliced lengthwise to reveal the root growing inside it.

FIGURE 7.12 A window plant. Note the transparent tips of the window leaves.

"Flower Pot" Leaves

Some of the leaves of *Dischidia* (figure 7.11), a greenhouse curiosity from tropical Australasia, develop into urnlike pouches that become the home of ant colonies. The ants carry in soil and add nitrogenous wastes, while moisture collects in the leaves through condensation of the water vapor coming through stomata from the mesophyll. This situation creates a good growing medium for roots, which develop adventitiously from the same node as the leaf and grow down into the soil contained in the urnlike pouch. In other words, this extraordinary plant not only reproduces itself by conventional means but also, with the aid of ants, provides its own fertilized growing medium and flower pots and then produces special roots that "take advantage" of the situation.

Window Leaves

In the Kalahari desert of South Africa, there are at least three plants belonging to the carpetweed family that have unique adaptations to living in dry, sandy areas. Their leaves, which are shaped like ice-cream cones, are about 3.75 centimeters (1.5 inches) long (figure 7.12) and are buried in the sand; only the dime-sized wide end of a leaf is exposed at the surface. This exposed end is covered with a thick epidermis and cuticle. It has a few stomata and is relatively transparent. Below the exposed end is a mass of tightly packed water-storage cells; these allow light coming through the "windows" to penetrate to the chloroplasts in the mesophyll, located all around the inside of the shell of the leaf. This arrangement, which keeps most of the plant buried and away from drying winds, allows the plant to thrive under circumstances that most other plants could not tolerate.

Reproductive Leaves

Some of the leaves of the walking fern are most un-usual in that they produce new plants at their tips. Occasionally, three generations of plants may be found linked together. The succulent leaves of air plants (figure 7.13) have little notches along the leaf margins in which tiny plantlets are produced, complete with roots and leaves, even after a leaf has been removed from the parent plant and pinned to a curtain or left on a counter. Each of the plantlets can develop into a mature plant if given the opportunity to do so.

Floral Leaves (Bracts)

Modified leaves known as **bracts** are found at the bases of flowers or flower stalks. In some plants such as the Christmas flower (poinsettia), the flowers themselves have no petals, but the brightly colored floral bracts that surround the small flowers make up for the absence of petals (figure 7.14). In other plants such as the dogwood, the tiny flowers in their buttonlike clusters do have inconspicuous petals, and the large white-to-pink bracts that surround the flower clusters, although they appear to the casual observer to be petals, are actually modified leaves.

FIGURE 7.13   A leaf of an air plant, showing plantlets being produced along the margins.

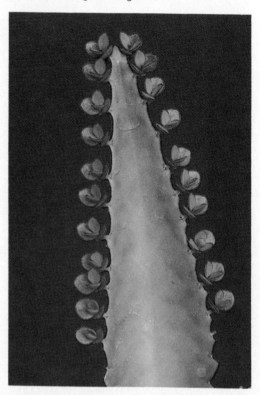

FIGURE 7.14   A poinsettia "flower." There are several flowers without petals in the center; the most conspicuous parts of the "flower" are the colored bracts (modified leaves) surrounding the true flowers.

FIGURE 7.15   Insect-trapping leaves. *A.* Pitcher plant
leaves.

A.

Insect-Trapping Leaves

Highly specialized *insect-trapping leaves* have in-
trigued humans for hundreds of years. Almost 200
species of flowering plants are known to have these
leaves. Such plants occur mainly in swampy areas
and bogs of tropical and temperate regions. In such
an environment, certain elements may be deficient
in the soil, or they may be in a form not readily
available to the plants. Some of the needed elements
are furnished when the soft parts of insects and other
small organisms trapped by the specialized leaves
are broken down and digested. All the plants contain chlorophyll and thus are able to make their own
food. It has been demonstrated that they can develop in apparently normal fashion without insects
if they are given the nutrients they need. The mechanisms of trapping the insects fall into four types,
represented by the following plants.

Pitcher Plants

Some of the leaves of *pitcher plants* (figure 7.15A)
have flattened blades and function like any other
leaves, but others are formed like vases of various
sizes and shapes, with or without umbrellalike flaps
over the open ends. Some Asian pitcher plants are
vines whose long petioles are twisted around
branches for support and whose pitchers are formed
at the tips of leaves. The plants give off a distinctive
odor that attracts insects; these, upon arrival, seek
out nectar-secreting glands located at the rim of the

FIGURE 7.15 *continued*   *B.* Sundew leaves.

B.

pitcher and usually fall into a pool of fluid at the bottom. If they try to climb out, they find the walls highly polished and slippery. In fact, in some species of pitcher plants the walls are coated with wax, and as the insects struggle up the surface, their feet become coated with the wax, which builds up until the victims seem to have acquired heavy clodlike boots. Most insects never make it up the walls, but even if they do, they still face a formidable barricade of stiff downward-pointing hairs near the rim. Eventually they drown, and their soft parts are digested by bacteria and by enzymes secreted by digestive glands near the bottom of the plant. In North America, the pitcher plants produce their pitchers in erect clusters on the ground, but the mechanisms for trapping insects are similar.

Merchandisers have been shameless in their indiscriminate collection of these plants for sale. Pitcher plants may become extinct in the wild. The cobra plant, a pitcher plant restricted to a few swampy areas in California and Oregon, has been placed on official rare species lists.

Sundews

The tiny plants called *sundews* (figure 7.15B) often do not measure more than 2.5 to 5.0 centimeters (1 to 2 inches) in diameter. The roundish to oval leaves are covered with about 150 to 200 upright glandular hairs that look like miniature clubs. There is a clear glistening drop of sticky fluid containing digestive enzymes at the tip of each hair. As the droplets sparkle in the sun, they may attract insects, which find themselves stuck if they alight. The hairs are exceptionally sensitive to contact, responding to weights of less than one-thousandth of a milligram, and bend inward, surrounding any trapped insect within the space of a few minutes. The digestive enzymes break down the soft parts of the insects, and after digestion has been completed (within a few days), the glandular hairs return to their original positions. If bits of nonliving debris happen to catch in the sticky fluid, the hairs barely respond, showing they can distinguish between protein and something "inedible." Some sundew owners regularly feed their plants tiny bits of hamburger and boiled egg white. Portuguese peasants use relatives of sundews with less specialized leaves in their homes as an effective

FIGURE 7.15 *continued*   *C.* A Venus flytrap plant.
D. Bladderwort leaves.

C.

D.

substitute for flypaper. In response to contact by living insects the edges of specialized leaves of similar plants called *butterworts* rapidly curl over and trap unwary victims.

Venus Flytraps

The *Venus flytrap* (figure 7.15C), which has leaves constructed along the lines of an old-fashioned steel trap, is found in nature only in wet areas of North Carolina and South Carolina. The two halves of the blade have the appearance of being hinged along the midrib, with stiff hairlike projections located along their margins. There are three tiny trigger hairs on the inner surface of each half. If two trigger hairs are touched simultaneously or if any one of them is touched twice within a few seconds, the blade halves suddenly snap together, trapping the insect or other small animal. As the organism struggles, the trap closes even more tightly. Digestive enzymes secreted by the leaf break down the soft parts of the insect, which are then absorbed. After digestion has been completed, the trap reopens, ready to repeat the process. As is the case with sundews, the traps do

not normally close for bits of debris that might accidentally fall on the leaf, presumably because nonliving material does not move about and stimulate the trigger hairs.

Bladderworts

Bladderworts (figure 7.15D) are plants with finely dissected leaves that are found submerged and floating in the shallow water along the margins of lakes and streams. Toward the bases of many of the leaves are tiny stomach-shaped bladders, each with a trapdoor over the opening at one end. The bladders, which are between 0.3 and 0.6 centimeter (0.125 to 0.25 inch) in diameter, trap aquatic insects and other small animals through a complex mechanism. Four curled but stiff hairs located toward one end of the trapdoor act as triggers when an insect touches one of them. The trapdoor springs open, and water rushes into the bladder. The stream of water propels the victim into the trap, and the door snaps shut behind it. The action takes place in less than one-hundredth of a second and makes a distinct popping sound, which can be heard with the aid of a sensitive underwater microphone. The trapped insect eventually dies, is broken down by bacteria, and the breakdown products are absorbed by cells in the walls of the bladder.

Science fiction writers have contributed to superstitions and beliefs that deep in the tropical jungles there are plants capable of trapping humans and other large animals. No such plants have been proved to exist, however. The largest pitcher plants known hold possibly 1 liter (roughly 1 quart) of fluid in their pitchers, and small frogs have been known to decompose in them, but the trapping of anything larger than a mouse or possibly a small rabbit seems very unlikely.

LEAF COLOR CHANGES IN AUTUMN

The chloroplasts of mature leaves contain several groups of pigments, such as green *chlorophylls,* and *carotenoids*—which include yellow *carotenes* and pale yellow *xanthophylls.* Each of these groups plays a role in photosynthesis. The chlorophylls are usually present in much larger quantities than the other pigments, and their intense green color masks or hides the presence of the carotenes and xanthophylls. In the fall, however, the chlorophylls break down, and other colors are revealed. The exact cause of the chlorophyll breakdown is not known, but it does appear to involve, among other factors, a gradual reduction in daylength. Water-soluble *anthocyanin* and *betacyanin* pigments may also accumulate in the vacuoles of the leaf cells in the fall. Anthocyanins, the more common of the two groups, are red if the cell sap is slightly acid and blue if it is slightly alkaline, and of intermediate shades if it is neutral. Betacyanins are usually red; they apparently are restricted to several plant families such as the cacti, the Goosefoot Family (to which beets belong), the Four-o'clock Family, and the Portulaca Family.

Some plants (e.g., birch trees) consistently exhibit a single shade of color in their fall leaves, but many (e.g., maple, ash, sumac) vary considerably from one locality to another or even from one leaf to another on the same tree, depending on the combinations of carotenes, xanthophylls, and other pigments present. Some of the most spectacular fall colors in North America occur in the Eastern Deciduous Forest, particularly in New England and the

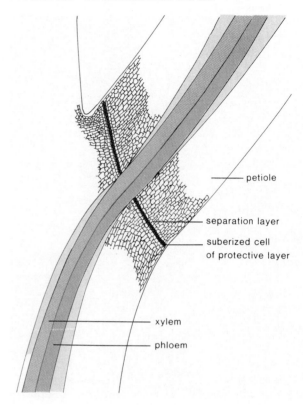

FIGURE 7.16   An abscission zone of a leaf.

petiole

separation layer

suberized cell of protective layer

xylem

phloem

upper reaches of the Mississippi Valley. In parts of Wisconsin and Minnesota one can observe the brilliant reds, oranges, and golds of maples, the deep maroons (and also yellows) of ashes, the bright yellows of aspen, and the seemingly glowing reds of sumacs and wahoos, all in a single locality. Some fall coloration is found almost anywhere in temperate zones where deciduous trees and shrubs exist.

ABSCISSION

Plants whose leaves drop seasonally are said to be **deciduous.** In temperate climates, new leaves are produced in the spring and are shed in the fall, but in the tropics the cycles coincide with wet and dry seasons rather than with temperature changes. Even so-called evergreen trees shed their leaves; they do so a few at a time, however, so that they never have the bare look of deciduous trees in their winter condition. The process by which the leaves are shed is called **abscission.**

Abscission occurs as a result of changes that take place in an *abscission zone* near the base of the petiole of each leaf (figure 7.16). Sometimes the abscission zone can be seen externally as a thin band of slightly different color on the petiole. Hormones that apparently inhibit the formation of the specialized layers of cells that facilitate abscission are produced in young leaves. As the leaf ages, hormonal changes take place, and at least two layers of cells become differentiated. Closest to the stem, the cells of the *protective layer,* which may be several cells deep, become coated and impregnated with fatty *suberin.* On the leaf side, a *separation layer* develops in which the cells swell, sometimes divide, and also become gelatinous. In response to any of several environmental changes (such as lowering temperatures, decreasing daylengths or light intensities, lack of adequate water, or damage to the leaf), the pectins in the middle lamella of the cells of the separation layer are broken down by enzymes. All that holds the leaf on to the stem at this point are some strands of xylem. Wind and rain then easily break the connecting strands, leaving tiny bundle scars within a leaf scar (see figure 5.1), and the leaf falls to the ground.

HUMAN AND ECOLOGICAL RELEVANCE OF LEAVES

Humans use shade trees and shrubs in landscaping for cooling as well as for aesthetic effects. The leaves of shade plants planted next to a dwelling can make a very significant difference in energy costs to the homeowner. Humans also use the leaves of cabbage, parsley, celery, lettuce, spinach, chard, and the petioles of rhubarb, to mention a few, as food. A large number of spices and flavorings are derived from leaves, including thyme, marjoram, oregano, tarragon, peppermint, spearmint, wintergreen, basil, dill, sage, and savory. Many dyes (e.g., a yellow dye from bearberry, a reddish dye from henna, and a pale blue dye from blue ash) are extracted from leaves (see appendix 3), although nearly all commercial dyes are now derived from coal tar. Many cordage fibers for ropes and twines come from leaves, with various species of *Agave* (century plants) accounting for about 80% of the world's production. Bowstring fibers are obtained from a relative of the common houseplant *Sansevieria,* and Manila hemp fibers, which are used both in fine-quality cordage and in textiles, are obtained from the leaves of a close relative of the banana. Panama hats are made from the leaves of the panama hat palm, and palms and grasses are used in the tropics as thatching material for huts and other buildings.

FIGURE 7.17   A pennyroyal plant.

In the high mountains of Chile and Peru, the leaves of the yareta plant are used for fuel. They produce a resin that causes the leaves to burn with an unusually hot flame. Leaves of many plants produce oils. Petitgrain oil from a variety of orange tree leaves and lavender, for example, are used for scenting soaps and colognes. Patchouli and lemongrass oils are used in perfumes, as is citronella oil, which was once the leading mosquito repellent before synthetic repellents gained favor. Eucalyptus oil, camphor, cajeput, and pennyroyal (figure 7.17) are all used medicinally.

Leaves are an important source of drugs used in medicine and also of narcotics and poisons. Cocaine, obtained from plants native to South America, is used medicinally as a local anesthetic, and its use as a narcotic has, in recent years, become a major problem in western cultures. Andean natives chew coca leaves while working and are reported capable of performing exceptional feats of labor with little or no food while under the influence of cocaine. Apparently the drug, which is addictive, anesthetizes the nerves that convey hunger pangs to the brain. Belladonna is a drug complex obtained from leaves of the deadly nightshade, a native of Europe. The plant has been used in medicine for centuries, and several drugs are now isolated from belladonna. Included among the isolates are atropine, which is used in shock treatments, to dilate eyes, to relieve pain locally, and to check secretions. Scopolamine, also a belladonna derivative, is used in tranquilizers and sleeping aids. Another European plant, the foxglove, is the source of digitalis. This drug has been used for centuries in regulating blood circulation and heartbeat.

At present, more than 940 million kilograms (2 billion pounds) of tobacco for smoking, chewing, and use as snuff are produced annually around the world. Also at present, about 125,000 Americans die annually from lung cancer, and almost all of them have a history of cigarette smoking. Cigarette smoking is also evidently a principal contributor to cardiovascular diseases. During recent years the increase in the use of chewing tobacco has seen a corresponding increase in the development of mouth and throat cancers. Federal law forbids concentrations of more than five parts per billion of nitrosamines (cancer-causing chemicals) in cured meats, but the levels of nitrosamines in the five most popular brands of chewing tobacco range from 9,600 to 289,000 parts per billion. Though humans have long used tobacco, it has only been in the last two decades that its health threat has become appreciated.

Similarly, marijuana, the controversial plant widely used as an intoxicant, has been utilized in various ways for thousands of years. The active principle, tetrahydrocannabinol (THC), although found in the leaves, is concentrated in hair secretions among the female flowers. In recent decades marijuana has found increasing acceptance in the western hemisphere; it appears, however, not to be the harmless intoxicant many thought it to be. Regular use of marijuana for a year has been shown to have the same effect on human lungs as smoking one and a half packs of cigarettes a day for 13 years. A decision to legalize its prescription in pill form for the alleviation of nausea caused by chemotherapy treatments has been criticized by some medical authorities.

The drug lobeline sulphate is obtained from the leaves of a close relative of garden lobelias. It is used in compounds taken by those who are trying to stop smoking. The leaves of several species of *Aloe,* especially *Aloe vera,* yield a juice that is used to treat various types of skin burns, including those accidentally received from X-ray equipment.

Several beverages are extracted by the brewing of leaves. Numerous teas have been obtained from a wide variety of plants, but most now in use come from a close relative of the garden camellia. Maté,

the popular South American tea, is brewed from the leaves and twigs of a relative of holly. The alcoholic beverages pulque and tequila find their origin in the mashed leaves of *Agave* plants, and absinthe liqueur receives its unique flavor from the leaves of wormwood, a relative of western sagebrush, and other flavorings such as anise.

Insecticides of various types are also derived from leaves. A type of rotenone and a substance related to nicotine are obtained from tropical plants; both are effective against a variety of insects. Mexico's cockroach plant has leaves that when dried are highly effective in killing cockroaches, flies, fleas, and lice. Water extracts of the leaves (as well as other parts) of India's Neem Tree are reported to control more than 100 species of insects, mites and nematodes, with little effect on useful predator insects.

Carnauba and caussu waxes are obtained from the leaves of tropical palms. Leaves are used extensively by florists in floral arrangements and bouquets, and their uses for other aesthetic purposes are legion. Leaves may find more extensive use in the future as a direct source of food. It has been shown that a curd obtained by coagulating juice squeezed from alfalfa leaves contains more than 40% protein, and juices from other leaves have yielded better than 50% protein. Experiments are under way to make the leaf curd palatable for human consumption. Additional information on past and present uses of leaves is given in chapter 22 and in appendices 2, 3, and 4. The scientific names of the plants that are the sources of the various substances discussed are given in appendix 1.

SUMMARY

Leaves are similar to solar panels in that they are covered with a transparent epidermis that admits light to the interior, and many twist on their petioles so that their flat surfaces are inclined to the sun throughout the day. Their lower and often their upper surfaces are dotted with stomata that permit air circulation and that function in other ways to facilitate photosynthesis, the primary function of leaves. Leaves also respire, accumulate wastes, and eliminate excess moisture via transpiration. The shapes, sizes, and textures of leaves vary greatly.

All leaves originate as primordia. Most leaves consist of a petiole and a blade. Some have paired stipules at the base of the petiole. Leaves of flowering plants may be simple or compound, and all are associated with leaf gaps. Both the venation and the compounding of the leaves may be either pinnate or palmate. In pinnate leaves, the main vein is the midrib in simple leaves and the rachis in compound leaves. A few leaves with evenly forking veins are said to have dichotomous venation.

Three groups of tissues occur in leaves. The epidermis is a single layer of cells covering the entire surface; it is coated with a thin cuticle, and may also have waxes, glands, hairs, and an occasional crystal within the cells. The lower epidermis usually contains numerous pores (stomata), each formed by a pair of guard cells that regulate both evaporation of water vapor from the leaf and gas exchange between the interior and the atmosphere. The mesophyll between the upper epidermis and the lower epidermis is divided into an upper palisade layer, which consists of rows of tightly packed parenchyma cells containing numerous chloroplasts, and a lower spongy layer in which the parenchyma cells are loosely arranged. The spongy layer contains a considerable amount of air space. Veins traverse the mesophyll.

Leaves may be specialized in many ways, each modification usually being associated with specific combinations of environmental factors. Plants in arid regions may have thick leaves, a reduction in the number of stomata, and water-storage leaves, or no leaves at all. Other specialized leaves include tendrils, which aid in the support of the plant; spines, which reduce leaf surface and aid in protection; food-storage leaves; flower pot leaves; window leaves; reproductive leaves; floral leaves (bracts); and several types of insect-trapping leaves.

Leaves drop from plants in autumn because of changes that occur in the abscission zone at the base of the petiole of each deciduous leaf. A protective layer of suberized cells and a separation layer of cells, which swell and become gelatinous, form adjacent to one another when hormones that inhibit the formation of these layers fall below a certain level in apparent response to environmental changes. Leaves change color in the autumn because the green chlorophyll pigments break down, revealing pigments of other colors. In addition, water-soluble anthocyanin and betacyanin pigments that may be red or blue in color, accumulate in cell vacuoles. The leaves of some plants consistently turn a certain color in the fall, whereas leaves of other plants vary in color, depending on their locality or even on their position on the tree.

REVIEW QUESTIONS

1. Leaves have no secondary xylem and phloem. Why not?
2. What are *bracts*?
3. Identify or define hydathode, transpiration, guard cells, mesophyll, venation, glands, and compound leaf.
4. How can one distinguish between the upper and lower epidermis in most leaves?
5. What is the function of *bundle sheaths*?
6. How do leaves in shaded areas differ from leaves in sunlit areas on the same plant?
7. What leaf modifications are associated with dry areas, wet areas (e.g., lakes), climbing, and reproduction?
8. Why do leaves turn different colors in the fall?

DISCUSSION QUESTIONS

1. Can you think of any advantages or disadvantages to a plant in having very tiny or very large leaves?
2. In chapter 3, it was noted that living cells are connected to one another by plasmodesmata that extend through tiny holes in the walls. If this is true, does a leaf really need veins for the acquisition and distribution of materials?
3. Is there any advantage to a leaf in having palisade mesophyll on the upper half and spongy mesophyll on the lower half?
4. Pines and other plants that grow in very cold climates have sunken stomata just like plants of the desert, yet the annual precipitation where they grow may be very abundant. Is there any advantage to them in having such a modification?
5. Since tendrils and spines can be either modified leaves or modified stems, how might you determine the origin of a given specimen?

ADDITIONAL READING

Addicott, F. T. 1982. *Abscission.* Berkeley, CA: University of California Press.

Cutter, E. G. 1971. *Plant anatomy: experiment and interpretation. Part II: Organs.* Reading, MA: Addison-Wesley Publishing Co.

Dale, J. E., and F. L. Milthorpe, eds. 1983. *The growth and functioning of leaves.* Fair Lawn, NY: Cambridge University Press.

Darwin, C. 1884. *Insectivorous plants.* New York: Murray.

Esau, K. 1977. *Anatomy of seed plants.* 2d ed. New York: John Wiley & Sons.

Foster, A. S., and E. M. Gifford, Jr. 1974. *Comparative morphology of vascular plants.* 2d ed. San Francisco, CA: W. H. Freeman and Co.

Leaves: One hundred seventy-seven photographs by Andreas Feininger. 1984. Mineola, NY: Dover Publications.

Lloyd, F. E. 1976. *Carnivorous plants.* New York: Dover Publications, Inc.

Maksymowych, R. 1973. *Analysis of leaf development.* Fair Lawn, NY: Cambridge University Press.

Martin, E. S., and M. E. Donkin. 1983. *Stomata.* Baltimore, MD: E. Arnold Publications.

Philipson, W. R., and E. E. Balfour. 1965. *Vascular patterns in dicotyledons.* Botanical Review 29: 382–404.

Prance, G. T. 1985. *Leaves: The formation, characteristics and uses of hundreds of leaves found in all parts of the world.* New York: Crown Publishers, Inc.

Steeves, T. A., and I. M. Sussex. 1972. *Patterns in plant development.* Englewood Cliffs, NJ: Prentice-Hall.

Telek, L., and H. D. Graham. 1983. *Leaf protein concentrates.* Westport, CT: AVI Publishing Co.

Williams, R. F. 1975. *The shoot apex and leaf growth.* Fair Lawn, NY: Cambridge University Press.

Overview

*I*n this chapter the structure and
parts of flowers are described,
and some modifications are
noted; a brief comparison between
dicots and monocots is made. The
nature and development of fruits is
discussed next, and fruit structure and
parthenocarpy are described. A listing
and discussion of the various types of
fleshy and dry fruits follows, and then
the agents of fruit and seed dispersal
are explored. The chapter concludes
with an examination of seed structure
and germination, and some
observations on seed dormancy,
longevity of seeds, stratification, and
vivipary.

8   Flowers, Fruits, and Seeds

Some Learning Goals

1. Know the parts of a typical flower and the function of each part.

2. Learn the features that distinguish monocots from dicots.

3. Understand the distinction between a fruit and a vegetable.

4. Know the regions of mature fruits.

5. Learn five types of fleshy and dry fruits, and know how simple, aggregate, and multiple fruits are derived from the flowers.

6. Learn the adaptations of fruits and seeds to the agents by which they are dispersed.

7. Diagram and label a mature dicot seed (e.g., bean) and a monocot seed (e.g., corn) in section to show the parts and regions.

8. Understand the changes that occur when a seed germinates and note the environmental conditions essential to germination.

9. Know the types of factors that control dormancy; learn how dormancy may be broken both naturally and artificially.

Outline

INTRODUCTION

Several years ago an Australian farmer glanced back at a field he was plowing and was startled to see what looked like flowers being tossed to the surface of the furrows. Closer inspection revealed that they were indeed flowers but both they and the plants to which they were attached were pale and devoid of chlorophyll. He had the foresight to report his find to a university, where botanists determined that the farmer had stumbled upon the first known underground flowering plant (figure 8.1). The plant proved to be an orchid that derived its nourishment from organic matter in the soil, and was pollinated by tiny flies that gained access to the below-ground flowers via mud cracks that developed in the dry season.

The underground-flowering orchid is but one of some 240,000 known species of flowering plants; additional unknown numbers of undescribed species occur in remote areas, particularly in the tropics. The flowering plants are vital to humanity in that they provide countless useful products, with just 11 species—10 of them members of the Grass Family—furnishing 80% of the world's food (see the amplification of this subject in chapters 21 and 22).

FIGURE 8.1   Flowers of an Australian orchid produced by plants that complete their entire life cycle below ground. (Drawing Courtesy Karen Hamilton)

FIGURE 8.2   A *Rafflesia* flower that is nearly 1 meter (3 feet 3 inches) in diameter. (Photograph © Kjell B. Sandved)

FIGURE 8.3 Male catkins of an alder. Each catkin consists of numerous tiny, inconspicuous, wind-pollinated flowers that have no petals.

FIGURE 8.4 Spanish moss hanging from a tree. Spanish moss is a nonparasitic flowering plant that goes through its life cycle suspended from other plants or objects such as wires. This plant should not be confused with *lichens* (see chapter 17), which may also hang suspended from other objects.

The flowers themselves range in size from the minute flowers of the duckweed *Wolffia columbiana,* whose entire speck of a plant body is only 0.5 to 0.7 millimeter (0.02 to 0.03 inch) wide, with flowers little more than 0.1 millimeter long, to the enormous *Rafflesia* flowers (figure 8.2) of Indonesia that are up to 1 meter (3 feet 3 inches) in diameter and may weigh 9 kilograms (20 pounds) each. Flowers may be any color or combination of colors of the rainbow, as well as black or white; they may have virtually any texture, from filmy and transparent to thick and leathery, spongy to sticky, hairy, prickly, or even dewy wet to the touch. Numerous flowers of trees, shrubs and garden weeds are quite inconspicuous and lack odor (figure 8.3), but many others are strikingly beautiful, particularly when examined with a dissecting microscope. Their fragrances, which can be exhilarating to seductive or even putrid, are the basis of international perfume and pet repellent industries.

The habitats of flowering plants are as varied as their form. In addition to the underground habitat we have noted, some go through their life cycles dangling from wires or other plants (figure 8.4). They also occur in both fresh and salt water, in the cracks and crevices of rocks, in deserts and jungles, in frigid arctic regions, and in areas where the temperatures regularly soar to 45° C (113° F) in the shade. In fact, they occur almost anywhere they can receive their basic needs of light, moisture, and a minimal supply of minerals. A species of chickweed survives at an altitude of 6,135 meters (over 20,000 feet) in the Himalaya Mountains, and fumitory plants in the same area flower even when the night temperatures plummet to −18° C (0° F).

Flowering plants can complete their life cycles in less than a month, or they may take as long as 150 years. Those that complete the cycle in a single season are called **annuals. Biennials** take two years to complete the cycle; **perennials,** however, may take several to many years to go from a germinated seed to a plant producing new seeds. Perennials may also produce flowers on new growth, which dies back each winter, while other parts of the plant persist indefinitely.

Flowering plants occur in two major classes, the Dicotyledonae and the Monocotyledonae, commonly referred to as **dicots** and **monocots,** respectively. Members of the two classes are distinguished from one another on the basis of features listed in table 8.1.

DIFFERENCES BETWEEN DICOTS AND MONOCOTS

Slightly less than three-fourths of all flowering plant species are dicots. Dicots include many annual plants and virtually all flowering trees and shrubs. Monocots, which are primarily herbaceous, include plants that produce bulbs (e.g., lilies), grasses and related plants, orchids, irises, and palms; they are believed to have developed from ancestors derived from primitive dicots.

TABLE 8.1
Some Differences between Dicots and Monocots

Dicots	Monocots
1. Seed with two cotyledons (seed leaves)	1. Seed with one cotyledon (seed leaf)
2. Flower parts mostly in mutliples of four or five	2. Flower parts in multiples of three
3. Leaf with a distinct network of veins	3. Leaf with more or less parallel veins
4. Vascular cambium, and frequently cork cambium, present	4. Vascular cambium and cork cambium absent
5. Vascular bundles in a ring	5. Vascular bundles scattered
6. Pollen grain mostly with three *apertures* (thin areas in the wall—see figure 21.4)	6. Pollen grain mostly with one *aperture*

STRUCTURE OF FLOWERS

Regardless of form, all flowers share certain basic features. Let us look briefly at a peach flower to familiarize ourselves with the parts of a typical flower (figure 8.5). Each flower, which begins as an embryonic *primordium* that develops into a *bud* (see chapter 5) occurs as a specialized branch at the tip of a stalk called a **peduncle.** The peduncle swells at its tip into a small pad known as the **receptacle.** The other parts of the flower, some of which are in whorls, are attached to this receptacle. The outermost whorl consists of five small green, somewhat leaflike **sepals.** The sepals of a flower are collectively referred to as the **calyx.** In many flowers, the calyx functions in protection, particularly while the flower is in the bud. The next whorl of flower parts consists of petals, which are known collectively as the **corolla.** When corollas are showy, they function in attracting pollinators such as bees, moths, or birds. Some may have special markings that are invisible to humans but can be seen by bees whose vision functions in the ultraviolet range of the light spectrum (see figure 23.19). The corolla is often missing in wind-pollinated plants (see figures 8.3 and 22.24). In peach flowers, the petals are distinct separate units, but in other flowers, such as petunias or jimson weeds, the petals are fused together so that the corolla consists of a single, flared, trumpetlike sheet of tissue (figure 8.6). Sepals also may be fused together. While peach

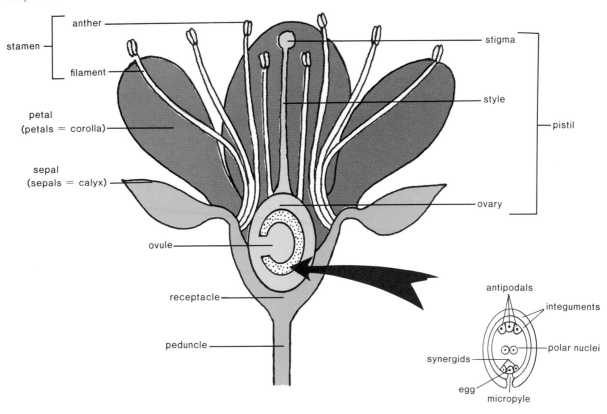

FIGURE 8.5   Parts of a typical flower. The interior structure of the ovule and the sexual process involved are discussed in chapter 21.

stamen
anther
filament

petal
(petals = corolla)

sepal
(sepals = calyx)

ovule

receptacle

peduncle

stigma

style

pistil

ovary

antipodals
integuments
polar nuclei
synergids
egg
micropyle

FIGURE 8.6   A jimson weed flower.

FIGURE 8.7   Types of inflorescences. Each ball represents a flower.

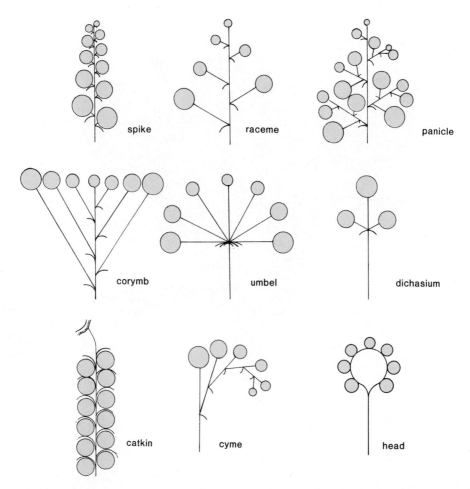

spike   raceme   panicle

corymb   umbel   dichasium

catkin   cyme   head

flowers are produced singly on their own peduncles, those of lilacs, grapes, bridal wreaths, and many others are produced in clusters called **inflorescences** (figure 8.7). In an inflorescence, the peduncle has numerous little stalks serving individual flowers. These additional little stalks, one to each flower, are called **pedicels.**

The **pistil,** a small greenish vaselike structure in the center of the peach flower, consists of three regions that merge with one another. At the top is a slight swelling called the **stigma,** which is connected by a slender stalklike **style** to the swollen base called the **ovary**. The ovary later develops into a *fruit*. In a peach flower the calyx and corolla are attached to the receptacle at the base of the ovary and the ovary is said to be **superior,** but in other flowers the receptacle grows up around the ovary so that the calyx and corolla appear to be attached at the top; such

flowers are said to have **inferior** ovaries (see figure 21.9 for variations in ovary positions). Within the ovary is a cavity containing an egg-shaped **ovule,** which is attached to the wall of the cavity by means of a short stalk. The ovule (the development of which is described in chapter 21) eventually becomes a **seed.**

Several to many **stamens** are attached to the receptacle around the base of the pistil. Each stamen consists of a semirigid but otherwise slender **filament** with a sac called an **anther** at the top. The development of **pollen grains** (see figure 21.4) in anthers is described in chapter 21 (see figures 21.3 and 21.5). In most flowers the release of pollen is facilitated by lengthwise slits that develop on the anthers, but pores that function in the same way develop on the anthers of members of the Heath Family and those of a few other groups.

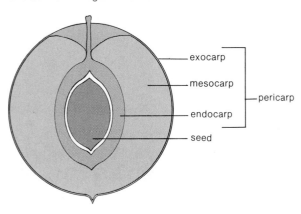

FIGURE 8.8   Regions of a mature fruit.

exocarp

mesocarp

pericarp

endocarp

seed

FRUITS

Introduction

In 1893 the United States Supreme Court, in the case of Nix vs. Hedden, ruled that a tomato was legally a vegetable rather than a fruit. This was in keeping with the general conception in the public's mind that fruits tend to be relatively sweet and in the nature of a dessert food, while vegetables tend to be more savory as salad or main course foods. Be that as it may, botanically speaking a **fruit** is any ovary that has developed and matured. It also usually contains seeds. By this definition, many so-called vegetables, including tomatoes, string beans, cucumbers, and squashes, are really fruits. On the other hand, vegetables can consist of leaves (e.g., lettuce, cabbage), stems (e.g., white potato), roots (e.g., sweet potato), stems and roots (e.g., beets), flower buds (e.g., globe artichoke), or other parts of the plant. All fruits arise from flowers and thus are found exclusively in the flowering plants. *Fertilization* (see chapter 21) usually indirectly determines whether or not the ovary or ovaries (and sometimes additional tissues such as the receptacle) of a flower will develop into a fruit. If at least a few of the ovules are not fertilized, the flower normally withers and drops without developing further. Pollen grains contain specific stimulants called *hormones* (discussed in detail in chapter 11) that may initiate fruit development, and sometimes a quantity of dead pollen is all that is needed to stimulate an ovary into becoming a fruit. It is the hormones produced by the developing seeds, however, that promote the greatest fruit growth. These hormones, in turn, stimulate the production of more fruit growth hormones by the ovary wall.

Fruit Regions

By the time an ovary has matured into a fruit, the bulk of it is divided into three regions, which are, however, sometimes difficult to distinguish from one another (figure 8.8). The skin forms the **exocarp,** while the inner boundary around the seed(s) forms the **endocarp.** The endocarp may be hard and stony (as in a peach pit around the seed). It also may be papery (as in apples), or it may not be distinct from the **mesocarp,** which is the name given to everything between the exocarp and the endocarp. The three regions are collectively called the **pericarp.** In dry fruits, the pericarp is often quite thin.

Some fruits consist of only the ovary and its seeds. Others have adjacent flower parts such as the receptacle or calyx fused to the ovary or different parts modified in various ways. Fruits may be either fleshy or dry at maturity, and they may split, exposing the seeds, or no split may occur. They may be derived from a single ovary or from more than one. Traditionally all these features have been used to classify fruits, but unfortunately not all fruits lend themselves to neat pigeonholing by such characteristics. Some of these problems are pointed out in the classification that follows.

Kinds of Fruits

Fleshy Fruits

Fruits whose mesocarp is at least partly fleshy at maturity are classified as fleshy fruits.

Simple Fleshy Fruits

Simple fleshy fruits are fruits that develop from a flower with a single pistil. The ovary may be superior or inferior, and it may be *simple* (derived from a single pistil) or it may consist of one or more segments called **carpels** and thus be *compound* (for a discussion of the derivation of carpels and compound ovaries see chapter 21). The ovary alone may develop into the fruit, or other parts of the flower may develop with it.

FIGURE 8.9 Representative drupes. *A.* Peach. *B.* Almond. *C.* Olive.

A.

C.

B.

Drupe A **drupe** is a simple fleshy fruit with a single seed enclosed by a hard, stony endocarp, or pit (figure 8.9). It usually develops from flowers with a superior ovary containing a single ovule. The mesocarp is not always obviously fleshy, however. In coconuts, for example, the *husk* (consisting of the mesocarp and the exocarp), which is usually removed before the rest of the fruit is sold in markets, is very fibrous (the fibers are used in making mats and brushes). The seed or "meat" of the coconut is hollow and contains a watery *endosperm* (see chapter 21) commonly but incorrectly referred to as "milk." It is surrounded by the thick, hard endocarp typical of drupes. Other examples of drupes include the stone fruits (e.g., apricots, cherries, peaches, plums), olives, and almonds. In almonds, the husk, which dries somewhat and splits at maturity, is removed before marketing, and it is the endocarp that we crack to obtain the seed.

FIGURE 8.10   Representative berries. *A.* Grapes.
B. Tomatoes.

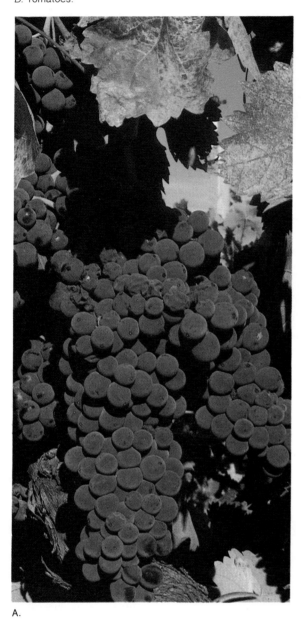

A.

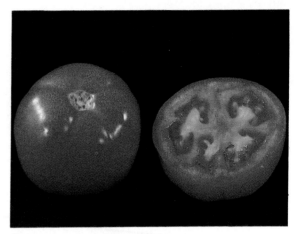

B.

Berry **Berries** usually develop from a compound ovary and commonly contain more than one seed. The entire pericarp is fleshy, and it is difficult to distinguish between the mesocarp and the endocarp (figure 8.10). Three types of berries may be recognized.

A *true berry* is a fruit with a thin skin and one in which the entire pericarp is relatively soft at maturity. Although most contain more than one seed, notable exceptions are dates and avocados, which have only one seed. Typical examples of true berries include tomatoes, grapes, persimmons, peppers, and eggplants. Some fruits that popularly include the word "berry"in their common name (e.g., strawberry, raspberry, blackberry) botanically are not berries at all.

Some berries are derived from flowers with inferior ovaries so that other parts of the flower also contribute to the flesh. They can usually be distinguished by the remnants of flower parts or their scars that persist at the tip. Examples of such berries include gooseberries, blueberries, cranberries, pomegranates, and bananas. In the cultivated banana, fruit development is *parthenocarpic* (i.e., the fruit develops without the ovules having been fertilized—see chapter 21), so there are no seeds. Several other species of bananas produce an abundance of seeds.

The *pepo* has a relatively thick rind. Fruits of members of the Pumpkin Family, including pumpkins, cucumbers, watermelons, squashes, and cantaloupes, are pepos.

FIGURE 8.11 Apples (representative of pomes). The bulk of the flesh is derived from the receptacle and the bases of the calyx, corolla, and stamens.

- endocarp
- seed
- vascular bundle in outer part of the ovary

The *hesperidium* is a berry with a leathery skin containing oils. Numerous outgrowths from the inner lining of the ovary wall become saclike and swollen with juice as the fruit develops. All members of the Citrus Family produce this type of fruit. Examples include oranges, lemons, limes, grapefruit, tangerines, and kumquats.

Pome **Pomes** are simple fleshy fruits, the bulk of whose flesh comes from the enlarged receptacle that grows up around the ovary. The endocarp around the seeds is papery or leathery. Examples include apples, pears, and quinces. In an apple, the ovary consists of the core and a little adjacent tissue. The remainder of the fruit has developed primarily from the receptacle (figure 8.11). Botany texts often refer to pomes, pepos, some berries, and other fruits derived from more than an ovary alone as *accessory fruits* or as fruits having *accessory tissue.*

Aggregate Fruits

An **aggregate fruit** is one that is derived from a single flower with several to many pistils. The individual pistils develop into tiny drupes or other fruitlets, but they mature as a clustered unit on a single receptacle (figure 8.12). Examples include raspberries, blackberries, and strawberries. In a strawberry, the cone-shaped receptacle becomes fleshy and red, while each pistil becomes a little *achene* (described in the following section on dry fruits) on its surface. In other words, the strawberry, while being an aggregate fruit, is also partly composed of accessory tissue.

A.

B.

Multiple Fruits

Multiple fruits are derived from several to many individual flowers in a single inflorescence. Each flower has its own receptacle, but as the flowers mature separately into fruitlets they develop together into a single larger fruit, as in aggregate fruits. Examples of multiple fruits include mulberries (figure 8.13), osage oranges, pineapples, and figs. Pineapples, like bananas, usually develop parthenocarpically, and there are no seeds. The individual flowers are fused together on a fleshy axis, and the fruitlets coalesce into a single fruit. Figs mature from a unique "outside in" inflorescence. The individual flowers of the

FIGURE 8.13   Inflorescence and fruit of a mulberry.

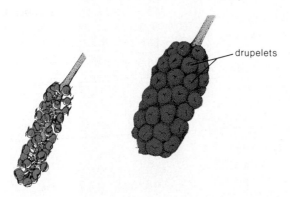

drupelets

FIGURE 8.14   Section through a developing fig.

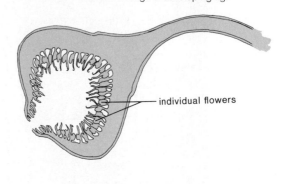

individual flowers

FIGURE 8.15   Follicles. *A.* Milkweed. *B. Magnolia.* The fruit of the magnolia is actually an aggregate fruit consisting of approximately 40 to 80 individual 1-seeded follicles on a common axis; the follicles and axis fall from the tree as a unit.

A.

B.

FIGURE 8.16 A legume of a garbanzo bean plant after it has split open. The two halves of the legume have curled on drying, releasing the seeds with force. (Courtesy Field Museum of Natural History, Chicago)

FIGURE 8.17 *A*. A silique after it has split open. The seeds are borne on a central, membranous partition. *B*. Silicles of a dollar plant. (*B*. Courtesy Field Museum of Natural History, Chicago)

A.

B.

inflorescence are enclosed by the common receptacle, which has an opening to the outside at the tip (figure 8.14). Some varieties develop parthenocarpically, but others are pollinated by tiny wasps that crawl in and out through the opening. Some multiple fruits, such as those of the sweet gum, are dry at maturity.

Dry Fruits

Fruits whose mesocarp is definitely dry at maturity are classified as *dry fruits*.

Dry Fruits That Split at Maturity

The fruits in this group are distinguished from one another by the manner in which they split.

Follicle The **follicle** splits along one side or seam only, exposing the seeds within (figure 8.15). Examples include larkspur, columbine, milkweed, and peony.

Legume The **legume** splits along two sides or seams (figure 8.16). Literally thousands of members of the Legume Family produce this type of fruit. Examples include peas, beans, garbanzo beans, lentils, carob, kudzu, and mesquite. Peanuts are also legumes, but they are atypical in that the fruits develop and mature underground. The seeds are usually released in nature by bacterial breakdown of the pericarp instead of through an active splitting action.

Silique **Siliques** also split along two sides or seams, but the seeds are borne on a central partition, which is exposed when the two halves of the fruit separate (figure 8.17A). Such fruits, when they are less than three times as long as they are wide, are called *silicles* (figure 8.17B). Siliques and silicles are typically produced by members of the Mustard Family, which includes broccoli, cabbage, radish, shepherd's purse, and watercress.

FIGURE 8.18   Capsules. *A.* Butterfly iris. *B. Bletilla* orchid.
C. Unicorn plant. *D.* Poppy. Note the row of pores through
which the seeds are released near the top.

A.

B.

C.

D.

Capsule **Capsules** are the most common of the dry fruits that split (figure 8.18). They consist of at least two carpels and split in a variety of ways. Some split along the partitions between the carpels, while others split through the cavities (*locules*) in the carpels. Still others form a cap toward one end that pops off and permits release of the seeds, or they form a row of pores through which the seeds are shaken out as the capsule rattles in the wind. Examples include irises, orchids, lilies, poppies, violets, and snapdragons.

Dry Fruits That Do Not Split at Maturity

In this type of dry fruit, the single seed is fused or attached to the pericarp to varying degrees.

Achene The single seed of the **achene** is attached to its surrounding pericarp only at its base. Thus the husk (pericarp) is relatively easily separated from the seed. Examples include sunflower "seeds" (the edible kernel plus the husk constitute the achene) (figure 8.19), buttercup, and buckwheat.

Nut **Nuts** are one-seeded fruits similar to achenes, but they are generally larger, and the pericarp is much harder and thicker. They develop with a cup, or cluster, of bracts at their base. Examples include acorns (figure 8.19), hazel nuts, hickory nuts, and chestnuts. Many nuts in the popular sense are not nuts, botanically speaking. We have already seen that peanuts are atypical legumes, and that coconuts and almonds are drupes. Walnuts and pecans are also drupes, whose "flesh" withers and dries after the seed matures. Brazil nuts are the seeds of a large capsule, and a cashew nut is the single seed of a peculiar drupe. It appears as a curved appendage at the end of a swollen pedicel, which is eaten raw in the tropics or made into preserves or wine. Pistachio nuts are also the seeds of drupes.

Grain (Caryopsis) The pericarp of the **grain** is tightly fused to the seed and cannot be separated from it (figure 8.19). All members of the Grass Family, including corn, wheat, rice, oats, and barley, produce grains (also called **caryopses**).

Samara In **samaras** the pericarp, which extends out in the form of a wing or membrane around the seed, aids in dispersal (figure 8.20). Samaras are produced in pairs in maples. In ashes, elms, and the tree of heaven they are produced singly.

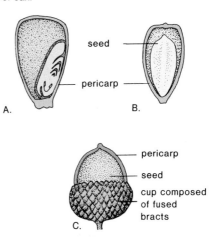

FIGURE 8.19 Dry fruits that do not split at maturity. *A.* Grain (caryopsis) of corn. *B.* Achene of sunflower. *C.* Nut (acorn) of oak.

FIGURE 8.20 Samaras. *Top:* Maple. *Center:* Elm. *Bottom:* Ash.

FIGURE 8.21   Schizocarps of carrots.

FIGURE 8.22   Types of seeds and fruits dispersed by wind.

Schizocarp   The twin fruit called a **schizocarp** (figure 8.21) is unique to the Parsley Family. Members of this family include parsley, carrots, anise, caraway, and dill. The twin fruits break into two one-seeded segments (fruitlets) upon drying.

FRUIT AND SEED DISPERSAL

Why are so many species of orchids rare, while dandelions, shepherd's purse, and other weeds occur all over the world? Why are some plants confined to single continents, mountain ranges, or small niches occupying less than a hectare (2.47 acres) of land? The answers to these questions involve many different factors, including climate, soil, the adaptability of the plant, and its means of seed dispersal. How fruits and seeds are transported from one place to another is the subject of the following sections. Other factors are discussed in chapters 13 and 23.

Dispersal by Wind

Fruits and seeds have a variety of adaptations for wind dispersal (figure 8.22). The samara of a maple has a curved wing that causes the fruit to spin as it is released from the tree. In a brisk wind, samaras may be carried up to 10 kilometers (6 miles) away from their source. In hop hornbeams, the seed is enclosed in an inflated sac that gives it some buoyancy in the wind. In some members of the Buttercup and Sunflower Families, the fruits have plumes, and in the Willow Family the fruits are surrounded by cottony or woolly hairs that aid in wind dispersal. In button snakeroots and Jerusalem sage, the fruits are too large to be airborne, but they are spherical enough to be rolled along the ground by the wind. Seeds themselves may be so tiny and light that they can be blown great distances by the wind. Orchids

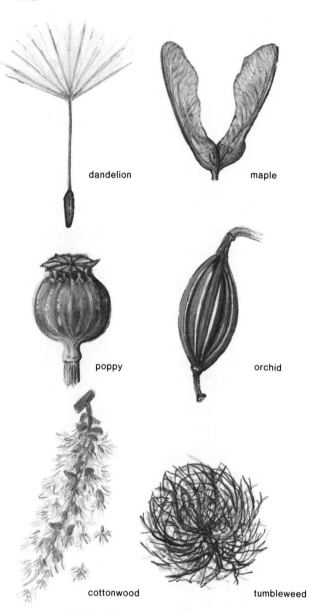

dandelion

maple

poppy

orchid

cottonwood

tumbleweed

and heaths, for example, produce seeds that are as fine as dust and equally light in weight. In trees such as catalpas and jacarandas, the seeds themselves are winged rather than the fruits, which remain on the branches and split, releasing their contents. Dandelion fruitlets have plumes that radiate out at the ends like tiny parachutes; these catch even a slight breeze. In tumble mustard and other tumbleweeds, the whole aboveground portion of the plant may break off and be blown away by the wind, releasing seeds as it bumps along.

FIGURE 8.23 Types of seeds and fruit dispersed by animals and birds.

mistletoe berries

cocklebur

bedstraw fruit

bur clover fruit

hoof of animal

capsule of
unicorn plant

to their feet. Other birds and mammals eat fruits whose seeds pass unharmed through their digestive tracts. In blackbirds, the seeds may remain in the tract as little as 15 minutes, but in mammals the period is more commonly about 24 hours. In the giant tortoises of the Galápagos Islands, seeds do not pass through the tract for two weeks or more, and the seeds usually will not germinate unless they have been subjected to such treatment (see chapter 11). Some seeds and fruits are gathered and stored by rodents such as squirrels and mice and then are abandoned. Blue jays, woodpeckers, and other birds carry away nuts and other fruits, which they may drop in flight and abandon.

Many fruits and seeds catch in or adhere to the fur or feathers of animals and birds. Some, such as those of bedstraw and bur clover, are covered with small hooks that catch in fur (or a hiker's socks). The large capsules of unicorn plants have two giant curved extensions about 15 centimeters (6 inches) long. These catch in the fetlock of a deer or other animal that happens to step on the fruit, and the seeds are scattered as the animal moves along. Twinflowers and flax have fruits with sticky appendages that adhere to fur on contact, and those of the puncture vine penetrate the skin and stick by means of vicious little thorns (figure 8.24). Bleeding hearts, trilliums, and several dozen other plants have

Dispersal by Animals

The adaptations of fruits and seeds for animal dispersal are legion. Birds, mammals, and ants all act as disseminating agents (figure 8.23). Shore birds may carry seeds great distances in mud that adheres

FIGURE 8.25   Seeds of the Pacific bleeding heart. The glistening appendages are *elaiosomes*, which are removed from the seeds by ants and used for food.

FIGURE 8.26   Sedge adaptation to water dispersal. *A.* A sedge fruit. The seed is enclosed within an inflated covering that enables it to float on water. *B.* A sedge plant.

A.   B.

appendages on their seeds that contain oils attractive to ants (figure 8.25). The Scandinavian scientist Sernander once estimated that more than 36,000 such seeds were carried by members of a single ant colony to their nest, where the ants stripped off the appendages for food but did not harm the seeds themselves.

Dispersal by Water

Some fruits are adapted to water dispersal by virtue of the fact that they contain trapped air. Many sedges, for example, have inflated sacs around the seeds (figure 8.26) that enable them to float. Others have waxy material on the surface of the seeds, which temporarily prevents them from absorbing water while they are floating. Sometimes a heavy downpour will create a torrent of water that dislodges

masses of vegetation along a stream bank, carrying whole plants and their fruits to new locations. Large raindrops themselves may splash seeds out of their opened capsules. Seeds and fruits of a few plants have thick spongy pericarps that absorb water very slowly. Such fruits are adapted to dispersal by ocean currents, even though salt water eventually may penetrate enough to kill the delicate embryos. Enough fruits are beached before this occurs to ensure the survival of the species. The best known of ocean dispersed plants is the coconut palm, whose large fruits apparently have been carried many hundreds of kilometers throughout the tropical seas of the world.

Other Dispersal Mechanisms

Fruits of many families mechanically eject seeds, sometimes with considerable force. For example, as the capsules of witch hazel dry, a splitting action occurs that may fling the seeds over 12 meters (40 feet) away. Seeds of some legumes and touch-me-nots also

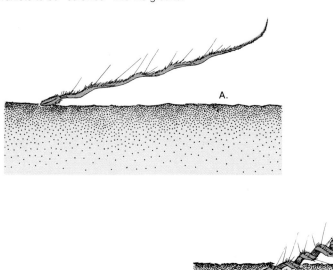

FIGURE 8.27 Filaree fruitlets. *A.* Under humid conditions. *B.* Under dry conditions. Alternate coiling and uncoiling causes the fruitlets to be "screwed" into the ground.

are explosively released, and those of dwarf mistletoes, when they are violently released in response to the heat of a warm-blooded animal coming close to the plants, can actually raise small welts on the skin. In manroots and a few other members of the Pumpkin Family, the seeds are squirted out of one end of the melonlike fruits like an erupting geyser. In filarees, each carpel of the fruit splits away and curls back from a central axis. Each fruitlet contains a single seed, which is pointed at its base. At the other end is a long, slender beak, which is sensitive to changes in humidity (figure 8.27). At night when the humidity increases, the beak is relatively straight, but in the sun it coils up like a corkscrew, literally drilling the pointed seed into the ground as it does so and effectively planting it in the process.

Humans are by far the most efficient transporters of fruits and seeds. Many noxious weeds and plant diseases, as well as valuable food and medicinal plants, have been carried from one continent to another by explorers and travellers over the past few hundred years in particular. Most countries now have strict regulations barring the importation of plant materials, except by special permit, and some plants

are not allowed across borders under any circumstances. Even within a country certain fruits and seeds that might carry diseases harmful to local agriculture may be barred from entry. In the United States, for example, Arizona, California, and Hawaii have border inspections in an attempt to prevent the importation of fruits such as citrus or popcorn across state lines, as such plants have in the past been carriers of diseases or pests that are presently under control.

SEEDS

Structure

If an ordinary kidney bean (a dicot) is examined closely, one can see a small white scar called a *hilum* on the concave side. This marks the point at which the ovule was attached to the ovary wall. The *micropyle* may be visible as a small pore adjacent to the hilum. If this bean is placed in water for an hour or two, it may swell enough to split the seedcoat. Once the seedcoat is removed, the bean can be seen

FIGURE 8.28   A common garden bean. *A.* Seed structure.
B. Germination and development of the seedling.

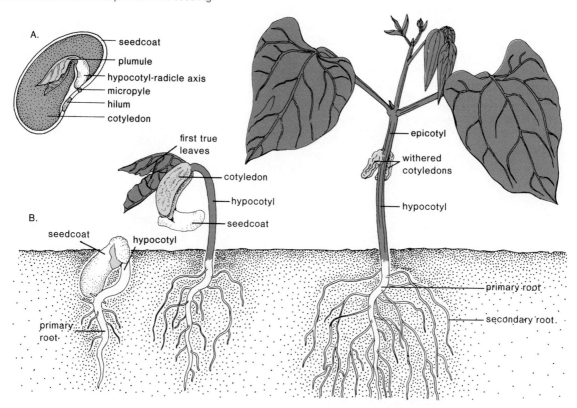

to consist of two halves, or **cotyledons,** with a tiny rudimentary bean plant between them toward one edge (figure 8.28). The cotyledons are food-storage organs, that also function as the first "seed leaves" of the seedling plant. They and the tiny, rudimentary bean plant to which they are attached constitute the *embryo.* Some seeds (e.g., those of grasses and all other monocots) have only one cotyledon.

The tiny plant bears undeveloped leaves and a meristem at one end. This embryo shoot is called a **plumule.** The cotyledons are attached just below the plumule. The stem part of the axis above the cotyledons, which at this stage is almost nonexistent, is referred to as the **epicotyl,** whereas the part below the point of attachment is the **hypocotyl.** In an embryo it is often difficult to tell where the stem ends and the root begins, but the tip that will develop into

a root is called a **radicle.** When a kidney bean germinates, the hypocotyl lengthens below a crook so that the cotyledons are pulled above the ground, but in lima beans and peas the hypocotyl remains short so that the cotyledons do not emerge above the surface (see chapter 11).

In other seeds, a plumule-radicle axis may be in the center instead of to one side, and the cotyledon(s) may not play a significant food-storage role. In corn, for example, the bulk of the food-storage tissue is *endosperm* (see chapter 21). Corn "seeds" (figure 8.29) also display other features not seen in beans. The plumule and the radicle are enclosed in tubular, sheathing structures called the **coleoptile** and the **coleorhiza,** respectively. These protect the delicate tissues within as the seeds germinate. Development of the coleoptile and coleorhiza ceases after they have attained several millimeters in length, and the plumule and radicle burst through the tips.

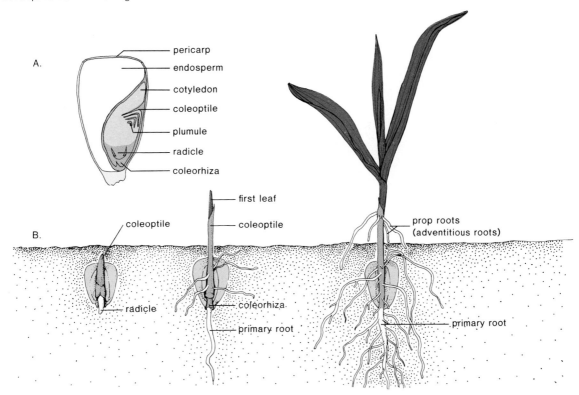

FIGURE 8.29   Corn. *A.* Grain structure. *B.* Germination and development of the seedling.

A.

pericarp
endosperm
cotyledon
coleoptile
plumule
radicle
coleorhiza

B.

coleoptile
first leaf
coleoptile
radicle
coleorhiza
primary root
prop roots
(adventitious roots)
primary root

Germination

Germination of a seed depends on the interplay of a number of factors, both internal and external. Many seeds need to undergo a period of **dormancy** before they will germinate (see the discussion of dormancy in chapter 11). Dormancy is brought about by either mechanical or physiological circumstances or both. In the Legume Family and others, the seeds may have seedcoats so thick or tough that they prevent the absorption of water or oxygen. Some seeds even have a one-way valve that lets moisture out but prevents its uptake (see figure 11.19). Dormancy in such seeds may sometimes be broken artificially by *scarification,* which involves nicking or slightly cracking the seeds or dipping them in a concentrated acid for a few seconds to a few minutes. In nature, such seeds may remain dormant until mechanical abrasion by rock particles in the soil, alternate thawing and freezing, or in some cases bacterial action, creates cracks in the seedcoat.

Dormancy may also be induced by growth-inhibiting substances present in the seedcoat, the interior of the seed, or tissues of the fruit surrounding it. Many desert plants have inhibitors in the seedcoat. These have to be washed away by soaking rains before germination will occur. The inhibitors function in survival of the species by preventing germination unless there has been sufficient rainfall for a seedling to become established. Apples, pears, citrus fruits, tomatoes, and other fleshy fruits contain inhibitors that prevent germination of the seeds within the fruits. Once the seeds are removed and washed, they germinate readily. The embryos of some seeds, such as those of the American holly, consist of only a few unspecialized cells when the fruit ripens. The seeds will not germinate after the fruit has dropped until the embryo has developed fully with the aid of food materials stored in its endosperm. Such a process of development is called *after-ripening.*

In many woody plants of temperate areas, germination stimulators need to be present to initiate growth. These normally do not develop unless the seeds encounter a wet period accompanied by cold temperatures. Usually this period needs to be a minimum of four to six weeks. The dormancy of such seeds can be broken artificially by placing them in a refrigerator, preferably in damp sand, for a few weeks. This technique is called *stratification.*

Even when mechanical and physiological barriers to germination are not present, a seed will not normally germinate unless environmental factors are favorable. Water and oxygen are essential to the completion of germination, and light or its absence also plays a role. Many seeds imbibe 10 or more times their total weight in water before the radicle emerges. Some seeds, such as those of castor beans (figure 8.30) and certain spurges, have appendages that function in water absorption, and thereby speed up the germination process. After water has been imbibed, enzymes begin to function in the protoplasm, which has now been rehydrated. Some enzymes convert stored proteins to amino acids, others convert fats and oils to soluble compounds, and still other enzymes aid in the conversion of starch to sugar. The soluble substances can then be conveyed to the embryo, and respiration, which in a dormant seed is almost imperceptible, can be greatly accelerated. Respiration releases the energy needed to initiate growth of the embryo, and a new plant begins to develop as mitosis and cell elongation take place.

Anaerobic respiration (see chapter 10) often initially furnishes the energy for the embryo, with aerobic respiration furnishing the energy as soon as the splitting of the seedcoat admits oxygen. If seeds are kept waterlogged after planting, oxygen available to them is greatly reduced and germination then may fail to be completed. Most seeds require temperatures within certain ranges to germinate. These usually need to be above freezing but below 45° C

(113° F). Germination percentages tend to be low approaching either extreme, however. Most crop plants have an optimum (ideal) germination temperature of between 20° C and 30° C (68° F to 86° F). The role of light in germination varies with the kinds of plants concerned. Seeds of some varieties of lettuce will not germinate in the dark (see discussion of *phytochrome* in chapter 11), while those of other seeds such as the California poppy germinate only in the dark. In lettuce seeds, the light apparently inactivates germination inhibitors, while in the California poppy it stimulates inhibitor formation.

Longevity

From time to time one reads or hears of seeds of wheat or other edible plants germinating after lying dormant in Egyptian pyramids or Native American tombs and caves for thousands of years, but none of these reports have been confirmed. In fact, there is evidence in a few instances that the seeds concerned were carried in recent times by rats or other rodents to their nests. However, reports of seeds of the aquatic lotus plant germinating after a little more than 1,000 years, and another documenting the germination of Arctic tundra lupine seeds that were frozen for an estimated 10,000 years have been confirmed.

Seeds remain viable (retain the capacity to germinate) for periods that vary greatly, depending on the species and the conditions of storage. Some seeds such as those of certain willows, cottonwoods, orchids, and tea remain viable for only a few days or weeks, regardless of how they are stored, but the period of viability of most seeds is extended by months or even years when they are stored at low temperatures and kept dry. Packets of vegetable and flower seeds sold in stores are dated, giving the buyer a rough idea of how long a significant number of the seeds might be expected to remain viable. Generally, seeds of Pumpkin Family members (e.g., squash, cantaloupe, cucumber) retain a relatively high percentage of viability for several years, whereas those of members of the Amaryllis Family (e.g., onion, leek, chives) retain a good percentage of viability for only two or three years. Properly stored wheat seeds have been reported to retain better than 30% viability for more than 30 years, and some weed seeds stored under conditions of low oxygen, high humidity, and cool temperatures have remained viable for even longer periods.

FIGURE 8.31 Young seedlings of red mangrove whose seeds have no dormant period and germinate while the fruit is still on the tree, growing up to 25 centimeters (10 inches) in length before falling and becoming planted in the mud below.

FIGURE 8.31 Young seedlings of red mangrove whose seeds have no dormant period and germinate while the fruit is still on the tree, growing up to 25 centimeters (10 inches) in length before falling and becoming planted in the mud below.

In 1879 William J. Beal, a botanist who pioneered in the development of hybrid corn, buried 20 pint-sized bottles of weed seeds on the campus of what is now Michigan State University in East Lansing, Michigan. Each bottle contained 1,000 seeds of 20 different species of weeds. Every five years a bottle of seeds was dug up and the seeds were planted, until the schedule was changed in 1920 to every ten years. When the first bottle was dug up in 1884, seeds of most of the weeds germinated; in 1960 seeds of evening primrose, curly dock, and moth mullein still germinated, and in 1980, 29 moth mullein seeds, 1 mullein seed, and 1 mallow seed germinated—101 years after they were placed in the bottles. It is of interest to note that a mallow seed previously had not germinated since 1899. Only 6 of the original bottles now remain; they are not scheduled to be unearthed until the year 2040. Recent evidence indicates that the timing of the digging up of the seed bottles did not take into account the fact that certain temperature patterns are critical to the germination of many weed seeds, and that if Beal's experiment had been conducted in another manner the germination results would have been quite different.

A few species of both dicots and monocots produce seeds that have no period of dormancy at all. In some instances, the embryo that develops from the zygote continues to grow without pause in a phenomenon known as *vivipary*. In the red mangrove, a tropical tree associated with coastal waters and estuaries, each fruit contains a single seed in which the embryo continues to grow while the fruit is still on the tree, reaching a length of 25 centimeters (10 inches) or more before the seedling becomes detached and essentially plants itself in the mud below (figure 8.31).

SUMMARY

Flowers occur in a wide variety of sizes, colors, textures, and habitats. Flowering plants that complete their life cycles in one growing season are annuals; biennials take two seasons and perennials may take several to many years to complete their cycles. Members of the two classes of flowering plants, Dicotyledonae (dicots) and Monocotyledonae (monocots) are distinguished from one another on the basis of differences in numbers of flower parts, numbers of cotyledons, venation, presence or absence of cambia, arrangement of vascular bundles, and number of apertures in the pollen grains.

A peach flower consists of a stalk (peduncle) with a receptacle to which are attached sepals (calyx), petals (corolla), stamens, and pistil. Some flowers have no petals; in others the petals may be fused together; and in yet others many flowers may be aggregated in an inflorescence. The pistil has three regions: the stigma, the style, and the ovary. The ovary, which may be superior with other flower parts attached to the receptacle below it or inferior with flower parts attached at the top, contains one or more ovules. Stamens consist of a pollen-bearing anther and a stalk or filament.

A fruit is a mature ovary; fruits are found exclusively in the flowering plants. Hormones produced by developing seeds promote the greatest fruit growth. Parthenocarpic fruits develop without fertilization occurring and are seedless. A mature fruit has an outer exocarp (skin), an endocarp around the seed(s), and a mesocarp ("flesh") between the exocarp and endocarp. The three regions may be fused together as a pericarp. At maturity fruits may be fleshy or dry, and the fruit parts may have been derived from the ovary alone or from adjacent flower parts as well.

Fleshy fruits may be simple fruits, which develop from a flower with a single pistil; aggregate fruits, which develop from a single flower having more than one pistil; or multiple fruits, which are derived from several to many individual flowers in a single inflorescence. Simple fleshy fruits include drupes, berries, and pomes. Drupes have a single seed enclosed by a hard endocarp. Berries, which develop from either a superior or an inferior compound ovary and have more than one seed are divided into true berries that have a thin skin and soft pericarp; pepos, which have a relatively thick rind present; and hesperidiums, which have a leathery skin containing oils. The flesh of pomes, for the most part, is derived from the enlarged receptacle that grows up around the ovary. Fruits derived from more than the ovary alone are sometimes referred to as accessory fruits. Some aggregate fruits (e.g., strawberries) are partly composed of accessory tissue. The individual fruitlets of a multiple fruit develop together into a single larger fruit.

Some fruits that are dry at maturity split as they mature. Among such fruits are follicles, which split along one side only; legumes, which split along two sides; siliques or silicles, which not only split along two sides but bear their seeds on a central partition between the two halves; and capsules, which split in various ways. Dry fruits that do not split at maturity include achenes, in which the single seed is attached to its surrounding pericarp at the base only; nuts, which resemble achenes but are larger and have a harder pericarp; grains (caryopses), which have the pericarp tightly fused to the seed; samaras, which have the pericarp extended in a wing; and schizocarps, which are twin fruits that break into two one-seeded fruitlets upon drying.

Some fruits and seeds have wings, plumes, and other adaptations for wind dispersal. Other fruits and seeds have appendages and other adaptations for animal and bird dispersal, and still others have adaptations for water dispersal. Some fruits eject seeds with force, and some have modifications that drill seeds into the ground. Humans disperse many seeds, and most countries and a few states have strict regulations governing the importation of plant materials, primarily to control the spread of pests and diseases.

A bean seed bears a scar called a hilum on the concave side and a pore called the micropyle adjacent to the hilum. Within the seedcoat are two cotyledons with a tiny rudimentary bean plant between them, the cotyledons and plant constituting the embryo. The embryo shoot is called a plumule; the part of the stem axis above the point of attachment of the cotyledons is called the epicotyl, and the part below is called the hypocotyl; the tip that develops into the root is called the radicle. In corn and other grain seeds the plumule and the radicle are protected by tubular sheathings called the coleoptile and the coleorhiza, respectively.

Germination of a seed depends on the interplay of a number of factors bringing about the cessation of dormancy. Dormancy may be sustained by mechanical circumstances such as tough seedcoats that need to be broken to admit water or valves that prevent the entry of water; scarification (nicking or softening the seedcoat) may break dormancy in such seeds. Dormancy may also be induced by growth or germination inhibitors, or immature embryos that must develop in a process of after-ripening before germination can occur. Germination stimulators that do not develop unless the seeds are subjected to a period of cold temperatures are necessary for the germination of seeds of some temperate woody plants; stratification (placing seeds in a moist cool environment for a few weeks) may break the dormancy of such seeds. Even when mechanical and physiological barriers to germination are not present, a seed will not germinate unless environmental factors including water, oxygen, light, and certain temperature ranges are favorable.

Seeds remain viable (capable of germinating) for periods that range from a few days to thousands of years, depending on the species and conditions of storage. The viability of most seeds is extended by storage at low temperatures under dry conditions, but some weed seeds have their viability extended by storage under humid, cool conditions that include little oxygen. William J. Beal buried bottles of weed seeds in 1879, and the bottles have been periodically dug up, one at a time, to check germination percentages. Seeds of three species still germinated 101 years after the bottles were buried. Some plants produce seeds that undergo no dormancy at all; the growth of the embryo while the seed and fruit are still on the plant is termed vivipary.

REVIEW QUESTIONS

1. Define calyx, corolla, receptacle, peduncle, pedicel, pistil, filament, and ovary.
2. Indicate the features by which dicots are distinguished from monocots.
3. What is the difference between a fruit and a vegetable?
4. What causes an ovary to develop into a fruit?
5. What are the various parts of a fruit?
6. Define parthenocarpy and give examples.
7. Distinguish among simple, aggregate, and multiple fruits.
8. Distinguish among achenes, grains, samaras, and nuts.
9. What adaptations do seeds and fruits have for dispersal by water and animals?
10. Define plumule, radicle, coleoptile, coleorhiza, hypocotyl, after-ripening, stratification, and vivipary.

DISCUSSION QUESTIONS

1. Most wind-pollinated flowering plants have inconspicuous, nonfragrant flowers. How might nature be affected if all flowers were that way?
2. Do you believe the botanical distinction between fruits and vegetables is a good one? If not, how would you change it?
3. In discussing pomes, it was observed that the bulk of the flesh in an apple comes from the receptacle. What could you do to prove or disprove this?
4. Seed and fruit dispersal is achieved with the aid of wind, water, animals, mechanical means, and humans. If you were "designing" a new plant, can you think of any new way in which it might be dispersed?
5. When volcanic activity or coral polyps cause new islands to appear in the oceans, they eventually acquire some vegetation. Would you expect the types of dispersal mechanisms for the flowering plants on these islands to be the same as they were for ancient continents?

ADDITIONAL READING

Bold, H. C. et al. 1987. *Morphology of plants and fungi,* 5th ed. New York: Harper & Row Publishers.

Cutter, E. G. 1971. *Plant anatomy: experiment and interpretation. Part II: Organs.* Reading, MA: Addison-Wesley Publishing Co.

Holm, E. 1979. *The biology of flowers.* New York: Penguin Books.

Pijl, L. van der. 1972. *Principles of dispersal in higher plants,* 2d ed. Berlin: Springler-Verlag.

Raven, P., R. F. Evert, and S. E. Eichhorn. 1986. *Biology of plants,* 4th ed. New York: Worth Publishers, Inc.

Ridley, H. N. 1930. *The dispersal of plants throughout the world.* Ashford, Kent, England: L. Reeve and Co.

Scagel, R. F. et al. 1984. *Plants: an evolutionary survey.* Belmont, CA: Wadsworth Publishing Co.

Overview

*T*his chapter begins by introducing molecular movement through a comparison between balls in motion in a room and molecular activity. This introduction is followed by a discussion of diffusion, osmosis, turgor, plasmolysis, imbibition, and active transport.

The chapter then takes up the entry of water into the plant, the movement of water through the plant, the evaporation of water into leaf air spaces, and transpiration. After a brief examination of soil horizons, the chapter concludes with a discussion of the development of soil, its texture, composition, structure, and its water.

9   Water in Plants; Soils

Some Learning Goals

1. In simple terms explain diffusion, osmosis, turgor, imbibition, and active transport.

2. Discuss the pressure-flow hypothesis and the cohesion-tension theory.

3. Know the pathway, movement, and utilization of water in plants.

4. Explain how a stoma opens and closes.

5. Understand how a good agricultural soil is developed from raw materials.

6. Contrast the various forms of soil water with regard to their specific location and availability to plants.

7. Compare sand, silt, and clay particles.

Outline

INTRODUCTION

Nearly everyone has had the experience of driving along a highway or city street when someone in the car says, "What's that smell?" Soon a bakery, or a paper mill, or perhaps a dead skunk comes into view, and the smell gets stronger. Then it fades away as the source is left behind. We take for granted the fact that there is a correlation between our proximity to an odor source and the intensity of the odor, but how and why does the odor reach us?

By way of an answer, let us imagine two adjacent rooms identical in height, width, and length, with no windows, doors, fixtures, or furniture. Now suppose we lift a small flap in the ceiling of one of them and drop in 100 tennis balls. These are ordinary tennis balls except for one extraordinary feature: they have perpetual motion motors in them that cause them to travel in any direction at 30 MPH. The tennis balls quickly make the room seem like a battlefield as they whiz around, bounce off the walls, floor, and ceiling, and also collide with one another, each time being deflected at a different angle. Shortly after they are introduced, the tennis balls will probably become randomly distributed throughout the room.

Now what would you expect to happen if we opened up a small hole in the wall between the two rooms? Eventually a tennis ball should bounce or travel at just the right angle to go through the hole into the other room. Theoretically, it could then bounce straight back into the first room, but it seems unlikely that it would. Reason tells us that long before it might happen to strike the exact angle it needed to return, several more balls will come in from the first room. Given enough time, some balls might indeed bounce back into the first room, but in the long run, each of the two rooms would end up with roughly 50 tennis balls, with an occasional ball going between the rooms in either direction.

The situation just described is analogous to what takes place in nature on a molecular level.

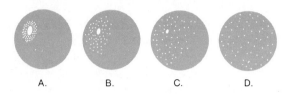

FIGURE 9.1 Diffusion of molecules from a crystal of dye placed on an agar plate. If the plate is left undisturbed, it may take years before the diffusion is complete.

MOLECULAR MOVEMENT

Molecules and ions are constantly in random motion. Visual evidence of this can be seen with an ordinary light microscope. If a drop of India ink is diluted with water and observed through a microscope under high power, the tiny carbon particles of the ink appear to be in constant motion. This motion, known as *Brownian movement,* is the result of the bombardment of the visible particles by invisible water molecules, which are in constant motion themselves.

Diffusion

The differing intensity of smells discussed earlier involves molecules behaving something like the tennis balls. Through their random motion, molecules tend to become distributed throughout the space available to them. Thus, if perfume molecules are kept in a bottle, they will become distributed throughout the bottle, but if the stopper is removed, they will eventually become dispersed throughout the room, even if there is no fan or other device to move the air. This movement of molecules or ions from a region of higher concentration to a region of lower concentration is called **diffusion** (figure 9.1). Molecules that are moving from a region of higher concentration to a region of lower concentration are said to be moving *along a diffusion gradient,* while molecules going in the opposite direction are said to be moving *against a diffusion gradient*. When the molecules, through their random movement in all directions, have become distributed throughout the space available, they are considered to be in a state of *equilibrium*. The rate of diffusion depends on several factors, including temperature and the density of the medium through which it is taking place.

Except within the area immediately surrounding the source, unaided diffusion requires a great deal of time because molecules and ions are infinitesimal. Something that is less than a forty-billionth of a millimeter in diameter is going to take a long time to move just 1 millimeter, even though the amount of movement may be great in proportion to the size of the particle concerned. In gases there is a great deal of space between the molecules and correspondingly less chance of the molecules bumping into each other and thus being slowed down. Accordingly, gas molecules occupy a space that becomes available to them relatively rapidly whereas liquids do so more slowly and solids are slower yet.

Large molecules move much more slowly than small molecules. If you added a tiny drop of a dye (which has relatively large molecules) to one end of a bathtub of water without disturbing the water in any way, it would take years for the dye molecules to diffuse throughout the tub and reach a state of equilibrium. In nature, however, wind and water currents distribute molecules much faster than they ever could be distributed by diffusion alone. Diffusion rates are also affected by the density of the molecules concerned and by other factors such as temperature.

Osmosis

Despite the fact that the protoplasts of living cells are bound by membranes, it is now well known that water (a **solvent**) moves freely from cell to cell. This has led scientists to believe that plasma, vacuolar, and other membranes have tiny holes or spaces in them, even though such holes or spaces are invisible to the instruments presently available, and it also has led to the construction of models of such membranes (see figure 3.7). Membranes through which different substances diffuse at different rates are described as **differentially permeable.** All plant cell membranes appear to be differentially permeable.

FIGURE 9.2   A simple osmometer made by tying a membrane over the mouth of a thistle tube.

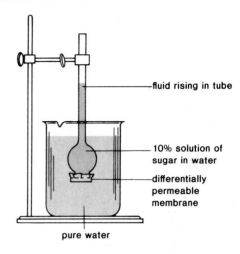

fluid rising in tube

10% solution of sugar in water

differentially permeable membrane

pure water

In plant cells **osmosis** is the *diffusion of water* (or any other solvent), *through a differentially permeable membrane*. A demonstration of osmosis can be made by tying a membrane over the mouth of a thistle tube that has been filled with a solution of 10% sugar in water (i.e., the solution consists of 10% sugar and 90% water). Fluid rises in the narrow part of the tube as osmosis occurs when the thistle tube is immersed in water (figure 9.2). Pressure (applied to the top of the tube) just sufficient to prevent the rise of the fluid is called **osmotic pressure** or more frequently **osmotic potential** of the solution. In other words osmotic pressure or osmotic potential is the pressure required to prevent osmosis from taking place.

Water enters a cell by osmosis until the osmotic potential is balanced by the resistance of the cell wall to expansion. Water gained by osmosis may keep a cell firm, or **turgid,** and the **turgor pressure** that develops against the walls as a result of water entering the vacuole of the cell is called *pressure potential*. The release of turgor pressure can be heard each time you bite into a crisp celery stick or the leaf of a young head of lettuce. When we soak carrot sticks, celery, and lettuce in pure water to make them crisp we are merely bringing about an increase in the *turgor* of the cells. The **water potential** (chemical potential of

FIGURE 9.3   A portion of a leaf of the water weed *Elodea*.
A. Normal cells. *B*. Plasmolyzed cells.

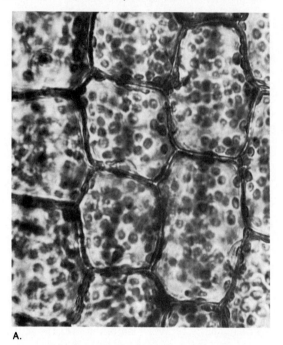

A.

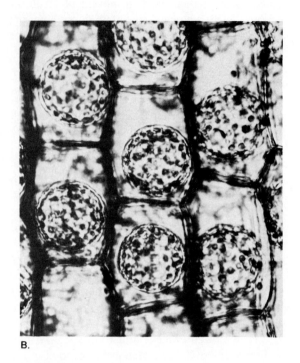

B.

water) of a plant cell is essentially its osmotic potential and pressure potential combined. If we have two adjacent cells of different water potentials, water will move from the cell having the higher water potential to the cell having the lower water potential.

Osmosis is the primary means by which plants obtain water from their environment. In land plants, water from the soil enters the cell walls and intercellular spaces of the epidermis and the root hairs and travels along the walls until it reaches the endodermis. Here it crosses the differentially permeable membranes and protoplasts of the endodermal cells on its way to the xylem, where it flows to the leaves, evaporates within the leaf air spaces, and diffuses out (*transpires*) through the stomata into the atmosphere. The movement of water occurs because there is a water potential gradient between relatively high soil water potential to successively lower water potentials in roots, stems, leaves, and the atmosphere.

Plasmolysis

If you place turgid carrot and celery sticks in a 10% solution of salt in water, they soon lose their rigidity and become limp enough to curl around your finger. The water potential inside the carrot cells is greater than the water potential outside, and so diffusion of water out of the cells into the salt solution takes place. If you were to examine such cells with the aid of a microscope, you would see that the vacuoles, which are largely water, had disappeared and that the protoplasm had shrunk away from the walls and was clumped in the middle of the cell. Such cells are said to be *plasmolyzed*. This loss of water through osmosis, which is accompanied by the shrinkage of protoplasm away from the cell wall, is called **plasmolysis** (figure 9.3). If plasmolyzed cells are placed in fresh water before permanent damage is done, water reenters the cell by osmosis, and the cells become turgid once more.

FIGURE 9.4   Pinto bean seeds before and after imbibition of water.

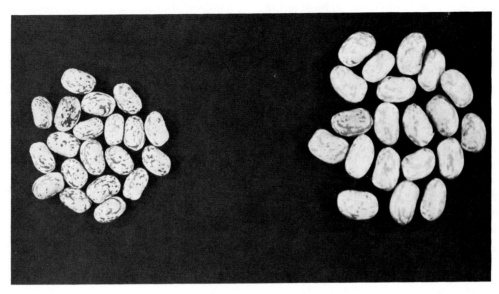

Imbibition

Osmosis is not the only force involved in the absorption of water by plants. *Colloidal* materials (i.e., materials that contain a permanent suspension of fine particles) and large molecules such as cellulose and starch usually develop electrical charges when they are wet, and they attract water molecules that adhere to the internal surfaces of the materials. Because water molecules are *polar* (polar molecules have slightly different electrical charges at each end due to their asymmetry—see the discussion in chapter 2) they can become both highly adhesive to large organic molecules like cellulose and cohesive with one another. This process, known as **imbibition**, results in the swelling of tissues, whether they are alive or dead, often to several times their original volume. Imbibition is the initial step in the germination of seeds (figure 9.4).

The physical forces developed during germination can be tremendous, even to the point of causing a seed to split a rock weighing several tons (see figure 23.8). It has been found, for example, that a pressure of 42.2 kilograms per square centimeter (600 pounds per square inch) is needed to break the seedcoat of a fresh walnut from within, and that water being imbibed by a cocklebur seed develops a force of up to 1,000 times that of normal atmospheric pressure. Yet when water and oxygen reach walnut and cocklebur seeds they germinate readily, as do seeds that fall into the crevices of rocks or have boulders roll over on them. The huge stone blocks used in the construction of the pyramids of Egypt are believed to have been quarried by hammering rounded wooden stakes into holes made in the face of the stone and then soaking the stakes with water. As the stakes swelled, the force created by imbibition was sufficient to split the rock.

Active Transport

Let us return for a moment to our two rooms with the tennis balls. Suppose that, in addition to the 100 tennis balls, we drop in 50 slightly underinflated basketballs, also extraordinary in having perpetual motion motors that propel them in any direction at 12 MPH. They should also become randomly distributed throughout the room shortly after they are introduced. Assume, however, that the hole in the wall (which is large enough for the passage of a tennis ball) is not quite large enough to allow a basketball to pass through freely. The basketballs will then remain in the first room. But if we installed a mechanical arm next to the hole in the second room, and if this arm could grab basketballs that come near the hole and squeeze them through, basketballs would be transported into both rooms *through the expenditure of energy.*

Plants expend energy, too. Plant cells generally contain a larger number of mineral molecules and ions than exist in the soil immediately adjacent to the root hairs. If it were not for the barriers imposed by the differentially permeable membranes, therefore, these molecules and ions would move from a region of higher concentration in the cells to a region of lower concentration in the soil. Most molecules needed by cells are polar, and those of solutes may set up an electrical gradient across a differentially permeable membrane of a living cell and require special proteins embedded in the membrane (see figure 3.7) to get across. The plants absorb and retain these substances against a diffusion or electrical gradient, however, *through the expenditure of energy.* This process is called **active transport.** The precise mechanism of active transport is not fully understood, but it apparently involves an enzyme and what has been referred to by some scientists as a proton "pump" in the plasma membrane of plant and fungal cells, whereas in animal cells the "pump" evidently involves sodium and potassium ions. Both "pumps" are energized by special energy-storing ATP molecules (discussed in chapter 10).

WATER AND ITS MOVEMENT THROUGH THE PLANT

If you were to cover the soil at the base of a plant with foil, place the pot where it receives light, and then put the potted plant under a glass bell jar, you would notice moisture accumulating on the inside of the jar within an hour or two. Because of the foil barrier, the water could not have come directly from the soil; it had to have come through the plant. More than 90% of the water entering a plant passes through and evaporates—primarily into leaf air spaces and then through the stomata into the atmosphere (see figure 9.6), with usually less than 5% of the water escaping through the cuticle. This process of water vapor loss from the internal leaf atmosphere is called **transpiration.**

The amount of water transpired by plants is greater than one might suspect. For example, mature corn plants each transpire about 15 liters (4 gallons) of water per week, while four-tenths of a hectare (1 acre) of corn may transpire more than 1,325,000 liters (350,000 gallons) in a 100-day growing season. A hardwood tree utilizes about 450 liters (120 gallons) of water while producing 0.45 kilogram (1 pound) of wood (or 1,800 liters while producing 0.45 kilogram of dry weight substance), and the 200,000 leaves of an average-sized birch tree will transpire from 750 to more than 3,785 liters (20 to 100 gallons) per day during the growing season. Humans recycle much of their water via the circulatory system, but if they were to have requirements similar to those of plants, each adult would have to drink well over 38 liters (10 gallons) per day.

Why is so much water involved in the normal processes of living plants? Water constitutes about 90% of the weight of young cells. The thousands of enzyme actions and other chemical activities of cells take place in water, and additional although relatively negligible amounts are used in the process of photosynthesis. The exposed surfaces of the chlorenchyma cells within the leaf have to be moist at all times, for it is through this film of water that the carbon dioxide molecules needed for the process of photosynthesis enter the cell from the air. Water is also needed for cell turgor, which gives rigidity to herbaceous plants. Consider also what it must be like in the mesophyll of a flattened leaf that is fully exposed to the midsummer sun in areas where the air temperature soars to well over 38° C (100° F) in the shade. If it were not for the evaporation of water molecules from the moist surfaces, which brings about some cooling, and reradiation of energy by the leaf, the intense heat could damage the plant. Sometimes the transpiration is so rapid that the loss of water begins to exceed the intake, and the stomata may close, thus preventing wilting (see also the role of *abscisic acid* relative to excessive water loss, discussed in chapter 11).

How does water travel through the roots from 3 to 6 meters (10 to 20 feet) or more beneath the surface and then up the trunk to the topmost leaves of a tree that is more than 90 meters (300 feet) tall? We know that continuous tubular pathways of xylem run throughout the plant, extending from the young roots up through the stem and branches to the tiny veinlets of the leaves. We also know that the water gets to the start of this "plumbing system" by osmosis following a water potential gradient. But water is raised through the columns apparently by a combination of factors, and the process has been the subject of much debate for the past 200 years.

One of the earliest explanations for the rise of water in a living plant was given by the English scientist Nehemiah Grew in 1682. He suggested that cells surrounding the xylem vessels and tracheids performed a pumping action that propelled the water along. This was questioned, however, when it was found that water will rise in lengths of dead stems also. Then the belief that capillary action moved the water became popular. It is well known that water will rise in a narrow tube and that the heights attained are inversely proportional to the diameter of the tube. It is also known that this rise occurs through the forces involved in the forming of a concave *meniscus* (curved surface) at the top of the water column (figure 9.5). But even though water can, indeed, rise 1 meter (3 feet) or more in a very narrow tube, air must be present above the column for the forces to work, and such is not the case in a plant. In fact, any air introduced into a water column in xylem interferes with the rise of water. Also, while capillarity might produce sufficient force to raise water a meter or two, the diameter of the tubes is not small enough to raise it more than that.

Others have pointed to *root pressure* as the means whereby water is moved through plants. When some plants are pruned after growth has begun in the spring, water will exude from the cut ends. This is the result of root pressure, which involves osmosis and perhaps other forces. Some plants do not "bleed" when they are pruned, however, and the force exerted by root pressure has been shown to be less than 30 grams per square centimeter (a few pounds per square inch), as a rule. This is considerably less than that needed to raise water to the tops of tall trees. Furthermore, root pressure seems to drop to negligible amounts in the summer, when the greatest amounts of water are moving through the plant.

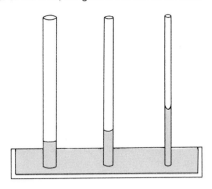

FIGURE 9.5 Capillarity in narrow tubes; the smaller the diameter of the tube, the greater the rise of the fluid.

The Cohesion-Tension Theory

The most satisfactory explanation of the rise of water in plants thus far suggested involves the **cohesion-tension theory.** It is known that water molecules adhere to capillary walls (e.g., those of xylem tracheids and vessels) and cohere to each other, permitting a certain amount of tension. It is possible, for example, to fill a small glass with water, place a thin smooth sheet of cardboard over the mouth, and invert the glass without the water spilling, because the adhesion of the water molecules to the cardboard and the cohesiveness of the water molecules to one another holds the cardboard against the rim of the glass.

When water evaporates from the mesophyll cells in a leaf and diffuses out of the stomata (*transpires*), the cells involved develop a lower water potential than the adjacent cells. Because the adjacent cells then have a correspondingly higher water potential, replacement water moves into the first cells by osmosis. This continues across rows of mesophyll cells until a small vein is reached. As indicated in earlier chapters, each small vein is connected to a larger vein, and the larger veins are connected to the main xylem in the stem, and that, in turn, is connected to the xylem in the roots which receive water, via osmosis, from the soil. As transpiration takes place it creates a "pull," or tension on water columns, drawing water from one molecule to another all the way through an entire span of xylem cells. The cohesion required to move water to the top of a tall tree is considerable, but the cohesive strength of the water columns is usually more than adequate.

Any breaking of the tension through the introduction of a gas bubble results in a temporary or permanent blocking of water transport. This seldom is a problem, however, because small bubbles may be redissolved and larger gas bubbles rarely block more than a few of the numerous capillary tubes of xylem at any time the tissue is functioning.

Most water and solutes reaching the root xylem can travel across the epidermis and cortex via the cell walls until it reaches the endodermis where it is then forced (by the *Casparian strips*—see chapter 6) to cross the protoplasts of the endodermal cells on its way to the vessels or tracheids of the xylem. If rapid transpiration is occurring, the soil around the root hairs may be so quickly drained that water will move from a considerable distance away via capillary pores in the soil. Unless the soil is very dry, however, the roots are more likely to grow rapidly toward the available water. In corn plants, for example, the main roots may grow at a rate of more than 6 centimeters (2.3 inches) a day. Solutes, as well as water, may move so rapidly during periods of rapid transpiration that there is little osmosis taking place across the endodermis. Scientists believe that at such times water may be pulled through the roots by *bulk flow,* which is the passive movement of a liquid from higher to lower potential by pressure or gravity or a combination of the two.

To recapitulate: "columns" of water molecules move through the plant from roots to leaves, and the abundant water of a normally moist soil supplies these "columns" as the water continues to enter the root by osmosis (figure 9.6); simply put, the difference between the water potentials of two areas (e.g., soil and the air around stomata) generates the force to transport water in a plant.

TRANSPORT OF FOOD SUBSTANCES IN SOLUTION

One of the most important functions of water in the plant involves the *translocation* (transportation) of food substances in solution by the phloem, a process that has only recently come to be better understood. Many of the studies that led to our present knowledge of the subject utilized aphids (small, sucking insects) and organic compounds designed as radioactive tracers.

Most aphids feed on phloem by inserting their tiny tubelike mouthparts (*stylets*) through the leaf or stem tissues until a sieve tube is reached and punctured. The turgor pressure of the sieve tube then

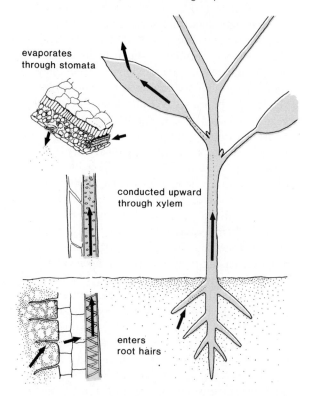

FIGURE 9.6 Pathway of water through a plant.

evaporates through stomata

conducted upward through xylem

enters root hairs

forces the fluid present in the tube through the aphid's digestive tract, and it emerges at the rear as a droplet of "honeydew." In the studies, research workers anaesthetized feeding aphids and cut their stylets so that much of the tiny tube remained where it had been inserted. The fluid exuding (sometimes for many hours) from the cut stylets was then collected and analyzed. In other studies, the pathway of food substances manufactured by the leaf could be traced by exposing a photosynthesizing leaf to carbon dioxide (a basic raw material of the process) that had been synthesized with radioactive carbon. The radioactive substances produced on photographic film an image corresponding to the food pathway. Data obtained from such studies revealed that food substances in solution are confined entirely to the sieve tubes while they are being transported. At one time it was believed that ordinary diffusion and cyclosis (see chapter 3) were responsible for the movement of the substances from one sieve-tube element to the next, but it is now known that the substances move through the phloem at approximately 100 centimeters (almost 40 inches) per hour—far too rapid a movement to be accounted for by diffusion and cyclosis alone.

FIGURE 9.7   A portion of a strip of lily leaf epidermis
showing stomata and surrounding epidermal cells.
(Photomicrograph by Carolina Biological Supply Company)

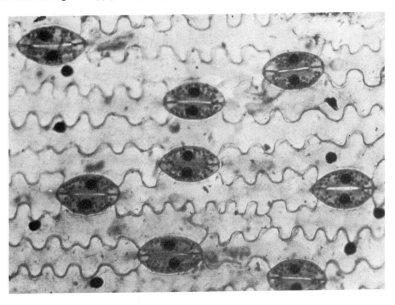

The Pressure-Flow Hypothesis

At present, the most widely accepted theory for movement of substances in the phloem is called the **pressure-flow hypothesis.** According to this theory, food substances in solution flow along concentration gradients between *sources* (food-producing and storage regions) and *sinks,* which are places where the food is utilized (e.g., the growing tips of stems and roots). Turgor pressure, which develops as osmosis occurs, is responsible for driving the fluid through the sieve tubes toward the sinks. According to the theory, water is taken up by osmosis at the source (e.g., mesophyll tissue of a leaf) and is given up at the sink. As the food substances in solution are actively removed at the sink, water also leaves the sink cells and the pressure in the adjacent sieve tubes is lowered, causing movement from the higher pressure at the source to the lower pressure at the sink. Most of the water diffuses back to the xylem, where it then returns to the source and is transpired or recirculated. The pressure-flow hypothesis nicely explains how nontoxic dyes applied to leaves or substances entering the sieve tubes, such as viruses introduced by aphids, are carried through the phloem.

REGULATION OF TRANSPIRATION

As previously indicated, the stomata *regulate transpiration and gas exchange.* The guard cells forming each stoma have relatively elastic walls with radially oriented microfibrils, making them analogous to pairs of sausage-shaped balloons that are joined at each end, each with a row of rubber bands around it; the part of the wall adjacent to the hole itself is considerably thicker than the remainder of the wall (figure 9.7). This thickness allows each stoma to be opened and closed by means of changes in the turgor of the guard cells. The stomatal pore is closed when turgor pressure is low, and open when turgor pressure is high. Changes in turgor pressures in the guard cells occur when the cells are exposed to light and/or low concentrations of carbon dioxide. Changes in turgor pressures also result from changes in solute concentrations brought about by osmosis and active transport between the epidermal cells and the guard cells (for example, when potassium and chloride ions are taken up from other epidermal cells or organic acids are produced from starch). Energy is expended by guard cells to take up potassium ions from

FIGURE 9.8 How a stoma opens and closes. *A.* A closed stoma. *B.* As turgor pressure in the guard cells increases, the thinner outer walls stretch more than the thicker inner walls, causing the stoma to open.

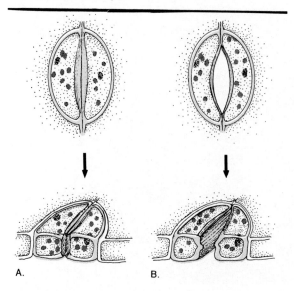

A.   B.

adjacent epidermal cells as the stomata open, and to release potassium ions as the stomata close. With an increase in potassium ions, the water potential in the guard cells is lowered, and the osmosis that takes place as a result brings in water that makes the cells turgid. Some believe that energy may also be expended to cause the potassium ions to leave, but whether the ions leave because of active transport or not, their departure also results in water leaving, making the cells less turgid, and closing the stomata (figure 9.8). Stomata will passively close whenever water stress occurs, but there is evidence that the hormone *abscisic acid* (see page 205) is produced in leaves subject to water stress and that this hormone causes membrane leakages, which induce a loss of potassium ions from the guard cells, causing them to close.

The stomata of most plants are open during the day and closed at night, but water is conserved in a number of desert plants whose stomata open only at night when there is less water stress on the plants. Such plants have a specialized form of photosynthesis (see *CAM photosynthesis* on page 187) since the carbon dioxide they require for typical photosynthesis is shut off during daylight hours by the closing of the stomata. Some desert plants are able to store carbon dioxide taken in at night in the form of special substances that can be converted back to carbon dioxide during the daytime when photosynthesis occurs. Other desert plants have their stomata below the surface of the leaf or stem in little sunken chambers. These chambers often are partially filled with epidermal hairs that further reduce water loss. Pine trees, which frequently occur in areas where the soil is frozen for part of the year and where limitations thus occur on available water, also have sunken stomata (see figure 7.8). A few tropical plants that occur in damp, humid areas (e.g., ruellias; see also figure 4.8B) actually have stomata that are raised above the surface of the leaf, whereas plants that occur in water generally have no stomata on submerged surfaces.

Although transpiration is affected by light and carbon dioxide concentration, several other factors play at least an indirect role. For example, the stomata may be regulated by the water gradient being affected by air currents sweeping away the water molecules as they emerge from the stomata, which speeds up the rates of evaporation within. Humidity plays an inverse role in transpiration rates: high humidity reduces transpiration, and low humidity accelerates it. A direct correlation also exists between temperature and the movement of the water molecules out of the leaf. The transpiration rate of a leaf at 30° C (86° F), for example, is about twice as great as it is for the same leaf at 20° C (68° F). The various adaptive modifications of leaves and their surfaces, which were mentioned in chapter 7, and the availability of water to the roots also may play important roles in influencing the amount of water transpired.

If a cool night follows a warm humid day, some plants may produce water droplets at the tips of veins in their leaves through structures called **hydathodes.** This loss of water in liquid form is called **guttation** (figure 9.9). Although the droplets resemble dew, the two should not be confused. Dew is water that is condensed from the air, whereas guttation water is literally forced out of the plant by root pressures. As the sun strikes the droplets in the morning, they dry up, leaving a residue of salts and organic substances, one of which is used in the manufacture of commercial flavor enhancers (e.g., the monosodium glutamate in products such as Accent). In the tropics, the amount of water produced by guttation can be considerable. In taro plants, used by the Polynesians to make poi, a single leaf may produce as much as a cupful of water overnight through guttation.

FIGURE 9.9   Droplets of guttation water at the tips of leaves of young barley plants.

SOILS

The soil is a dynamic, complex, constantly changing part of the earth's crust that extends from a few centimeters deep in some places to hundreds of meters deep in others. It is essential not only to our existence but, at least indirectly, to the existence of most living organisms. It has a pronounced effect on the plants that grow in it, and they, in turn, have an effect on it.

In many parts of North America if one were to dig up a shovelful of soil from one's backyard and examine it, one would probably find a mixture of ingredients, including several grades of sand, rocks and pebbles, powdery silt, humus (a dark substance made up of incompletely decomposed organic materials), dead leaves and twigs, clods consisting of soil particles held together by clay and organic matter, plant roots, and small animals such as ants, pillbugs, millipedes, and earthworms. Also present, but not visible, would be hundreds of thousands of microorganisms, particularly bacteria, and, of course, air and water.

The soil became what it is today through the interaction of a number of factors: climate, parent material, topography of the area, living organisms, and time. Because there are thousands of ways in which these factors may interact, there are many thousands of different soils. The solid portion of a soil consists of mineral matter and organic matter. Pore spaces, shared by variable amounts of water and air, occur between the solid particles. The smaller pores often contain water, and the larger ones usually contain air. The sizes of the pores and the connections between them largely determine how well the soil is aerated.

If one were to dig down 1 or 2 meters (3 to 6 feet) in an undisturbed area, a soil profile of three intergrading regions that soil scientists refer to as

FIGURE 9.10   A Miami loam soil profile. (After Foth, H. D., L. V. Withee, H. S. Jacobs, and S. J. Thien. 1982. *Laboratory Manual for Introductory Soil Science,* 6th ed. Wm. C. Brown Publishers, Dubuque, Iowa)

DM

1

2

3

4

5

6

7

8

9

10

11

A₁ horizon (dark loam)

A₂ horizon (light-colored loam)

B horizon (clay loam)

C horizon (parent material)

horizons would probably be exposed (figure 9.10). The horizons show the soil in different stages of development, and the composition varies accordingly. The upper layer, usually extending down 10 to 20 decimeters (4 to 8 inches), is called the A horizon, or *topsoil*. It is usually subdivided into a darker upper portion called the A₁ horizon and a lighter lower portion called the A₂ horizon. The A₁ portion in particular contains more organic matter than the layers below. The next 0.3 to 0.6 meter (1 to 2 feet) is called the B horizon, or *subsoil*. It usually contains more clay and is lighter in color than the topsoil. The C horizon at the bottom may vary from about 10 centimeters (4 inches) to several meters (6 to 10 feet or more) in depth; it may even be absent. It is commonly referred to as the *soil parent material* and extends down to *bedrock*.

Parent Material

The first step in the development of soil is the formation of *parent material*. This material accumulates through the weathering of rocks, which originate from volcanic activity (*igneous rocks*), from fragments of rocks deposited by glaciers, water, or wind (*sedimentary rocks*), or from changes in either igneous or sedimentary rocks brought about by great pressures, or heat, or both (*metamorphic rocks*).

Climate

Climate varies greatly throughout the globe, and its role in the weathering of rocks varies correspondingly. In desert areas, for example, there is little weathering by rain, and soils are poorly developed; whereas in areas of high rainfall, well-developed soils are common. In areas where there are great temperature ranges, rocks may split or crack as their outer surfaces expand or contract at different rates from the material beneath the surface. Low temperatures cause water in rock crevices to freeze, which, in turn, also causes cracks and splits.

Living Organisms and Organic Composition

There are numerous organisms of many kinds in the soil, in addition to plant roots. In the upper 30 centimeters (1 foot) of a good agricultural soil, living organisms constitute about one-thousandth of the total weight of the soil. This may not sound significant, but it amounts to approximately 6.73 metric tons per hectare (3 tons per acre). Bacteria and fungi present in the soil decompose all types of organic matter that accumulates when leaves fall and roots and animals die (this process is further discussed under the section on composting in chapter 15). Roots and all other living organisms present produce carbon dioxide, which combines with water to form an acid, thereby increasing the rate at which minerals dissolve. Ants and other insects, earthworms, burrowing animals, and birds all cultivate the soil through their activities and add to its organic increment either through wastes that they deposit or through the decomposition of their bodies when they die.

The total organic composition of a soil varies greatly. An average topsoil might consist of about 25% air, 25% water, 48% minerals, and 2% organic matter; however, soils in low wet areas, where a lack of oxygen prevents microorganisms from their normal activities, may contain as much as 90% organic matter. Except in legumes and a few other plants, almost all of the nitrogen utilized by growing plants, as well as much of the phosphorus and sulphur, comes from decomposing organic matter. In addition, as organic matter breaks down, it produces acids that, in turn, decompose minerals. Other roles of organic matter in the soil are discussed in the section on soil structure.

Topography of the Area

If the topography of an area is steep, soil may wash away or erode through the action of wind, water, and ice as soon as it is weathered from the parent material. It has been estimated that more than 20 metric tons of topsoil per hectare (8.2 tons per acre) is washed away annually from some prime croplands in the central United States. If an area is flat and poorly drained, however, pools and ponds may appear in slight depressions when it rains. If these bodies of water cannot drain relatively quickly, the activities of organisms in the soil are interrupted and the development of the soil is arrested. The ideal topography for the development of soil is one that permits drainage without erosion.

TABLE 9.1
Soil Mineral Components as Classified by the
United States Department of Agriculture

Mineral	Diameter (range in mm)	Comments
Stones	> 76 mm	do not support plant growth but affect permeability and erosion of the soil
Gravel	76 mm–2.0 mm	
Very coarse sand	2.0 mm–1.0 mm	
Coarse sand	1.0 mm–0.5 mm	
Medium sand	0.5 mm–0.25 mm	make soil feel gritty
Fine sand	0.25 mm–0.10 mm	
Very fine sand	0.10 mm–0.05 mm	
Silt	0.05 mm–0.002 mm	a lens or microscope needed to see any but the coarsest silt particles
Clay	< 0.002 mm	may absorb water, swell, and later shrink, causing the soil to crack as it dries

Soil Texture and Mineral Composition

Soil *texture* refers to the sizes of the individual soil particles (see table 9.1). The principal texture designations are *sand, silt,* and *clay.* Sands and gravels are usually composed of many small particles bound together chemically or by a cementing matrix material. Silt consists of particles that are mostly too small to be seen without a lens or a microscope. Clay particles are so tiny that even a powerful light microscope would not render them visible, although they can be seen with an electron microscope. Individual clay particles are called *micelles;* they are somewhat sheetlike, negatively charged, and held together by chemical bonds. They attract, exchange, or retain positively charged ions. Clay is plastic in nature because the water that tightly adheres to the surface of the particles acts both as a binding agent and a lubricant. Physically, clay is a **colloid** (that is, a suspension of particles that are larger than molecules but that do not settle out of a fluid medium).

"Light" soils have high sand and low clay content. "Heavy" soils have high clay content. Coarse-textured soils do not retain large amounts of water, whereas clay soils permit very little water to pass through.

Over half of the composition by weight of mineral matter is oxygen. Other common elements present include hydrogen, silicon, aluminum, iron, potassium, calcium, magnesium, and sodium, but soil inherits hundreds of different mineral combinations from parent material.

More plant nutrients are stored in the form of ions by clay and organic matter than by sand and silt. This is because the clay and organic particles, being smaller than those of sand and silt, have a greater total surface area per unit volume of material to which ions can become attached.

Soil Structure

Soil structure refers to the arrangement of the soil particles into groups of particles called aggregates. Aggregates in sands and gravels show little cohesion, but most agricultural soils have aggregates that stick together. Structure develops when colloidal particles clump together as a result of the activities of soil organisms, freezing and thawing, and so on. If the individual granules do not become coated with organic matter, they may continue to clump until they become clods.

Good agricultural soils are granular soils with pore spaces that occupy between 40% and 60% of the total volume of the soil. The pores contain air and water, and their sizes are more important than their total volume. Clay soils, for example, have more actual pore space than sandy soils, but the pores are so small that water and air are restricted in their movement through the soil. When the pores are full of water, air is kept out, which results in insufficient oxygen for root growth. Sandy soils have large pores that drain by gravity soon after they are filled. The water is replaced by air, but too much air speeds up nitrogen release by microorganisms. The plants cannot use nitrogen this quickly, and much of it is lost as a result.

Water itself can be harmful, because too much of it depletes mineral nutrients and slows the mineralization process under the anaerobic (absence of free oxygen) conditions, slows the release of nitrogen, and interferes with plant growth. Water-logged conditions also accelerate the breakdown of nitrates to the extent that virtually all the nitrate present in the soil is lost in as little as half an hour. Countless numbers of houseplant enthusiasts and outdoor gardeners have stunted or killed the very plants they were trying to foster by "drowning" the objects of their affection with too much water or too frequent waterings. Generally, one should refrain from watering until the surface of the soil appears dry.

Water in the soil occurs in three forms. **Hygroscopic water** is physically bound to the soil particles and is unavailable to plants. **Gravitational water** drains out of the pore spaces after a rain. If drainage is poor, it is this water that interferes with normal plant growth. Plants depend mainly on the third type, **capillary water,** for their needs. The structure and organic matter of the soil—which enable the soil to hold water against the force of gravity—the density and type of vegetational cover, and the location of underground water tables largely determine the amount of capillary water available to the plant. The ancient Incas of Peru knew that water would rise just so far in some areas. Where the water table was close to the surface, they removed the upper 0.6 meter (2 feet) of soil and planted their crops down in the hollowed-out areas so that the roots would be able to reach the capillary water. In some areas having sandy soils and low annual precipitation, soils can be compacted by a heavy roller to create finer capillaries to raise water from below. This technique is effective only if the available water is within 1.5 to 3.0 meters (5 to 10 feet) of the surface and if the soils do not contain much silt or clay.

After rain or irrigation, water in the soil drains away by gravity. The water remaining after such draining is referred to as the *field capacity* of the soil. Field capacity is mainly governed by the texture of the soil, but the structure and organic content also influence it to a certain extent. Plants readily absorb water from the soil when it is at, or near, field capacity; but as the soil dries, the film of water around each soil particle becomes thinner, and

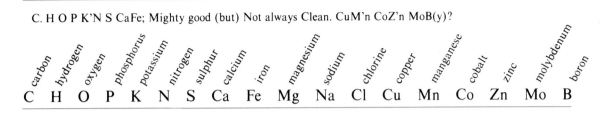

C. H O P K'N S CaFe; Mighty good (but) Not always Clean. CuM'n CoZ'n MoB(y)?

carbon · hydrogen · oxygen · phosphorus · potassium · nitrogen · sulphur · calcium · iron · magnesium · sodium · chlorine · copper · manganese · cobalt · zinc · molybdenum · boron

C   H   O   P   K   N   S   Ca   Fe   Mg   Na   Cl   Cu   Mn   Co   Zn   Mo   B

the force exerted by the plants to remove the water from the soil increases. If water is not added to the soil, a point is eventually reached at which the rate of absorption of water by the plant is insufficient for its needs, and the plant wilts permanently. The soil is then said to be at the *permanent wilting point*. In clay soils the permanent wilting point is reached when water content drops below 15%, while in sandy soils the permanent wilting point may be as low as 4%. *Available water* is soil water between field capacity and the permanent wilting point.

The pH (acidity or alkalinity) of a soil affects both the soil and the plants growing in it in various ways. A soil that is unusually acid or alkaline may be toxic to the roots, but these conditions do not normally affect plants directly nearly as much as they affect nutrient availability. For example, minerals such as copper, iron, and manganese become less available to plants as alkalinity increases, whereas highly acid soils inhibit the growth of nitrogen-fixing bacteria. Acid soils tend to be common in areas of high precipitation where significant amounts of bases are leached from the topsoil. It is a common agricultural practice to counteract soil acidity by adding compounds of calcium or magnesium in a process known as *liming*. Alkaline soils can be made more acid by the addition of sulphur, which is converted by bacteria to sulphuric acid; the addition of some nitrogenous fertilizers may have the same effect.

Mineral Requirements for Growth

Growth phenomena are not controlled by internal means alone. Light, temperature, soil structure and minerals, and other external factors all play a role. In fact, growth depends on a complex, interrelated combination of chemical and physical forces, both internal and external, all of which are in delicate balance with one another.

Plants may take up many elements from the soil, but besides the carbon, hydrogen, and oxygen obtained from carbon dioxide and water, only 15 elements are essential to most plants as building blocks for numerous compounds they synthesize. Sodium, a comparatively abundant element, is apparently required by relatively few plants. The **essential elements** can be remembered by a sentence (figure 9.11) that includes the symbols of the elements involved.

Macronutrients and Micronutrients

The mineral elements are usually put into two categories: (1) *macronutrients*, which are used by plants in greater amounts and constitute from about 0.5% to 3.0% of the dry weight of the plant; (2) *micronutrients*, which are needed by the plant in very small amounts, often constituting only a few

FIGURE 9.12 Tobacco plants grown in media deficient in various elements to show deficiency symptoms. *A.* A normal plant that has been supplied with all elements. The other plants have been grown in media deficient in specific nutrients, as follows: *B.* Nitrogen. *C.* Phosphorus. *D.* Potassium. *E.* Boron. *F.* Calcium. *G.* Magnesium. (USDA Photo)

parts per million of the dry weight. The macronutrients (other than carbon, hydrogen, and oxygen) are nitrogen, potassium, calcium, phosphorus, magnesium, and sulphur, with the first 4 usually making up about 99% of the nutrient total. The remaining elements, the micronutrients, are present in amounts ranging from bare traces—as in the case of sodium and cobalt, neither of which may actually be essential for some plants—to 1,500 parts per million of iron and manganese, and up to 10,000 parts per million of chlorine. In addition to these widely required elements, specific organisms may require others. For example, certain algae apparently also require the elements vanadium, silicon, or iodine, while some ferns utilize aluminum. Several loco weeds absorb and accumulate selenium in amounts constituting up to 5 micrograms per gram of dry weight. Selenium, which is often fatally poisonous to livestock, appears to enhance the growth of these plants by reducing toxic effects of phosphates, but there is no direct evidence it is essential to them.

When any of these elements is deficient in the soil, the plants exhibit characteristic symptoms of the deficiency, which disappear after the problem has been corrected (figure 9.12). Table 9.2 shows some of the uses of essential elements in plants and describes the symptoms of deficiency for each element.

Uses of Essential Elements in Plants

Element	Some Functions	Deficiency Symptoms
Nitrogen	part of proteins, nucleic acids, chlorophyll	relatively uniform loss of color in leaves, occurring first on the oldest ones
Potassium	activates enzymes; concentrates in meristems	yellowing of leaves, beginning at margins and continuing toward center; lower leaves mottled and often brown at tip
Calcium	essential part of middle lamella; involved in movement of substances through cell membranes	terminal bud often dead, young leaves often appearing hooked at tip; tips and margins of leaves withered; roots dead or dying
Phosphorus	necessary for respiration and cell division; high-energy cell compounds	plants stunted; leaves darker green than normal; lower leaves often purplish between veins
Magnesium	part of the chlorophyll molecule; activates enzymes	veins of leaves green but yellow between them with dead spots appearing suddenly; leaf margins curling
Sulphur	part of some amino acids	leaves pale green with dead spots; veins lighter in color than the rest of the leaf area
Iron	needed to make chlorophyll and in respiration	larger veins remaining green while rest of leaf yellows*
Manganese	activates some enzymes	dead spots scattered over leaf surface; all veins and veinlets remain green; effects confined to youngest leaves
Boron	influences utilization of calcium ions, but functions unknown	petioles and stems brittle; bases of young leaves break down

*NOTE: The symptoms of iron deficiency may be caused by several factors such as overwatering, cold temperatures, and nematodes (small roundworms) in the roots. The iron may be relatively abundant in the soil, but its uptake may be prevented or sharply reduced by these environmental conditions. Iron becomes more soluble under acid conditions—so much so that it can produce toxic conditions for most plants. Acid soil plants, on the other hand, have a much higher iron requirement than plants that require more nonacid conditions; thus, azaleas and other plants having high iron requirements can achieve normal growth in a nonacid soil.

It should be noted that all the micronutrients are harmful to plants when supplied in excessive quantities. Copper will kill algae in concentrations of one part per million, and boron has been used in weed killers. Even macronutrients are harmful if present in heavy amounts, although nonessential elements sometimes can counteract their toxicity.

SUMMARY

Molecules and ions are in constant random motion and tend to distribute themselves evenly in the space available to them. In doing so they move from a region of higher concentration to a region of lower concentration by simple diffusion along a diffusion gradient, although they may also move against a diffusion gradient. Evenly distributed molecules are said to be in a state of equilibrium; diffusion rates are affected by temperature, the size of the molecules, the density of the molecules, and other factors such as wind and currents.

Osmosis is the diffusion of water through a differentially permeable membrane. It takes place in response to concentration differences of dissolved substances. Osmotic pressure or potential is the pressure required to prevent osmosis from taking place. The pressure that develops in a cell as a result of water entering it is called turgor. Water moves from a region of higher water potential (osmotic potential and pressure potential combined) to a region of lower water potential when osmosis is occurring.

Osmosis is the primary means by which plants obtain water from their environment. Shrinkage of the protoplasm away from the cell wall as a result of osmosis taking place when the water potential inside the cell is greater than outside is called plasmolysis. Imbibition is the attraction and adhesion of water molecules to the internal surfaces of materials; it results in swelling and is the initial step in the germination of seeds. Active transport is the expenditure of energy by a cell that results in molecules or ions entering the cell against a diffusion gradient.

Most of the water that enters a plant passes through it and evaporates into the atmosphere through stomata by a process called transpiration. The water retained by the plant is used in photosynthesis and other metabolic activities of the cell; it is involved in the translocation of food substances in solution, and according to the pressure-flow hypothesis, such substances flow along concentration gradients between their sources and sinks (places where the food is utilized). According to the cohesion-tension theory, water rises through plants because of a combination of factors including the adhesion of water molecules to the walls of the capillary conducting elements of the xylem, cohesion of the water molecules, and tension on the water columns created by the pull developed by transpiration.

Transpiration is regulated by the stomata, which open and close through changes in turgor pressure of the guard cells. These changes, which involve potassium ions, result from osmosis and active transport between the guard cells and the adjacent epidermal cells. Aquatic, desert, tropical, and some cold zone plants have modifications of stomata or specialized forms of photosynthesis that adapt them to their particular environments. Guttation is the loss of water in liquid form as it is forced out of the tips of leaf veins by root pressures through structures called hydathodes.

Soils contain a mixture of ingredients, including sands, rocks, silt, humus, dead organic matter, plant roots, small animals and microorganisms, plus air and water within pore spaces of various sizes.

A vertical column of soil exhibits intergrading regions called horizons; these show the soil in various stages of development. The A horizon (topsoil) is divided into two layers; the upper A_1 horizon is darker and contains more organic matter than the lower A_2 horizon. The B horizon (subsoil) usually contains more clay and is lighter in color than the A horizon. The C horizon (bottom portion) is referred to as the soil parent material. Parent material is developed through the weathering of rocks. The extent of weathering varies with the climate.

Living organisms in the soil decompose organic matter, producing carbon dioxide that combines with water to form an acid that increases the rate at which minerals dissolve and are decomposed themselves. Animals also cultivate the soil. The total amount of organic matter in a soil can vary greatly; most important plant nutrients come from decomposing organic matter. The topography of an area may cause the soil to erode as soon as it is produced or it may arrest its development.

Soil texture pertains to sizes of soil particles. Sands and gravels usually consist of many small particles bound together. Silt particles are too small to be seen without magnification, and clay particles (micelles) are visible only with the aid of an electron microscope. More than half of the composition by weight of mineral matter is oxygen. Soil structure refers to the arrangement of the soil particles into aggregates. Good soils are highly granular and have pore spaces that constitute about half the total volume, with the sizes of the pores being more important to aeration and water retention than the numbers present. Water in the soil occurs as hygroscopic water, which is chemically bound to the soil particles and unavailable to plants; gravitational water, which drains out of pores; and capillary water, which is water held against the force of gravity. The field capacity of the soil is the amount of water that remains after the rest of the water has drained away by gravity. Soil reaches the permanent wilting point when plants wilt permanently because they can no longer extract enough water from the soil for their needs. Available water is soil water between field capacity and the permanent wilting point.

Growth phenomena are controlled by both internal and external means and by chemical and physical forces in balance with one another. Besides carbon, hydrogen, and oxygen, 15 of the elements taken up are essential to most plants. When any of the 18 elements are deficient in the soil, characteristic deficiency symptoms appear in the plants.

REVIEW QUESTIONS

1. Distinguish among diffusion, osmosis, active transport, plasmolysis, and imbibition.
2. Why do living plants need a great deal of water for their activities?
3. Explain how a tall tree gets water to its tips without the aid of mechanical pumps.
4. What is the difference between transpiration and guttation?
5. Explain the pressure-flow hypothesis.
6. What is soil parent material? Where does it come from, and how does it become soil?
7. What is the difference between soil *texture* and soil *structure*?
8. What types of soil water are recognized? What is *available* soil water?
9. What are macronutrients? List them.
10. When nutrients are deficient in the soil, how are the deficiencies manifested in plants?

DISCUSSION QUESTIONS

1. Why is salted meat less likely to spoil than unsalted meat?
2. Why would dye molecules in a bathtub of water take a long time to diffuse completely throughout the tub, but perfume molecules released in a closed room take considerably less time to do the same thing?
3. Why does osmosis not cause submerged water plants to swell up and burst?
4. Some bodies of water such as the Dead Sea have considerably higher salt concentrations than those of the human body. If you were swimming in such water, would you expect your cells to become plasmolyzed? Why?
5. Persons associated with commercial nurseries and greenhouses often sterilize their soil by heating it to get rid of pests, but then they have to wait for a short time after it has cooled to use it. Why?

ADDITIONAL READING

Crafts, A. S., and C. E. Crisp. 1971. *Phloem transport in plants.* NewYork: W. H. Freeman and Co.

Donahue, R. L. et al. 1983. *Our soils and their management,* 5th ed. Danville, IL: Interstate Printers & Publishers, Inc.

Donahue, R., and J. Miller. 1983. *Soils: an introduction to soils and plant growth,* 5th ed. Englewood Cliffs, NJ: Prentice-Hall, Inc.

Epstein, E. 1972. *Mineral nutrition of plants: principles and perspectives.* New York: John Wiley & Sons, Inc.

Galston, A. W., P. J. Davies, and R. L. Satter. 1980. *The life of the green plant,* 3d ed. Englewood Cliffs, NJ: Prentice-Hall, Inc.

Janick, J. et al. 1981. *Plant science,* 3d ed. New York: W. H. Freeman & Co.

Kozlowski, T. T., ed. 1968–1978. *Water deficits and plant growth,* 5 vols. Orlando, FL: Academic Press.

Luettge, U., and N. Higinbotham. 1979. *Transport in plants.* New York: Springer-Verlag.

Milburn, J. A. 1979. *Water flow in plants.* White Plains, NY: Longman, Inc.

Morrill, L. G. et al. 1982. *Organic compounds in soils: sorption, degradation and persistence.* Stoneham, MA: Butterworth's.

Rutter, A. J. 1972. *Transpiration.* Oxford Biology Readers, No. 24. Burlington, NC: Carolina Biological Supply Co.

Salisbury, F. B., and C. W. Ross, eds. 1985. *Plant physiology,* 3d ed. Belmont, CA: Wadsworth Publishing Co., Inc.

Singer, M. J. 1986. *Soils.* New York: Macmillan Publishing Co.

Sutcliffe, J. 1968. *Plants and water.* New York: St. Martin's Press, Inc.

Overview

Photosynthesis and respiration are each presented at three different levels: the essence of the process is examined, the major steps are briefly introduced, and the processes are then examined in greater detail. One or two levels may be sufficient for some readers, whereas others will want to explore all three. The chapter discusses the importance of the main features of each process and summarizes the light reactions, the dark reactions, glycolysis, the Krebs cycle, and the electron transport system. It concludes with a tabular comparison between photosynthesis and respiration, and makes a few brief observations on digestion and assimilation.

10   *Metabolism in Plants*

Some Learning Goals

1. Contrast the generalized equations of photosynthesis and respiration.

2. Understand what occurs in the light and dark reactions of photosynthesis, and know the principal products of the reactions.

3. Explain what occurs in glycolysis, the Krebs cycle, and the electron transport chain of respiration.

4. Distinguish between aerobic respiration and fermentation.

5. Compare digestion and assimilation.

Outline

INTRODUCTION

A number of years ago I had a student who told me her father was an alchemist, and that he was on the verge of discovering how to transform lead into gold. The belief that an element could be transformed into another element became widespread in medieval times, but with the development of modern science and technology, such romantic yet totally unscientific ideas have been discarded by educated people in general and relegated to the history books. Yet some amazing molecular "transformations" of another kind take place all around us every day in green plants and have been taking place since long before humans appeared on this planet. By far the most important of these transformations, *photosynthesis,* involves little more than fresh air, water, a green pigment, and light; yet parts of the water and air are converted in cells to a sugar without the aid of any cumbersome machinery. In addition, 24 hours a day, as long as any cell remains alive, stored energy is released from sugar by another process, *respiration.* These two processes take up the bulk of our discussion in this chapter.

PHOTOSYNTHESIS

Oil and coal today provide about 90% of the energy needed to power trains, trucks, ships, airplanes, factories, and a myriad of electrically energized appliances, computers, and communications systems. The energy within that oil and coal was originally "captured" from the sun by plants growing millions of years ago and transformed into fossil fuels by geological forces.

The energy needs of transportation, industry, and the modern household seem insignificant, however, when compared with the combined energy requirements of all living organisms. Every living cell requires energy just to remain alive, and additional energy is needed for the cell to reproduce, grow, or do physical work as part of an organism. In addition, oxygen is vital to nearly all life in processes releasing that stored energy for use. Photosynthesis, at least indirectly, is not only the principal means of enabling a civilized society to function normally but also the sole means of sustaining life—except for a few bacteria that derive their energy from sulphur salts and other inorganic compounds. This unique manufacturing process of green plants furnishes raw material, energy, and oxygen. In photosynthesis, energy from the sun is harnessed and packed into sugar molecules made from water and carbon dioxide with the aid of chlorophyll; oxygen is given off as a by-product of the process.

It has been estimated that all of the world's green organisms (including those in the oceans) together produce between 100 billion and 200 billion metric tons (110 billion and 220 billion tons) of sugar each year. In addition, much of the sugar produced is used in the formation of wood, fibers, and other structural materials. The immediate products of photosynthesis may be converted to storage forms of carbohydrates such as starch and other polysaccharides (chains of simple sugars) and disaccharides (e.g., sucrose), and others that the digestive activities of all living organisms break down to smaller molecules. Photosynthetic sugars are also involved in the production of amino acids for proteins and a host of other cell constituents. In fact, photosynthesis produces more than 94% of the dry weight substance of green organisms, with the remainder coming from the soil or dissolved matter.

The capacity of plants to satisfy humanity's energy needs probably will determine the ultimate size of human populations. Already in some quarters of the globe, starvation is commonplace where insufficient food is being produced to maintain the minimal energy requirements of overpopulated lands. While the Western world is still spending large sums of money on weight reduction as a result of the consumption of too much food, its populations are rapidly approaching the point at which they must either stabilize or they, too, will exceed the capacity of the plants to sustain them.

The food energy produced by plants is not necessarily the final limiting factor for human populations, for pollution eventually could interfere with the principal by-product of photosynthesis to the extent of diluting the oxygen content of the atmosphere.

A great deal of photosynthesis occurs in organisms present in the oceans. Consequently, it is estimated that more than 50% of the oxygen in the atmosphere originates in the seas. Laboratory tests have shown, however, that just one of the thousands of pollutants, DDT—in concentrations as low as 20 parts per billion—is capable of stopping many of the delicate marine algae from carrying on photosynthesis. The concentration of such substances in ocean waters at present is thousands of times less than that (except in estuaries where DDT concentrations of up to 7 or more parts per billion have been reported). But, while the use of DDT and related chemicals has been curtailed in the United States in recent years, other countries are still using it, and residues are still washing off into rivers and on into the oceans. If the concentrations eventually were to build up to the point of stopping algal growth and reproduction, it is conceivable that the oxygen consumption of an overpopulated world could eventually exceed the capacity of the remaining plant life to replenish it, with suffocation of nearly all animal life the grisly result. In actuality, however, starvation would be far more likely to occur first.

Note to the Reader: *Photosynthesis is undoubtedly the most important life process on earth. It is also a complex process that can be summarized briefly or examined in detail. The following is a discussion of the subject at three different levels: (1) the essence of photosynthesis; (2) a brief introduction to the major steps of photosynthesis; and (3) a closer look at photosynthesis. The process of respiration, which is discussed after photosynthesis, is treated in similar fashion. The third level, in particular, contains detail that either may or may not be discussed in your course.*

1. The Essence of Photosynthesis

Photosynthesis is an energy-storing process that takes place in leaves and other green parts of plants in the presence of light. The light energy is stored in a simple sugar molecule that is produced from carbon dioxide present in the air and water absorbed by the plant. When the carbon dioxide and the water are combined and form a sugar molecule in a chloroplast, oxygen gas is released as a by-product. The oxygen diffuses out into the atmosphere. The process is summed up in the following equation:

Photosynthesis takes place in chloroplasts (see figures 3.4, 3.5, and 3.11). The principals in the process are: *carbon dioxide, water, light,* and *chlorophyll;* a brief examination of each follows.

Carbon Dioxide

Our atmosphere consists of approximately 79% nitrogen and about 20% oxygen. The remaining 1% is made up of a mixture of less common gases, including 0.035% carbon dioxide and a little hydrogen, helium, argon, and neon. The 0.035% carbon dioxide in the air surrounding the leaves reaches the chloroplasts in the mesophyll cells by diffusing through the stomata, which are well adapted to the absorption of gases, and by going into solution in the thin film of water on the outside walls of the cells. By diffusion, it then passes through the walls into the cytoplasm and reaches the chloroplasts.

The amount of carbon dioxide constantly being taken from the atmosphere during daylight hours by all green plants is enormous. Just four-tenths of a hectare (1 acre) of corn (10,000 plants) accumulates more than 2,500 kilograms (5,512 pounds) of carbon from the atmosphere during a growing season. Over 10 metric tons (11 tons) of carbon dioxide are needed to furnish this much carbon. It also has been calculated that the total present atmospheric supply of carbon dioxide (more than 2.2 billion metric tons or about 50 metric tons over each hectare of the earth's surface) would be completely used up in about 22 years if it were not constantly being replenished. A large reservoir of carbon dioxide in the oceans maintains the atmosphere's percentage of the gas at an ambient 0.03, while plant and animal respiration, the burning of material containing carbon, volcanoes, and similar sources replace it at roughly the same rate at which it is removed during photosynthesis.

There is some evidence that atmospheric carbon dioxide concentration is increasing; in fact, a 1970 study by scientists at the Massachusetts Institute of Technology on critical environmental problems predicted a 20% increase by the year 2000. This could actually benefit plants, providing the level does not

$$6\ CO_2 + 6\ H_2O + \text{light energy} \xrightarrow[\text{enzymes}]{\text{chlorophyll}} C_6H_{12}O_6 + 6\ O_2$$

carbon dioxide   water   glucose   oxygen

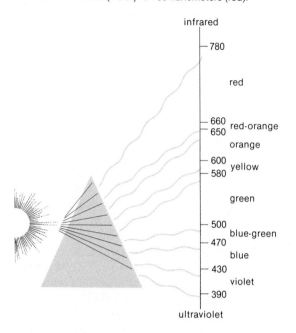

FIGURE 10.1 Visible light that is passed through a prism is broken up into individual colors with wavelengths ranging from 390 nanometers (violet) to 780 nanometers (red).

infrared

— 780

red

— 660
— 650   red-orange

orange

— 600   yellow
— 580

green

— 500
— 470   blue-green

blue

— 430
violet

— 390

ultraviolet

climb beyond 0.1% to 0.2% (which is extremely unlikely). Increases in yields of between 100% and 200% have been obtained by "fertilizing" plants with carbon dioxide, and some large commercial greenhouses have pipes over plant beds to supplement their natural supply. The changes in climate that would result from an increase in atmospheric carbon dioxide, however, could well offset positive benefits.

Water

Less than 1% of all the water absorbed by plants is used in photosynthesis; most of the remainder is transpired and incorporated into protoplasm and other materials. The water used is the source of the oxygen released as a by-product of photosynthesis, even though carbon dioxide also contains oxygen. This has been demonstrated by using isotopes of oxygen to make both the carbon dioxide and the water used in photosynthesis. When the isotope is used only in the water, it appears in the oxygen gas released. If, however, it is used only in the carbon dioxide, it is confined to the sugar produced and

never appears in the oxygen gas. Thus it has been shown clearly that the water is the sole source of this important by-product. If it is in short supply, water may indirectly become a limiting factor in photosynthesis; in this case, the stomata usually close and sharply reduce the carbon dioxide supply.

Light

Light exhibits properties of both waves and particles. Energy reaches the earth from the sun in waves of different lengths, the longest waves being radio waves, and the shortest being X rays. About 40% of the radiant energy we receive is in the form of visible light. If this visible light is passed through a glass prism, it splits up into its component colors. Reds are on the longer wave end and violets on the shorter wave end, with yellows, greens, and blues between (figure 10.1). Although nearly all of the visible light colors can be used in photosynthesis, those in the blue-violet and red-orange wavelengths are used most extensively; many in the green range are reflected. Leaves commonly absorb about 80% of the visible light that reaches them.

The intensity of light varies with the time of day, season of the year, altitude, latitude, and atmospheric conditions. On a clear summer day at noon in a temperate zone, sunlight attains an intensity in the vicinity of 10,000 footcandles. (A footcandle is the light cast by a standard candle at a distance of 1 foot.) In contrast, consider that a good reading lamp produces only about 25 footcandles. Plants vary considerably in the light intensities they need for photosynthesis to occur at optimal (ideal) rates, and other factors such as the amount of carbon dioxide available are related. For example, an increase in photosynthesis will not occur in some plants receiving more than 3,000 footcandles of light unless supplemental carbon dioxide is provided. With supplements, however, rates of photosynthesis will continue to increase up to about 5,000 footcandles. Herbaceous plants on a forest floor can survive with less than 2% of full daylight, and some mosses are reported to thrive on intensities as low as 0.05% to 0.01%. Most land plants that naturally grow in the open need at least 30% of full daylight to thrive, and the optimal amount for some species of trees approaches full daylight. Shade plants often do well in 10% of full daylight, and light that is too intense may change the way in which some of the metabolic activities of their cells takes place. For example, in high light intensities *photorespiration,* a special type of respiration that uses oxygen and releases carbon dioxide but differs from common aerobic respiration in its chemical pathways (see page 187), and *photooxidation,* which involves the destruction of chlorophyll by light, may occur. Although carbon dioxide is released in this process, most of the chemical steps involved are quite different from those of normal respiration. High light intensities may also cause an increase in transpiration, resulting in the closing of stomata. A sharp reduction in the available carbon dioxide supply inevitably follows.

Chlorophyll

There are several different types of chlorophyll molecules, all of which contain the element magnesium. They are otherwise similar in structure to hemoglobin, the red pigment in blood. Each molecule has a long lipid tail, which is believed to anchor the chlorophyll molecule in the lipid layers of the grana.

Chloroplasts of most plants contain two kinds of chlorophyll associated with the thylakoid membranes. Chlorophyll *a* is bluish green in color and has the formula $C_{55}H_{72}O_5N_4Mg$. Chlorophyll *b* is yellowish green in color and has the formula $C_{55}H_{70}O_6N_4Mg$. Usually a chloroplast has about three times more chlorophyll *a* than *b*. When a molecule of chlorophyll *b* absorbs light, it transfers the energy to a molecule of chlorophyll *a*. Chlorophyll *b,* then, is an **accessory pigment** that makes it possible for photosynthesis to take place over a broader spectrum of light than would be possible with chlorophyll *a* alone. Other such pigments include carotenoids (yellowish to orange pigments found in all plants), phycobilins (blue or red pigments found in blue-green bacteria and red algae), and several other types of chlorophyll. Chlorophylls *c* and possibly *d* and *e* take the place of chlorophyll *b* in certain algae, and several other photosynthetic pigments are found in bacteria. The various chlorophylls are all closely related and differ from one another only slightly in the structure of their molecules.

In chloroplasts, about 250 to 400 pigment molecules are grouped as a light-harvesting complex called a **photosynthetic unit,** with countless numbers of these units in each granum. Two types of these photosynthetic units function together in the chloroplasts of green plants, bringing about the first phase of photosynthesis, the *light reactions* (discussed in the next section).

2. Introduction to the Major Steps of Photosynthesis

The process of photosynthesis takes place in two successive major steps called the **light reactions** and the **dark reactions.**

The Light Reactions

In the 1930s the English biochemist, Robin Hill, discovered that a solution of fragmented and whole chloroplasts, isolated from leaves that had been ground up and centrifuged, could briefly produce oxygen if an electron acceptor was present to receive electrons from water. In 1951 it was shown that *nicotinamide adenine dinucleotide phosphate,* usually abbreviated to NADP (a substance derived from the B vitamin *niacin*), was a natural electron acceptor in this reaction which, in honor of its discoverer, has become known as the *Hill reaction.* The Hill reaction and other research gave insight into a major phase of photosynthesis, called the *light reactions,* which involve light striking chlorophyll molecules that are embedded in the thylakoids of the grana in the chloroplasts (see figure 3.11). The light initiates reactions that result in the conversion of some of the light energy to chemical energy. In the process, (1) water molecules are split apart, producing hydrogen ions and electrons, and oxygen gas is released; (2) energy-storing molecules commonly known as **ATP** are created;[1] and (3) the hydrogen from the split water molecules is involved in the creation of $NADPH_2$ (more precisely shown as $NADPH + H^+$), which carries hydrogen and which is used in the second major phase of photosynthesis, the *dark reactions.*

The Dark Reactions

The *dark reactions* are a series of reactions that take place outside of the grana in the stroma of the chloroplast (see page 42), if the products of the light reactions are available, regardless of whether or not light is present. The term *dark,* therefore, may be slightly misleading. The dark reactions may develop in different ways, depending on the particular kind of plant involved. In 1961, Dr. Melvin Calvin of the University of California received a Nobel Prize for

---

1. ATP stands for adenosine triphosphate; there are millions of such molecules in living cells. When an ATP molecule releases its energy, it gives up one of its three phosphate "groups" and becomes ADP, or adenosine diphosphate. An ADP molecule becomes an ATP molecule again when it regains a third phosphate "group" and stores energy. ATP is an important participant in many reactions involving the transfer of energy.

FIGURE 10.2   A simple summary of photosynthetic reactions.

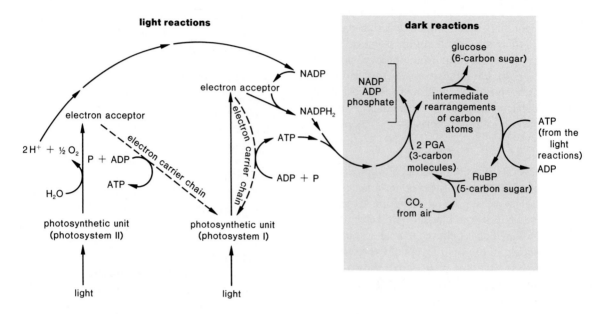

discovering how the most widespread type of dark reactions occur, and this process is now often referred to as the **Calvin cycle.**

In this cycle, carbon dioxide from the air is combined with a 5-carbon sugar, RuBP (ribulose bisphosphate), and then the combined molecules are converted, through several steps, to 6-carbon sugars such as glucose. Energy and other items involved in these steps are furnished by the ATP molecules and $NADPH_2$ produced during the light reactions. Some of the 6-carbon sugars that are produced during the dark reactions are recycled, while others are stored as starch or other polysaccharides (simple sugars strung together in chains). A simple summary of the photosynthetic reactions is portrayed in figure 10.2. More detailed diagrams of light and dark reactions are shown in figures 10.5 and 10.6.

Two molecules of a 3-carbon compound are shown as the first stable substance produced when carbon dioxide from the air and RuBP are combined and then converted during the dark reactions (figure 10.2). Many tropical plants produce a 4-carbon compound at this point instead. This 4-carbon pathway is discussed, along with another variation found mostly in desert plants, in the next section.

3. A Closer Look at Photosynthesis

Although many details of photosynthesis need to be further elucidated, a great deal has been learned about this vital process since 1772 when the English naturalist, Joseph Priestly (1733–1804), received a medal for demonstrating that vegetation "restored" oxygen so that a mouse could live in air that had been "used up" by a burning candle. Seven years later the Dutch physician, Jan Ingen-Housz (1730–1799), confirmed Priestly's observations, and showed that the air was "restored" only when green parts of plants were in the presence of sunlight. In 1782 Jean Senebier (1742–1809), a Swiss pastor, discovered that the photosynthetic process required carbon dioxide, and in 1796 Ingen-Housz showed that carbon went into the nutrition of the plant. The final component of the overall photosynthetic reaction was explained in 1804 by another Swiss, Nicholas Theodore de Saussure (1767–1845), who showed that water was involved in the process. A little of what is currently known about the details of photosynthesis is given in the following modest amplification of the preceding "second level" outline. Those who wish more information are referred to the reading list at the end of the chapter.

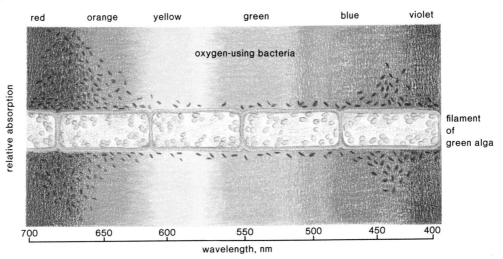

FIGURE 10.3   Engelmann's experiment. A tiny spectrum of light was focused on a microscope slide with a row of algal cells suspended in water containing bacteria that move toward an oxygen source. The bacteria assembled mostly in the areas exposed to the red and blue portions of the spectrum.

The Light Reactions Reexamined

We previously noted that light exhibits the dual characteristic of waves and particles. In the seventeenth century when Sir Isaac Newton first used a prism to produce a spectrum of colors from visible white light, he postulated that light consisted of a series of discrete particles he called "corpuscles." His theory only partially explained light phenomena, however, and by the middle of the nineteenth century James Maxwell and others showed that light and all other parts of the vast electromagnetic spectrum travel in waves.

Before the beginning of the twentieth century, the wave theory of light also became inadequate to explain certain attributes of light, particularly in relation to the fact that a *photoelectric effect* can be produced in all metals. When a metal is exposed to radiation of a critical wavelength, either visible or invisible, it becomes positively charged because the radiation forces electrons out of the metal atoms. The ability of light to force electrons from metal atoms depends on its particular wavelength and not on its intensity or brightness. The shorter the wavelength, the greater the energy and vice versa.

Einstein returned to the particulate nature of light in 1905 and proposed that light is composed of packets of energy he called **photons,** but the wave theory of light was not discarded. Instead, physicists gradually came to regard both theories of light as necessary for a complete description, and both waves and particles (photons) are now almost universally recognized as aspects of light. We also now recognize that the energy of a photon is not the same for all kinds of light; those of longer wavelengths have lower energy, and those of shorter wavelengths have proportionately higher energy.

As noted earlier, chlorophylls, the principal pigments of photosynthetic systems, absorb light primarily in the violet to blue and also in the red wavelengths; they reflect green wavelengths, which is why leaves appear green. This was first ingeniously demonstrated in 1882 by T. W. Engelmann. He focused a tiny spectrum of light on a filament (single row of cells) of a freshwater alga; the alga had been mounted in a drop of water on a microscope slide containing bacteria that move toward an oxygen source. As shown in figure 10.3, the bacteria assembled in greatest numbers along the algal filament in the blue and red portions of the spectrum; this showed that oxygen production is directly related to the light the chlorophyll absorbs. Each pigment has its own distinctive pattern of light

FIGURE 10.4   The absorption spectrum of chlorophyll *a*.

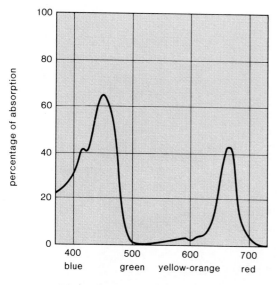

wavelength of visible radiation (in nanometers)

Photosystems

The two types of photosynthetic units present in most chloroplasts collectively comprise **photosystems** designated as *photosystem I* and *photosystem II* (figure 10.5). Each photosynthetic unit of photosystem I consists of 200 or more molecules of chlorophyll *a*, small amounts of chlorophyll *b*, carotenoid pigment with protein attached, and 1 special *reaction center* molecule of chlorophyll *a* called P_{700}. Although all pigments in a photosystem can absorb photons, the reaction center molecule is the only one that can actually use the light energy; the remaining photosystem pigments are called *antenna pigments* because together they function somewhat like an antenna in gathering and passing light to the reaction center molecule. In addition, there are iron-sulfur proteins that are the primary electron acceptors for photosystem I; i.e., iron-sulfur proteins first receive electrons from P_{700}. A photosynthetic unit of photosystem II consists of chlorophyll *a*, β-carotene attached to protein, a little chlorophyll *b*, and 1 special reaction center molecule of chlorophyll *a* called P_{680}. The letter *P* stands for pigment and the numbers *700* and *680* of the reaction center molecules of chlorophyll *a* refer to peaks in the absorption spectra of light with wavelengths of 700 and 680 nanometers, respectively. These peaks differ slightly from those of the otherwise identical chlorophyll *a* molecules of the photosynthetic units. Also present in photosystem II is a primary electron acceptor, believed to be a molecule called *pheophytin* or Pheo.

When a photon of light strikes a P_{680} molecule of a photosystem II photosynthetic unit, the energy boosts an electron out of place to a higher energy level. This is referred to as "exciting" an electron. The excited electron is picked up by the primary acceptor molecule[3] Pheo, and is then passed to another acceptor, Q. The electrons lost by the P_{680} molecule are replaced by electrons extracted from a water molecule through a process that is not yet fully understood. As these water molecules are split, a molecule of oxygen and 4 protons are also produced. This splitting of water molecules, which requires manganese atoms and is mediated by an enzyme on the *inside* of the thylakoid membrane, is called *photolysis*. The location of the enzyme results in what

absorption; this is referred to as the *absorption spectrum* of the pigment (figure 10.4). When a pigment absorbs light, the energy levels of some of the pigment's electrons are raised. When this occurs, the energy may be emitted immediately as light (a phenomenon called *fluorescence*). In chlorophyll, this is characteristically in the red part of the visible light spectrum, so that an extract of chlorophyll placed in light will appear red. The absorbed energy may also be emitted as light after a delay (a phenomenon called *phosphorescence*); it may otherwise be converted to heat, or it may be stored in a chemical bond as it is in photosynthesis.

As indicated at the beginning of our discussion, photosynthesis is a very complex process. It involves several types of chemical reactions, including *oxidation-reduction reactions,*[2] which are also vital to the process of respiration (discussed later in this chapter).

---

2. Oxidation is the removal of electrons from a compound; reduction is the addition of electrons. When an electron is removed a proton may follow, with the result that hydrogen is often removed during oxidation and added during reduction. Oxygen is usually the oxidizing agent (i.e., the final acceptor of the electron), but oxidations can occur without oxygen being involved.

---

3. Except for the high speed at which it usually carries out its function, an acceptor molecule operates something like a pickup order telephoned to a clerk in a store in the sense that an item is picked up from a source, temporarily held, and then transferred elsewhere.

FIGURE 10.5   A summary of the light reactions of
photosynthesis.

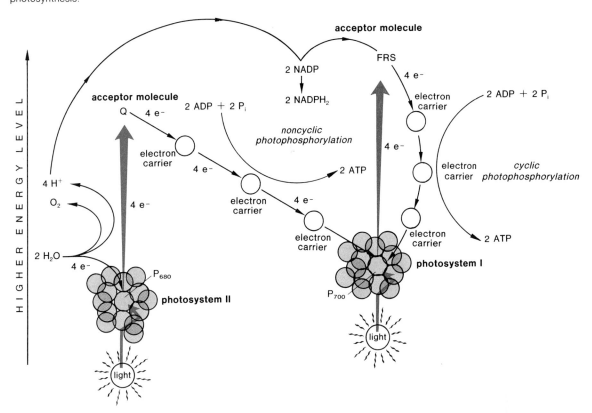

is believed to be a proton gradient being formed
across the membrane. The movement of protons
across the membrane is thought to be a source of
energy for the synthesis of ATP; it has been described as similar to the movement of molecules
during osmosis, and has been called *chemiosmosis*
or the "Mitchell Theory," after its author, Peter
Mitchell. In this concept, protons move across a thylakoid membrane through protein channels called
ATPase or *coupling factor.* Coupled to the proton
movement, ADP and phosphate combine, producing
ATP. This light-powered production of ATP is called
photophosphorylation.

Let us return to the acceptor molecule Q. It releases the excited electron to an electron transport
system that functions something like a high-speed
bucket brigade in moving electrons from H_2O to a
temporary high energy-storage molecule, *nicotinamide adenine dinucleotide phosphate*
($NADPH + H^+$), an electron acceptor for photosystem I. The electron transport chain consists of
iron-containing pigments called *cytochromes* and

other electron transfer molecules plus a copper-
containing protein, *plastocyanin.* While electrons
pass along the electron transport system and protons
are being moved through the coupling factor, ATP
molecules are being formed from ADP and phosphate in the process of *photophosphorylation,* as
noted above.

A somewhat similar series of events occurs in
photosystem I. When a photon of light strikes a P_{700}
molecule in a photosynthetic unit, the energy excites
an electron, which is transferred to an acceptor molecule believed to be an iron-sulfur protein designated Fe-S. This then passes the electron to another
iron-sulfur protein, *ferredoxin,* which in turn releases it to $NADP^+$. The $NADP^+$ is reduced to
$NADPH + H^+$. The electrons removed from the P_{700}
molecule are replaced by electrons from photosystem II. This overall movement of electrons from
water to photosystem II to photosystem I to $NADP^+$
is called *noncyclic* electron flow, because it goes in
one direction only. The production of ATP that correspondingly occurs is designated as *noncyclic photophosphorylation.*

FIGURE 10.6   The Calvin cycle (dark reactions) of photosynthesis.

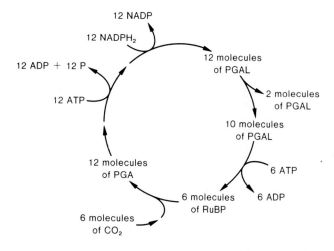

Photosystem I can also work independently of photosystem II, and when it does, the electrons boosted from P_{700} reaction center molecules (of photosystem I) are passed from an acceptor molecule called P_{430} to the electron transport chain between the two photosystems (instead of to ferredoxin and $NADP^+$) and back into the reaction center of photosystem I. This is *cyclic* electron flow. ATP generated by cyclic electron flow is called *cyclic photophosphorylation,* but no $NADPH+H^+$ or oxygen are produced nor are water molecules split. A summary of the light reactions in the most widespread form of photosynthesis is shown in figure 10.5.

The Dark Reactions Reexamined

We have seen how ATP and $NADPH+H^+$ are synthesized during the light reactions. Both of these substances play key roles in the synthesis of carbohydrate from carbon dioxide of the atmosphere, which reaches the interior of chlorenchyma tissues via stomata, during the *dark reactions.* As indicated earlier, the dark reactions are really a whole series of reactions, each mediated by an enzyme in this major phase of photosynthesis. The reactions take place in the stroma of the chloroplasts and, as long as the products of the light reactions are present, they need no light to occur although they normally take place during daylight hours.[4]

----

4. Note: Some of the enzymes involved in the dark reactions may actually need light for their activation (that is, conversion to a form in which they can actively catalyze reactions of the dark reactions).

The Calvin Cycle

There are three known mechanisms through which carbon dioxide is converted to carbohydrate during the dark reactions. The most widespread is the *3-carbon pathway* or *Calvin cycle* (named after Nobel laureate Melvin Calvin who, with his associates at the University of California at Berkeley, discovered and unraveled some of its details in the 1950s) (figure 10.6).

The main steps of the Calvin cycle (dark reactions) are as follows:

1. Six molecules of carbon dioxide from the air combine with 6 molecules of ribulose 1,5-bisphosphate (RuBP, the 5-carbon sugar continually being formed while photosynthesis is occurring), with the aid of the enzyme RuBP carboxylase.
2. The resulting 6 6-carbon molecules are immediately split into 12 3-carbon molecules known as 3-phosphoglycerate (PGA).
3. The $NADPH+H^+$ (which has been temporarily holding the hydrogen ions and electrons released during the light reactions) and ATP (also from the light reactions) supply energy to convert the PGA to 12 molecules of glyceraldehyde 3-phosphate (a 3-carbon sugar-phosphate), the stable end products of photosynthesis.
4. Ten of the 12 glyceraldehyde 3-phosphate molecules are restructured and become 6 5-carbon molecules of RuBP, the sugar with which the Calvin cycle was initiated.

The net gain of 2 glyceraldehyde 3-phosphate molecules can result either in an increase in the carbohydrate content of the plant (glucose, starch, cellulose, or related substances), or they can be used in pathways that lead to the net gain of lipids and amino acids.

The 4-Carbon Pathway

Plants of at least 100 genera that occur primarily in the tropics produce a 4-carbon compound, *oxaloacetic acid* instead of 3-carbon PGA during the initial steps of the dark reactions. Oxaloacetic acid is produced when a 3-carbon compound, phosphoenolpyruvate (PEP), and carbon dioxide are combined with the aid of an enzyme, phosphoenolpyruvate carboxylase, in mesophyll cells. Depending on the species, the PEP may then be converted to aspartic, malic, or other 4-carbon acids. These organic acids, which are not substitutes for

FIGURE 10.7 A portion of a cross section of a leaf of a C_4 plant. Compare this with the leaf of a typical C_3 plant, as illustrated in figure 7.4. (G. S. Ellmore)

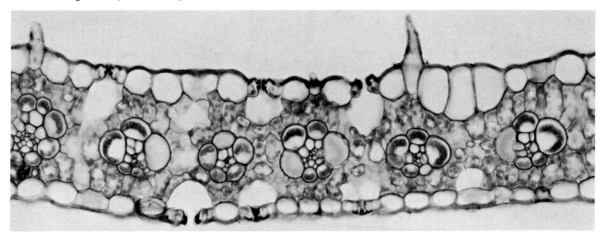

PGA, migrate to bundle sheath cells surrounding the vascular bundles (veins) of leaves where they are converted to pyruvic acid and carbon dioxide. The pyruvic acid returns to the mesophyll cells and interacts with ATP, forming additional PEP molecules; the carbon dioxide combines with RuBP in the bundle sheath cells and is converted to PGA and related molecules in the Calvin cycle. Because the PEP system produces 4-carbon compounds, plants having this system are called 4-carbon, or C_4, plants, to distinguish them from plants that have only the 3-carbon, or C_3, system.

C_4 plants have several characteristic features. (1) Unlike C_3 plants, which have one form of chloroplast and can run only the Calvin cycle, C_4 plants possess two forms of chloroplasts and have an alternate pathway for carbon dioxide utilization. There are large chloroplasts, often with few to no grana, in the bundle sheath cells surrounding the veins. These large chloroplasts contain numerous starch grains and the grana, when present, are poorly developed. The much smaller, very numerous chloroplasts of the mesophyll cells usually lack starch grains and have well-developed grana (figure 10.7). (2) High concentrations of PEP carboxylase are found in the mesophyll cells. This is significant because PEP carboxylase permits the conversion of carbon dioxide to carbohydrate at much lower concentrations than does RuBP carboxylase (found only in the bundle sheath cells), the corresponding enzyme in the Calvin cycle. (3) The optimum temperatures for C_4 photosynthesis are much higher than those for C_3 photosynthesis, thus allowing C_4 plants to thrive under conditions that would adversely affect, or even kill, C_3 plants.

Obviously the C_4 pathway furnishes carbon dioxide to the Calvin cycle in a more roundabout way than does the C_3 pathway, but it should be noted that during *photorespiration* (see page 181), which is in competition with the Calvin cycle and takes place in light all the time the Calvin cycle is functioning, RuBP reacts with oxygen and eventually releases carbon dioxide, whereas in photosynthesis RuBP and carbon dioxide are involved in the synthesis of carbohydrate. In C_4 plants both photorespiration (which reduces the efficiency of carbon fixation) and the Calvin cycle take place almost exclusively in the bundle sheath cells, but high concentrations of carbon dioxide and low concentrations of oxygen limit the amount of photorespiration that can occur. As a result, C_4 plants have photosynthetic rates that are two to three times higher than those of C_3 plants. The C_4 pathway in the mesophyll cells results in carbon dioxide being picked up even at low concentrations (via PEP carboxylase) and in concentrating carbon dioxide in the vicinity of RuBP carboxylase in the bundle sheath cells so that RuBP will react with carbon dioxide rather than oxygen.

CAM Photosynthesis

Crassulacean acid metabolism (CAM) photosynthesis is found in plants of at least 20 families, including cacti, stonecrops, orchids, bromeliads, and many succulents growing in regions of high light intensity. A few succulents do not have CAM photosynthesis, however, and several non-succulent plants do. Plants with CAM photosynthesis typically do not have a well-defined palisade mesophyll in the leaves, and, in contrast to the chloroplasts of the bundle sheath cells of C_4 plants those of CAM photosynthesis plants resemble the mesophyll cell chloroplasts of C_3 plants.

CAM photosynthesis is similar to C_4 photosynthesis in that 4-carbon compounds are produced during the dark reactions. In these plants, however, malic acid accumulates in the chlorenchyma tissues at night and is converted back to carbon dioxide during the day. The enzyme PEP carboxylase is responsible for converting the carbon dioxide to malic acid at night. During daylight the malic acid diffuses out of the cell vacuoles in which it was stored and is converted back to carbon dioxide for use in the Calvin cycle. A much larger amount of carbon dioxide can be converted to carbohydrate each day than would otherwise be possible, since the stomata of such plants generally admit very little carbon dioxide during the day because they are closed (closed stomata greatly reduce water loss). This arrangement permits the plants to function well under conditions of both limited water supply and high light intensity.

RESPIRATION

The solar energy that is converted to chemical energy by the process of photosynthesis is stored in various organic compounds—for example, wood or coal. If the organic compounds are burned, the energy is released very rapidly in the form of heat and light, and much of the usable energy is lost. Living organisms, however, "burn" their energy-containing compounds in numerous small steps that release tiny amounts of immediately usable energy, usually storing the released energy in ATP molecules and thus permitting the available energy to be used more efficiently.

1. The Essence of Respiration

Respiration is an energy-releasing process that takes place in all living cells 24 hours a day, regardless of whether or not photosynthesis happens to be occurring simultaneously in the same cells. It is initiated in the cytoplasm and completed in the mitochondria. The energy is released from simple sugar molecules that are broken down during a series of steps controlled by enzymes. No oxygen is needed to initiate the process, but in **aerobic respiration** (the most widespread form of respiration) the process cannot

be completed without oxygen gas; carbon dioxide and water are by-products. Aerobic respiration is summed up in the following equation:

$$C_6H_{12}O_6 + 6\ O_2 \xrightarrow{\text{enzymes}} 6\ CO_2 + 6\ H_2O + \text{energy}$$

glucose   oxygen   carbon   water
                             dioxide

Anaerobic respiration and **fermentation** are two forms of respiration carried on by certain bacteria and other organisms in the absence of oxygen gas. These forms of respiration release much less energy than aerobic respiration. The two forms differ from one another in the manner in which hydrogen released from the glucose is combined with other substances (see discussion in a later section). Fermentation is very important industrially, particularly in the brewing industry. Two well-known forms of fermentation are illustrated by the following equations:

$$C_6H_{12}O_6 \xrightarrow{\text{enzymes}} 2\ C_2H_5OH + 2\ CO_2 + \text{energy}$$

glucose   ethyl   carbon
                      alcohol   dioxide

$$C_6H_{12}O_6 \xrightarrow{\text{enzymes}} 2\ CH_3CHOHCOOH + \text{energy}$$

glucose   lactic acid

The relatively small amount of energy released during these forms of respiration is stored for the most part in 2 ATP molecules.

2. Introduction to the Major Steps of Respiration

Glycolysis

In all forms of respiration, the first major phase takes place in the cytoplasm and requires no oxygen gas (O_2). This phase, called **glycolysis** involves three main steps and several smaller ones, each controlled by an enzyme. During the process a small amount of energy is released, and some hydrogen atoms are removed from compounds derived from a glucose molecule. The essence of this complex series of steps is as follows:

1. In a series of reactions the glucose molecule becomes a fructose molecule carrying 2 phosphates.
2. This sugar (fructose) molecule is split into 2 3-carbon fragments.
3. Some hydrogen, energy, and water are removed from these 3-carbon fragments, leaving **pyruvic acid** (figure 10.8).

FIGURE 10.8   A summary of respiration.

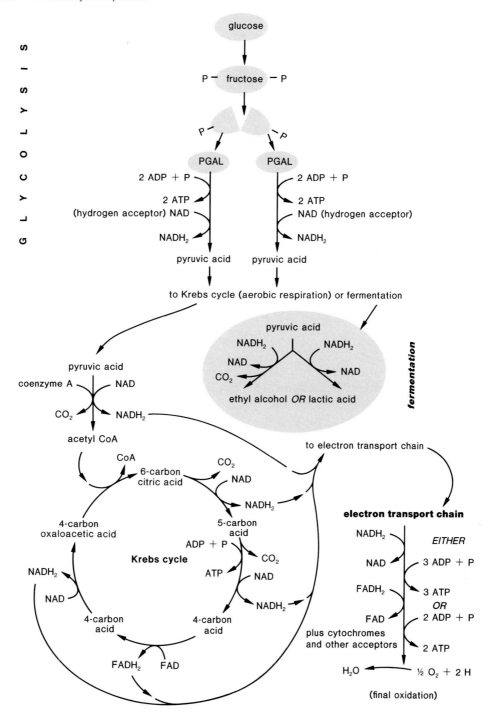

The energy needed to start the process of glycolysis is supplied by 2 ATP molecules. By the time pyruvic acid has been formed, however, 4 ATP molecules have been produced from the energy released along the way, for a net gain of 2 ATP molecules; a great deal of the energy originally in the glucose molecule remains in the pyruvic acid. The hydrogen ions and electrons released during the process are picked up and temporarily held by a hydrogen acceptor molecule called *nicotinamide adenine dinucleotide* (NAD). What happens to them next depends on the kind of respiration involved: *aerobic respiration, true anaerobic respiration,* or *fermentation.*

Aerobic Respiration

In **aerobic respiration** (the most common type of respiration), glycolysis is followed by two major stages: the *Krebs cycle* and the *electron transport chain.* Both stages occur in the mitochondria and involve many smaller steps, each of which is controlled by enzymes (figure 10.8).

The Krebs Cycle

The **Krebs cycle** was named after the British biochemist Hans Krebs, who received a Nobel Prize in 1953 for his unraveling of many of the complex reactions that take place in respiration. This cycle is sometimes referred to as the *citric acid* (or *organic acid*) *cycle* because of the important role played by several organic acids during the process.

Before entering the Krebs cycle, which takes place in the *matrix* located between the cristae membranes of mitochondria, carbon dioxide is released from the pyruvic acid produced by glycolysis; what remains is restructured to a 2-carbon acetyl group. This acetyl group combines with an acceptor molecule called *coenzyme A* (CoA). This combination (called *acetyl CoA*) then enters the Krebs cycle, which is a series of chemical reactions that are catalyzed by enzymes. During glycolysis little of the energy originally trapped in the glucose molecule is released but, as the Krebs cycle proceeds, small amounts of energy and hydrogen are successively removed from a series of organic acids until the remaining energy has been transferred to compounds such as $NADH_2$ (*nicotinamide adenine dinucleotide,* a temporary energy-storing compound—see discussion in level three), ATP, and $FADH_2$ (*flavin adenine dinucleotide*—also discussed in level three). Carbon dioxide is produced as a by-product while the cycle is proceeding.

The Electron Transport Chain

Although all of the energy originally in the glucose molecule has been changed to other forms by the end of the Krebs cycle, much of it remains in the hydrogen ions and electrons that were passed to the electron carriers NAD and FAD, which became $NADH_2$ and $FADH_2$, respectively. The hydrogen ions and electrons now are passed along an *electron transport chain*—a chain of special acceptor molecules, which include iron-containing proteins called *cytochromes*—releasing energy in small increments at each step, with water being formed as the hydrogen ions and electrons finally combine with oxygen from the air. ATP is produced from ADP as the energy is released. Thus, to summarize the final step in aerobic respiration (see figure 10.8), water is produced by oxygen acting as a "wastebasket" for hydrogen.

By the time the process is complete, the energy locked in a molecule of glucose has been released, with about 39% of it being stored in ATP molecules. This stored energy is then available for use in the synthesis of other molecules and for growth, active transport, and a host of other metabolic processes. Aerobic respiration produces a net gain of 36 ATP molecules from 1 glucose molecule, using up 6 molecules of oxygen and producing 6 molecules of carbon dioxide and a net total of 6 molecules of water. For each mole (180 grams) of glucose aerobically respired, 686 kilocalories (kcal) of energy is released, with about 39% of it being stored in ATP molecules and the remainder being released as heat.

Anaerobic Respiration and Fermentation

Many bacteria carry on both fermentation and true anaerobic respiration simultaneously, making it difficult to distinguish between the two processes. Some texts use the terms *anaerobic respiration* and *fermentation* interchangeably to designate respiration occurring in the presence of little or no oxygen gas.

In many instances in the biological realm, glucose molecules may undergo glycolysis without sufficient oxygen being available to permit the completion of aerobic respiration. In such cases, the hydrogen released during glycolysis is simply transferred from the hydrogen acceptor molecules back to the pyruvic acid, after it has been formed, creating ethyl alcohol, lactic acid, or similar substances. A little energy is released during either fermentation or true anaerobic respiration, but most of it remains locked up in the alcohol, lactic acid, or other compounds produced.

In true anaerobic respiration, the hydrogen removed from the glucose molecule during glycolysis is combined with an inorganic ion, as is the case when sulphur bacteria (discussed in chapter 15) convert sulphate (SO_4) to sulphur or another sulphur compound or when certain cellulose bacteria produce methane gas (CH_4) by combining the hydrogen with carbon dioxide.

Oxygen gas is not required to make these compounds, but few organisms can live long without oxygen, and many that carry on fermentation can also respire aerobically. If oxygen becomes available, the remaining energy can be released by breaking these compounds down further. During anaerobic respiration or fermentation only about 7% of the total energy in the glucose molecule is removed, and so much of that energy goes into the making of the alcohol or the lactic acid or is dissipated as heat that there is a net gain of only 2 ATP molecules (compared with 36 ATP molecules produced in aerobic respiration).

Living cells can tolerate only certain concentrations of alcohol. In media in which yeasts are fermenting sugars, for example, once the alcohol concentration builds up beyond 12%, the cells die and fermentation ceases. This is why most wines have an alcohol concentration of about 12%.

Factors Affecting the Rate of Respiration

Temperature

Temperature plays a major role in the rate at which the various respiratory reactions occur. For example, when air temperatures rise from 20° C (68° F) to 30° C (86° F), the respiration rates of plants double and sometimes even triple. The faster respiration occurs, the faster the energy is released from sugar molecules, with an accompanying decrease in weight. In growing plants, this weight loss is more than offset by the production of new sugar by photosynthesis. In harvested fruits, seeds, and vegetables, however, respiration continues without sugar replacement; some water loss also occurs. Respiring cells primarily convert energy stored as starch or sugar to ATP, but much of the energy is lost in the form of heat, with only 39% being stored as ATP. Most fresh foods are kept under refrigeration, not only to lower the respiration rate and retard water loss, but also to dissipate the heat. Keeping the temperatures down is also important to prevent the growth and reproduction of food-spoiling molds and bacteria, which may prefer warmer temperatures.

Heat inactivates most enzymes at temperatures above 35° C (95° F), but a few organisms such as various algae in the hot springs of Yellowstone National Park and similar places have adapted in such a way that they are able to thrive at temperatures of up to 60° C (140° F)—heat that would kill other organisms of comparable size almost instantly.

Water

Water inside the cells and their organelles acts as a medium in which the enzymatic reactions can take place. Living cells often have a water content in excess of 90%, but the cells of mature seeds may have a water content of less than 10%. When water content becomes this low, respiration does not cease completely, but it continues at a drastically reduced rate, resulting in only very tiny amounts of heat being released and of carbon dioxide being given off. Seeds may remain viable (capable of germinating) for many years if stored under dry conditions. If they come in contact with water, however, they swell by imbibition. Respiration rates then increase rapidly. If the wet seeds happen to be in an unrefrigerated storage bin, the temperature may increase to the point of killing the seeds. In fact, if fungi and bacteria begin to grow on the seeds, temperatures from their respiration can become so high that spontaneous combustion can sometimes occur.

Oxygen

If, due to flooding, the roots of trees and houseplants have their oxygen supply sharply reduced, their respiration rates may be decreased and their growth retarded; they may even die if the condition persists too long. In the case of stored foods, however, reducing the oxygen in the storage areas to bring about lower rates of respiration may be desirable, and consequently, it has become a common commercial practice to reduce the oxygen present in warehouses where crops are stored to as little as 1% to 3% by pumping in nitrogen gas, while maintaining low temperatures and humidity.

3. A Closer Look at Respiration

Respiration, like photosynthesis, is a very complex process, and, as with photosynthesis, it is beyond the scope of this book to explore the subject in great detail. The following amplification of information already discussed is modest, and those who wish further information are referred to the reading list at the end of the chapter.

Glycolysis Reexamined

As previously discussed, this initial phase of all forms of respiration brings about the conversion of each 6-carbon glucose molecule to 2 3-carbon pyruvic acid molecules via three main steps, each mediated by an enzyme. The three main steps are (1) *phosphorylation,* whereby the 6-carbon sugar is prepared for the reaction by the addition of phosphate; (2) *sugar cleavage,* which involves the splitting of the sugar molecule into 2 3-carbon fragments; and (3) *pyruvic acid formation,* which involves the oxidation of the sugar fragments.

Energy needed to initiate the process is furnished by an ATP molecule which also furnishes the phosphate group for the phosphorylation of the sugar, glucose, to yield glucose 6-phosphate. Another ATP, with the aid of an appropriate enzyme, yields fructose 1,6-bisphosphate. As a result of the cleavage of the fructose 1,6-bisphosphate, 2 different 3-carbon molecules are produced, but ultimately only 2 glyceraldehyde 3-phosphate (GA3P) molecules remain. Through several small steps mediated by enzymes the phosphate groups are removed from the GA3P molecules; a controlled energy release takes place during the process, with a net energy gain of 2 ATP molecules. Thus hydrogen, protons, and electrons are also removed from the GA3P; these are picked up by a hydrogen acceptor molecule, nicotinamide adenine dinucleotide (NAD^+). Two $NADH_2$ (more precisely written $NADH+H^+$) molecules also are produced from each glucose molecule during glycolysis, which requires no molecular oxygen.

The glucose molecule is prepared for the reaction by the addition of 2 phosphate groups furnished by ATP molecules. This restructured glucose molecule (now fructose 1,6-bisphosphate) is split into 2 3-carbon molecules, both of which eventually become glyceraldehyde 3-phosphate (GA3P). After removal of the phosphate groups and water 2 pyruvic acid molecules, which still contain a considerable amount of energy, remain. Glycolysis, which requires no oxygen gas, is summarized in figure 10.8.

The Krebs Cycle Reexamined

Before a pyruvic acid molecule enters the Krebs cycle, (which also takes place in the mitochondria) a molecule of carbon dioxide is removed and a molecule of $NADH_2$ is produced, leaving an acetyl fragment. The 2-carbon fragment is then bonded to a large molecule called *coenzyme A.* Coenzyme A consists of a combination of the B vitamin pantothenic acid and a nucleotide. Pantothenic acid is one of several B vitamins essential to respiration in both plants and animals; others include thiamine (vitamin B_1), pyridoxine, niacin, and riboflavin. The bonded acetyl fragment and coenzyme A molecule is referred to as *acetyl CoA.* The following equation summarizes the fate of the 2 pyruvic acid molecules produced during glycolysis:

$$2 \text{ pyruvic acid} + 2 \text{ CoA} + 2 \text{ NAD} \longrightarrow$$

$$2 \text{ acetyl CoA} + 2 \text{ NADH}_2 + 2 \text{ CO}_2$$

In addition to pyruvic acid, fats and amino acids can also be converted to acetyl CoA and enter the process at this point.

The $NADH_2$ molecules donate their hydrogen to an electron transport chain (discussed in the next section), and the acetyl CoA enters the Krebs cycle (figure 10.8).

In the Krebs cycle, acetyl CoA is first combined with oxaloacetic acid, a 4-carbon compound, producing citric acid, a 6-carbon compound. The cycle then progresses in circular fashion. A carbon dioxide is soon removed, producing a 5-carbon compound. Then another carbon dioxide is removed, producing a 4-carbon compound. This 4-carbon compound, through additional steps, is converted back to oxaloacetic acid, the substance with which the cycle began, and the cycle is repeated. Each full cycle uses up an acetyl group and regenerates an oxaloacetic acid molecule for the next turn of the cycle. Some hydrogen is removed during the process and is picked up by a flavoprotein hydrogen acceptor molecule, flavin adenine dinucleotide (FAD^+), and also by NAD. One molecule of ATP, 3 molecules of $NADH_2$, and 1 molecule of $FADH_2$ are produced for each turn of the cycle. The Krebs cycle may be summarized as follows:

$$\text{oxaloacetic acid} + \text{acetyl CoA} + \text{ADP} + \text{P} + 3 \text{ NAD}^+ + \text{FAD}^+ \longrightarrow$$

$$\text{oxaloacetic acid} + \text{CoA} + \text{ATP} + 3 \text{ NADH}_2 + \text{FADH}_2 + 2 \text{ CO}_2$$

The Electron Transport Chain and Oxidative Phosphorylation

After completion of the Krebs cycle, the glucose molecule has been totally dismantled and some of its energy has been transferred to ATP molecules. A considerable portion of the energy was transferred to NAD^+ and FAD^+ when they were used to pick up H^+ and electrons from the glucose molecule as it was broken down during glycolysis and the Krebs cycle. This energy is released as the electrons are passed along an *electron transport chain*. This chain, like the electron transport chain of photosynthesis, functions something like a high-speed bucket brigade in passing along electrons from their source to their destination. Several of the electron carriers in the transport chain are cytochromes. They are very specific and, as electrons flow along the chain, they can transfer their electrons only to other specific acceptors. When the electrons reach the end of the chain, they are picked up by oxygen, and combine with hydrogen ions, forming water.

Part of the energy that is released during the movement of electrons along the electron transport chain can be used to make ATP. If hydrogen ion and electron transport begins with $NADH_2$ that was produced inside the mitochondria, (that is during the conversion of pyruvic acid to acetyl CoA and during the Krebs cycle) sufficient energy is produced to yield 3 ATP molecules from each $NADH_2$ molecule. Similarly, if hydrogen ion and electron transport begins either with $FADH_2$ or with $NADH_2$ produced outside the mitochondria (that is, during glycolysis), 2 ATP molecules are produced.

Essentially the same *chemiosmotic* concept that was applied earlier to proton movement across thylakoid membranes and photophosphorylation (see page 185) needs to be mentioned at this point. In recent years there has been some evidence to support the theory, proposed in the 1960s by British biochemist Peter Mitchell, concerning the electron transport chain. This theory holds that oxidative phosphorylation is energized by a gradient of protons (H^+) that flow by **chemiosmosis** across the inner membrane of a mitochondrion. Mitchell, who received a Nobel Prize for his work in 1978, surmised that protons are "pumped" from the matrix of the mitochondria to the region between the two membranes (see figure 3.7) as electrons flow from their source in $NADH_2$ molecules along the electron transport chain, which is located in the inner membrane. The protons are believed to "diffuse" back into the matrix via channels provided by an enzyme complex, releasing energy that is used to synthesize ATP.

If we retrace our steps through the entire process of aerobic respiration, we find that glycolysis yields 4 molecules of ATP and 2 molecules of $NADH_2$ (from which 4 more molecules of ATP are formed), for a total of 8 ATP from the conversion of glucose to 2 pyruvic acid molecules. However, 2 ATP are used in the process leaving a net gain of 6 ATP.

When 2 pyruvic acid molecules are converted to 2 acetyl CoA in the mitochondria, 2 more $NADH_2$ molecules (which are converted to 6 molecules of ATP) are produced. The 2 acetyl CoA molecules metabolized in the Krebs cycle yield 2 molecules of ATP, 2 molecules of $FADH_2$ (from which 4 ATP are formed), and 6 molecules of $NADH_2$ (which cause the formation of 18 molecules of ATP), making a Krebs cycle total of 24 ATP. Thus a grand total of 36 ATP is produced for the aerobic respiration of 1 glucose molecule. The 36 ATP molecules represent about 39% of the energy originally present in the glucose molecule. The remaining energy is lost as heat or is unavailable.

A condensed comparison between photosynthesis and respiration is shown in table 10.1.

TABLE 10.1
Summary Comparison of Photosynthesis and Respiration

Photosynthesis	Respiration
1. Stores energy in sugar molecules	1. Releases energy from sugar molecules
2. Uses carbon dioxide and water	2. Releases carbon dioxide and water
3. Increases weight	3. Decreases weight
4. Occurs only in light	4. Occurs in either light or darkness
5. Occurs only in cells containing chlorophyll	5. Occurs in all living cells
6. Occurs in chloroplasts (in eukaryotic cells)	6. Occurs in cytoplasm (glycolysis) and mitochondria (aerobic respiration)
7. Produces oxygen in green plants	7. Utilizes oxygen (aerobic respiration)
8. Produces ATP with energy from light	8. Produces ATP with energy released from sugar

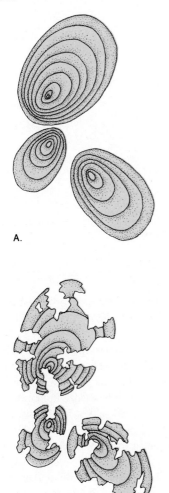

A.

B.

DIGESTION AND ASSIMILATION

Sugars produced through photosynthesis may undergo many transformations. Some sugars are used directly in respiration, but others not needed for that purpose may be transformed into lipids, proteins, or other carbohydrates. Among the most important carbohydrates produced from simple sugars are sucrose, starch, and cellulose. When photosynthesis is occurring in a leaf, sugar may be produced faster than it can be used or transported away to other parts of the plant. When this happens, the excess sugar may be converted to starch, temporarily stored in the chloroplasts, and then later changed back to a soluble form that is transported out of the leaf. The conversion of starch and other insoluble carbohydrates to soluble forms is called **digestion** (figure 10.9). The process is nearly always one of **hydrolysis,** in which water is taken up and, with the aid of enzymes, the links of the chains of simple sugars that comprise the molecules of starch and similar carbohydrates are broken. The disaccharide malt sugar (maltose), for example, is transformed to 2 molecules of glucose, with the aid of an enzyme, by the addition of 1 molecule of water as follows:

$$C_{12}H_{22}O_{11} + H_2O \xrightarrow[\text{(enzyme)}]{\text{maltase}} 2\ C_6H_{12}O_6$$

Fats are broken down to their component fatty acids and glycerol, and proteins to their amino acid building blocks in similar fashion. Digestion is carried on in any cell where there may be stored food, with very little energy being released in the process. In animals, special digestive organs also play a role in digestion, but plants have no such additional "help" in the process. In both plants and animals, however, digestion within cells is similar and is a normal part of metabolism.

Digestion in plants takes place within the cells where the carbohydrates, fats or proteins are stored (except in insect-trapping plants), whereas in animals digestion usually occurs outside of the cells in the digestive tract. Apart from the location, however, the process is essentially similar in plants and animals.

Much of the organic matter produced through photosynthesis is eventually used in the building of protoplasm and cell walls. This conversion process is called *assimilation*. Energy released through respiration is used in each building step, not only in the formation of fats, proteins, and complex carbohydrates but also in the manufacture of such products as resins, gums and oils, drugs, pigments, acids, and tannins. The capacity to assimilate is a fundamental attribute of living organisms, but much is still not known about the details of the process.

SUMMARY

Photosynthesis is an energy-storing process that combines carbon dioxide and water in the presence of light with the aid of chlorophyll; oxygen is given off as a by-product. Virtually all life depends on photosynthesis, which takes place in the chloroplasts of both terrestrial and aquatic organisms. The carbon dioxide for the process occurs as a constant 0.03% of the atmosphere; less than 1% of water absorbed by a plant is used in photosynthesis, which utilizes light of primarily blue-violet and red-orange wavelengths. All types of chlorophyll contain a magnesium atom. Chlorophyll *a* is the principal chlorophyll pigment present in most leaves; chlorophylls *b, c, d,* and *e,* carotenoids, phycobilins, and other pigments make it possible for photosynthesis to occur over a broader spectrum of light than is possible with chlorophyll *a* alone. In chloroplasts various pigment molecules are grouped into photosynthetic units that function in the light reactions, the first major phase of photosynthesis.

During the light reactions of photosynthesis, water molecules are split and oxygen gas is released, and hydrogen ions and electrons are released from water and transferred to create $NADPH_2$ and energy-storing ATP. In the dark reactions, carbon dioxide is combined with RuBP, and this combination is converted to 6-carbon sugars. The ATP and $NADPH_2$, produced during the light reactions, furnish the energy and materials needed for the dark reactions. The first substance produced in the dark reactions is usually a 3-carbon compound, but some tropical and desert plants produce a 4-carbon compound instead.

A closer look at photosynthesis reveals that there are two types of photosynthetic units, collectively comprising photosystems I and II, which are in most chloroplasts. Each photosystem contains a reaction center molecule of chlorophyll *a* that reacts to the energy of a photon of light by boosting an electron out of place to a higher energy level. In photosystem II, this "excited" electron is picked up by Pheo and

then Q, an electron acceptor molecule, which releases it to an electron transport system consisting of cytochromes and other molecules; these pass it along to a special transfer molecule. The excited electron leaves a "hole" that is filled by an electron from an electron donor. The donor's losses are replaced with electrons from water molecules that are split, releasing oxygen. Electron passage along the transport chain and proton movement across thylakoid membranes are both involved in ATP production. A similar series of events occurs in photosystem I, where the excited electron is either returned to the special reactive molecule (cyclic electron flow) or passed on to $NADP^+$, which forms $NADPH_2$ with electrons from the original splitting of water molecules in photosystem II (noncyclic electron flow).

In the dark reactions, which take place in the stroma of chloroplasts, the combination of RuBP and carbon dioxide is initially split into PGA molecules; then GA3P is formed with hydrogen ions and electrons from $NADPH_2$ and energy from ATP. Ten of the 12 GA3P molecules are converted back to RuBP; the remaining 2 GA3P molecules are the basis for an increase in the carbohydrate content or net gain in lipids or amino acids of a plant. In plants that have a 4-carbon pathway (C_4 plants), 4-carbon oxaloacetic acid is initially produced instead of 3-carbon PGA in the dark reactions. C_4 plants have large chloroplasts, which contain RuBP carboxylase, in the bundle sheaths, and small chloroplasts, which contain higher concentrations of PEP carboxylase than those of C_3 plants, in the mesophyll of their leaves. The high PEP carboxylase concentrations permit the conversion of carbon dioxide to carbohydrate at much lower concentrations than is possible in C_3 plants. In desert plants that have CAM photosynthesis, the 4-carbon compounds that accumulate in the chlorenchyma tissues at night are converted back to carbon dioxide during the day, permitting regular photosynthesis to occur; this process is advantageous to plants whose stomata are closed during the day and thus admit little carbon dioxide.

Respiration is an energy-releasing process that takes place in the mitochondria and cytoplasm of cells. The energy is released from simple sugar and acid molecules with the aid of enzymes. In aerobic respiration, oxygen is needed to release all the stored energy; carbon dioxide and water are by-products of the process. Anaerobic respiration and fermentation do not require oxygen gas, and much less energy is released. The remaining energy is in the ethyl alcohol, lactic acid, or other such substances produced. Some of the released energy is stored in ATP molecules. Temperatures, available water, and the oxygen content of the environment all affect respiration rates.

During glycolysis, which is common to all forms of respiration and requires no molecular oxygen, 2 phosphates are added to a 6-carbon sugar molecule and the prepared molecule is split into 2 3-carbon sugars (GA3P). Some hydrogen, energy, and water are removed from the GA3P, eventually producing pyruvic acid. There is a net gain of 2 ATP molecules. Hydrogen ions and electrons released during glycolysis are picked up by NAD, a hydrogen acceptor molecule, which becomes $NADH_2$.

In aerobic respiration, the pyruvic acid loses carbon dioxide and is restructured and combined with coenzyme A, becoming acetyl CoA. This enters the Krebs cycle, which involves successive enzyme-catalyzed reactions of a series of organic acids, and during which energy, carbon dioxide, and hydrogen are removed. $NADH_2$ passes the hydrogen gained during glycolysis and the Krebs cycle along an electron transport chain; during the process small increments of energy are released and partially stored in ATP molecules, and the hydrogen is combined with oxygen gas, forming water in the final step of aerobic respiration.

Anaerobic respiration differs from fermentation in that hydrogen removed from glucose during glycolysis is combined with an inorganic ion in the anaerobic process, whereas in fermentation, the hydrogen is simply combined with the pyruvic acid or one of its derivatives; both processes occur in the absence of oxygen gas, with only about 7% of the total energy in the glucose molecule being released, for a net gain of 2 ATP molecules.

A closer look at respiration reveals that 2 molecules of $NADH_2$ and 2 ATP molecules are gained during glycolysis when 2 3-carbon pyruvic acid molecules are produced from a single glucose molecule. Another molecule of $NADH_2$ is produced when the pyruvic acid is restructured and combined with acetyl CoA prior to entry into the Krebs cycle. In the Krebs cycle, acetyl CoA combines with oxaloacetic acid, a 4-carbon compound, to produce first a 6-carbon compound, next a 5-carbon compound, and then several 4-carbon compounds. The last 4-carbon compound, after several steps, is oxaloacetic acid. Some hydrogen removed during the process is picked up by the hydrogen acceptor molecules FAD^+ and NAD^+; 1 molecule of ATP, 3 molecules of $NADH_2$, and 1 molecule of $FADH_2$ are produced during one complete cycle. Energy associated with electrons and/or with hydrogen picked up by NAD^+ and FAD^+ is gradually released as the electrons are passed along the electron transport chain, and some of this energy is transferred to ATP molecules during oxidative phosphorylation. Energy used in ATP synthesis is believed to be derived from a gradient of protons formed across the inner membrane of a mitochondrion while electrons are moving in the electron transport chain by what has been termed "chemiosmosis." Altogether, 38 ATP molecules are produced during the complete aerobic respiration of 1 glucose molecule, but 2 are used to prime the process so that there is a net gain of 36 ATP molecules.

Digestion takes place within plant cells with the aid of enzymes. During digestion large insoluble molecules are broken down by hydrolysis to smaller soluble forms that can be transported to other parts of the plant. The conversion of the sugar produced by photosynthesis to fats, proteins, complex carbohydrates, and other substances is termed assimilation.

REVIEW QUESTIONS

1. What happens in the light reactions of photosynthesis?
2. What roles do water, light, carbon dioxide, and chlorophyll play in photosynthesis?
3. What is glycolysis?
4. What are the differences among aerobic respiration, anaerobic respiration, and fermentation?
5. How do temperature, water, and oxygen affect respiration?
6. Explain digestion and assimilation.

DISCUSSION QUESTIONS

1. Since plants apparently benefit from small increases in the carbon dioxide available to them, would you suppose you would be doing your favorite houseplant a favor by breathing repeatedly on it?
2. If the dark reactions of photosynthesis take place in the light as well as in the dark, how did they come to be so named?

ADDITIONAL READING

Clayton, R. K. 1981. *Photosynthesis: physical mechanisms and chemical patterns.* Fair Lawn, NY: Cambridge University Press.

Danks, S. M. et al. 1984. *Photosynthetic systems: structure, function and assembly.* New York: Wiley-InterScience.

Douce, R. and D. A. Day, eds. 1985. *Higher plant cell respiration.* New York: Springer-Verlag.

Fong, F. K., ed. 1982. *Light reaction path of photosynthesis.* New York: Springer-Verlag.

Galston, A. W., P. J. Davies, and R. L. Satter. 1980. *The life of the green plant,* 3d ed. Englewood Cliffs, NJ: Prentice-Hall.

Hall, D. O., and K. K. Rao. 1981. *Photosynthesis,* 3d ed. Baltimore: University Park Press.

Kluge, M., and I. P. Ting. 1978. *Crassulacean acid metabolism.* Ecological Studies, Vol. 30. New York: Springer-Verlag.

Lehninger, A. L. 1975. *Biochemistry: the molecular basis of all structure and function,* 2d ed. New York: Worth Publishers, Inc.

Levine, R. P. 1969. "The mechanisms of photosynthesis." *Scientific American* 221:58–70.

Öpik, H. 1982. *The respiration of higher plants.* Baltimore: University Park Press.

Walker, D. 1979. *Energy, plants and man: an introduction to photosynthesis in C_3, C_4 and CAM plants.* Philadelphia: International Ideas, Inc.

Zelitch, I. 1971. *Photosynthesis, photorespiration and plant productivity.* New York: Academic Press.

Overview

*T*his chapter introduces growth phenomena with a discussion of the distinctions among growth, differentiation, and development. This is followed by a discussion of plant hormones (auxins, gibberellins, cytokinins, abscisic acid, ethylene) and their roles in plant growth and development. The chapter explores plant movements, including spiraling, twining, contraction, nastic, tropisms, turgor, taxes, and miscellaneous other movements. The discovery and functions of phytochrome and photoperiodism are briefly surveyed. The chapter concludes with a discussion of the relationship of temperature to growth and dormancy.

11 Growth

Some Learning Goals

1. Know distinctions among growth, differentiation, development, hormones, and vitamins.

2. Identify the types of plant hormones and describe the major functions of each; discuss commercial applications for each.

3. Distinguish among the various types of plant movements and know the forces behind them.

4. Understand photoperiodism and make distinctions among short-day, long-day, intermediate-day, and day-neutral plants.

5. Explain what phytochrome is and how it functions.

6. Discuss the role of temperature in plant growth.

7. Learn dormancy and stratification, and give examples.

Outline

INTRODUCTION

The word **growth** is used in a variety of contexts. If, for example, you observe a rubber balloon being inflated with gas, you may refer to its gradual increase in diameter as its growth in size. Or if there is a leak in the roof you might remark that the puddle beneath is growing bigger. In the biological realm, however, growth is always associated with cells. It may be defined simply as an "increase in mass due to the division and enlargement of cells," and may be applied to an organism as a whole or to any of its parts. The growth of some plants or parts of plants may be *determinate* in that they attain a certain size, stop growing, and then break down and die. Other plant parts (see meristems in chapter 4) may exhibit *indeterminate* growth and continue to be active for thousands of years. Many plants go through a regular sequence of growth rates, with an initial period of rapid growth followed by little if any increase in volume, and eventually death after complete cessation of growth and breakdown of tissues.

As mentioned at the beginning of chapter 3, all living organisms begin as a single cell, which usually divides and keeps dividing until an entity consisting of possibly billions of cells is formed. As the various cells mature, they usually become larger and then *differentiate*—that is, they assume different forms adapted to specific functions such as conduction, support, or secretion of special substances. If **differentiation** did not occur, the result would be a shapeless blob with no distinct tissues and little, if any, coordination.

The term **development** is applied to the process of growth and differentiation of cells into tissues, organs, and organisms. What brings about growth and differentiation, and how do they occur? We already know they are controlled by the genes of the chromosomes, which not only dictate the kinds of cells that will develop but also their proportionate numbers and the ultimate size of the organism. The sizes of the cells themselves are limited by several factors. They are dependent on molecular diffusion within for the transport of oxygen and food and for the removal of wastes, since they have no circulatory organs of their own. As a spherical cell enlarges, the total area of its surface, through which substances enter and leave, does not increase at the same rate as its volume. For example when a cell's surface area has increased fourfold, its volume has increased roughly six times, obviously putting a great strain on its capacity to admit and exchange materials through its surface. Most plant cells have become adapted to the problem through various changes in shape (e.g., development of a convoluted outer surface; flattening of the cell) that increase the surface area, and also through the development of large vacuoles, with the cytoplasm becoming a thin layer adjacent to the cell wall. These adaptations alone, however, do not increase the strength of the bounding membranes, and most cells divide before their enlargement becomes a significant problem.

Each metabolic step within cells is under the control of *enzymes* (discussed in chapter 2), the synthesis and development of which is dictated by *genes* (see chapter 13). In addition, many growth phenomena are influenced by **hormones.** These are chemical substances produced, usually in minute quantities, in one part of an organism and transported to another part where they have specific effects, such as causing stems to bend or initiating flowering. They should not be confused with enzymes, which are also produced in minute quantities. As previously noted, enzymes are organic catalysts that speed up chemical reactions in the cells where they are produced. They generally are not transported from one part of an organism to another, and their chemical structure is different from that of hormones.

Vitamins, which function in activating enzymes, are organic molecules of varied structure that are synthesized in the membranes and cytoplasm of cells. They are essential in relatively small amounts for the normal growth and development of all organisms. Most vitamins essential to all living organisms are produced by plants, although a few are also synthesized by animals. At least one vitamin (Vitamin A) does not, however, occur in plants, but carotene pigments found in chloroplasts (see chapter 10) act as *precursors* (simple molecules that are converted by living organisms to more complex building blocks) for Vitamin A in animals. Because some of their effects are similar to those of hormones, vitamins and hormones are sometimes difficult to distinguish from one another. To avoid having to distinguish between the two, the term *growth-regulating substances* for both vitamins and hormones is gaining acceptance.

FIGURE 11.1 Went's experiment with oat coleoptiles.
A. An intact oat coleoptile. *B.* The tip of a coleoptile was cut
off, placed on a small agar block, and left for an hour or two.
C. When this agar block was then placed squarely on a
decapitated coleoptile, growth was vertical. *D.* When the agar
block was placed off center so that only half the coleoptile tip
was in contact with it, the tip bent away from the block. This
demonstrated that a substance affecting growth diffused from
the coleoptile tip into the agar, and from the agar to the
decapitated coleoptile.

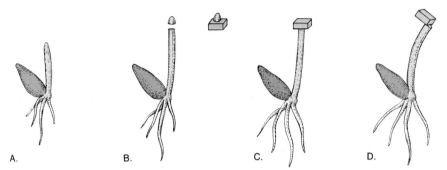

A.   B.   C.   D.

PLANT HORMONES

In 1881 Charles Darwin and his son Francis were
fascinated by the fact that *coleoptiles* (more or less
tubular, closed sheaths protecting the emerging
stems of germinating seeds of the Grass Family)
bend toward a light source. They noticed that the
bending did not occur if they placed opaque little
caps over the coleoptile tips and concluded that this
part of the plant must be sensitive to light. In 1926
a young Dutch plant physiologist named Frits Went
followed up the investigations of the Darwins and
others by cutting off the tips of oat coleoptiles and
placing the tips, cut surface down, on flat portions
of **agar** (a substance obtained from marine algae; it
is similar to gelatin in consistency). After a few hours
he removed the tips and cut the agar into little square
blocks, which he then placed on decapitated coleoptiles. He found that if he placed a little block squarely
over the cut top of a coleoptile, it grew straight up,
but if he placed the block off center, the tip bent away
from the side on which the block had been placed
(figure 11.1). He also found that the more coleoptile
tips he placed on the agar to begin with, the more
pronounced was the response. His experiments demonstrated conclusively that something produced in
the coleoptile tips moved out into the agar and that
this substance was responsible for the bending observed. Went named the substance **auxin** (from the
Greek word *auxein,* meaning "to increase"), a term
now in widespread use.

Auxins were the first plant hormones to be discovered, with several others coming to light later. At
least three major groups apparently promote the
growth of plants. Some may also have an inhibitory
effect, depending on the concentration. Other hormones and growth regulatory substances are known
only for their growth-inhibiting features. All are
produced in minute quantities and must be replenished, as they are constantly being used up during
metabolic activities of cells. The following are the
major known groups of hormones: *auxins, gibberellins, cytokinins, abscisic acid,* and *ethylene.*

Auxins

Auxins are synthesized by living protoplasm from
an amino acid in both plant and animal cells. Production of auxins occurs mainly in apical meristems, buds, young leaves, and other active young
parts of plants. It is sometimes difficult to predict
how cells will respond to auxins because responses
vary according to the concentration, location, and
other factors. For example, if an auxin of specific
concentration promotes shoot growth in a certain
plant, the same auxin of identical concentration will
inhibit its root growth. At appropriate concentrations auxins normally stimulate the enlargement of
cells and increase the flexibility of cell walls. They
also may have many other effects, including trig-
gering the production of other hormones (especially
ethylene), causing the Golgi bodies (dictyosomes)
to increase the rates of secretion, playing a role in
controlling some phases of respiration, and are involved in numerous developmental aspects of growth.
These aspects include flower initiation, sex deter-
mination, growth rate, fruit growth and ripening,
abscission, rooting, aging, dormancy, and apical

FIGURE 11.2 Effect of auxin applied to the base of a
Gardenia cutting four weeks after application. *Left:* A treated
cutting. *Right:* An untreated cutting.

dominance (see the discussion of apical dominance on page 205). Many monocots are less sensitive than dicots to auxins, and shoots are less sensitive than roots, but higher concentrations will kill almost any living plant tissues. The effects of auxins combined with those of other hormones produce many of the growth phenomena.

Movement of auxins from the cells where they originate requires the expenditure of energy stored in ATP molecules. The migration is relatively slow (about 1 centimeter per hour) although the active transport mechanism involved carries the hormone up to ten times faster than it would be by simple diffusion alone (see chapter 9). The movement is *polar,* which refers to the flow of auxins away from their source, usually downward from a stem tip. They do not seem to be carried through the sieve-tube "plumbing" of the phloem but proceed from cell to cell, particularly through parenchyma cells surrounding vascular bundles. It was once believed that there were several naturally occurring auxins, but

until recently scientists generally had become convinced that *indoleacetic acid,* usually referred to as IAA, was the only active auxin, and that other organic acids that act like auxins are actually converted to IAA in the plants where they perform their normal functions. Recent evidence now indicates that plants do actually produce two other hormones that bring about many of the same responses as IAA. One, called *phenylacetic acid* (usually abbreviated to PAA), is often more abundant but less active than IAA. The other, called *4-chloro-indoleacetic acid* (abbreviated to 4-chloroIAA), is found in germinating seeds of legumes.

IAA and a number of organic acids that regulate growth have been produced synthetically. They are in widespread use in agriculture and horticulture. At the proper concentrations, IAA and other growth regulators stimulate the formation of roots on almost any plant organ. Accordingly, as noted in appendix 4, nurseries apply these substances in paste or powder form to the bases of cut pieces of stems to stimulate a much more vigorous production of roots than would occur naturally (figure 11.2).

Orchardists spray fruit trees with growth regulators to promote uniform flowering and fruit set, and they later spray the fruit to prevent the formation of abscission layers and subsequent premature fruit drop. A substantial saving in labor expenses results from being able to go through an orchard or a pineapple plantation and pick all of the fruit at one time. If growth regulators are applied to flowers before pollination occurs, seedless fruit (see chapter 8) can be formed and developed. Some orchardists use growth regulators for fruit thinning; that is, they control the number of fruits that will mature in order not to have too many small ones, and some even control the shapes of plants for sales appeal.

Other types of synthetic auxins such as 2,4-D, 2,4,5-T, and 2,4,5-TP have been used as herbicides (plant killers). Weeds were controlled by hand labor and by the use of caustic or otherwise poisonous substances until it was discovered that these hormone-like compounds, when sprayed at low concentrations, kill some weeds. Broad-leaved plants like dandelions, plantains, and others (for reasons that are not clearly understood) are more susceptible to low concentrations of 2,4-D and 2,4,5-T than narrow-leaved plants like grasses. Such growth regulators, therefore, have been used on lawns to kill weeds without noticeably affecting the grass.

Thus far there is little evidence of 2,4-D having directly adverse effects on humans and other animal life, but such is not the case with 2,4,5-T, which was banned in 1979 by the Environmental Protection Agency for most uses in the United States. The controversy over the use of 2,4,5-T began in the 1960s during the Vietnam War when Agent Orange, a 1-to-1 mixture of 2,4-D and 2,4,5-T, was used to defoliate jungles. Subsequent tests and experiments with 2,4,5-T have produced leukemia, miscarriages, birth defects, and liver and lung diseases in laboratory animals. The diseases and defects are apparently caused by TCDD (2,3,7,8-tetrachlorodibenzoparadioxin), a dioxin contaminant that evidently is unavoidably produced in minute amounts during the manufacture of 2,4,5-T. TCDD kills laboratory animals in doses as low as a few parts per billion.

Gibberellins

In 1926 a Japanese scientist named E. Kurosawa reported the discovery of a new substance that was causing what was referred to as *bakane* or "foolish seedling" effect on rice. The stems of rice seedlings infected with a certain fungus grew twice as long as those of uninfected plants, but the stems were weakened so that they eventually collapsed and died. Kurosawa found that extracts of the fungus applied to uninfected plants brought about the same growth stimulations as the fungal disease, but it was not until 9 years later that other Japanese scientists were able to purify the substance, which he named **gibberellin,** after the scientific name of the fungus that produced it (*Gibberella fujikuroi*). Today more than 60 different gibberellins have been isolated from seeds (especially those of dicots), or from fungi. Gibberellins probably also occur in algae, mosses, and ferns, but none are known in bacteria. Acetyl coenzyme A, the coenzyme vital to the process of respiration (see chapter 10), functions as a precursor in the synthesis of gibberellins. Interest in their hormonal properties is high, for their ability to increase growth in plants is far more impressive than that of auxin alone, although traces of auxin apparently need to be present for gibberellins to produce their maximum effects. Most dicots and a few monocots grow faster with an application of gibberellins, but coniferous trees like pines and firs generally show little, if any, response. If a gibberellin, at the appropriate concentration, is applied to a cabbage, the plant may become 2 meters (6 feet) tall, and bush bean plants can become pole beans with a single application. There is not usually, however, a corresponding stimulation of root growth. Many genetically dwarf plants grow as tall as their normal counterparts, after an application of the appropriate gibberellin.

Gibberellins not only dramatically increase stem growth, but they are involved in nearly all of the same regulatory processes in plant development as auxins (figure 11.3). In some kinds of plants, flowering can be brought about by applications of gibberellins, and the dormancy of buds and seeds can be broken. Some gibberellins appear to lower the threshold of growth; that is, plants may start growing at lower temperatures than usual after an application of gibberellin. For example, an application to a lawn could cause it to turn green 2 or 3 weeks earlier in the spring. Gibberellins have been used experimentally to increase yields of sugar cane and hops and have revolutionized the production of seedless grapes through increasing the size of the fruit and in lengthening fruit internodes—which results in slightly wider spaces between grapes in the bunches. Better air circulation between grapes reduces their susceptibility to fungus diseases, and the need for hand-thinning is eliminated. They have been used in navel orange orchards to delay the aging of the fruit's skin, and have been found to increase the length and crispness of celery petioles (the parts that are most in demand as a raw food).

Field experiments have shown that if gibberellins are applied to plants at the same time selected herbicides are applied, the gibberellins may reverse the effects of the herbicides. The relatively high cost of gibberellins has limited the extent of their use in horticulture and agriculture.

Cytokinins

By the close of the nineteenth century there was conjecture concerning the existence of a chemical that regulates cell division in plants. The conjecture gave way in 1913 to the discovery, by G. Haberlandt, of an unidentified substance in the phloem of various plants that stimulated cell division and initiated the production of cork cambium. In the 1950s several substances that promote cell division were found in coconut "milk" (a liquid *endosperm*—see chapter 21), but it was not until 1964 that the identity of such a substance was determined in kernels of corn. These various stimulants to cell division came to be known as **cytokinins.** The several cytokinins now known differ somewhat in their molecular structure and possibly also in origin, but they are all basically adenine-like. You will recall that adenine is a building block of one of the four nucleotides found in DNA (see chapter 2), although none of the cytokinins appears to be derived from

FIGURE 11.3   Effect of gibberellins on growth. *Left*: An untreated *Chrysanthemum* plant. *Right*. A similar *Chrysanthemum* plant that has been treated with gibberellins. (USDA Photo)

DNA. Some do, however, occur in certain forms of RNA. Cytokinins are found most abundantly in meristems and other developing tissues, particularly those of young fruit. Cytokinins participate in the enlarging of cells, the differentiation of tissues, the development of chloroplasts, the stimulation of cotyledon growth, the delay of aging in leaves, and in many of the growth phenomena also brought about by auxins and gibberellins. Most have not yet been approved for general agricultural use, but experiments have shown that they do prolong the life of vegetables in storage. Related synthetic compounds have been used extensively to regulate the height of ornamental shrubs, and to keep harvested lettuce and mushrooms fresh. They have also been used to shorten the straw length in wheat so as to minimize the chances of the plants blowing over in the wind, and to lengthen the life of cut flowers.

FIGURE 11.4   The trunk of this Jeffrey pine tree, which would normally be single, is forked because of the earlier removal of the terminal bud.

Apical Dominance

Apical dominance is the suppression of the growth of the axillary or lateral buds brought about by an auxin inhibitor in a terminal bud. It is strong in trees with conical shapes and little branching toward the top (e.g., many pines, spruces, firs) and weak in trees with stronger branching above (e.g., elms, ashes, willows). When a terminal bud is deliberately or accidentally removed the nearest axillary buds begin to grow (figure 11.4). Presumably the removal of the terminal bud also results in the removal of the inhibitor (assumed to be an auxin), but this presumption has not yet been confirmed, and some studies actually point to a cytokinin deficiency in axillary buds as playing a greater role than terminal bud auxin inhibitors in apical dominance. Further studies are needed before definitive statements pertaining to the precise causes of apical dominance can be made.

Experiments with living pith cells of tobacco plants have shown that such cells will enlarge when supplied with auxin and nutrients, but they will not divide unless small amounts of cytokinin are added. By varying the amounts of cytokinin it is possible to stimulate the pith cells to differentiate into roots or into buds from which stems will develop. This last regulatory effect can be used to offset apical dominance. The extent of the suppression depends on how close the axillary buds are to the terminal bud and on the amount of inhibitor involved. Experiments with pea plants have shown that the axillary buds will begin to grow as little as 4 hours after a terminal bud has been removed. If cytokinins are applied in appropriate concentration to axillary buds, however, they will begin to grow even in the presence of a terminal bud.

Abscisic Acid

In 1963 three groups of investigators independently discovered another growth-inhibiting hormone. In the United States F. T. Addicott and his associates, in Wales a group led by P. F. Wareing, and R. F. M. Van Steveninck's group in New Zealand and England discovered what in 1967 was officially called **abscisic acid** (ABA). It is synthesized in plastids, apparently from carotenoid pigments. It is found in many plant materials but is particularly common in fleshy fruits, where it evidently prevents seeds from germinating while they are still on the plant. ABA almost universally inhibits cell growth. In addition it was once generally believed to promote the formation of abscission layers in leaves and fruits (see chapter 7), but that belief has become quite controversial, with some claiming it plays a major role in abscission and others suggesting ethylene (discussed next) is far more important in abscission, and that ABA has little, if any, influence on the process.

When ABA is applied to active plant buds, the leaf primordia become bud scales and the bud goes into a dormant winter condition. Since nursery plants that are being shipped stand to suffer less damage in a dormant condition than in an active one, the application of ABA and similar growth inhibitors to such nursery stock has great practical value. The inhibiting effects of ABA can be reversed by the application of gibberellins.

FIGURE 11.5   Effect of ethylene on holly twigs. Two similar twigs were placed under glass jars for a week; a ripe apple was at the same time placed under the jar on the right. Ethylene produced by the apple caused abscission of the leaves.

ABA apparently helps leaves respond to excessive water loss. When the leaves wilt, ABA is produced in amounts several times greater than usual. This interferes with the transport or retention of potassium ions in the guard cells, causing the stomata to close. When the uptake of water again becomes sufficient for the leaf's needs, the ABA is broken down metabolically and the stomata reopen.

Ethylene

In 1934, R. Gane discovered that **ethylene** was produced naturally by fruits although it had been known for some time that artificially adding this simple gas to green fruits hastens ripening. It is now known that ethylene is produced not only by fruits but also by flowers, seeds, leaves, and even roots. Several fungi and a few bacteria are also known to produce it, and its regulatory effects make it a hormone in the broad sense of the word.

The production of ethylene by plant tissues varies considerably under different conditions. A surge of ethylene lasting for several hours becomes evident after various tissues, including those of fruits, are bruised or cut, and applications of auxin can cause an increase in ethylene production of two to ten times. As pea seeds germinate, the seedlings produce a surge of ethylene when they meet interference with their growth through the soil. This apparently causes the stem tip to form a tighter crook, which may aid the seedling in pushing to the surface. Ethylene probably is also involved in *thigmomorphogenesis* (a reduction in the elongation of stems that are subject to mechanical stresses such as those imposed by wind or contact).

Ethylene apparently can trigger its own production. If minute amounts are introduced to the tissues that produce the gas, a tremendous response by the tissues often results. These tissues may then produce so much ethylene that it can adversely affect the part concerned (e.g., flowers may fade in the presence of excessive amounts, and leaves may abscise—see figure 11.5).

In ancient China, growers ripened fruits in rooms where incense was being burned, and citrus growers ripened their fruits in rooms equipped with a kerosene stove. In houses where artificial gas is used, occupants often experience great difficulty in growing houseplants, and greenhouse heaters using such fuel create a dilemma for owners attempting to promote plant growth. In the days before electric street lights became commonplace, gas lights were used, and in some cities in Germany leaves fell from the shade trees if they were located near a gas light that leaked. In all of these instances, minute amounts of ethylene gas resulting from the fuel combustion or leakage brought about the results, both good and bad. In fact, as little as 1 part of ethylene per 10 million parts of air may be sufficient to trigger responses.

Today, commercial use of ethylene is extensive. It has been used for many years to ripen harvested green fruits such as bananas, mangoes, and honeydew melons and to cause citrus fruits to color up before marketing. (Some growers still use natural ethylene to ripen pears and peaches by wrapping each fruit individually in tissue paper. The paper retards the escape of ethylene from the fruit and hastens ripening.) "Resting" potato tubers will sprout following brief applications of ethylene, and seeds may be stimulated to germinate if given a short exposure to the gas just before sprouting, although treatment after sprouting inhibits growth. Ethylene is used in Hawaii to promote flowering in pineapples, and it causes members of the Pumpkin Family to produce more female flowers and thus more fruit. Because trees grown in containers in nurseries are usually crowded, they are often tall and spindly, but applications of small amounts of ethylene to the trunks while they are enclosed in plastic tubes causes a marked thickening, making the trees sturdier and less likely to break.

SENESCENCE

The breakdown of cell components and membranes that eventually leads to the death of the cell is called **senescence.** As mentioned in chapter 7, the leaves of deciduous trees and shrubs senesce and drop through a process of abscission every year. Even evergreen species often retain their leaves for only 2 or 3 years (with the notable exception of bristlecone pines, which hold their leaves for up to 30 years), and the above-ground parts of many herbaceous perennials senesce and die at the close of each growing season. Why do plant parts senesce? Some studies have suggested that certain plants produce a senescence "factor" that behaves like, or is actually, a hormone, but we are not yet certain of the precise mechanisms involved. We do know, however, that both ABA and ethylene promote senescence, whereas auxins, gibberellins, and cytokinins delay senescence in a number of plants that have been studied. Other factors such as nitrogen deficiency and drought also hasten senescence.

PLANT MOVEMENTS

We noted in chapter 2 that all living organisms exhibit movement but that most plant movements are relatively slow and imperceptible unless seen in time-lapse photography or demonstrated experimentally. We now want to examine how and why plant movements occur.

Growth Movements

Growth movements result from varying growth rates in different parts of an organ. They are principally related to young parts of a plant and as a rule are quite slow, usually taking at least 2 hours to become apparent although the plant may have begun microscopic changes within minutes of receiving a stimulus. The stimulus may be either internal or external.

Movements Resulting Primarily from Internal Stimuli
Helical (Spiraling) Movements
Charles Darwin once attached a tiny sliver of glass to the tip of a plant growing in a pot. Then he suspended a piece of paper blackened with carbon over the tip, and as the plant grew he raised the paper just enough to allow the tip to touch the paper without hurting the plant. He found that the growing point traced a spiral pattern in the blackened paper. We now know that such helical movements of growth are common to many plants (figure 11.6).

FIGURE 11.6   Charles Darwin's demonstration of spiraling growth.

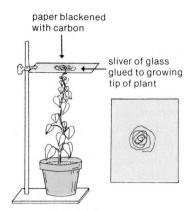

paper blackened
with carbon

sliver of glass
glued to growing
tip of plant

tip spirals as it grows,
tracing its pattern on the paper

FIGURE 11.7   Typical twining of a squash tendril.

Nodding Movements

Members of the legume family such as garden beans, whose ethylene production upon germinating causes the formation of a thickened crook in the hypocotyl, exhibit a slow oscillating movement (that is, the bent hypocotyl nods from side to side like an upside down pendulum) as the seedling pushes up through the soil. This nodding movement apparently facilitates the progress of the growing plant tip through the soil.

Twining Movements

Although twining movements are primarily internally stimulated, external forces such as gravity and contact may also play a role. These movements occur when cells in the stems of climbing plants such as morning glory elongate to differing extents, causing visible spiraling in growth (in contrast with the spiraling movements previously mentioned, which are not visible to the eye). Tendril twining, which is initiated by contact, results from an elongation of cells on one side of the stem and a shrinkage of cells on the opposite side, followed by differences in growth rates (figure 11.7). Some tendrils are stimulated to coil by auxin, while others are stimulated by ethylene.

Contraction Movements

In chapter 6 we noted that the bulbs of a number of dicots and monocots have contractile roots that pull them deeper into the ground. In lilies, for example, seeds germinating at the surface ultimately produce bulbs that end up 10 to 15 centimeters (4 to 6 inches) below ground level because of the activities of contractile roots. The aerial roots of some banyan trees straighten out by contraction after the roots have made contact with the ground. The "shrinking" of roots has been shown to take place at the rate of 2.2 millimeters (0.1 inch) a day in sorrel. There is some evidence that temperature fluctuations at the surface determine how long the contracting will continue. When the bulb gets deep enough that the differences between daytime and nighttime temperatures are slight, the contractions cease.

FIGURE 11.8   This painted lady (stonecrop) plant grew in
the direction of light it received from only one side over a
period of several weeks.

Nastic Movements

When flattened plant organs such as leaves or flower
petals first expand from buds, they characteristically alternate in bending down and then up as the
cells in the upper and lower parts of the leaf alternate in enlarging faster than those in the opposite
parts. Such nondirectional movements (i.e., movements that do not result in an organ being oriented
toward or away from the direction of a stimulus) are
called **nastic.** In prayer plants (see figure 11.11) and
others where the leaves or flowers fold up and reopen daily, turgor movements are also involved, as
are the external stimuli of light and temperature.
These movements are controlled by a biological
"clock" and are referred to as *circadian rhythms* (see
the section on *turgor movements* on page 211). In a
few instances nastic movements apparently are initiated by external stimuli alone.

Movements Resulting from External Stimuli

Permanent, directed movements resulting from external stimuli are commonly referred to as **tropisms.**

Phototropism

The main shoots of most plants growing in the open
tend to develop vertically, although the branches
often grow horizontally. If a box is placed over a plant
growing vertically and a hole is cut to admit light
from one side, the tip of the plant will begin to bend
toward the light within a few hours. If the box is
later removed, a compensating bend develops,
causing the tip to grow vertically again. Such a
growth movement toward light is called a *positive
phototropism* (figure 11.8), whereas a similar
bending away from light is called a *negative phototropism.* The shoot tips of most plants are positively phototropic, while roots are either insensitive
to light or negatively phototropic.

Leaves often twist on their petioles and become
oriented at right angles to a light source in response
to the illumination. In fact, many plants have solar-
tracking leaves with the blades constantly oriented
at right angles to the sun throughout the day. Some
have referred to solar-tracking movements as *heli-
otropisms* but, unlike phototropic responses of stems
and roots, growth is not involved. Strictly speaking,
the twisting of petioles that facilitates this type of
movement should be called *phototorsion,* because
motor cells (in *pulvini*—see *turgor movements* on
page 211) at the junction of the blade and the petiole control the movement. If one views a tree from
directly overhead or observes a vine growing on a
fence, it is perhaps surprising to note how little
overlap of the leaves occurs; each leaf is placed so
that it receives the maximum amount of light available.

FIGURE 11.9 This *Coleus* plant was placed on its side the day before the photograph was taken. The stem bent upward within 24 hours of the pot being tipped over.

We have already noted that if the tip of a coleoptile is covered or removed, the structure will not bend toward light, and that auxin is produced in the tip (see figure 11.1). We have also noted that auxin promotes the elongation of cells, at least in certain concentrations. For some time it was believed that stem tips bent toward light because auxin was destroyed or inactivated on the exposed side, leaving more growth-promoting hormone on the side away from the light and thus causing the cells there to elongate more and produce a bend. Careful experiments have shown, however, that stem tips growing in the open have the same total amount of auxin present as stem tips from the same species that have received light from one side. Other experiments indicate that the auxin migrates away from the light, accumulating in greater amounts on the opposite side, thus promoting greater elongation of cells on the "dark" side. Apparently an active transport system enables the auxin to migrate against a diffusion gradient.

Different intensities of light may bring about different phototropic responses. In Bermuda grass, for example, the stems tend to grow upright in the shade and parallel with the ground in the sun. In other plants, such as the European rock rose, which grows among rocks or on walls, the flowers are positively phototropic, but once they are fertilized they become negatively phototropic. As the pedicels (stalks) elongate, the developing fruits are buried in cracks and crevices, where the seeds then germinate. Phototropic responses are not confined to flowering plants. A number of mushrooms, for example, show marked positive responses to light, and certain fungi that grow on horse dung manifest striking phototropic movements, which are discussed in chapter 17.

Gravitropism

Growth responses to the stimulus of gravity are called **gravitropisms.** The primary roots of plants are positively gravitropic, whereas shoots forming the main axis of plants are negatively gravitropic (figure 11.9). As indicated in chapter 6, there is evidence that plant organs perceive gravity through the

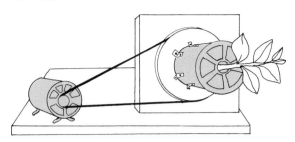

FIGURE 11.10   A clinostat.

movement of starch grains in special cells of the root cap. These special cells, called *statocytes,* are also found in coleoptile tips and in the endodermis. When a potted plant is placed on its side, the starch grains in the statocytes will, within a few minutes, begin to float or tumble down until they come to rest on the side of the cells closest to the gravity stimulus. In roots, the cells on the side opposite the stimulus begin elongating within an hour or two, bringing about a downward bend, while the opposite occurs in stems. Some have suggested that mitochondria and Golgi bodies (dictyosomes) also respond to gravity, but precisely how either they or the starch grains bring about the response is the subject of much conjecture; it may be that auxin and ABA are agents involved in modifying cell elongation so as to produce these gravitropic bendings. The stimulus of gravity can be negated by rotating a plant placed in a horizontal position. A simple device called a *clinostat* uses a motor and a wheel to rotate a potted plant slowly about a horizontal axis (figure 11.10). As the plant rotates, both the stem and roots continue to grow horizontally instead of exhibiting characteristic gravitropic responses. Obviously neither the starch grains in the statocytes nor other organelles can settle while the plant is moving, and this apparently prevents transport of auxin, which would bring about cell elongations that produce curvatures of root and stem.

Other Tropisms

A plant or plant part response to contact with a solid object is called a *thigmotropism.* One of the most common thigmotropic responses is seen in the coiling of tendrils and in the twining of climbing plant stems (see figures 7.9 and 11.7). Such responses can be relatively rapid, with some tendrils wrapping around a support two or more times within an hour. The coiling results from cells in contact becoming slightly shorter while those on the opposite side elongate.

It is well known that roots will enter cracked water pipes and sewers. In fact, roots have been known to grow upward for considerable distances in response to water leaks. Some have called such growth movements *hydrotropisms,* but most plant physiologists today doubt that responses to water and several other "stimuli" are true tropisms. Other external "stimuli" that produce tropic responses designated as tropic by some scientists include chemicals (*chemotropism*), temperature (*thermotropism*), wounding (*traumotropism*), electricity (*electrotropism*), dark (*skototropism*), and oxygen (*aerotropism*). It has been noted that greater concentrations of roots occur on the north and south sides of wheat seedlings, and it has been suggested that magnetic forces may be involved; the term *geomagnetotropism* has been proposed for this phenomenon. Some of these tropic or tropic-like responses have been induced primarily by artificial means, but others occur regularly in nature. For example, germinating pollen grains produce a long tube that follows a diffusion gradient of a chemical released within a flower; this is considered a chemotropism. Thermotropic responses to cold temperatures may be seen in the shoots of many common weeds, which grow horizontally when cold temperatures prevail and return to erect growth when temperatures become warmer.

Turgor Movements

Turgor movements result from changes in internal water pressures. The cells concerned may be in normal parenchyma tissue of the cortex, or they may be in special swellings called *pulvini* located at the bases of leaves or leaflets. Some turgor movements may be quite dramatic, taking place in a fraction of a second. Others may require up to 45 minutes to become visible.

Sleep Movements

About 90% of the members of the legume family, which includes peas and beans, and members of several other flowering plant families exhibit movements in which either leaves or petals fold as though

A.

B.

"going to sleep" (figure 11.11). The folding and unfolding usually takes place in regular daily cycles, with folding most frequently taking place at dusk and unfolding occurring in the morning. Such cycles, which have been more extensively documented in the Animal Kingdom, have come to be known as **circadian rhythms.** They appear to be controlled internally by the plants, although they are also geared to changing daylengths and seasons. These cycles do not, however, generally accelerate when temperatures increase. The actual timing of circadian rhythms varies with the species, although most plants exhibiting sleep movements do so at dusk and at dawn. The flowers of several species, however, open their flowers at different hours of the night.

About 200 years ago, the famous Swedish botanist Linnaeus planted wedge-shaped segments of a circular garden with plants that exhibited sleep movements. The plants were arranged in successive order of their sleep movements throughout a 24-hour day. One could tell the approximate time simply by noting which part of the garden was "asleep" and which was "awake." A few others copied the "garden clock" idea, but the expense and difficulty of obtaining all the plants from different parts of the world and replanting them each year proved to be too great for the practice to be continued.

The movements are produced by turgor changes caused by the passage of water in and out of cells at the bases of the leaves or leaflets. The function of these movements is not clear; the rhythms are also not confined to sleep movements. Species of certain algae called *dinoflagellates* (see chapter 16) inhabiting warmer ocean waters glow in the dark through *bioluminescence,* a process by which chemical energy is reconverted to light energy. One species always glows brightly within 2 or 3 minutes of midnight, even if it is maintained in culture in continuous dim light. This particular dinoflagellate also glows when culture containers in which it is suspended are jarred, and it displays two other types of circadian rhythms. One rhythm involves cell division, with peak mitotic activity occurring just before dawn, and the other rhythm pertains to photosynthesis, which reaches a maximum around noon.

Another type of rhythm is seen in the giant bamboos of Asia, which send up huge flowering stalks every 33 or 66 years, even if the plants have been transplanted to other continents and are growing under different conditions. These flowering stalks use up all the energy reserves of the bamboo, and they die shortly after appearing. This has especially been a problem in cities where nearly all of the bamboo plantings were propagated from a single source, and then all died simultaneously after flowering.

FIGURE 11.12   Water conservation movement in a grass leaf when insufficient water to maintain normal turgor is available. *A.* The leaf when adequate water is available. *B.* The leaf after it has rolled up. *C.* Enlargement of a cross section of a rolled leaf showing the large, thin-walled *bulliform cells* (arrows) which partially collapse under dry conditions and thus bring about the rolling of the leaf blade.

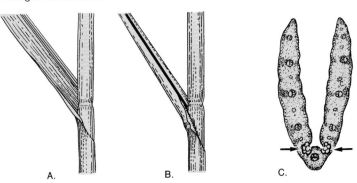

A.   B.   C.

Water Conservation Movements

Leaves of many grasses have special thin-walled cells (*bulliform cells*) below parallel, lengthwise grooves in their surfaces. During periods when sufficient water is not available, these cells lose their turgor and the leaf rolls up or folds (figure 11.12). Experiments with certain prairie grasses have shown that the rolling effectively reduces transpiration to as low as between 5% and 10% of normal.

Contact Movements

The sudden movements of bladderworts (discussed in chapter 7), involve turgor changes apparently triggered by electric charges released upon contact or as a result of variations in light or temperature. The springing of the trap of the Venus flytrap (also discussed in chapter 7) was thought to be brought about in similar fashion but recent research has shown that the trap closes when its outer epidermal cells expand rapidly, and it reopens when the inner epidermal cells expand in the same way. About one-third of the ATP available in the cells is used in each movement, so that repeated stimulation of the trap by touching the trigger hairs readily fatigues the trap if sufficient time for ATP replenishment is not given between stimulations.

The sensitive plant (*Mimosa pudica*) has well-developed swellings (*pulvini*) at the bases of its many leaflets and a large pulvinus at the base of each leaf petiole (figure 11.13). When the leaf is stimulated

FIGURE 11.13   A longitudinal section through the pulvinus of a sensitive plant (*Mimosa pudica*).

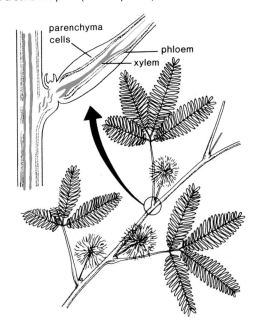

parenchyma cells

phloem

xylem

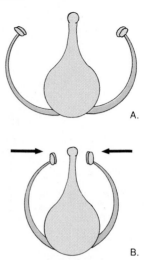

by touch, heat, or wind, there is a type of chain reaction in which potassium ions migrate from one half of each pulvinus to the other half. This is followed by a rapid shuttling of water from half of the pulvinar parenchyma cells to those of the other half. The result is sufficient loss of turgor to bring about a folding of both the leaflets and the leaf as a whole (figure 11.14). If a part of the stem of a sensitive plant (*Mimosa pudica*) is cut off and then immediately reattached with a water-filled piece of rubber tubing, it can be shown that something is transmitted from above the cut through the water. Within a few minutes after a leaf above is stimulated to fold, a leaf or two below the cut will also fold. Although potassium ions have been shown to leave cells in pulvini that are losing turgor, it is not known if the reaction is transmitted across water in the rubber tubing by these ions, by electrical charges, or by something else. A normal response to a stimulus by a sensitive plant (*Mimosa pudica*) takes from a few seconds to less than 2 minutes.

The redwood sorrel, whose leaflets also have pulvini, flourishes on the floor of redwood forests where the light is only one-two hundredth that of full sunlight. Occasionally, light of sufficient intensity to damage delicate leaves may temporarily penetrate the overhead canopy. When this occurs, the leaflets begin to fold downward within 10 seconds, and are completely folded in about 6 minutes, unfolding once again when the brighter light is gone.

Contact movements are not confined to leaves. Many flowers exhibit movements of stamens and other parts, most such movements facilitating pollination. The pollen-receptive stigmas found in flowers of the African sausage tree have two lobes. These lobes fold inward, enclosing pollen grains that have landed on them. The stigmas of monkey flowers and Asian cone flowers exhibit similar turgor movements upon contact, while the pollen-bearing stamens of barberry and moss rose flowers "snap" inward suddenly upon contact, dusting visiting insects with pollen (figure 11.15). Orchid flowers exhibit some of the most spectacular of all contact movements, including those by which little sacs of pollen are forcibly attached to the bodies of visiting insects; in a few instances the benefactors even receive dunkings in small buckets of fluid from which they escape via a trap door (see figure 23.25 for a discussion of a few examples).

Taxes

The *taxis,* a type of movement that involves either the entire plant or its reproductive cells (see chapters 15–21), does not occur among flowering plants. In response to a stimulus, the cell or organism, either propelled or pushed by flagella (whiplike appendages) or cilia (shorter whiplike appendages), moves toward or away from the source of the stimulus.

Stimuli include chemicals, light, oxygen, and gravitational fields. In ferns, for example, the female reproductive structures produce a chemical that prompts a *chemotactic* response in the male reproductive cells (sperms)—that is, the sperms swim toward the source of the chemical. Certain one-celled algae manifest *phototactic* responses, swimming either toward or away from a light source. Other similar organisms exhibit *aerotactic* movements in response to changes in oxygen concentrations.

Miscellaneous Movements

Slime molds (discussed in chapter 17) inhabit damp logs and debris. During their active stages their protoplasm, which has no rigid cell walls, "flows" over dead leaves and other substrates like thin, slowly moving gelatin. Certain blue-green bacteria (e.g., *Oscillatoria*) wave slowly back and forth or slide up and down against each other in *gliding movements,* and diatoms (one-celled algae with thin glass shells) give the appearance of being jet-propelled. It is not certain just how these movements occur, but there is evidence that submicroscopic fibrils produce rhythmic waves that bring about the motion. In the case of diatoms, some materials are apparently forced out of the cells, and the friction set up by the process may propel the organisms through the water in which they are found.

Dehydration movements do not involve living cells or hormones, the forces being purely physical. They are caused by imbibition or by the drying out of tissues or membranes. The individual fruitlets of filarees and other members of the Geranium Family have long, stiff, pointed extensions that are sensitive to changes in humidity. As humidity decreases during the day, they coil up, and then at night, when humidity increases, the coils relax. This alternating coiling and uncoiling results in the pointed fruitlets planting themselves into the ground in a corkscrew-like fashion (see figure 8.27).

A number of fruits that are podlike and dry at maturity split in various ways with explosive force, flinging seeds as far as 12 meters (39 feet) from the plant. Examples of such fruits include those of the garbanzo bean, witch hazel, vetches, and Mexican poppies. In ice plants and stonecrops, the converse is true; rain or dew causes the mature fruits to open due to the swelling of membranes. Pressures inside the squirting cucumber build up to the point where, upon abscission of the fruit, the seeds are expelled from the stalk in distances of up to 10 or more meters (33 feet). Dwarf mistletoes, which are parasitic on coniferous trees, produce tiny sticky fruits that are explosively released and adhere to the trunks and branches of trees in the vicinity.

PHYTOCHROME

Phytochrome is an extraordinary pale blue proteinaceous pigment that apparently occurs in all higher plants and is associated with the absorption of light. It was first isolated in 1959, although its presence was suspected several years before. It was not visible, however, and a special pigment-analysis instrument had to be constructed to detect it. Only minute amounts are produced, mostly in meristematic tissues. It occurs in two stable forms, either of which can be converted to the other: P_{red} or P_r is a form that absorbs red light; $P_{far-red}$ or P_{fr} is a related form that absorbs the far-red light found at the edge of the visible light spectrum. When either form absorbs light, it is converted to the other form, so that P_r becomes P_{fr} when it absorbs red light, and P_{fr} becomes P_r when it absorbs far-red light. P_r is stable indefinitely in the dark. The normal effect of light in nature is to cause more P_r to become P_{fr} than vice versa. P_{fr} converts back to P_r in the dark over a period of several hours, or else becomes inactivated, but its conversion in the presence of appropriate light is instantaneous (figure 11.16).

FIGURE 11.16   Phytochrome interconversions.

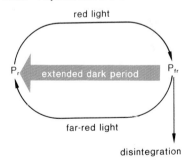

FIGURE 11.17   Photoperiodism. Five cockscomb plants exposed to different daylengths. The plant on the left received 8 hours of light per day. The plant on the right received 16 hours of light per day. The plants between (from left to right) received 10, 12, and 14 hours of light, respectively. The plants that received 14 or more hours of light did not flower. (USDA Photo)

Phytochrome pigments play a role in a great variety of plant responses: they are involved with several aspects of plant development, *photoperiodism* (discussed in the next section), changes in plastids, the production of anthocyanin pigments, and with the detection of shading by other plants. One of the most studied phenomena associated with phytochrome pigments involves the germination of seeds. Some seeds, for example, do not germinate in darkness, but the red part of the spectrum in sunlight converts P_r to P_{fr}, which in turn somehow unblocks the germinating mechanism. If such seeds (e.g., those of "Grand Rapids" lettuce) are given suitable moisture and oxygen conditions and then exposed only to red light, they will germinate readily, but if they are given only far-red light they will not. Furthermore, if they are given alternating flashes of red and far-red light, the type of light in the final flash determines whether or not they will germinate. If the last flash, which may be of only a few seconds duration, is red light, the seeds will germinate, but if it is far-red light they will not.

When a seedling first emerges from the soil, light changes the P_r to P_{fr}, signaling the cells producing ethylene to reduce production. This in turn permits the crooks in the hypocotyls to relax, and the plant straightens up. Elongation of stems appears to be inhibited by P_{fr}. The mechanism is so sensitive that as little as 3 seconds of clear moonlight may produce a shortening of internodes of **etiolated** (spindly and pale from having been grown in the dark) oat seedlings. When stems of a tree are shaded, however, more far-red light reaches them, causing the P_{fr} to be converted to P_r and thus lowering the inhibition so that the stems grow longer and out from under the shade.

PHOTOPERIODISM

In the early 1900s, two American plant physiologists, W. W. Garner and H. A. Allard, became curious about a variety of tobacco plant growing near Washington, D.C. The plants grew 3 to 4 meters (10 to 13 feet) tall during the summer, but they failed to flower unless they were brought into a greenhouse for the winter. If the plants were started in pots in the fall in a greenhouse, they grew only about a meter (3 feet) tall before flowering. Garner and Allard decided to see what would happen if they kept the tobacco plants, as well as some soybean seedlings that would not bloom in Washington before September, in complete darkness from 4 P.M. until 9 A.M., thus allowing them only 7 hours of daylight. They found that all the plants flowered, and further investigations showed that the length of day (actually the length of the night) was directly related to the onset of flowering in many plants. They published the results of their investigations in 1920 and later called the phenomenon they had discovered **photoperiodism** (figure 11.17).

The critical length of day (i.e., the maximum or minimum length of day) for the initiation of flowering is frequently about 12 to 14 hours, although it can vary considerably. Plants that will not flower unless the daylength is shorter than the critical length (e.g., in the fall or spring) are called **short-day plants.** They include asters, chrysanthemums, dahlias, goldenrods, poinsettias, ragweeds, sorghums, salvias, strawberries, and violets. Plants that will not flower unless periods of light are longer than the critical length are called **long-day plants.** These include garden beets, larkspur, lettuce, potatoes, spinach, and wheat. Such plants usually flower in the summer but will also flower when left under continuous artificial illumination. Accordingly, leafy vegetables such as lettuce and spinach need to be harvested in the spring and grown again in the fall in temperate latitudes if *bolting* (producing flowering stalks) is to be avoided. Potato breeders in the United States grow their plants in northern states where the long summer days initiate flowering, but since the potatoes themselves are produced independently of flowering, the plants may be grown for crop purposes at any latitude where other conditions are favorable.

Indian grass and several other grasses have two critical photoperiods; they will not flower if the days are too short and they also will not flower if the days are too long. Such species are referred to as **intermediate-day plants.** Other plants, particularly those of tropical origin, will flower under any length of day, providing, of course, they have received the minimum amount of light necessary for normal growth. Such plants are called **day-neutral plants** and include garden beans, calendulas, carnations, cyclamens, cotton, nasturtiums, roses, snapdragons, sunflowers, and tomatoes, as well as many common weeds such as dandelions.

With some plants, small differences in daylength may be critically important. Some varieties of soybeans, for example, will not flower when days are 14 hours long but will flower if the daylength is increased to 14½ hours. This difference could amount to less than 320 kilometers (200 miles) of latitude, so that certain varieties grown in the southern states might not produce fruit in the northern states and vice versa.

Commercial florists and nurserymen have made extensive use of photoperiods, manipulating with artificial light the flowering times of poinsettias, some lilies, and other plants so as to have them flower at times of the biggest demand, such as Christmas or Mother's Day. The light intensities used to lengthen the days artificially can be very low—often less than 10 footcandles (i.e., one-thousandth the intensity of full sunlight)—and can come from incandescent bulbs, which have more red wavelengths than fluorescent lamps and therefore have more effect on phytochromes, which are involved in photoperiodism.

A number of vegetative activities of plants are also affected by photoperiods. In the shortening days of the fall, for example, many woody plants will begin undergoing the changes that lead to the dormancy of buds, regardless of whether or not "Indian summer" temperatures may be prevailing. In the spring, certain seeds respond to photoperiods in their germination, both long-day and short-day species having been discovered. Plants that produce tubers (e.g., Jerusalem artichoke) may develop them only under short-day conditions, even though they are long-day plants with respect to flowering. Usually these photoperiodically controlled responses are valuable to the plants in that they prepare them for changes in the seasons and thus ensure their survival and perpetuation.

FLORIGEN

After photoperiodism was discovered scientists conducted experiments to determine which part of a plant was sensitive to daylength. They soon found it was mainly the leaf, although they also noted that buds of long-day plants exhibited similar responses. If they completely enclosed the leaf of a short-day plant in black paper for all but 8 hours of a 24-hour day while exposing the rest of the plant to long days, flowering was initiated, but if they treated a long-day plant the same way, flowering did not occur

FIGURE 11.18   An experiment illustrating the effect of
subjecting one leaf of a short-day plant to short days while
the rest of the plant is exposed to long days. *A.* The short-
day plant exposed to long days. No flowers were produced.
B. The same plant exposed to long days while one leaf was
covered with black paper 16 hours a day for a few weeks.
The plant flowered, presumably because some substance
that initiates flowering was produced in the shaded leaf and
then diffused or was carried to the stem tip where flower
buds are produced.

A.

B.

(figure 11.18). This suggested that something in the leaf was being carried to an area where flowers were initiated. Added support for this theory came from showing that when part of a plant was exposed to the appropriate daylength to initiate flowering and then grafted to a plant that had not been so exposed, something would cross the graft so that both parts would flower. Further evidence was obtained in the 1950s through experiments in which the leaves of some plants were removed immediately after exposure to critical photoperiods, while the leaves of similar plants were removed several hours after exposure. The plants whose leaves were removed immediately after exposure did not flower, but those whose leaves were removed later flowered almost as well as others whose leaves were not removed. This indicated something initiating flowering moved out of the leaves, but its departure was prevented if the leaves were removed before it had a chance to do so.

After the existence of phytochrome had been demonstrated it was theorized that it might be involved in photoperiodism, with P_{fr} inhibiting flowering in short-day plants and promoting flowering in long-day plants. Presumably the P_{fr} would accumulate in the light in short-day plants and revert back to P_r or be broken down during dark periods. After the length of the night increased sufficiently, all or most of the P_{fr} would disappear and flowering then would cease to be inhibited, with the converse taking place for long-day plants. P_{fr} has been demonstrated to disappear in many plants in as little as 3 or 4 hours of darkness, however, indicating that phytochrome interconversions alone cannot explain photoperiodism. On the basis of the evidence it was theorized that plants produce one or more flowering hormones called **florigen,** which may then be transported to the apical meristems where flower buds are initiated. Despite all the circumstantial evidence, however, florigen has not yet actually been isolated from a plant, nor has it otherwise been proved to exist. This has led to speculation that photoperiods may bring about a shunting of nutrients that initiate flowering, but others feel the evidence for florigen is still compelling, or that there may be no simple explanation for the phenomenon of photoperiodism.

TEMPERATURE AND GROWTH

Each species of plant has an optimum temperature for growth, although the optimum may vary throughout the life of the plant, and a minimum temperature below which growth will not occur. Each species also has a maximum temperature above which injury may result (or, at least, growth will cease).

Dr. Frits Went, the discoverer of auxin, experimented with the growth of tomato plants under conditions in which night temperatures were lowered to 17° C (63° F) while day temperatures were maintained at 25° C (77° F). These plants were found to grow better than plants kept at a steady temperature around the clock. Went applied the term *thermoperiodism* to this phenomenon. It has since been shown that the optimal thermoperiod may change with the growth stage of the plant. Thus, young pepper plants have been shown to develop best when night temperatures are approximately 25° C (77° F), but as they grow older the optimum night temperature for their development drops more than 11° C (20° F). Lower night temperatures often result in a higher sugar content in plants and may also produce greater root growth, although some plants such as peas seem to be unaffected in this regard.

The growth of many field crops is roughly proportional to prevailing temperatures—making it possible to predict harvest times—although the number of days until maturity varies considerably with the locale. In 1855 the Swiss botanist A. P. de Candolle established a basis for the harvest dates of crops. His method, based on summing the temperature means for each day, makes it possible to follow a crop's progress with some precision. This method has been refined by multiplying the temperature means by the number of hours of daylight. In Hawaii a high degree of precision has been attained in predicting the date of the pineapple harvest by measuring the rate of growth within the temperature range at which the plants grow and multiplying the number of hours per day at each temperature. These methods, which work best for crops that are not closely regulated by photoperiods, do not, however, take into account the requirements of some plants for different optimum temperatures at different stages of their development.

DORMANCY AND QUIESCENCE

Dormancy may be defined as a period of growth inactivity in seeds, buds, bulbs, and other plant organs even when environmental requirements of temperature, water, or daylength are met. **Quiescence** is a state in which a seed is unable to germinate unless environmental conditions normally required for growth are present. We have already discussed the formation of abscission layers (see chapter 7) in the fall as plants change from an active condition to either a dormant or a quiescent one. In temperate and cold climates, most plants and their seeds go through a dormant period lasting from a few days to several months, and in each case the dormancy nearly always has some survival value for the species. Many of the stone fruits (for example, cherries, peaches, and plums) need at least several weeks of rest at temperatures below 7° C (45° F) before the buds will develop into flowers or new growth, and in some instances the buds will not develop before the tree has encountered a period of freezing temperatures. This dormancy adapts the plants to the conditions prevailing in the cooler temperate zones of the world. In response to prevailing conditions, seeds of desert plants germinate only after appreciable rain has fallen, and seeds of plants native to vernal pools and other wet areas may germinate only if under water.

The change from dormancy to a state in which germination will occur in seeds is sometimes controlled by a variety of factors referred to as *after-ripening*, depending on the species concerned. Some plants become sensitive to photoperiods only after they have been subjected to a period of chilling, either while they are still seeds or shortly after germination. Growers often place moistened seeds of such plants in a refrigerator for several weeks. This process is known as *stratification* and is one of several practices employed to break dormancy. In some biennial plants, the cold treatment can be replaced by an application of gibberellins. Natural stratification occurs when winter grains are sown in the fall, and it has been shown that the extent of flowering and yield in such grains is profoundly reduced in the absence of a period of natural chilling.

Some plants have their dormancies broken only after they have received a series of similar stimulations, such that a single stimulation, for example an unseasonal rainstorm, will not trigger growth. Other plants, such as poor man's pepper, require both certain temperatures and photoperiods for germination, and some weeds require specific thermoperiods plus some mechanical abrasion of the seedcoat to break dormancy. Still other seeds may require specific enzyme action or a combination of conditions acquired only by passing through the digestive tracts of birds, bats, or other animals, before they will germinate.

When a mechanical tomato harvester was being developed in California during the early 1960s, tomato breeders wanted to have fruits that would be able to stand up to the bumping they would receive in the machinery, and so they tried to produce suitable varieties. Their search for new genes to use in their breeding programs led them to the wild tomatoes of the Galápagos Islands. When they brought the seeds back to California, however, they experienced difficulty in getting them to germinate, even after subjecting them to a wide variety of stratification treatments. Then they remembered that tomatoes are eaten by the giant tortoises of the Galápagos Islands. After feeding some to giant tortoises in a zoo, seeds that passed through these animals were found to germinate normally. Apparently something in the tortoises' digestive tracts broke the dormancy of the seeds.

Germination is prevented by a thick or restrictive seedcoat in the seeds of plants such as cocklebur, and the seeds of some legumes are almost

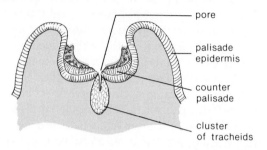

pore

palisade epidermis

counter palisade

cluster of tracheids

stonelike because of the exceptional hardness of the outer covering or seedcoat. Growers often nick or file seedcoats to allow water and gases to enter. In nature, various decay organisms corrode the seedcoats, or the seeds themselves may secrete enzymes that aid in this process. The seeds of certain lupines have remarkable valves that restrict the uptake of water, preventing germination before the seedcoat is eroded by other means (figure 11.19). Many plants have seeds that will not germinate until they have been exposed to fire, and one such Australian legume with a thick seedcoat will not readily germinate until heat from a fire causes a plug in the seed to pop out.

Dormancy may be restricted to certain parts of the plant rather than occurring in the plant as a whole. In coffee, for example, only the flower buds may show dormancy. In other woody plants the onset of dormancy progresses from buds toward the base of the branch and on to the tip. Some seeds go through two cycles of dormancy before full growth occurs. In wake-robins, the radicle (embryo root) of the seed emerges and becomes established after being subjected to a period of cold, but the epicotyl does not emerge until a second season of cold has been encountered.

In many seeds, growth inhibitors such as ABA are present in the seedcoat, often in abundance. Before germination can occur, these inhibitors may need to be leached out by rain, broken down by enzymes, or have their effects overcome by other hormones. Studies have shown that gibberellins, which have the capacity to break dormancy when applied artificially, are produced by the germinating seeds of both cone-bearing and flowering plants. Ethylene, which is also effective in breaking dormancy, is produced by various germinating seeds, including those of oats, peanuts, and clover. Cytokinins do not generally appear to be involved in breaking dormancy, but a few instances are known of their doing so artificially. Although auxins are formed during the early stages of germination, they probably are not involved in the control of dormancy.

Many aspects of growth discussed in this chapter are applied to, or are involved in, vegetative propagation, which is surveyed in appendix 4.

SUMMARY

Growth is defined as an "increase in volume due to the division and enlargement of cells." As cells mature they differentiate into forms adapted to specific functions. Development is a change in form as a result of growth and differentiation combined. Cells themselves assume different shapes and forms that adapt them to the problem involved in their total volume increasing at a greater rate than their total surface area.

Many growth phenomena are influenced by hormones, which are chemical substances produced in minute amounts in one part of an organism and transported to another part, where they have specific effects on growth, flowering, and other plant activities. Vitamins are organic molecules that function in activating enzymes; they are sometimes difficult to distinguish from hormones.

Darwin and his son noted in 1881 that coleoptiles bend toward a light source. Frits Went, in 1926, followed up on the Darwinian observations and demonstrated that something in coleoptile tips moved out into agar when decapitated tips were placed on it. He called the substance auxin and showed that it could cause coleoptiles to bend. Since Went's experiments, three groups of plant hormones that promote the growth of plants have been found; others that have been discovered are known only for their inhibitory effects. Auxins stimulate the enlargement of cells and are involved in many other growth phenomena. 2,4-D and 2,4,5-T are hormonelike compounds that have been used as weed killers and defoliants. 2,4,5-T has a highly toxic contaminant, dioxin, that has caused disease and defects in laboratory animals; it is now banned for most uses in the United States. Gibberellins promote stem growth without corresponding root growth; cytokinins promote cell division, and can be used to stimulate the growth of axillary buds. Abscisic acid causes buds to become dormant and apparently helps leaves respond to excessive loss of water; ethylene gas hastens ripening of fruits and is used commercially to ripen green fruits. Senescence is the breakdown of cell parts that leads to the death of the cell.

Plant movements such as helical (spiraling) movements, nodding movements, twining movements, contraction movements, and some nastic movements are growth movements that result primarily from internal stimuli. Tropisms are permanent, directed growth movements that result from external stimuli such as light (phototropism), gravity (gravitropism), contact (thigmotropism), chemicals (chemotropism), etc. Turgor movements result from changes in internal water pressures; they may be very rapid or take up to 45 minutes to become visible. Turgor movements include sleep movements in which leaves or flowers fold daily in what is known as a circadian rhythm. Other types of plant movements include water conservation movements, contact movements, taxes, and dehydration movements.

Phytochrome is a pale blue pigment that occurs in all higher plants and plays a role in many different plant responses. It occurs in two forms, each of which can be converted to the other by the absorption of light. Daylight generally results in more P_r being converted to P_{fr} than vice versa. P_r absorbs red light and P_{fr} absorbs the far-red light found at the edge of the visible light spectrum. P_{fr} will convert back to P_r over a period of several hours in the dark; P_r is stable indefinitely in the dark. The P_r–P_{fr} mechanism is so sensitive that as little as 3 seconds of clear moonlight may produce enough P_{fr} to result in a shortening of the internodes of etiolated (spindly and pale from having been grown in the dark) oat seedlings.

Photoperiodism is a response of plants to the duration of day or night. Short-day plants will not flower unless the daylength is shorter than a critical daylength. Long-day plants will not flower unless the daylength is longer than a critical daylength. Intermediate-day plants have two critical photoperiods and will not flower if the day is either too short or too long; day-neutral plants will flower under any daylength providing they have received a minimum amount of light.

The precise mechanism of photoperiodism is not yet fully understood. Phytochrome interconversions may play a role in photoperiodism but cannot be used alone to explain it. A flowering hormone called florigen is believed to be produced in leaves and to be involved in the initiation of flowering, but it has not yet been isolated from plants.

Each species of plant has an optimum temperature for growth. Dormancy is a period of growth inactivity in seeds, buds, bulbs, and other plant organs even when appropriate environmental conditions are met. Quiescence is a state in which a seed is unable to germinate unless appropriate environmental conditions exist. The change from dormancy to a state in which germination will occur in seeds is controlled by a variety of factors, including temperature, moisture, photoperiod, thickness of seedcoat, enzymes, and growth inhibitors. Stratification is the breaking down of dormancy in seeds by keeping them refrigerated and moist for several weeks.

REVIEW QUESTIONS

1. What are the differences among hormones, enzymes, and vitamins?
2. Auxins, gibberellins, and cytokinins all promote growth. How does one distinguish among them?
3. How do hormonal herbicides function?
4. List several commercial applications of ethylene gas.
5. What are statocytes, and what role do they play in plant growth?
6. Distinguish among internally and externally induced growth movements, turgor movements, and dehydration movements.
7. How do day-neutral and intermediate-day photoperiod plants differ in their requirements?
8. What is the difference between dormancy and quiescence? How may dormancy be artificially broken?
9. What are the requirements for a seed to germinate?
10. How does phytochrome differ from other plant pigments?

DISCUSSION QUESTIONS

1. If green plants require light in order to produce food needed for growth, why are seedlings that are germinated in the dark taller than those of the same age grown in the light?
2. If you remove a terminal bud from a stem, will growth in length stop altogether? Explain.
3. Would it be technically correct to say that some plants go to sleep at night? Explain.
4. Many plants produce seeds that require a period of dormancy before they will germinate. Of what value is the dormancy to the plant? Where might you expect to find plants with seeds that do not undergo dormancy?
5. The rapid turgor movements of the sensitive plant (*Mimosa pudica*) are rare in the Plant Kingdom. If all plants evolved such movements, would the phenomenon have possible survival value or other value to the plants? Explain.

ADDITIONAL READING

Briggs, W. R., R. L. Jones, and V. Walbot, eds. 1986. *Annual review of plant physiology.* Vol. 37. Palo Alto, CA: Annual Reviews, Inc.

Chadwick, C. M., and D. R. Garrod. 1987. *Hormones, receptors and cellular interactions in plants.* New York: Cambridge University Press.

Hillman, W. S. 1979. *Photoperiodism in plants and animals.* Oxford Biology Readers, No. 107. Burlington, NC: Carolina Biological Supply Company.

Kendrich, R. E. and B. Frankland. 1983. *Phytochrome and plant growth: studies in biology, no. 68,* 2d ed. London: Edward Arnold Publishers.

Nickell, L. G. 1982. *Plant growth regulators—agricultural uses.* New York: Springer-Verlag.

Ray, P. M., T. A. Steeves, and S. A. Fultz. 1983. *Botany.* Philadelphia: Saunders College Publishing.

Salisbury, F. B., and C. W. Ross. 1985. *Plant physiology,* 3d ed. Belmont, CA: Wadsworth, Inc.

Thimann, K. V. 1977. *Hormone action in the whole life of plants.* Amherst: University of Massachusetts Press.

Torrey, J. G. 1985. "The development of plant biotechnology." *American Scientist* 73: 354–63.

Wareing, P. F., and I. D. Phillipps. 1981. *The control of growth and differentiation in plants,* 3d ed. Elmsford, NY: Pergamon Press, Inc.

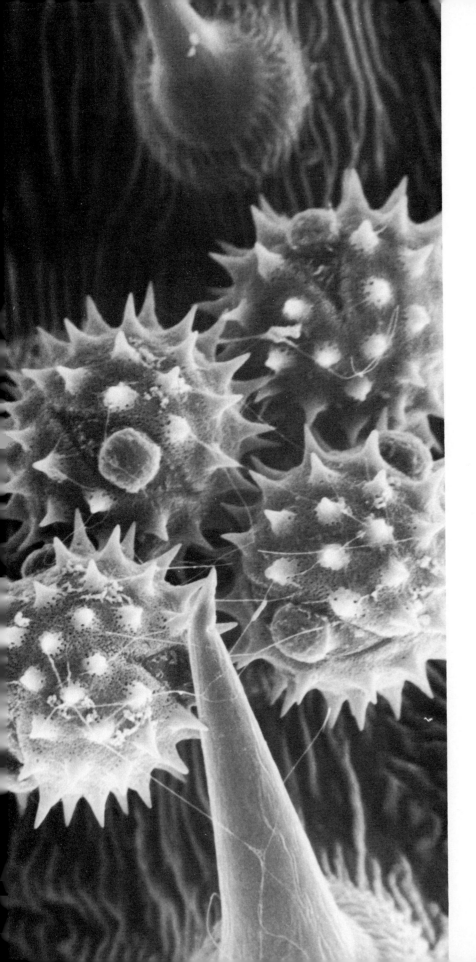

Overview

*A*fter a comparison of asexual and sexual reproduction, this chapter explores the necessity for halving the chromosome number at some stage in the life cycle of sexually reproducing organisms. The phases of meiosis are systematically covered. The chapter also presents a discussion of Alternation of Generations and general rules for understanding the process.

Scanning electron micrograph courtesy of Robert L. Carr

12 Meiosis and Alternation of Generations

Some Learning Goals

1. Know the phases of meiosis and briefly describe what occurs in each of them.

2. Understand clearly what features meiosis and mitosis have in common and how they differ.

3. Explain the significance of crossing over to offspring.

4. In Alternation of Generations, indicate at what point each of the following occurs: a change from n to $2n$; a change from $2n$ to n; initiation of the gametophyte generation.

5. Relate meiosis and Alternation of Generations to the process of DNA replication discussed in chapter 2.

Outline

INTRODUCTION

As plants grow, the number of cells multiplies through the process of cell division discussed in chapter 3. Mitosis ensures, in very precise fashion, that the number of chromosomes and the nature of the DNA in the daughter cells will be identical with those of the parent cell. But living organisms do not grow indefinitely, although a few may remain alive for a long time. In order to perpetuate the species, they must reproduce or they will become extinct. This reproduction may take place through natural vegetative propagation (see appendix 4) or by means of special cells called *vegetative spores* (see chapters 17–21), which are produced by mitosis. Such processes, in which the cells involved are identical in their chromosomes with the cells from which they arise, are forms of **asexual reproduction** (the prefix *a* of *asexual* means "without," so asexual reproduction simply means reproduction without sex). However, nearly all plants also undergo **sexual reproduction,** which in flowering and cone-bearing plants ultimately results in the formation of **seeds.** In sexual reproduction, special sex cells called **gametes** are produced. Two gametes, called **egg** and **sperm** in higher plants and animals, form a single cell called a **zygote** when they fuse together. The zygote is the first cell of a new individual (figure 12.1).

If gametes had the same number of chromosomes as all the other cells of the organism, then the zygote, since it results from the union of 2 gametes, would have double the number of its parents' chromosomes. If such zygotes were then to develop into mature organisms, each of the new organisms' cells would have double the original number of chromosomes, and when they produced gametes, the zygotes resulting from the union of this next generation of gametes would have four times the number of chromosomes of the grandparents. If that sort of thing were to continue for just 20 generations, a species having 16 chromosomes in each cell to begin with would end up with no fewer than 8,388,608 chromosomes per nucleus! The problem does not arise, however, because at a certain point in the life cycle of all sexually reproducing organisms a unique process called **meiosis** occurs. This process brings about the development of gametes that have only half the number of chromosomes of any body developing from the zygote, so that when the gametes form zygotes as they unite in pairs, the original chromosome number is restored. Since some aspects of meiosis are similar to those of mitosis (see figure 12.2), it is strongly recommended that you review mitosis (chapter 3) before going through the phases of meiosis outlined in the following section.

When an interphase cell undergoes meiosis, 4 cells are produced as a result of two successive divisions, which take place without a "resting period" between them. Because of the remarkable events that occur during the process, the 4 cells, depending on the organism involved, are rarely, if ever, identical in all respects with the original cell or with each other.

FIGURE 12.1   Contrast between asexual and sexual reproduction in a strawberry plant. More detailed life cycles are shown in figure 12.6.

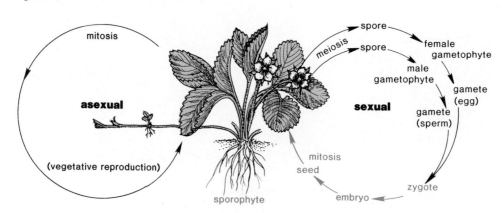

THE PHASES OF MEIOSIS

As in mitosis, a doubling of the DNA of each chromosome takes place before the process of meiosis actually begins, so that all the chromosomes each initially have 2 identical parallel strands held together by a centromere. Like mitosis, meiosis is a continuous process that has been divided into arbitrary phases for convenience. Bear in mind, however, that the lines between one phase and another are indistinct. Four main phases are recognized in each of the two divisions, and the first phase is further subdivided. Since the main phases of the first division bear the same names as those of the second division, it has become customary to designate first division phases with a Roman numeral I and those of the second division with a Roman numeral II. The number of the chromosomes is reduced to half during the first division, but no further reduction in number takes place throughout the second division. Accordingly, some have referred to division I as a *reduction division* and to division II as an *equational division.*

Meiosis generally takes much longer than mitosis to be completed. In lilies, for example, mitosis runs its full course in about 24 hours, whereas meiosis takes about 2 weeks. In other organisms, meiosis may take months or even years to be completed. After the process has begun in human females, for example, it sometimes takes as long as 50 years to reach its conclusion since it remains in a state of arrest for most of that time.

Division I (Meiosis I or Reduction Division)

Prophase I
The main features of prophase I are: (1) The chromosomes pair up as they become shorter and thicker, and their 2-stranded nature becomes apparent. (2) The nuclear envelope and the nucleolus disappear. (3) The pairs of chromosomes often exchange parts with each other and then separate (figure 12.2; see figure 12.4A, B).

As with prophase of mitosis, the beginning of this phase is marked by the appearance of chromosomes as faint threads in the nucleus. These threads gradually coil like a spring so that they become shorter in length, and the coil obviously becomes thicker than the thread alone. As the chromosomes become shorter and thicker they align themselves in pairs, and eventually 2 parallel threads, the **chromatids,** can be distinguished for each chromosome—each pair of chromosomes, therefore, has 4 chromatids. One of the chromosomes of each pair came from the male parent and the other from the female parent. They are identical with each other in length, in the amount of DNA present, and in having a centromere at the same precise location, but their genes may control contrasting characteristics (see chapter 13). Such chromosomes are called *homologues,* and associated pairs of **homologous chromosomes** are referred to as **bivalents.** The 4 chromatids of a bivalent each have their own centromere (figure 12.2; see figure 12.4B).

As prophase I progresses, parts of the chromatids of the homologous chromosomes may break and be exchanged with each other. Depending on the length of the chromosomes, this exchange may occur at one to several points throughout the length of the paired homologous chromosomes, or if the chromosomes are short, it may not occur at all. The evidence for this exchange of parts appears a little later in prophase I, when the chromosomes of each bivalent appear to push each other apart. It can then be seen that adjacent chromatids have crossed over each other at one to several points, and remain attached at these points. This phenomenon of **crossing over** brings about an exchange of some of the DNA contributed by the two parents and is the basis for some of the variability seen in the offspring. The crossovers are called **chiasmata,** and their relative positions provide a basis for the study of where various

FIGURE 12.2   Contrast between mitosis and meiosis.

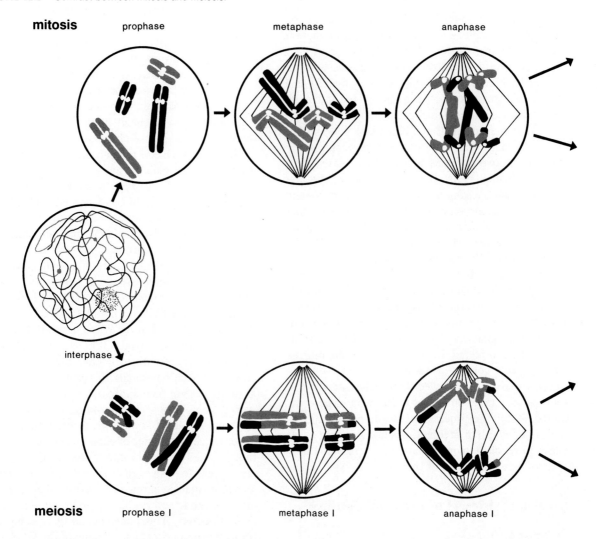

genes are located on the chromosomes (figure 12.3). After the chiasmata have appeared, the chromatids slowly tear apart, each remaining the same length as it was originally and with the same amount of DNA but now possessing "traded" material. The process is something like exchanging three fenders and the front bumper of a red car that has chrome bumpers with an otherwise identical blue car that has white bumpers. When the exchange is completed, the cars will still be structurally identical, but each will have a different combination of fender colors and bumpers.

By the end of prophase I, the nuclear envelope has disintegrated, the nucleolus has disappeared, and spindle fibers are beginning to form. As in mitosis, some spindle fibers are attached to the centromeres of the chromosomes, whereas others extend from pole to pole.

Metaphase I

The main features of metaphase I are: (1) The chromosomes line up at the equator of the cell in pairs. (2) The spindle becomes conspicuous and complete (figure 12.2; figure 12.4C).

telophase

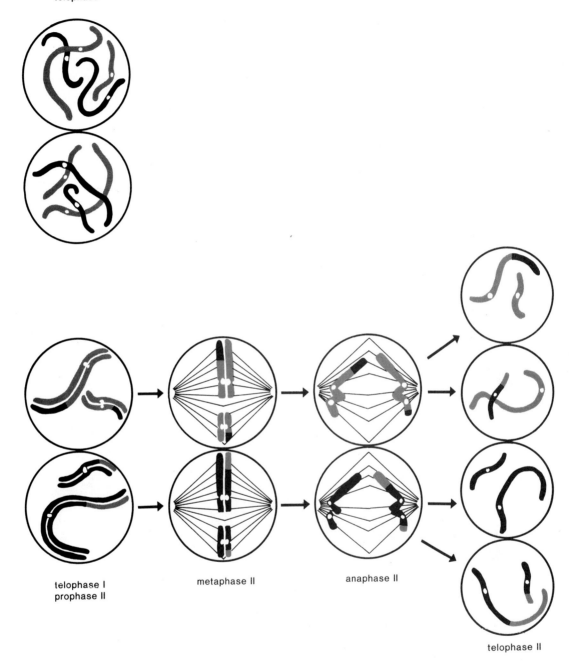

telophase I
prophase II

metaphase II

anaphase II

telophase II

FIGURE 12.3   Chiasmata in chromosomes of crested
wheat grass at prophase I of meiosis. (Photomicrograph by
William Tai)

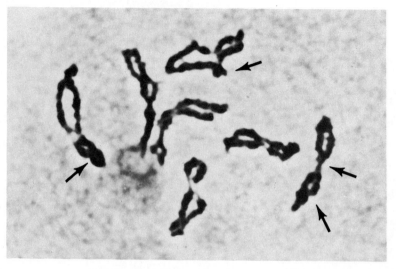

FIGURE 12.4   Various phases of meiosis in crested wheat
grass. (Photomicrographs by William Tai)

A. **Early Prophase I.**
Chromosomes begin to coil and
contract and appear as threads.
B. **Prophase I.** Homologous
chromosomes become aligned in
pairs, and some crossing-over is
visible.
C. **Metaphase I.** Bivalents become
aligned along the equator.
D. **Anaphase I.** Homologous
chromosomes separate and migrate
to opposite poles.
E. **Late Telophase I.**
Chromosomes are at the poles, and
the original cell becomes two cells.

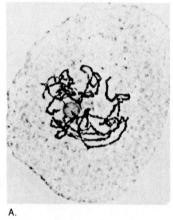

A.

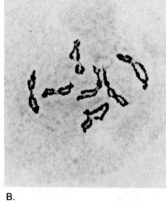

B.

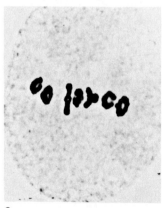

C.

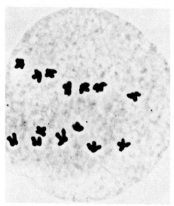

D.

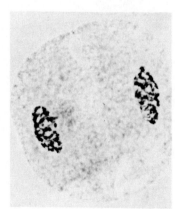

E.

This phase resembles metaphase of mitosis except that when the chromosomes move to the equator, the chromosomes become lined up along it in pairs, with those of homologous chromosomes directly opposite one another on each side of the invisible equator (which is actually an invisible, roughly circular plate) (figure 12.5). Also, as indicated in our discussion of mitosis in chapter 3, the 2 chromatids of a chromosome are joined together at their centromeres and function as a single unit at this stage.

Anaphase I

The main feature of anaphase I is: One whole chromosome (consisting of 2 chromatids) from each pair migrates to a pole (figure 12.2; figure 12.4D).

In anaphase of mitosis, the 2 chromatids of each chromosome, which are being held together at their centromeres, separate as the centromeres become detached from each other, and a chromatid of each pair migrates to an opposite pole. Anaphase I of meiosis is fundamentally different in that the chromatids of each chromosome remain cohered at their centromeres and do *not* separate from one another. Instead, a whole, 2-stranded chromosome from each pair migrates to an opposite pole. Thus, when the chromosomes arrive at the poles, they still consist of 2 chromatids, but only half the total number of chromosomes is at each pole. If a particular chromosome was involved in crossing over in prophase I, 1 of the chromatids will consist of a mixture of chromatid material from a homologous chromosome and some of its original DNA.

F. **Prophase II.** The chromosomes coil and contract again; because of crossing over, the chromatids are no longer identical with each other.
G. **Metaphase II.** The chromosomes of each cell become aligned along their respective equators.
H. **Anaphase II.** The chromatids separate completely and migrate to the poles.
I. **Telophase II.** The four groups of chromatids (now called chromosomes again) are at the poles; new cell walls begin to form.
J. **Formation of cell walls.** Cell walls form; the four cells will become pollen grains.

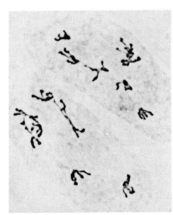

F.

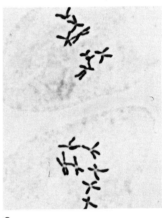

G.

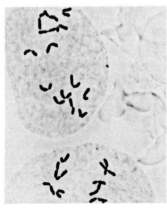

H.

I.

J.

FIGURE 12.5 Diagram of homologous chromosomes at metaphase I of meiosis.

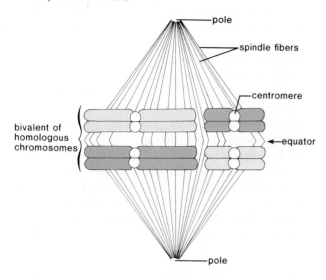

Telophase I

Depending upon the species, the chromosomes may now either partially revert back to interphase by becoming longer and thinner, or they may proceed directly to division II (figure 12.2; figure 12.4E). Normally no new nuclear envelopes are organized around the chromosomes, but the nucleoli usually do reappear. This phase has been completed when the original cell has become 2 cells or 2 nuclei.

Division II (Meiosis II or Equational Division)

In the second division of meiosis, the events of the various phases are similar to those of mitosis except that there is no duplication of DNA during any part of the interphase that may occur between the two divisions. Also, the number of chromosomes in each cell at the end of meiosis is half the number that existed at the beginning of meiosis. In some organisms where there is no interphase, the process proceeds almost directly from telophase I to metaphase II.

Prophase II

The main feature of prophase II is: The chromosomes of both groups become shorter and thicker simultaneously, and their 2-stranded nature once more becomes apparent (figure 12.2; figure 12.4F).

Metaphase II

The main features of metaphase II are: (1) The centromeres of the chromosomes line up along the equator. (2) New spindles become conspicuous and complete.

The 2 spindles may be formed either at right angles to the ones that had formed in metaphase I or at the same angle. A spindle fiber is attached to each centromere, and other spindle fibers extend from pole to pole (figure 12.2; figure 12.4G).

Anaphase II

The main feature of anaphase II is: The centromeres and chromatids of each chromosome separate and migrate to opposite poles (figure 12.2; figure 12.4H).

Telophase II

The main features of telophase II are: (1) The chromatids (now called chromosomes again) lengthen and become thinner. (2) New nuclear envelopes and nucleoli become organized for each group of chromosomes (figure 12.2; figure 12.4I).

This phase is accompanied by the formation of new cell walls between each of the four groups of chromosomes. The set of chromosomes present in each of the 4 cells formed by the end of telophase II constitutes half the original number, and if crossing over and exchange of material between chromatids has occurred, none of the 4 cells will have exactly the same combination of DNA. In many organisms, these 4 cells are called *meiospores;* they develop into various structures or bodies from which gametes are produced by mitosis.

ALTERNATION OF GENERATIONS

As we have noted, meiosis occurs at some point in the life cycle of all organisms that reproduce sexually. The chromosomes of each cell that result from the process constitute a complete set, since one member of every original pair ends up in each cell. The original chromosomal complement, consisting of two complete sets of chromosomes, is restored when gametes unite, forming a *zygote.* Any cell having one set of chromosomes is said to be **haploid,** and any cell with two sets of chromosomes is said to be **diploid.** By the time meiosis is complete 4 *haploid* cells have been produced from 1 *diploid* cell. The gametes of any organism are haploid, whereas a zygote of the same organism is diploid. This holds true regardless of the number of chromosomes peculiar to a given organism. Thus, we can state that an organism having *n* (a specified quantity) chromosomes in its haploid cells will have twice as many, or 2*n* (twice a specified quantity), chromosomes in its diploid cells.

In most animals the only haploid cells (i.e., cells with *n* or a single set of chromosomes) are the gametes (egg and sperm) and the cells that become the gametes. In plants and other green organisms, however, this is generally not so. In a complete life cycle involving sexual reproduction, there is an alternation between a diploid (2*n*) **sporophyte** phase and a haploid (*n*) **gametophyte** phase. This is commonly referred to as **Alternation of Generations.** The diploid body itself is called a **sporophyte.** It develops from a zygote and eventually produces **spore mother cells (meiocytes),** each of which undergoes meiosis, producing 4 **spores.** The haploid bodies that develop from these spores are called **gametophytes.** These eventually form sex organs in which gametes are produced by mitosis.

As we shall see in chapters to follow, the gametophytes of many primitive forms constitute a large part of the visible organism, but as we progress up through the Plant Kingdom to more complex plants, they become proportionately reduced in size until they may be only microscopic. The switch from one generation to the other takes place as spores are produced when spore mother cells undergo *meiosis,* and again when a zygote is produced through fusion of gametes or **fertilization** (also called **syngamy**). Although the sporophyte generation of some primitive organisms may consist of a single cell (the zygote), the basic plan of Alternation of Generations can be seen in the Protistan, Fungal, and Plant Kingdoms. It becomes most conspicuous, however, in the Plant Kingdom, and it differs from one organism to the next in the forms of the various bodies and cells. If you will note, along with the accompanying diagram (figure 12.6), the following six rules pertaining to Alternation of Generations in the majority of plants and other green organisms, you should have little trouble following the life cycle of any sexually reproducing organism discussed in this book.

1. The first cell of any *gametophyte generation* is normally a *spore* (*sexual spore* or *meiospore*), and the last cell is normally a *gamete.*
2. Any cell of a gametophyte generation is usually *haploid* (*n*).
3. The first cell of any *sporophyte generation* is normally a *zygote,* and the last cell is normally a *spore mother cell* (*meiocyte*).
4. Any cell of a sporophyte generation is usually *diploid* (2*n*).
5. The change from a sporophyte to a gametophyte generation usually occurs as a result of *meiosis.*
6. The change from a gametophyte to a sporophyte generation usually occurs as a result of *fertilization* (fusion of gametes), which is also called *syngamy.*

The word *generation* as used in *alternation of generations* simply means *phase of a life cycle* and should not be confused with the more widespread use of the word pertaining to time or offspring.

SUMMARY

Reproduction may take place through natural vegetative propagation or by spores (asexual reproduction) or by sexual processes (sexual reproduction). In sexual reproduction 2 gametes unite, forming a zygote, which is the first cell of a new individual. The process of meiosis ensures that gametes will have half the chromosome number of the zygotes, and also usually ensures that offspring will not be identical with the parents in every respect.

Meiosis takes place by means of two successive divisions, each of which, like mitosis, is divided into arbitrary phases even though the process is continuous. In prophase I the chromosomes pair up, often exchange parts, and then separate. The similar chromosomes of each pair are referred to as being homologous, and associated pairs are referred to as bivalents. Exchange of parts of chromatids in bivalents may be affected by the parts initially crossing over one another and forming chiasmata, and then tearing apart again. In metaphase I the chromosomes become aligned at the equator in pairs and in anaphase I whole chromosomes from each pair migrate to opposite poles. In telophase I the chromosomes either partially revert to their interphase state

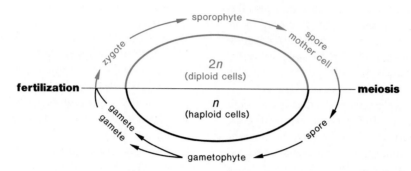

or initiate the second division, which is essentially similar to mitosis. In prophase II the chromosomes of each of the two groups become shorter and thicker again, with both groups becoming aligned at their respective equators in metaphase II. In anaphase II the chromatids of each chromosome separate and migrate to opposite poles, and in telophase II the chromatids (now called chromosomes again) lengthen, and new nuclear envelopes and nucleoli appear for all four groups; new cell walls are produced between each of the four groups. If the chromatids have exchanged parts earlier, none of the four groups may have identical combinations of DNA, and each group has half the original number of chromosomes. Each of the 4 cells constitutes a spore (sexual spore or meiospore), which may develop into a body or structure within which gametes may be produced by mitosis.

Any cell having one set of chromosomes is said to be haploid or to have n chromosomes; any cell having two sets of chromosomes is said to be diploid or to have $2n$ chromosomes. In the life cycle of an organism that undergoes sexual reproduction, there is an alternation between a haploid phase and a diploid phase. The haploid body is called a gametophyte and the diploid body is called a sporophyte.

The change from the haploid phase to the diploid phase occurs when 2 gametes (each n) unite, forming a zygote ($2n$) in the process of fertilization (syngamy). The change from the diploid to the haploid phase occurs as a result of a spore mother cell (meiocyte) undergoing meiosis, when a $2n$ cell becomes 4 n cells. This switching of phases is referred to as Alternation of Generations.

REVIEW QUESTIONS

1. Indicate when during meiosis each of the following events occurs: (a) crossing over, (b) chromatids separating at their centromeres and migrating to opposite poles, (c) chromosomes aligning themselves in pairs at the equator.
2. Is there any difference between mitosis and the second division phases of meiosis?
3. What is the significance of meiosis with respect to sexual reproduction?
4. What is meant by saying that the cells of a sporophyte phase are diploid or $2n$?
5. At what stage of a life cycle does the chromosome number of cells switch from n to $2n$?

DISCUSSION QUESTIONS

1. If an organism reproduces very freely by asexual means, is there any advantage to its also reproducing sexually?
2. Would it make any difference if the events of the second division of meiosis occurred before the events of the first division?
3. Would it be correct to say that a bivalent has four times the amount of DNA present in the chromosomes of a cell in anaphase II?
4. Would you assume that the length of a chromosome might have anything to do with the number of crossovers it might form with its homologue?
5. Two mitotic divisions may take place in as little as a few hours, whereas meiosis may take much longer. Can you suggest any reasons for this?

ADDITIONAL READING

DeRobertis, E. D. P., and E. M. F. DeRobertis, Jr. 1980. *Cell and molecular biology,* 7th ed. Philadelphia: W. B. Saunders Co.

Dyson, R. D. 1985. *Cell biology: a molecular approach,* 3d ed. Newton, MA: Allyn and Bacon.

Karp, G. 1984. *Cell biology,* 2d ed. Hightstown, NJ: McGraw-Hill Book Co.

Raven, P. H., R. F. Evert, and S. E. Eichorn. 1986. *Biology of plants,* 4th ed. New York: Worth Publishers, Inc.

Wolfe, S. L. 1981. *Biology of the cell,* 2d ed. Belmont, CA: Wadsworth Publishing Co.

Overview

*I*n this chapter the story of Karpechenko's crossing of a radish and a cabbage introduces the topic of genetics. This is followed by a brief history of Mendel and his classic experiments with pea plants. Each of Mendel's principles or laws is discussed, and examples of monohybrid and dihybrid crosses are given. The chapter then introduces the backcross, linkage, chromosomal mapping, and the Hardy–Weinberg law. A few basic principles of plant breeding conclude this section.

The discussion of evolution begins with an abbreviated history of Charles Darwin and the principles or tenets involved in the theory of evolution through natural selection. Mechanisms of evolution, including mutations, hybridization and introgression, polyploidy, apomixis, and reproductive isolation are discussed. The chapter concludes with an examination of the role of isolation and mutation in modifying Darwinism and with an examination of the evidence for evolution.

13 Genetics and Evolution

Some Learning Goals

1. Understand the significance of Mendel's experiments with peas.

2. Learn dominance, phenotype, genotype, homozygous, heterozygous, monohybrid cross, dihybrid cross, backcross, linkage, chromosomal mapping, the Hardy–Weinberg law.

3. Give the ratios of the offspring in the first two generations from a monohybrid and a dihybrid cross. Describe the genotypes involved.

4. Understand how introgressive hybridization may lead to the development of new species.

5. Know the contributions of Charles Darwin to theories of organic evolution and the tenets of natural selection as he understood them.

6. Explain the significance of mutation and reproductive isolation to evolution.

7. Give reasons, past and present, for the controversy over evolutionary theory.

Outline

Introduction

When a plant is being grown as a crop, the more parts of the plant that are edible or otherwise useful, the more efficient the use of the land on which the crop is grown. Some parts of many crop plants are not used, however, except possibly as ingredients in compost. In lettuce, for example, the roots are not useful for either food or fuel. In tomatoes and grapes, only the fruits are edible; in potatoes, only the tubers are used, and in cotton the fibers and the seeds constitute the useful parts. In 1928 the Russian biologist G. D. Karpechenko reported on an experiment in which he tried to cross two food plants in the cabbage family—the radish, with an edible root, and the cabbage, with edible leaves. One might assume that such a cross, if successful, would produce an ideal crop plant. Karpechenko found that these two plants, although related, did not cross readily, but he did succeed in obtaining a few seeds from such crosses. When these seeds were planted, the results were dramatic. Huge plants grew and produced fertile seeds of their own. From a crop point of view, however, the cross was a dismal failure because these large plants had leaves mostly like those of a radish and roots like those of a cabbage—they were virtually useless!

Such crossing experiments are a part of **genetics,** the science that deals with heredity, or natural inheritance. Genetics is one of the youngest of the biological sciences, having developed after the details of meiosis and mitosis became understood at the turn of the century. It has since become a vast and very important field of study, having the potential not only for improving food production in a hungry world but also for solving various problems related to the inheritable human diseases, the effects of radiation, the control of population growth, and many other matters of direct human interest and concern.

Mendel and His Work

Genetics as a science originated about a generation before its significance became appreciated in the scientific community as a whole. An Austrian monk, Gregor Mendel (figure 13.1), taught between 1853 and 1868 in what is currently the Czechoslovakian

FIGURE 13.1   Gregor Mendel. (Courtesy of Hunt Institute for Botanical Documentation, Carnegie Mellon University, Pittsburgh, PA)

city of Brno. In the monastery there, he carried on a wide range of studies and experiments in physics, mathematics, and natural history. He also became an authority on bees and meteorology and kept notes on experiments involving two dozen different kinds of plants. Today he is best known for the studies he conducted with a number of varieties of pea plants.

Mendel published the results of his studies on pea plants in a biological journal in 1866, and he sent copies of his paper to leading European and American libraries. He also sent copies to at least two eminent botanists of the time. His work, nevertheless, was completely disregarded until 1900, when three botanists (Eric von Tschermak of Austria, Carl Correns of Germany, and Hugo de Vries of Holland), working independently in their own countries, reached the same conclusions as Mendel. Each, as a result of library research, also came across Mendel's original paper.

Peas, unlike most flowering plants, are self-pollinating and self-fertile. In other words, a pollen grain of a pea flower is capable of germinating on its own stigma, fertilizing one of its own ovules, and developing a fertile pea seed. Although Mendel did not know about mitosis and meiosis, he had observed that if a pea plant grew to a certain height,

FIGURE 13.2   A cross between a tall variety and a dwarf variety of peas.

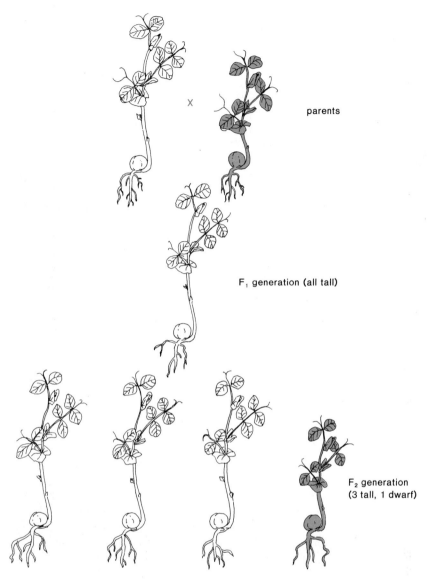

parents

F₁ generation (all tall)

F₂ generation
(3 tall, 1 dwarf)

its offspring, if grown under the same conditions, would always reach approximately the same height, generation after generation. He wondered if he crossed a tall plant with a short one if the offspring would be intermediate in height. To make such a cross he would need to prevent self-pollination by reaching into a flower of one plant and removing the stamens before the pollen had matured. Then he could take pollen from another plant and apply it by hand to the stigma of the first flower. He also needed to do something to prevent insects from accidently bringing pollen from other flowers after the cross had been made (e.g., cover the experimental flowers with a cloth or paper bag).

When Mendel made such crosses, the results were astonishing. All of the offspring were tall. There were no short or intermediate plants. He found, however, that if he allowed the offspring plants to pollinate themselves, their offspring, in turn, occurred in a ratio of approximately three tall plants to one short plant (figure 13.2).

Mendel then tried crosses between peas with smooth seeds and those with wrinkled seeds. He also crossed green-seeded plants with yellow-seeded ones, and ultimately used seven different pairs of contrasting characteristics in all.

TABLE 13.1
Summary of Mendel's Data

Parents	F$_1$	F$_2$
Yellow $\times$ green seeds	all yellow	6,022 yellow: 2,001 green
Smooth $\times$ wrinkled seeds	all smooth	5,474 smooth: 1,850 wrinkled
Green pod $\times$ yellow pod	all green	428 green: 152 yellow
Long stem $\times$ short stem	all long	787 long: 277 short
Axial flowers $\times$ terminal flowers	all axial	651 axial: 207 terminal
Inflated pods $\times$ constricted pods	all inflated	882 inflated: 299 constricted
Red flowers $\times$ white flowers	all red	705 red: 224 white

The original plants involved in making the crosses were referred to as the *parents,* or the *parental generation,* and their offspring were referred to as the *first filial generation (filial* is defined as "of or relating to a son or daughter"), usually abbreviated to F$_1$. The offspring of F$_1$ plants were called the *second filial generation,* or F$_2$. Mendel counted the number of offspring in each cross, something none of his predecessors had done. In addition, he studied the inheritance of single pairs of contrasting characteristics and succeeded where others had failed because he did not try to interpret inheritance in terms of whole organisms. Besides being astute and meticulous in his work, it was also fortunate that he chose the plants and the characteristics he did, because we presently know that many other characteristics are controlled by complex factors that do not give the simple, clear-cut results Mendel obtained. Mendel was also exceptionally fortunate in choosing peas for his experiments because, although he was unaware of it, peas have just 7 chromosomes in each set, and each of the pairs of characteristics he studied is carried on different homologous chromosomes. A summary of the data accumulated by Mendel in his work with peas is provided in table 13.1.

As Mendel began to collect data in his experiments, he realized that there must be something inside the pea plants that made them have yellow or green seeds and red or white flowers. He referred to this unknown agent as a *factor.* He also deduced that each plant must have two such factors for each characteristic, since even though all the offspring of an F$_1$ generation appeared more or less identical, the F$_2$ generation revealed some plants with one characteristic and others with the contrasting characteristic. These discoveries and deductions came to be a principle, or law, known as *the law of unit characters.* Stated simply, this law says that "factors,

which always occur in pairs, control the inheritance of various characteristics." These factors later became known as **genes** (discussed in chapter 3). We know now that there may be hundreds or even thousands of pairs of genes on each pair of homologous chromosomes (see chapter 12).

From his data and observations Mendel also deduced that one factor, or gene, of a pair may overcome or conceal the expression of the other. For example, in the F$_1$ generation of a cross involving yellow-seeded and green-seeded parents, all the plants had yellow seeds, yet both kinds of parents obviously had to contribute something to the cross since some of the F$_2$ plants had yellow seeds whereas others had green seeds. This deduction led to another principle known as the *law of dominance.* This principle says that "in any given pair of factors or genes, one may suppress or mask the expression of the other." The expressed factor is referred to as the **dominant** while the one not expressed is referred to as the **recessive**. In the yellow-seeded $\times$ green-seeded cross, the factor yellow is dominant and the factor green is recessive.

Mendel's crosses also made it clear that a plant's having yellow seeds did not indicate that both members of the pair of factors, or genes, controlling seed color were present, because the dominant could suppress or mask the presence of the recessive one. To distinguish between the appearance of an organism and its genetic components we use the terms **phenotype** and **genotype**. Phenotype refers to the physical appearance of the organism, and genotype refers to its genetic makeup. We customarily describe phenotypes with words and designate genotypes with letters. In Mendel's crosses, for example, seven pairs of phenotypes are shown as parents—these phenotypes include yellow seeds, green seeds, smooth seeds, wrinkled seeds, green pods, yellow pods, etc. In designating the genotypes of each of these plants, a capital letter is used to indicate the dominant

FIGURE 13.3   Absence of dominance (also referred to as *incomplete dominance* or *co-dominance*). A cross between a red-flowered and a white-flowered variety of snapdragons.

parents

red   white

F₁ (all pink)

pink

F₂

red   pink   pink   white

member of a pair of genes; the same letter in lowercase is used to indicate the corresponding recessive. In the yellow-seeded × green-seeded pea cross, for example, yellow is dominant. If a plant has a yellow phenotype, its genotype could be either **YY** or **Yy.** The green genotype would be **yy.** The plant is said to be genotypically homozygous if both members of a pair of factors, or genes, are the same (e.g., **YY** or **yy**) and heterozygous if the pair includes both types (e.g., **Yy**).

Although the principle of dominance was amply demonstrated by Mendel's crosses, we know now that in many other instances neither member of a pair of factors, or genes, completely dominates the other. In other words, in some species there is an absence of dominance. In snapdragons, for example, the F₁ plants of a cross between a red-flowered parent and a white-flowered parent are all pink-flowered because the characteristics of both factors blend, yielding an intermediate condition. When an F₂ generation is produced from such a cross, the flowers of the offspring appear in a ratio of 1 red: 2 pink: 1 white. This lack of dominance is also referred to as *incomplete dominance* or *co-dominance* (figure 13.3).

You will recall that the cells contributed by the parents in sexual reproduction (egg and sperm in higher organisms) are called gametes and that the offspring develop from *zygotes* formed when the gametes unite. You will also recall that gametes have only half the number of chromosomes present in the body cells of the organism. Each member of a pair of factors, or genes, controlling a given characteristic is located at a specific point on the chromosome, and the other member of the pair is found at the same precise point on a corresponding (homologous) chromosome in the cell. When meiosis occurs, the members of pairs of factors become separated and are not matched up again until a zygote is formed. Mendel called this separation of paired factors, or genes, into separate cells the *law of segregation.*

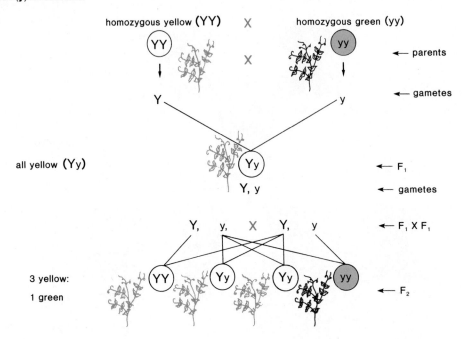

The Monohybrid Cross

With these first three laws in mind, let us examine a cross between a homozygous yellow-seeded pea plant and a homozygous green-seeded pea plant (figure 13.4). The genotype of the homozygous dominant parent (yellow-seeded) is **YY;** the genotype of the homozygous recessive parent (green-seeded) is **yy.** You will recall from chapter 12 that gametes have only one member of a pair of homologous chromosomes. The gametes of the yellow-seeded parent (**YY**), therefore, will be **Y;** the gametes of the green-seeded parent (**yy**) will be **y.** No matter which egg unites with any sperm of the other parent, all the zygotes of this cross will be heterozygous, with the genotype **Yy.** This means that all the individuals of the F_1 generation will be yellow-seeded phenotypes, and all will be, like the zygotes, heterozygous. A cross of this type, involving two parents differing by only a single genetic trait, is called a **monohybrid cross.**

It is customary when making such crosses to try to produce an F_2 generation by crossing members of the F_1 generation with one another. Two kinds of gametes are produced in equal numbers by each F_1 parent. All of the F_1 plants involved are yellow-seeded phenotypes, and their genotypes are all **Yy.**

When gametes are produced, half will be **Y** and half will be **y.** Thus, a **Y** egg may unite with either a **Y** sperm nucleus, producing a **YY** zygote, or, purely at random, the same egg may unite with a **y** sperm nucleus, producing a **Yy** zygote. The same type of random unions occur with **y** eggs so that either **Yy** or **yy** zygotes are produced. If all the offspring of a large number of such crosses are counted, it will be seen that they have been produced in a ratio of 1 **YY:** 2 **Yy:** 1 **yy** genotypes and the ratio will be approximately 3 yellow-seeded phenotypes to 1 green-seeded phenotype. These 4 equally possible F_2 offspring are shown at the bottom of figure 13.4.

The Dihybrid Cross

Thus far, the crosses we have been discussing have involved single pairs of factors, or genes. But, as we have noted, homologous chromosomes may contain hundreds or even thousands of pairs of genes. If crosses are made involving parents that are homozygous for two or more pairs of genes, the nature of meiosis dictates that the genes will not necessarily be inherited together. In fact, the two or more sets of genes are separated when meiosis occurs and, if they are on separate chromosomes, they are combined again in the formation of zygotes. Another law of Mendel's, the *law of independent assortment,* recognizes this and states that the "factors (genes) controlling two or more contrasting pairs of characteristics segregate independently and the gametes combine at random." This law is illustrated in **dihybrid crosses,** which involve two pairs of genes located on separate pairs of chromosomes.

Let us assume that the factors for yellow seeds and green seeds are located on one pair of chromosomes while those for tall plants and dwarf plants are located on another pair. You will recall that tallness in peas is dominant and dwarfness is recessive. The phenotypes of the homozygous dominant parent will be yellow and tall while those of the recessive parent will be green and dwarf; their corresponding genotypes will be **YYTT** and **yytt.** All the gametes from the dominant parent will be **YT**; all those from the recessive parent will be **yt.** The dihybrid genotypes of the zygotes will all be **YyTt** and the phenotypes of the F_1 generation will all be yellow and tall. If these dihybrid members of the F_1 generation are then crossed with one another to produce an F_2 generation, 4 kinds of gametes will be produced in equal numbers by each parent because of the independent segregation that occurs during meiosis. These gamete genotypes are **YT, Yt, yT,** and **yt.** Since any 1 of the 4 kinds of gametes can unite randomly with any of the 4 kinds of gametes of the other parent, 16 combinations are possible.

In order to avoid the confusion of having to remember all possible combinations of gametes, a *Punnett square* is used to determine the genotypes of the zygotes. The Punnett square diagram looks somewhat like a checkerboard, with one set of gametes across the top and the other set down one side (figure 13.5). If the genotypes of the zygotes are added up, it can be seen that there are nine different kinds in a ratio of 1 **YYTT**: 2 **YYTt**: 2 **YyTT**: 4 **YyTt**: 1 **YYtt**: 2 **Yytt**: 1 **yyTT**: 2 **yyTt**: 1 **yytt.** Four kinds of phenotypes are produced from these genotypes in a ratio of 9 yellow, tall: 3 yellow, dwarf: 3 green, tall: 1 green, dwarf. These ratios are expected when large numbers of individuals are involved. With small numbers, chance alone may not produce the expected ratios. Human populations, for example, are divided relatively equally into males and females, but in smaller groups such as families, the general population ratio of one male to one female may not be evident.

The Backcross

If a scientific theory is valid, one should be able to test it experimentally. Mendel himself tested his predictions by means of **backcrosses.** He had found in pea flowers that red color is dominant and that white is recessive (as contrasted with snapdragons in which color dominance is lacking), so that all the F_1 offspring of a cross between a homozygous red-flowered parent and a homozygous white-flowered parent would be red. The genotypes of such a cross would be **RR** (red) × **rr** (white), and the F_1 generation would be all **Rr.** The gametes of such F_1 offspring would be either **R** or **r,** with each type being produced in equal numbers. Thus, if pollen from an F_1 hybrid is placed on the stigma of a white-flowered plant (which, since it shows the recessive trait, has to be homozygous), there is one chance in two that a dominant (**R**) gamete will form a zygote with a recessive (**r**) one and an equal chance that two recessive (**r**) gametes will form a zygote, with the offspring being produced in equal numbers of heterozygous red (**Rr**) and white (**rr**). Such backcrosses involving the homozygous recessive parent and the F_1 offspring are routinely made today, and invariably they yield the same results, namely, a 1:1 ratio of phenotypes. The same 1:1 ratios are produced even when there is no dominance. In snapdragons, for example, when the pink-flowered F_1 offspring are backcrossed with a red-flowered parent, a ratio of 1 red: 1 pink results. If the white parent is used, the backcross produces a ratio of 1 white: 1 pink.

Linkage

There may be thousands of genes on each chromosome, and the closer the genes are to one another on any given chromosome, the more likely they are to be inherited together. The tendency of genes on the same chromosome to be inherited together is called **linkage.**

In 1906, just a few years after the basic details of meiosis became known, W. Bateson and R. C. Punnett (after whom the Punnett square is named) of Cambridge University in England became the first to report on linkage in sweet peas. They knew from earlier work with sweet peas that purple flower color was dominant and red was recessive; they also knew that pollen grains with a shape that was oblong in outline ("long pollen") were dominant and that

FIGURE 13.5 A dihybrid cross between a yellow-seeded,
tall pea plant and a green-seeded, dwarf pea plant.

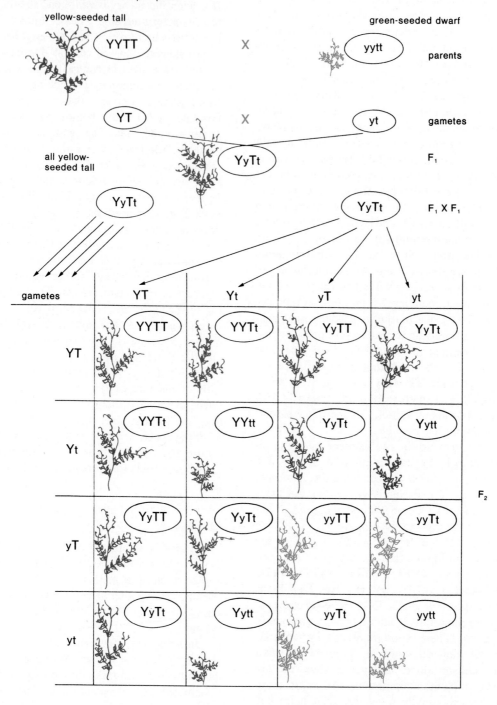

spherical pollen grains ("round pollen") were re-cessive. When they crossed a homozygous purple, long plant with a homozygous red, round plant, all the F_1 offspring were purple, long, as expected. When they crossed the F_1 plants with one another to obtain an F_2 generation, however, the F_2 offspring pheno-types were produced in numbers that differed mark-edly from the expected 9:3:3:1 ratio. Puzzled by these results they tried a backcross, crossing F_1 plants (heterozygous purple, long) with the recessive parent (homozygous red, round). Again, the expected 1:1:1:1 ratio was not produced. Instead they ob-tained a ratio of 7 purple, long: 1 purple, round: 1 red, long: 7 red, round. Bateson and Punnett could not explain adequately what they had observed. In 1910, however, T. H. Morgan, who had observed similarly puzzling ratios in his experiments with fruit flies, theorized that linkage and crossing over were responsible, an explanation that is still considered valid today.

If the genes for purple and red, and long and round, were on separate pairs of chromosomes, the genotypes of the F_1 would all be **PpLl** and the homo-zygous recessive would be **ppll**. This should have produced 1 **PpLl**: 1 **Ppll**: 1 **ppLl**: 1 **ppll** instead of the 7:1:1:7 ratio actually obtained. Apparently the F_1 parent produced many more **PL** and **pl** gametes than **Pl** and **pL** gametes. If we designate the parents in the backcross $\frac{PL}{pl}$ and $\frac{pl}{pl}$ to indicate that **P** and **L** are on one chromosome and **p** and **l** are on another, we would expect the F_1 parent $\frac{PL}{pl}$ to have produced just two kinds of gametes, **PL** and **pl,** and the back-cross to have produced only two kinds of pheno-types, purple, long and red, round in equal numbers. But some purple, round and red, long phenotypes also were produced. The only plausible explanation for this seems to be that some *crossing over* (see pro-phase I of meiosis in chapter 12) occurs, with **P** and **L** genes occasionally being switched with **p** and **l** genes.

Chromosomal Mapping

As a result of thousands of experiments with many organisms, particularly those with fruit flies and with corn, it is currently known that crossing over be-tween particular pairs of linked genes occurs regu-larly. The frequency of crossing over varies widely, depending on the genes involved. We must conclude, therefore, that each gene has a specific location (locus) on a chromosome. Indeed, by calculating percentages of crossing over through counting large numbers of offspring phenotypes, we can effectively map the relative locations of genes on chromosomes. If, for example, we assume that the space within which 1% crossing over occurs is one unit of map distance between linked genes, then 7% crossing over between genes A and B would mean that they are 7 units of map distance apart on their chromosome.

A ◄——— 7 units ———► B

If we find there is 10% crossing over between A and a third linked gene C, and that B and C are shown by crossover percentages to be 17 units apart, we can deduce that A lies between B and C on its chro-mosome.

C ◄——— 10 units ———► A ◄——— 7 units ———► B
17 units

However, if there is only 3% crossing over between B and C, we would have to conclude that B lies be-tween A and C.

C ◄——— 10 units ———► A
◄ 3 units ►B ◄——— 7 units ——►

In Bateson and Punnett's backcross, 2 of every 16 F_2 offspring were produced through recombina-tions due to crossing over, this two-sixteenths being 12.5%. We can deduce from this that the genes con-trolling flower color and pollen shape in sweet peas are 12.5 map units apart on the chromosomes. The

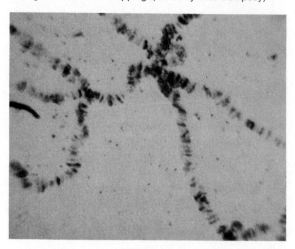

locations of many hundreds of pairs of genes have been determined and mapped in corn and in fruit flies (figure 13.6), and the mapping of chromosomes in other organisms is currently being undertaken at institutions around the world.

The Hardy–Weinberg Law

Before Mendel's work became known, most biologists believed that inherited characteristics were a blending of those furnished by the parents, and it was difficult to understand why unusual characteristics did not eventually become so diluted that they essentially disappeared. After Mendel, biologists asked why dominant genes did not eventually completely eliminate recessive ones in breeding populations. G. H. Hardy, a British mathematician, and W. Weinberg, a German physician, pointed out the reason in 1908, and their observation became known as the *Hardy–Weinberg Law.*

The Hardy–Weinberg law, which essentially specifies the criteria for genetic equilibrium in large populations, states that "the proportions of dominant genes to recessive genes in a normally interbreeding population will be retained from generation to generation unless there are forces that change those proportions." In small populations, there can be a random loss of individual genotypes through,

for example, a failure to breed, or if breeding is not totally random with respect to genotype. This in itself can lead to loss of a gene. The proportions can also be upset by a particular gene mutating to another form and then mutating back again to its original form at a slower rate. But *selection* is by far the most significant cause of exceptions to the Hardy–Weinberg law. The manner in which selection operates is discussed in the section on evolution, which follows.

PLANT BREEDING

Most of the varieties of crop, ornamental, and other economically important plants available today have been developed from wild ancestral parents through controlled plant breeding. Such breeding appears to possess significant potential for reducing world hunger and improving renewable natural resources in the future. This, in turn, has led a number of field botanists to team up with anthropologists, medical doctors, and interpreters to explore for new sources of wild plants. One activity of such teams involves the interviewing of native medicine men and indigenous tribesmen in the tropics of the Americas and Africa in attempts to glean information about their utilization of local plants. Tasks of this kind have taken on a measure of urgency as undescribed species of tropical plants and other organisms are being lost daily to large-scale clearing of forests and savannahs, and the destruction is eliminating unknown numbers of plants with possible potential for food, medicines, fiber, and other uses. When the teams have sifted through all the gathered information, they hope to save significant material of new *gene pools* (a gene pool consists of all the genes of all the individuals in a population) for use in breeding better-yielding crop plants with improved disease resistance. In the case of food plants, they also hope to improve the quality and quantity of protein produced by the plants. There are several ways of improving existing varieties and of developing new ones. The basic methods include *hybridization, polyploidy, mutation,* and *tissue culture.*

Hybridization

When different varieties (or in some cases, species) are crossed, the F_1 generation often displays *hybrid vigor,* with the plants being bigger or better-yielding than either parent. The effect usually disappears in the next generation, however, so that F_1

seeds need to be planted each season to produce predictable results. Most corn grown in North America today comes from F_1 hybrid seed. When a new variety is being developed through hybridization, the experimental work is carried through at least two, and more often up to seven or more generations before development is complete. During the experiments, backcrossing is used to reinforce in the offspring desirable characteristics of one parent, with *inbreeding* (brought about by repeated self-pollination) and managed *outcrossing* (cross-pollination between individuals of the same variety or species) normally playing significant roles. Repeated inbreeding develops homozygous purebred strains, and as undesirable recessive genes come together, yields and vigor may decline. Through careful selection, however, desirable dominant and recessive genes are retained. Then pure strains are crossed, often resulting in vigorous F_1 hybrids; these hybrids, in turn, may be crossed with one another to produce even more vigorous double hybrids.

Since cultivation of wheat began thousands of years ago in the Mediterranean region, literally hundreds of different varieties have been developed. One widely used group of bread wheat varieties apparently originated from a hybrid formed between a species that had 28 chromosomes ($n = 14$) and a grass that had 14 chromosomes ($n = 7$). The hybrid was sterile because the cells contained non-homologous chromosomes that could not pair properly during meiosis, but a doubling of the chromosomes followed, presumably when chromosomes failed to separate during mitosis or meiosis, resulting in a fertile hybrid with 42 ($n = 21$) chromosomes. Many varieties were bred to counteract the susceptibility of wheat to black stem rust, a fungus that has cost farmers millions of dollars in losses. New resistant varieties constantly must be developed as older varieties lose their resistance to mutant rust strains that seem to evolve almost as fast as resistant wheats can be produced.

Norman Borlaug and the Green Revolution

Norman Borlaug (figure 13.7), who is known as the "Father of the Green Revolution," was awarded a Nobel Prize in 1970 for developing new strains of wheat in Mexico. The new high-yielding strains, which were produced from crosses with a dwarf variety from Japan, were widely planted. Prior to the beginning of the project in 1944 Mexico had imported wheat for many years, but the new varieties

FIGURE 13.7 Norman Borlaug, the "Father of the Green Revolution." (USDA Photo)

were so successful that within 20 years Mexico had quadrupled its wheat production and had sufficient to export to other countries. India and Pakistan have experienced similar gains.

Although the "Green Revolution" has been a dramatic success, the success has not been unqualified. The high-yielding crops generally perform well only in areas where irrigation water is available and require extensive applications of fertilizer, most of which are produced from fossil fuels. In 1973 when OPEC brought about a major increase in the cost of fossil fuels, the agricultural areas of the world discovered that the increased costs suddenly changed what had been profits into deficits. In addition, the agricultural methods employed had not taken into account the longer term effects of farming without appropriate ecological considerations (see the "Humans in the Ecosystem" section in chapter 23), and the growing areas had become degraded with pesticides, herbicides, salts, and topsoil erosion. Since 1973, numerous small farmers have been forced out of business, and there is much debate over how to proceed with food and energy production in the coming decades.

Polyploidy

We noted in chapter 12 that any cell that has 1 set of chromosomes (e.g., a gamete) is *haploid*, and that any cell that has 2 sets of chromosomes (e.g., a zygote) is *diploid*. But many plants, including about half of all the flowering plants, have more than 2 sets of chromosomes in their cells and are therefore **polyploid.** Polyploids that have 3, 4, 6, or 8 sets of chromosomes are said to be *triploid, tetraploid, hexaploid,* or *octoploid,* respectively. Polyploidy can arise in various ways. Hybridization followed by a doubling of the chromosome number, as exemplified by the bread wheat mentioned above, results in a polyploid called an *alloploid*. The chromosome sets can also be doubled when pairs of chromosomes fail to separate during mitosis or meiosis. The resulting polyploid is called an *autoploid;* a diploid plant could become a tetraploid by either alloploidy or autoploidy.

Since tetraploids are often bigger or more vigorous than their diploid counterparts, plant breeders sometimes try to develop tetraploids artificially by treating seeds with *colchicine,* an alkaloidal drug derived from the corms of the autumn crocus. The colchicine interferes with spindle formation during mitosis so that pairs of chromosomes do not separate during anaphase, and if the cell with the double set of chromosomes survives and becomes functional, a tetraploid plant may be the ultimate result. Developing tetraploids with colchicine usually involves treating seeds with just enough colchicine (usually 0.1% to 0.3%) to kill most of them, and then checking for the small percentage (usually about 5%) of surviving seedlings whose cells have twice the original number of chromosomes. Some tetraploids are artificially induced without the use of colchicine. In tomatoes, for example, wounding of the plant causes it to produce *callus* tissue whose cells may have double the basic complement of chromosomes. In corn, tetraploids may be produced by applying heat to seeds. See the additional discussion of polyploidy under the "Role of Hybridization in Evolution" section later in this chapter.

Mutation

Mutations (see the "Mechanisms of Evolution" section later in this chapter), which involve a change in a gene or chromosome and occur naturally in all living organisms, can be artificially induced by radiation and chemicals. Like those occurring naturally, artificially induced mutations are largely harmful, and the expense of screening large populations for a rare desirable mutant form generally is not practical. A few useful modifications have, however, been obtained through induced mutation. These include higher yielding strains of *Penicillium* mold (see chapter 17), blight resistant varieties of oats, and an improved variety of grape whose individual fruits previously were bunched too compactly.

Tissue Culture

Tissue culture, which is discussed in appendix 4, offers considerable potential for crop improvement in that selection for desirable traits can be accomplished in large populations of cells cultured in a few glass containers instead of having to deal with populations of plants on large acreages of land. Manipulation of the cells often involves subjecting them to a variety of stresses such as herbicide and other poisons, heat, cold, disease bacteria and fungi, radiation, etc., and then isolating cells that survive from among those that are susceptible. Tissue culture also lends itself to various *genetic engineering* techniques (see the section on "Genetic Engineering" in chapter 15), including the transfer of organelles from one cell to another, and the hybridization of cells through *protoplast fusion.* This latter technique involves the digestion of cell walls with enzymes, leaving naked protoplasts. A small percentage of the protoplasts actually undergo fusion in the culture, resulting in "hybrid" cells with genes from both parent cells. The hybrid cells are isolated from those with identical genotypes by culturing them on a medium on which only the hybrid cells can grow.

EVOLUTION

Introduction

Few historical events since 1859 have had a greater impact on society in general and the biological sciences in particular than Charles Darwin's book, *On the Origin of Species by Means of Natural Selection, or the Preservation of Favoured Races in the Struggle for Life.* Darwin's theory of evolution through "natural selection" has stimulated an enormous amount of thinking and research, and has provided an explanation based on natural laws for the diversity of life around us. Initially, the theory caused a great deal of controversy. Although Darwin had a life-long belief in a Divine Creator, he also believed that the Creator had used natural laws to develop gradually the great diversity of all living things over aeons of time. Most of his contemporaries, however, were guided by a literal interpretation of the biblical account of creation, and were convinced that all living things had been created in six days and had existed unchanged since the beginning. Although the controversy has subsided, some disagreement still exists today. When the popular *Scofield Reference Bible* was first published in 1909, it included in the margin opposite the account of creation "4004 B.C.," a date arrived at by the seventeenth-century Irish archbishop James Ussher who based his calculation on faulty interpretation of biblical genealogies. Ussher's date has been deleted from the editions published since 1967; the editors have observed that little evidence exists for fixing dates of biblical events prior to 2100 B.C. Many *scientific creationists,* who since the 1970s increasingly have sought to have a non-evolutionary interpretation of the living world included in public school biology textbooks, do not necessarily believe the earth was created in 4004 B.C., nor do they refuse to recognize the existence of minor variations in living organisms. The majority, however, believe the earth is not more than 30,000 years old, and reject the foundations of evolution as incompatible with a literal interpretation of the biblical account of creation. In doing so, scientific creationists reject most of the evidence for evolution, including that accumulated since Darwin.

Part of the remaining disagreement also stems from a failure of people in diverse fields to define terms and to distinguish between fact and theory in evolutionary matters. Evolution itself, for example, has been broadly defined by some as simply being synonymous with change. We are told that anything from cars to computers to cultures is evolving and that even our thought processes evolve. Thus, to distinguish between change in inanimate or intangible entities and progressive change in living organisms over time, we need to refer to the latter as *organic evolution.* Even then we do not pin down the subject entirely, for there are variations in organic evolutionary theory. That which pervades and unifies most biological thought today, however, finds its origin in the observations of Charles Darwin (figure 13.8) and of a contemporary of his, Alfred Wallace, who independently arrived at the same conclusions as Darwin.

Charles Darwin

Charles Darwin was born in Shrewsbury, England, in 1809. His father, Dr. Robert Darwin, was a relatively wealthy, successful country physician. In 1825, at the age of 16, young Darwin was sent to the University of Edinburgh Medical School to follow in his father's footsteps. He did not do well in his studies, however, and dropped out after two years. The following year he went to Cambridge University to study for the ministry but did not excel academically in this field either. Part of the reason

FIGURE 13.8   Charles Darwin. (Courtesy National Library of Medicine)

for his poor scholastic showing apparently was due to the fact that he spent much time in the countryside collecting beetles, rocks, and other specimens or talking at length with his Cambridge biology professors, who held him in high regard. He barely attained what today could be called a C average, but he managed to graduate in 1831 with a theology degree. He was then 22 years old but was still not really sure what he wanted to do with his life. While he was pondering such matters, King William the IV of England commissioned a sailing vessel, the H. M. S. *Beagle,* to undertake a voyage around the world for the purpose of charting coastlines, particularly those of South America. Young Darwin was recommended for the (unpaid) post of naturalist on the voyage by his Cambridge biology professors. His father was opposed to the idea, but his uncle sided with him and he accepted the position.

The voyage, which began December 27, 1831, and took five years, gave Darwin an opportunity to collect specimens on both sides of South America, as well as in the Galápagos Islands and along the coasts of Australia and New Zealand. He kept a daily journal and spent countless hours collecting specimens. This gave him ample opportunity to think about the forms and distribution of the myriad new organisms he encountered. His thoughts slowly led to the development of ideas that later blossomed into his theory of evolution through natural selection.

Upon his return to England in 1836, Darwin obtained the financial support of his father and his cousin Emma, whom he married. Although still young, he retired to the country and began working on his collections and journal. He also carried on a voluminous correspondence with other biologists and made extensive investigations into pollinating mechanisms, earthworm ecology, geographical distributions of plants and animals, and several other areas of natural history. Throughout all of his activities he was guided by a concept that he had adapted from an essay written by Malthus in 1798 on human populations and food supplies. Darwin realized that although human beings might artificially improve or increase their food supply through selective breeding and cultivation, plants and animals could not do so and were therefore vulnerable to a process of selection in nature, which would explain changes in natural populations. He was reluctant to publish his ideas, however, and did not begin putting together his book on the origin of species until 1856.

Meanwhile, an English naturalist by the name of Alfred R. Wallace (1823–1913), who made major contributions to our knowledge of animal geography, independently concluded that natural selection contributed to the origin of new species and in 1858 sent Darwin a brief essay on the topic. Urged by friends, Darwin published a statement of his views jointly with Wallace, and in 1859 Darwin's classic book *On the Origin of Species by Means of Natural Selection* was published.

Tenets of Natural Selection

In essence, Darwin's theory of evolution through natural selection is based on four principles, or tenets:

1. *Overproduction.* Many living organisms produce enormous quantities of reproductive cells or offspring. For example, a single maple tree produces thousands of seeds each year, most being capable of becoming a new tree. Some fungi produce trillions of spores, each with the potential to become a new fungus.

2. *Struggle for Existence.* All the germinating seeds, spores, and other reproductive cells of living organisms "compete" for available moisture, light, nutrients, and space. The amounts of these elements available in nature are not sufficient to support all of these organisms, and as a result many die.

3. *Inheritance and Accumulation of Favorable Variations.* All living organisms vary. Those hereditary variations with survival value are inherited from generation to generation and accumulate in time, while other variations not important to the survival of the species are gradually eliminated.

4. *Survival and Reproduction of the Fittest.* Those forms of organisms best adapted to the environment have the best chance to survive and reproduce, while others less well adapted may succumb. A tree with thicker bark, for example, has a variation favorable to surviving cold temperatures; it hence may survive to reproductive age and therefore its offspring will bear the inherited features of the thicker-barked tree.

One of the criticisms of Darwin's theory, published in 1859, was that it did not explain how hereditary variations originated and developed. It should be remembered, however, that Mendel's findings were not published until 1866, and the details of mitosis and meiosis did not become known until 1900–1906. Today, with our far greater knowledge of how variations occur and are inherited, we are more able to understand the mechanisms of evolution in populations than was possible in Darwin's time.

Populations of organisms from aardvarks to zinnias are composed of a few to billions of individuals, yet even within very large populations it is virtually impossible to find two that are identical down to the last molecule. As humans, we are well aware of this within our own species, but variation exists in all living organisms, even though in simpler one-celled forms the differences may be much less obvious. Whether the differences are obvious or subtle, however, if the environment or other factors favor certain hereditary variations and bring about the elimination of others, the general characteristics of a population will eventually change.

In some instances, the environment may alter the phenotype slightly without affecting the genetic constitution of an organism. Plants that grow relatively tall at sea level may become dwarfed when they are transplanted to cooler areas in the mountains or drier areas near deserts, yet they are capable of breeding freely with plants at the original location if they are returned to that area. If one applies fertilizers to plants, stimulates their growth with hormones, or prunes them, the changes are not passed on to the offspring, for no permanent change occurs in a given population unless there is *inheritable variation*. The changes in transplanted, fertilized, or pruned plants are not transmitted to the offspring because the gametes of those plants will carry the same genetic information they would have carried if the transplanting, fertilizing, or pruning had not occurred.

Before the details of genetics became known, many prominent biologists believed that hereditary changes in populations over long periods of time (evolution) occurred as a result of the inheritance of acquired characteristics. One of the more prominent supporters of this widespread idea was Jean-Baptiste Lamarck (1744–1829). He believed that giraffes attained their long necks over many generations as a result of the gradual increase in neck length as shorter-necked animals stretched to reach leaves on the branches of trees. Slight stretching of the neck was supposed to have been passed on to the offspring as it occurred, and eventually numerous tiny increases due to individual stretching added up to the present great neck-length of giraffes. If such were the case, one should be able to demonstrate it experimentally, and indeed, attempts were made, but all failed. For example, one such attempt involved the surgical removal of the tails of mice for many successive generations, but the mice of the last generation had tails of exactly the same average length as those of the first generation, because the repeated removal of the tails in no way affected the hereditary characteristics carried in the genes within the cells. Likewise, fruit trees that are pruned annually never produce even slightly dwarfed tree seeds as a result, even after many generations.

Mechanisms of Evolution

Despite the experiments just discussed, short-tailed mice and dwarf fruit trees do occasionally occur, but for reasons quite different from those proposed by Lamarck and his contemporaries. They come about

FIGURE 13.9 Some types of chromosomal changes that can occur. *A.* **Deletion.** Part of a chromosome breaks off and is lost. *B.* **Translocation.** Part of one chromosome becomes attached to another. *C.* **Inversion.** Part of a chromosome breaks off, becomes inverted from its original position, and then is reattached.

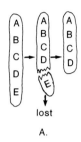

A.

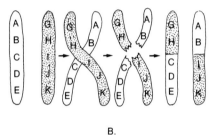

B.

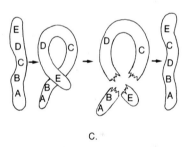

C.

as a result of a sudden change in a gene or chromosome. Such a change is called a **mutation,** a term introduced in 1901 by the Dutch botanist Hugo de Vries. Changes within chromosomes may occur in several ways. A part of a chromosome may break off and be lost (*deletion*), or a piece of a chromosome may become attached to another (*translocation*). In some instances a part of a chromosome may break off and then become reattached in an inverted position (*inversion*) (figure 13.9). A mutation of a gene itself may involve a change in one or more nucleotide pairs (see chapter 3). Mutation rates vary considerably from gene to gene, but mutations occur constantly in all living organisms at an average roughly estimated to be about 1 mutant gene for every 200,000 produced. *Mutator genes* that increase the rate of mutation in other genes have been discovered, but generally the mutation rate for a specific gene remains relatively constant unless changes in the environment (e.g., increase in cosmic radiation) occur. Most mutations are harmful, many times to the point of killing the cell. About 1% or fewer are either harmless or produce a characteristic that may help the species survive changes in its environment. Modern theories of evolution extend Darwin's natural selection theory by recognizing that mutation and recombination of genes through hybridization are fundamental to bringing about the changes that result in new species. They also include in the process *reproductive isolation* (discussed in a later section).

Role of Hybridization in Evolution

Hybridization (the production of offspring by two parents that differ in one or more characteristics) seldom occurs naturally, but when it does occur the hybrids may be significant or important in evolutionary change, depending on how the characteristics of the parents were combined. If, for example, the environment changes (e.g., average temperatures drop or annual precipitation increases), hybrids may have gene combinations that are better suited to the new environment than those of either parent. If two species hybridize occasionally, *introgression* (backcrossing between the hybrids and the parents) may occur. If the backcrossing occurs repeatedly, some characteristics of the parents may disappear eventually from the population if the new combinations of genes in the offspring happen to be better suited to the environment than those of the parents, and as natural selection favors the offspring. Both parents and hybrids may, however, also evolve in other ways.

The phenomenon of *polyploidy* occurs occasionally in nature when during mitosis a new cell wall fails to develop between two daughter cells even though the chromosomes have divided. As observed during our discussion of plant breeding, this situation can result in a cell with twice the original number of chromosomes. If mitosis were to occur

normally after such a cell was formed and the cell divided repeatedly until a complete organism resulted, that organism would have double the original number of chromosomes in all its cells. The hybrids resulting from a cross between two species are often sterile because the chromosomes do not pair up properly in meiosis. If polyploidy does occur in such a hybrid, however, the extra set of chromosomes present from each parent offers an opportunity for any one chromosome to pair with its homologue in meiosis, and thus possibly the problem of sterility may be overcome. This type of polyploidy apparently occurred frequently in the past (in terms of geological time), and it is believed that more than 100,000 species of today's flowering plants originated in this fashion.

Sterile hybrids also may reproduce by asexual means. Asexual propagation, called *apomixis*, may include the development and production of seeds without fertilization, as well as other forms of vegetative reproduction discussed in chapters 4–7 and in appendix 4. When species reproduce principally by apomixis, but sometimes also hybridize so that new combinations of genes are occasionally produced, they can be highly successful in nature. Dandelions and wild blackberries, for example, reproduce apomictically, as well as by other means, and are among the most successful plants known.

Reproductive Isolation

If new genes are produced in a freely interbreeding population they will gradually be spread throughout the population and the nature of the whole population will change in time. If some barrier divides the population, however, the two new populations eventually may become distinct from each other, sometimes in a relatively short period of time. The log of a Portuguese sailing vessel of more than 600 years ago indicates that, for unknown reasons, some rabbits from Portugal were released on one of the Canary Islands during a visit. When twentieth-century biologists examined the island descendants of those rabbits, which in Portugal forage during the day, they found them to be smaller than their continental ancestors, to be nocturnal in foraging habits, and to have larger eyes. In addition, attempts at breeding them back to their European ancestors failed because in the short space of 600 years a new species of rabbit had evolved. How do two populations of organisms that initially have the same gene pool come to have gene pools different enough to prevent their interbreeding?

Geographic or other isolation of the two populations from each other prevents the flow of genes between the two populations, and random mutations (which are rarely identical) then spread only throughout the population in which they arise. In time the genetic changes may become so great that even when the isolation is removed, gene flow between the two populations no longer can occur. In the United States there are two closely related species of small trees or shrubs called *redbuds* that look very much alike. The eastern redbud occurs on the borders of streams, mostly east of the Mississippi River between the Canadian border and Florida; the western redbud is native to California, Utah, Nevada, and Arizona. The two species can be artificially hybridized, but each is so adapted to its own wild habitat and associated climate that specimens of either species transplanted to the other's wild habitat soon succumb.

Several other factors contribute to the development of new species from geographically isolated populations with a common ancestry. When separation first occurs it is most unlikely that both populations will have genes that are identical in all respects, and a small population will have only a small percentage of the genetic variation present throughout the original population. In addition, geographically isolated populations normally will be subjected to selection pressures from numerous subtle to conspicuous differences in environment. The Canary Island rabbits, for example, initially probably found it difficult to compete with other animals for food during the day, and as mutations for improved night vision occurred those acquiring the genes were able to survive while those without them perished.

Isolation leading to the development of new species is not limited to physical barriers such as mountains or oceans, or to climate. Soils may play a role, or a mutant form within a population may flower at a different time, preventing exchange of genes between it and nonmutant forms. In the temperate deciduous woods of eastern North America and in the Columbia River Basin in the Pacific Northwest there are many populations of early spring-flowering herbs called Dutchman's breeches

(discussed further in chapter 14). These have delicate, highly dissected leaves, which are nearly indistinguishable from those of related squirrel corn plants often found growing right among the Dutchman's breeches in eastern North America (see figure 14.1). So closely do the plants resemble each other vegetatively that early botanists and many lay persons assumed they were the same species. It is believed, however, that at some point in geological history a mutation or mutations occurred that caused some plants to begin flowering after other plants had already set seed. As a result, one group became reproductively isolated from the other group (figure 13.10). In due course, other mutations affecting the form and fragrance of the flowers and the shape and pigmentation of food-storage bulblets beneath the surface also occurred, but the plants continued to occupy the same habitats. In short, we now have two closely related but distinct species growing together without interbreeding because it is no longer possible for them to do so.

Other isolating mechanisms may be mechanical. In orchids, for example, the pollen is usually produced in little sacs called *pollinia* (see figure 23.25) that stick to the heads or bodies of visiting insects. If pollination is to occur, the pollinia must be inserted within a concave stigma. Each species of orchid is constructed so that it is highly unlikely that a pollinium of one species will be inserted within the stigma of another species. As a result many species of related orchids can be *sympatric* (occupy the same territorial range) without genes being exchanged. In fact, four closely related sympatric species of Peruvian *Catasetum* orchids, which can easily be artificially hybridized, have no known natural hybrids, despite their being pollinated by a single species of bee. Microscopic examination of the pollinators has shown that the pollinium of one species is attached to the insect's head, another is attached to the insect's back, a third to the abdomen, and the fourth only to the left front leg, and even after visits to hundreds of flowers none of the pollinia is misplaced! Even if pollen from one species is placed on or within the stigma of another species, however,

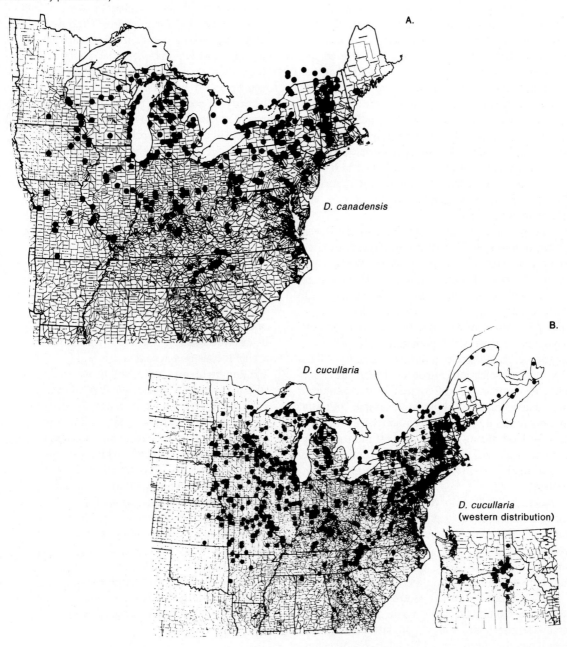

FIGURE 13.10   Geographical distribution of two species of
Dicentra. A. Distribution of *D. canadensis* (squirrel corn).
B. Distribution of *D. cucullaria* (Dutchman's breeches). (After
Stern, K. R. 1961. "*Revision of Dicentra* (Fumariaceae)."
Brittonia 13(1):1–57. © The New York Botanical Garden.
Reproduced by permission.)

fertilization frequently does not follow because the sperm is chemically or mechanically prevented from reaching the egg. Other isolating mechanisms include the failure of hybrids to survive or breed, and failure of embryos to develop.

Rates of Evolution

Darwin believed that evolution by natural selection was a slow and gradual process. A number of contemporary biologists, however, favor variations of *punctuated equilibrium* theories, which hold that major changes have taken place in spurts of maybe 100,000 years or less, followed by periods of millions of years during which changes, if any, have been minor. They base their theories on fossil records, which have large "chains" of missing organisms. Although missing link fossils are occasionally discovered, the record does little to support Darwin's concept of gradual, long-term change, even though it is widely acknowledged that possibly as few as one organism in a million ever became a fossil. Others opposed to theories of evolution through sudden change argue that because such a tiny percentage of organisms becomes fossilized, and usually only the harder parts of organisms (e.g., bones, teeth) are preserved, drawing definite conclusions from fossil evidence about evolution through either gradual or sudden change is not warranted. As indicated in chapter 19, the conditions necessary for an organism to become a fossil are very specialized and limited in occurrence, and probably also were in the past. This makes it quite improbable that large numbers of missing link fossils will ever be found. Proponents of evolution through periods of rapid change argue that under conditions of changing climates or other situations exerting strong selection pressures on forms with adaptive mutations, new species of organisms could arise in less than 100 generations, making 100,000 years ample time for considerable evolution to occur. Since it is not possible to prove or disprove the various theories experimentally, the debate on evolution rates will undoubtedly continue indefinitely, until or unless new evidence convincingly supports one theory more than another.

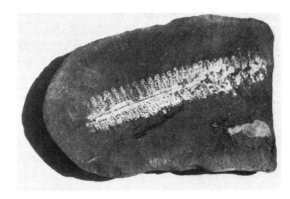

FIGURE 13.11   A fossil fern.

Discussion

If we now know and understand some of the mechanisms of evolution, why does there yet remain any controversy over the broad subject itself? Obviously ignorance is a factor, but it is not the sole reason. Science deals with tangible facts and evidence that can be experimentally tested; beliefs stemming from metaphysics and religion are outside the realm of science to prove or disprove. Evidence in support of evolution is drawn from several additional areas, including the form and ecology of living organisms, and the way they are related to each other today. Other evidence comes from the structure and relationships of proteins and other molecules, and from fossils, which are remnants of previously living forms (figure 13.11). The simplest fossils are generally found in the oldest geological strata, while more complex forms tend to be found in younger strata. Still further evidence is drawn from the geographical distribution of organisms (see figure 13.10). Many groups are confined to a single continent or island. In some instances where similar organisms occur on more than one land mass, there is evidence that the land masses concerned were once linked together, which would have permitted terrestrial migration. Still other conclusions are drawn from the physiology and chemistry of the organisms, and in animals, particularly, from the presence of vestigial organs (structures that are apparently now functionless but are presumed to represent previously functional organs in predecessor species). Among the best known such structures in humans are the appendix and wisdom teeth.

Most scientists who have studied the evidence feel that some form of evolutionary process is the only plausible explanation for the unity of life at the molecular and cellular level and the extraordinary diversity of the organisms now around us. There is little unanimity of thought, however, as to precisely how evolution proceeded in the past. One authority will be convinced that a certain group evolved from another, while other equally eminent authorities will maintain that the exact reverse occurred. Part of the reason for such paradoxes is that the historical record is quite incomplete. Although the fossil evidence for the evolution of a few organisms such as certain mollusks and the horse is fairly substantial, other fossil evidence is very fragmentary. As mentioned in the previous section, possibly fewer than one in each million organisms that once existed ever became a fossil, and there are probably no fossils at all of many herbaceous and soft-bodied organisms. With such evidence, scientists can deal only in probabilities or possibilities, and it is inevitable that various, sometimes conflicting, interpretations result. Objective scientists freely acknowledge that numerous problems concerning the interpretation of the geological past and the pathways of organic evolution exist, but they ask their detractors to suggest more plausible alternatives. At this point, some apply the tenets of religious faith, which like history, is not subject to scientific experimentation. Some see no conflict between science and religion, and others are convinced that the two are mutually exclusive. Depending on individual points of view, an impasse may result, with persons of different persuasions—including, unfortunately, some scientists—becoming dogmatic on the topic. Virtually no objective thinkers will deny, however, the extraordinary impact theories of evolution have had on modern peoples and on their concepts of the living world around them.

SUMMARY

Karpechenko crossed a radish and a cabbage in an attempt to produce a better food crop, but the results were unsatisfactory. Such crossing experiments are a part of genetics, the science that deals with heredity or natural inheritance. Genetics as a science originated with Gregor Mendel, an Austrian monk, who performed experiments with peas. Mendel made crosses with varieties of peas that had single pairs of contrasting characteristics, each pair being carried on a separate pair of homologous chromosomes. The original plants involved in making the crosses were referred to as the parents, or parental generation, while their offspring were designated as the F_1 generation and the offspring of crosses between members of the F_1 generation as the F_2 generation. F_1 plants were also called hybrids. Mendel differed from his predecessors in counting the number of all offspring and in being meticulous. He also was exceptionally fortunate in the characteristics he chose to study. In the pea crosses he made, he found that the F_1 hybrids all resembled their parents in appearance, but the F_2 offspring were produced in a ratio of 3 plants resembling one parent to 1 plant resembling the other parent. He referred to agents for the characteristics inside the plants as "factors" and also deduced that each plant must have two factors for each characteristic. From this came the principle or law of unit characters, which states that "factors, which always occur in pairs, control the inheritance of various characteristics." The factors later became known as genes.

Mendel further deduced that one member of a pair of factors may suppress or mask the expression of the other member. The suppressing member was called a "dominant" and its counterpart a "recessive." To distinguish between the appearance of an organism and its genetic makeup, different terms are applied: "phenotype" for appearance and "genotype" for genetic makeup. Phenotypes are described with words while genotypes are designated with letters; capital letters are used for dominants and lowercase letters for recessives. If both members of a pair of factors, or genes, are identical, the plant is said to be homozygous; if the members of a pair of genes are contrasting, the plant is said to be heterozygous (for those factors). Sometimes the F_1 generation may phenotypically be more or less intermediate between the parents as a result of absence of dominance (incomplete dominance).

In a monohybrid cross the offspring are produced in a ratio of 3:1, when dominants and recessives are involved. In a dihybrid cross involving dominance and two pairs of genes carried on separate pairs of chromosomes, the offspring usually are produced in a phenotypic ratio of 9:3:3:1.

Backcrosses are routinely made to check the results of genetic experiments. Such crosses involve $F_1 \times$ the recessive parent and normally produce a 1:1 (monohybrid) ratio or a 1:1:1:1 (dihybrid) ratio of phenotypes.

There are often thousands of genes on the same chromosome; genes that are inherited together are said to be linked. Bateson and Punnett discovered that flower color and pollen grain shape in sweet peas did not produce an expected 9:3:3:1 ratio of phenotypes in the F_2 generation. Morgan postulated that the 7:2:2:7 ratio they obtained instead was the result of linkage and occasional crossing over. Calculating crossover percentages has enabled workers to determine the relative position of genes on chromosomes, a process known as chromosomal mapping. The closer the genes are to each other on a chromosome, the less likely they are to be involved in a crossover, and vice versa.

Before Mendel's work became known, most biologists did not understand why recessive characteristics in a population did not eventually disappear. Hardy and Weinberg explained, in what is known as the Hardy–Weinberg law, that the proportions of dominant genes to recessive genes in a normally interbreeding population will be retained from generation to generation unless there are forces that change those proportions. The most significant cause for exceptions to the Hardy–Weinberg law is selection.

Plant breeding offers considerable potential for alleviating world hunger problems in the future. Improvement of crops involves various techniques, including hybridization, polyploidy, mutation, and tissue culture. The use of tissue culture for crop improvement can entail selection of cells that have been subjected to stresses, and protoplast fusion whereby two cells unite after their walls have been broken down by enzymes.

The theory of organic evolution received its greatest impetus from Charles Darwin's *On the Origin of Species*. Darwin's theory is based on four principles, or tenets: (1) overproduction (many organisms produce enormous numbers of offspring); (2) "struggle for existence" (all living organisms "compete" for available nutrients, etc., but there is not enough for all and as a result many die); (3) "survival and reproduction of the fittest" (those organisms best adapted to the environment have the best chance to survive); and (4) variation and inheritance (those hereditary variations with survival value are inherited from generation to generation and accumulate in time, while other variations are eliminated).

Before 1900 many biologists believed that evolution occurred as a result of the inheritance of acquired characteristics. Lamarck believed giraffes had attained their long necks by constantly stretching for food and that the slight stretching in each generation had been passed on until in time the numerous small stretchings had added up to the present long necks. This theory was discredited experimentally by surgically shortening tails of mice for many generations and by pruning trees; none of the acquired characteristics were passed on.

Mechanisms of evolution include mutations in chromosomes or genes, hybridization (particularly introgression, the backcrossing of hybrids to parents in a natural population), polyploidy (having additional sets of chromosomes as a result of a cell's failure to produce a cell wall during mitosis and having the new cell develop into a new organism—those developing from sterile hybrids are especially significant), apomixis (asexual reproduction), and reproductive isolation.

If new genes are produced in a freely interbreeding population, they will gradually be spread throughout the population, and the nature of the whole population will change in time. However, if a population is divided by a barrier such as a mountain range or an ocean, genes occurring in the one population will not spread throughout the isolated population as before. In time, because of the isolation, each new population may develop into separate species incapable of breeding with one another.

Darwin believed that evolution through natural selection was a gradual process over great periods of time. Some contemporary biologists believe that evolution has taken place in spurts between long periods of little change, based on evidence from the fossil record. Interpretation of the fossil record is, however, controversial, and will be debated indefinitely.

Additional evidence in support of evolution is drawn from the form, the ecology and relationships of living organisms, the molecular structure, the geographical distribution, fossils, and vestigial organs (structures that are apparently now functionless but are presumed to represent previously functional organs in predecessors). Most scientists feel some form of evolution is the only plausible explanation for the unity of life at the molecular and cellular level and the great diversity of life, but there is little agreement among them as to precisely how evolution proceeded in the past. This is partly due to the inadequacy of the fossil record, which is incomplete because proportionately so few organisms ever became fossils, and the fact that history is not subject to scientific experimentation. There are a variety of opinions and convictions on origins, but few advocates can deny the major impact that theories of evolution have had on modern peoples and on their concepts of life.

REVIEW QUESTIONS

1. When did genetics originate as a science?
2. Why did Mendel succeed where others had failed?
3. Who were Tschermak, Correns, and de Vries?
4. Define hybrid, F_1 generation, phenotype, genotype, homozygous, heterozygous, dominant, and recessive.
5. What are Mendel's laws?

6. In a variety of flowering plants, blue flowers and dwarfness are controlled by dominant genes; white flowers and tallness are the recessives. The gene pairs are carried on separate pairs of chromosomes. If a homozygous blue dwarf plant is crossed with a homozygous white tall plant,
   a. What are the phenotypes and genotypes of the F_1 generation?
   b. How many different *kinds* of genotypes could occur in the F_2 generation?
   c. What phenotypic ratio would you expect to occur in the F_2 generation?
   d. What genotypes could result from the cross **BBDD** $\times$ **BbDd**?
   e. What phenotypes could result from the cross **bbDd** $\times$ **BBDd**?
7. In summer squashes, disk-shaped fruit is dominant over spherical fruit and white fruit color is dominant over yellow fruit color. If a homozygous white, disk-shaped variety is crossed with a homozygous yellow, spherical variety, how many different homozygous genotypes are possible in the F_2 generation?
8. A cross is made between a white-flowered plant and a red-flowered plant of the same variety. The F_2 generation consisted of 32 white-flowered plants, 64 pink-flowered plants, and 29 red-flowered plants. Explain.
9. In tomatoes, smooth skin is dominant and fuzzy skin is recessive; also, tallness is dominant and dwarfism is recessive. A homozygous smooth, tall variety is crossed with a homozygous fuzzy, dwarf variety, and the F_1 is backcrossed to a homozygous fuzzy, dwarf variety. The offspring of the backcross are 95 smooth, tall; 3 fuzzy, tall; 4 smooth, dwarf; and 96 fuzzy, dwarf. Is linkage involved? If so, what is the percentage of crossing over?
10. In squashes, white color, which is controlled by a single gene, is dominant, and yellow is recessive. Give the phenotypes and the genotypes for each of the following crosses: Ww $\times$ Ww; Ww $\times$ ww; WW $\times$ ww.
11. If red flowers are dominant and white flowers are recessive in an *annual* variety of *self-pollinating* beans, assume that you plant one heterozygous bean seed on an island, and that it germinates and thrives. What will the colors and ratios of the phenotypes be in the F_4 generation?

12. In snapdragons, red flower color shows incomplete (lack of) dominance over white flower color. If you wanted to produce seeds that would produce only pink-flowered plants when sown, how would you go about it?

13. If one plant is homozygous for three dominant characteristics and another plant is correspondingly recessive for the same characteristics, what proportion of the F_2 generation will resemble each parent if the two plants are crossed?

14. What would happen in the production of gametes if a species had an odd number of chromosomes in each of its cells?

15. What is meant when genes are said to be linked?

16. What is the difference between organic evolution and other forms of evolution?

17. How did Darwin's theory of evolution differ from Lamarck's?

18. What basic modifications have been made in evolutionary theory since Darwin's time?

19. Why is a hybrid often sterile?

20. What is meant by a mechanism of evolution?

21. How are polyploids produced?

22. Why is reproductive isolation necessary for evolution to occur?

23. What is chromosomal mapping and how is it done?

24. What is the Hardy–Weinberg law?

25. What evidence is there to support modern concepts of evolution? Are there any problems with the evidence?

DISCUSSION QUESTIONS

1. Mendel is said to have been exceptionally lucky in his discoveries. Do you think most scientists who make significant discoveries are lucky?

2. The peas Mendel worked with were largely self-fertile. Of what advantage or disadvantage to a species is such a phenomenon?

3. One of Darwin's tenets was that there is a struggle for existence among living organisms. Do plants struggle with one another to survive? If so, how do they do it?

4. Some populations change noticeably in form within 100 years. If only 1 gene in every 200,000 per cell mutates, and if most mutations are harmful, how is such change possible?

5. Do you think there might be viable alternatives to organic evolutionary theory to account for the diversity of life around us? Explain.

ADDITIONAL READING

Ayala, F. J. 1983. *Genetic variation and evolution.* Carolina Biology Readers, Head, J. J., ed. Burlington, NC: Carolina Biological Supply Company.

Crow, J. F. 1983. *Genetics notes: an introduction to genetics,* 8th ed. Minneapolis: Burgess Publishing Co.

Darwin, C. 1975. *On the origin of species.* Facsimile 1st ed. of 1859. New York: Cambridge University Press.

Ehrlich, P. R., R. W. Holm, and D. R. Parnell. 1975. *The process of evolution,* 2d ed. New York: McGraw-Hill Book Co.

Grant, V. 1981. *Plant speciation,* 2d ed. New York: Columbia University Press.

Janick, J., R. W. Schery, F. W. Woods, and V. W. Ruttan. 1981. *Plant science,* 3d ed. San Francisco: W. H. Freeman and Company.

Kerkut, G. A. 1960. *Implications of evolution.* Oxford and New York: Pergamon Press.

Kiger, A., and J. Kiger. 1983. *Modern genetics,* 2d ed. Menlo Park, CA: Benjamin-Cummings Publishing Co.

Olby, R. C. 1985. *The origins of Mendelism,* 2d ed. Chicago: University of Chicago Press.

Stebbins, G. L. 1977. *Processes of organic evolution,* 3d ed. Englewood Cliffs, NJ: Prentice-Hall.

Strickberger, M. W. 1985. *Genetics,* 3d ed. New York: Macmillan Publishing Co., Inc.

Wagner, R. P., B. H. Judd, B. G. Sanders, and R. H. Richardson. 1980. *Introduction to modern genetics.* New York: John Wiley & Sons, Inc.

Overview

*T*his chapter begins with a discussion of the problems involved in the use of common names for plants, as illustrated by a survey of such names for two related species of American spring-flowering perennials. It continues with a brief historical account of the events that led to the development and acceptance of Linnaeus's Binomial System of Nomenclature. The history of the development of a five-kingdom classification is presented, along with a list of the divisions and classes included in the kingdoms covered in this text. The chapter concludes with a synoptic key to the kingdoms and divisions of organisms.

Robert A. Schlising

14 *Plant Names and Classification*

Some Learning Goals

1. Understand several problems associated with the application of common names to organisms.

2. Know what the Binomial System of Nomenclature is, how it developed, and how it is currently used.

3. Learn several reasons for recognizing more than two kingdoms of living organisms.

4. Understand the bases for Whittaker's five-kingdom system.

5. Construct a dichotomous key to ten different objects (e.g., pencils, golf balls).

Outline

INTRODUCTION

Biologists sometimes are thought by the general public to be slightly pompous when they refer to peanuts as *Arachis hypogaea* or an African hoopoe bird as *Upupa epops*. They are not, however, merely "showing off" or trying to be difficult. Rather, they are identifying organisms by their *scientific names* through a system that has evolved over the last few centuries. Circumstances have dictated the necessity of distinguishing among the 10 million existing types of organisms, as well as those that have become extinct, in a manner that will identify them anywhere, regardless of the languages spoken locally.

At present, all living organisms are given a single two-part Latin scientific name, and most also have *common names*. Only one correct scientific name applies to all individuals of a specific kind of organism, no matter where they may be found, but many common names may be given to the same organism, and one common name may be applied to a number of different organisms. For example, the scientific name *Dicentra cucullaria* has been given to a pretty spring-flowering plant native to eastern North America and the Columbia River basin in the West (see distribution map in figure 13.9). Its unique flower shape, reminiscent of the baggy pants of a traditional Dutch costume, has resulted in its being given the common name Dutchman's breeches, but it also has the common names of little-boys' breeches, kitten's breeches, breeches-flower, boys-and-girls, Indian boys-and-girls, monkshood, white eardrops, soldier's cap, colicweed, little blue stagger, white hearts, butterfly banners, rice roots, and meadow bells. In addition to these English common names, the plant has Indian names, and in the Canadian province of Quebec it has French names. Often growing with it is a related plant, with similar leaves but flowers of a slightly different shape, having the scientific name *Dicentra canadensis*. Because of the similarities between the two plants and their close association in the woods where they occur, some people in the past assumed that they were merely two different forms of the same plant. This is reflected in the fact that some of the common names of *Dicentra cucullaria* (e.g., colicweed, Indian boys-and-girls, little blue stagger) have also been applied to *Dicentra canadensis* (figure 14.1), in addition to the names squirrel corn, turkey corn, turkey pea, wild hyacinth, fumitory, staggerweed, and trembling

FIGURE 14.1 *A. Dicentra cucullaria. B. Dicentra canadensis.*

A.

B.

stagger. But the problem of the diversity and the overlapping of common names for these two plants does not stop there, for the names monkshood, soldier's cap, rice roots, meadow bells, and turkey corn have also been applied to completely unrelated plants.

In Europe, with its many languages, common names can become very numerous indeed. The widespread weed with the scientific name *Plantago major,* for example, is often called broad-leaved plantain in English, but it also has no fewer than 45 other English names, 11 French names, 75 Dutch names, 106 German names, and possibly as many as several hundred more names in other languages, with literally dozens of these common names also applying to quite different plants. If it were not for the early recognition by biologists and others of the urgent need for worldwide uniformity in naming and classifying all organisms, utter chaos eventually might have prevailed in communications concerning them.

DEVELOPMENT OF THE BINOMIAL SYSTEM OF NOMENCLATURE

As was indicated in chapter 1, the first person of note to attempt to organize and classify plants was Theophrastus, the brilliant student of Aristotle and Plato. His third-century B.C. classification of nearly 500 plants into trees, shrubs, and herbs, along with his distinctions between plants on the basis of leaf characteristics, was used for hundreds of years. Not until the thirteenth century A.D. was a distinction between monocots and dicots made on the basis of stem structure. After this, herbalists added considerably to Theophrastus's list, and by the beginning of the eighteenth century, details of fruit and flower structure were utilized in classification schemes in addition to form and habit. By this time European scholarly institutions were bulging with thousands of plants that explorers were bringing back from around the world, and confusion over scientific and common names was multiplying. The use of Latin in schools and universities had become widespread, and it was then customary to use descriptive Latin phrase names for both plants and animals. All organisms were grouped into **genera** (singular: **genus**), with the first word of the Latin phrase indicating the particular genus to which the organisms belonged. For example, all known mints were given phrase names beginning with the word *Mentha,* the name of the genus. Likewise, the phrases for lupines began

with *Lupinus,* and those for poplars began with *Populus.* A complete phrase name for spearmint read *Mentha floribus spicatis, foliis oblongis serratis.* Roughly translated, it means "Mentha with flowers in a spike (an elongated but compact flower cluster); leaves oblong, saw-toothed." A phrase name for the closely related peppermint read *Mentha floribus capitatus, foliis lanceolatis serratis subpetiolatis,* meaning "Mentha with flowers in a head; leaves lance-shaped, saw-toothed, with very short petioles."

It was at this point in history that the Swedish botanist, Carolus Linnaeus (1707–1778), entered the scene and established the basis for the present-day naming and classification of organisms. Linnaeus, who was nicknamed the "Little Botanist" at school, inherited his passion for plants from his father, who was a minister and an amateur gardener. He is said to have been much impressed at the age of four by his father's remarks about the uses of neighborhood plants in his home community of Råshult, located in southern Sweden. After a brief tenure as a student at the University of Lund, he spent most of his time making excursions to Lapland, Holland, France, and Germany. Eventually he became the professor of botany and medicine at Uppsala, where he inspired large numbers of students, 23 of whom became professors themselves. It is said that he frequently led large field trips into the countryside, accompanied by a musical band.

When Linnaeus began his work, he set out to classify all known plants and animals according to their genera. In 1753, he published a two volume work entitled *Species Plantarum,* which was later to become the most important of all works on plant names and classification. In this work, he not only included a referenced list of all the Latin phrase names previously given to the plants, but he also changed some of the phrases, when necessary, to reflect relationships, placing one to many specific kinds of organisms called **species**[1] in each genus. He limited each Latin phrase to a maximum of 12 words, and in the margin next to the phrase he listed a single word, which, when combined with the generic name,

---

1. Note that *species* is like the word *sheep* in that it is spelled and pronounced the same way in either singular or plural usage. There is no such thing as a plant or animal "specie." Since Linnaeus's time, a species has been defined as "a population of individuals capable of freely interbreeding in nature but not generally interbreeding with members of another species."

FIGURE 14.2 A page from *Species Plantarum* by Linnaeus. (Courtesy of the National Library of Medicine)

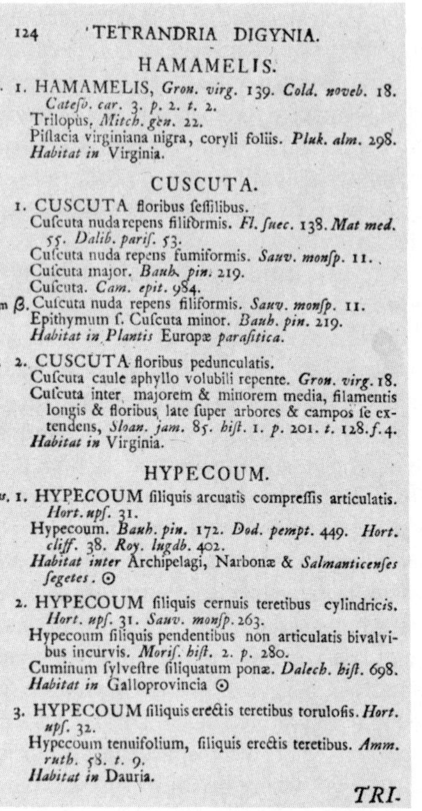

124 TETRANDRIA DIGYNIA.

HAMAMELIS.

virginiana. 1. HAMAMELIS, *Gron. virg.* 139. *Cold. noveb.* 18. *Catefb. car.* 3. *p.* 2. *t.* 2.
Trilopus, *Mitch. gen.* 22.
Piftacia virginiana nigra, coryli foliis. *Pluk. alm.* 298.
Habitat in Virginia.

CUSCUTA.

europæa. 1. CUSCUTA floribus feffilibus.
Cufcuta nuda repens filiformis. *Fl. fuec.* 138. *Mat med.* 55. *Dalib. parif.* 53.
Cufcuta nuda repens fumiformis. *Sauv. monfp.* 11.
Cufcuta major. *Bauh. pin.* 219.
Cufcuta. *Cam. epit.* 984.
epithymum β. Cufcuta nuda repens filiformis. *Sauv. monfp.* 11.
Epithymum f. Cufcuta minor. *Bauh. pin.* 219.
Habitat in Plantis Europæ *parafitica.*

americana. 2. CUSCUTA floribus pedunculatis.
Cufcuta caule aphyllo volubili repente. *Gron. virg.* 18.
Cufcuta inter majorem & minorem media, filamentis longis & floribus late fuper arbores & campos fe extendens, *Sloan. jam.* 85. *hift.* 1. *p.* 201. *t.* 128. *f.* 4.
Habitat in Virginia.

HYPECOUM.

procumbens. 1. HYPECOUM filiquis arcuatis compreffis articulatis. *Hort. upf.* 31.
Hypecoum. *Bauh. pin.* 172. *Dod. pempt.* 449. *Hort. cliff.* 38. *Roy. lugdb.* 402.
Habitat inter Archipelagi, Narbonæ & *Salmanticenfes fegetes.* ⊙

pendulum. 2. HYPECOUM filiquis cernuis teretibus cylindricis. *Hort. upf.* 31. *Sauv. monfp.* 263.
Hypecoum filiquis pendentibus non articulatis bivalvibus incurvis. *Morif. hift.* 2. *p.* 280.
Cuminum fylveftre filiquatum ponæ. *Dalech. hift.* 698.
Habitat in Galloprovincia ⊙

erectum. 3. HYPECOUM filiquis erectis teretibus torulofis. *Hort. upf.* 32.
Hypecoum tenuifolium, filiquis erectis teretibus. *Amm. ruth.* 58. *t.* 9.
Habitat in Dauria.

TRI-

formed a convenient abbreviated designation for the species. The word in the margin for spearmint was *spicata,* and the word for peppermint was *piperita.* Thus, the abbreviated name for spearmint was *Mentha spicata* and for peppermint *Mentha piperita.* Although Linnaeus originally considered the phrase names the official names of the plants, he and those who followed him eventually replaced all the phrase names with abbreviated ones. Because of their two parts, these names became known as *binomials* and the method of naming became known as the Binomial System of Nomenclature. Today all organisms are named according to this system, which in current use also includes the authority for the name, either in abbreviated form or in full, after the Latin name. Thus, for example, the full scientific name for spearmint is now written *Mentha spicata* L., the L. standing for Linnaeus, and the full scientific name for the common dandelion is written *Taraxacum officinale* Wiggers, because Fredericus Henricus Wiggers was the first, in his *Flora of Holstein* (published in 1780), to describe the species.

In addition to establishing the Binomial System of Nomenclature, Linnaeus sought to do more than merely publish a long list of plants. After all, such lists have limited usefulness if they cannot be used to identify the plants concerned. Linnaeus organized all known plants into 24 **classes,** which were based primarily on the number of stamens (pollen-bearing structures) in flowers. For example, all plants having 5 stamens per flower were placed in one class, while all having 6 stamens were placed in another. Plants and other organisms that have no flowers at all (e.g., mosses, fungi) were assigned to a class of their own. His arrangement was for convenience and was artificial to the extent that it did not necessarily reflect natural relationships, but for the first time it became possible for a worker to identify an unknown plant. This arrangement was used in *Species Plantarum* and other works by Linnaeus (figure 14.2).

In 1867, more than 100 years after *Species Plantarum* was published, about 150 European and American botanists met in Paris to try to standardize rules governing the naming and the classifying of plants. They agreed to use the works of Linnaeus as the starting point for all scientific names of plants and decided that his binomials, or the earliest ones published after him, would have priority over all others. International congresses of botanists have met at varying intervals since then and have revised and expanded these rules. Today the modified rules comprise what is known as the *International Code of Botanical Nomenclature,* which is a single book with a common index to its English, French, and German translations of the various rules and recommendations. It now specifically recognizes Linnaeus's *Species Plantarum* as the starting point for scientific names of plants and spells out details of naming and classifying, which are followed by botanists of all nationalities. A similar code for animals has been developed and established by zoologists. It, too, uses a work of Linnaeus as the starting point for scientific names.

DEVELOPMENT OF THE KINGDOM CONCEPT

If you were to ask the average person the differences between plants and animals, you would probably be told that animals move and plants do not, and that animals eat plants or other animals for food whereas plants produce their own food through photosynthesis. A distinction between plants and animals in general has probably always existed in the minds of

TABLE 14.1
Four Classifications of Organisms into Kingdoms

Two Kingdoms (Traditional)	Three Kingdoms (Hogg and Haeckel)	Four Kingdoms (Copeland)	Five Kingdoms (Whittaker)	
		Monera Bacteria	**Monera** Bacteria	cells prokaryotic
	Protista Bacteria Algae Slime molds Flagellate fungi True fungi Protozoa Sponges	**Protista** Algae Slime molds Flagellate fungi True fungi Protozoa Sponges	**Protista** Algae Slime molds[1] Protozoa Sponges	cells eukaryotic
			Fungi Flagellate fungi True fungi	mostly absorb food in solution
Plantae Bacteria Algae Slime molds Flagellate fungi True fungi Bryophytes Vascular plants	**Plantae** Bryophytes Vascular plants	**Plantae** Bryophytes Vascular plants	**Plantae** Bryophytes Vascular plants	produce food via photosynthesis
Animalia Protozoa Sponges Multicellular animals	**Animalia** Multicellular animals	**Animalia** Multicellular animals	**Animalia** Multicellular animals	ingest solid food

1. Slime molds are treated as a subkingdom of Kingdom Fungi in this text.

intelligent beings, and it was natural when classification schemes were first developed that all living organisms would be placed, according to the highest category of kingdom, in either the *Plant Kingdom* or the *Animal Kingdom*. While this distinction still works well for the more complex plants and animals, it breaks down for some of the so-called simpler organisms. For example, there are more than 300 species of single-celled organisms called *euglenoids* (see figure 16.6) inhabiting a variety of freshwater habitats. These little creatures have *flagella* (see chapter 15) that pull them through the water, and they also can ingest food particles through a groove called a *gullet*. Both of these features would be considered animallike, and so euglenoids are treated as animals in some textbooks. Many of these organisms also have chloroplasts, however, and if light is present, they can carry on photosynthesis efficiently enough to eliminate the need for ingesting food. Accordingly, they have often been treated as plants in botany books. A similar problem has existed with slime molds, which resemble masses of protoplasm slowly flowing or creeping across dead leaves and debris, feeding on bacteria and other substances as they go, and looking somewhat like amoebae, which botanists and zoologists alike have traditionally called

animals. When slime molds reproduce, however, dramatic changes take place. They become stationary and develop reproductive bodies that are distinctly funguslike, causing some to consider them fungi.

In an attempt to overcome this problem, the biologists J. Hogg and E. Haeckel proposed a third kingdom in the 1860s. All organisms that did not develop complex tissues (e.g., algae, fungi, and sponges) were placed in a third kingdom called *Protoctista,* or *Protista.* This third kingdom included such a heterogeneous variety of organisms, however, that in 1938 another biologist by the name of H. F. Copeland proposed it again be divided. He assigned the name *Monera* to all single-celled protists with prokaryotic cells (see chapters 3 and 15), leaving the algae, fungi, and single-celled organisms with eukaryotic cells in the Kingdom Protoctista. Although many biologists considered Copeland's four-kingdom system of classification a definite improvement, it too had problems—particularly the fact that basic differences in the mode of nutrition existed among organisms within the Kingdom Protoctista. As a result, a sizable number of biologists today favor variations of a five-kingdom system proposed by R. H. Whittaker in 1969 (table 14.1). In Whittaker's system,

Classification of Organisms in Five Kingdoms

Kingdom Monera
Subkingdom Archaebacteriophytineae (archaebacteria)
  Division Archaebacteriophyta (methane, salt, and
    sulfolobus bacteria)
Subkingdom Eubacteriophytineae (true bacteria)
  Division Eubacteriophyta
    Class Eubacteriae (unpigmented, purple, and green
     sulfur bacteria)
    Class Cyanobacteriae (blue-green bacteria)
    Class Prochlorobacteriae (prochlorobacteria)

Kingdom Protista
  Division Chrysophyta
    Class Xanthophyceae (yellow-green algae)
    Class Chrysophyceae (golden-brown algae)
    Class Bacillariophyceae (diatoms)
    Class Cryptophyceae (cryptophytes)
  Division Pyrrophyta (dinoflagellates)
  Division Euglenophyta (euglenoids)
  Division Chlorophyta
    Class Chlorophyceae (green algae)
    Class Charophyceae (stoneworts)
  Division Phaeophyta (brown algae)
  Division Rhodophyta (red algae)
  [Phylum Protozoa—protozoans]
  [Phylum Porifera—sponges]

Kingdom Fungi
Subkingdom Myxomycotineae
  Division Acrasiomycota (cellular slime molds)
  Division Myxomycota (true slime molds)
Subkingdom Mastigomycotineae
  Division Chytridiomycota (chytrids)
  Division Oomycota (water molds)
Subkingdom Eumycotineae
  Division Zygomycota (coenocytic fungi)
  Division Eumycota (septate fungi)
    Class Ascomycetes (cup fungi)
    Class Basidiomycetes (club fungi)
    Class Deuteromycetes (imperfect fungi)
    [Class Lichenes (lichens)]

Kingdom Plantae
  Division Bryophyta
    Class Hepaticae (liverworts)
    Class Anthocerotae (hornworts)
    Class Musci (mosses)
  Division Psilophyta (whisk ferns)
  Division Lycophyta (club mosses)
  Division Sphenophyta (horsetails)
  Division Pterophyta (ferns)
  Division Cycadophyta (cycads)
  Division Ginkgophyta (ginkgoes)
  Division Coniferophyta (conifers)
  Division Gnetophyta (*Gnetum, Ephedra, Welwitschia*)
  Division Anthophyta (flowering plants)
    Class Dicotyledonae (dicots)
    Class Monocotyledonae (monocots)

(Division Tracheophyta)

Kingdom Animalia (multicellular animals)

---

three kingdoms of more complex organisms based on three basic forms of nutrition (photosynthesis, ingestion of solid food, and absorption of food in solution) are recognized, along with two kingdoms of protists, which are distinguished on the basis of differences in cellular structure. Even the five-kingdom arrangement is not without its critics and problems, however. Carl Woese, a microbiologist and leading authority on bacteria, has presented cogent arguments for dividing the Kingdom Monera into two kingdoms, based on some previously unrecognized fundamental differences between two major groups of bacteria (see chapter 15). The slime molds, which have no cell walls during their active state but do develop walls when they reproduce, still do not fit well into the five-kingdom system, and other organisms traditionally regarded as fungi may be more closely related to protists. Much information still needs to be accumulated to understand natural relationships, and until this occurs, taxonomic categories will remain subject to different interpretations.

CLASSIFICATION OF MAJOR GROUPS

Since Linnaeus's time, a number of classification categories have been added between the levels of kingdom and genus. Genera are now grouped into **families,** families into **orders,** orders into **classes,** classes into **divisions,**[2] and divisions into **kingdoms.** Depending on which system of classification is used, there may be between 12 and nearly 30 divisions of "plants" recognized.

The classification of the major groups of living organisms shown in table 14.2 utilizes the modification of Whittaker's five-kingdom system. It is the classification followed throughout this book. Hypothetical derivations and relationships are indicated in figure 14.3.

---

2. *Division* is equivalent to the term *phylum,* which is used in animal classification. Although use of the word phylum for all living organisms has been proposed, the *International Code of Botanical Nomenclature* presently permits only the use of the word *division* for plants.

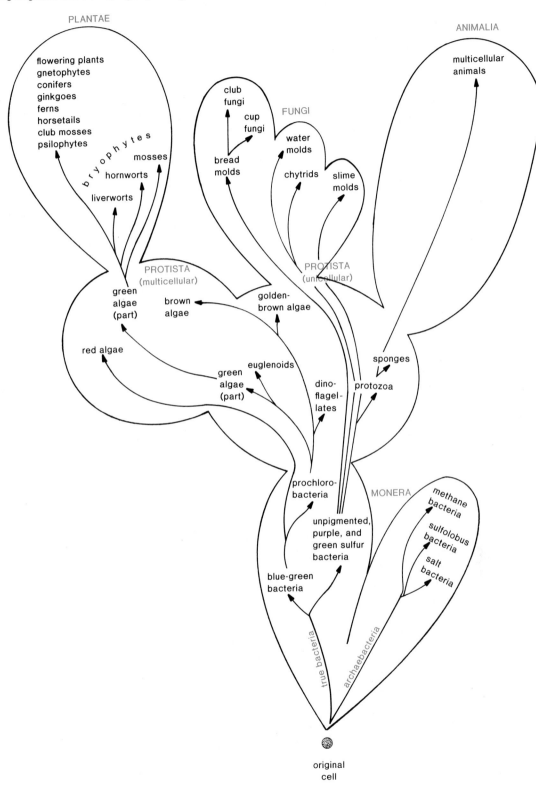

FIGURE 14.3 Hypothetical derivations and relationships among kingdoms and the major groups of organisms.

Viruses, which have neither cellular structure nor most of the other attributes of living organisms, do not fit into this classification but are discussed in chapter 15 after the bacteria. Lichens, each of which is a combination of an alga and a fungus in close association, are discussed with the fungi in chapter 17.

If we were to give a complete classification of an onion according to this particular arrangement, it would look like this:

Kingdom: Plantae
  Division: Anthophyta
    Class: Monocotyledonae
      Order: Liliales
        Family: Amaryllidaceae
          Genus: *Allium*
            Species: *Allium cepa* L.

The second part of the scientific name, the *specific epithet,* is followed by the name of the author, usually in abbreviated form. The author is the person (or persons) who originally named the plant or placed the species in a particular genus. It is customary to give binomials in italics (or to underline the words). In the case of *Allium cepa,* the L. following the name stands for Linnaeus, since this plant was one of the 7,300 species he included in his *Species Plantarum.*

In addition to these categories of classification, various "in-between" categories such as *subdivision, subclass,* and *suborder* have been used, and species themselves are sometimes further divided into *subspecies, varieties,* and *forms.* Some authors also recognize "super" categories such as *superkingdom* and *superorder.* One popular classification arrangement recognizes only two divisions in Kingdom Plantae (Division Bryophyta for nonvascular plants and Division Tracheophyta for all vascular plants) with a number of subdivisions. These subdivisions are treated as full divisions in this and a number of other texts.

Since human judgment has to enter into the compilation of classifications, there undoubtedly will never be complete agreement on them, but they are nevertheless useful, indeed necessary, for making some order of the diversity of plant life around us.

Those who specialize in identifying, naming, and classifying organisms and trying to sort out natural relationships are called *taxonomists* (see chapter 1). One of the activities of taxonomists is the construction of keys to help others identify organisms with which they may not be familiar. Most such keys are *dichotomous,* that is, they give the reader pairs of statements based on features of the organisms. By carefully examining an organism and choosing from each pair the statements that most closely apply to the organism, we can arrive at an identification. If, for example, we agree with the first of the two statements given for each number in the key, then we either use the identification at the end of the line or proceed to the next indented statement beneath. If we disagree with the first of the paired statements, then we look for the matching alternate statement and proceed from there. The following synoptic key, which deals with major groups of organisms, illustrates natural relationships. It covers only general features and does not indicate various details or occasional exceptions that would appear in keys for lesser groups. All organisms in any given group are presumed to be more closely related to each other than to organisms in another group.

SYNOPTIC KEY TO MAJOR GROUPS OF ORGANISMS
(exclusive of Kingdom Animalia)

1. Unicellular, prokaryotic organisms with cell walls...**Kingdom Monera**
  2. Cell walls without muramic acid..*Subkingdom Archaebacteriophytineae*
  2. Cell walls with muramic acid...*Subkingdom Eubacteriophytineae*
1. Unicellular, colonial, filamentous, or multicellular eukaryotic organisms, with or without cell walls.
    3. Organisms whose female (and usually male) reproductive structures consist of a single cell or with sterile cells surrounding the one-celled reproductive structures; zygotes not developing into embryos.
      4. Organisms unicellular or filamentous
        5. Cells with plastids...**Kingdom Protista**

6. Plastids having yellow, brown, or orange pigments more conspicuous than the chlorophyll pigments.
    7. Food reserves oils or carbohydrates other than starch; two flagella both located at one end of the cell... Division Chrysophyta
    7. Food reserve starch; cells with a flagellum at one end and another at right angles to it in a central groove .... Division Pyrrophyta
6. Plastids with chlorophyll pigments more conspicuous than other pigments.
    8. Cells flexible; carbohydrate food reserve *paramylon* ....
    .... Division Euglenophyta[3]
    8. Cells not flexible; carbohydrate food reserve starch...................................
    .... Division Chlorophyta (in part)
5. Cells without plastids...**Kingdom Fungi**
    9. Vegetative bodies amoebalike.............*Subkingdom Myxomycotineae*
    9. Vegetative bodies not amoebalike.
      10. At least asexual reproductive cells with flagella; aquatic organisms................................. *Subkingdom Mastigomycotineae*
      11. Vegetative bodies consisting primarily of a single spherical cell, often with rhizoids.........................Division Chytridiomycota
      11. Vegetative bodies consisting of branching, filamentous cells with numerous nuclei.............................. Division Oomycota
      10. All reproductive cells without flagella; terrestrial organisms........
      ....*Subkingdom Eumycotineae*
      12. Filaments of the vegetative bodies branched and with numerous nuclei, but not partitioned into individual cells
      ....Division Zygomycota
      12. Filaments of the vegetative bodies partitioned into individual cells, each with one to several nuclei ....
      ....Division Eumycota
4. Organisms multicellular, not filamentous
      13. Organisms with accessory pigments essentially similar to those of higher plants; carbohydrate food reserve starch..............Division Chlorophyta (in part)
      13. Organisms with some accessory pigments differing from those of higher plants; food reserves carbohydrates other than ordinary starch.
        14. Organisms brownish in color due to presence of brown pigments; carbohydrate food reserve *laminarin*...........................Division Phaeophyta
        14. Organisms reddish in color due to presence of red pigments; carbohydrate food reserve *floridean starch*......................Division Rhodophyta

---

3. Only about a third of the euglenoids develop chloroplasts.

3. Organisms with multicellular reproductive structures; zygotes developing into embryos.............
....**Kingdom Plantae**
    15. Plants without true xylem or phloem .... Division Bryophyta
    15. Plants with true xylem and phloem.
        16. Plants with true leaves absent; *enations*[4] present; stems branching dichotomously........
        .... Division Psilophyta
        16. Plants with true leaves present; enations absent; stems branching in various ways or
        unbranched;
            17. Plants with leaves having a single vein (*microphylls*).
                18. Stems not ribbed and not containing silica; leaves photosynthetic ....
                .... Division Lycophyta
                18. Stems ribbed and containing silica; leaves reduced to scales and
                nonphotosynthetic.. Division Sphenophyta
            17. Plants with leaves usually having more than one vein (*megaphylls*).
                19. Plants reproducing by means of spores produced on the leaves.................
                .... Division Pterophyta
                19. Plants reproducing by means of seeds developed from ovules.
                    20. Male reproductive cells (sperms) *motile* (i.e., capable of "swim-
                    ming").
                        21. Leaves pinnate and large, resembling those of palms...................
                        ....Division Cycadophyta
                        21. Leaves smaller and fan-shaped, with numerous dichotomously
                        forking veins... Division Ginkgophyta
                  20. Male reproductive cells nonmotile.
                        22. Plants without flowers; seeds produced in cones or conelike
                        structures.
                          23. Wood containing no vessels; resin canals present.............
                        ....Division Coniferophyta
                          23. Wood containing vessels; resin canals absent..................
                        .... Division Gnetophyta
                    22. Flowering plants....................................Division Anthophyta

Once again, it should be emphasized that a key of this type, while calling attention to basic differences between major groups of organisms, may also oversimplify the differences, partly because it does not list all the characteristics of each member of a given group. Moreover, it does not necessarily identify exceptions to the rule or some of the intermediate forms that frequently are transferred back and forth as research uncovers new evidence or as new lines of speculation develop. Keys at this level of classification also are not completely practical in that one sometimes needs to have specific stages in the life cycles of the plants available to be able to arrive at an identification. Furthermore chemical tests that are not easily performed may be required, or details that may be difficult to see with a light microscope may be called for. Nevertheless, such a synoptic key is generally preferable to a simple listing of groups and their characteristics for purposes of distinguishing among them.

---

4. See page 382 for a discussion of *enations*.

SUMMARY

All organisms are identified by a two-part Latin name (binomial), regardless of the language spoken in the area where the organisms occur; most also have one to many common names. The multiplicity of common names led to an early recognition by biologists of the need for worldwide uniformity in naming and classifying organisms.

Theophrastus (third century B.C.) was the first person of note to attempt to organize and classify plants. In the thirteenth century a distinction was made between monocots and dicots, and by the beginning of the eighteenth century herbalists had added many plants to Theophrastus's original list, with details of fruit and flower structure in addition to form and habit being utilized in classification schemes. By this time it had become customary in Europe to use descriptive Latin phrase names for all organisms, with the first word of the phrase indicating the genus (classification group) of each organism. Linnaeus compiled a comprehensive list of all known Latin phrase names for plants according to their genera in a work entitled *Species Plantarum,* published in 1753. In this work he placed one to many specific kinds of related organisms (species) in each genus; in the margin next to the Latin phrase he listed a single word, which, when combined with the name of the genus, formed an abbreviation for each species. This two-part Latin name (binomial), plus the authority for it, eventually replaced the phrase name, and the method of naming plants became known as the Binomial System of Nomenclature, a system still in use today. Linnaeus also organized all known plants into 24 classes, based primarily on the number of stamens in flowers.

In 1867, European and American botanists met in Paris and agreed to use Linnaeus's 1753 publication and binomials as the starting point for all scientific names of plants. The rules for naming and classifying plants drawn up at that meeting have periodically been revised and expanded, and today constitute the *International Code of Botanical Nomenclature,* a compendium of rules and recommendations now followed by botanists of all nationalities. A similar code governs the naming and classifying of animals.

When classification schemes were first developed, all living organisms were placed in either the Plant Kingdom or the Animal Kingdom. Many microscopic organisms have characteristics of both plants and animals, however, and in the 1860s Hogg and Haeckel proposed a third kingdom (Protista) for all organisms that did not develop complex tissues. In 1938 H. F. Copeland divided the Protista into two kingdoms, with organisms having prokaryotic cells being placed in the Kingdom Monera, and those with eukaryotic cells being left in the Kingdom Protoctista. Since there are differences in the mode of nutrition of organisms in Copeland's Kingdom Protoctista, many biologists now favor five kingdoms, as proposed by R. H. Whittaker in 1969. In Whittaker's system there are three kingdoms based on forms of nutrition (photosynthesis, ingestion of solid food, absorption of food in solution) and two kingdoms of protists based on differences in cellular structure. Slime molds, which have no cell walls at certain stages and do have walls at others, still do not fit well into the five-kingdom system, but this system appears to be the most satisfactory arrangement so far proposed.

Since Linnaeus's time several classifications have been added between the level of kingdom and genus. Genera are grouped into families, families into orders, orders into classes, classes into divisions, and divisions into kingdoms. All forms of bacteria are placed in the Kingdom Monera; all algae, euglenoids, and dinoflagellates are placed in the Kingdom Protista; slime molds, flagellate fungi, and true fungi are placed in the Kingdom Fungi; liverworts, hornworts, mosses, and all vascular plants are placed in the Kingdom Plantae. Viruses, which lack most attributes of living organisms and are not classified within the five kingdoms, are discussed in chapter 15; lichens, which are a combination of an alga and a fungus, are discussed in chapter 17.

The second part of a binomial, the specific epithet, is followed by the name of the author (the one who named or classified the species), usually in abbreviated form. Subcategories (e.g., subspecies, suborder) are sometimes used in classification. Taxonomists may construct keys to aid in the identification of organisms. Most keys are dichotomous, that is, one arrives at an identification by making choices between pairs of statements given throughout the key. The key given to major groups of organisms deals only in generalities and does not indicate occasional exceptions or allow for intermediate forms that may be transferred back and forth between groups as new knowledge about them is gained.

REVIEW QUESTIONS

1. What are the advantages and drawbacks of using scientific names as compared with common names?
2. What is the *International Code of Botanical Nomenclature?*
3. What is meant by the term *binomial?*
4. Which divisions include organisms found mainly in water?
5. List some characteristics by which the following may be distinguished from one another, and then construct a simple dichotomous key that could aid someone from another planet in identifying them: blondes, oranges, hammers, bicycles, rosebushes, automobiles, brunettes, cats, rats, raspberry bushes, dogs, wrenches, peaches.

DISCUSSION QUESTIONS

1. Since phrase names are generally more descriptive of organisms than binomials, do you think they should be revived?
2. After Linnaeus organized all the genera known to him into classes, a number of other categories of classification between the level of kingdom and genus were added. Are these other categories really useful?
3. Do you think biologists are justified in dividing the traditional Plant and Animal Kingdoms into five kingdoms, and would it make any significant difference to the world at large if they had not? Explain.

ADDITIONAL READING

Bell, P. R., and C. L. F. Woodcock. 1972. *The diversity of green plants,* 2d ed. Reading, MA: Addison-Wesley Publishing Co.

Bold, H. C., C. J. Alexopoulos, and T. Delevoryas. 1987. *Morphology of plants and fungi,* 5th ed. New York: Harper & Row, Publishers, Inc.

Cronquist, A. 1968. *The evolution and classification of flowering plants.* Boston: Houghton Mifflin Co.

Ornduff, R., ed. 1967. *Papers on plant systematics.* Boston: Little, Brown and Co.

Radford, A. E. 1986. *Fundamentals of plant systematics.* New York: Harper & Row, Publishers, Inc.

Scagel, R. F., R. J. Bandoni, J. R. Maze, G. E. Rouse, W. B. Schofield, and J. R. Stein. 1984. *Plants: an evolutionary survey.* Belmont, CA: Wadsworth Publishing Co.

Stearn, W. T. 1957. *An introduction to the Species Plantarum and cognate botanical works of Carl Linnaeus.* Printed for the Ray Society and sold by Bernard Quaritch, London.

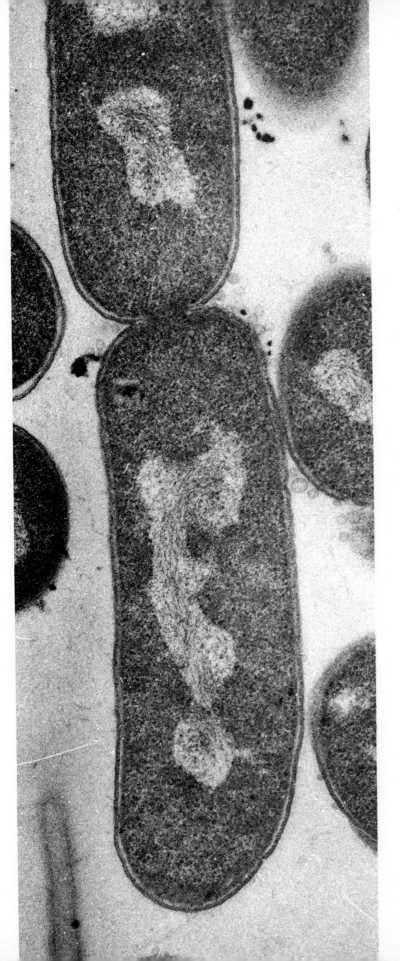

Overview

*T*he chapter introduces Kingdom Monera with an overview of its features, and describes the luminescent bacteria associated with flashlight fish as an introduction to bacteria. A brief investigation of features of bacteria and their reproduction is followed by a discussion of archaebacteria and true bacteria, with representatives and human relevance of heterotrophic and autotrophic true bacteria being surveyed; sections on composting, disease transmission, bacterial diseases, and Koch's Postulates are included. Blue-green bacteria and prochlorobacteria are explored next, and human relevance of blue-green bacteria is examined. The chapter continues with an overview of the nature, reproduction, and human relevance of viruses, and concludes with an introduction to genetic engineering.

Courtesy of R. G. E. Murray

15   Kingdom Monera and Viruses

Some Learning Goals

1. Know the basic forms of bacteria. Explain how a prokaryotic cell differs from a eukaryotic cell and why prokaryotic organisms are difficult to classify.

2. Learn the forms of nutrition in bacteria.

3. Know at least ten bacteria useful to humans, and understand how they are useful.

4. Understand the various ways in which disease bacteria are transmitted; describe how each type of disease bacterium functions in causing disease.

5. Learn Koch's Postulates.

6. Know how the major groups of bacteria and viruses differ from one another in form, pigmentation, and reproduction.

7. Understand what genetic engineering is and explain how it is accomplished.

Outline

FEATURES OF KINGDOM MONERA

The members of Kingdom Monera all have **prokaryotic** cells. Such cells have no nuclear envelopes, but each has a long strand of DNA. In addition, there may be small DNA fragments called *plasmids,* ribosomes, and membranes present in most bacterial cells. Membrane-bound organelles such as plastids, mitochondria, Golgi bodies, and endoplasmic reticulum are lacking (see figure 15.15). In a number of members of Kingdom Monera, cells may occur in a common matrix of gelatinous material as variously shaped colonies, or in the form of chains or **filaments** (threads), but each cell is completely independent—protoplasmic connections between them are absent in virtually all species. Some species are **motile** (capable of independent movement), usually by means of simple **flagella** (whiplike tails that are often, but not always, in pairs) that propel or pull a cell through the water. Several filamentous species exhibit a gliding motion in which the filaments glide back and forth independently or against each other, but most species are nonmotile. Nutrition is primarily by the absorption of food in solution through the cell wall, but some bacteria are *chemosynthetic* (capable of obtaining their energy through chemical reactions involving various compounds or elements) whereas a few true bacteria, including blue-green bacteria and prochlorobacteria, exhibit forms of photosynthesis. Reproduction is predominantly asexual, by means of **fission,** a form of cell division that does not involve mitosis since there are no nuclei or other organelles. The strand of DNA duplicates itself, however, and the DNA is distributed between the two new cells originating from the parent cell. Sexual reproduction is unknown, but genetic recombination occurs in several groups. The genetic recombination is facilitated by means of *pili* (minute tubes) that form connections between cells.

INTRODUCTION TO THE BACTERIA

The astonishing variety and beauty of undersea life, with representatives from all five kingdoms, has attracted the attention of snorkelers and scuba divers in ever-increasing numbers, particularly since Jacques Cousteau and other marine biologists have added to our knowledge of this vast domain with their sophisticated investigations and color cinematography. The shallow tropical waters of the Great Barrier Reef of Australia, the Virgin Islands National Park, Hanauma Bay of Hawaii, and countless other reefs teem with brilliantly colored

FIGURE 15.1   A flashlight fish. (Photo by David Powell, Steinhart Aquarium, San Francisco, California)

marine life that may command the rapt attention of even the most blase of observers. Scuba divers who explore the oceans at night in areas such as Indonesia's Banda Sea, the region just north of Darwin, Australia, or Israel's Gulf of Elat are frequently rewarded with a spectacular sight not seen by those who explore by day. They find themselves in the midst of a shimmering, constantly changing light display.

Countless numbers of lights, bright enough to read by, flicker, dart, glide slowly, or even congregate in glimmering groups. The lights come from large pouches beneath the eyes of small fish, which can turn them off at will by covering them with special folds of tissue. Unlike their deep-sea relatives, however, these *flashlight fish,* as they are called, do not produce the light within their own bodies. Instead, it comes from thousands of microscopic, luminescent *bacteria,* which enjoy a *symbiotic,* or specifically a *mutualistic* relationship with the fish (figure 15.1). **Symbiosis** is simply an intimate association between two dissimilar organisms. **Mutualism** is a form of symbiosis that benefits both organisms. In this instance, the fish furnish some food and oxygen to the bacteria from numerous blood vessels in the pouches, while the bacteria produce the light used by the fish not only to search for food at night but also to attract live food and to confuse predators by flashing the lights on and off while swimming in erratic patterns.

Luminescent bacteria are but minor representatives of the most numerous and geologically ancient of the earth's living organisms. Fossils of bacteria have been found in geological strata that are estimated to be 3.5 billion years old. Since fossils of the first organisms with eukaryotic cells date back

FIGURE 15.2 Duplication of DNA in a bacterial cell. The duplication begins (A) at a single point or bubble (upper right) that expands (B) until the two new rings of DNA separate (C). (Electron micrographs courtesy Tsuyoski Kakefuda)

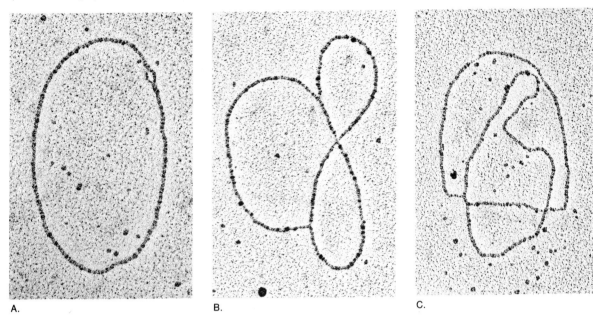

A.　　　　　　　　　　B.　　　　　　　　　　C.

"only" 1.3 billion years, bacteria are almost three times as old as any other organism known to have existed on this planet.

The approximately 2,500 species of bacteria recognized today occur in astronomical numbers in almost every conceivable natural habitat. Each gram of garden soil, for example, is estimated to contain as many as 2 billion bacteria. They are found on and in all plants and animals, in all types of soils, throughout both fresh and salt waters, in polar ice caps and bubbling hot springs, in coal and petroleum, throughout the atmosphere, in bottles of India ink, and almost anywhere else one may care to look for them. Just how many species there are is uncertain, since it is difficult to classify simple one-celled organisms on the basis of visible features alone. For example, one disease-causing bacterium with the scientific name of *Streptococcus pneumoniae* occurs in 77 known "strains," all of which look alike but each of which causes a different form of pneumonia or related disease. Should each strain then be called a species? *Microbiologists,* who study microscopic organisms of all kinds, are not sure, but they tend to recognize clusters of strains based on what they do rather than how they look.

Although bacteria that cause disease and spoilage receive most of the publicity, more than 90% of bacterial species are either harmless or useful to

humans. Bacteria are used in industry for a wide variety of purposes and are so important ecologically that life today would be vastly different without the activities of these tiny organisms. Both industrial uses and the role of bacteria in ecology are discussed in later sections.

Cellular Detail and Reproduction

As indicated in the introduction, bacterial cells are *prokaryotic.* Although such cells have no nuclear envelopes or organelles, folds of the plasma and other membranes apparently perform some of the functions of the organelles of eukaryotic cells. Ribosomes that are about half the size of those in the cytoplasm of eukaryotic cells are also present. Bacterial cells have a *nucleoid,* which is a single, long, very condensed DNA molecule in the form of a ring; it is usually attached at one point to the plasma membrane. In addition, up to 30 or 40 small, circular DNA molecules called *plasmids* may be present. The plasmids replicate independently of the large DNA molecule or chromosome, and the entire complement of plasmids often consists of copies of one or at most very few different plasmids. The chromosome and sometimes all or part of the plasmids replicate before the cell divides (figure 15.2).

FIGURE 15.3   A dividing bacterial cell. The new wall is growing inward. X48,750. (Courtesy R. G. E. Murray)

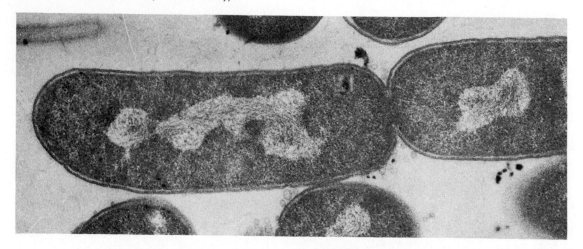

Mitosis as seen in eukaryotic cells does not occur. Instead, there is an internal reorganization of material during which the two DNA molecules migrate to opposite ends of the cell. Then, at approximately the middle of the cell, the plasma membrane and cell wall, as a result of forming a transverse wall, divide the cell in two. Finally, the two new cells separate and enlarge to their original size, or in some bacteria the cells remain attached to each other in chains. This simple form of asexual reproduction, *fission,* is found so universally in bacteria that they have been called *schizomycetes,* which means "fission fungi" (figure 15.3). Bacteria are not, however, currently believed to be closely related to the fungi, as was the case when the name was originally applied.

Under ideal conditions of moisture, food supply, and temperature, a bacterium may undergo fission every 10 to 20 minutes. If a single bacterium of average size were to duplicate itself every 20 minutes for only 36 hours, a mass of bacteria numbering 322,981,536,679,200,000,000,000,000,000,000 and weighing 126,464,618,590 metric tons (137,438,953,472 tons) would be produced. If this mass were not compacted, its volume would exceed that of the earth. Of course, bacteria do not continue to reproduce at their maximum rate for very long because several factors prevent it. These factors include the exhaustion of food supplies and the accumulation of toxic wastes.

FIGURE 15.4   Conjugation in bacteria. Part of the DNA strand of the donor cell migrates through the hollow, tubelike pilus to the recipient cell, where it is incorporated into the DNA of its new cell. Shorter pili cover the surfaces of both cells.

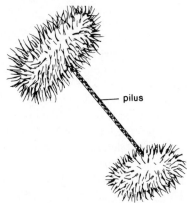

pilus

Bacteria do not produce gametes or zygotes, nor do they undergo meiosis. At least three forms of genetic recombination occur, however. In *conjugation,* part of the DNA strand or chromosome is transferred from a donor cell to a recipient cell through a tiny, hollow pilus while the two cells are in contact (figure 15.4). Once in the recipient cell the DNA fragment becomes part of the new cell. Bacterial cells that are capable of conjugating have different "sexual" forms of DNA, and only cells of different mating types may undergo conjugation. In *transformation,* a living cell picks up fragments of DNA released by dead cells into the medium in which it occurs and incorporates them into its own cell. In *transduction,* fragments of DNA are carried from one cell to another by viruses.

FIGURE 15.5   Three basic forms of bacteria. *A*. Cocci. *B*. Bacilli. *C*. Spirilli. (Photomicrographs by Carolina Biological Supply Company)

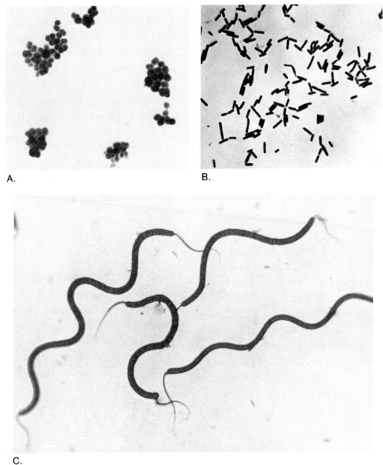

A.

B.

C.

Size, Form, and Classification

Except for the blue-green bacteria, which are discussed separately later in this chapter, and a few other giant bacteria, which attain lengths of up to 60 micrometers, most are less than 2 or 3 micrometers in diameter, and a few of the smaller species approach 0.15 micrometer. The latter are so small that 6,500 of them arranged in a row would not quite extend across the head of a pin. Because of their small size, bacteria are not visible individually to the unaided eye and are studied in the laboratory with electron microscopes or with the highest magnifications of light microscopes. The thousands of different kinds occur primarily in three forms: *cocci* (singular: *coccus*), which are spherical or sometimes elliptical; *bacilli* (singular: *bacillus*), which are rod-shaped or cylindrical; and *spirilli* (singular: *spirillum*), which are in the form of a helix, or spiral (figure 15.5). Further classification is based on numerous visible features including, for example, presence of pigments, development of slimy or gummy capsulelike sheaths around cells, presence of hairlike or budlike appendages, development of internal thicker-walled *endospores,* and gliding movements by which threadlike groups of bacteria appear to slide back and forth. Some bacteria have slender flagella, usually about 5 to 10 micrometers in length, which propel them through fluid media. Others have somewhat shorter tubelike *pili* (singular: *pilus*), which resemble flagella but do not function in locomotion. These apparently enable bacteria to attach themselves to surfaces or to each other.

Bacteria are also grouped into two large categories based on their reaction to a dye. After a heat-fixed smear of cells has been stained blue-black with a violet dye, a dilute iodine solution, alcohol or ace-tone is added. When so treated some species rapidly lose their color (but absorb pink safranin dye), and are called *gram-negative;* others, called *gram-positive,* retain most of the blue-black color. Gram's stain, named after Christian Gram who discovered it in 1884, has many variations, but all the variations produce similar results. Other characteristics used in classifying bacteria are indicated in the discussions that follow.

SUBKINGDOM ARCHAEBACTERIOPHYTINEAE— THE ARCHAEBACTERIA

The **archaebacteria** represent one of two quite distinct lines of the most primitive known living organisms. They differ fundamentally from the other line of bacteria, the *eubacteria,* in having a unique sequence of bases in their RNA molecules, in lacking muramic acid in their walls, in having distinctive lipids, and in their metabolism. These basic differences have led University of Illinois microbiologist Carl Woese and his colleagues, who have conducted considerable research on the archaebacteria, to suggest that the organisms belong in a kingdom of their own. At present, three distinct groups of bacteria are included in the archaebacteria.

The Methane Bacteria

The *methane bacteria,* which comprise the lion's share of the archaebacteria, are killed by oxygen and are active only under anaerobic conditions found in swamps, ocean and lake sediments, hot springs, animal intestinal tracts, sewage treatment plants, and other areas not open to the air. In their metabolic activities they derive their energy from the generation of *methane* gas from carbon dioxide and hydrogen. Methane or "marsh gas" is a principal component of natural gas and may have been a major part of the earth's atmosphere in early geologic times. It still is present in the atmospheres of the planets Jupiter, Saturn, Uranus, and Neptune, and is the main ingredient of "firedamp" that causes serious explosions in mines. Methane will burn when it constitutes only 5% to 6% of the air, and a flitting, dancing light called *ignis fatuus* or "will-o'-the-wisp" that is occasionally seen at night over swamps and marshy places is said to be due to the spontaneous combustion of the gas.

The Salt Bacteria

Air travelers to the West Coast and other areas where there are shallow salt evaporation ponds are frequently struck by how red the ponds appear as they fly over them. The red color is due to a distinctive group of archaebacteria, the *salt bacteria,* whose metabolism enables them to thrive under conditions of extreme salinity that instantly kills other living cells. The bacteria carry on a simple form of photosynthesis with the aid of a membrane-bound red pigment called *bacterial rhodopsin.* The concentration of salt inside the cells is much lower than in their surroundings, but their metabolism is so closely tied to their environment that the bacteria die if placed in waters with lower salt concentrations.

The Sulfolobus Bacteria

A third group of archaebacteria, the *sulfolobus bacteria,* is found in sulphur hot springs. The extraordinary metabolism of these bacteria permits them to thrive not only at very high temperatures—mostly in the vicinity of 80° C (176° F), with some doing very well at only 10° C below boiling point—but their environment is also exceptionally acidic, the hot springs often having a pH of less than 2 (you will recall from chapter 2 that the neutral point on the 14 point pH scale is 7). One genus of bacteria (*Thermoplasma*) in this group is bound by only a plasma membrane and has no cell wall. It is found solely in the embers of coal tailings. Another genus (*Thermoproteus*) appears to be confined to the geothermal areas of Iceland.

HUMAN RELEVANCE OF THE ARCHAEBACTERIA

In the future, methane bacteria may be used on a large scale to furnish energy for engine fuels and for heating, cooking, and light, since the methane gas they produce can be substituted for natural gas. Methane has an octane number of 130, and has been used as a motor fuel in Italy for over 40 years. It is clean, nonpolluting, safe and nontoxic, and prolongs the life of automobile engines, also making them easier to start. The methane is given off by the bacteria as they "digest" organic wastes in the absence of oxygen. Nine kilograms (20 pounds) of horse manure or 4.5 kilograms (10 pounds) of pig manure fed daily into a methane digester will produce all the gas needed for the average American adult's cooking

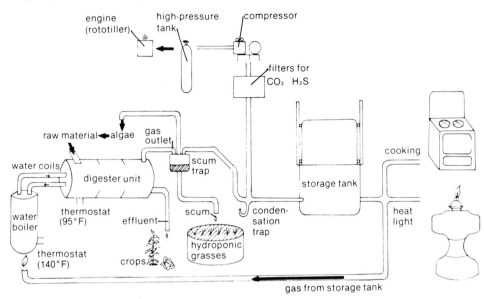

FIGURE 15.6   Integrated organic digester operation for the production of methane gas. (Reprinted from *Producing Your Own Power* © 1974 by Rodale Press, Inc. Permission granted by Rodale Press, Inc., Emmaus PA, 18049)

needs. Considerably less green plant material is required to produce the same amount. The digester basically consists of an airtight drum connected by a pipe to a storage tank with a means of drawing off the sludge left after the gas has been produced (figure 15.6). The sludge itself makes an excellent fertilizer, although many sludges that originate from municipal and large-scale agricultural plants carry with them toxic levels of metals. France had over 1,000 methane plants in operation by the mid 1950s, and in India, where cows produce over 812 million metric tons (800 million tons) of manure per year, many villages satisfy their fuel needs by means of manure-fed methane gas digesters, which are now being employed on a small scale in rural areas in the United States. Methane is the primary source of hydrogen in the commercial production of ammonia.

SUBKINGDOM EUBACTERIOPHYTINEAE— THE TRUE BACTERIA

Division Eubacteriophyta

Class Eubacteriae—The Unpigmented, Purple, and Green Sulfur Bacteria

The **eubacteria** or *true bacteria* have *muramic acid* in their cell walls and differ fundamentally from archaebacteria in several other ways (see the introduction to *archaebacteria*). They constitute quite a

heterogeneous assemblage, with most being **heterotrophic** (organisms that cannot synthesize their own food and therefore depend on other organisms for it); the majority of heterotrophic bacteria are *saprobes* (living organisms that obtain their food from nonliving organic matter). Saprobic bacteria, along with fungi, are primarily responsible for decay and for recycling all types of organic matter in the soil. Some of their recycling activities are discussed in the section on the nitrogen cycle in chapter 23. Other heterotrophic bacteria are *parasites* (living organisms that depend on other living organisms for their food); several parasites that cause important human diseases are discussed later in the chapter.

Autotrophic Bacteria

A few groups of true bacteria are similar to green plants in being **autotrophic;** that is, they are capable of synthesizing organic compounds from simple inorganic substances. Some carry on photosynthesis without, however, producing oxygen as a by-product. Included in the photosynthetic bacteria are the *purple sulfur bacteria,* the *purple nonsulfur bacteria,* and the *green sulfur bacteria.* The *blue-green bacteria* and the *prochlorobacteria,* which do produce oxygen as a product of their photosynthesis, are discussed separately later in this chapter. Cells of the first two groups appear purplish, or occasionally red to brown, because of the presence of a mixture of greenish, yellow, and red pigments. Their greenish *bacteriochlorophyll* pigments (there are several closely related ones) are very similar to the chlorophyll *a* of higher plants.

The pigments of bacteria are located in thylakoids or small spherical bodies, no plastids being present. In their photosynthesis, the purple bacteria substitute hydrogen sulfide, the bad-smelling gas given off by rotten eggs, for the water used by higher plants, so that the generalized equation for the process is as shown below:

The equation for the purple nonsulfur bacteria is similar, but hydrogen from organic molecules is used instead of hydrogen sulfide. Green sulfur bacteria do use hydrogen sulfide, but their chlorophyll, called *chlorobium chlorophyll,* differs significantly in its chemistry from the chlorophylls of higher plants.

Other groups of bacteria are *chemoautotrophic;* that is, they are capable of obtaining the energy they require from various compounds or elements through chemical reactions involving the oxidation of reduced inorganic groups such as NH_3, H_2S, and Fe^{++}, or the oxidation of hydrogen gas (see the footnote on oxidation-reduction reactions in chapter 10). Chemoautotrophic bacteria include *iron bacteria,* which transform soluble compounds of iron into insoluble substances that accumulate as deposits (e.g., in water pipes); *sulfur bacteria,* which can convert hydrogen sulfide gas to elemental sulfur and sulfur to sulphate; and *hydrogen bacteria,* which flourish in soils where they utilize molecular hydrogen derived from the activities of anaerobic and nitrogen-fixing bacteria. (See chapter 23 for a discussion of nitrifying and nitrogen-fixing bacteria and their roles in the nitrogen cycle.)

HUMAN RELEVANCE OF THE UNPIGMENTED, PURPLE, AND GREEN SULFUR BACTERIA

Compost and Composting

Long before the existence of bacteria was known or even suspected, primitive agriculturists made rough piles of weeds, garbage, manure, and other wastes.

They watched as the piles became reduced to a fraction of their original volume and changed into *compost,* a dark fluffy material that conditions and enriches soil when mixed with it. Today the compost pile is at the heart of the activities of numerous organic gardeners and farmers. With many communities instituting ordinances forbidding the burning of leaves and refuse, and with space for waste disposal becoming scarce, a significant number of cities and towns have turned to the composting of street leaves and other materials on a large scale. They have found that they not only save space but are also producing a useful, ecologically compatible product (figure 15.7).

Because bacteria are ubiquitous, a single leaf or nonliving tissues in a pile of any size will eventually be decomposed. Ideally, however, a compost pile is 2 meters wide, 1.5 meters deep, and at a minimum, 2 meters long (6 feet wide, 3 feet deep, and at least 6 feet long). If the pile is of lesser length or breadth, or greater depth, the conditions that favor the breakdown of materials (created by the bacteria themselves) develop at a slower pace.

Any accumulation of household garbage, leaves, weeds, grass clippings, and/or manure may be heaped together, and it has been demonstrated that no so-called starter culture of microorganisms is needed to initiate decomposition. Once the ever-present decay organisms have material to decompose, their numbers increase rapidly, and much heat is generated. In fact, the temperature in the center of the pile often rises to 70° C (158° F). As it rises, populations of organisms adapted to higher temperatures replace those not as well adapted. The remains of the latter are then added to the compost. The high temperatures also kill many weed seeds and most disease organisms.

The proportion of carbon to nitrogen present in a compost pile largely determines the pace at which the microbial activities proceed, a ratio of 30 carbon to 1 nitrogen being optimal. Microbial growth stops if the proportion of carbon gets much greater, and nitrogen is lost in the form of ammonia if the ratio drops lower.

$$CO_2 + 2\,H_2S \xrightarrow[\text{light}]{\text{bacteriochlorophyll}} (CH_2O) + H_2O + 2\,S$$

carbon dioxide   hydrogen sulfide   carbohydrate   water   sulfur

If the materials are kept moist and turned occasionally to aerate the pile, composting may be completed in as little as two weeks. Shredding the materials exposes much more surface area for the microorganisms to work on and speeds up the process. Composters generally avoid or keep to a minimum a few materials such as *Eucalyptus* and walnut leaves, which contain substances that inhibit the growth of other plants, when the compost is intended for use as a soil conditioner or crop fertilizer. Bamboo and some types of fern fronds are particularly resistant to decay bacteria and will decompose much more slowly than other materials. Domestic cat manure may contain stages of a parasite that can infect humans, although the organism should be killed if proper composting techniques are employed. Generally, however, almost any organic materials are suitable.

While the value of compost to the soil is indisputable, its value as a fertilizer has certain limitations. The nitrogen content of good compost is about 2% to 3%, phosphorus 0.5% to 1%, and potash 1% to 2%, as compared with five to ten times those amounts in chemical fertilizers, which also require less labor to produce. Nevertheless, with the spiraling costs of chemical fertilizers, an increasing awareness of the problems accompanying their use, and greater public enlightenment concerning the accumulation and disposal of solid wastes, compost is undoubtedly destined to play a major role in pertinent agricultural practices of the future.

True Bacteria and Disease

It has been calculated that plant diseases cause American farmers losses in excess of $4 billion per year. Although many plant diseases are caused by fungi, a number involve bacteria. The latter include diseases of pears, potatoes, tomatoes, squash, melons, carrots, citrus, cabbage, and cotton. Bacteria also cause huge losses of foodstuffs after they have been harvested or processed. They are even better known, however, as the culprits in many serious diseases of animals and humans.

Since bacteria are so tiny and often incapable of independent movement, how are they transmitted, and what is it about their activities that enables them to fell creatures billions of times their size? We have learned much about the activities of bacteria in the past century and presently know that they can gain access to human tissues in a variety of ways.

FIGURE 15.9   Botulism bacteria. (Photomicrograph Courtesy Robert McNulty)

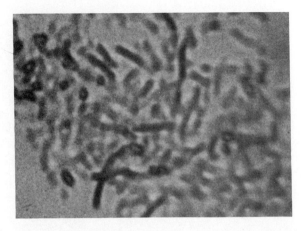

Modes of Access of Disease Bacteria

Access from the Air

Every time a person coughs, sneezes, or just speaks loudly, an invisible cloud of tiny saliva droplets is produced (figure 15.8). Each droplet contains bacteria or other microorganisms and a tiny amount of protein. The moisture usually evaporates almost immediately, leaving minute protein flakes to which live bacteria adhere. As normal breathing takes place, these soon may find their way into the respiratory tracts of other humans or animals, particularly if they have been confined within the air of a room. Fortunately, the natural resistance of most of those who acquire the bacteria in this way prevents the bacteria from multiplying to the point of causing a disease. When the resistance is not there, however, a number of diseases, including diphtheria, whooping cough, some forms of meningitis and pneumonia, and "strep throat" can develop. Psittacosis, a disease carried by birds, is caused by the inhalation of *chlamydias,* which are exceptionally minute organisms unable to manufacture their own ATP molecules. They are apparently "energy parasites," depending on their host cells for the energy needed to carry on their own functions. Some chlamydias are transmitted sexually, and in recent years have become a widespread human problem.

Access through Contamination of Food and Drink

When sewers were commonly open and food processing was done under less sanitary conditions than exist at present, a number of bacterial diseases, including cholera, dysentery, *Staphylococcus* and *Salmonella* "food poisoning," and typhoid and paratyphoid fevers often reached epidemic proportions. Although *Staphylococcus* food poisoning, which is seldom fatal, is still fairly common, civilized countries rarely see epidemics caused by other bacteria unless a natural disaster such as a flood or a typhoon disrupts normal sewage disposal. The diseases are more often spread by "carriers" who handle food, or by houseflies. *Salmonella* bacteria often multiply in raw shellfish to the extent that governments of the United States and other countries bordered by ocean waters place a ban on their harvesting at certain times to reduce the possibility of human contamination from consuming them without sufficient cooking. These bacteria multiply in the intestinal tract or spread from there to other parts of the body.

Some bacteria (e.g., those responsible for Legionnaire's disease, which killed 34 members of the American Legion attending a convention in a Philadelphia hotel in 1976 and infects an estimated 50,000 Americans annually) are very common on algae in freshwater streams, lakes, and reservoirs. They usually pass through the human digestive tract harmlessly without multiplying, but on rare occasions, something unknown triggers their reproduction, often with fatal results.

The most deadly of all known biological toxins is produced by a bacterium with the scientific name of *Clostridium botulinum* (figure 15.9). The name comes from *botulus,* the Latin word for sausage, since the bacterium was first discovered after people had died from eating some contaminated sausage at a picnic. Unlike salmonella "food poisoning," **botulism** is not an infection but is poisoning from a substance produced by bacteria that can grow and multiply anaerobically (in the absence of oxygen) in improperly processed or stored foods. Home-canned beans, beets, corn, and asparagus in particular have been known to permit the development of botulism bacteria. Just 1 gram (0.035 ounce) of the toxin is sufficient to kill 14 million adults, and a little more than half a kilogram (1.1 pound) could eliminate the human race!

The bacteria, which are present in most soils, produce unusually heat-resistant spores and are likely to be present on any soil-contaminated foods. They may not be destroyed during canning unless the food is heated for 30 minutes at 80° C (176° F), or boiled for 10 to 15 minutes. They ordinarily will not grow in foods that have been preserved in brines containing at least 10% salt (sodium chloride) or in fairly acid media such as the juices produced by most stone fruits. The toxin is absorbed directly from the stomach and the intestines, affecting nerves and muscles and causing paralysis. While some antidotes are available, they are ineffective after symptoms have become advanced. Deaths reported from botulism in the United States reached a peak of about 25 per year in the 1930s but have declined to 5 or 6 per year since then, although there is recent evidence that deaths of infants due to hitherto unknown causes are actually due to botulism. For reasons that are not clear, botulism bacteria, which pass harmlessly through the digestive tracts of humans over the age of 6 months, may germinate in the intestines of infants between the ages of 1 to 6 months. Between 1976 and 1980, 170 cases of sudden infant death reported worldwide appear to have been due to botulism, and the evidence now suggests that botulism is responsible for about 5% of the 8,000 cases of sudden infant death that occur each year in the United States.

Access through Direct Contact

Bacteria responsible for diseases such as syphilis, gonorrhea, anthrax, and brucellosis enter the body through the skin or mucous membranes (i.e., those membranes lining tracts with openings to the exterior of the body). Both syphilis and gonorrhea are transmitted through sexual intercourse or other forms of direct contact, rarely through the use of public washroom towels or toilets. In 1976, gonorrhea accounted for the largest number of reported cases of any communicable disease in the United States, and currently more than 3 million Americans become infected every year. The symptoms of both syphilis and gonorrhea, which include persistent sores or discharges from the genitalia, sometimes disappear after a few weeks, leading victims to believe the body has healed itself. More often than not, however, symptoms reappear in different parts of the body at a later date. Both diseases are curable when treated promptly, but it is very important that such treatment be sought, since failure to do so can lead to sterility, blindness, and even death.

Anthrax, which is primarily a disease of cattle and other farm animals in addition to wild animals, is sometimes transmitted to humans, particularly workers in the wool and hide industries. Like syphilis and gonorrhea, it can be effectively eliminated if treated early enough but may be fatal if allowed to progress. Brucellosis, another disease of farm animals, is occasionally transmitted to humans through direct contact or through the consumption of contaminated milk. It is sometimes called "undulant fever" because of a daily rise and fall of temperature apparently associated with the release of toxins by the bacteria.

Access through Wounds

When one steps on a dirty nail or is wounded in such a way that dirt is forced into body tissues, tetanus ("lockjaw") bacteria, which are common soil organisms, may gain access to dead or damaged cells. There they can multiply and produce a deadly toxin so powerful that 0.00025 gram (0.00000088175 ounce) is sufficient to kill an adult. In contrast, about 150 times that amount of strychnine is needed to achieve the same result. Control of tetanus through immunization is very effective and has become widespread. Several related bacteria that gain access to the body in the same way are responsible for potentially fatal gas gangrene, which in the past was feared on the battlefields but is now controlled through the use of antibiotics and aseptic techniques.

FIGURE 15.11   Robert Koch. (Courtesy National Library of Medicine)

Access through Bites of Insects and Other Organisms

Bubonic plague (the "Black Death") and tularemia are two bacterial diseases transmitted by fleas, deerflies, ticks, or lice that have been parasitizing infected animals, particularly rodents.

Rat fleas, found on infected rats that inhabit dumps, sewers, barnyards, and ships (figure 15.10), acquire the bacteria for plague and then pass them on to humans through their bites. The disease has been found in ground squirrels and other rodents in the United States, particularly in the West, since 1900. In the past, bubonic plague has spread with great speed and reached devastating epidemic proportions. In 1665 in London, hundreds of thousands perished from the disease, and between 1347 and 1349 it is believed to have killed one-fourth of the entire population of Europe (about 25 million persons). Today plague is rare in North America, but it still occasionally manifests itself in port cities of Asia, Europe, and South America. Control depends on control of rats and fleas, which are virtually impossible to eradicate entirely, and the use of vaccines, which produce immunity for about 6 to 12 months.

Tularemia is primarily a disease of animals, but it may be transmitted to humans through the bites of infected ticks or deerflies. It is an occupational disease of meat handlers and is fatal in 5% to 8% of the cases. Ticks, lice, and fleas may also transmit rickettsias, which cause typhus and spotted fevers. Rickettsias are extremely tiny bacteria that live inside of eukaryotic cells. Pleuropneumonia-like organisms (PPLOs) are also minute bacteria that may be transmitted by various means. These have no cell walls and therefore are quite plastic. They are found in many plants, in hot springs, and in the moist surfaces of the respiratory and intestinal tracts of animals and humans. They are responsible for a form of human pneumonia and for numerous plant diseases. They are the only prokaryotes known that, as a group, are resistant to penicillin.

Koch's Postulates

Since there are so many bacteria present everywhere, how can one be certain that a given bacterium obtained from an infected person is actually the organism responsible for the observed disease? A German physician, Robert Koch (figure 15.11), became known during the latter half of the nineteenth century for his investigations of anthrax and tuberculosis. As a result of his work, Koch formulated rules for proving that a particular microorganism is the cause of a particular disease. His rules, with minor modifications, are still followed today. They have come to be known as *Koch's Postulates,* and their essence is as follows:

FIGURE 15.12   A tomato hornworm before, and 3 days after spraying with *Bacillus thuringiensis*.

A.

B.

1. The microorganism must be present in all cases of the disease.
2. The microorganism must be isolated from the victim in pure culture (i.e., in a culture containing only that single kind of organism).
3. When the microorganism from the pure culture is injected into a susceptible host organism, it must produce the disease in the host.
4. The microorganism must be isolated from the experimentally infected host and grown in pure culture for comparison with that of the original culture.

True Bacteria Useful to Humans

For many years, insect pests of food plants have been controlled primarily through the use of toxic sprays. Residues of the sprays remaining on the fruits and vegetables have accumulated in human tissues, however, often with adverse effects, while at the same time many organisms have become immune or resistant to the toxins. In addition, the sprays kill useful organisms, and precipitation runoff washes the toxins into streams, lakes, and oceans, harming or killing aquatic creatures. As an awareness of the undesirable effects of the use of toxic sprays has developed, alternative means of controlling crop predators have been sought. Today many harmful pests and even weeds can be limited to inconsequential numbers through the use of "biological controls" (see appendix 2).

Three such controls involve bacteria that have been registered for use by the United States Department of Agriculture. One, with the scientific name of *Bacillus thuringiensis* (often referred to as BT), has been remarkably effective against a wide range of caterpillars and worms, including peach tree borers, European corn borers, bollworms, cabbage worms and loopers, tomato and fruit hornworms (figure 15.12), tent caterpillars, fall webworms, leaf miners, alfalfa caterpillars, leaf rollers, gypsy moth larvae, and cankerworms. The bacteria, which are easily mass-produced by commercial companies, are sold in the form of a stable wettable dust containing millions of spores. When the spores are sprayed on food plants, they are harmless to humans, birds, animals, earthworms, or any living creatures other than moth or butterfly larvae. When a caterpillar ingests any tissue with BT spores on it, the bacteria quickly become active and multiply within the digestive tract, soon paralyzing the gut. The caterpillar stops feeding within two or three hours and slowly turns black, dropping off the plant in two to four days.

In 1983, a variety of *Bacillus thuringiensis* (var. *israelensis*) was introduced into the commercial market for the control of mosquitoes. Called BTI, the bacterium attacks only mosquito larvae ("wigglers") and one or two other pests. Six years of experiments performed on more than 70 species of fish, snails, shrimp, and insects demonstrated no adverse effects on either plants or animals other than mosquitoes (and a couple of lesser pests) even at dosages 100 times more powerful than those needed to kill mosquito larvae. Use of this bacterium to control mosquitoes in the future may constitute a significant step in lessening the ecological damage and disruption that so frequently accompanies the use of toxic chemicals for pest control.

Another bacterium, *Bacillus popilliae,* is also marketed in a powder form. When it is applied to soil where grubs of the highly destructive Japanese beetle are present, it causes what is known as "milky spore disease" in the grubs, which die in a few days. The spores are carried throughout the topsoil by rainwater, foraging grubs, and by organisms that feed on the grubs.

Bacteria play a major role in the dairy industry. Milk, which is composed of proteins, carbohydrates, fats, minerals, vitamins, and about 87% water, has exceptional nutritive value for animals and is also an excellent medium for the growth of many kinds of bacteria. Milk is sterile when secreted within the udder of the cow, but it picks up bacteria as it leaves the cow's body. It spoils rapidly if it is not obtained and stored under strictly sanitary conditions. Even after it has been pasteurized and refrigerated, the numbers of bacteria in it will increase the longer it stands. If milk is left in open containers in household refrigerators, for example, bacterial growth will not be kept in check for much longer than 24 hours.

Except for milk itself, either alone or in mixtures (e.g., ice cream), all dairy products are manufactured from raw or pasteurized milk by the controlled introduction of various bacteria. Such products include buttermilk, acidophilus milk, yogurt, sour cream, kefir, and cheese. Whey, the watery part of the milk separated from curd during cheesemaking, is utilized, along with starches and molasses, for the commercial production of lactic acid. Lactic acid is used extensively in the manufacture of textile and laundry products, in the leather tanning industry, as a solvent in lacquers, and in the treatment of calcium and iron deficiencies in humans.

In their metabolism of various sugars, proteins, and other organic substances, bacteria also produce waste products that have important industrial uses. Such products are often produced in large quantities by culturing bacteria in huge vats. These products include acetone, used in the manufacture of photographic film, explosives, and as a solvent (e.g., nail polish remover); butyl alcohol, another solvent used in the manufacture of synthetic lacquers; dextran, used as a food stabilizer and as a blood plasma substitute; sorbose, used in the manufacture of ascorbic acid (vitamin C); and citric acid, used as a lemon flavoring for foods. Some vitamin and medicinal preparations also involve bacterial synthesis. Bacteria are now used to break down oil molecules when spills have occurred, and are increasingly being used to neutralize toxic wastes. With the probability of the development of bacteria specifically engineered (see the *genetic engineering* section that follows) to deal with a host of pollution problems, their use for this purpose may become widespread in the future.

Bacteria are used in the curing of vanilla pods, cocoa beans, coffee, and black tea and in the production of vinegar, sauerkraut, and dill pickles. Fibers for linen cloth are separated from flax stems by bacteria, and green plant materials are fermented in silos to produce ensilage for cattle feed. In recent years, the production of several important amino acids by bacteria has been exploited commercially. More than 6,800 metric tons (7,500 tons) of one amino acid, glutamic acid, are produced in the United States each year. This is in demand as a flavor-enhancing agent in the form of monosodium glutamate.

Class Cyanobacteriae—The Blue-Green Bacteria

Introduction

In the past, algae as a group have been distinguished from other organisms in being photosynthetic, in having relatively simple structures, and for those reproducing sexually, in having sex organs consisting of a single cell. As more has become known about cellular details, however, it is apparent that the differences between the **blue-green bacteria** (formerly known as the blue-green algae) and true algae are quite basic. Blue-green bacteria have prokaryotic cells, like all other organisms comprising the Kingdom Monera, whereas all algae that are assigned to the Kingdom Protista have eukaryotic cells. In fact, blue-green bacteria are so much like other true bacteria (figure 15.13) that most biologists now have abandoned the reference to them as algae and consider them true bacteria. The principal distinctions between organisms traditionally regarded as bacteria and blue-green bacteria are: blue-green bacteria possess chlorophyll *a,* which is found in higher plants, and oxygen is produced when they

FIGURE 15.13 Similarity of form between various unpigmented bacteria and blue-green bacteria. (After Pelczar, M. J., Jr., and R. D. Reid. 1972. *Microbiology,* 3d ed. Redrawn by permission of the McGraw-Hill Book Co.)

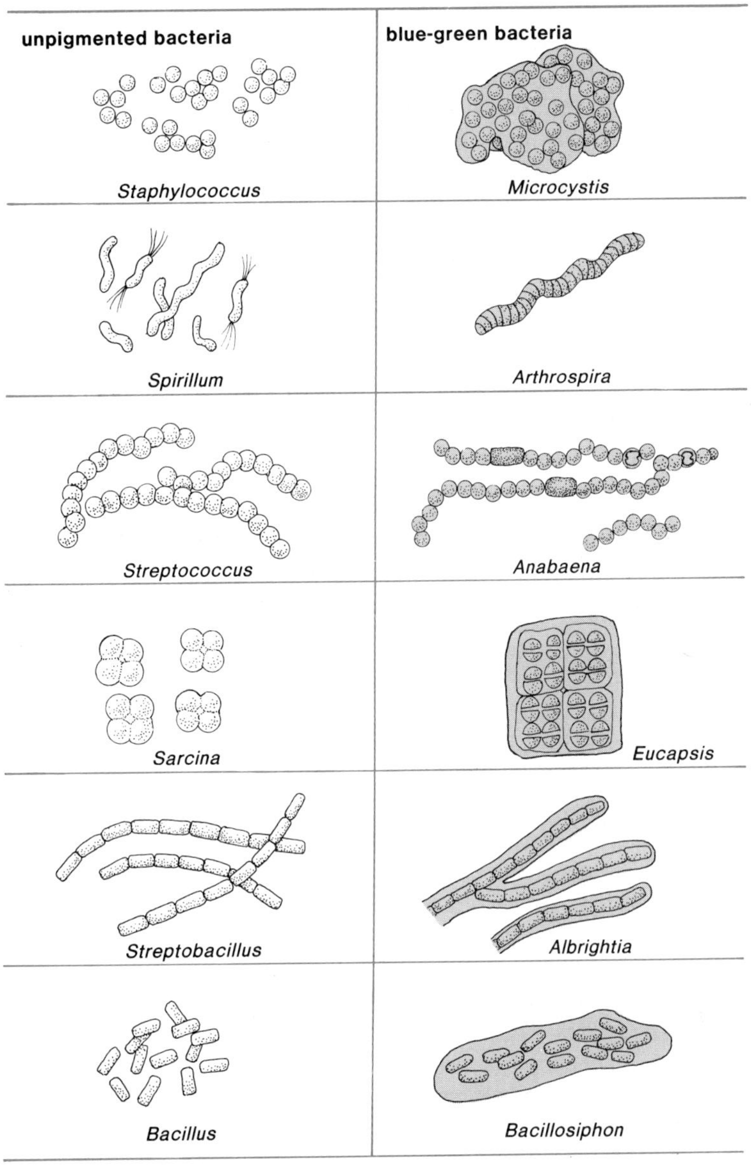

unpigmented bacteria	blue-green bacteria
Staphylococcus	Microcystis
Spirillum	Arthrospira
Streptococcus	Anabaena
Sarcina	Eucapsis
Streptobacillus	Albrightia
Bacillus	Bacillosiphon

undergo photosynthesis; blue-green bacteria also possess blue *phycocyanin* and red *phycoerythrin* phycobilin pigments. Blue-green bacteria are the only organisms that can both fix nitrogen and produce oxygen—a paradox, since nitrogen-fixation is essentially an anaerobic process. Except for the recently discovered prochlorobacteria, other bacteria capable of carrying on photosynthesis do not produce oxygen, and they do not have chlorophyll *a*.

Distribution

Blue-green bacteria are found in almost as diverse a variety of habitats as other true bacteria. They are common in temporary pools or ditches, particularly if the water is polluted. They are not found in acidic waters, but they occur abundantly in other fresh and marine waters around the globe, from the frozen lakes of Antarctica to warm tropical seas, and in hot springs whose water temperatures approach 85° C (185° F). In Yellowstone National Park, a different species of blue-green bacterium is found in each temperature range of hot springs. There the bacteria precipitate chalky, insoluble carbonate deposits, which become a rocklike substance called *travertine*. The deposits accumulate at the rate of up to 2 to 4 millimeters per week, with other blue-green bacteria often forming brilliantly colored streaks in the travertine. Blue-green bacteria are frequently the first photosynthetic organisms to appear on bare lava after a volcanic eruption, and they also occur in the tiny fissures of desert rocks. Some are found in jungle soils, or on the shells of turtles and snails, while others live symbiotically in various types of organisms, including amoebae, protozoans, diatoms, certain sea anemones and their relatives, some fungi, and in the roots of tropical palmlike plants called *cycads* (see chapter 20). They also flourish in tiny pools of water formed at the bases of the leaves of tropical grasses and other plants, and they are well-known components of "compound" organisms called *lichens,* which consist of a fungus and a photosynthetic partner (see chapter 17).

Form, Metabolism, and Reproduction

The cells of blue-green bacteria often occur in chains or hairlike filaments, which are sometimes branched. Several species occur in irregular, spherical, or platelike colonies, the individual cells being held together by the gelatinous sheaths they secrete (figure 15.14). The sheaths may be colorless or pigmented with various shades of yellow, red, brown, green, blue, violet, or blue-black, which makes some colonies quite striking in appearance. The cells themselves appear blue-green in color in about half of the approximately 1,500 known species. This color, which is due to the presence of chlorophyll *a* and phycocyanin, is often masked by the presence of other pigments. Several yellow or orange carotenoid pigments similar to those of higher plants are usually present, and varying amounts of phycoerythrin (a red phycobilin pigment in a form unique to the blue-green bacteria), may give the cells a distinct red color. The periodic appearances of large numbers of blue-green bacteria with considerable amounts of phycoerythrin are believed to have given the Red Sea its name.

The organisms produce a nitrogenous food reserve called *cyanophycin.* The production of such food reserves is atypical for prokaryotic organisms. Blue-green bacteria also produce and store carbohydrates and lipids. Flagella are unknown in the blue-green bacteria, but some of these algae are nevertheless capable of movement. *Oscillatoria* filaments (figure 15.14), for example, seem to rotate on axes and move in a gliding fashion, apparently by the twisting of minute fibrils inside the cell walls while secreting mucilage that creates friction. New cells are formed through fission, while new colonies or filaments may arise through *fragmentation* (breaking up) of older ones.

In the common genera *Nostoc* and *Anabaena* (figure 15.14), which form chains of cells, fragmentation often occurs at special larger cells called **heterocysts** produced at intervals in the chains. Heterocysts are the sites of fixation of nitrogen from the air. Members of these two genera also may produce thick-walled cells called *akinetes.* These cells can resist freezing and other adverse conditions. This feature permits the species to survive cold or dry conditions, germinating when favorable conditions return, and becoming new chains or filaments. Some akinetes have been known to germinate after lying dormant for more than 80 years. Blue-green bacteria do not produce gametes or zygotes and do not undergo meiosis; genetic recombination has, however, been reported—apparently taking place in similar fashion to that reported for other bacteria—but its occurrence evidently is rare.

Blue-Green Bacteria, Chloroplasts, and Oxygen

Blue-green bacteria that occur symbiotically in other organisms commonly lack a cell wall and appear to function essentially as chloroplasts. When a eukaryotic cell containing chloroplasts divides, the chloroplasts divide at the same time. The cells of blue-green bacteria occurring within the cells of other organisms divide in similar fashion, leading to speculation that chloroplasts originated as blue-green bacteria or *prochlorobacteria* (see the discussion of prochlorobacteria which follows the blue-green bacteria) living within other cells. Fossils of photosynthetic organisms believed to be 3.5 billion years old, and closely resembling present-day blue-green bacteria, have been found in Australia, although it apparently was not until half a billion years later that these organisms began producing oxygen as a byproduct of photosynthesis. The oxygen slowly began to accumulate, becoming substantial about 1 billion

FIGURE 15.14   Representative blue-green bacteria.

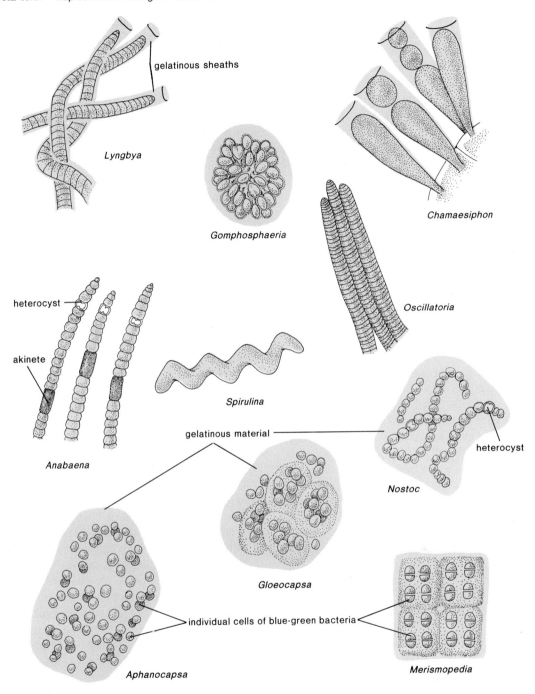

gelatinous sheaths

Lyngbya

Gomphosphaeria

Chamaesiphon

Oscillatoria

heterocyst

akinete

Anabaena

Spirulina

gelatinous material

heterocyst

Nostoc

Gloeocapsa

individual cells of blue-green bacteria

Aphanocapsa

Merismopedia

years ago. At the same time the oxygen was accumulating, other photosynthetic organisms appeared, and forms of aerobic respiration (see chapter 10) developed. Within the last half billion years *ozone,* a by-product of oxygen, accumulated in sufficient quantity to become an effective shield against most of the harmful ultra-violet radiation originating with the sun, and photosynthetic organisms, which had been protected from the radiation by their watery environments, were then able to survive on land. Thus, blue-green bacteria appear to have played a fundamental role in almost the entire history of living organisms.

HUMAN RELEVANCE OF THE BLUE-GREEN BACTERIA

Blue-green bacteria are included among the many aquatic and photosynthetic organisms at the bottom of various food chains. They store energy in the sugar they produce through photosynthesis. Small fish and crustaceans feed on them, only to be eaten by larger fish, which then become food for other aquatic organisms or humans.

During warmer months various algae and blue-green bacteria may become abundant in bodies of fresh water, especially if the water is polluted. A floating scum or mat called a *bloom* may cover 2,000 or more square kilometers (800 square miles) of the surfaces of larger lakes in late summer. Such blooms often impart a "fishy," or otherwise objectionable, odor or taste to the water, and toxic substances may be produced when the blue-green bacteria die and are decomposed. There have been occasional reports of cattle, hogs, sheep, rabbits, dogs, and even poultry having been poisoned through drinking water that had an algal-bacterial bloom. While the blue-green bacteria are alive, the oxygen content of the water increases temporarily, but later the decay bacteria decomposing their dead bodies may deplete the oxygen to the extent of killing fish and other organisms.

One species of blue-green bacteria, which occurs in abundance on the coral beaches of the Gulf of Mannar on the southeast coast of India, is palatable to horses but is toxic to them and frequently causes their death. Many poisonous fish that are immune to the toxins of certain blue-green bacteria become poisonous to their predators only after feeding on the bacteria. Most blue-green bacteria are not palatable to humans, but one exception is a species of *Spirulina,* which has been used for food by the natives of the Lake Chad region of central Africa for many years. A few colorless forms are mild parasites in humans and animals.

In human water supplies, blue-green bacteria frequently clog filters, corrode steel and concrete, cause natural softening of water, and produce undesirable odors or coloration in the water. Many communities control the blue-green bacteria in reservoirs through the addition of very dilute amounts of copper sulphate.

More than 40 species of blue-green bacteria are known to fix nitrogen from the air at roughly the same rates as the nitrogen-fixing bacteria of leguminous plants (see chapter 23). They may be more important than originally thought in this regard. In Southeast Asia, so much usable nitrogen is produced in the rice fields by naturally occurring blue-green bacteria that rice is often grown for many years on the same land without the addition of fertilizer.

Class Prochlorobacteriae—The Prochlorobacteria

In 1976 R. A. Lewin of the Scripps Institute of Oceanography announced the discovery of unicellular, prokaryotic organisms with bright green cells that were living on marine animals called *sea squirts.* These new organisms proved to have the chlorophylls *a* and *b* of higher plants, but no trace of the phycobilin accessory pigments associated with blue-green bacteria. Instead, their accessory pigments were confined to the carotenoid pigments found in higher plants. Also, unlike the single membrane thylakoids of blue-green bacteria, those of the new organisms were double. Lewin considered the pigment differences between blue-green bacteria and these bright green cells to be basic enough to warrant recognition at the division level, and he proposed a new division to be known as the *Prochlorophyta.* Many microbiologists are reluctant to recognize these organisms as belonging to a separate Moneran division because the prokaryotic cell structure and chemistry is similar to that of blue-green and other true bacteria (figure 15.15), but others agree with Lewin's assessment of the significance of the pigment system. While the pigment system is, indeed, significant, they are treated here as a class of true bacteria because their remaining structure and features are essentially indistinguishable from those of other true bacteria.

The discovery of prochlorobacteria adds weight to the theory that chloroplasts may have originated from such cells living within the cells of other organisms, especially since the pigments involved are identical with those of higher plants.

FIGURE 15.15 A section through a cell of the prochlorobacterium *Prochloron*, which lives on the surface of sea squirts (marine animals). Note the absence of a nucleus and other organelles, and the concentric layers of membranes that perform some of the functions of organelles. ×6,000. (Electron micrograph courtesy Jean Whatley)

VIRUSES

Introduction

Smallpox is a communicable disease that apparently was widespread for thousands of years, having often been fatal in the past, periodically killing countless numbers of individuals. As it developed in its victims, it manifested itself as blisterlike lesions on the skin, which later frequently became permanent pits or depressions. In 1901 an outbreak of smallpox in New York caused 720 deaths. Since that time, however, it has been eliminated in the United States through vaccination, and the World Health Organization believes it has now been eradicated from the remainder of the world.

Vaccination involves the introduction of a weakened form of disease agent to the body. The body's natural defenses, in fighting the agent, build up an immunity to the disease. The first known vaccinations were performed by two Englishmen, Benjamin Jesty, a farmer, and Edward Jenner, a country physician. They both had noticed that farmhands working with cows having cowpox, a comparatively mild disease related to smallpox, did not contract smallpox itself during the devastating epidemics of the disease that occasionally occurred in Europe. In 1796 Jenner scratched the skin of a boy with fluid he had obtained from a cowpox blister on a milkmaid's hand. Six weeks later, he deliberately inoculated the boy with fluid from a blister of a smallpox victim, but the boy developed no symptoms of the disease. Jenner thus had performed a successful vaccination against a dread disease more than 50 years before Louis Pasteur developed the germ theory of disease.

Size and Structure

We know now that smallpox was caused by agents considerably smaller in size than bacteria. During Pasteur's time virtually all infectious agents, including bacteria, protozoans, and yeast, were called **viruses.** One of Pasteur's associates, Charles Chamberland, discovered that porcelain filters would not allow bacteria to pass through but would permit the passage of an unseen agent that caused rabies, another serious disease of both animals and humans. Such agents of disease that could pass through filters became known as *filterable viruses*. Today we know that not only smallpox and rabies are caused by these filterable viruses but also measles, mumps, chicken pox, polio, yellow fever, influenza, fever blisters, warts, and the common cold. The word *filterable* is no longer used, and only those organisms that have certain unique features are called viruses. These features, which include a complete lack of cellular structure, make viruses quite different from anything else in the five kingdoms of living organisms currently recognized. In fact, there is some question as to whether or not viruses are even living. In 1946 Wendell Stanley, an American chemist, received a Nobel Prize for demonstrating that a virus causing tobacco mosaic, a common plant disease, could be isolated, purified, and crystalized and that the crystals could be stored indefinitely but would always produce the disease in healthy plants at any time they were placed in contact with them. We also know that viruses do not grow by increasing in size, nor do they respond to external stimuli. They cannot move on their own, and they cannot carry on independent metabolism.

FIGURE 15.16   Papavoviruses in a human wart. (Electron micrograph Courtesy Richard S. Demaree, Jr.)

Viruses are about the size of large molecules, varying in diameter from about 15 to 300 nanometers (figure 15.16). Thousands of the smallest ones could fit inside a single bacterium of average size. Basically they consist of a nucleic acid core surrounded by a protein coat. The architecture of the protein coats varies considerably, but many have 20 sides and resemble tiny geodesic domes, whereas others have distinguishable head and tail regions. The nucleic acid core consists of either DNA or RNA—never both. Viruses have been classified in several ways. Originally they were grouped according to their hosts and the types of tissues or organs they affected. Now they are separated first according to the DNA or RNA in their cores. Then they are grouped according to size and shape, the nature of their protein coats, and the number of identical structural units in their cores.

FIGURE 15.17   Phage viruses. Inset drawing shows some of the detail. (Electron micrograph by Carolina Biological Supply Company)

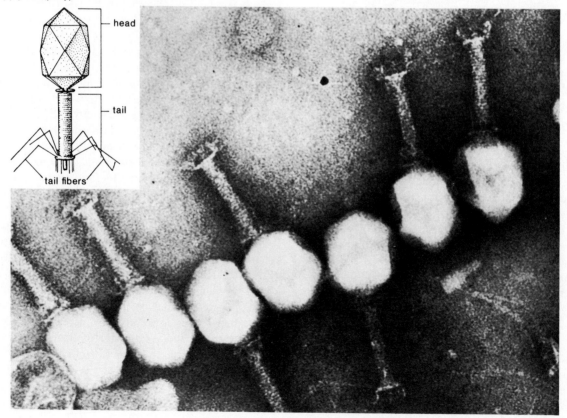

Bacteriophages

Viruses that attack bacteria have been studied extensively. These are called **bacteriophages** or simply **phages** (figure 15.17). Some resemble the space exploration vehicles portrayed in the literature and films of science fiction. They consist of a head on top of a thin cylindrical core, which is surrounded by a sheathing coat. Toward the base of the core are six spiderlike fibril "legs," which anchor the virus in place.

Viral Reproduction

Viruses can replicate (reproduce themselves) only at the expense of their host cells. In doing so, they first become attached to a susceptible cell by their "legs." Then they penetrate to the interior, some types leaving their coats on the outside. Inside the cell, their DNA or RNA directs the synthesis of new virus molecules, which are then assembled into complete viruses. These are released from the host cell, usually as it dies (figure 15.18).

Some viruses (e.g., those causing influenza) can mutate (see chapter 13) very rapidly, allowing them to attack organisms that previously had been immune to them. As a result, new vaccines have to be developed constantly to combat new strains of viruses as they appear. Some viruses greatly affect the metabolism of their host cells. For example, in botulism bacteria, the botulism toxins are produced only if specific phages are present and active. Evidence is mounting that many forms of cancer, which usually involves abnormal cell growth, are caused by viruses. Scientists also suspect that all living organisms carry viruses in an inactive form in their cells, and they are trying to discover what causes the inactive viruses suddenly to become active.

Cells of higher animals that are invaded by viruses produce a protein called *interferon,* which is released into the fluid around the cells or into the bloodstream. Minute amounts of interferon in contact with cells cause the cells to produce a protective protein that prevents or inhibits the propagation of numerous types of viruses within the protected cells and also inhibits the capacity of viruses causing tumors to transform normal cells into tumor cells. Because of these properties of interferon, considerable research is currently being conducted on methods of producing it in much greater quantities than has formerly been possible, for use in controlling cancers and many other viral diseases. Especially promising is the use of bacteria as hosts for donor DNA.

FIGURE 15.18 Stages in the development of a phage virus within a bacillus bacterium. *A.* The virus becomes attached to the bacterium. *B.* The DNA of the virus enters the cell. *C.* Various components of the virus are synthesized from the DNA of the bacterium. *D.* The viral components are assembled into units. *E.* The assembled viruses are released as the bacterial wall breaks down. (After Pelczar, M. J., Jr., and R. D. Reid. 1972. *Microbiology,* 3d ed. Redrawn by permission of the McGraw-Hill Book Co.)

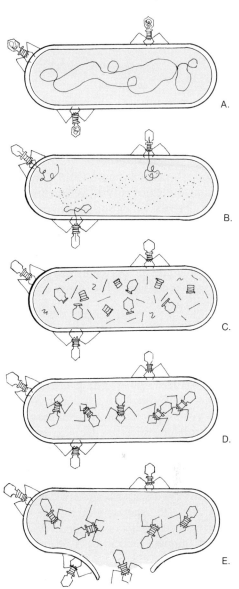

A.

B.

C.

D.

E.

FIGURE 15.19   Splicing genes to bacterial plasmids.

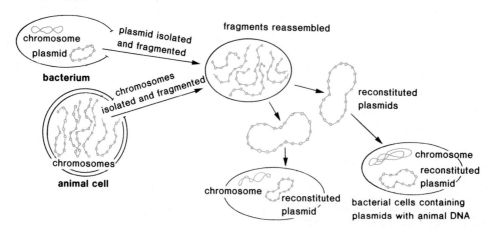

This process, in effect, turns the bacteria into interferon synthesis centers (figure 15.19), and in 1981, scientists at the University of Washington and the Genetech Corporation of San Francisco announced that they had succeeded in producing a form of interferon by splicing interferon genes into yeast cells. Because yeast cells are larger than bacteria, the process can potentially produce much larger quantities of interferon at considerably lower cost than is possible using bacteria.

HUMAN RELEVANCE OF VIRUSES

The economic impact of viruses in both developing and industrialized countries is enormous. The annual loss in work time due to common cold and influenza viruses alone amounts to millions of hours. While discomfort, adverse effects on employment, and even deaths due to viral diseases such as chicken pox, measles, German measles, mumps, and yellow fever have declined dramatically since immunizations against the diseases became widespread, they still take their toll. Another viral disease, infectious hepatitis, periodically also still claims victims. AIDS (Acquired Immune Deficiency Syndrome), a fatal disease caused by one or more cancer-related viruses, was first reported in the early 1980s. Since its initial appearance increasing research is being applied to the development of a vaccine with the use of a genetically engineered virus. If successful, the vaccine will cause the human body to develop a defense against AIDS viruses without the disease itself developing.

The production of vaccines for a number of other diseases is undertaken by a flourishing worldwide multimillion dollar industry. Other mass-produced viruses are used to infect ticks, insects, and other disease organisms of both animals and plants. Some viruses cause great losses when they infest creamery vats during the manufacture of dairy products or culture vats during the production of antibiotics. One phage attacks nitrogen-fixing bacteria in the roots of leguminous plants, while other phages attack diphtheria and tuberculosis bacteria.

GENETIC ENGINEERING

Controversy has arisen over the successful artificial addition of DNA to certain bacteria. Some fear that such tinkering by scientists could accidentally create uncontrollable monsters, while others hail such **genetic engineering** (now increasingly referred to as *biotechnology*) as the forerunner of eventual suppression or elimination of human defects and diseases.

The process of genetic engineering, or gene "splicing," begins with the isolation of pure bacterial DNA, which, as indicated in chapter 3, consists of two strands wound in a double helix. This is done by breaking up the bacterial cells and extracting the DNA, which is centrifuged to separate the heavier chromosomes or nucleoids from the plasmids. Next, the linkages between adjoining nucleotides of plasmids are broken down by special bacterial enzymes known as *restriction enzymes*, which break a circular plasmid at a specific nucleotide sequence, leaving a linear strand with "sticky" ends where the nucleotides are unpaired. Recombinant DNA is made by mixing, from two different sources, large numbers of DNA strand segments broken by the same restriction enzyme so that the "sticky" ends are complementary to each other. When so mixed, a fragment from one source may associate with a fragment from the other source, and with the aid of a *repair enzyme,* the two fragments may link—in much the same fashion as recombination occurs during crossing over in meiosis—yielding a circular recombinant DNA molecule with parts of both original molecules. The recombinant plasmids are then inserted into bacterial cells where they replicate, thus introducing new traits into a particular strain of bacteria (figure 15.19).

Bacteria are easily propagated, and it is thus relatively simple to obtain large amounts of DNA with enormous potential for scientific studies. If, for example, genes for desirable traits in plants (e.g., superior nutritional quality of fruits or vegetables, superior metabolic efficiency, superior capacity for nitrogen fixation) could be isolated from selected plants and incorporated into bacterial plasmids, the bacteria could produce large amounts of recombinant DNA for introduction into plants that, in turn, could be propagated asexually if the desirable trait had been successfully incorporated. Viruses, some of which naturally infect plants, are also easily propagated and are subject to similar engineering techniques. It is possible that viruses will play a major role in the genetic manipulation of higher plants in the future.

Bacteria have been used for many years in the production of industrial chemicals, but genetic engineering may greatly increase such use in the future. More complex organisms have more complex genetic complements, however, and much additional research is needed before genetic engineering becomes commonplace in the envisioned elimination of diseases and defects in higher plants and animals. Pioneering experiments with higher animals and plants have already been performed. In 1980, for example, J. Gordon of Yale University squirted thousands of recombinant plasmids into newly fertilized mouse eggs that were in a dish, and then implanted the eggs in the oviducts of about 10 female mice. The mice produced 78 offspring, 2 of which proved to have traces of the new DNA in their cells. In 1981 scientists of the University of Wisconsin and the United States Department of Agriculture reported transferring a gene from a French bean seed into a cell of sunflower, and called the tissue that

developed after the transfer "sunbean." The gene, which directs major protein production, is stable in the sunflower cell, and high levels of bean protein production are anticipated when technology is developed to regenerate a sunflower plant from the "sunbean" cells (see the section on mericloning and tissue culture in appendix 4).

In 1986 scientists at Oxford University in England were given permission to field test a genetically engineered virus specific for the control of the pine beauty moth, a major pest of European coniferous forests. The virus was produced with a single defective gene that prevents its duplication in its victims but at the same time effectively brings about their demise. Other genetically engineered organisms developed during 1985 and 1986 include the so called "ice-minus" bacteria that produce a protein coat, which deters the formation of ice crystals and reduces frost damage to crops. Although this organism has enormous potential for extending growing seasons and ranges for a variety of crops, its developers have experienced strong opposition to field testing, even though they have received permission to do so. Much of the opposition is based on a fear of the unknown, and the belief that the release of genetically engineered organisms outside the laboratory could be potentially disastrous if unanticipated effects such as uncontrollable growth or virulent mutations were to occur.

The United States Department of Agriculture in 1986 withdrew and then reinstated the first license for a genetically engineered vaccine that contained a herpes virus that could not duplicate itself. Prior to the promise of the vaccine there has been no cure for herpes, and a vaccine to prevent it would be a major step in the anticipated eventual suppression or elimination of numerous diseases. Also in 1986, the Environmental Protection Agency granted the Mycogen Corporation of San Diego, California, permission to conduct small-scale field tests of a genetically engineered pesticide. Other research projects in progress, and involving genetically engineered organisms, include the production of an enzyme that would greatly reduce the cost and improve the quality of certain expensive antibiotics; the development of an anti-inflammatory agent previously produced in only trace amounts; the development of a bacterium, to which human genes have been introduced, for the production of insulin to be used by diabetics; the production of a human growth hormone; and the production of a yeast that would enhance fermentation processes and improve wine flavors.

SUMMARY

Kingdom Monera consists of prokaryotic organisms, which are grouped into archaebacteria (methane, salt, and sulfolobus bacteria) and eubacteria (unpigmented, purple, green sulfur, blue-green bacteria, and prochlorobacteria). The bacteria occur as single cells, in colonies, or in the form of chains or filaments. Some cells may be motile (capable of movement) by means of whiplike flagella, or they may exhibit a gliding motion, but most are nonmotile. Nutrition is primarily by absorption of food in solution through the cell wall, but some members of the kingdom are photosynthetic or chemosynthetic (capable of obtaining their energy through chemical reactions with various compounds or elements). Reproduction is asexual by means of fission, a form of cell division that does not involve mitosis; some genetic recombination occurs in several groups by means of pili (minute tubes) that form connections between cells.

Luminescent bacteria are minor representatives of the most numerous of earth's living organisms. Bacteria are mostly less than 2 or 3 micrometers in diameter and are not visible individually except with the aid of a microscope. They occur primarily in the form of spheres (cocci), rods (bacilli), and spirals (spirilli) and are further classified on the basis of sheaths, appendages, and motion. Depending on their reaction to the blue-black Gram's stain, they are also classified as gram-positive or gram-negative.

Prokaryotic cells have no nuclear envelopes or organelles, but membranes present apparently perform some of the functions of organelles. Each cell has a nucleoid consisting of 1 long DNA molecule and sometimes up to 30 or 40 small, circular DNA molecules called plasmids, which replicate independently of the large DNA molecule or chromosome. Mitosis does not occur, but a transverse wall forms near the middle of the cell and creates two cells by fission after the DNA has replicated. Meiosis also does not take place, and gametes and zygotes are not produced. Genetic recombination occurs through conjugation (during which part of the DNA strand is transferred from a donor cell to a recipient cell through a hollow tubelike pilus). A living cell may also pick up and incorporate fragments of DNA released by dead cells in a process called transformation, or fragments of DNA may be carried from one cell to another by viruses in a process called transduction.

Heterotrophic bacteria (organisms that depend on other organisms for food) are mostly saprobes (organisms that obtain their food from nonliving organic matter), but some are parasites. Autotrophic bacteria are photosynthetic but do not produce oxygen; chemoautotrophic bacteria obtain their energy through oxidation of reduced inorganic groups.

Any nonliving tissues will eventually be decomposed to compost by bacteria and fungi, but piles of moderate size favor the sequence of bacteria usually involved in composting, which is increasingly being done by communities where ordinances against burning of leaves have been instituted. The proportion of carbon to nitrogen in a compost pile largely determines the pace at which decomposition proceeds, with a ratio of 30 carbon to 1 nitrogen being optimal. Compost is definitely good for the soil but has limited value as a fertilizer.

Bacteria cause huge losses through plant diseases and food spoilage and many serious diseases in animals and humans. They gain access to their hosts from the air (e.g., diphtheria, whooping cough, some pneumonia forms), through contamination of food and drink (e.g., cholera, dysentery, salmonella food poisoning), through consumption of bacterial toxins (e.g., botulism), through direct contact (e.g., syphilis, gonorrhea, anthrax, brucellosis), through wounds (e.g., tetanus, gas gangrene), and through the bites of insects and other organisms (e.g., bubonic plague).

Koch formulated postulates (rules) for proving that a particular microorganism is the cause of a particular disease: (1) the microorganism must be present in all cases of the disease; (2) the microorganism must be isolated from the victim in pure culture; (3) when the microorganism from the pure culture is injected into a susceptible host it must produce the disease in the host; (4) the microorganism must be isolated from the experimentally infected host and grown in pure culture for comparison with that of the original culture.

Bacteria useful to humans include *Bacillus thuringiensis,* which attacks a wide range of caterpillar pests; *Bacillus thuringiensis* var. *israelensis,* which is highly destructive of mosquito larvae, and *Bacillus popilliae,* which attacks grubs of Japanese beetles. Bacteria also play a major role in the manufacture of dairy products such as yogurt, sour cream, kefir, and cheese. They are used in the manufacture of industrial chemicals, vitamins, flavorings, food stabilizers, and a blood plasma substitute; they play a role in the curing of vanilla, cocoa beans, coffee, and tea, and in the production of vinegar and sauerkraut; and they aid in the extraction of linen fibers from flax stems, in the production of ensilage for cattle feed, and in the production of several important amino acids.

Blue-green bacteria occur in fresh and marine water, in hot springs, on bare lava and desert rocks, on the shells of turtles and snails, and symbiotically within other organisms. The cells may occur singly, in colonies within a gelatinous matrix, or as single or branched filaments. They are similar to the cells of other true bacteria, but are distinguished from them in having chlorophyll *a,* in producing oxygen when they undergo photosynthesis, and in having blue phycocyanin and red phycoerythrin phycobilin pigments. They produce cyanophycin, a nitrogenous food reserve. There are no flagella, but some species have gliding movements. Fragmentation may occur at heterocysts, which are special larger cells produced at intervals in the chains of some species. Thick-walled cells called akinetes may also be produced; akinetes can lie dormant for many years and are resistant to cold and drought. Blue-green bacteria cells may have been the origin of chloroplasts, since they divide as chloroplasts do when their host cells divide.

Blue-green bacteria may become very abundant in bodies of polluted fresh water. Toxic substances are produced when the bacteria die and are decomposed. There are reports of various animals dying after drinking water in which blue-green bacteria have become abundant. At least 40 species of blue-green bacteria are known to fix nitrogen.

Prochlorobacteria are similar in form to blue-green bacteria but have pigmentation similar to that of higher plants and lack phycobilins.

Vaccination, which involves introducing to the body a weakened disease agent that in turn causes the body to build up an immunity to the disease while fighting the agent, was first performed by Jesty and Jenner in connection with smallpox. Smallpox, a disease caused by a virus, is now believed to have been eradicated.

Viruses, which have no cellular structure, are about the size of large molecules. Some can be isolated, purified, and crystallized, yet remain virulent. They cannot grow or increase in size, and cannot be reproduced outside of a living cell, upon whose DNA they depend for duplication. Such attributes have led to speculation that they are not living organisms. Viruses consist of a core of nucleic acid surrounded by a protein coat. They are now classified on the basis of the DNA or RNA in their core, their size and shape, the number of identical structural units in their cores, and the nature of their protein coats.

Bacteriophages are viruses that attack bacteria. In replicating they become attached to a susceptible cell, which they penetrate, with their DNA or RNA directing the synthesis of new virus molecules from host material; the assembled new viruses are released when the host cell dies.

Cells of higher animals being invaded by viruses produce interferon, a protein that causes cells to produce a protective substance that inhibits duplication of viruses, and also inhibits the capacity of viruses to transform normal cells into tumor cells.

Viral diseases such as chicken pox, measles, mumps, and yellow fever have declined since immunizations against the diseases have become widespread. Mass-produced viruses are used to infect insects and other pests of both plants and animals. Some viruses cause losses in creamery vats.

Genetic engineering involves the alteration of a cell's DNA by artificially mixing large numbers of enzymatically broken DNA strands from two different sources. Fragments from one source may associate with those from another source, creating recombinant DNA with parts of both original molecules. The recombinant plasmids are inserted into live bacterial cells where they replicate. Since bacteria are easy to propagate, large amounts of recombinant DNA may be obtained in this fashion. Such recombinant DNA may be used in the future for improving crops and possibly eliminating diseases of both plants and animals.

REVIEW QUESTIONS

1. What is symbiosis? Give examples other than those mentioned in the text.
2. Why are bacteria not classified on the basis of visible features alone?
3. How do bacteria exchange DNA?
4. How does fission differ from mitosis?
5. Is photosynthesis the same in bacteria as it is in higher plants? Explain.
6. How do chemoautotrophic bacteria differ from photosynthetic bacteria?
7. What is the difference between nitrification and nitrogen fixation?
8. If decay bacteria use nitrogen, how does composting accumulate any nitrogen?
9. How are disease bacteria transmitted?
10. What are Koch's Postulates?
11. Why are many bacteria considered "useful"?
12. What do blue-green bacteria and other bacteria have in common? How do they differ?
13. How do blue-green bacteria survive freezing and desiccation?
14. What is an algal-bacterial bloom?
15. How do viruses differ from bacteria?
16. What is a vaccination?
17. What is a phage?
18. How do viruses multiply?

DISCUSSION QUESTIONS

1. If a virulent phage were to eliminate all the bacteria in North America for one year, how would our lives be affected?
2. What would be the feasibility and the advantages or disadvantages of using only blue-green bacteria and other nitrogen-fixing bacteria for our agricultural nitrogen needs?
3. Methane gas produced by bacteria is proving to be sufficient to meet all the fuel needs of villages in India. Do you think we could produce and use methane in a similar fashion in the United States?
4. If blue-green bacteria are capable only of asexual reproduction, does this mean that species of these organisms can never change in form?
5. As long as viruses can multiply, why should there be any question as to whether or not they are living?

ADDITIONAL READING

Brock, T. D. 1984. *Biology of microorganisms,* 4th ed. Englewood Cliffs, NJ: Prentice-Hall, Inc.

Carr, N. G., and B. A. Whitton, eds. 1982. *The biology of Cyanobacteria.* Berkeley, CA: University of California Press.

Chakrabarty, A. M. 1978. *Genetic engineering.* Boca Raton, FL: CRC Press.

Gibbs, A. J., and B. D. Harrison. 1979. *Plant virology: the principles.* New York: John Wiley & Sons, Inc.

Holt, J. G., and N. R. Krieg, eds. 1984. *Bergey's manual of determinative bacteriology,* vol. 1. Baltimore: Williams and Wilkins Co.

Luria, S. E., et al. 1978. *General virology,* 3d ed. New York: John Wiley & Sons, Inc.

Margulis, L. 1982. *Early life.* Portola Valley, CA: Jones and Bartlett.

Old, R. W., and S. B. Primrose. 1982. *Principles of gene manipulation: an introduction to genetic engineering,* 2d ed. Berkeley: University of California Press.

Palmer, C. M. 1962. *Algae in water supplies.* Public Health Service Publication No. 657. Washington, DC: U.S. Government Printing Office.

Pelczar, M. J., Jr., and E. C. S. Chan. 1981. *Elements of microbiology.* New York: McGraw-Hill Book Co.

Stanier, R. Y., E. A. Adelberg, and J. Ingraham. 1986. *The microbial world,* 5th ed. Englewood Cliffs, NJ: Prentice-Hall.

Starr, M. P., H. Stolp, H. G. Trüper, A. Balows, and H. G. Schlegel, eds. 1981. *The prokaryotes.* 2 vols. New York: Springer-Verlag.

Stoner, C. H., ed. 1975. *Producing your own power.* New York: Random House Publishers.

Tortora, G. J., B. R. Funke, and C. L. Case. 1982. *Microbiology: an introduction.* Menlo Park, CA: Benjamin/Cummings.

Woese, C. R. 1981. "Archaebacteria." *Scientific American* 244(6): 98–122.

Overview

Beginning with a summary of the features of members of Kingdom Protista, this chapter discusses the divisions of algae. Initially, the Chrysophyta (yellow-green algae, golden-brown algae, diatoms, and cryptophytes) are discussed. Next, mention is made of the Division Pyrrophyta (dinoflagellates) and the role of members of the division in red tides and bioluminescence. After a brief overview of the Euglenophyta (euglenoids), the Chlorophyta (green algae), asexual and sexual reproduction in *Chlamydomonas*, *Ulothrix*, *Spirogyra*, and *Oedogonium* are explored, and mention is made of *Chlorella*, desmids, *Acetabularia*, *Volvox*, and *Ulva*. Next the Phaeophyta (brown algae) and Rhodophyta (red algae) are covered, with a discussion of the life cycle of a representative of each division being included. This is followed by a table summarizing divisional differences with respect to food reserves, special pigments, and flagella. The chapter concludes with a digest of the human and ecological relevance of the algae.

16 Kingdom Protista

FEATURES OF KINGDOM PROTISTA

In contrast with Kingdom Monera, whose members have prokaryotic cells, all members of Kingdom Protista have eukaryotic cells. The organisms comprising this kingdom are very diverse and heterogenous, but none have the distinctive combinations of characteristics possessed by members of Kingdoms Plantae, Fungi, or Animalia. Many, including the euglenoids, protozoans, and some algae, consist of a single cell, whereas other algae are multicellular or occur as colonies or filaments. Nutrition is equally varied, with the algae being photosynthetic, the protozoans ingesting their food, and the euglenoids either carrying on photosynthesis or ingesting their food.

Individual life cycles vary considerably, but reproduction is generally by cell division and sexual processes. Many protists are motile, usually by means of flagella, but others are nonmotile.

INTRODUCTION TO THE ALGAE

Children fortunate enough to have lived within access of ocean beaches where seaweeds are cast ashore by the surf, have often enjoyed stamping on the bladders ("floats") of kelps to make them "pop." Some have collected and pressed beautiful, feathery red seaweeds, and others who have waded around the shores of freshwater lakes or in slow-moving streams have encountered slimy-feeling pond scums. All who have kept tropical fish in glass tanks have sooner or later had to scrape a brownish or greenish film from the inner surfaces of the tank, and those who have lived in homes with their own swimming pools have learned that the white plaster of the pool soon acquires colored patches on its surface if chemicals are not regularly added to prevent them from appearing.

Algae, which have no flowers or leaves (although some of the seaweeds do have flattened, leaf-like blades), are involved in our everyday lives in more ways than most people realize (see the discussion on the human and ecological relevance of the algae, which begins on page 321), and seaweeds, pond scums, fish tank films, and colored patches in swimming pools are but a few of the numerous kinds of algae all assigned to the Kingdom Protista. Based on the form of their reproductive cells and combinations of pigments and food reserves, the algae are grouped into several major divisions.

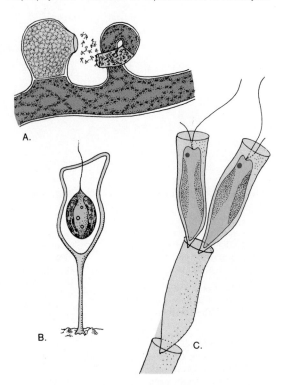

FIGURE 16.1   Some representatives of the Division Chrysophyta. A. *Vaucheria*. B. *Stipitococcus*. C. *Dinobryon*.

DIVISION CHRYSOPHYTA— THE GOLDEN-BROWN ALGAE

If the roughly 6,000 members of this division were not primarily microscopic, many undoubtedly would become collectors' items in the art and antique shops of the world because of their exquisite form and ornamentation (figures 16.1, 16.2, and 16.3). They are grouped into four classes: *Yellow-green algae* (Xanthophyceae), *golden-brown algae* (Chrysophyceae), *diatoms* (Bacillariophyceae), and *cryptophytes* (Cryptophyceae). Superficially the organisms of each class may appear unrelated to each other, but they do have several features in common, including food reserves, specialized pigments, and other cell characteristics. Some members of each class produce a unique "resting" cell called a *statospore* (figure 16.2). These cells resemble miniature glass apothecary bottles, complete with plugs that dissolve or "uncork," releasing the protoplast inside. Many statospores are striking in form, with finely sculptured ornamentations on the surface.

FIGURE 16.2   Statospores, which resemble apothecary bottles, are formed by many of the golden-brown algae. (After G. M. Smith. 1950. *The Freshwater Algae of the United States*. 2d ed. Redrawn by permission of the McGraw-Hill Book Co.)

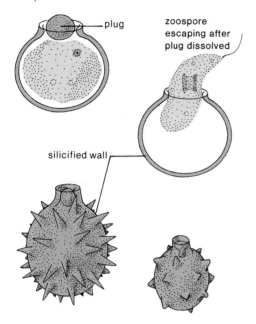

FIGURE 16.3   A diatom. (Courtesy J. D. Pickett-Heaps)

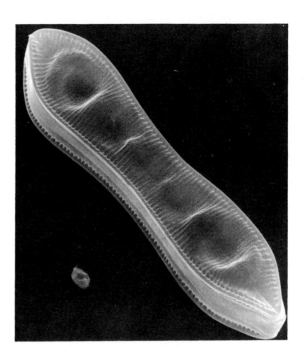

Diatoms

Diatoms (figure 16.3) are the best known and economically the most important members of the division. These mostly unicellular algae occur in astronomical numbers in both fresh and salt water but are particularly numerous in colder marine habitats. They also usually dominate the algal flora on damp cliffs, the bark of trees, bare soil, or the sides of buildings. More than 5,600 living species are recognized, with almost as many more known only as fossils. Some can withstand extreme drought, and one species is known to have become active after lying dormant for 48 years in dry soil.

Diatoms look like little glass pillboxes, with half of the rigid crystal-clear wall fitting inside the other overlapping half. Many marine diatoms are circular in outline when viewed from the top, whereas freshwater species tend to resemble the outline of a kayak when viewed in the same manner. The walls contain as much as 95% silica, an ingredient of glass, deposited in an organic framework of pectin or other substances. They usually have exquisitely fine grooves and pores that actually are exceptionally minute passageways connecting the protoplasm with the watery environment outside the shell. These fine grooves and pores are so uniformly spaced they have been used to test the resolution of microscope lenses.

Each diatom may have one, two, or many chloroplasts per cell. In addition to chlorophyll *a,* the accessory pigments chlorophyll c_1 and chlorophyll c_2 are typically present. The chloroplasts usually are golden-brown in color because of the dominance of **fucoxanthin,** a brownish pigment also found in the brown algae. Food reserves are oils, fats, or the carbohydrate *chrysolaminarin.*

Many freshwater diatoms move backward and forward with somewhat jerky motions. It is believed that the movements occur in response to external stimuli. Extremely tiny fibrils are thought to take up water as they are discharged into a lengthwise groove or pores through which moving cytoplasm protrudes. When they come in contact with a surface, they stick and contract, moving the cell as the caterpillar treads on a tractor do and leaving a trail similar in some respects to that of a snail.

Reproduction in diatoms is unique (figure 16.4). After a protoplast, which is diploid, has undergone mitosis and division, the two halves of the cell separate, with a daughter protoplast remaining in each half. Then a new half wall fitting inside the old half is formed. This occurs for a number of generations, with the result that some of the cells become progressively smaller. Eventually, however, a protoplast undergoes meiosis, producing four gametes, which

FIGURE 16.4   Reproduction in diatoms.

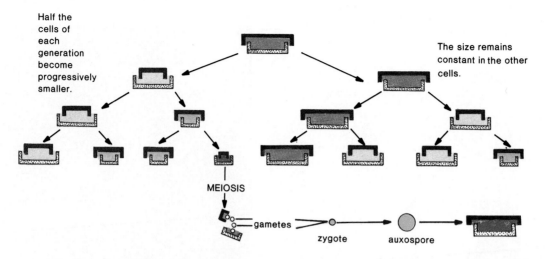

Half the
cells of
each
generation
become
progressively
smaller.

The size remains
constant in the other
cells.

MEIOSIS

gametes — zygote — auxospore

then escape. These fuse with other gametes, becoming zygotes called *auxospores*. Auxospores are like any other zygotes except that they rapidly increase in size before their rigid "pillbox" walls are formed around them, thus restoring the original size of the diatom species.

DIVISION PYRROPHYTA— THE DINOFLAGELLATES

Occasionally visitors to an ocean beach in midsummer may notice a distinct reddish tint to the water, usually as a result of a phenomenon known as a **red tide.** Some, upon observing such an event, have dipped a cup of seawater and saved it for examination with a microscope. (A few drops of formaldehyde or other similar substance added to the cup preserves the material indefinitely).

Red tides are caused by the sudden and not fully understood multiplication of unicellular organisms called **dinoflagellates** (figure 16.5), over 3,000 species of which are presently known. The cup of seawater usually contains a large number of these dinoflagellates, which are the best known representatives of the Division Pyrrophyta. Some resemble armor-plated spaceships, whereas others may be smooth or have fine lengthwise ribs. The "armor" plates, which are located just inside the plasma membrane, are composed of cellulose and vary in thickness. They have two flagella that are distinctively arranged, usually being attached near each other in two adjacent and often intersecting grooves. One, which acts as a rudder, trails behind the cell, while the other gives the cell a spinning motion as

FIGURE 16.5   Dinoflagellates. *A. Ceratium. B. Gonyaulax.*

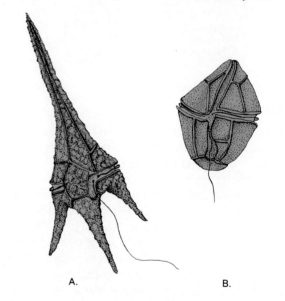

A.   B.

it undulates in its groove like a tiny snake. The latter groove encircles the cell at right angles to the first groove. Most species have two or more disc-shaped chloroplasts, which contain distinctive brown pigments in addition to various other pigments, including chlorophylls *a* and c_2. About 45% of the species are, however, non-photosynthetic, and some ingest food particles, whether or not chlorophyll is present. Some have an *eyespot* (a pigmented organelle that is sensitive to light), and all have a unique nucleus in which the chromosomes remain condensed and clearly visible throughout the life of the cell. The chromosomes contain a disproportionately large amount of DNA—as much as 40 times that

306   *Chapter 16*

found in human cells. The food reserve is starch, which in dinoflagellates is stored outside the chloroplasts.

Dinoflagellates occur in most types of fresh and salt water, but those that cause red tides have received the most publicity because they also produce powerful toxins that kill fish and shellfish. In fact, the toxins, which also paralyze the human nervous system, have been studied for possible use in chemical warfare. The havoc to the fishing industry caused by major red tides is so great that several laboratories have been conducting research with dinoflagellates and their marine habitats to try to find a way of preventing such destruction in the future.

Some dinoflagellates are luminescent and play a major role in the phenomenon of *bioluminescence* (see chapter 11), often seen in tropical oceans. At night, if dinoflagellates are numerous and the water is agitated, the ocean appears to glow and sparkle, especially as the waves break.

Reproduction is by cell division. Sexual reproduction appears to be rare.

DIVISION EUGLENOPHYTA— THE EUGLENOIDS

Barnyard pools and sewage treatment ponds often develop a rich green bloom of algae. A superficial examination of water from such a pool with the aid of a microscope usually reveals large numbers of active green cells, and a closer inspection probably will reveal one to several of the more than 750 species of **euglenoids,** of which *Euglena* is a common example (figure 16.6).

A *Euglena* cell, which is spindle-shaped and has no rigid wall, can be seen to change shape even as the organism moves along. Just beneath the plasma membrane are fine strips that spiral around the cell parallel to one another. The strips and the plasma membrane are devoid of cellulose and together are called a *pellicle*. A single flagellum, which has numerous tiny hairs along one side, pulls the cell through the water. A second very short flagellum is present within a *reservoir* at the base of the functional flagellum. Other features of *Euglena* include the presence of a *gullet,* or groove, through which food can be ingested, and in about a third of the 500 species there are several to many mostly disc-shaped chloroplasts present. A red eyespot is located in the cytoplasm near the base of the flagella, and a carbohydrate food reserve called *paramylon* normally is present in the form of small whitish bodies of various shapes.

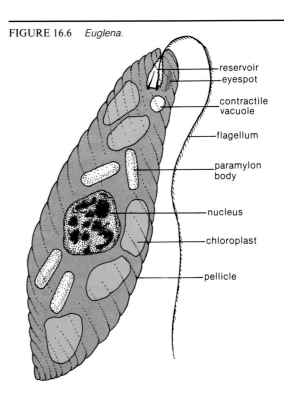

FIGURE 16.6 *Euglena.*

reservoir
eyespot
contractile vacuole
flagellum
paramylon body
nucleus
chloroplast
pellicle

Reproduction is by cell division. The cell starts to divide at the flagellar end and eventually splits lengthwise, forming two complete cells. Sexual reproduction has been hinted at but has never been confirmed.

Some species of *Euglena* can live in the dark if appropriate food and vitamins are present. Others are known to reproduce faster than their chloroplasts under certain circumstances, so that some chloroplast-free cells are formed. As long as a suitable environment is provided, these cells also can survive indefinitely. *Euglena*'s clear manifestation of both photosynthetic capacity and animallike methods of obtaining nutriment caused it in the past, when only two kingdoms were recognized, to be treated as a plant in botany texts and as an animal in zoology texts.

DIVISION CHLOROPHYTA— THE GREEN ALGAE

The approximately 7,500 species of organisms comprising the largest of the algal divisions, Division Chlorophyta, are commonly known as the **green algae.** They are a very diverse and cosmopolitan group. Some consist of microscopic unicellular forms or threadlike filaments, while others form platelike colonies, netlike tubes, or hollow balls. Some green

FIGURE 16.7 Representative green algae. *A. Volvox*. The cells form hollow, spherical colonies that spin on their axes as the flagella of each cell beat in such a way that the motion is coordinated. New colonies are formed within the old ones. *B. Scenedesmus*. The outer cells of each four-celled colony have spinelike extensions of the cell wall; the function of the spinelike extensions is unknown. *C. Pediastrum*. The cells form flattened, platelike, circular colonies. *D. Zygnema* (Portion of a filament). Each cell of this green alga contains two somewhat starlike chloroplasts. (*A.*, *B.*, and *D.*, by Carolina Biological Supply Company. *C.* by © J. Robert Waaland/University of Washington/BPS)

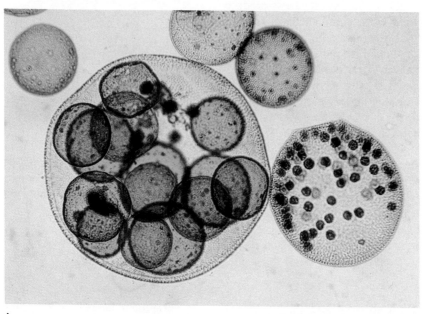

A.

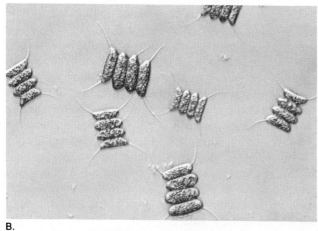

B.

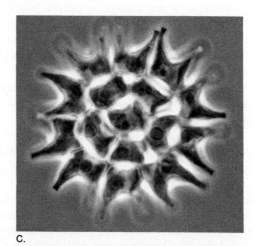

C.

D.

algae are seaweeds, resembling lettuce leaves or green ropes up to several meters long. A number of unicellular forms grow in greenish patches or streaks on the bark of trees or help camouflage jungle animals, growing in large numbers while attached to their fur. Still others thrive in snowbanks, live in flatworms and sponges, or are found on the backs of turtles. They are the most common member in lichen "partnerships" (see chapter 17). The greatest variety, however, is found in freshwater ponds, lakes, and streams, with additional diversity evident in the oceans, where they are an important part of the *plankton* (free-floating, mostly microscopic organisms), and thus of food chains (figure 16.7).

The cells of green algae resemble those of higher plants in having the same kinds of chlorophylls (*a* and *b*) and other pigments in their chloroplasts. The green algae also are believed to have been ancestral to the higher plants and, like the higher plants they store their food within the chloroplasts in the form of starch. Although most green algae have a single nucleus in their cells, one group, the *bryopsids,* has multinucleate cells. Most green algae can undergo both asexual and sexual reproduction (see chapter 12). The manner in which they do so is illustrative of the forms of reproduction found in most of the organisms discussed in the chapters to follow, and so several different representative green algae will be examined in some detail here.

Chlamydomonas

A lively little alga, *Chlamydomonas* (figure 16.8), is a common inhabitant of quiet freshwater pools. It has an ancient history among eukaryotic organisms, with fossil relatives occurring in rock formations reported to be nearly 1 billion years old.

Chlamydomonas is unicellular, with a somewhat oval cell surrounded by a cellulose wall. A pair of whiplike flagella at one end enable the cell to move very rapidly. The flagella are, however, difficult to see with ordinary light microscopes, and the cell itself is usually less than 25 micrometers (one tenthousandth of an inch) long. Near the base of the flagella are two or more vacuoles, which can contract and expand. They apparently regulate the water content of the cell. A dominant feature of each *Chlamydomonas* is a single, usually cup-shaped chloroplast, which at least partially hides the centrally located nucleus. Located in the chloroplast are one or two roundish glistening *pyrenoids,* which are proteinaceous structures thought to contain enzymes associated with the synthesis of starch. Most species also have a red eyespot on the chloroplast

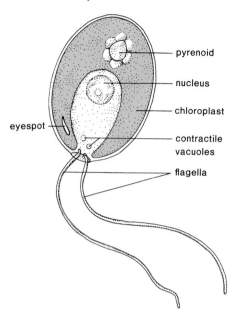

FIGURE 16.8   *Chlamydomonas.*

near the base of the flagella. Although it is merely part of an organelle within a single cell and nothing like an eye in structure, it has been demonstrated to be sensitive to light.

Asexual Reproduction

Before a *Chlamydomonas* reproduces asexually, the cell's flagella degenerate and drop off or are reabsorbed. Then the nucleus divides by mitosis, and the entire protoplasm becomes 2 cells within the cellulose wall. These 2 daughter cells, which develop flagella, escape and swim away as the parent cell wall breaks down. Once they have grown to their full size, they may repeat the process. Sometimes mitosis occurs more than once, so that 4, 8, or up to 32 little cells with flagella are produced inside the parent cell. Occasionally flagella do not develop, and the cells remain together in a colony. When growth conditions change, however, each cell of the colony may develop flagella and swim away. This type of reproduction brings about no changes in the number of chromosomes present in the nucleus, and all the cells remain *haploid* (see chapter 12).

Sexual Reproduction

Under certain environmental conditions, unknown forces within the organisms may bring about a congregating of many cells in populations of *Chlamydomonas.* Careful study of such events has revealed that pairs of cells appear to be attracted to each other by their flagella and function as gametes, which sometimes are of two types. The cell walls break down as the protoplasts slowly emerge and mate,

FIGURE 16.9   Sexual life cycle of *Chlamydomonas*.

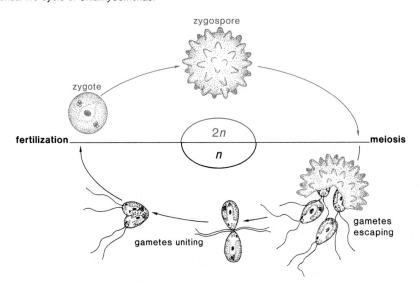

zygospore

zygote

fertilization

2*n*

n

meiosis

gametes
escaping

gametes uniting

fusing together and forming zygotes. A new wall, often relatively thick and ornamented with little bumps, forms around each zygote. This may remain dormant for several days, weeks, or even months, but under favorable conditions a dramatic change occurs. The protoplast, which is now *diploid,* undergoes meiosis, producing 4 haploid **zoospores** (motile cells that never unite with other cells; many different kinds of algae produce zoospores). When the old zygote wall breaks down, the zoospores swim away and grow to full-sized *Chlamydomonas* cells (figure 16.9).

Ulothrix

Often an examination of dead twigs, rocks, and other debris in cold freshwater ponds, lakes, and streams reveals a threadlike alga called *Ulothrix* (figure 16.10), whose name is derived from the Greek words *oulos* (woolly) and *thrix* (hair); some species of *Ulothrix* are marine. Each alga consists of a single row of cylindrical cells attached end to end and forming a thread or **filament.** The basal cell of each filament is slightly longer than the other cells and functions as an attachment cell, or **holdfast.** The nucleus of each cell is surrounded by a wide chloroplast that is shaped like a partial to nearly complete bracelet. Each chloroplast contains one to several pyrenoids. Filaments increase in length when individual cells divide, and any cell except the holdfast may do so.

Asexual Reproduction

The protoplast of any cell except the holdfast appears to clump and condense inside the rigid cell wall, divide by mitosis, and become *zoospores*. The zoo-

spores of *Ulothrix* are quite similar to *Chlamydomonas* cells in that they have contractile vacuoles and an eyespot, but they have four flagella instead of two.

Frequently a protoplast divides one to several times before becoming zoospores, but after zoospores are formed they usually escape from the parent cell through a pore in the wall. After swimming about for a few hours to several days, they settle on submerged objects, shed their flagella, and divide. One of the first two daughter cells becomes a holdfast, while the other continues to divide and forms a new filament. In some instances, the protoplasts do not produce flagella after they have condensed and divided like developing zoospores, but they are otherwise capable of germinating and producing new filaments. Such cells, which are released when the parent cell wall breaks down, are called *aplanospores.*

Sexual Reproduction

Both asexual and sexual reproduction start out in the same way. The protoplast of any cell except the holdfast appears to condense and then divides. Up to 64 zoosporelike cells, each with two flagella, may be produced. When these cells escape from the parent cell walls, however, they function as gametes, uniting in pairs with gametes from other filaments and forming zygotes. The zygotes form thick walls and become dormant. Eventually their protoplasts undergo meiosis, giving rise to zoospores, which then can become new filaments.

Although the gametes of *Ulothrix* come from different filaments, they are identical in size and appearance. Sexual reproduction involving such gametes is called **isogamy.** It is found only in simpler

FIGURE 16.10   Life cycle of *Ulothrix*.

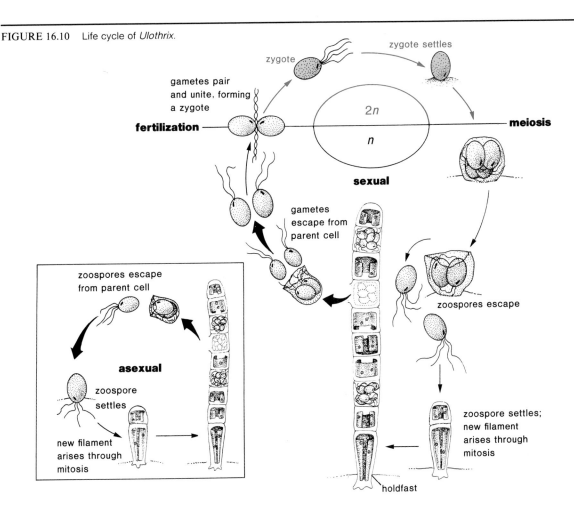

organisms. As in *Chlamydomonas,* the zygotes of *Ulothrix* are the only diploid (2*n*) cells in the life cycle. All the other cells are haploid (*n*).

Spirogyra

One's initial reaction to a first encounter with "watersilk," as *Spirogyra* is called, may be less than ecstatic because of the slimy feel of the watery sheaths surrounding its filaments. The microscope, however, reveals a beautiful alga with some of the most striking chloroplasts known.

These common freshwater algae, which form unbranched filaments of cylindrical cells, are found most frequently floating in masses at the surface of quiet waters. The cells each contain one or more long, slightly frilly, ribbon-shaped chloroplasts that look as though they had been spirally wrapped around an invisible pole occupying most of the cell's interior. Some species of *Spirogyra* have as many as 16 of these chloroplasts in each cell. Every one of these elegant green ribbons has pyrenoids at regular intervals along it.

Asexual Reproduction

Unlike the two algae just discussed, *Spirogyra* does not form any zoospores or other cells with flagella. Any cell is capable of dividing, but the only asexual reproduction resulting in new filaments is brought about through the breakup or *fragmentation* of existing filaments. This frequently occurs as a result of a storm or other disturbance.

Sexual Reproduction

In colonies of *Spirogyra,* the filaments usually are produced so close to each other that they may actually be touching. When sexual reproduction begins, the individual cells of adjacent filaments form little dome-shaped bumps, or **papillae** (singular: **papilla**), opposite each other. As these papillae grow, they force the filaments apart slightly, and then they fuse at their tips, forming small cylindrical **conjugation tubes** between each pair of cells. The condensed protoplasts then function as gametes. Usually

FIGURE 16.11 *Spirogyra* (watersilk). *A.* A portion of a vegetative filament showing the ribbonlike chloroplasts spirally arranged in each cell. The centrally located darker object in each cell is a nucleus. *B.* Papillae have grown out from opposite cells of two closely adjacent filaments and formed conjugation tubes. *C.* The condensed protoplasts in the cells on the left are functioning as male gametes and migrating through the conjugation tubes to the stationary female gametes in the cells on the right. *D.* Zygotes have been produced in some of the cells on the right as a result of fusion of gametes. (Photomicrographs by Carolina Biological Supply Company)

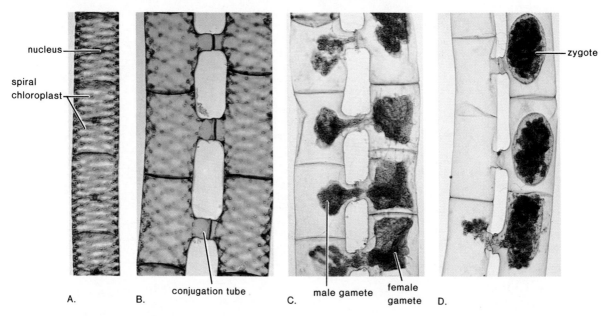

nucleus

spiral chloroplast

A.

B. conjugation tube

C. male gamete female gamete

D. zygote

those of one filament will seem to flow or crawl like amoebae through the conjugation tubes to the adjacent cells, where each fuses with the stationary gamete, forming a zygote. Each moving protoplast is considered a male gamete, while the stationary ones function as female gametes (figure 16.11).

The zygotes usually develop thick walls and remain dormant for some time, often over the winter. Thick-walled zygotes are characteristic of most freshwater green algae. Eventually their protoplasts undergo meiosis, producing four haploid cells. Three of these disintegrate, and a single new *Spirogyra* filament grows from the interior of the old zygote shell. The type of sexual reproduction shown by *Spirogyra* is called **conjugation.**

Oedogonium

Aquatic flowering plants and other algae often provide surfaces to which *Oedogonium* (pronounced ee-doh-goh'-nee-um), a filamentous green alga, may attach itself. It is, however, in no way parasitic. Algae and plants that attach themselves to other organisms in such a manner are called **epiphytes.** The basal cells of the unbranched filaments form holdfasts, and

the terminal cell of each filament is rounded. The remaining cells are cylindrical and attached end to end. The name *Oedogonium* comes from words meaning "swollen reproductive cell" and is quite apt, as the female reproductive cells do indeed bulge noticeably. Each cell contains a large netlike chloroplast, which rolls and forms a tube something like a loose wickerwork basket around and toward the periphery of each protoplast. There are pyrenoids at a number of the intersections of the net.

Asexual Reproduction

Akinetes (thick-walled overwintering cells) may occasionally be formed, but more commonly zoospores are produced singly in cells at the tips of the filaments (figure 16.12). Unlike the zoospores of most other algae, those of *Oedogonium* have about 120 small flagella forming a fringe around the cell toward one end. The zoospores look like tiny balding faceless heads, and their rapid rotating movements can be quite entertaining when viewed through a microscope. After they escape from their filaments, they eventually settle and form new filaments in the same manner as *Ulothrix.*

FIGURE 16.12   Life cycle of *Oedogonium*.

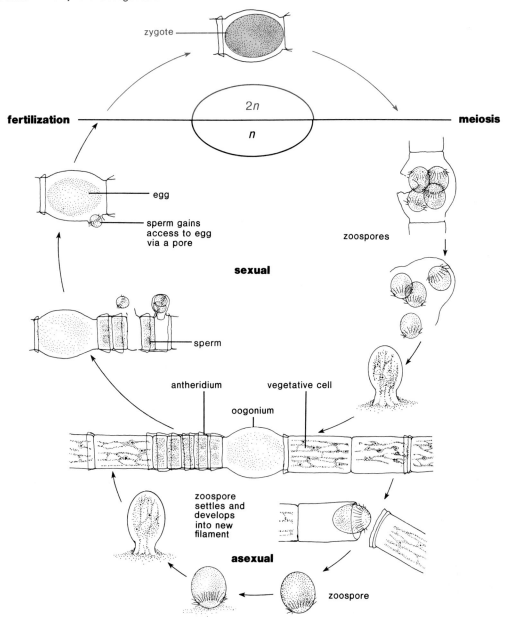

zygote

2n

n

fertilization

meiosis

egg

sperm gains
access to egg
via a pore

zoospores

sexual

sperm

antheridium

vegetative cell

oogonium

zoospore
settles and
develops
into new
filament

asexual

zoospore

Sexual Reproduction

Oedogonium shows more sexual specialization than any of the other three green algae previously discussed. Short boxlike cells called **antheridia** (singular: **antheridium**) are formed in the filaments alongside the ordinary vegetative cells. A pair of male gametes, or **sperms,** is produced in each antheridium. The sperms resemble the zoospores but are smaller. Certain cells become swollen and round to elliptical in outline. These cells, called **oogonia** (singular: **oogonium**), each contain a single female gamete or **egg,** which occupies nearly all of the cell.

As the egg matures, a pore develops on the side of the oogonium. When sperms escape from the antheridia, they are attracted to the oogonia by a substance released by the eggs. One sperm eventually enters the oogonium through the pore and unites with the egg, forming a zygote. In some species of *Oedogonium,* oogonia are produced only on female filaments and antheridia only on male filaments. Sometimes the male filaments are dwarf. They attach themselves to the female filaments, and then both produce hormones that influence the other's development.

FIGURE 16.14   A portion of a water net (*Hydrodictyon*), a
green alga whose cells form a network in the shape of a tube.
(Photomicrograph by Carolina Biological Supply Company)

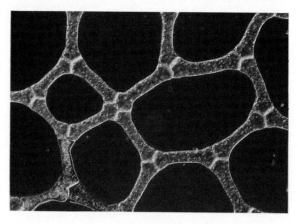

Zygotes may remain dormant for a year or more, but they eventually undergo meiosis, producing four zoospores, each of which is capable of developing into a new filament. The sexual reproduction exhibited by *Oedogonium,* in which one gamete is *motile* (capable of spontaneous movement) while the other gamete is larger and stationary, is called **oogamy.**

Other Green Algae

As indicated earlier, the green algae constitute a very diverse group, with an extraordinary variety of forms and chloroplast shapes. Each species obviously has to reproduce in order to perpetuate itself. Most undergo both sexual and asexual reproduction, but a few do not. For example, the worldwide algae that make parts of some tree trunks appear as though they had received a light brushing or spattering of green paint usually are unicellular or colonial forms that reproduce only asexually. These and *Chlorella,* another widespread green alga composed of tiny spherical cells, reproduce by forming either daughter cells or aplanospores through mitosis. The daughter cells often remain together in packets, while the aplanospores of *Chlorella,* which are formed inside the parent cells, grow to full size as the parent cell wall breaks down.

Chlorella is very easy to culture and is a favorite organism of research scientists. It has been used in many major investigations of photosynthesis and respiration, and in the future it may not only become important in human nutrition (see the section on human and ecological relevance of the algae, which begins on page 321), but it could also play a key role in long-range space exploration. Present exploration is severely limited by the weight of oxygen tanks and food supplies needed on a spacecraft, and so scientists have turned to *Chlorella* and similar algae as portable oxygen generators and food sources. Future spacecraft may be equipped with tanks of such algae. These would carry on photosynthesis, using available light and carbon dioxide given off by the astronauts, while furnishing them with oxygen. As the algae multiplied, the excess could either be eaten or fed to freshwater shrimp, which could, in turn, become food for the astronauts. Still other algae and bacteria could recycle other human wastes. Such a self-perpetuating "closed system," as it is called, has already been successfully tested with mice and other animals. Many algae, however, are known to produce traces of deadly carbon monoxide gas, and until this problem is resolved, humans will not be subjected to such research.

Desmids (figure 16.13), whose 2,500 species of crescent-shaped, elliptical, and star-shaped cells are mostly free-floating and unicellular, reproduce by conjugation. In the beautiful *water nets* (*Hydrodictyon*) (figure 16.14), sexual reproduction is isogamous, with up to 100,000 flagellated gametes being produced in a single cell. Sexual reproduction is also

FIGURE 16.15   Mermaid's wineglass (*Acetabularia*).

FIGURE 16.16   Sea lettuce (*Ulva*).

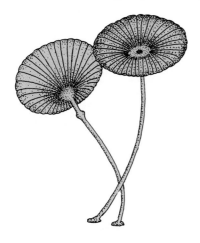

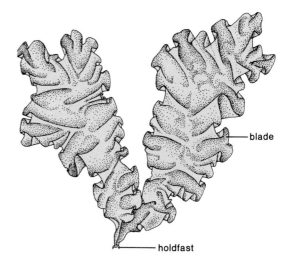

blade

holdfast

isogamous in the *mermaid's wineglass* (*Acetabularia*), a marine alga consisting of a single huge cell shaped like a delicate mushroom (figure 16.15). Each cell is up to 5 centimeters (2 inches) long. This alga has been used in classic experiments demonstrating the influence of the nucleus on the form of the cell. If the "cap" of an alga is removed and the nucleus is replaced with a nucleus taken from another species, the base regenerates a cap identical with the previous one. If this new cap is also removed, however, the next cap that develops shows form characteristics of both species. If the intermediate cap is then removed, the next cap that develops is identical with that of the species from which the nucleus originally came. Clearly the original nucleus directed development of cytoplasmic substances regulating cap form, and when these are gone, the replacement nucleus exerts its own influence.

Volvox (see figure 16.7A) is representative of a line of green algae that forms colonies, apparently by means of single cells similar to those of *Chlamydomonas,* held together in a secretion of gelatinous material. In some colonies, the cells are actually connected to one another by cytoplasmic strands. The flagella of individual cells beat separately but pull the whole colony along. A *Volvox* colony may consist of several hundred to many thousands of cells arranged so that they resemble a hollow ball, which appears to spin on its axis as it moves. Reproduction may be either asexual or sexual, with smaller daughter colonies being formed inside the parent colony. The daughter colonies are released when the parent colony breaks apart.

Sea lettuce (*Ulva*) is a multicellular seaweed with flattened, crinkly-edged green blades that may attain lengths of 1 meter (3 feet) or more (figure 16.16). The blades, which may be either haploid or diploid, are anchored to rocks by means of a holdfast at the base. Diploid blades produce spores that develop into haploid blades bearing gametangia. The gametes from the haploid blades fuse in pairs, forming zygotes, each of which can potentially grow into a new diploid blade. Except for the reproductive structures, the haploid and diploid blades of sea lettuce are indistinguishable from one another.

Cladophora is a branched, filamentous green alga whose species are represented in both fresh and marine waters. Unlike other green algae, the cells of *Cladophora* and its relatives are mostly multinucleate.

Stoneworts, which are aquatic and loosely resemble small calcium-encrusted horsetail plants (see chapter 19), are often included with the green algae. Their sexual reproduction is oogamous, and they are sometimes placed in their own division (Charophyta) because the antheridia in which the sperms are produced are multicellular and because other features of both their vegetative growth and reproduction are more complex than those of any of the other green algae. Some botanists have even considered them more closely related to mosses than to algae.

A.

B.

DIVISION PHAEOPHYTA— THE BROWN ALGAE

Most seaweeds that are brown to olive green in color are assigned to this division, which includes between 1,500 and 2,000 species of **brown algae** of varying size. Many are relatively large, and none are unicellular or colonial. Only 4 of the 260 known genera occur in fresh water, the vast majority being found in colder ocean waters, usually in shallower areas, although the giant kelp (figure 16.17) may be found in water up to 30 meters (100 feet) or more deep. One giant kelp measured 274 meters (710 feet) in length, which is believed to be a record for any single living organism.

Many have a body that is differentiated into a *holdfast,* a *stipe,* and flattened leaflike *blades* (figure 16.18). The **holdfast** is a tough, sinewy structure resembling a mass of intertwined roots. It holds the seaweed to rocks so tenaciously that even the heaviest pounding of surf will not readily dislodge it. The **stipe** is an often hollow stalk, which has a meristem either at its base or at the blade junctions. Since the meristem produces new tissue at the base, the oldest parts of the blades are at the tips. The **blades,** which like most of the rest of the body are photosynthetic, may have gas-filled floats called *bladders* toward their bases. Analysis of the bladder contents of some kelps has revealed the presence of more than 10% carbon monoxide, the deadly gas also produced by internal combustion engines.

FIGURE 16.18 Parts of the brown alga *Nereocystis*, a kelp. The heaviest pounding surf seldom dislodges these brown algae from rocks.

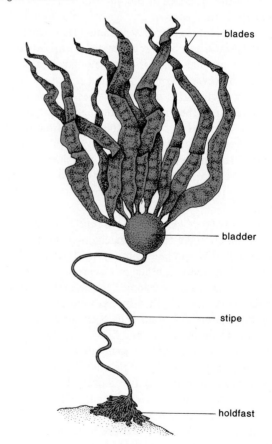

blades

bladder

stipe

holdfast

FIGURE 16.19 *Sargassum*, a floating brown alga from which the Sargasso Sea got its name. It is also found in other marine waters. (Courtesy I. A. Abbott)

The color of the brown algae, which can vary from light yellow-brown to almost black, reflects the presence of varying amounts of the brown pigment *fucoxanthin,* in addition to chlorophylls *a* and *c* and several other pigments in the chloroplasts. The main food reserve is a carbohydrate called *laminarin.* **Algin** or *alginic acid* (see table 16.2 and the discussion under human and ecological relevance of the algae, page 321) occurs on or in the cell walls, and can represent as much as 40% of the dry weight of some kelps. Reproductive cells are unusual in that the two flagella are inserted laterally (i.e., on the side) instead of at the ends. The only motile cells in the brown algae are the reproductive cells.

In some localities off the coast of British Columbia and Washington, herring deposit spawn (eggs) in layers up to 2.5 centimeters (1 inch) thick on both surfaces of giant kelp blades in late spring. In the past, and to a limited extent at present, native North Americans have harvested these spawn-covered blades, sun-dried them, and used them for winter or feast food. Even today some school children are given small pieces of the dried or preserved material for lunchbox snacks.

The Sargasso Sea gets its name from a brown rockweed, *Sargassum,* which is washed up in large quantities on the shores along the Gulf of Mexico after tropical storms (figure 16.19). A species occurring in the Pacific Ocean has been used, in chopped form, as a poultice on cuts received from coral by native Hawaiians. This and several other brown algae reproduce asexually by fragmentation, while some produce aplanospores. Sexual reproduction takes several different forms, depending on the species, with the conspicuous phases of the life cycles usually being diploid.

In the common rockweed *Fucus* (figure 16.20), separate male and female **thalli** (singular: **thallus,** the term for a body that is usually flattened, multicellular, and not organized into leaves, stems, and roots) are produced. Somewhat puffy fertile areas called *receptacles* develop at the tips of the branches of the thallus. The surface of each receptacle is dotted with pores (visible to the naked eye) that open into special, spherical hollow chambers called *conceptacles.* Within the conceptacles **gametangia** (cells or structures in which gametes are produced) are formed. Eight eggs are produced in each *oogonium* (female gametangium) as a result of a single diploid nucleus undergoing meiosis followed by mitosis. Meiosis also occurs in each *antheridium* (male gametangium), but three mitotic divisions follow meiosis so that 64 sperms are produced. Eventually both eggs and sperms are released into the water, where fertilization takes place, and the zygotes develop into mature thalli, completing the life cycle.

DIVISION RHODOPHYTA— THE RED ALGAE

Like the brown algae, most of the more than 5,000 species of **red algae** are seaweeds (figure 16.21) that tend, however, to occur in warmer and deeper waters than their brown counterparts. Some grow attached to rocks in intertidal zones, where they may be exposed at low tide. Others grow at depths of up to 175 meters (575 feet), where light barely reaches them. A few are unicellular, but most are filamentous. The filaments frequently are so tightly packed that the plants appear to have flattened blades or to form branching segments. Some develop as beautiful

FIGURE 16.20   Life cycle of the common rockweed *Fucus.*

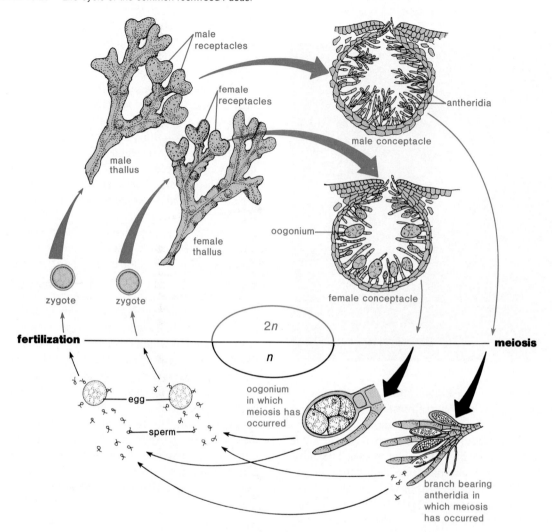

feathery structures that have the appearance of delicate works of art.[1] None match the large kelps in size, the largest species seldom exceeding a meter (3 feet) in height.

The red algae have relatively complex life cycles, often involving three different types of body structures. Meiosis usually occurs on a special body called a *tetrasporophyte,* while gametes are produced on separate male and female bodies. All of the reproductive cells are nonmotile and are carried passively by water currents. Zygotes may migrate from one cell to another through special tubes, which form bizarre loops in some species.

1. Most seaweeds produce their own "glue" and are easy to mount on paper for display. Fresh specimens can be laid directly on clean, high rag-content paper, covered with a layer or two of cheesecloth, and pressed between sheets of blotting paper for a day or two until they are dry. Feathery types will make better specimens if they are placed in a shallow pan of water so that their delicate structures float out, as they do in the ocean. Paper should be slid under them and then carefully lifted so that the seaweed spreads out naturally on the surface. Once the specimens are dry, the cheesecloth (which keeps them from sticking to blotting paper) is removed and they remain glued to the paper. They can then be displayed or stored indefinitely. Green and brown seaweeds can be treated in similar fashion.

Certain marine algae, which form crusty growths or jointed-appearing upright structures on rocks, accumulate calcium salts as they grow and often contribute to the development of coral reefs. These coralline algae need no special treatment to be displayed, although some may lose their natural pinkish or purplish color when they die.

A.

B.

C.

D.

FIGURE 16.22 Life cycle of the red alga *Polysiphonia*.

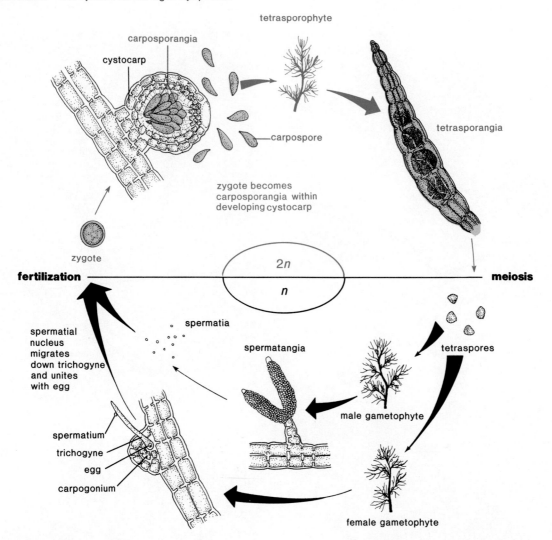

tetrasporophyte

carposporangia

cystocarp

carpospore

tetrasporangia

zygote becomes
carposporangia within
developing cystocarp

zygote

fertilization   **meiosis**

2*n*

n

spermatia

spermatangia

tetraspores

spermatial
nucleus
migrates
down trichogyne
and unites
with egg

spermatium

trichogyne

egg

carpogonium

male gametophyte

female gametophyte

In *Polysiphonia* (figure 16.22), a feathery red alga that is widespread in marine waters, the three types of thalli (male gametophyte, female gametophyte, and tetrasporophyte) all outwardly resemble one another in being about 2 to 15 centimeters (1 to 6 inches) tall, and in being branched into many fine, threadlike segments. The male sex structures, called *spermatangia,* somewhat resemble dense clusters of tiny grapes on slender branches of the male gametophyte thallus. Each spermatangium, as it is released from its branch, functions as a nonmotile male gamete or *spermatium.* The female sex structures, called *carpogonia,* are produced on the female gametophyte thallus. Each carpogonium consists of a single cell that looks something like a microscopic bottle with a long neck called the *trichogyne.* A

single nucleus at the base of the carpogonium functions as the female gamete, or *egg.* Since the spermatia have no flagella they cannot move of their own accord, but currents may carry them considerable distances. If a spermatium should brush against a trichogyne it may become attached, with the walls between the spermatium and the trichogyne breaking down and permitting the nucleus of the spermatium to migrate to the egg nucleus, with which it unites, forming a zygote.

Next, the zygote develops into a cluster of club-like *carposporangia* toward the base of an urn-shaped body (the *cystocarp*) formed around the carposporangia by the female gametophyte thallus. Diploid asexual spores, called *carpospores,* are produced by the carposporangia; these, when released, are carried away by ocean currents. When a carpospore lodges in a suitable location (e.g., rock

Comparison of the Divisions of Eukaryotic Algae

Division	Food Reserves	Special Pigments[1]	Flagella
Chrysophyta 6,000 species of yellow-green algae, golden-brown algae, diatoms, and cryptophytes	oils, fats, chrysolaminarin	chlorophyll c_1, chlorophyll c_2	2, 1, or none, usually unequal, at or near apex
Pyrrophyta 3,000 species of dinoflagellates	starch	chlorophyll c_2	2; 1 trailing, 1 girdling; few with none
Euglenophyta 750 species of euglenoids	paramylon	chlorophyll b	2, 1, or 3, at or near apex
Chlorophyta 7,500 species of green algae	starch	chlorophyll b	2, 4, or more at apex, or none
Phaeophyta 1,500 to 2,000 species of brown algae	laminarin, mannitol	chlorophyll c	2, lateral and unequal
Rhodophyta More than 5,000 species of red algae	floridean starch, mannitol	chlorophyll d (some species), phycobilins	none

1. **Note:** All divisions of eukaryotic algae have chlorophyll *a* and yellow to orange pigments called *carotenes* and *xanthophylls*, although the combinations and specific types of the latter vary with the divisions. The fucoxanthin of the golden-brown algae and the brown algae is a xanthophyll.

crevice or hull of a ship) it germinates and grows into a *tetrasporophyte,* which closely resembles the gametophyte thalli. Tetrasporangia are formed along the branches of the tetrasporophytes; each tetrasporangium undergoes meiosis, giving rise to four haploid *tetraspores.* When tetraspores germinate they develop into male or female gametophytes, and the life cycle is complete.

The red to purplish colors of most red algae are due to the presence of varying amounts of red and blue accessory pigments called *phycobilins,* which are closely related to those found in the blue-green bacteria and which have led to the belief that the red algae may have been derived from the blue-green bacteria. Several other pigments, including chlorophyll *a* and sometimes chlorophyll *d,* are also present in the chloroplasts. The principal reserve food is a carbohydrate called *floridean starch.* A number of red algae also produce **agar** and other important gelatinous substances discussed in the next section.

A comparative summary of the food reserves, special pigments, and flagella of the divisions of eukaryotic algae is given in table 16.1.

HUMAN AND ECOLOGICAL RELEVANCE OF THE ALGAE

We have already noted that numbers of blue-green bacteria are at the bottom of aquatic food chains, but members of all the protistan algal divisions play similar roles. Diatoms, for example, are consumed by fish that feed on plankton. The oils produced by diatoms are converted into cod-liver and other liver oils, which are rich sources of vitamins for humans.

Diatoms also have other extensive and more direct industrial uses. In the past, they have apparently been at least as numerous as they are now. As billions upon billions of them have reproduced and died, their microscopic glassy shells have accumulated on the ocean floor, forming deposits of *diatomaceous earth.* These deposits have accumulated to depths of hundreds of meters in some parts of the world and are quarried in several areas where past geological activity has raised them above sea level. At Lompoc, California, beds of diatomaceous earth are more than 200 meters (650 feet) deep, while in the Santa Maria oil fields of California, deposits reach a depth of 1,000 meters (3,280 feet).

This light, porous, powdery-looking material contains about 6 billion diatom shells per liter (1.057 quarts), yet a liter weighs only eight-tenths of a kilogram (1.76 pounds). It also has an exceptionally high melting point of 1,750° C (3,182° F) and is insoluble in most acids and other liquids. These properties make it ideal for a variety of industrial and domestic uses, including many types of filtration. The sugar industry uses diatomaceous earth in sugar refining, and its use for swimming pool filters is widespread. It is also used in silver and other metal polishes, in toothpaste, and in the manufacture of light-reflecting paint used on highway markers and signs and on the automobile license plates of some states. It is packed as insulation around blast furnaces and boilers. It is also manufactured into the construction panels used in prefabricated housing.

TABLE 16.2
Some Uses of Algin

Food
1. As a thickening agent in toppings, pastry fillings, meringues, potato salad, canned foods, gravies, dry mixes, bakery jellies, icings, dietetic foods, flavored syrups, candies, puddings.
2. As an emulsifier and suspension agent in soft drinks and concentrates, salad dressings, barbecue sauces, frozen food batters.
3. As a stabilizer in chocolate drinks, eggnog, ice cream, sherbets, sour cream, coffee creamers, party dips, buttermilk, dairy toppings, milkshakes, marshmallows.

Paper
1. Provides better ink and varnish holdout on paper surfaces; provides uniformity of ink acceptance, reduction in coating weight and improved holdout of oil, wax, and solvents in paperboard products. Makes improved coating for frozen food cartons.
2. Used for coating greaseproof papers.

Textiles
1. Thickens print paste and improves dye dispersal. Reduces weaving time and eliminates damage to printing rolls or screens.

Pharmaceuticals and Cosmetics
1. As a thickening agent in weight control products, cough syrups, suppositories, ointments, toothpastes, shampoos, eye makeup.
2. As a smoothing agent in lotions, creams, lubricating jellies. Binder in manufacture of pills. Blood anticoagulant.
3. As a suspension agent for liquid vitamins, mineral oil emulsions, antibiotics. Gelling agent for facial beauty masks, dental impression compounds.

Industrial Uses
1. Used in manufacture of acidic cleaners, films, seed coverings, welding rod flux, ceramic glazes, boiler compounds (prevents minerals from precipitating on tubes), leather finishes, sizing, various rubber compounds (e.g., automobile tires, electric insulation, foam cushions, baby pants), cleaners, polishes, latex paints, adhesives, tapes, patching plaster, crack fillers, wall joint cement, fiberglass battery plates, insecticides, resins, tungsten filaments for light bulbs, digestible surgical gut (disappears by time incision is healed), oil well-drilling mud. Used in clarification of beet sugar. Mixed with alfalfa and grain meals in dairy and poultry feeds.

Brewing
1. Helps create creamier beer foam with smaller, longer-lasting bubbles.

A few green algae have occasionally been used for food, but members of this division have generally been used less by humans than those of other divisions. With dwindling world food supplies, this could soon change. Sea lettuce has been used for food on a limited scale in Asian countries for some time, and several countries are experimenting with the suitability of plankton for human consumption. Except for vitamin C, *Chlorella* contains most of the vitamins needed in human nutrition, and since it is so easy to culture, it may become an important protein source in many parts of the globe. *Chlorella* has also been investigated as a potential oxygen source for atomic submarines, in addition to its possible use in space exploration.

Commercially produced ice cream, salad dressing, beer, jelly beans, latex paint, penicillin suspensions, paper, textiles, toothpaste, ceramics, and floor polish today all share a common ingredient, *algin,* produced by the giant kelps and other brown algae. It is now used in so many products (table 16.2) that one might wonder how the world used to get along without it.

Algin has the unique ability to regulate water "behavior" in a wide variety of products. It can, for example, control the development of ice crystals in frozen foods, regulate the penetration of water into a porous surface, and generally stabilize any kind of suspension such as an ordinary milkshake or other thick fluid containing water. It is produced by several kinds of seaweeds, but a major source is the giant kelp found in the cooler ocean waters of the world, usually just offshore where there are strong currents (see figure 16.17A). This large seaweed, which has been known to attain a length of 92 meters (300 feet), sometimes grows at the rate of 3 to 6 decimeters (1 to 2 feet) per day. Specially equipped oceangoing vessels (figure 16.23) harvest the kelp by mowing off the top meter (3 feet) of growth, taking the chopped material aboard and then transferring it to processing centers onshore where it is extracted and refined.

Brown algae also produce a number of other useful substances. Many seaweeds, but particularly kelps, build up concentrations of iodine to as much as 20,000 times that of the surrounding seawater. Although it is cheaper to obtain iodine from other sources in the United States, in other parts of the world dried kelp has been used in the treatment of goiter, which results from iodine deficiency. Kelps are relatively high in nitrogen and potassium and have been used as fertilizer for many years. They have also been used as livestock feed in northern Europe and elsewhere. In the Orient, many marine algae are used for food in soups, confections, meat dishes, vegetable dishes, and beverages. In Japan, acetic acid is produced through fermentation of seaweeds.

During the Irish famine of 1845–1846, *dulse,* a red seaweed, became an important substitute for the potato crop that had been destroyed by blight. Dulse also occurs along both the Atlantic and Pacific coasts of North America, where some is still gathered for food. Another red seaweed, *purple laver* ("nori"), which occurs in both American and Asian waters, is used extensively for food, particularly in the Orient. In Japan, it is cultured on nets or bamboo stakes set out in shallow marine bays (figure 16.24). It is harvested when the thin, crinkly, gelatinous blades are several centimeters (2 to 3 inches) in diameter and is used in meat and macaroni dishes, soups, and dry spiced delicacies. *Irish moss* is another important edible red alga. It is also used in bulking laxatives, cosmetics, and pharmaceutical preparations. Blancmange is a dessert made from Irish moss and milk. *Carrageenan* is a mucilaginous substance extracted from Irish moss and used as a thickening agent in chocolate milk and other dairy products. *Funori,* obtained from yet another red alga, is used as a laundry starch, as an adhesive in hair dressings, and in some water-based paints.

Agar

One of the most important of all algal substances is **agar,** produced most abundantly by the red alga *Gelidium.* This substance, which has the consistency of gelatin, is used (with nutrients added) around the world in laboratories and medical institutions as a culture medium for the growth of bacteria. When various nutrients are added to it, it can also be used as a culture medium for the growth of both plant and animal cells. Full-sized plants have been induced to develop from pollen grains sown on nutrient agar. Orchid tissues are cultured commercially on it and induced to grow into full-sized plants (see the section on mericloning in appendix 4), and its use in making the capsules containing drugs and vitamins is now worldwide. It is also used as an agent in bakery products to retain moistness, as a base for cosmetics, and as an agent in gelatin desserts to promote rapid setting.

Current research involving red algae and other seaweeds indicates they contain a number of substances of potential medicinal value. More than 20 seaweeds have been used in preparations designed for the expulsion of digestive tract worms, control of diarrhea, and the treatment of cancer. Some have shown considerable potential as antibiotics and insecticides. Chemical relatives of DDT have been found to be produced by certain red algae. The sea hare and other marine animals feed on such algae, and it is possible that such animals may degrade (break down) the DDT-like compounds to simpler substances—a feat unknown among land animals.

OTHER MEMBERS OF KINGDOM PROTISTA

Protozoans (Phylum Protozoa) and *sponges* (Phylum Porifera) are included within Kingdom Protista; they have, however, traditionally been regarded as animals and are not covered in this book. A number of biologists regard the *slime molds* (Myxomycotineae) as protists, but they are included in Kingdom Fungi in this text, for reasons given in chapter 17.

SUMMARY

Kingdom Protista includes organisms that all have eukaryotic cells but are otherwise diverse. Members may be unicellular or multicellular, and occur as either colonies or filaments. Modes of nutrition include photosynthesis, ingestion of food, or a combination of both. Some members are nonmotile but most are motile or at least have stages that are motile by means of flagella or by amoeboid movements.

The golden-brown algae are grouped into four classes that include the yellow-green algae, the true golden-brown algae, the diatoms, and the cryptophytes. Some members of each class produce statospores. Diatoms, which have a glassy shell that consists of two "halves" that fit together like a pillbox, are extremely abundant, particularly in colder marine waters. The shells are usually etched with fine grooves and pores through which the cytoplasm is in contact with the environment. Diatoms are often golden-brown in color due to the presence of the brownish pigment fucoxanthin in the one to many chloroplasts occurring in each cell. Diatoms move in caterpillar fashion by contact of the cytoplasm with a surface as it protrudes through the pores. In asexual reproduction the two "halves" of a cell separate after mitosis of the protoplast, and a new "half" forms within each original portion. An auxospore (zygote) is produced through a sexual process involving the fusion of gametes.

Dinoflagellates are unicellular organisms with two flagella. Some are known to cause red tides and when present in large numbers the paralyzing toxins they release into the water can kill fish and shellfish. Dinoflagellates exhibit bioluminescence (emission of light) when disturbed in tropic waters.

Euglenoids have no rigid cell wall, only one functional flagellum, a gullet, and a carbohydrate food reserve called paramylon. Reproduction is by cell division; sexual reproduction has not been confirmed.

Green algae occur in a wide variety of aquatic habitats. Their cells have the same pigments and reserve food (starch) as those of higher plants. The unicellular green alga, *Chlamydomonas,* has a pair of flagella that enable the cell to move rapidly. Within the cell there is a chloroplast containing one or two pyrenoids, two or more vacuoles, and often a red eyespot. Asexual reproduction is by mitosis; sexual reproduction is by union of like gametes (i.e., it is isogamous).

Ulothrix is a filamentous green alga that attaches to various objects by means of a basal cell called a holdfast; each cell contains a bracelet-shaped chloroplast. Asexual reproduction is by zoospores, sexual reproduction is isogamous.

Spirogyra is a floating, filamentous green alga with spiral, ribbonlike chloroplasts. It does not produce flagellated cells of any kind. Asexual reproduction is by fragmentation (breaking of filaments with each fragment adding new cells by mitosis). Sexual reproduction is by conjugation.

Oedogonium is a filamentous green alga that is epiphytic (attached in a nonparasitic manner) on aquatic plants and other algae; it has cylindrical, netted chloroplasts. In asexual reproduction, zoospores are produced. In sexual reproduction short cells of a filament function as antheridia in which sperms are produced, and large swollen cells function as oogonia in each of which a single egg is produced. A sperm enters an oogonium through a pore and fertilizes the egg. This union of a small motile gamete and a large nonmotile gamete is called oogamy.

Other green algae include *Chlorella,* a tiny unicellular alga that is easy to culture and has been used in experiments to produce oxygen for space vehicles. Desmids are free-floating unicellular algae that reproduce by conjugation; mermaid's wineglass (*Acetabularia*) produces huge mushroomlike cells; *Volvox* is a colonial alga that forms motile hollow balls of hundreds to thousands of cells; sea lettuce (*Ulva*) has blades that are anchored to rocks by means of a holdfast. *Cladophora* has multinucleate cells. Stoneworts are sometimes included with the green algae or retained in a separate division (Charophyta) because they have multicellular antheridia and other features more complex than those of the green algae.

The brown algae include the largest of seaweeds. Many are differentiated into a stalk (stipe), flattened blades that may have bladders toward their bases, and a tough, sinewy holdfast that holds the seaweed to the rocks. Fucoxanthin is largely responsible for the color of brown algae whose main carbohydrate food reserve is laminarin. Some produce algin (alginic acid), a useful gelatinous substance. The reproductive cells have lateral flagella. In the common rockweed, *Fucus,* eggs (from oogonia) and sperms (from antheridia) are released into the water where fertilization occurs, and the zygotes develop into mature thalli.

The red algae, whose members tend to be smaller than those of the brown algae, have relatively complex life cycles involving three different types of thalli and nonmotile gametes. In *Polysiphonia* and other representatives, the three thalli are the tetrasporophyte, the male gametophyte, and the female gametophyte.

The colors of red algae are partially due to the presence of red and blue phycobilins, which are related to those of the blue-green bacteria; chlorophyll *d* may be present in the chloroplasts. The main carbohydrate food reserve is floridean starch. Some red algae produce agar, an economically important gelatinous substance.

Algae are important in aquatic food chains and in numerous other ways. Diatom shells have accumulated for thousands of years on ocean floors and make up diatomaceous earth that is used for filtering, polishes, insulation, and reflectorized paint. *Chlorella* is a potentially important food and oxygen source. Algin is used as a stabilizer and thickening agent in hundreds of food products, paints, medicines, papers, ceramics, cosmetics, beer, etc. Brown algae are a source of fertilizer and iodine, and some serve as food for both livestock and humans. Red algae are a source of agar, which is used as a culture medium for bacteria and other organisms or tissues; some are also used for human food; and substances produced by red algae (e.g., carrageenan, funori) are used as thickening agents, as laundry starch, and as adhesives. Some red algae have potential medicinal value.

REVIEW QUESTIONS

1. How do cells of diatoms differ from those of other organisms?
2. What forms of sexual and asexual reproduction occur in the green algae?
3. Which groups of algae produce the following important products: (1) agar, (2) algin, (3) nerve poisons, (4) abrasives for polishes?
4. How would you distinguish *Chlamydomonas* from *Euglena?*
5. *Spirogyra, Ulothrix,* and *Oedogonium* all form filaments. How can you tell them apart?
6. In the green algae studied, where in the life cycles does the chromosome number change from *n* to *2n* and vice versa?
7. Where and how is algin obtained?
8. Is there any difference in structure between the holdfasts of microscopic green algae and brown algae?
9. Why are some green algae red and some red algae green?
10. Which divisions of algae have only unicellular representatives?

DISCUSSION QUESTIONS

1. Some algae are attached to solid objects or other organisms, while others are free-floating. What are the advantages and disadvantages of each type of growth?
2. The variety and sizes of algae found in the oceans are considerably greater than those of freshwater forms. Can you suggest reasons?
3. Seaweeds that grow in intertidal zones, where they may be exposed between tides, are often more gelatinous than their continually submerged counterparts. Explain.
4. Why do some algae grow on one side of a tree and not all around the trunk?
5. Should the bladders, or floats, of some of the kelps give these algae any advantage over those algae that do not possess such structures?

ADDITIONAL READING

Abbott, I. A. 1984. *Limu: an ethnobotanical study of some edible Hawaiian seaweeds,* 3d ed. Lawai, HI: Pacific Tropical Botanical Garden.

Abbott, I. A., and E. Y. Dawson. 1978. *How to know the seaweeds,* 2d ed. Pictured Key Nature Series. Dubuque, IA: Wm. C. Brown Publishers.

Abbott, I. A., and G. J. Hollenberg. 1976. *Marine algae of California.* Stanford: Stanford University Press.

Bold, H. C., and M. J. Wynne. 1985. *Introduction to the algae,* 2d ed. Englewood Cliffs, NJ: Prentice-Hall.

Buetow, D. E., ed. 1968. *The biology of Euglena: General biology and ultrastructure.* 2 vols. New York: Academic Press.

Chapman, V. J., and D. J. Chapman. 1980. *Seaweeds and their uses,* 3d ed. London: Chapman & Hall.

Crowder, W. "Marvels of the mycetozoa." 1926. *National Geographic Magazine* 49(4):421–44.

Fritsch, F. E. 1935, 1945. *Structure and reproduction of the algae.* 2 vols. New York: Cambridge University Press.

Gibor, A. 1966. "*Acetabularia:* a useful giant cell." *Scientific American* 215: 118–24.

Irvine, D., and D. M. John, eds. 1985. *Systematics of the green algae.* New York: Academic Press, Inc.

Jackson, D. F., ed. 1964. *Algae and man.* Proceedings of the NATO Advanced Study Institute. New York: Plenum Press.

Kelco Company. 1968. *Search in the aquasphere. The story of Kelco.* San Diego, CA: Kelco Co.

Lee, R. E. 1986. *Phycology.* New York: Cambridge University Press.

Palmer, C. M. 1962. *Algae in water supplies.* U.S. Department of Health, Education and Welfare. Public Health Service. Washington, DC: Superintendent of Government Documents.

Pickett-Heaps, J. D. 1975. *Green algae: structure, reproduction and evolution in selected genera.* Sunderland, MA: Sinauer Associates.

Prescott, G. W. 1979. *How to know the freshwater algae,* 3d ed. Pictured Key Nature Series. Dubuque, IA: Wm. C. Brown Publishers.

Round, F. E. 1984. *The ecology of algae.* New York: Cambridge University Press.

Scagel, R. F., R. J. Bandoni, J. R. Maze, G. E. Rouse, W. B. Schofield, and J. R. Stein. 1984. *Plants: an evolutionary survey.* Belmont, CA: Wadsworth Publishing Co.

Smith, G. M. 1950. *Freshwater algae of the United States,* 2d ed. New York: McGraw-Hill Book Co.

Turner, N. J. 1974. *Food plants of British Columbia Indians. Part 1: Coastal peoples.* Victoria: British Columbia Provincial Museum.

Overview

*T*he chapter opens with a summary of the features of Kingdom Fungi and a review of how the kingdom came to be recognized. The kingdom is treated here as being composed of three independent subkingdoms whose members do not appear to have been derived from members of the other subkingdoms. One subkingdom includes the slime molds, another the chytrids and water molds, and the third the true fungi. Selected members of each subkingdom and division are discussed, and representative life cycles are presented, with discussions of their economic importance. Among the topics or fungi discussed are slime molds, nematode-trapping fungi, *Pilobolus,* truffles, morels, ergot, yeasts, stinkhorns, puffballs, bracket fungi, bird's-nest fungi, smuts, rusts, poisonous and hallucinogenic fungi, Black Forest mushrooms, mushroom culture, antibiotics, industrial products obtained from fungi, and fungi in nature. The chapter concludes with a description and discussion of various forms of lichens. Natural dyeing is discussed in a footnote, and the economic importance of lichens is reviewed.

17 Kingdom Fungi and Lichens

FEATURES OF KINGDOM FUNGI

In the past the true fungi, slime molds, and bacteria were all placed in a single division of the Plant Kingdom. Once the fundamental differences between prokaryotic and eukaryotic cells became known, however, the bacteria were placed in the prokaryotic Kingdom Monera. Then it became increasingly apparent that the metabolism, reproduction, and general lines of diversity of fungi were different from those of plants, with the fungi evidently having been independently derived from ancestral unicellular organisms. Accordingly, the true fungi were placed in their own kingdom.

The slime molds, with their absence of cell walls, the amoeboid movement of their *plasmodia* and *myxamoebae* (discussed in the section on slime molds), and their mode of nutrition (ingestion of food particles), did not appear related to the true fungi. All true fungi are filamentous or unicellular *saprobic* (see chapter 15) or parasitic decomposers that absorb their food in solution through their cell walls. Partly because of these differences the slime molds were assigned to the heterogeneous Kingdom Protista by several respected contemporary biologists. However, although slime molds are obviously not closely related to the true fungi, their derivation and relationship to members of Kingdom Protista is unclear. In this text they are assigned to a subkingdom of Kingdom Fungi because their formation of *sporangia* and *spores* as a means of reproduction is more funguslike than it is either protist- or plantlike. They also share a primary food reserve (*glycogen*) with the fungi (and incidentally with animals) and, with the increasing awareness that a number of protists also absorb their food in solution, the distinction between Kingdoms Protista and Fungi on the basis of nutrition alone becomes problematical. The treatment of Kingdom Fungi in this chapter recognizes three subkingdoms for three distinct and apparently independently derived groups of organisms: the *slime molds,* the *chytrids and water molds,* and the *true fungi.*

The *slime molds* have no cell walls in their active state, and consist of multinucleate masses of protoplasm called **plasmodia** (singular: **plasmodium**). The plasmodia flow or creep over damp leaves or debris and become converted to stationary, variously-shaped **sporangia** containing tiny globular **spores** when environmental changes occur. The

aquatic *chytrids and water molds,* which apparently have been derived independently from protists, range in form from single spherical cells to branching, threadlike **hyphae** (singular: **hypha**) with numerous nuclei. They produce motile cells with flagella at various stages of their life cycles. The filamentous **true fungi** do not have motile cells and produce threadlike hyphae, which grow at their tips, usually branching and often anastomosing (forming a network) into a mass known as a **mycelium.** Structures such as mushrooms are formed from hyphae tightly interwoven and packed together. All fungi except the water molds (Division Oomycota) have cell walls that consist primarily of *chitin,* a material also found in the shells of arthropods (e.g., insects, crabs). Fungi exhibit a variety of forms of sexual reproduction. The food substances, which most fungi absorb through their cell walls, are often broken down with the aid of enzymes secreted to the outside by the cells. Because of the great variety of form and reproduction throughout Kingdom Fungi, a neat pigeonholing of all the members into distinct groups is difficult; broad groups can, however, be recognized.

INTRODUCTION TO THE FUNGI

To many, the word fungus evokes visions of mushrooms or some sort of creeping, insidious growth. While mushrooms are indeed fungi, and while many fungi do give the appearance of creeping along the ground, the fungi assume a great variety of forms. There are about 100,000 known species of mushrooms, rusts, smuts, mildews, molds, stinkhorns, puffballs, truffles, and other organisms assigned to Kingdom Fungi, and more than 1,000 new species are described each year. Possibly as many as 200,000 species still await to be described or discovered. The hyphae of fungi (figure 17.1) grow so rapidly if appropriate food is available that a single day's growth could extend more than a kilometer (0.6 mile) if all the hyphae produced were laid end to end. Some fungi thrive in freezers if the temperature is not lower than −5° C (23° F), while others freely reproduce under temperature conditions of 55° C (131° F) or higher. Along with bacteria, fungi are vital to the natural recycling of dead organic material, but they also cause huge economic losses through food spoilage and disease (see the sections on the human and ecological relevance of fungi later in the chapter). Scientists who study fungi are known as *mycologists,* and consumers of fungi are called *mycophagists* (from the Greek word *myketos,* a fungus).

SUBKINGDOM MYXOMYCOTINEAE—
THE SLIME MOLDS

At the Chicago World's Fair in 1933 at the "Believe It or Not" pavilion there was an exhibit of "hair growing on wood." The "hairs," while indeed superficially resembling short human hair, were actually the reproductive structures of a species of **slime mold** (figure 17.2). These curious organisms are totally without chlorophyll and thus are incapable of producing their own food. As indicated in the introduction, they are a bit of a puzzle to biologists because they are distinctly animallike during much of their life cycle but just as distinctly funguslike when they reproduce.

The tiny roundish *spores* of the more than 500 species of slime molds average only 10 to 12 micrometers in diameter and thus are individually invisible to the naked eye. Nevertheless, they are present nearly everywhere and are especially abundant in airborne dusts. If one places almost any dead leaf or piece of bark on a food source such as a few dry oatmeal flakes in a covered dish and adds a few drops of water, the chances are that any slime mold spores present will germinate. In some instances, they will germinate in as little as 15 minutes, but germination usually takes several hours or longer. Within a few

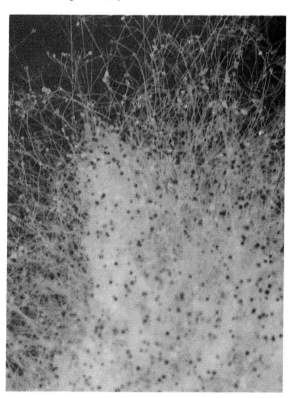

FIGURE 17.1   Typical fungal hyphae, as seen with the aid of a dissecting microscope.

FIGURE 17.2   Reproductive bodies (sporangia) of the slime mold *Stemonitis.*

FIGURE 17.3 A plasmodium of the slime mold *Physarum*.

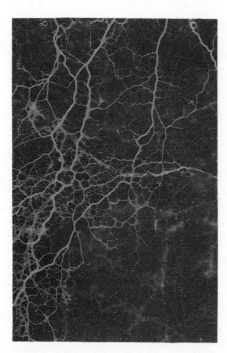

FIGURE 17.4 Common slime mold sporangia. *A. Arcyria. B. Lycogala.* (*B*, Courtesy L. L. Steimley)

A.

B.

days a curious glistening mass of active slime mold material somewhat resembling the netted venation of a leaf may appear. This material, whose "veins" tend to merge into the shape of a fan at its leading edges, is the slime mold **plasmodium** (figure 17.3). Inspection of a plasmodium with a hand lens or dissecting microscope reveals no cell walls. The protoplasm in the veins, particularly that located toward the center, flows very rapidly and rhythmically. Brief pauses in the protoplasmic movement occur at regular intervals, and the flow reverses after each pause.

Plasmodia are often white, but they also may be brilliantly colored in shades of yellow, orange, blue, violet, or black. A few are colorless and essentially transparent. They are found on damp forest debris, under logs, on old shelf or bracket fungi, sometimes on older mushrooms, and in other moist places where dead organic matter is present. They tend to creep forward at a rate of up to 2.5 centimeters (1 inch) or more per hour, often against slow moisture seepage, feeding on bacteria and other organic particles as they go. They contain many diploid nuclei, all of which divide simultaneously and

frequently as growth occurs. With an adequate food supply, a plasmodium may increase to 25 times its original size in just one week.

When the food supply dwindles or other features of the environment such as moisture or light change significantly, dramatic events occur. Usually the plasmodium heaps up into many separate small *sporangia* (figure 17.4), each containing thousands of minute one-celled *spores*. The sporangia frequently are globe-shaped, but in some species they occur as long or wide stationary bodies of variable shape. Depending on the species, sporangia may either be on slender stalks or without stalks. In other species, the whole plasmodium may form a single spore-bearing body that develops in a variety of shapes. In still others, combinations of body forms may occur. The spores are often dispersed throughout a jumbled mass of threads called a *capillitium*.

FIGURE 17.5   Life cycle of a slime mold. (After
Alexopoulos, C. J.; Mims, E. W. 1979. *Introductory Mycology*,
3d ed. Copyright 1979 John Wiley & Sons, Inc., New York.
Redrawn by permission of John Wiley & Sons, Inc.)

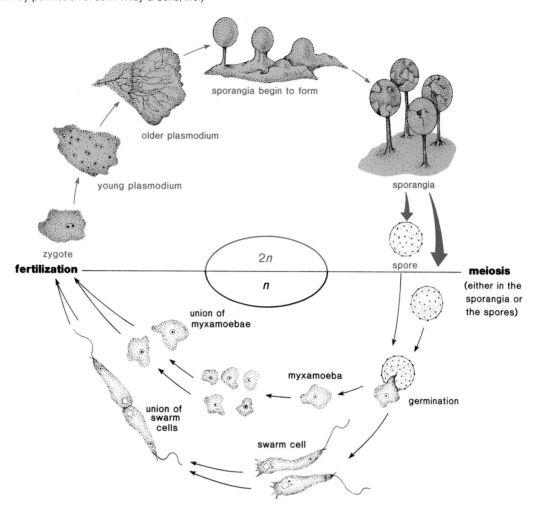

As a spore is formed, a single nucleus and a little cytoplasm become surrounded by a wall. Meiosis takes place in the spore, and three of the four resulting nuclei degenerate. When the spore germinates, one or more amoebalike cells called *myxamoebae* emerge. Sometimes these have flagella, in which case they are called *swarm cells*. Either form may become like the other through the development or loss of flagella. Myxamoebae or swarm cells at first feed on bacteria and other food particles. Sooner or later they function as gametes, fusing in pairs and forming zygotes. A new plasmodium usually develops from the zygote, although occasionally zygotes or small plasmodia may fuse and form larger plasmodia (figure 17.5).

Two or more divisions of slime molds are recognized. The majority of species have typical plasmodia and follow the patterns just discussed. About two dozen species that evidently are not closely related to the other slime molds are called *cellular slime molds* because they do not produce true plasmodia. Instead, individual amoebalike cells feed independently, dividing and producing new separate cells from time to time. When a population reaches a certain size, they stop feeding and clump together, forming a mass called a *pseudoplasmodium*. The pseudoplasmodium looks and crawls like a garden slug. It eventually becomes stationary and is transformed into a sporangiumlike mass of spores.

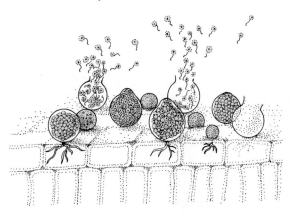

FIGURE 17.6   Chytrids on a dead leaf that is under water.

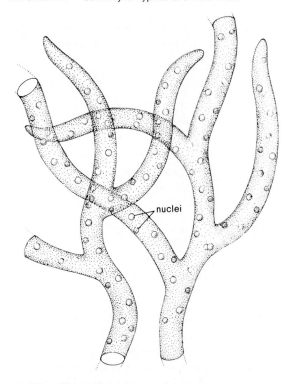

FIGURE 17.7   Coenocytic hyphae of a water mold.

nuclei

HUMAN AND ECOLOGICAL RELEVANCE OF THE SLIME MOLDS

The slime molds perform an ecological role similar to that of bacteria in breaking down organic particles to simpler substances, and also reduce bacterial populations in their forest and woodland habitats. Except for one atypical species that occasionally attacks cabbages, and another that causes powdery scab of potatoes and a disease of watercress, they are of little economic significance.

SUBKINGDOM MASTIGOMYCOTINEAE

Division Chytridiomycota—The Chytrids

If you were to immerse dead leaves or flowers, old onion bulb scales, dead beetle wing covers, or other organic material in water that has been mixed with a little soil, it is very probable that within a day or two, thousands of microscopic *chytrids* (pronounced kitt-ridds) would appear on the surfaces of the immersed objects (figure 17.6). These simplest of fungi include many parasites of aquatic flowering plants and algae and also numerous saprophytic species. A chytrid consists mostly of a spherical cell with colorless branching threads called **rhizoids** at one end. The rhizoids anchor the fungus to its food source. Many chytrid species reproduce only asexually, producing zoospores within a spherical cell. The zoospores, which each have a single flagellum, settle upon release and grow into new chytrids. Some species undergo sexual reproduction by means of the fusion of two motile gametes with haploid nuclei or by the union of two nonmotile cells whose diploid zygote nuclei undergo meiosis. The origin of chytrids is unknown, but the presence of flagella on the motile cells has led some authorities to suggest they may have originated from the protozoa; some species have cellulose present in their cell walls, and they do not appear to be directly related to the other fungi.

Division Oomycota—The Water Molds

Water mold fungi are familiar to those who have kept tropical fish aquaria or have seen salmon at the end of a spawning run. They appear as cottony growths wherever cuts and bruises have occurred on a fish's body, and they frequently attack the eyes of ailing fish. They also grow on dead insects and readily develop on marijuana seeds placed in water.

The hyphae of water molds may branch repeatedly and form large mycelia. No individual cells are formed during growth, however, and large numbers of nuclei are scattered throughout the mycelium. Such multinucleate mycelia without crosswalls are said to be **coenocytic** (pronounced see-no-sitt-ik) (figure 17.7). In asexual reproduction, tips of certain hyphae are separated from the remainder of the mycelium through the formation of crosswalls. Numerous biflagellated zoospores are produced in these special tips. The zoospores eventually give rise to new water mold mycelia after emerging through a terminal pore. Sexual reproduction involves *oogonia* and *antheridia,* which arise on side

FIGURE 17.8   Life cycle of the water mold *Saprolegnia*.

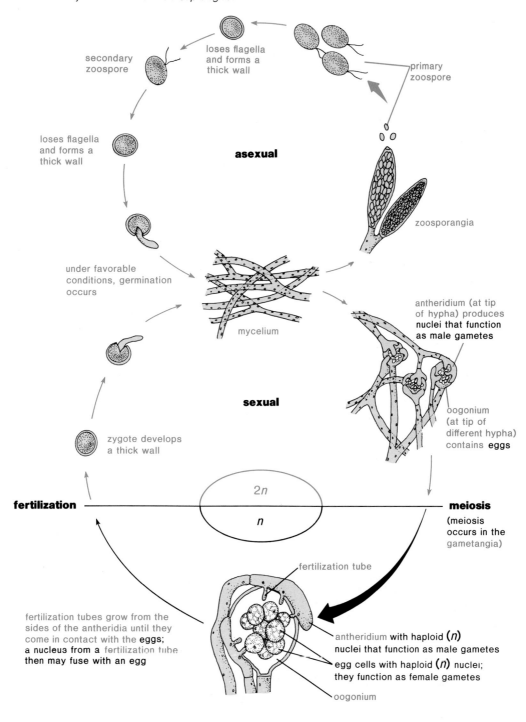

secondary
zoospore

loses flagella
and forms a
thick wall

primary
zoospore

loses flagella
and forms a
thick wall

asexual

zoosporangia

under favorable
conditions, germination
occurs

mycelium

antheridium (at tip
of hypha) produces
**nuclei that function
as male gametes**

sexual

oogonium
(at tip of
different hypha)
contains **eggs**

zygote develops
a thick wall

fertilization

2*n*

n

meiosis

(meiosis
occurs in the
gametangia)

fertilization tube

fertilization tubes grow from the
sides of the antheridia until they
come in contact with the **eggs**;
a nucleus from a fertilization tube
then may fuse with an egg

antheridium with haploid (*n*)
nuclei that function as male gametes

egg cells with haploid (*n*) nuclei;
they function as female gametes

oogonium

branches under the influence of hormones produced by the fungus. The mycelium is diploid, and meiosis takes place in the oogonia and antheridia. Zygotes formed in the oogonia eventually give rise to new mycelia (figure 17.8).

Two very important members of this fungal division are responsible for serious diseases of higher plants. Neither grows under water, although dew or rainwater is necessary for their reproduction. One, called *downy mildew of grapes,* develops and undergoes its life cycle on grape leaves, usually killing the leaves and consequently the vine if not controlled. This disease seriously threatened the French wine industry in the latter part of the nineteenth century after it had been introduced into the vineyards on imported American cuttings. It was controlled within a few years when it was discovered that Bordeaux mixture, a combination of copper sulphate and lime, inhibited the growth of downy mildew. This unappetizing-looking mixture, which is the first substance known to have been used as a fungicide, was originally sprayed on grape vines to discourage passersby from picking the grapes.

The other important disease-causing member of the Oomycota is called *late blight of potato.* This virulent organism plagued potato farmers to such an extent in the past that at times it altered the course of history. In the summer of 1846, the entire potato crop of Ireland was wiped out in one week, bringing about a famine during which a million persons starved to death and many more emigrated to the United States. The fungus produces its mycelium inside the leaves, and so it is relatively immune to spraying once its spores land on the surface and send hyphae into the tissues beneath. Development of the fungus is checked, however, if the leaves are sprayed with Bordeaux mixture or other fungicides before the spores have germinated.

The exceptional resonance of the violins made by Stradivari is believed to have been due, in part, to the growth of a water mold in the raw soundboard wood.

SUBKINGDOM EUMYCOTINEAE— TRUE FUNGI

Division Zygomycota—The Coenocytic Fungi

Although black bread molds are the best known members of this division, they are not the only fungi that grow on bread; in fact, so many organisms can contribute to bread spoilage that nearly all commercially baked goods have, in the past, had chemicals such as calcium propionate added to the dough to retard the growth of such organisms. There is now a trend toward eliminating preservatives from bread, pies, and other bakery items since the chemicals that have made the goods a less suitable medium for the growth of fungi apparently are unhealthy for humans after prolonged use, and alternative ways of retarding fungal growth and spoilage are being sought.

Rhizopus (figure 17.9), a well-known representative black bread mold of this division, has spores that are exceedingly common everywhere. They have been found in the air above the North Pole, over jungles, on the inside and outside of buildings, in soils, clothing, automobiles, and hundreds of kilometers out to sea, easily carried there by prevailing winds and breezes. When a spore lands in a suitable growing area, it germinates and soon produces an extensive mycelium, which, like that of the water molds, is *coenocytic* (not partitioned into individual cells) and contains numerous haploid nuclei. After the mycelium has developed, certain hyphae called **sporangiophores** grow upright and produce globe-shaped *sporangia* at their tips. Numerous black spores are formed within each sporangium. When these spores are released through the breakdown of the sporangium wall, they may blow away and repeat the cycle.

Such reproduction, involving no union of gametes, is asexual, but black bread molds also reproduce sexually in a unique fashion. Although there is no visible difference in form, black bread mold mycelia occur in two different mating strains. When a hypha of one strain encounters a hypha of the other, swellings called *progametangia* develop opposite each other on the hyphae. These protuberances grow toward each other until they touch. A cross-wall is formed a short distance behind each tip, and the two *gametangia* merge, becoming a single large multinucleate cell in which the nuclei of the two strains fuse in pairs. A thick ornamented wall then develops around this cell with its numerous diploid nuclei. This structure, called a *zygospore,* is the characteristic sexual spore of members of this division; it may lie dormant for months. Eventually the zygospore cracks open, and one or more sporangiophores with sporangia at their tips emerge. Meiosis apparently takes place just before this occurs.

FIGURE 17.9 Life cycle of the black bread mold *Rhizopus*.

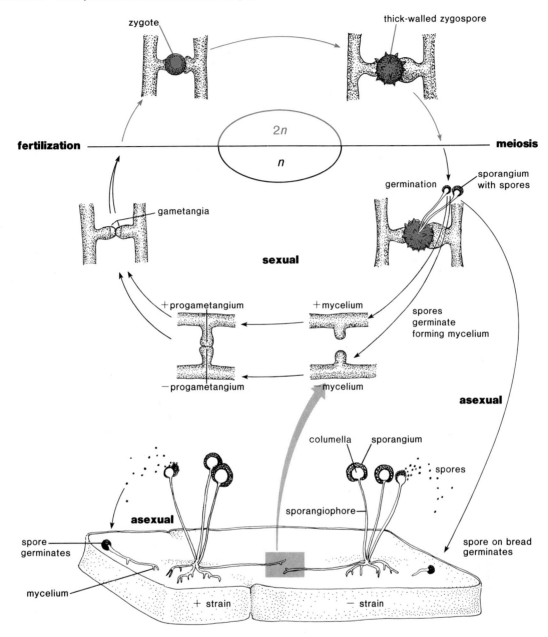

No members of this division produce zoospores, but some produce the spores externally on hyphae instead of forming them in a sporangium. One interesting group of such fungi in this class parasitizes protozoans and other small animals in various ways. Some develop an unbranched body inside their victim and slowly absorb nutrients until the host dies. They then produce a chain of spores that may stick to or be eaten by another victim. Others capture their prey on sticky hyphae to which passing amoebae adhere. One soil-dwelling group captures *nematodes* (eelworms) in hyphal rings or loops akin to the tripknots mischievous children sometimes tie in long grasses.

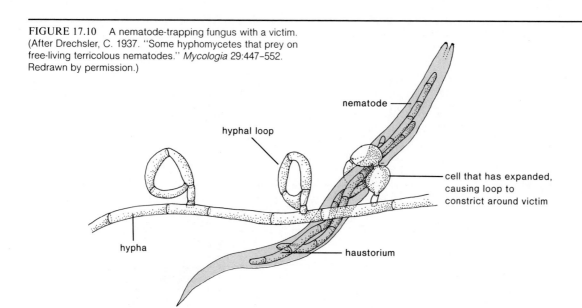

FIGURE 17.10   A nematode-trapping fungus with a victim. (After Drechsler, C. 1937. "Some hyphomycetes that prey on free-living terricolous nematodes." *Mycologia* 29:447–552. Redrawn by permission.)

nematode

hyphal loop

cell that has expanded, causing loop to constrict around victim

hypha

haustorium

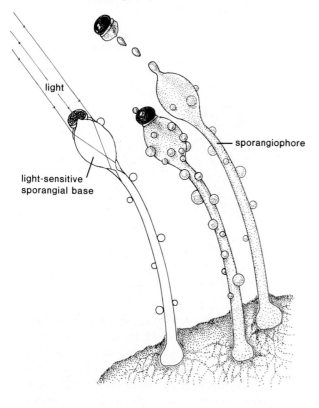

FIGURE 17.11   The dung fungus *Pilobolus,* which releases its spores with force toward a light source. (After A. H. Reginald Buller. 1934. *Researches on Fungi,* vol. 6. Longman Group Ltd., Harlow, England.)

light

light-sensitive sporangial base

sporangiophore

In some, the rings are a little smaller than the circumference of a nematode, which tapers at both ends. When a nematode randomly happens to stick its head through such a loop, it frequently tries to struggle forward instead of backing out and becomes trapped (figure 17.10). The fungus then produces small rhizoidlike outgrowths called *haustoria* which grow into the worm's body and digest it. More specialized species in this group have loops that constrict around the worm less than a tenth of a second after contact. The loops are spaced so that sometimes the tail of the worm gets caught in a second loop while the worm is thrashing around after being caught by the head. A number of these worm-trapping fungi can easily be grown on agar media, but they do not form loops unless an eelworm is placed in the dish. Evidently the eelworms produce a substance that promotes the development of the loops.

One dung-inhabiting genus of fungi in this class has the scientific name of *Pilobolus,* derived from two Greek words meaning "cap-thrower." The name is quite appropriate as the mature sporangia are forcibly "shot" a distance of up to 8 meters (26 feet), where they adhere to grass or other vegetation (figure 17.11). When the vegetation is ingested by animals, the spores pass unharmed through the digestive tract and germinate in the dung. These fungi release their sporangia with force precisely in the direction of light, to which they are very sensitive. This action can be demonstrated by placing some horse dung in a glass dish that has a lid and then covering it with black paper. If an opening is cut in the paper (the pattern of the opening can be elaborately frilly or angular if desired), and the dish is set where it will receive sufficient light for a few days, any *Pilobolus* sporangia that have been produced and forcibly discharged will form a black pattern on the glass corresponding closely to the cut-out area.

HUMAN AND ECOLOGICAL RELEVANCE OF THE COENOCYTIC FUNGI

One species of bread mold is used in Indonesia and adjacent areas to produce a food called *tempeh*. Basically, tempeh consists of boiled skinless soybeans that have been inoculated with a bread mold and allowed to sit for 24 hours. The mycelium that develops holds the soybeans together and produces enzymes that increase the content level of several of the B vitamins. The tempeh is fried, roasted, or diced for soup and is prepared fresh daily. Other bread mold species are used with soybeans to make a Chinese cheese called *sufu*.

At least two species of coenocytic fungi are now used commercially to carry out important steps in the manufacture of birth control pills and anesthetics, while others are utilized in the production of industrial alcohols and as a meat tenderizer. Yet another species produces a yellow agent used for coloring margarine.

Division Eumycota

Class Ascomycetes—The Cup or Sac Fungi

Should you happen one summer to find yourself in the south of France, you might be startled to see men and women with coils of rope slung around their shoulders pushing pigs in wheelbarrows. They are not on their way to a hog-hanging festival but are instead wheeling the pigs into the woods to help them find *truffles*. These are gourmet "mushrooms" that grow between 2.5 and 15 centimeters (1 and 6 inches) beneath the surface of the ground, usually near oak trees. They are somewhat prunelike in appearance and may be more than 10 centimeters (4 inches) in diameter, although most are smaller. They give off a tantalizing aroma that pigs can detect from more than 15 meters (50 feet) away, and the animals, in an attempt to get to them, strain at the ropes that are tied around their necks. The owners dig up the truffles, which in 1986 were selling in the United States for $186 a pound, and reward the pigs with

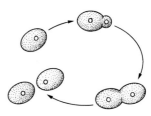

FIGURE 17.12 How a yeast cell reproduces asexually by budding.

acorns or other less interesting food. Trained dogs, which also can be used for other purposes, are now increasingly being substituted for pigs in truffle-hunting.

Truffles are the reproductive bodies of representatives of a large and varied class of true fungi called **cup** or **sac fungi.** More than 20,000 species, including yeasts, powdery mildews, brown fruit rots, ergot, morels, microscopic parasites of insects, canned fruit fungi, and many others, have thus far been described. Most produce mycelia superficially similar to those of the water molds and coenocytic true fungi, but the hyphae are divided into individual cylindrical cells. Each cross-wall has one or more pores through which the single to many tiny nuclei can pass. Little bodies that serve as plugs are located near the pores. They seal off individual cells if they become damaged.

Asexual reproduction is by means of **conidia** (singular: **conidium**). Conidia are spores that are produced externally—outside of a sporangium—either singly or in chains at the tips of hyphae called *conidiophores* (see figure 17.37). In the case of yeasts, asexual reproduction is by **budding** (figure 17.12). When a yeast cell buds, a small protuberance appears to balloon out slowly from the cell and become pinched off as it grows to full size.

Sexual reproduction always involves the formation of sacs called **asci** (singular: **ascus**). When hyphae of two different "sexes" become closely associated in the more complex cup or sac fungi, male *antheridia* may be formed on one and female *ascogonia* on the other. Male nuclei migrate into an ascogonium and pair but do not fuse with the female nuclei present. Then new hyphae (*ascogenous hyphae*), whose cells each contain one male and one female nucleus, grow from the ascogonium, the cells dividing in a unique way so that each cell has one

FIGURE 17.13 Life cycle of a cup fungus. When hyphae of two different strains of a cup fungus become closely associated, male *antheridia* may be formed on one and female *ascogonia* on the other. Male nuclei migrate into an ascogonium and pair but do not fuse with the female nuclei present. Then new hyphae (*ascogenous hyphae*), whose cells each contain a pair of nuclei, grow from the ascogonium. In a process involving fusion of the pairs of nuclei (followed by meiosis), fingerlike *asci*, each containing four or eight haploid nuclei, are formed in a layer called a *hymenium* lining an *ascocarp*. The haploid nuclei become *ascospores*, which are discharged into the air. They are potentially capable of initiating new mycelia and repeating the process.

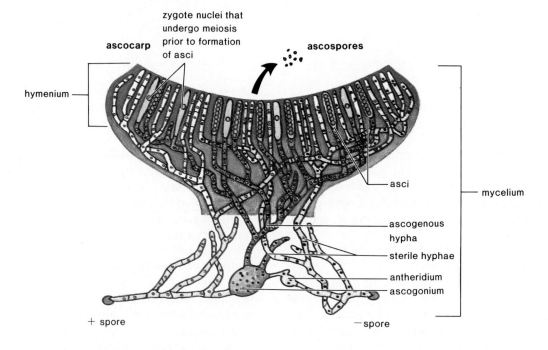

FIGURE 17.14 An ascocarp of a cup fungus.

of each kind of nucleus. At maturity the pairs of nuclei in the cells at the tips of each ascogenous hypha unite, and tubular asci develop in a layer (referred to as the *hymenium*) at the surface of a structure called an *ascocarp*. The diploid zygote nucleus undergoes meiosis, and the resulting four haploid nuclei usually divide once more by mitosis so that there is a row of eight nuclei in each ascus. These nuclei become *ascospores* as they are walled off from one another with a little cytoplasm (figure 17.13). Thousands of asci are often packed together in an ascocarp, which may be cup-shaped (figure 17.14), completely enclosed, or flask-shaped with a little opening at the top. Truffles are really enclosed ascocarps. Cup-shaped ascocarps may be several centimeters (2 to 3 inches) in diameter and may be brilliantly colored on the inside. When ascospores are mature, they are often released with force from the asci. If an open ascocarp is jarred at the right time, it may appear to belch fine puffs of smoke consisting of thousands of ascospores.

When an ascospore lands in a suitable area, it may germinate, producing a new mycelium, and then repeat the process. In many instances, however, a number of asexual generations involving conidia are produced between the sexual cycles.

HUMAN AND ECOLOGICAL RELEVANCE OF THE CUP OR SAC FUNGI

Morels, which have been called by some "the world's most delicious mushrooms," and *truffles* have been prized as food for centuries. Neither is a true mushroom; mushrooms are included in the *club fungi* discussed in the next section. Wealthy Romans and Greeks insisted on preparing them personally according to various recipes they had concocted, and morels and truffles are still prized as gourmet food today. Prior to 1982, numerous unsuccessful attempts were made to cultivate them under controlled conditions, but morels now can be mass-produced commercially and may be marketed in the near future. Morels (figure 17.15) are the size of a medium to smallish mushroom, tan in color, and have a wrinkled, slightly conelike top on a stalk that resembles a miniature tree trunk. The numerous depressions between the wrinkles each contain thousands of asci. Although morels are perfectly edible by themselves, there are unconfirmed reports that some persons have become ill after consuming them with alcohol; caution in this regard is advised.

A related "mushroom" called a *false morel* or *beefsteak morel* is considered a delicacy by many but has caused death in others. This points up the peculiar unpredictability of the interaction between the human metabolism and certain fungi. For some, the fungi are quite safe to eat; for others they are not—a literal example of "one man's meat being another man's poison."

When rye, and to a lesser extent, other grains come into flower in a field, they may become infected with *ergot* fungus (figure 17.16). This fungus seldom causes serious damage to the crop, but as it develops in the maturing grain, it produces several powerful drugs. If the infected grain is harvested and milled, a disease called **ergotism** may occur in those who eat the contaminated bread. The disease can affect the central nervous system, often causing hysteria, convulsions, and sometimes death. Another

FIGURE 17.15   A morel. (Courtesy Mary Steinke)

FIGURE 17.16   Ergotized rye. The prominent darker objects scattered throughout the ears of rye are kernels infected with ergot fungus. (USDA Photo)

form of ergotism causes gangrene of the limbs; the affected limb may drop off as a result. It can also be a serious problem for cattle grazing in infected fields. They frequently abort their fetal calves and may succumb themselves.

Ergotism was common in Europe in the Middle Ages. Known then as *St. Anthony's Fire,* it killed 40,000 people in an epidemic in 994 A.D. In 1722, the cavalry of Czar Peter the Great was felled by ergotism just as the czar was about to conquer Turkey. The conquest never took place. As recently as 1951, an outbreak hospitalized 150 victims in a French village; 5 persons died and 30 became temporarily insane, imagining they were being chased by snakes and demons. Ergotism has been suggested as a cause, in the past, for believing people to be witches.

In small controlled doses, ergot drugs have proved medically useful. They stimulate contraction of the uterus to initiate childbirth and have been used in abortions and in the treatment of migraine headaches. Ergot is also an initial source for the manufacture of the hallucinogenic drug lysergic acid diethylamide, popularly known as LSD.

Fungi in this class play a basic role in the preparation of baked goods and alcohol. Enzymes produced by yeast aid in fermentation (see chapter 10), producing ethyl alcohol and carbon dioxide in the process. The carbon dioxide produced in bread dough forms bubbles of gas, which are trapped, causing the dough to rise and giving bread its porous texture. Part of the flavor of individual wines, beers, ciders, sake, and other alcoholic beverages is imparted by the species, or strain, of yeast used to ferment the fruits or grains.

Yeasts are at least indirectly involved in the manufacture of a number of other important products. Ephedrine, a drug also produced by leafless western desert shrubs, is obtained commercially from certain yeasts. The drug is used in nose drops and in the treatment of asthma. Yeasts are also a rich source of B vitamins and are used in the production of glycerol for explosives. Ethyl alcohol is used in industry as a solvent and in the manufacture of synthetic rubber, acetic acid, and vitamin D. Yeast contains about 50% protein and makes a nourishing feed for livestock. Its enzyme, invertase, is used to soften the centers of chocolate candies after the chocolate coating has been applied.

Several very important plant diseases are found in this class of fungi. *Dutch elm disease,* a disease originally described in Holland, has devastated the once numerous and stately elm trees in many towns

FIGURE 17.17 Peach leaves infected with peach leaf curl.

and cities in the midwestern and eastern United States and has now spread to the west and south. When Dutch elm disease was first discovered, limited control of the disease was achieved by spraying the trees with DDT to kill the elm bark beetles that spread the disease from tree to tree. DDT killed many useful organisms as well and its use for such purposes was eventually banned. Other sprays have since been used, again with limited success, and biological controls are now being sought. One such control was reported in 1980 by G. Strobel of Montana State University. He injected pseudomonad bacteria into 20 diseased trees and saved 7 of them when the bacteria multiplied and killed the Dutch elm disease fungus. Whether or not this control technique will meet with widespread success remains to be seen, but it is the type of control that is much to be preferred to poisonous sprays that upset delicate ecological balances (see appendix 2, which deals with biological controls).

Chestnut blight has virtually eliminated the once numerous American chestnut trees from the eastern deciduous forests. Attempts to control the disease have met with very limited success thus far. Much better success has been obtained in controlling *peach leaf curl,* a disease that attacks the leaves of some stone fruits, especially peach trees (figure 17.17). Sprays that contain copper or zinc salts apparently inhibit the germination of spores of peach leaf curl, and have been effective in preventing serious damage when applied to trees before the dormant buds swell and open in the spring.

FIGURE 17.18 A common stinkhorn. Note the slimy mass of spores on the cap. (Courtesy Leland Shanor)

Class Basidiomycetes—The Club Fungi

At the end of my first year in college I did odd jobs during the summer, including various types of yard work. On one occasion while cleaning dead leaves and debris from a shaded garden area I came across two or three "growths," about the width of a pencil and the length of a finger, rising above the surface of the ground. On closer inspection I observed that these "growths" had the consistency and appearance of a sponge and the tips were partially covered with a slimy and evil-smelling substance. The "growths" turned out to be fungi called *stinkhorns* (figure 17.18), whose odor attracts flies, which disseminate the sticky spores that adhere upon contact to their bodies.

Stinkhorns are interesting but relatively unimportant representatives of another large class of true fungi, the **club fungi**. Other members of this class include mushrooms (figure 17.19*A*) or toadstools (the only distinctions between mushrooms and toadstools are based on folklore or tradition—botanically there is no difference), puffballs, earth stars (figure 17.19*B*), shelf or bracket fungi (figure 17.19*C*), rusts, smuts, jelly fungi, and bird's-nest

FIGURE 17.19 *A.* Common woods mushrooms. *B.* Earth stars. *C.* A bracket fungus. (*B,* Courtesy Perry J. Reynolds)

A.

B.

C.

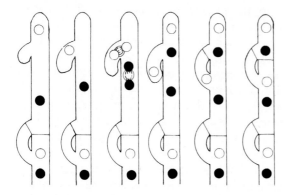

fungi. They are called club fungi because in sexual reproduction they produce their spores at the tips of swollen hyphae that often resemble baseball bats or clubs. These swollen hyphal tips are called **basidia** (singular: **basidium**). The hyphae, like those of the cup or sac fungi, are divided into individual cells; these cells, however, have either a single nucleus or, in some stages, two nuclei. The crosswalls have a central pore that is surrounded by a swelling, and both the pore and swelling are covered by a cap. This cap, with some exceptions, blocks passage of nuclei between cells but allows cytoplasm and small organelles to pass through.

Asexual reproduction is much less frequent in club fungi than in the other classes of fungi. When it does occur, it is mainly by means of conidia, although a few species produce buds similar to those of yeasts, and others have hyphae that fragment into individual cells, each functioning like a spore and forming a new mycelium after germination.

Sexual reproduction in many mushrooms begins in the same way as it does for members of the two classes previously discussed. When a spore lands in a suitable place—often an area with good organic material and humus in the soil—it germinates and produces a mycelium just beneath the surface. The

mycelium soon becomes divided into cells, each containing a single haploid nucleus. Such a mycelium is said to be **monokaryotic.** Monokaryotic mycelia of club fungi often occur in four mating types, usually designated simply as types 1, 2, 3, and 4. Only types 1 and 3, or types 2 and 4, can mate with each other. If the growth of the hyphae of compatible mating types happens to bring them close together, cells of each mycelium may unite, initiating a new mycelium in which each cell has two nuclei. Such a mycelium is said to be **dikaryotic.** Dikaryotic mycelia usually have little walled-off bypass loops called **clamp connections** between cells on the surface of the hyphae (figure 17.20). The clamp connections develop as a result of a unique type of mitosis that ensures that each cell will have one nucleus of each original mating type within it. When a cell at the tip of a dikaryotic hypha is ready to divide, a small looping protuberance appears on the side of the cell and starts to grow backward. As it does so, one nucleus migrates into it, and both nuclei then undergo mitosis simultaneously. The protuberance fuses back to the cell, and a transverse wall forms across it between the daughter nuclei. Another crosswall forms in the original cell just above the point at which the loop or clamp connection has united with it. This latter crosswall forms between the two daughter nuclei of the second original nucleus. When the process has been completed, there are two cells, each with two nuclei, and a walled-off clamp connection to one side.

After developing for a while, the dikaryotic mycelium may become very dense and form a compact, solid-looking mass called a *button.* This pushes above the surface and expands into a *basidiocarp,* commonly called a **mushroom** (figure 17.21). Most mushrooms have an expanded umbrellalike *cap* and a *stalk.* Some have a ring called an *annulus* on the stalk. It is the remnant of a membrane that extended from the cap to the stalk and tore as the cap expanded. Some mushrooms, such as the notorious *death angel mushroom,* also have a cup called a *volva* at the base (figure 17.22). Thin, fleshy-looking plates called **gills** radiate from the stalk on the underside of the cap. Microscopic examination of a gill reveals the compacted hyphae of which it is composed and innumerable **basidia** oriented at right angles to the flat surfaces of the gill.

FIGURE 17.21   Life cycle of a typical mushroom.

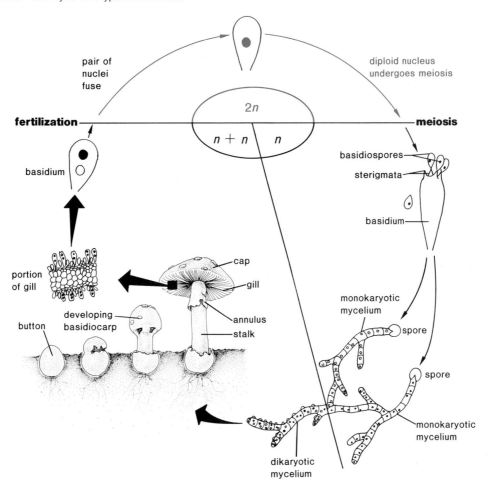

As each basidium matures, the two nuclei unite and then undergo meiosis. The four nuclei that result from meiosis become **basidiospores** when they migrate through four (in a few species, two) tiny pegs at the tip of the basidium and form their own walls around them. The tiny pegs, called *sterigmata* (singular: *sterigma*), serve as stalks for the basidiospores. One large mushroom may produce several billion basidiospores within a few days. These are forcibly discharged into the air between the gills, and they blow away with the slightest draft. If you remove the stalk and place a mushroom cap gill-side down on a piece of paper, covering it with a dish to eliminate air currents, the spores will fall and adhere to the paper in a pattern perfectly reflecting the

FIGURE 17.22   A death angel mushroom. Note the egglike volva at the base. (Courtesy Dr. T. Duffy)

FIGURE 17.23   A spore print.

FIGURE 17.25   Oyster mushrooms. These are shelf, or bracket, fungi with gills; many bracket fungi have pores instead of gills. (Courtesy Perry J. Reynolds)

FIGURE 17.24   A bolete—a mushroom that has pores instead of gills beneath its cap. (Courtesy H. S. Thiers)

arrangement of the gills. The dish and cap can be removed in a day or two, and the *spore print* (figure 17.23) can be made more or less permanent with the application of a little clear varnish or shellac. Professional mycologists use such spore prints as an aid to identification, employing white paper for dark-colored spores and black paper for white or light-colored spores.

In nature, some of the basidiospores eventually repeat the reproductive cycle. Often a dikaryotic mycelium radiates out in an ever-expanding circle from its starting point, periodically producing bas-idiocarps in so-called *fairy rings*. If conditions are favorable, the mycelium continues to grow at the edges for many years while dying in the center as food resources are depleted. Some mycelia have been known to grow in this fashion for over 500 years.

Some mushrooms produce their spores on the surfaces of thousands of tiny pores instead of on gills (figure 17.24).

Shelf or *bracket fungi* (figure 17.25) grow out horizontally from the bark or dead wood to which they have become attached, some adding a new layer of growth each year. Perennial species can become large enough and so securely attached that they can support the weight of an adult.

Other members of this class produce spores within parchmentlike membranes, forming some-what ball-like basidiocarps called *puffballs*. Puff-balls, which are generally edible, range in diameter from a few millimeters to 1.2 meters (0.125 inch to 4 feet) (figure 17.26). They have no stalks and rest in contact with the ground. Literally trillions of spores may be produced by a large puffball. These are released through a pore at the top or from random locations when the outer membrane breaks down. *Earth stars* (see figure 17.19B), which are similar to puffballs, differ from them in having a ring of appendages at the base that look like a set of woody petals around a flower.

FIGURE 17.26   Giant puffballs. The largest weighed 6.12 kilograms (13.5 pounds). (Courtesy Louise White, Redding, California, *Record-Searchlight*)

FIGURE 17.27   A bird's-nest fungus.

Bird's-nest fungi (figure 17.27), which grow on wood or manure, form nestlike cavities in which "eggs" containing basidiospores are produced. In some species, each "egg" has a sticky thread attached to it. When raindrops fall in the nests, the eggs may be splashed out, and as they fly through the air the sticky threads catch on nearby vegetation, whipping the eggs around it (figure 17.28). When animals graze on the vegetation, the spores pass unharmed through the intestinal tract.

Smuts are parasitic club fungi that do considerable damage to corn, wheat, and other grain crops. In corn smut (figure 17.29), the mycelium grows between the cells of the host. The hyphae absorb nourishment from these cells and also secrete substances that stimulate them to divide and enlarge, forming tumors on the surfaces of the corn kernels. These eventually break open, revealing millions of sooty black spores, which are blown away by the wind. Some smuts affect only the flowering heads or grains, while others infect the whole plant.

Rusts, which also are parasites, attack a wide variety of plants. Some rusts grow and reproduce on only one species of flowering or cone-bearing plant. Others, however, require two or more different hosts to complete their life cycles. *Black stem rust of wheat,* which has reduced wheat yields by millions of bushels in a single year in the United States alone,

FIGURE 17.28   How the "eggs" in a bird's-nest fungus are dispersed. (After Brodie, Harold J. 1951. "The splash-cup dispersal mechanism in plants." *Canadian Journal of Botany* 29:224–234. Reproduced by permission of the National Research Council of Canada from the *Canadian Journal of Botany*, vol. 29, pp. 224–234, 1951.)

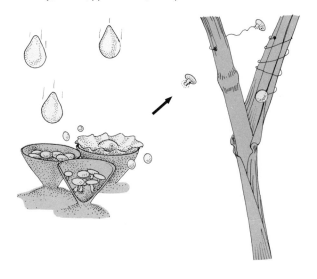

FIGURE 17.29   Corn smut fungus on ears of corn. (USDA Photo)

has plagued farmers ever since wheat was first cultivated thousands of years ago. More than 300 races of black stem rust are now known. This rust requires both common barberry plants and wheat to complete its life cycle (figure 17.30).

Since two hosts are necessary for black stem rust of wheat to complete its life cycle, partial control of the disease has been accomplished through an attempt to eradicate common barberry bushes. An estimated 600 million such plants have been destroyed in the United States since 1918. Producing rust-resistant strains of wheat has also helped, but even as new strains are developed, the rusts themselves hybridize or mutate (see chapter 13), producing new races capable of attacking previously immune varieties of cereals—a striking example of evolution in action.

Another serious rust with two hosts is the *white pine blister rust,* which has caused huge losses of valuable timber trees in both the eastern and western United States. Basidiospores infect the pine trees;

FIGURE 17.30   Life cycle of black stem rust of wheat.

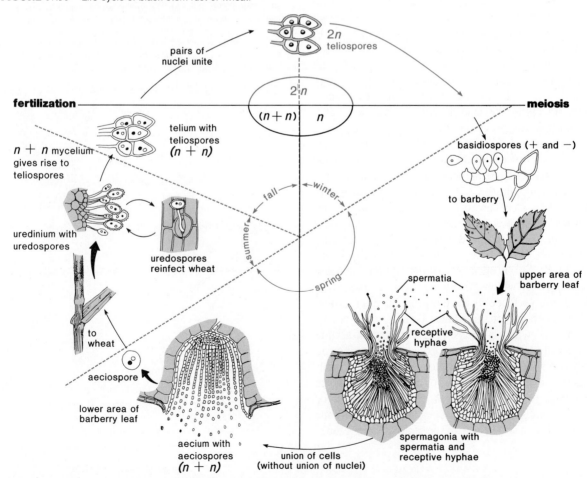

when the basidiospores germinate, other types of spores are produced. These different spores in turn infect currant and gooseberry bushes, and the spores formed on the currants and gooseberries eventually give rise to new basidiospores, completing the cycle. The U.S. Forest Service had a program of gooseberry bush eradication in operation for many years in an attempt to alleviate the problem. Spraying programs have been implemented more recently and rust-resistant trees are being developed as alternatives, with some success. Other rusts with two hosts include *apple rust* (alternate host: cedar trees), *poplar leaf spot* (alternate host: larch or tamarack trees), and *corn rust* (alternate host: sorrel).

HUMAN AND ECOLOGICAL RELEVANCE OF THE CLUB FUNGI

Of the approximately 25,000 described species of club fungi, fewer than 75 are known to be poisonous. Many of the latter are not readily distinguishable by amateurs from edible species, and some edible forms such as the inky cap and shaggy mane mushrooms, which cause no problems by themselves, may make one very ill if consumed with alcohol. Normally few of the poisonous forms are fatal, but unfortunately some—such as the death angel, which causes 90% of the fatalities attributed to mushroom poisoning—are relatively common. Poisoning from death angels and similar species usually takes from 6 to 24 hours to manifest itself, and until recently, successful treatment was impossible by the time the intense stomachache, blurred vision, violent vomiting, and other symptoms occurred. Hope for survival of the victims now lies with the administration of a drug known as thioctic acid. Even thioctic acid does not always prevent a fatality and either the drug or the mushroom poison usually leave the patient with hypoglycemia (a blood sugar deficiency). Some wild mushroom lovers have fed a part of their collection to a dog or cat, and after an hour or two when nothing happened to the animal, they have eaten the mushrooms themselves. Later both they and the animals succumbed. Others have mistakenly believed that the toxic substances are destroyed by cooking. Records show, however, that before the discovery of thioctic acid as an antidote, death ensued in 50% to 90% of those who had eaten just one or two of the deadly mushrooms, cooked or raw, and that even as little as 1 cubic centimeter (less than a 0.5 cubic inch) can be fatal.

FIGURE 17.31   Teonanacatl mushrooms. (Drug Enforcement Administration)

A number of widespread but very unreliable beliefs exist concerning distinctions between edible and poisonous mushrooms. One holds that a silver coin placed in the cooking pan will turn black while the mushrooms are cooking if any poison is present. Some members of both edible and poisonous species will turn such a coin black, and some will not. Another superstition holds that edible species can be peeled while poisonous ones cannot. Again this is a fallacy, for death angels peel quite easily. Still other erroneous beliefs maintain that poisonous mushrooms appear only in the fall or the early spring, or that all mushrooms eaten by snails and beetles are edible, or that all purplish-colored mushrooms are poisonous, or that all mushrooms growing in grassy areas are edible. Again, these notions simply are not supported by the facts. Wild mushrooms can be eaten with confidence *only* if they have been identified by a knowledgeable authority. It is foolhardy for anyone to do otherwise.

Some poisonous mushrooms cause hallucinations in those that eat them. During the Mayan civilization in Central America a number of *teonanacatl* ("God's flesh") sacred mushrooms (figure 17.31) were used in religious ceremonies. The consumption of these mushrooms, which has continued to the present among native groups in Mexico and Central America, results in sharply focused, vividly

FIGURE 17.32 Fly agaric mushrooms. (Courtesy Perry J. Reynolds)

colored visions. Similar use of the striking, fly agaric mushroom (figure 17.32) in Russia, and for a while in the Indus valley of India, dates back to many centuries B.C. Users appear to go into a state of intoxication. Related species that occur in the United States have not produced the same effects but have, instead, caused the user to become nauseated and to vomit. In Siberia, users have noted that the intoxicating principle is passed out in the urine; as a result, some persons have adopted the practice of drinking the urine of fly agaric users. Reindeer, incidentally, are reported to be obsessed with both fly agaric and human urine.

More than 90% of a mushroom is water, and mushrooms generally contain smaller quantities of nutritionally valuable substances than most foods. An apparent exception is the legendary Black Forest (*Shiitake*) mushroom grown for centuries in China and Japan on oak logs and now cultured in the northwestern United States. It has more than double the protein of ordinary, commercially grown mushrooms and is very rich in calcium, phosphorous, and iron. It also contains significant amounts of B vitamins, vitamin D_2 (ergosterol), and vitamin C and has excellent flavor. Ancient Chinese royalty believed that eating Black Forest mushrooms would promote healthful vigor and retard the aging process. It appears there could be some substance to this

belief, since recent research suggests that mushrooms in general may be one of the richer sources of RNA, which has been shown to retard the aging process in cells through its metabolic conversion to ATP. Lentinacin, an agent capable of lowering human cholesterol levels, has been obtained from Black Forest mushrooms, and purified extracts from spores of the mushrooms have demonstrated antiviral activity against influenza and polio viruses in laboratory animals. The extracts evidently induce the formation of *interferon* (see chapter 15), a virus-combating substance produced by animal cells.

Many types of mushrooms have been cultivated for food since ancient times. In the second century B.C., a Greek doctor by the name of Nicandros taught people how to grow mushrooms underneath fig trees on soil fertilized with manure. Andrea Cesalpino, a noted Italian botanist and physician of the sixteenth century, cultivated mushrooms by scattering pulverized poplar bark on very rich soil near poplar trees with which the mushrooms were associated. In more recent times, Italians have cultivated mushrooms on waste material from olives, on coffee grounds, on remnants of oak leaves after tannins have been extracted for leather tanning, and on laurel berries. Today jelly fungi and various mushrooms are cultivated in the Orient on media composed of wood and manure.

In Geneva, Switzerland, there is a special market for wild mushrooms where more than 50 species are sold under the supervision of a state mycologist. Although wild mushrooms are consumed by many in the United States, few such species are sold in markets and only one species of mushroom (*Agaricus campestris*) is cultivated to any extent. It has been grown in basements, caves, and abandoned mines, but contrary to popular belief, light does not affect its growth at all. Large-scale operations generally use windowless warehouses with stacked rows of shallow planting beds because temperatures, humidity, and other climatic factors are easier to control in such buildings (figure 17.33). The mushrooms are grown on compost made from straw and on horse or chicken manure. Prior to use, the compost is pasteurized for a week to destroy insects and their eggs. Then it is inoculated with spawn, which is compact mycelium grown from germinated basidiospores sown on bran of wheat or other grains. After inoculation, the spawn is covered with a thin layer of soil. The moisture content of the compost is controlled with regular, light waterings, and a humidity of about 75% is maintained. The mycelium grows in temperatures ranging from just above freezing to 33° C (91.4° F), but commercial growers try to keep the temperatures between 9° C and 13° C (48° F to 55° F) because the mushrooms are less subject to disease or insect attacks and are also firmer when grown under cool conditions. The mushroom buttons appear within a week to ten days after spawn is planted and continue appearing for six months or more. About 1 kilogram (2.2 pounds) of mushrooms is obtained from each 10 square decimeters (slightly over 1 square foot) of growing area. An estimated 59,000 metric tons (65,000 tons) of mushrooms are produced annually in the United States.

As indicated in the chapter introduction, bacteria and fungi of all divisions and classes are the most important organisms involved in natural recycling processes. Fungi are constantly breaking down dead wood and debris and returning the components to the soil where they can be recycled. Sometimes, as we have seen, they can be very destructive from a human viewpoint, attacking everything from living plants and harvested or processed food to shoe leather, paper, cloth, construction timbers, paint, petroleum products, upholstery, and even glass, particularly in warmer climates. Others play a nondestructive ecological role. For example, several species of fungi are found in ant and termite

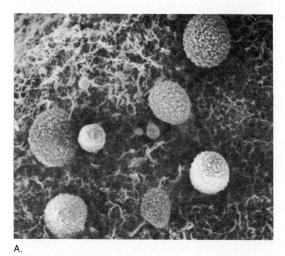

A.

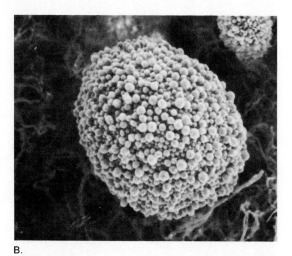

B.

FIGURE 17.35   A *Penicillium* colony. The tiny droplets of fluid on the surface contain penicillin.

nests. The insects cultivate their fungus gardens by bringing in bits of leaves, other plant debris, caterpillar droppings, and their own feces to form a rich growing medium (figure 17.34). They also lick the hyphae and constantly probe them, using them for food as they grow. Some wood-boring beetles have pouches in their bodies that function as fungus spore containers. The spores germinate in the wood tunnels, their mycelia forming a lining that produces yeastlike cells on which the beetles and their larvae feed. Many higher plants have **mycorrhizal fungi** (see chapter 6) associated with their roots. These fungi greatly increase the absorptive surface area around the roots, and may be far more important than root hairs in this regard, particularly in mature roots.

Class Deuteromycetes—The Imperfect Fungi

Any fungus for which a sexual stage has not been observed is classified as an **imperfect fungus.** Many otherwise unrelated fungi are grouped together in this artificial class, which includes several well-known disease organisms as well as fungi important in disease control and food processing. If a member of this group is studied further and a sexual stage is discovered, the fungus is reassigned to its appropriate class.

All imperfect fungi reproduce by means of conidia. In one such group, which is found growing on dead leaves and debris at the bottom of streams with rapidly moving water, the spores often have four long extensions on them. These arms are arranged in such a way that three may catch like a tripod on a flat surface, keeping the fungus from being washed downstream.

HUMAN AND ECOLOGICAL RELEVANCE OF THE IMPERFECT FUNGI

Among the best known of the imperfect fungi are the *Penicillium* molds (figure 17.35), which produce *penicillin,* the well-known and widely used *antibiotic* (a substance produced by a living organism that interferes with the normal metabolism of another living organism). Sir Alexander Fleming, a British scientist, noticed in 1929 that certain bacteria would not grow in the vicinity of the mycelium of a *Penicillium* mold and gave the name penicillin

FIGURE 17.36   Blue cheese.

to the element in the mold that prevented the bacterial growth. He apparently did not grasp the significance of his findings and the findings did not particularly excite the medical profession until the outbreak of World War II some ten years later. At that time, a team of British and American scientists at Northern Regional Laboratories in Peoria, Illinois, set out to see if they could coax *Penicillium* molds into producing more of this antibacterial substance, primarily because war casualties created a need for greater quantities of more effective medicines that could keep wounds from becoming infected. They began with cultures from the original mold observed by Fleming, but the amounts of penicillin it produced were so small that it was very expensive to obtain significant quantities for human use. They appealed to the general public, asking them to send in any material they found with a greenish or bluish mold on it. They received huge quantities of moldy trash from all over the United States. The breakthrough in the research came, however, when a different species of *Penicillium* mold—one that yielded 25 times the penicillin produced by the original culture—was found on a moldy cantaloupe from a local market. The scientists set to work germinating individual spores of this new mold on culture media, and by careful selection they eventually were able to isolate a strain that produced more than 80 times the original quantity of penicillin. Later, when this strain was subjected to X-radiation, still other forms were produced that

upped the penicillin output to 225 times that of Fleming's mold. Today most of the penicillin produced around the world comes from descendants of that cantaloupe mold. Literally hundreds of other antibiotics effective in combating human and animal diseases have been discovered since the close of World War II, and the production of these drugs is a vast worldwide industry.

Penicillium molds are also useful for purposes other than the manufacture of penicillin. Some are introduced into the milk of cows, sheep, and goats at stages in the production of "smelly" cheeses such as blue (figure 17.36), Camembert, Roquefort, Gorgonzola, and Stilton. The molds produce enzymes that break down proteins and fats in the milk, giving the cheeses their characteristic flavors.

Since the early 1980s, organ transplants have been significantly facilitated by the discovery and production of a "wonder drug" called *cyclosporine* from an imperfect fungus found in soil. The drug suppresses immune reactions that cause rejection of transplanted organs without risking the development of leukemia and other undesirable side effects associated with drugs previously used.

Aspergillus is a genus of imperfect fungi whose species produce dark brown to blackish or yellow spores. It is closely related to the *Penicillium* molds, and is extensively used in industry. One or more species are used commercially for the production from sugar of citric acid, a substance for flavoring foods and for the manufacture of effervescent salts that were originally obtained from oranges. Citric acid is also used in the manufacture of inks and in medicines, and it is even used as a chicle substitute in some chewing gums. These same fungi also produce gallic acid used in photographic developers, dyes, and indelible black ink. Other species are used in the production of artificial flavoring and perfume stabilizers, chlorine, alcohols, and several acids. Further uses are in the manufacture of plastics, toothpaste, and soap, and in the silvering of mirrors.

One species of *Aspergillus* is used in the Orient and elsewhere to make soy sauce, or *shoyu,* by fermenting soybeans with the fungus. A Japanese food called *miso* is made by fermenting soybeans, salt, and rice with the same fungus. More than one-half million tons of miso are consumed annually.

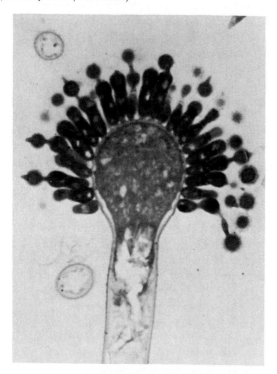

FIGURE 17.37 Tip of a conidiophore of *Aspergillus* as seen with the aid of a microscope. Conidia are radiating out around the swollen tip. Several species of *Aspergillus* are responsible for human respiratory and skin diseases called *aspergilloses.* (Courtesy Turtox/Cambosco)

A number of diseases of both man and animals are caused by *Aspergillus* species. The diseases, called *aspergilloses,* attack the respiratory tract after the spores have been inhaled (figure 17.37). One type thrives on and in human ears.

Other diseases caused by different genera of imperfect fungi include those responsible for the widespread problem of "athlete's foot" and also ringworm, for white piedra (a mild disease of beards and mustaches), and for tropical diseases of the hands and feet that cause the limbs to swell in grotesque fashion. One serious disease called "valley fever," found primarily in the drier regions of the southwestern United States, is usually initiated by the inhalation of dust-borne spores of an imperfect fungus that produces lesions in the upper respiratory tract and lungs. The disease may spread elsewhere in the body, with sometimes fatal results.

Two imperfect fungi show promise as biological controls of pest organisms. One has already been used with some success in controlling scale insects in Florida and other warm, humid regions. Another may be used to combat water hyacinths, which have caused serious clogging of waterways in areas of the world with mild to tropical climates.

LICHENS

A student who was interested in natural dyeing came to me a few years ago and asked if she could experiment with the dye potential of local plants as a special project. In the course of her experimentation, she obtained beautiful shades of yellow, brown, and green from two dozen common local plants, using simple "recipes."[1] During the following summer she extended the project to include lichens growing on the trees and rocks at a camp where she served as a counselor. The rich colors she obtained from the lichens were even more spectacular, which is perhaps not surprising since these organisms were in the past a major source of dyes and still are used in a minor way in commercial dyeing.

Lichens have traditionally been referred to as prime examples of symbiotic relationships. Each consists of a fungus and an alga intimately associated in a spongy body called a **thallus.** The thallus can range in diameter from less than 1 millimeter to more than 2 meters (0.04 inch to 6.5 feet). The alga supplies the food for both organisms, while the fungus protects the alga from harmful light intensities, produces a substance that accelerates photosynthesis in the alga, and absorbs and retains water and minerals for both organisms. The evidence suggests, however, that it would probably be more correct to say that the fungus parasitizes the alga in a controlled fashion, actually destroying algal cells in some instances.

There are about 25,000 known species of lichens. The algal component is either a green alga or a blue-green bacterium, and a few lichens have two species of algae present. Three genera of green algae and one genus of blue-green bacterium are involved in 90% of all lichen species, and one species

1. Most natural dye "recipes" call for simmering at least a liter (approximately 1 quart) or 2 of loosely packed fresh or dry material covered with water in a large enamel kettle until most of the coloring appears to be in the water. This usually takes from 1 to several hours. Next the solid waste is strained and discarded. A *mordant* (substance that helps fibers take up dye permanently) is then often added to the liquid in amounts varying from 1 teaspoon to one-half cup. Commonly used mordants include alum, detergent ammonia, copper sulphate, tin (stannous chloride), and white vinegar. White wool or other fibers are washed in warm water and detergent, rinsed in warm water, and left to soak in hot water for up to an hour. The fibers are quickly transferred to the hot dyeing liquid and simmered for an hour or left overnight. As a final step, the dyed material is rinsed in warm water until no more dye diffuses out, washed again with detergent, and dried. Reaction of the dyed material to light can be tested by placing it in direct sunlight for several days. (See appendix 3 for a number of sources of natural dyes.)

FIGURE 17.38   A section through a foliose lichen. (G. S. Ellmore)

ascocarp (apothecium)

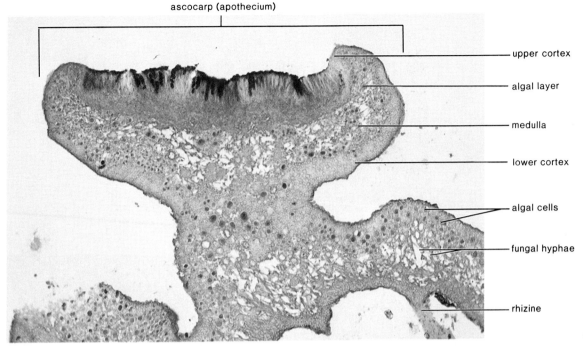

- upper cortex
- algal layer
- medulla
- lower cortex
- algal cells
- fungal hyphae
- rhizine

of alga may be found in many different lichens. Each lichen, however, has its own unique species of fungus. With the exception of about 20 tropical species of lichens that have a club fungus, and one species (associated with bald cypress trees) that has a bacterial component, all lichens have members of the cup or sac fungi for their fungal components. It is possible to isolate and culture the components separately. When this is done the fungus takes on a very different, compact but indefinite shape, and the algae grow faster than they do when they are part of a lichen. The fungal component is very rarely found growing independently in nature, whereas the algal component may occasionally do so. Lichen species, therefore, are identified according to the fungus present.

Lichens grow very slowly, at a maximum rate of 1 centimeter (0.4 inch) and a minimum of as little as 0.1 millimeter (.004 inch) per year. They are capable of living to an age of 4,500 or more years, and are tolerant of environmental conditions that kill most other forms of life. They are found on bare rocks in the blazing sun or bitter cold in deserts, in both arctic and antarctic regions, on trees, and just below the permanent snow line of high mountains where nothing else will grow. One species grows completely submerged on ocean rocks. They even attach themselves to such man-made substances as glass, concrete, and asbestos. Part of the reason for this wide range of tolerance is the presence in the

lichens' thalli of gelatinous substances that enable them to withstand periods of rapid drying alternating with wet spells. While they are dry their water content may drop to as low as 2% of their dry weight, and the upper part of the thallus becomes opaque enough to exclude much of the light that falls on them. In this state most environmental extremes do not affect them at all as they temporarily become dormant and do not carry on photosynthesis. One environmental factor from modern civilization does, however, have a marked effect on them. They are exceptionally sensitive to pollution, particularly that of sulphur dioxide. In fact, their sensitivity to this air pollutant is so great that it has been possible to calculate the amount of sulphur dioxide present in the air solely by mapping the occurrence or disappearance of certain lichens in a given area. Such studies have shown that some species of lichens have disappeared entirely from industrial areas and are now facing extinction due to air pollution. They are also very sensitive to nuclear radiation, and have been used on at least one occasion in monitoring radioactive contamination when a satellite fell to Earth in a remote area.

The lichen thallus generally consists of three or four layers of cells or hyphae (figure 17.38). At the surface is a protective layer constituting the *upper*

FIGURE 17.39   Three types of lichen thalli. *A.* Crustose
lichens on the surface of a rock. *B.* A foliose lichen. *C.* A
fruticose lichen.

A.

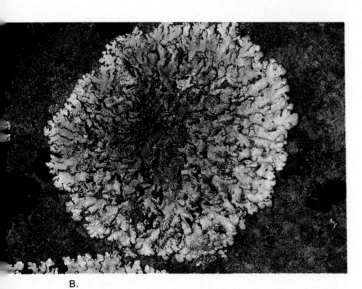

B.

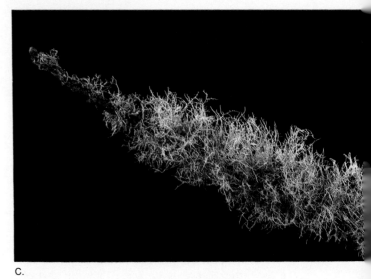

C.

cortex. The hyphae are so compressed they resemble parenchyma cells, and it is here that gelatinous substances are present that aid in the layer's
protective function. Below the upper cortex is an
algal layer in which algal cells are scattered among
strands of hyphae. Next is a *medulla* consisting of
loosely packed hyphae, which occupies at least half
the volume of the thallus. A number of substances
produced by the lichen are stored here. A fourth
layer, called the *lower cortex,* is frequently but not
always present. It resembles the upper cortex but is
usually thinner and is often covered with anchoring
strands of hyphae called *rhizines.*

Lichens have been loosely grouped into three
major growth forms, which have no basis in their
natural relationships but are convenient as a first step
in their identification (figure 17.39). *Crustose lichens* are attached to or embedded in their substrate
over their entire lower surface. They often form
brightly colored crusty patches on bare rocks and
tree bark. The hyphae of some that grow on sedimentary rocks penetrate as much as 1 centimeter
(0.4 inch) into the rock. Others grow just beneath
the cuticle of the leaves of tropical hardwood trees,
with no apparent harm to the leaves. *Foliose lichens*
have somewhat leaflike thalli, which often overlap
one another. They are weakly attached to the substrate. The edges are frequently crinkly or divided

into lobes. *Fruticose lichens* may resemble miniature shrubs, or they may hang down in festoons from branches. Their thalli, which are usually branched, are basically cylindrical in form and are attached at one point.

It should be stressed that while lichens may be attached to trees or other plants, the majority in no way parasitize them, although a few have been shown to derive supplementary nutrition by partial parasitism of the cortex of their hosts.

Although the fungal component of a lichen reproduces sexually, lichens are naturally dispersed in nature primarily by asexual means. In about a third of the species, small powdery clusters of hyphae and algae called **soredia** (singular: **soredium**) are formed and cut off from the thallus in a set pattern as it grows. Rain, wind, running water, and animals act as agents of dispersal. In other lichens, specialized parts of the thallus may break off or be separated by decay. Sexual reproduction is similar to that of the cup or sac fungi, except that the ascocarps produce spores continuously for many years. No one has ever observed the initiation of a new thallus in nature. It is believed a new thallus forms when a wind-disseminated ascospore germinates next to a suitable free-living alga and surrounds the alga with intertwining hyphae. Some of the hyphae penetrate the alga and parasitize it without, however, killing the alga or preventing it from reproducing itself within the new thallus. Lichen algae reproduce by mitosis and simple cell division.

HUMAN AND ECOLOGICAL RELEVANCE OF LICHENS

Lichens provide food for many lower animals as well as large mammals. Reindeer and caribou can survive exclusively on fruticose lichens in Lapland when other food is unavailable, while North African sheep graze on a crustose lichen. By themselves lichens do not make good food for human consumption, but they have been used as a food supplement (e.g., in soups) in parts of Europe. Most have acids that make them unpalatable, and some (e.g., rock tripe) have had harsh laxative effects on those who have tried them.

More than half of the lichens investigated have antibiotic properties. One lichen substance has been used in Europe in combination with another antibiotic in the treatment of tuberculosis. Europeans have also used lichen antibiotics to produce salves that have been effective in treating cuts and skin diseases.

Lichens were used for dyes by the Greeks and Romans, and a lichen dye industry persisted for many centuries in Europe. Native Americans and others also used lichens for dyes. Coal tar dyes have now largely replaced those of lichens, but lichens are still used in the manufacture of Scottish tweeds and in the production of litmus paper. Litmus paper is used in elementary chemistry laboratories as an acid-alkaline indicator, turning red under acid conditions and blue under alkaline ones.

Soaps are scented with extracts from lichens, and such extracts are still used in the manufacture of some European perfumes. Because of their resemblance to miniature trees and shrubs, some fruticose lichens are used by toy makers for the scenery of model railroads and car tracks. The importance of lichens in initiating soil formation is discussed in chapter 23.

SUMMARY

Members of Kingdom Fungi were grouped in the past with the slime molds and bacteria in a single division of the Plant Kingdom, but bacteria later were placed in Kingdom Monera because of their prokaryotic cells, and the slime molds were placed in Kingdom Protista because of the nature of their plasmodia, their reproduction, and other considerations. The reproduction, metabolism, and lines of diversity of the fungi were sufficiently different from those of members of the Plant Kingdom to warrant their being placed in a kingdom of their own. Kingdom Fungi is treated as consisting of three subkingdoms of organisms that appear to have been derived independently of each other.

Subkingdom Myxomycotineae includes the slime molds, which are animallike in their vegetative state; they consist of a multinucleate mass of protoplasm called a plasmodium that flows over damp surfaces, ingesting food particles encountered. They are funguslike when they reproduce, forming stationary sporangia containing spores, from which myxamoebae or swarm spores emerge upon germination. Myxamoebae or swarm spores function as gametes, with new plasmodia developing from the zygotes. Some members of this division are called cellular slime molds because they do not produce true plasmodia; instead, independent cells clump together, becoming a pseudoplasmodium that crawls like a slug and eventually is converted into a stationary, sporangiumlike mass of spores.

Subkingdom Mastigomycotineae includes two divisions of chytrids and water molds. Chytrids are simple fungi that consist mostly of spherical cells with rhizoids at one end. Reproduction is primarily asexual, by means of zoospores. Water molds include fungi that cause diseases of fish and other aquatic organisms, and grow on dead insects and submerged organic material. They also cause two major diseases of flowering plants (downy mildew of grapes, which seriously threatened the French wine industry until it was controlled with Bordeaux mixture; and late blight of potato, which ruined the Irish economy in the mid-nineteenth century and brought about starvation and death of large numbers of Irish peasants). Water molds have coenocytic mycelia (i.e., the hyphae have no cross-walls). They reproduce asexually by means of zoospores, and sexually by means of gametes produced in oogonia and antheridia produced in response to hormones secreted by the fungi.

Subkingdom Eumycotineae includes the true fungi, which are assigned to two divisions, the second division including three classes of fungi and one of lichens. True fungi are all filamentous or unicellular with chitinous walls. True fungi are nonmotile and most produce spores. Sexual reproduction is varied. They absorb their food in solution through their walls. About two-thirds of the possibly 300,000 species of fungi still await description or discovery. Filamentous fungi produce threads called hyphae that collectively constitute a mycelium. Fungi may thrive at temperatures below freezing or well above human comfort levels. They are major natural decomposers and cause huge economic losses through food spoilage and diseases.

The Division Zygomycota (coenocytic fungi)—represented by *Rhizopus,* which has exceptionally widespread spores—is characterized by coenocytic mycelia that produce upright hyphae (sporangiophores) with sporangia containing numerous spores at their tips (in asexual reproduction), and sexually by gametangia produced on hyphae of opposite strains that merge, creating a single large cell in which nuclei fuse in pairs. This large cell becomes a zygospore in which meiosis occurs prior to its cracking open and sporangiophores and sporangia emerging. Some spores, called conidia, are produced externally on hyphae in this division, but conidia production is more common in the other fungal division. Some conidia-producing zygomycetes trap eelworms and digest them. Another, *Pilobolus,* releases its sporangia with force toward a light source. Bread molds are used to make tempeh and sufu, and are also for various industrial purposes.

Members of the Division Eumycota do not have coenocytic hyphae, their filaments being partitioned into cells. Members of the Class Ascomycetes (cup or sac fungi) include truffles, yeasts, powdery mildews, brown fruit rots, ergot, morels, insect parasites, and others. The cross-walls of their hyphae have pores. Asexual reproduction is by conidia or budding. Sexual reproduction involves the formation of sacs called asci following union of male and female structures. In some groups ascogenous hyphae develop after the union, and these, in turn, develop into asci contained in cup-shaped or enclosed ascocarps. Ascospores are produced, following meiosis, in the asci. Upon release, each ascospore may germinate, forming a new mycelium.

Morels are prized edible members of the Class Ascomycetes. Ergotism is a disease produced when rye infected with ergot is ingested; the disease has killed thousands in the past. Yeasts produce carbon dioxide and alcohol used in baking and brewing, the drug ephedrine, and proteins and vitamins used for human and livestock consumption. Dutch elm disease, chestnut blight, and peach leaf curl are caused by fungi in this class.

Stinkhorns are relatively unimportant representatives of the Class Basidiomycetes (club fungi), which includes mushrooms, bracket fungi, rusts, smuts, jelly fungi, and bird's-nest fungi. The hyphae are divided into cells that have either one nucleus or two. Asexual reproduction, which is less common than in other classes, is usually by means of conidia. Sexual reproduction involves the union of cells of two mating types of monokaryotic hyphae (hyphae whose cells have a single nucleus); this initiates the development of a dikaryotic mycelium whose cells each have two nuclei, and usually have small external protuberances called clamp connections between each cell. A dikaryotic mycelium may become very dense and push above the surface, forming a button that expands into a basidiocarp or mushroom. The expanded cap of the mushroom is usually elevated on a stalk that may have a ring called an annulus around it and sometimes a cup, the volva, at the base. Gills radiate from the stalk under the cap. The gills consist of hyphae that produce club-shaped basidia, in each of which fusion of two nuclei occurs, followed by meiosis; four basidiospores usually develop on the outside of each basidium. The basidiospores may germinate and repeat the cycle. Puffballs are stalkless club fungi that may each produce trillions of spores. Shelf fungi may add a new layer of growth each year. Bird's-nest fungi produce their basidiospores in "eggs" that have a sticky thread attached.

When splashed by rain the eggs may be displaced and become attached to vegetation ingested by animals; the eggs pass through the intestinal tract of the animals unharmed. Smuts infest cereals; rusts, which may have more than one host, infest cereals, white pines, apple trees, and other important plants. Rusts may be partially controlled by eradicating alternate hosts.

Fewer than 75 species of the approximately 25,000 described club fungi are known to be poisonous, with the death angel mushroom responsible for about 90% of the fatalities attributed to mushroom poisoning. Thioctic acid has been found to be an antidote to mushroom poisoning. Many unreliable beliefs are associated with attempts to distinguish between poisonous and edible mushrooms; the only safe way, however, is to have a mushroom identified by an authority. Some mushrooms have been used for intoxication and hallucinatory purposes. Mushrooms are not high in nutritional value, with the exception of the Black Forest mushroom, which also has produced substances capable of lowering cholesterol levels and inhibiting viruses. Many types of mushrooms have been cultivated for food but only one species is commercially cultivated in the United States.

Fungi and bacteria play a major role in natural recycling processes. They are also destructive of many economically important substances; they are cultivated by ants and termites in their nests, and form mycorrhizal associations with the roots of higher plants.

Members of the Class Deuteromycetes (imperfect fungi) all lack known sexual reproductive phases but otherwise resemble members of the other classes of true fungi. Among the best known members of the class are the *Penicillium* molds, which produce penicillin, and *Aspergillus,* some species of which are used commercially in the production of a wide variety of products, including shoyu and miso. Aspergillus and other imperfect fungi also cause diseases in humans and animals, including athlete's foot and ringworm.

Lichens, whose bodies (thalli) consist of an alga and a fungus in intimate symbiotic relationship, are used commercially as a source of dyes. The algal component is usually a green alga or a blue-green bacterium, which may occur in many different species of lichens, but each lichen species has its own unique fungal component. Except for a few tropical species with club fungi and one with a bacterium, all lichens have a cup or sac fungus for this component, and it is by the fungus that lichens are identified.

Lichens grow very slowly; they can stand great temperature ranges and may live to be thousands of years old. They are, however, exceptionally sensitive to industrial pollution. The lichen thallus usually consists of several layers of hyphae, including a protective layer (the upper cortex), an algal layer containing algal cells, a medulla consisting of loosely packed hyphae, and a lower cortex, which is often covered with anchoring rhizines, but which may not always be present.

Lichens occur in three growth forms: crustose lichens, which are attached to their substrate over their whole lower surface; foliose lichens, which are somewhat leaflike and weakly attached; and fruticose lichens, which are erect or pendent. Lichens reproduce sexually but are naturally dispersed primarily by asexual means. Some produce small powdery clusters called soredia, which become detached and are dispersed by rain, wind, water, and animals.

Lichens provide food for animals, especially in cold areas; most are unpalatable to humans. Many lichens have antibiotic properties; some lichens are used for perfume, litmus paper, dyeing Scottish tweeds, and as scenery for model railroads.

REVIEW QUESTIONS

1. Why are fungi placed in a kingdom separate from protists and bacteria?
2. How do the cells of fungi differ from those of other organisms?
3. What does *coenocytic* mean? To which groups of fungi does it apply?
4. What means of asexual reproduction do the fungi exhibit?
5. If you were looking at hyphae of a bread mold, a cup fungus, and a club fungus through a microscope, could you distinguish among the three? How?
6. Which fungi produce zoospores?
7. Define plasmodium, conidia, haustoria, zygospore, sporangiophore, ascus, mycelium, ascocarp, monokaryotic, dikaryotic, and clamp connection.
8. Is there any one best way to tell a poisonous mushroom from an edible one?
9. What is ergotism?
10. Discuss the human relevance of the cup or sac fungi.
11. What is a spore print?

12. To which division and/or class of fungi do each of the following belong: truffles, puffballs, LSD fungi, stinkhorns, peach leaf curl, late blight of potato, bird's-nest fungi, jelly fungi, downy mildew of grape, death angel, shelf or bracket fungi?
13. What is a fairy ring, and why does it form?
14. How do rusts and smuts attack their hosts?
15. Under what conditions do commercially produced mushrooms grow best?
16. Why are some fungi referred to as being "imperfect"?
17. Discuss the economic importance of imperfect fungi.
18. What is the relationship between the alga and the fungus in a lichen?
19. What makes up the basic structure of a typical lichen?
20. How is it possible for lichens to live to such great ages?

DISCUSSION QUESTIONS

1. Since fungi produce so many trillions of spores, why is it we are not overrun by them?
2. Slime molds were once treated as animals in elementary zoology books. Do you think their being assigned to a different kingdom and specifically Kingdom Fungi is justified?
3. Why are fungus diseases much more common now in the United States than they were at the time of the Declaration of Independence?
4. If all rusts had only one host, would they (theoretically, at least) be easier to control?
5. Since most lichens are not good for food and are not very important economically, should we be as concerned about some of them becoming extinct as we are about Africa's big game animals disappearing?

ADDITIONAL READING

Ainsworth, G. C., and Bisby. 1983. *Dictionary of the fungi (including the lichens),* 7th ed. Forestburgh, NY: Lubrecht and Cramer.

Alexopoulos, C. J., and C. W. Mims. 1979. *Introductory mycology,* 3d ed. New York: John Wiley & Sons.

Baldwin, R. S. 1981. *The fungus fighters: two women scientists and their discovery.* Ithaca, NY: Cornell University Press.

Batra, L. R., ed. 1967. "Insect-fungus symbiosis: nutrition, mutualism and commensalism." *Scientific American* 217(5): 112–20.

Bessey, E. A. 1973. *Morphology and taxonomy of the fungi,* reprint of 1950 ed. New York: Hafner Press.

Burnett, J. H., and A. P. Trinci, eds. 1980. *Fungal walls and hyphal growth.* New York: Cambridge University Press.

Christensen, C. M. 1965. *The molds and man: an introduction to the fungi,* 3d ed. Minneapolis: University of Minnesota Press.

Christensen, C. M. 1975. *Molds, mushrooms and mycotoxins.* Minneapolis: University of Minnesota Press.

Christensen, C. M. 1981. *Edible mushrooms,* rev. ed. Minneapolis, University of Minnesota Press.

Deacon, J. W. 1984. *Introduction to modern mycology,* 2d ed. Palo Alto, CA: Blackwell Scientific Publications.

Hale, M. E., Jr. 1980. *The biology of lichens,* 3d ed. Baltimore: E. Arnold Publications, Ltd.

Hale, M. E., Jr. 1979. *How to know the lichens,* 2d ed. Dubuque, IA: Wm. C. Brown Publishers.

Kramer, J. 1972. *Natural dyes: plants and processes.* New York: Charles Scribner's Sons.

Lamb, I. M. "Lichens." *Scientific American,* October 1959, pp. 2–9.

Lawrey, J. D., and M. Hale. 1984. *Biology of lichenized fungi.* Westport, CT: Praeger Publications.

Martin, G. W., and C. J. Alexopoulos. 1969. *Myxomycetes.* Iowa City, IA: University of Iowa Press.

Miller, O. K., Jr., and D. F. Farr. 1985. *An index of the common fungi of North America.* Forestburgh, N.Y.: Lubrecht and Cramer.

Moore-Landecker, E. 1982. *Fundamentals of the fungi,* 2d ed. Englewood Cliffs, NJ: Prentice-Hall.

Rinaldi, A., and V. Tyndalo. 1974. *The complete book of mushrooms.* New York: Crown Publishers.

Smith, A. H., and N. Weber. 1980. *The mushroom field hunter's guide,* enlarged ed. Ann Arbor: University of Michigan Press.

Webster, J. 1980. *Introduction to fungi,* 2d ed. New York: Cambridge University Press.

Overview

*T*his chapter opens with a discussion of the features of the Plant Kingdom and its members, and then introduces bryophytes by noting their past use as bandages. Following a discussion of the habitats and general life history of bryophytes, liverworts are examined, with an emphasis on *Marchantia*. Next, the chapter examines hornworts and their life cycles, and then discusses mosses in greater detail. The chapter concludes with observations on the human and ecological relevance of bryophytes.

18 *Introduction to the Plant Kingdom: Bryophytes*

Some Learning Goals

1. Know the features that distinguish the Plant Kingdom from other kingdoms.

2. Understand how bryophytes as a group differ from other plants.

3. Learn the basic differences between thalloid liverworts and "leafy" liverworts.

4. Explain how a liverwort thallus can be distinguished from that of a hornwort.

5. Know the structures involved in the life cycle of a moss and where meiosis and fertilization occur.

6. Learn which features liverworts, hornworts, and mosses have in common, and understand how their sporophytes differ.

7. Learn five uses of bryophytes by humans.

Outline

FEATURES OF THE PLANT KINGDOM

Most older botany texts and a few newer ones apply the term *plants* to many of the simpler organisms mentioned in previous chapters. Today, however, the majority of botanists confine this term to mosses, ferns, cone-bearing plants, flowering plants, and various close relatives of each of these groups. These groups are the subjects of the next several chapters.

Plants have several major pigments (e.g., chlorophyll *a,* chlorophyll *b,* carotenoids) similar to those of green algae (see chapter 16). Plants and green algae also have in common starch as the primary food reserve, cellulose in the cell walls, and the development of phragmoplasts and cell plates (see chapter 3) when their cells divide. Phragmoplasts and cell plates are unknown in other organisms. These features shared by plants and green algae suggest they were derived from a common ancestor.

Fossil records indicate that plants first appeared about 400 million years ago, so that any hypothetical ancestor probably made the transition even earlier from an aquatic environment to the land. By the time plants became established on land, several features had developed that prevented them from drying out. Most plants have a fatty cuticle on their surfaces and also a jacket consisting of a sterile layer of cells surrounding both the gametangia and the spore-producing cells (sporangia). In addition, the zygote develops into an embryo within tissues that originally surround the egg.

Unlike members of Kingdoms Monera, Protista, and Fungi, which (with the exception of stoneworts) have unicellular gametangia, all plants produce their gametes in multicellular gametangia. Members of the Plant Kingdom exhibit greater structural organization and diversity than those of other kingdoms, with corresponding specialization in the tissues involved in photosynthesis, conduction, support, anchorage, and protection. Reproduction is primarily sexual, with sporophyte phases of the life cycles assuming the predominant roles in the more advanced members.

The multicellular embryos of plants, mentioned earlier, are unknown in the kingdoms studied in the preceding chapters. This was recognized in past classifications when all organisms were placed in either the Plant or Animal Kingdoms. The Plant Kingdom was then divided into two subkingdoms, with the monerans, protists, and fungi being placed in Subkingdom Thallophyta (thallus plants), and all other organisms except animals being placed in Subkingdom Embryophyta (embryo plants).

One division of essentially nonvascular plants, the *bryophytes,* and several divisions of vascular plants are included in Kingdom Plantae. The bryophytes are discussed in this chapter and the vascular plants are discussed throughout the remaining chapters of the book.

INTRODUCTION TO THE BRYOPHYTES

On one occasion in the midst of heavy battles in France during World War I, nurses at what is now referred to as a M.A.S.H. unit ran out of bandages for the wounded soldiers. In desperation they substituted some soft green plant material they found growing in the water at the edge of a nearby lake. To their surprise, the material proved to be not only an adequate substitute for the bandages, but they noticed that wounds to which the material was applied did not become infected as frequently as those bound with conventional cotton bandages.

The material the nurses used was a species of *Sphagnum* moss (bog or peat moss), which has since been experimentally demonstrated to have antiseptic properties. This moss, which has specialized water-absorbing "leaves" (see figure 18.10), has been used as a packing material in the past, and is still widely used as a soil conditioner. It is but one of about 23,000 species of **bryophytes,** which include *liverworts, hornworts,* and *mosses,* many of which frequently make a soft and cool-looking green covering on damp banks, trees, and logs that are shaded for at least a part of the day (figure 18.1). Some that can withstand long periods of desiccation are also found on bare rocks in the scorching sun (figure 18.2), and others occur on frozen alpine slopes. The bryophytes also frequent habitats ranging from sea level near ocean beaches up to altitudes of 5,486 or more meters (18,000 or more feet) on high mountains.

A number of bryophytes are very specific indeed as to their growth sites. For example, some species are found only on the antlers and bones of dead reindeer. Others are confined to the dung of herbivorous animals, while still others can grow only on the dung of carnivores. A few found in the tropics are attached exclusively to large insects. The pygmy mosses, which appear annually on bare soil after rains, are only 1 to 2 millimeters (0.04 to 0.08 inch)

FIGURE 18.1   Mosses and ferns growing on a tree trunk.

FIGURE 18.2   *Grimmia*, a rock moss that survives on bare
rocks, often in scorching sun.

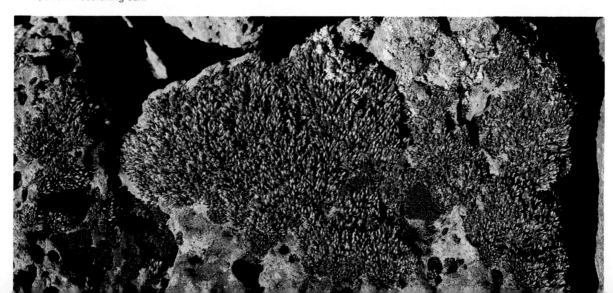

FIGURE 18.3 Lenslike cells of the protonema of the luminous moss. The cells concentrate light on the chloroplasts at their bases. (After von Denffer, D., Schumacher, W., Magdefrau, K. and Ehrendorfer, F. 1976. Strasburger's *Textbook of Botany,* 30th German ed., translated by Bell, P. and D. Coombe, 1976. Redrawn by permission of Gustav Fischer Verlag, Stuttgart, and Longman Group, London and New York.)

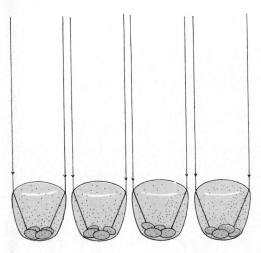

tall and can go through their whole life cycle in a few weeks. The widespread peat mosses, which are very important ecologically in bogs and in their transformation to dry land (see chapter 23), sometimes form floating mats over water and keep conditions so acid that the growth of bacteria and fungi is inhibited. Dead organisms deposited in such waters or bogs are often preserved for hundreds or even thousands of years. The luminous moss, which appears to glow in reflected light with an eerie golden-green radiance, is found in caves near the entrances and in other dark, damp places. The upper surfaces of its cells are curved slightly, and each cell functions as a tiny magnifying glass, concentrating the dim light on the chloroplasts at the base and thus adapting the plant to photosynthesis under conditions that would not support the survival of other organisms (figure 18.3).

Water, usually in the form of dew or rain, is essential for the reproduction of all bryophytes. None of the bryophytes have true xylem or phloem, but many mosses have special water-conducting cells called *hydroids* in the centers of their stems and a few have food-conducting cells called *leptoids* surrounding the hydroids. Neither type of cell is as efficient in conduction as xylem and phloem elements, however, and most water is absorbed directly through the surface. The absence of xylem or phloem makes

most bryophytes soft and pliable, and it is not surprising, therefore, that birds often use them to line their nests. In one study in the Appalachian Mountains of Virginia in 1975, for example, David Breil and Susan Moyle found at least 65 species of bryophytes used by a dozen species of native birds in the construction of their nests.

Bryophytes and *ferns* (discussed in chapter 19) exhibit an *alternation of generations* (see chapter 12) more conspicuously than do most organisms. In mosses the "leafy" plant is a major part of the *gametophyte* generation, which produces the *gametes.* Periodically the *sporophyte* generation, which produces the *spores,* grows from the tip of a "leafy" gametophyte, usually in the form of a tiny capped urn at the end of a slender stalk.

The bryophytes, which have similarities in their life cycles, chromosomes, and ecology, are divided into three distinct classes on the basis of structure and reproductive features. None of the bryophytes shows any close relationship to other living plants, and the fossil record provides little evidence of the relationship of the three classes to one another. This has led to speculation that each class may have arisen independently from ancestral green algae, and has resulted in some authorities elevating the classes to the rank of division.

DIVISION BRYOPHYTA

Class Hepaticae—Liverworts

The word *wort* simply means plant or herb. In medieval times, when the Doctrine of Signatures was promulgated (see chapter 1), the herbalists of the day thought that some of the bryophytes—specifically those that have flattened bodies with lobes shaped somewhat like those of the human liver (see figure 18.4)—were useful for treating ailments of that organ. No such efficacy has been demonstrated, but the name **liverwort** is still universally used today.

Structure and Form

The flattened lobed bodies of the liverworts just mentioned are called **thalli** (singular: **thallus**), but although thalloid liverworts are widespread, they actually constitute only about 20% of the approximately 8,000 known species. The remainder are "leafy" and superficially look more like true mosses. Liverworts differ from mosses (discussed in a later section) in several details and are considered less complex than mosses. Their thalli or "leafy" stages (gametophytes), which develop almost directly from spores, have comparatively smooth upper surfaces with various markings and pores, and lower surfaces with numerous one-celled *rhizoids*. The rhizoids resemble tiny roots and function in anchorage. Many liverworts have cells with special thickenings in the walls at the corners, and all have prostrate, as opposed to upright, forms of growth.

Probably the best known of the non-"leafy" or *thalloid* liverworts is one with the scientific name of *Marchantia* (named in honor of a French botanist, N. Marchant)(figure 18.4). The most widespread species of *Marchantia* is often found on damp soil, particularly after a fire has burned over the area. The thallus, which varies from a thickness of about 30 cells in the center to 10 at the margin, forks dichotomously as it grows. Each branch has a notch at the apex and a central lengthwise groove that extends behind the notch. Meristematic cells at these notches add new growth to the thalli. Older tissues at the rear decay as the new growth is added. The upper surface of the thallus is divided into diamond-shaped or polygonal segments, the segment lines marking the limits of the chambers below. Each segment has a small bordered pore opening into the interior.

FIGURE 18.4 *Marchantia* (a thalloid liverwort). *A.* A thallus with male gametophores. *B.* A thallus with female gametophores.

A.

B.

FIGURE 18.5 A section through a portion of a *Marchantia* thallus.

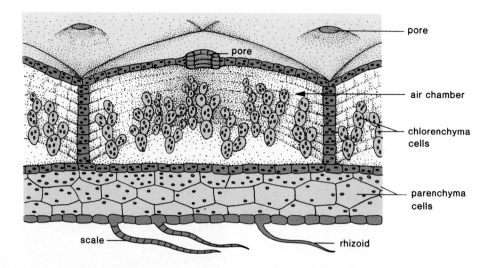

Examining a section through a thallus with a microscope is something like looking at a series of covered prickly pear cactus gardens sitting on a decorative rock wall (figure 18.5). The "rock wall," which may make up the bulk of the thallus, consists of parenchyma cells that contain few, if any, chloroplasts. The tissue apparently functions in the storage of substances manufactured by other cells. The bottom layer of cells constitutes an epidermis from which rhizoids and scales arise. The "cactus gardens" consist of upright branching rows of chlorenchyma cells in an air space. The individual "gardens" are enclosed by vertical walls and covered by a slightly dome-shaped layer of epidermal cells. In the center of each "roof," the conspicuous pore looks something like a short, suspended, open-ended barrel. The pore remains open at all times.

Marchantia—Asexual Reproduction

Marchantia reproduces asexually by means of **gemmae** (singular: **gemma**). These are groups of cells that form tiny lens-shaped pieces of tissue that become detached from the thallus. They are produced in small *gemmae cups* formed at intervals on the upper surface of the liverwort gametophyte (figure 18.6). Raindrops may splash the gemmae as much as 1 meter (3 feet) away. Each gemma is inhibited from developing further while it is in the cup (by lunularic acid), but after separation from the cup is capable of growing into a new thallus. In addition, parts of an older thallus may die, isolating patches of active tissue that may then continue to grow independently.

Marchantia—Sexual Reproduction

Marchantia develops sexual reproductive structures that are more specialized than those found in other liverworts. The male and female structures are produced on separate gametophytes, with both male and female gametangia being formed on **gametophores** (figure 18.6). These are umbrellalike structures rising on slender stalks from the central grooves of the thallus. The top of the male gametophore, or **antheridiophore,** is disclike with a scalloped margin, whereas that of the female gametophore, or **archegoniophore,** resembles the hub and spokes of a wagon wheel. **Antheridia** (club-shaped male sex structures each containing numerous sperms) are produced in rows just beneath the upper surface of the antheridiophore, whereas **archegonia** (flasklike female sex structures each containing a single egg), which are also produced in rows, hang neck downward beneath the spokes of the archegoniophore. Raindrops sometimes splash the released sperms more than one-half meter (1.5 feet) away. Fertilization may occur before the stalks of the archegoniophores have finished growing.

FIGURE 18.6   Life cycle of the thalloid liverwort
Marchantia.

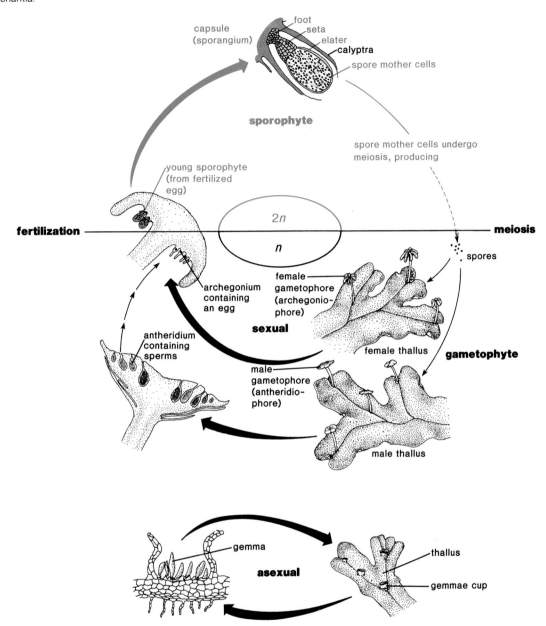

FIGURE 18.7 A longitudinal section through a sporophyte of *Marchantia*.

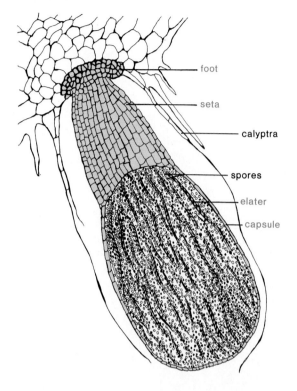

FIGURE 18.8 A "leafy" liverwort.

After fertilization, the zygote develops into a multicellular **embryo** (an immature *sporophyte*). The sporophyte (the diploid, spore-producing phase), is anchored in the tissues of the archegoniophore by means of a knoblike **foot,** and hangs suspended by a short thick stalk called the **seta.** The main part of the sporophyte, in which different types of tissues develop, is called a **capsule.** Stomata are typically absent from liverwort sporophytes. One type of tissue in the capsule consists of **spore mother cells** that undergo meiosis, producing haploid *spores,* whereas the cells of another type of tissue do not undergo meiosis but remain diploid and develop instead into long, pointed **elaters** (figure 18.7). Elaters, which have spiral thickenings and are sensitive to changes in humidity, twist and untwist rapidly in *Marchantia* species, thereby aiding in the dispersal of spores. In the sporophytes of other liverworts the elaters may aid spore dispersal through a snapping action or by sudden expansion. The young sporophyte is protected until it is mature by a caplike tissue, the **calyptra,** which grows out from the gametophyte, and by other membranes covering the

capsule. At maturity the capsule splits, and the spores are carried away by air currents. Under favorable conditions the spores germinate, producing new gametophytes.

Other thalloid forms such as the floating or amphibious liverworts do not produce gametophores. Instead the archegonia and antheridia develop within the thallus beneath the central grooves, where the sporophytes also are formed. The spores are liberated from the submerged sporophytes as the thallus decays.

Leafy Liverworts

"Leafy" liverworts (figure 18.8), which are often found abundantly in tropical jungles and in fog belts, always have two rows of partially overlapping "leaves" whose cells contain distinctive oil bodies. The "leaves" have no midribs, and unlike the "leaves" of mosses, they often have folds and lobes. In the tropics, the lobes form little water pockets in which tiny animals are nearly always present. It has been suggested that these water pockets may function like the pitchers of pitcher plants (see chapter 7). A third row of "underleaves," which are smaller in size than the other "leaves" and not visible from the top, frequently is present on the underside of "leafy" liverworts. A few rhizoids, which anchor the plants, develop from the stemlike axis at the base of the "underleaves."

Members of all classes of bryophytes often have mycorrhizal fungi (see chapters 6 and 17) associated with their rhizoids. In some instances the fungi are apparently at least partially parasitic. One species of liverwort, which lives underneath mosses, is completely colorless and depends on its fungal associate for its existence.

The archegonia and antheridia of the "leafy" liverworts are produced in cuplike structures composed of a few modified "leaves," either in the axils of "leaves" or on separate branches. At maturity, the sporophyte capsule may be pushed out from among the "leaves" by elongation of the seta. When a spore germinates, it produces a **protonema,** which consists of a few photosynthetic cells in the form of a short filament. The protonema soon develops into a new gametophyte plant.

Class Anthocerotae—Hornworts

Structure and Form

Hornworts, which received their common name from the resemblance of their mature sporophytes to miniature cattle horns, have rounded gametophytes superficially similar to those of thalloid liverworts, but they appear to be only distantly related to the liverworts. They seldom exceed 2 centimeters (0.4 inch) in diameter and are found mostly on moist earth in shaded areas, although some occur on trees. There are about 100 species distributed around the world, but they are uncommon in arctic regions. They differ from liverworts and mosses in several respects. The differences include the presence in hornworts of usually only one large chloroplast in each cell (a few species have up to eight), and the occurrence of pyrenoids similar to those of green algae on each chloroplast. The thalli have pores and cavities filled with mucilage, in contrast with the pores and cavities of thalloid liverworts, which are filled with air. Nitrogen-fixing blue-green bacteria often grow in the mucilage. Rhizoids anchor the plants to the surface.

Asexual Reproduction

Asexual reproduction in hornworts occurs through fragmentation or separation of lobes from the main part of the thallus. In a few instances, tiny tubers capable of becoming new gametophytes are formed.

Sexual Reproduction

In sexual reproduction, archegonia and antheridia are produced in rows just beneath the upper surfaces of the gametophytes. As is the case with both mosses and liverworts, some species of hornworts have *unisexual* plants whereas other species are *bisexual.* Sporophytes of hornworts have numerous

FIGURE 18.9   *A.* Hornworts. *B.* Detail of a hornwort sporophyte.

A.

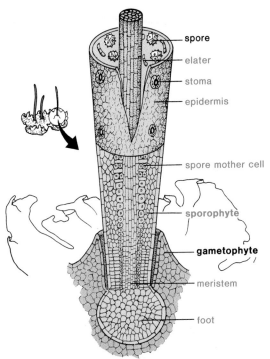

B.

stomata, and are most distinctive. They have no setae (stalks) and look like tiny green broom handles or horns rising through a basal sheath from a foot beneath the surface of the thallus (figure 18.9). A meristem above the foot continually increases the length of the sporophyte from the base when conditions are favorable. As growth occurs, spore mother cells surrounding a central rodlike axis in the sporophyte undergo meiosis, producing spores. The tip of the sporophyte horn splits into two or three ribbonlike segments, releasing the spores, and the segments continue to peel back as long as the meristem is producing new tissue at the base.

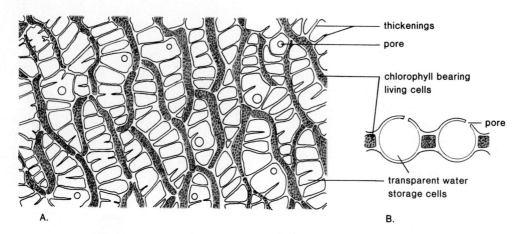

FIGURE 18.10 Enlargement of a portion of a peat moss (*Sphagnum*) leaf. *A.* Surface view. *B.* Further enlarged cross section of living and dead cells.

Class Musci—Mosses

Structure, Form, and Subclasses

Many different organisms have been called **mosses.** In fact, almost any greenish covering or growth on tree trunks and forest floors has probably been called "moss" at one time or another. Examples include lichens (e.g., reindeer moss), red algae (e.g., Irish moss), flowering plants (e.g., Spanish moss), and club mosses. The latter superficially resemble large true mosses but have xylem and phloem and thus are vascular plants. About 15,000 species of mosses are currently known. These are divided into three different subclasses, commonly called *peat mosses, true mosses,* and *rock mosses.* As observed in the following discussion, mosses are distinct, both in form and reproduction, and possibly in origin, from any other groups of organisms.

The "leaves" of moss gametophytes differ from the leaves of more complex plants in having no mesophyll tissue, stomata, or veins; moss "leaf" cells also are all haploid. The blades are nearly always merely one cell thick, except at the *midrib,* which runs lengthwise down the middle, and they are never lobed or divided, nor do they have a petiole. The midrib, which is absent in some genera, occasionally projects beyond the tip in the form of a hair or spine. The "leaf" cells usually contain numerous lens-shaped chloroplasts except at the midrib. The

"leaves" of peat mosses, however, have large transparent water-storage cells (without chloroplasts) that adapt them to water absorption and storage, with small green photosynthetic cells sandwiched between the large cells (figure 18.10). The "leaves," which initially are formed in three ranks, usually end up appearing to be arranged in a spiral or alternately on an axis that twists as it grows. The axis is somewhat stemlike but also has no xylem or phloem, although hydroids at the center often appear in the form of a strand that differs in appearance from that of the surrounding tissue. At the base are rootlike rhizoids, consisting of several rows of colorless cells. Their primary function is to anchor the plant. Some water absorbed by rhizoids rises up the central strand, but most water used by the plant apparently travels up the outside of the plant by means of capillarity. The closely packed habit of many mosses and the fact that they rarely extend more than a few centimeters (2 to 3 inches) into the air favors such outside movements of water, which is absorbed directly through the plant surfaces.

Sexual Reproduction

Sexual reproduction in mosses is initiated with the formation of multicellular gametangia, usually at the apices of the "leafy" shoots (see figure 18.13), although they frequently form on special separate branches. Both male and female sex structures are often produced on the same plant, but in some species they occur on separate plants. The *archegonia,* which are the female gametangia, are somewhat cylindrical and project upward from the base of the

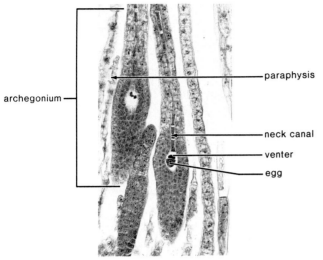

FIGURE 18.11   A longitudinal section through the tip of a female gametophyte of the moss *Mnium*. (Photomicrograph by G. S. Ellmore)

archegonium —

— paraphysis

— neck canal

— venter

— egg

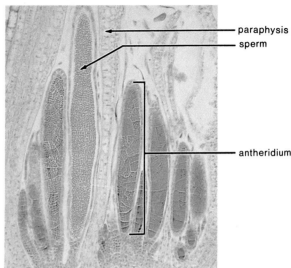

FIGURE 18.12   A longitudinal section through the tip of a male gametophyte of the moss *Mnium*. (Photomicrograph by G. S. Ellmore)

— paraphysis
— sperm

— antheridium

expanded gametophyte tip (figure 18.11). Certain cells break down in a slightly swollen central region of an archegonium, leaving a cavity called a **venter** in which a single *egg* cell is produced. The part of the archegonium above the venter is called the *neck*. It may taper toward the tip and contains a narrow *canal*. The canal is at first plugged with cells, but these break down as the archegonium matures, leaving an opening to the outside at the top. A number of archegonia usually are produced at the same time, with sterile hairlike, multicellular filaments called *paraphyses* (singular: *paraphysis*) scattered among them.

Male gametangia, which also have paraphyses among them, are sausage-shaped to roundish with walls that are one cell thick. These *antheridia* (figure 18.12) are borne on a short stalk. A mass of tissue inside each antheridium develops into numerous coiled or comma-shaped *sperm* cells. This mass of sperms is forced out of the top of the antheridium when it absorbs water and swells. After release, the sperm mass breaks up into individual cells, each with a pair of flagella. It is believed that the breakup of the sperm mass is aided, in some cases, by fats produced by the moss, while in other instances rain splash is responsible.

Archegonia release sugars, proteins, acids, or other substances that attract the sperm, and eventually a sperm, after swimming down the neck of an archegonium, unites with the egg, forming a *zygote* (figure 18.13). The zygote usually grows rapidly into a spindle-shaped *embryo*, which breaks down the cells at the base of the archegonium and becomes firmly established in the tissues of the stem by means of a swollen knob called a *foot*. As the embryo grows, cells around the venter divide, thereby accommodating its increasing size. The length of the embryo soon exceeds the length of the venter cavity. The top of the venter is split off and is left sitting like a pixie cap on top of the embryo. By this time, the embryo is a developing *sporophyte* whose cells are all diploid. The pixie cap, called a *calyptra,* remains until the sporophyte is mature. In one genus with the common name of "extinguisher mosses," the calyptra looks just like a little candlesnuffer, and in the haircap mosses it resembles a miniature, pointed, goatskin cap such as might be worn by a Shakespearean actor.

The cells of the sporophyte become photosynthetic as it develops, remaining so until maturity. The sporophyte, however, depends on the gametophyte to varying degrees for some of its carbohydrate needs as well as for at least a part of its water and minerals, which are absorbed through the foot.

FIGURE 18.13   Life cycle of a moss.

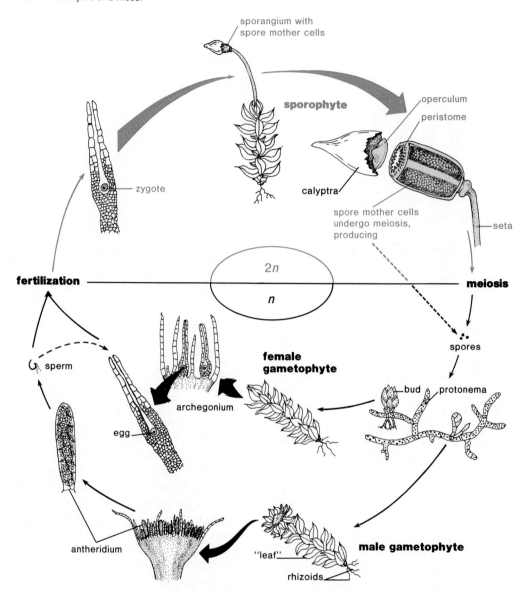

The mature sporophyte, which is initially green and photosynthetic, consists of a *capsule* located at the tip of a slender stalk, the *seta*. Depending on the species, the seta may be less than 1 millimeter (0.04 inch) long, or it may be up to 15 centimeters (6 inches) long. Most, however, are less than 5 centimeters (2 inches) long. The capsule, which may resemble a tiny apple, a pear, an urn, a box, or a wingtip fuel tank of an airplane, usually has from 3 or 4 to over 200 stomata at or near its surface. Unless extremely dry conditions prevail, the stomata normally remain open until the capsule begins to age, and then they close permanently. The free end of the capsule is usually protected by a little rimmed lid, the **operculum,** which falls off at maturity.

As the capsule matures, *spore mother cells* inside undergo meiosis, producing *spores*. These spores, often numbering in the millions, are released from the free end of the capsule after the operculum has fallen off, usually through a structure called a *peristome*. Most peristomes consist of a circular row, or often two rows, of narrowly triangular and membranous teeth arranged around the rim of the capsule, each row having 16 teeth. The teeth are frequently colored orange or red and are often beautifully sculptured with bars and fringes. They are sensitive to changes in humidity and open or close in response to such changes. In a few species of mosses, the peristome is a cone-shaped structure with pores through which the spores are released. Certain rock mosses have neither a peristome nor an operculum. The spores in these mosses are released when the capsule splits longitudinally along four lines. In the dung mosses, a putrid odor is given off when the spores are ready for release. Some of the spores adhere to the legs and bodies of flies, which are attracted by the odor, and are disseminated as the insects clean themselves. Most moss spores are, however, simply blown away by the wind, and if they fall in a suitable damp location, they usually germinate relatively quickly.

In the majority of mosses, fine green tubular threads consisting of single rows of cells with chloroplasts first emerge from the spores. These soon branch and grow, forming an algalike *protonema*. The protonema can be distinguished from a filamentous green alga by the crosswalls of its cells, which are usually formed at oblique angles instead of right angles, and by the chloroplasts, which are all lens-shaped. If light and other conditions are favorable, tiny "leafy" buds appear at intervals along the protonemal filaments after about two to four weeks of growth. These "leafy" buds develop rhizoids at the base and grow into new "leafy" gametophytes, thus completing the cycle.

Asexual Reproduction

Some reproduction in mosses does not depend on such a sexual cycle and its alternation of generations. It has been demonstrated under laboratory conditions that cells of archegonia and antheridia, paraphyses, leaves, stems, and rhizoids all can develop protonemata. Two American biologists, Norton Miller and Howe Ambrose, made a collection of bryophyte gametophyte fragments from a snowbed in the Canadian high arctic and found that 12% of their samples resumed growth in various ways when cultured in the laboratory. On this basis, they calculated that at their study site there were over 4,000 bryophyte fragments per cubic meter (1.3 cubic yards) of snow capable of becoming new plants. They suggested that wind dispersal by fragments may be routine in arctic regions. Such dispersal and vegetative reproduction also occur widely in more temperate areas.

HUMAN AND ECOLOGICAL RELEVANCE OF BRYOPHYTES

Certain bryophytes and lichens may be pioneers on bare rock after volcanic eruptions or other geological upheavals, and after the retreat of glaciers. They slowly convert the rock to soil, which can then be inhabited or utilized by other organisms (see chapter 23). Mosses, in particular, retain moisture, slowly releasing it into the soil. They reduce flooding and erosion and contribute to humus formation. They are indicators of the presence or absence of calcium in the soil, and of higher than usual soil salinity or acidity. The presence of one genus in a dry area indicates that running water occurs there during a part of the year. A few mosses are occasionally a problem in water reservoirs, where they may plug entrances to pipes. A few bryophytes are reported to be grazed, along with lichens, by foraging mammals in arctic regions, but bryophytes are not generally edible.

Some mosses have been used for packing dishes and stuffing furniture, and Native Americans are reported to have used mosses under splints when setting broken limbs. By far the most important bryophytes to humans, however, are the peat mosses. When allowed to absorb water, 1 kilogram (2.2 pounds) of dry peat moss will take up 25 kilograms (55 pounds) of water. Its extraordinary absorptive capacity has made it very useful as a soil conditioner in nurseries and as a component of potting mixtures. Live shellfish and other organisms are shipped in it, and its natural acidity, which inhibits bacterial and fungal growth, gives it antiseptic properties. This latter feature combined with its absorbency, which is greater than that of cotton, has made it a useful poultice material for application to wounds. It was used for this purpose during the Crimean War of 1854–1856 and, as indicated in the chapter introduction, on an emergency basis during World War I. Extensive peat deposits have been formed from the remains of peat mosses that flourished in past eras. Peat, like the undecomposed peat mosses, is used around the world as a soil conditioner and as a fuel. It is also used in the curing of malt for Scotch whiskey.

See appendix 1 for the scientific names of all the bryophytes discussed.

SUMMARY

Members of the Plant Kingdom have a cuticle and produce their gametes in multicellular organs surrounded by a sterile jacket of cells; their zygotes develop into embryos, and tissues specialized for photosynthesis, conduction, support, anchorage, protection, and reproduction are produced. Sporophyte phases of the life cycles are predominant in more advanced members. Cell plates and phragmoplasts are formed when plant cells divide; outside of the Plant Kingdom cell plates occur only in certain green algae. The similarity in pigments, food reserve (starch), and occurrence of cell plates suggests a common ancestor for the green algae and plants, which, according to fossil records, may have originated more than 400 million years ago. The Plant Kingdom includes one division of bryophytes and several divisions of vascular plants.

Bryophytes (liverworts, hornworts, mosses) are found in habitats ranging from bare rocks in hot sun to frozen slopes, and from sea level to high altitudes. Some bryophytes, such as those found on bones or dung or insects are very specific in their habitats. The luminous moss is adapted to existing under conditions of very low light. Water is essential to bryophyte reproduction. Despite the presence of primitive conducting cells in many mosses, most water is absorbed directly through the plant surfaces.

The best known liverworts have gametophytes with flattened, dichotomously forking thalli, but about 80% of the liverwort species are "leafy." Liverworts have distinct upper and lower surfaces, with one-celled rhizoids that function in anchorage on the lower surface.

The thallus of the well-known liverwort *Marchantia* has a central lengthwise groove along its upper surface, which is divided into polygonal segments that each contain a pore. Beneath each pore is a chamber containing upright branching rows of chlorenchyma cells arising from brick-shaped parenchyma cells in wall-like layers at their bases. Rhizoids and scales arise from the bottom row of parenchyma cells. *Marchantia* reproduces asexually by means of pieces of tissue called gemmae that are splashed by rain out of cups in which they are formed, and also by fragmentation of the thallus. *Marchantia* reproduces sexually by means of flasklike archegonia beneath the "spokes" of an umbrella-shaped archegoniophore that arises from the thallus, and club-shaped antheridia in the disclike tops of antheridiophores, which also arise from

the thallus. Each archegonium contains an egg and each antheridium contains many sperms. The zygote formed from the union of egg and sperm develops into a sporophyte that is anchored to the archegoniophore by a knoblike foot, from which is suspended a capsule connected to the foot by a stalk, the seta. The capsule contains spore mother cells, which undergo meiosis and produce spores, and pointed elaters, which do not undergo meiosis. The elaters, which twist, snap about, or suddenly grow larger with changes in humidity, aid in spore dispersal. The calyptra, which grows out from the gametophyte, and other membranes protect the spores until they are released as the capsule splits; the spores may then develop into new gametophytes.

"Leafy" liverworts have two rows of overlapping "leaves" and frequently a third row of "underleaves" not visible from above. The "leaves" often have lobes that retain rainwater.

Hornworts, which have one chloroplast with pyrenoids in each cell, resemble liverworts in their gametophytes, but their sporophytes are hornlike and have a meristem above the foot. Hornwort thalli have pores and cavities filled with mucilage in which blue-green bacteria often grow. Asexual reproduction involves the separation of lobes from the main part of the thallus. Archegonia and antheridia are produced in rows beneath the upper surface of a thallus. The tip of the hornlike sporophyte splits vertically, releasing the spores.

A moss gametophyte consists of an axis to which "leaves" are attached, with rhizoids at the base. The "leaves" are haploid and have no mesophyll, stomata, or veins. Water is primarily absorbed directly through the plant surfaces.

Multicellular archegonia and antheridia are produced at the tips of "leafy" shoots. Each archegonium has a cavity, the venter, containing a single egg and a neck through which a sperm gains access to the egg. Sperms are produced in sausage-shaped antheridia. After fertilization the zygote grows into an embryo, which is attached to the gametophyte by means of a foot embedded in its tissues. The embryo develops into a sporophyte consisting of a capsule and a seta, with a caplike calyptra partially covering the capsule. Spore mother cells in the capsule undergo meiosis, producing spores that are released through the teeth of the peristome, a structure at the tip of the capsule. The peristome is initially protected by a lid, the operculum, which falls off when the spores are mature. When moss spores germinate, they develop into threadlike protonemata on which "leafy" buds appear. The buds grow into new gametophytes. Protonemata have been demonstrated to develop from cells of various other moss tissues besides spores.

Mosses may be pioneers, along with lichens, on bare rocks. They are indicators of soil calcium, salinity, and acidity. Mosses are not generally edible, although a few are grazed in arctic regions. Some mosses are used for packing material, but the most significant use is that of peat mosses for soil conditioners. Peat mosses can absorb and retain large amounts of water, and their natural acidity gives them antiseptic properties. Peat deposits, from peat mosses that flourished in past eras, are used for fuel and also as a soil conditioner.

REVIEW QUESTIONS

1. What basic features distinguish members of the Plant Kingdom from those of other kingdoms?
2. What features distinguish the bryophytes discussed in this chapter from other plants?
3. How could you tell a hornwort thallus from that of a thalloid liverwort?
4. Contrast the sporophytes of mosses, liverworts, and hornworts.
5. What is a protonema? Do all bryophytes have them? How would you tell a protonema from a green alga?
6. What adaptations do bryophytes have for their particular habitats?
7. Which parts of the life cycles of bryophytes have haploid (n) cells? Which parts have diploid ($2n$) cells?
8. Why is a bryophyte "leaf" technically not the same as a flowering plant leaf?
9. Define calyptra, operculum, capsule, peristome, paraphysis, foot, seta, archegoniophore, thallus, underleaves, and elaters.
10. If you were to single out one bryophyte as being the most important member of its division from a human viewpoint, which would you choose? Why?

DISCUSSION QUESTIONS

1. Very few fossils of bryophytes have been found. Suggest reasons for this.
2. Do the multicellular sex organs of plants give them any advantages over other organisms?
3. After reading about the characteristics and uses of peat mosses, can you suggest some possible new uses for these plants?
4. Some bryophytes produce unisexual gametophytes, while others produce bisexual gametophytes. Should one type have any survival or adaptive advantage of the other? Explain.

ADDITIONAL READING

Clarke, G. C. S., and J. G. Duckett, eds. 1980. *Bryophyte systematics, No. 14.* New York: Academic Press, Inc.

Conard, H. S., and P. L. Redfearn, Jr. 1979. *How to know the mosses and liverworts,* 2d ed. Pictured Key Nature Series. Dubuque, IA: Wm. C. Brown Publishers.

Grout, A. J. 1903–1910. *Mosses with hand-lens and microscope.* Published in five parts by the author in Brooklyn, NY.

Richardson, D. H. S. 1981. *The biology of mosses.* New York: Halsted Press.

Scagel, R. F., et al. 1984. *Plants: An evolutionary survey.* Belmont, CA: Wadsworth Publishing Company.

Schofield, W. B. 1985. *Introduction to bryology.* New York: Macmillan.

Schultze-Motel, W., ed. *Advances in bryology: 1984.* Vol. 2. Forestburgh, NY: Lubrecht and Cramer.

Schuster, R. M. 1966–1974. *The Hepaticae and Anthocerotae of North America.* 4 vols. New York: Columbia University Press.

Schuster, R. M. 1983, 1984. *New manual of bryology.* 2 vols. Hattori Botanical Laboratory, Nichinan, Miyazaki, Japan.

Smith, A. J. E., ed. 1982. *Bryophyte ecology.* New York: Methuen, Inc.

Vitt, D. H., and S. R. Gradstein. 1985. *Compendium of bryology.* Forestburgh, NY: Lubrecht and Cramer.

Watson, E. V. 1971. *The structure and life of bryophytes,* 3d ed. Atlantic Highlands, NJ: Humanities Press International, Inc.

Overview

*T*his chapter opens with a brief review of the features that distinguish the major groups of seedless vascular plants from one another and from the bryophytes and then discusses representatives of each division. Included in the discussion are whisk ferns (*Psilotum*), club mosses (*Lycopodium, Selaginella*), quillworts (*Isoetes*), horsetails (*Equisetum*), and ferns. A digest of the human and ecological relevance of each group is given, and life cycles of representatives are illustrated. The chapter concludes with an examination of fossils, and each type is briefly described. A table showing the geologic time scale is provided.

19 Introduction to Vascular Plants: Ferns and Their Relatives

Some Learning Goals

1. Know the basic structural differences between bryophytes and vascular plants.

2. Distinguish the four divisions of seedless vascular plants from one another.

3. Understand the differences in the life cycles of ground pines (*Lycopodium*) and spike mosses (*Selaginella*).

4. Learn the structural features of horsetail (*Equisetum*) sporophytes.

5. Know all the structures involved in Alternation of Generations in a fern.

6. Learn 10 important uses of seedless vascular plants.

7. Explain what a fossil is and distinguish among the various types of fossils.

Outline

INTRODUCTION

Our survey of all organisms traditionally regarded as plants thus far has taken us from simple one-celled prokaryotic organisms, through more specialized eukaryotic protists and fungi, and on to the bryophytes. We now take up several divisions of plants whose members have developed internal conducting tissues but have not produced seeds.

As indicated in the introduction to the Plant Kingdom in the preceding chapter, the hypothetical ancestor of the bryophytes and the vascular plants was probably a multicellular green alga that became established on the land over 400 million years ago. The shift to terrestrial habitats was accompanied by the appearance of features such as sterile jackets of cells around gametangia, development of embryos within protective tissues, and cutin—all significant in preventing the drying out of vital parts. Any internal conducting tissues possessed by the bryophytes was, however, negligible and water was absorbed directly through the aboveground parts of the plants. In addition, external water was essential to the reproductive processes.

Then internal conducting tissues (xylem and phloem) began to evolve, true leaves appeared, and roots that function in anchorage and absorption developed. At the same time gametophytes became progressively smaller and more dependent on and protected by proportionately larger sporophytes. Unlike conifers and flowering plants, the primitive vascular plants discussed in this chapter do not produce seeds. They include the ferns and a number of their relatives often referred to as "fern allies."

Four divisions of seedless vascular plants are recognized: Division Psilophyta (whisk ferns), whose members are unique among vascular plants in having sporophytes that have neither true leaves nor roots and consist of stems and rhizomes that fork evenly; Division Lycophyta (club mosses and quillworts), whose stems are covered with mostly photosynthetic **microphylls** (leaves with a single vein whose trace is not associated with a leaf gap—see chapter 7); Division Sphenophyta (horsetails and scouring rushes), whose sporophytes have ribbed stems containing silica deposits and whorled scalelike microphylls that lack chlorophyll; and Division Pterophyta (ferns), whose sporophytes have **megaphylls** (leaves with more than one vein and a leaf trace that is associated with a leaf gap—see chapter 7) that are often large and much divided. Fossils, discussed at the end of the chapter, give us clues to the ancestry of some of these plants and also to the derivations of seed plants discussed in the chapters to follow.

FIGURE 19.1   A *Psilotum* sporophyte.

DIVISION PSILOPHYTA— THE WHISK FERNS

There is nothing very fernlike in the appearance of **whisk ferns,** more commonly referred to by their scientific name of *Psilotum,* but they do loosely resemble small green whisk brooms. Exactly where these plants fit in the Plant Kingdom is not clear. Traditionally they have been associated with a number of extinct plants called *psilophytes,* which flourished perhaps 400 million years ago, but there is a general lack of fossil evidence to substantiate the link. Some have classified them with the true ferns because their gametophytes share features with those of true ferns. By most criteria presently in use they are, however, among the simplest of all living vascular plants and thus will be the first discussed.

Structure and Form

Psilotum sporophytes consist of evenly forking stems that commonly approach 30 centimeters (1 foot) in height, occasionally growing to a height of 1 meter (3 feet) or more (figure 19.1). The visible stems arise from short branching rhizomes just beneath the surface of the ground. Tiny green, flattened protuberances called **enations** are spirally arranged at intervals along the stems, but *Psilotum* is unique

FIGURE 19.2 Life cycle of *Psilotum*.

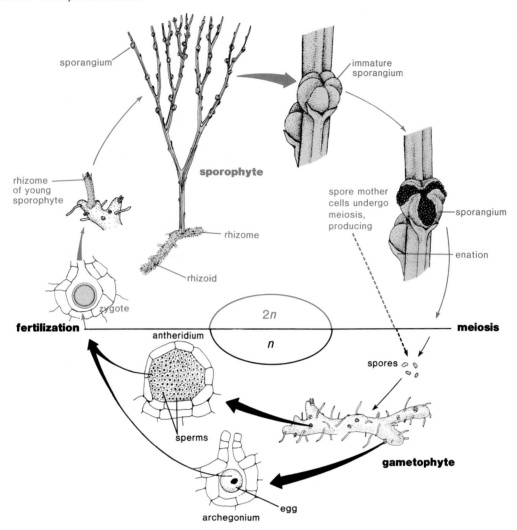

among living vascular plants in having neither leaves nor roots. Photosynthesis takes place in the outer cells of the stem, which contains a central cylinder of xylem surrounded by phloem. The xylem is star-shaped in cross section. The rhizomes, which have short rhizoids over their surfaces, perform the functions of roots with the aid of mycorrhizal fungi, some of which establish themselves in the cells of the cortex just beneath the epidermis.

Reproduction

Very short stubby branches appear at times along the upper parts of the angular stems. These are terminated by small sporangia fused together in threes and resembling miniature yellow pumpkins. Spores released from the sporangia germinate slowly in the soil, on the bark of tree ferns, or in similar habitats.

The gametophytes that arise from the spores are found beneath the surface of the soil and are easily overlooked as they are colorless, about 2 millimeters (0.08 inch) in diameter, and seldom more than 6 millimeters (0.25 inch) long. These gametophytes, which are cylindrical and may fork like the stems, sometimes resemble tiny dog bones. They are *saprophytic,* obtaining their nutrients through association with fungi. Archegonia and antheridia are produced randomly over the surface of the same gametophyte. After a sperm unites with an egg in an archegonium, the zygote develops a foot and a rhizome. As soon as the rhizome establishes itself with the aid of mycorrhizal fungi, upright stems are produced and the rhizome separates from the foot (figure 19.2).

Whisk ferns grow naturally primarily in tropical and subtropical regions. In the United States, they are found in Florida, Louisiana, Texas, Arizona, and Hawaii. They are cultivated, primarily as a botanical curiosity, in Japan and in greenhouses around the world. A close relative, *Tmesipteris,* with leaflike appendages, occurs in Australia and the South Pacific.

Fossil plants resembling whisk ferns in many respects have been found in Silurian geological formations (see table 19.1). As intimated in the introduction to the whisk ferns, these geological formations are estimated to be as much as 400 million years old. One group of these fossil plants, exemplified by *Rhynia* (see figure 19.22), had naked stems and spindle-shaped terminal sporangia. A second group of fossils, represented by *Zosterophyllum,* had somewhat rounded sporangia produced along the upper parts of naked stems. *Zosterophyllum* and its relatives first appeared during the Devonian period. They are considered to be likely ancestors of the club mosses discussed in the next section.

Whisk ferns are of little economic importance. Their spores, which have a slightly oily feel, were once used by Hawaiian men to reduce loincloth irritations of the skin. Hawaiians also made a laxative liquid by boiling whisk ferns in water.

DIVISION LYCOPHYTA—THE CLUB MOSSES AND QUILLWORTS

On one occasion during my youth, my father was interviewed by a newspaper reporter in a large South American city. At the end of the interview, which took place in a hotel room with the rest of the family present, the reporter set up a press camera and said he wanted to take a picture of the whole family. After posing us, he took a vial from his pocket and poured a little powder along a metal bar attached to one end of a T-shaped device that otherwise looked like a flashlight. He then told us to smile, and a moment later there was a blinding flash accompanied by a large billow of smoke. I learned later that the flash powder (forerunner of flashbulbs and strobe lights) he had used contained millions of spores of primitive vascular plants called **club mosses.**

The approximately 950 known species of club mosses superficially look enough like large true mosses to have caused the great Swedish botanist Linnaeus to lump both together in a single class. Once details of the structure and the form of club mosses were known, however, it became obvious that they are quite unrelated to true mosses.

FIGURE 19.3   A species of *Lycopodium* native to northern and eastern North America.

Two major and two minor genera of club mosses exist today (several others became extinct about 270 million years ago). The sporophytes of all have leaves called **microphylls** which are usually (but not always) quite small (see chapter 7); they also have true stems and true roots.

Lycopodium—Ground Pines

Structure and Form

The two major genera of club mosses with living members are *Lycopodium* with about 50 species, and *Selaginella,* with over 700 species. *Lycopodium* plants are frequently called *ground pines,* partly because they often grow on forest floors and resemble little Christmas trees, complete with "cones" that are usually upright or, in a few species, hang down (figure 19.3).

They are distributed throughout the world, although they are more abundant in the tropics and wetter temperate areas. Ground pine sporophytes have either simple or branched stems, mostly less than 30 centimeters (1 foot) tall although tropical species may attain heights of 1.5 meters (5 feet) or more. The upright, or sometimes pendent, stems develop from rhizomes, which also branch. Their leaves may be whorled or in a tight spiral and rarely exceed 1 centimeter (0.4 inch) in length. Adventitious roots arise along the rhizomes. Many of the epidermal cells of the roots produce root hairs (see chapter 5).

FIGURE 19.4   Life cycle of the ground pine *Lycopodium*.

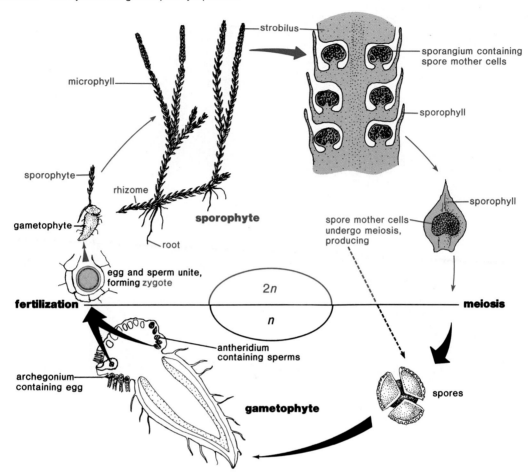

Reproduction

At maturity, some species of ground pines develop sporangia shaped like kidney beans on short stalks in the axils of several of the leaves. Such sporangium-bearing leaves are called **sporophylls.** In other species, the sporophylls are reduced in size, lack chlorophyll, and are clustered in terminal cone-like structures called **strobili** (singular: **strobilus**).

Spore mother cells in the sporangia undergo meiosis, producing spores that are carried away by air currents upon their release. Those of some species germinate in a few days if they land in a suitable location. Other species produce spores that may not germinate for up to several years. Upon germination, the spores produce independent gametophytes that occur in various shapes, some resembling tiny carrots, with most or all of the body developing in the ground. Other gametophytes develop primarily on the surface, assisted by mycorrhizal fungi. Surface gametophytes, or the parts of them exposed to light, are green. All types produce both antheridia and archegonia on the same gametophyte, which, in some species, may live for several years. Zygotes develop embryos with a foot, stem, and leaves, and these develop into mature sporophytes (figure 19.4). If the gametophyte is underground, chlorophyll does not develop in the young sporophyte until it emerges into the light. Several sporophytes may be produced from a single gametophyte. A number of ground pines also reproduce asexually by means of small *bulbils* (bulb produced in the axils of leaves), each of which is capable of developing into a new sporophyte.

FIGURE 19.5   *Selaginella*, a spike moss.

Selaginella—Spike Mosses

Structure and Form

Members of *Selaginella,* the other major genus of living club mosses, are sometimes called *spike mosses* (figure 19.5). The approximately 700 species are widely scattered around the world in wetter areas, but they are especially abundant in the tropics. One or two are common "weeds" in greenhouses. They tend to branch more freely than ground pines, from which they differ in several respects. The two most obvious differences are (1) their leaves each have a tiny extra appendage, or tongue, called a **ligule,** on the upper surface near the base, and (2) they produce two different kinds of spores and gametophytes—an advanced feature referred to as **heterospory.** The seed-bearing coniferous and flowering plants discussed in chapters to follow are all heterosporous.

Reproduction

Sexual reproduction of spike mosses initially is similar to that of ground pines. Sporangia are produced on sporophylls called either *microsporophylls* or *megasporophylls* (figure 19.6). **Microsporophylls** bear *microsporangia* containing numerous *microspore mother cells* that undergo meiosis, producing tiny **microspores,** whereas the *megasporangia* of megasporophylls usually contain a *megaspore mother cell* that becomes four comparatively large **megaspores** upon undergoing meiosis. Each microspore has the potential to develop into a male gametophyte, which consists simply of a somewhat

FIGURE 19.6   Life cycle of the spike moss *Selaginella*.

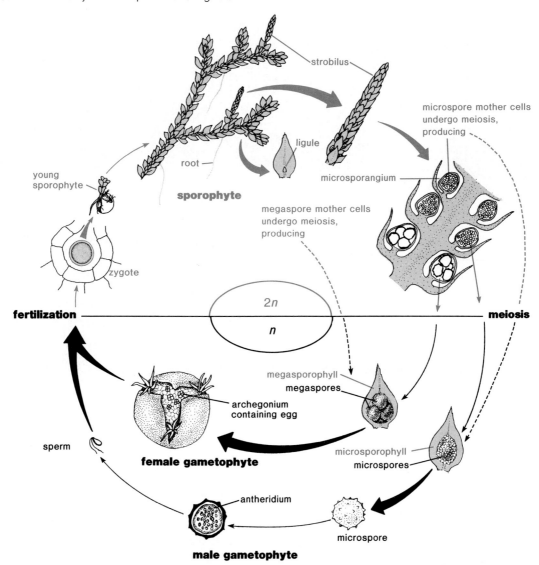

spherical antheridium surrounded by a sterile jacket of cells. Either 128 or 256 sperm cells with flagella are produced in each antheridium. A megaspore develops into a female gametophyte, which is also relatively simple in structure. By the time this gametophyte is mature, however, it consists of many cells that have been produced inside the megaspore.

As it increases in size, it eventually ruptures its thickened spore wall, and several archegonia are produced in the exposed seams. The development of both the male and the female gametophytes often occurs, or is at least initiated, before the spores are released from their sporangia. Fertilization and development of new sporophytes are similar to those of ground pines.

FIGURE 19.7   Quillwort (*Isoetes*) sporophytes. (Courtesy Robert A. Schlising)

Isoetes—Quillworts

Structure and Form

There are about 60 species of **quillworts** (all in the genus *Isoetes*) (figure 19.7). Most are found in areas where they are at least partially submerged in water for part of the year. Their leaves (microphylls), which are slightly spoon-shaped at the base, are reminiscent of green porcupine quills, although they are not as stiff and rigid as their animal counterparts. They are arranged in a tight spiral on a stubby stem that resembles the corm of a gladiola or a crocus. Ligules similar to those of spike mosses occur toward the leaf bases. The corms have a vascular cambium and may live for many years. Wading birds and muskrats often include the corms in their diet. The plants are generally less than 10 centimeters (4 inches) tall, but one species has leaves that reach a length of 0.6 meter (2 feet).

Reproduction

Reproduction is similar to that of spike mosses, with both types of sporangia being produced at the bases of the leaves (figure 19.8). Up to 1 million microspores may occur in a single microsporangium.

Some of the ancient and extinct relatives of the club mosses were large and treelike, reaching heights of 30 meters (100 feet) and having trunks up to 1 meter (3 feet 3 inches) in diameter. They were dominant members of the forests and swamps of the Carboniferous period, which reached its peak some 325 million years ago (table 19.1). Quillworts first appeared in the Cretaceous period, about 130 million years ago. Their fossils reveal that even the oldest were remarkably similar to their present-day relatives (figure 19.9). A knowledge of the life histories and ancestry of various members of this division enhances our understanding of the development and diversity of the Plant Kingdom as a whole.

FIGURE 19.8   Life cycle of a quillwort (*Isoetes*).

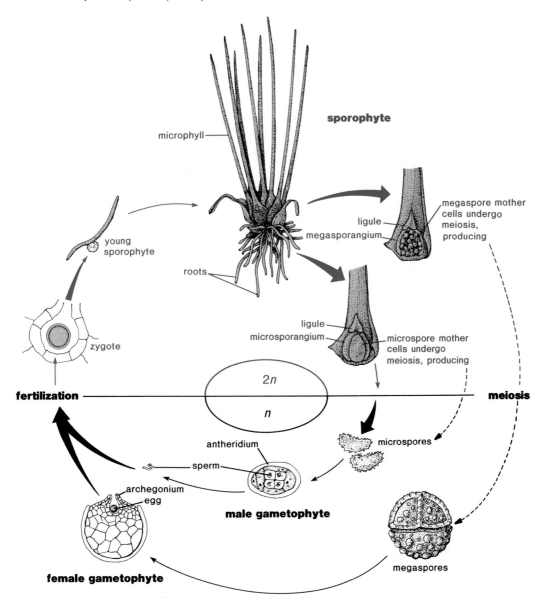

TABLE 19.1
Geologic Time Scale

Era	Period	Age	Earliest Evidence of Plants	Duration (millions of years)	Million Years before Present
Cenozoic	Quaternary	Age		63	
	Tertiary	of			
		Modern			
		Seed			
		Plants			65
Mesozoic	Cretaceous		grasses	65	
	Jurassic	Age	flowering plants	45	130
	Triassic	of		45	175
Paleozoic	Permian	Ancient		50	220
	Carboniferous	Seed	ginkgoes	80	270
		Plants	conifers		
		Age of			
	Devonian	Spore-bearing	seed ferns	50	350
		Plants	lycopods		
			horsetails		
	Silurian	Age	bryophytes	40	400
	Ordovician	of		60	440
	Cambrian	Bacteria		100	500
	Precambrian	and	marine algae	1,700?	600
		Algae			2,300?

FIGURE 19.9 A small portion of the surface of the fossil lycopod, *Lepidodendron,* showing leaf (microphyll) bases similar to those seen in modern lycopods. (Courtesy University of Illinois Paleobotany Laboratory)

HUMAN AND ECOLOGICAL RELEVANCE OF CLUB MOSSES AND QUILLWORTS

Like the whisk ferns, club mosses and quillworts are of little economic importance today. As was mentioned at the beginning of the discussion, large numbers of club moss spores produce a flash of light when ignited. This characteristic was exploited at one time in the manufacture of theatrical explosives and photographic flashlight powders. Druggists also used to mix spore powder with pills and tablets to prevent them from sticking to one another. The spore powder itself has been used for centuries in folk medicine, particularly for the treatment of urinary disorders and stomach upsets. Members of some Native American tribes used it as a talcum powder for babies, snuffed it to arrest nosebleeds, and applied it following childbirth to staunch hemorrhaging.

FIGURE 19.10 Horsetails (*Equisetum*). *A.* An unbranched species. *B.* A branched species.

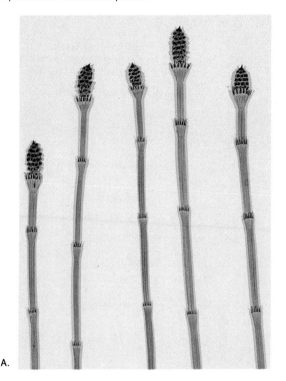

A.

B.

Sometimes it was also used to stop bleeding from wounds. Club moss extracts have been used in several countries in the past to reduce fevers, but medicinal use has now largely been abandoned, partly because of undesirable side effects.

Native Americans of Washington, Oregon, and British Columbia used to experience a mild form of intoxication after chewing parts of one local species of club moss. It is reported that they became unconscious if they chewed too much.

A species of spike moss native to Mexico and to the southwestern United States is sometimes sold in stores as a novelty. Known as the "resurrection plant," it shrivels and rolls up in a ball when dry, appearing to be completely dead, but quickly unfolds and turns green when sprinkled with water. Other spike mosses have been placed on shelves indoors without water for nearly three years and then have resumed growth when given water. Club mosses and spike mosses have been used ornamentally indoors and outdoors as ground covers. Some club mosses are spray painted and used as Christmas ornaments or in floral wreaths. Several species of *Lycopodium* have been exploited to the extent that they are now on rare and endangered or threatened species lists; they should no longer be collected.

Quillwort corms have been eaten by domestic and wild animals, waterfowl, and humans.

DIVISION SPHENOPHYTA— THE HORSETAILS AND SCOURING RUSHES

Many backpackers and campers have become aware of a unique use of a relatively common and widespread genus of plants (*Equisetum*) known as **horsetails** or **scouring rushes.** Significant deposits of silica accumulate on the inner walls of the epidermal cells of the stems, which make excellent scouring material for dirty metal pots and pans. Native Americans were aware of this scouring property of the plants and used them extensively.

Structure and Form

About 25 species of horsetails (the name usually applied to branching forms, which look a little like a horse's tail) and scouring rushes (unbranched forms) (figure 19.10) are scattered throughout all continents including Australia, where they are weeds. They usually grow less than 1.3 meters (4 feet) tall, but some in the tropics and coastal redwood forests of California exceed 4.6 meters (15 feet) in height. Where branches occur, they are normally in whorls at regular intervals along the jointed stems. Both branched and unbranched species have tiny scale-like leaves (microphylls) in whorls at the nodes.

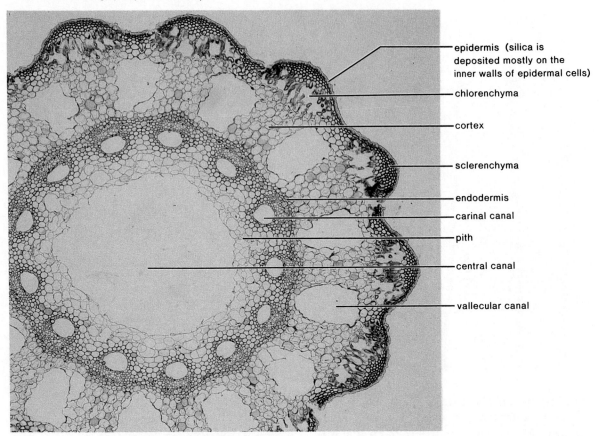

epidermis (silica is deposited mostly on the inner walls of epidermal cells)

chlorenchyma

cortex

sclerenchyma

endodermis

carinal canal

pith

central canal

vallecular canal

These leaves are fused together at their bases, forming a collar. They are green when they first appear, but they soon wither and bleach, and virtually all photosynthesis occurs in the stems. The stems are distinctly ribbed and exhibit well-defined nodes and internodes, with numerous stomata in the grooves between the ribs. When a stem is viewed in cross section (figure 19.11), it can be seen that the pith in the center breaks down at maturity, leaving a hollow central canal. Outside of the pith are two rings of smaller canals. The inner ring consists of canals that are aligned opposite the ribs of the stem. These *carinal canals* function in water conduction. Each has a patch of xylem and phloem to the outside. The canals of the outer ring, called *vallecular canals,* contain air. They are larger than the carinal canals and are aligned opposite the "valleys" between the ribs.

The stems arise from horizontal rhizomes, which, like the aerial stems, are ribbed and have regular nodes and internodes. In some species, the rhizomes may form extensive branching systems as much as 2 meters (6.5 feet) below the surface. They also have adventitious roots. Species of horsetails may be distinguished from one another on the basis of internal stem structure as well as by means of external features.

Reproduction

If stems or rhizomes are broken up by a disturbance such as a storm or foraging animals, the segments are potentially capable of growing into new sporophyte plants. Reproduction in most species usually involves a sexual process, however. In some species,

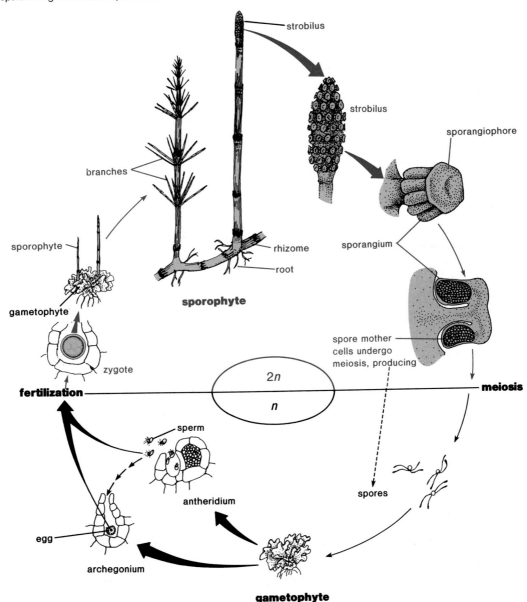

special cream- to buff-colored nonphotosynthetic stems are produced from rhizomes in the spring (figure 19.12). **Strobili** (singular: **strobilus**) in the shape of small cones are produced at the tips of these special stems, or at the tips of regular photosynthetic stems in other species. The strobili are usually about 2 to 4 centimeters (0.75 to 1.5 inches) long, the exterior having dovetailing hexagonal plates that resemble the surface of a honeycomb. Each hexagon marks the top of a sporangiophore, which will have five to ten elongate sporangia attached at the rim.

The sporangia, which surround the stalk of the sporangiophore, point inward toward the center of the strobilus to which the stalks of each sporangiophore are attached. The sporangia are not visible until maturity. At that time, the sporangiophores separate slightly, revealing the sporangia beneath, and the spores are released.

Distinctive-appearing spores are produced when the spore mother cells in the sporangia undergo meiosis. The spores, which are green, have at one

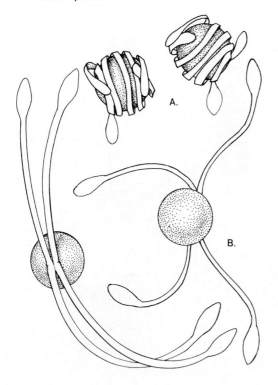

pole four ribbonlike appendages that are slightly expanded at the tips (figure 19.13). The appendages are called **elaters** (not related to the elaters of liverworts). The elaters are very sensitive to changes in humidity, aiding spores in their dispersal. While spores are being carried aloft by a breeze, the elaters are more or less extended like wings. If a humid air pocket is encountered above a damp area, the elaters coil around the spore, causing it to drop in an area where the probability of available moisture needed for germination is increased.

Upon germination, which usually occurs within a week, spores produce pinhead-sized, lobed, cushionlike green gametophytes that seldom grow to more than 8 millimeters (0.36 inch) in diameter. Rhizoids anchor them to the surface. At first, about half of the gametophytes are male with antheridia, and the other half are female with archegonia. After a month or two, however, the female gametophytes become bisexual in most species, producing only antheridia from then on. When water contacts mature antheridia, the sperms produced within appear to be explosively ejected by sudden changes in water pressures. Several eggs on a female or bisexual gametophyte may be fertilized, and the development of more than one sporophyte is common.

Horsetails and scouring rushes belong to the single remaining genus (*Equisetum*) of a division of several different orders that flourished in the Carboniferous period some 300 million years ago. At that time, some members of this division, like the ancient club mosses, were treelike (figure 19.14) and reached heights of over 15 meters (50 feet), whereas others may have been vinelike in appearance. A number were similar to present-day horsetails in having leaves in whorls (figure 19.15), jointed stems with internal canals, and sporangiophores with sporangia hanging down from the rims.

HUMAN AND ECOLOGICAL RELEVANCE OF HORSETAILS AND SCOURING RUSHES

The seventh-century Romans are reported to have eaten the young strobili of field horsetails boiled like asparagus or fried after coating with flour. Members of a number of Native American tribes peeled off the silica-impregnated outer parts of young stems and ate the inner portions either raw or boiled, while others cooked parts of the rhizomes of the giant horsetail for food. Cows, goats, muskrats, bears, and geese also have been known to eat horsetails. Hopi Indians ground dried stems of another species to flour, which they mixed with cornmeal to make bread or mush. Field horsetails have occasionally been reported to be poisonous to horses, however, and they are not recommended for human consumption unless they are cooked in three or four changes of water to eliminate or reduce their acid and drug content.

Native Americans and Asians had several other uses for these plants. One tribe drank the water from the carinal canals of the stem, and another thought the shoots were "good for the blood." It is not known if there was any basis for this latter idea, but there is an old unconfirmed report that field horsetail consumption "produces a decided increase in blood corpuscles." At least one or two species are known to have a mild diuretic effect (a *diuretic* is a substance that increases the flow of urine), and they have been used in the past in folk medicinal treatment of urinary and bladder disorders. Some have also been used as an antacid or an *astringent* (an astringent is a substance that arrests discharges, particularly of blood). One species was used in the treatment of gonorrhea, and others were used for tuberculosis.

At least two Native American tribes burned the stems and used the ashes to alleviate sore mouths or applied the ashes to severe burns. Members of another tribe ate the strobili of a widespread scouring

FIGURE 19.14   Reconstruction of a coal age (Carboniferous) forest. (Photo by Field Museum of Natural History, Chicago)

rush to cure diarrhea, and still others boiled stems in water to make a hairwash for the control of lice, fleas, and mites.

At one time, the use of scouring rush stems for scouring and sharpening was widespread. They were used not only for cleaning pots and pans but also for polishing brass, hardwood furniture, and flooring and for honing mussel shells to a fine edge. Scouring rushes are still in limited use for these purposes today.

Some species of horsetails accumulate certain minerals in addition to silica. Veins of such minerals have been located beneath populations of horsetails by analyzing the plants' mineral contents. This process of analysis involves a chemical treatment of the tissues followed by the use of X-ray equipment.

For the scientific names of species discussed, see appendix 1.

In the geological past, the giant horsetails and club mosses were a significant part of the vegetation growing in vast swampy areas. In some instances, the swamps were stagnant and slowly sinking, permitting the gradual accumulation of plant remains, which, because of the lack of oxygen in the water, were not readily attacked by decay bacteria. Such

FIGURE 19.15   Reconstruction of the fossil giant horsetail, *Calamites*. (Courtesy Field Museum of Natural History, Chicago)

FIGURE 19.16 Common ferns with typically dissected fronds. *A.* Cinnamon fern. The cinnamon-colored fertile fronds in the center are non-photosynthetic and produce large numbers of sporangia. *B.* Maidenhair fern. *C.* Holly fern. *D.* 'Ama'uma'u, a small Hawaiian tree fern. (*A, B, C,* Courtesy Perry J. Reynolds)

A.

B.

C.

D.

circumstances, over aeons of time, were ideal for the formation of coal. Today, if a lump of coal is sectioned thinly enough and examined under the microscope, bits of tissue and spores of plants that were living hundreds of millions of years ago can still be seen. One soft coal known as *cannel* consists primarily of the spores of giant horsetails and club mosses that through the ages were reduced to carbon.

DIVISION PTEROPHYTA— THE FERNS

If one were to take a worldwide poll of opinion concerning ornamental plants, ferns undoubtedly would be at or near the top of the popularity list. Their leaves display such an infinite and aesthetically pleasing variety of form that Thoreau was once moved to state, "God made ferns to show what He could do with leaves."

FIGURE 19.17   Life cycle of a fern.

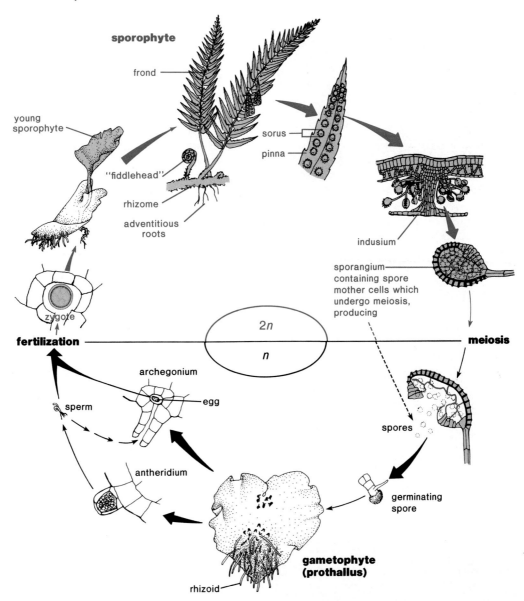

sporophyte

frond

young sporophyte

sorus

pinna

"fiddlehead"

rhizome

adventitious roots

indusium

sporangium containing spore mother cells which undergo meiosis, producing

zygote

2n

n

fertilization

meiosis

archegonium

egg

spores

sperm

germinating spore

antheridium

gametophyte (prothallus)

rhizoid

Structure and Form

The approximately 11,000 known species of ferns vary in size from the tiny floating forms less than 1 centimeter (0.4 inch) in diameter to the giant tropical tree ferns up to 25 meters (82 feet) tall. Although fern leaves, commonly referred to as **fronds,** are typically dissected and feathery in appearance (figure 19.16), some are undivided, pleated, or tonguelike, and others resemble a four-leaf clover or grow in such a way as to form "nests." The latter ferns often accumulate enough humus in the tropics to provide food and shelter for huge earthworms that are up to 0.6 meter (2 feet) long. Since external water is essential to their sexual reproduction, ferns are most abundant in wetter tropical and temperate habitats, but a few are adapted to survival in drier areas.

Reproduction

As in all the plants discussed in this chapter, the sporophyte is the conspicuous phase of the life cycle (figure 19.17). It consists of the fronds, a stem in the form of a rhizome, and adventitious roots that arise along the rhizome. The fronds, regardless of their ultimate form, usually first appear tightly coiled

FIGURE 19.18   Croziers ("fiddleheads") of a tropical tree fern.

at their tips. These **croziers,** or "fiddleheads" (figure 19.18), then unroll and expand, revealing the blades. At maturity, the blades are frequently divided into segments called **pinnae** (singular: **pinna**), which are attached to a midrib, or rachis. A stalk, or petiole, is usually present at the base.

When the fronds have expanded, small, often circular, rust-colored patches of powdery looking material may appear on the lower surfaces of some or all of the blades. Because these patches appear similar to fungal rusts, some fern owners have thought that their plants were diseased and have turned to sprays to deal with the "problem," or have even carefully scraped off the patches. The development of the patches is normal and healthy, however, and examination with a hand lens or dissecting microscope will reveal that they are actually clusters of *sporangia* (figure 19.19). The sporangia may be scattered evenly over the lower surfaces of the fronds, but more often they are confined to the margins; very frequently they are found in numerous discrete clusters called **sori** (singular: **sorus**). In many species of ferns, the sori are protected by individual flaps of tissue called **indusia** (singular: **indusium**) while they are developing. As the sporangia mature, the indusium, which often slightly resembles a semi-transparent umbrella attached by its base to the frond surface, shrivels and exposes the sporangia beneath.

Most of the sporangia, which are microscopic and stalked, look something like tiny transparent baby rattles with a conspicuous row of heavy-walled brownish cells along the edge. This row of cells, somewhat resembling a millipede, is called an **annulus.** It functions in catapulting spores out of the sporangium with a distinct snapping action influenced by moisture changes in the cells (figure 19.20).

Spore mother cells undergo meiosis in the sporangia, usually producing either 48 or 64 spores per sporangium. Sporangia of some of the primitive adder's tongue ferns may have up to 15,000 spores, however, and the number of sporangia is often so great that it has been estimated that a single beech fern plant will produce a total of 50 million spores. In certain aquatic or amphibious ferns, two kinds of spores are produced; single large megaspores are produced in some sporangia, whereas numerous tiny microspores are produced in others. The vast majority of fern species, however, produce only one kind of spore.

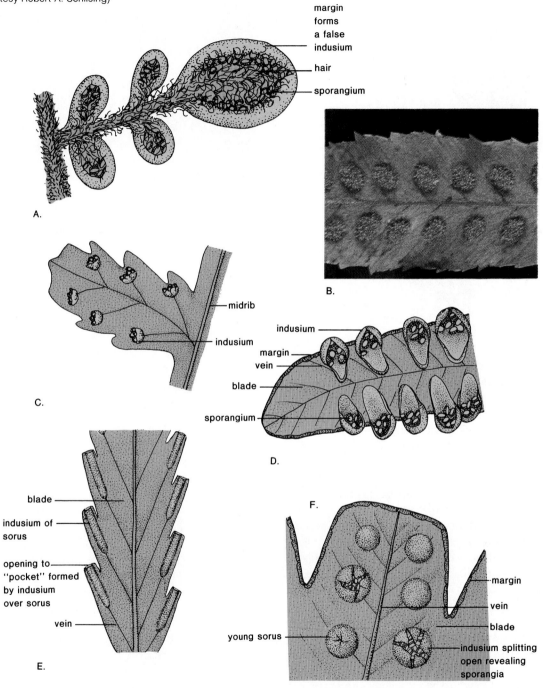

FIGURE 19.19 Pinnae of fern fronds, showing some types of arrangements of sporangia. *A. Cheilanthes. B. Polypodium. C. Cystopteris. D. Cibotium. E. Davallia. F. Cyathea.* (*B.* Courtesy Robert A. Schlising)

margin forms a false indusium

hair

sporangium

A.

B.

midrib

indusium

C.

indusium

margin

vein

blade

sporangium

D.

F.

blade

indusium of sorus

opening to "pocket" formed by indusium over sorus

vein

E.

margin

vein

blade

young sorus

indusium splitting open revealing sporangia

FIGURE 19.20   Release of spores from a fern sporangium. *A.* An intact sporangium. *B.* Spores being ejected as the sporangium splits; the annulus first draws back and then snaps forward.

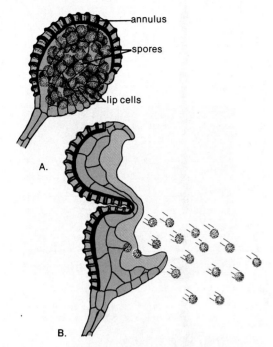

FIGURE 19.21   A fern prothallus.

FIGURE 19.22   Artist's restoration of ancient preferns. *A. Rhynia. B. Psilophyton.*

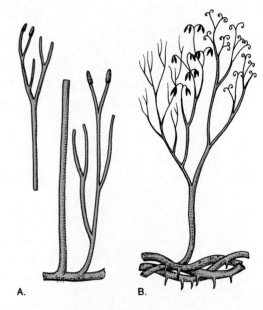

After the spores have been dispersed, initially by being flung out of their sporangia and then carried by wind, relatively few end up in habitats suitable for their survival. Such habitats include shady wet ledges and rock crevices or moist soil (the spores can also easily be induced to germinate on damp clay flowerpots in the home or greenhouse). Those that do come to rest in favorable locations germinate, producing little "Irish valentines," or **prothalli** (singular: **prothallus**), as the green heart-shaped gametophytes of these and other seedless vascular plants are more properly called (figure 19.21). These structures, which often curl slightly at their edges, may be 5 to 6 millimeters (0.25 inch) in diameter and are readily seen without a microscope. They are only one cell thick, except toward the middle where they are slightly thicker. Rhizoids are produced in this central area on the lower surface, with antheridia interspersed among them. Archegonia are also produced, usually closer to the notch of the heart-shaped gametophyte. The archegonia are somewhat flask-shaped, with curving necks that protrude above the surface, whereas the antheridia are more spherical and often elevated above the surface on short stalks. From 32 up to several hundred sperms with several to numerous flagella may be produced in a single antheridium.

Only one zygote develops through an embryo stage into a young sporophyte on any prothallus, regardless of the number of eggs that may be fertilized. This sporophyte usually has smaller, simpler fronds during its first season of growth, but typical full-sized fronds grow from the persisting rhizomes in succeeding years.

Fossil Relatives of Ferns

Fossil remains of ferns and plants thought to be ancestors of ferns abound in ancient deposits. The ancestors of ferns, sometimes referred to as *preferns* (figure 19.22) are found in Devonian formations (see table 19.1) estimated to be 375 million years old.

Most resemble ferns in habit but otherwise are more like whisk ferns than true ferns in that they lack broad fronds. Well-preserved fossils of tree ferns related to present-day large tropical ferns are found in geological strata dating back to between 250 and 320 million years ago (figure 19.23). These became so abundant during the latter part of the Carboniferous period some 300 million years ago that this stage in geological history used to be referred to as the "Age of Ferns." The twentieth-century discovery that a number of these fernlike plants produced seeds on their fronds raised questions, however, as to the true nature of the fossils, and the name "Age of Ferns" was dropped.

HUMAN AND ECOLOGICAL RELEVANCE OF FERNS

In some areas of the United States, it is unusual to find a house without at least a Boston fern growing within. Ferns make ideal houseplants because many of them are adapted to growing in low light, and most are not as susceptible as other plants to aphids, mites, mealybugs, and similar pests. Outdoors, ferns are equally popular as ornamentals. Invariably some eventually become subjects for the landscape artist's brush or the natural history photographer's camera.

A number of ferns also have practical value. Commercial growers of the brilliantly colored anthuriums in Hawaii and elsewhere have found that native tree ferns provide the ideal amount of shade and other environmental conditions needed for bringing their flowers to perfection. Tree fern rhizome or root "bark" and rhizome "bark" of certain other species such as the royal fern (which produces "osmunda bark") has long been a favorite medium of orchid, bromeliad, and staghorn fern growers (figure 19.24). Its texture is eminently suited to the growth of the orchids' aerial roots, and as the "bark" slowly decomposes, rainwater trickling over its surface picks up nutrients that are particularly appropriate for these plants. The demand for fern "bark" for orchids has exceeded the supply for a number of years, and it has become very expensive.

FIGURE 19.23　Fossil fern.

FIGURE 19.24　An orchid plant growing on bark.

A.

B.

As the young "fiddleheads" of many species of ferns unroll, a dense covering of hairs is visible on the petiole and rachis. In the past, the silky hairs of some of the larger tropical tree ferns (figure 19.25) were stripped and used for upholstery, pillow, and mattress stuffing. During the late 1800s over 1,900 metric tons (2,094 tons) of this material was shipped from Hawaii to the mainland, and if it were not for the eventual substitution of alternative materials, these magnificent plants might have been totally destroyed. Some tropical hummingbirds use these hairs along with scales of other ferns to line their nests, and at one time Polynesians used them in a form of embalming for their dead. The trunks of tree ferns have been used in the construction of small houses in the tropics. Parts of one Hawaiian species of small tree fern and the fronds of an Asian fern yield red pigments used for dyeing cloth.

The bracken fern, which is distributed worldwide, has been used and even cultivated for human food for many years, particularly in Japan and New Zealand, even though it has also long been known to be mildly poisonous to livestock. Recent research in both Europe and Japan has shown conclusively that bracken fronds fed to experimental animals produce intestinal tumors, and because of this the consumption of these fronds for food should be actively discouraged.

In the past, indigenous peoples of many areas where ferns occur used the rhizomes and young fronds of several species of ferns for food. Native Americans often baked the rhizomes of sword ferns, lady ferns, and others in stone-lined pits, removed the outer layers, and ate the starchy inner material. Similarly, Native Hawaiians ate the starchy core of their tree ferns as emergency food. In Asia, the oriental water fern is still sometimes grown for food in rice paddies and used as a raw or cooked vegetable. In Malaya, a relative of the lady fern is frequently used as a vegetable.

Uses of ferns in folk medicine abound. They have been used in the treatment of diarrhea, dysentery, rickets, diabetes, fevers, eye diseases, burns, wounds, eczema and other skin problems, leprosy, coughs, stings and insect bites, as a poison antidote, for labor pains, constipation, dandruff, and a host of other maladies. The male fern, which is more common in Europe than the United States, contains a drug that is effective in expelling intestinal worms (e.g., tapeworms). Its use for this purpose dates back to ancient times, and it is still occasionally used, although synthetic medicines have now largely replaced it. The licorice fern, which was used by Native Americans of the Pacific Northwest in the treatment of sore throats and coughs, was also used as a flavoring agent and a sugar substitute. Members of a tribe in California chewed stalks of goldback ferns to quell toothaches and snuffed a liquid made from the fronds of the bird's-foot fern to arrest nosebleeds.

The fronds of bracken and other ferns have been used in the past for thatching houses. Anyone who has placed such fronds in compost piles knows that they break down much more slowly than the leaves of other plants. Bracken fronds are still occasionally used as an overnight bedding base by fishermen and hunters. A substance extracted from these fronds has been used in the preparation of chamois leather, and the rhizome is used in northern Europe in the brewing of ale.

The chain fern, which has large fronds up to 2 meters (6.5 feet) long, has two flexible leathery strands in the petiole and rachis of each frond. Native Americans and others have gathered the fronds for many years to strip these strands for use in basketry and weaving. They do so by gently cracking the long axis with stones to expose the strands, which are then easily removed. The glossy black petioles of the five-finger fern have also been long used in intricate basketry patterns by Native Americans. In Southeast Asia, the climbing fern, whose rachis may attain lengths of up to 12 meters (40 feet), is still a favorite material (when available) for the weaving of baskets.

One of the floating water ferns, which forms tiny plants not much bigger than duckweeds, is spread over wide areas where the climate is relatively mild. It sometimes forms such dense floating mats that it is believed to suffocate mosquito larvae, which periodically need to reach the surface for air. Thus it is known as the mosquito fern in some parts of the United States. This same fern frequently has bluegreen bacteria living symbiotically in cavities between cells. The blue-green bacteria fix nitrogen, and it has been shown experimentally that plants without these organisms do not grow as well as those with them.

The scientific names of ferns discussed are given in appendix 1.

FOSSILS

Several references to **fossils** have been made in this and preceding chapters, and other references occur in the chapters that follow. We have become increasingly aware of worldwide limitations of energy sources, and we now frequently hear references to "fossil fuels." Exactly what is a fossil, and how did it come into existence?

Originally the word *fossil* was applied to anything unusual found in rocks, but its use is now more or less confined to any recognizable prehistoric object of an organic nature or its impression, which has been preserved from past geological ages in the earth's crust (table 19.1). Such a definition includes teeth marks, borings, impressions of footprints and tracks, dung deposits, and deposits of chemicals which are evidence of the activities of algae and bacteria. Age itself does not make a fossil. Human remains thousands of years old that have been found in caves and tombs are not regarded as fossils, but it is conceivable that other remains of the same age could be preserved in rocks and regarded as fossils.

Fossils are formed in a number of different ways. The conditions for their formation almost invariably include the accumulation of sediments in an area where plants, animals, or other organisms are present or to which they have been transported. Such areas are found in swamps, oceans, lakes, or other bodies of water. Hard parts such as wood or bones are more likely to be preserved than soft parts. Other environmental factors such as the presence of salt or antiseptic chemicals or stagnation resulting in low oxygen content of the water, all favor fossil formation. Another factor enhancing the chances of organic material becoming fossilized is quick burial. Such burial might result if an organism was unable to avoid a rain of ash from a volcanic eruption or was trapped in a cave or sudden sandstorm. Other forms of quick burial occur when an organism falls into quicksand, a body of water, or asphalt. Descriptions of several common types of fossils follow.

Molds, Casts, Compressions, and Imprints

After silt or other sediment has buried an object and hardened into rock, the organic material may be slowly washed away by water percolating through the pores of the rock. This leaves a space that may then be filled in with silica deposits. If only a space

FIGURE 19.26   A compression fossil.

is left, it is called a *mold;* if it is filled in with silica, it is called a *cast*. Artificial casts of plaster or wax compounds may be made from molds of the original objects. When objects such as leaves are buried by layers of sediment, the sheer weight of the overlying material may compress them to as little as 5% of their original thickness. When this occurs, all that is left is a thin film of organic material and an outline showing some surface details. Virtually no preservation of cells or other internal structure takes place. Such fossils, which are very common, are called *compressions* (figure 19.26). The image of a compression, like the details of a foot or a hand pressed into wet cement, is called an *imprint*. Coal is a special type of compression, often involving different plant parts that are thought to have been subjected to enormous pressures after the fallen plants slowly accumulated in a swamp.

Petrifactions

Petrifactions (figure 19.27) are uncompressed rocklike materials in which the original cell structure has been preserved. About 20 different mineral substances, including silica and the salts of several metals, are known to bring about petrifaction. At one time it was believed that the process occurred through the replacement of the plant parts by minerals in solution, one molecule at a time. It is now believed, however, that chemicals in solution infiltrate the cells and cell walls, where they crystallize and harden, preserving the original material permanently. Petrifactions can be studied by cutting thin sections with a diamond or carborundum saw and polishing the material with extremely fine grit powder until it is thin enough for light to pass through. Then it can be examined with a compound microscope. Another simpler method of studying petrifactions involves etching the cut surface with a dilute acid and then applying a plastic or similar film. As soon as the film hardens, it can be peeled off. Such peels display lifelike microscopic details of the surface with which they were in contact. They are commonly used by *paleobotanists* (botanists who study fossil plant materials) in their research.

Coprolites

Coprolites are the fossilized dungs of prehistoric animals and humans. They may contain pollen grains and other plant and animal parts that provide clues to the food and feeding habits of past organisms and cultures.

Unaltered Fossils

Some plants or animals fell into bodies of oil or water that, because of substances present or a nearly total lack of oxygen, did not permit decay to occur. Some animals died in snowfields, and their bodies were permanently frozen. In such instances, preservation in the unaltered state occurred. Unaltered fossils, particularly frozen ones, are very rare.

SUMMARY

Four divisions of seedless vascular plants are recognized: Members of Division Psilophyta (whisk ferns) have neither leaves nor roots, but have stems that fork evenly; members of Division Lycophyta (club mosses and quillworts) have mostly photosynthetic microphylls (leaves with one vein whose traces have no leaf gaps); members of Division Sphenophyta (horsetails and scouring rushes) have ribbed stems and scalelike nonphotosynthetic leaves; members of Division Pterophyta (ferns) have megaphylls (leaves with more than one vein and traces associated with leaf gaps).

Whisk ferns are the simplest of all living vascular plants, consisting of evenly forking green stems that have small protuberances called enations, but no leaves; roots are also lacking. The stem contains a central cylinder of xylem and phloem. Spores produced in sporangia formed along the stems germinate into tiny saprophytic gametophytes over the surface of which archegonia and antheridia are scattered. After fertilization the zygote develops a foot and a rhizome. Upright stems are produced when the rhizome separates from the foot. An Australian relative of whisk ferns, *Tmesipteris,* has leaf-like appendages. Fossil plants resembling whisk ferns have been found in Silurian geological formations.

Club mosses presently have living members of two genera. Ground pines (*Lycopodium*) develop sporangia in the axils of sporophylls (sporangium-bearing leaves); spore mother cells in the sporangia undergo meiosis, producing spores that germinate into gametophytes containing both archegonia and antheridia. Several sporophytes may be produced from one gametophyte. Spike mosses (*Selaginella*) have a ligule (tiny tonguelike appendage) on each microphyll; they are also heterosporous, producing microsporophylls with microsporangia in which microspores are formed, and megasporophylls with megasporangia in which fewer, larger megaspores are formed. The microspores develop into male gametophytes with antheridia and the megaspores develop into female gametophytes with archegonia.

Quillworts have quill-like microphylls that arise from a cormlike base. The corms have a cambium that remains active for many years.

Some fossil relatives of club mosses were large dominant members of the forests and swamps of the Carboniferous period.

Club moss spores have been used for flash powder, medicinal purposes, as talcum powder, and to staunch bleeding. The plants themselves have been used as ornamentals, novelty items, Christmas ornaments, and for intoxicating purposes.

Horsetails and scouring rushes (*Equisetum*) accumulate deposits of silica in their epidermal cells and have made good scouring material for Native Americans and others. They occur in both unbranched and branched forms. The stems are jointed and ribbed and have tiny scalelike leaves in whorls at each joint; they are also hollow in the center and contain two rings of canals. Carinal canals (inner ring) conduct water; vallecular canals (outer ring) contain air. The stems arise from rhizomes that branch extensively below the surface of the ground. In some species, nonphotosynthetic stems with conelike strobili are produced in the spring; in other species strobili are produced on photosynthetic stems. The strobili are composed of sporangiophores that support sporangia in which spore mother cells undergo meiosis, producing spores with ribbonlike elaters that are sensitive to changes in humidity. Equal numbers of male and female gametophytes are produced, but female gametophytes may become bisexual after a month or two. The development of more than one sporophyte from a gametophyte is common.

Ancient relatives of horsetails were the size of trees when they flourished in the Carboniferous period of 300 million years ago.

Horsetails have been used for food after the parts containing silica were removed, but they are not recommended for human consumption. They have also been used medicinally as a diuretic, as an astringent, and in the treatment of venereal disease and tuberculosis; other uses include as a hairwash, a mineral indicator, and a metal polish. Cannel coal consists primarily of spores of giant horsetails that were reduced to carbon.

Fern leaves (fronds) are typically dissected and feathery in appearance but vary greatly in form. They usually first appear as croziers (tightly coiled immature fronds) that unroll and expand. Rust-colored patches of sporangia appear on the lower surfaces of fronds. The sporangia may be arranged in various ways but commonly occur in discrete clusters called sori, which may be protected by flaps of tissue called indusia. Each sporangium has a row of heavy-walled cells, the annulus, that functions in catapulting mature spores out of the sporangium after they have developed from spore mother cells following meiosis. Heart-shaped gametophytes called prothalli develop after spores germinate; the prothalli contain both archegonia and antheridia; only one zygote develops into a sporophyte. Preferns, the ancestors of ferns, are found in Devonian deposits estimated to be 375 million years old, and ferns (especially tree ferns) became so abundant during the latter part of the Carboniferous period that this era was once referred to as the "Age of Ferns," but the discovery of seeds on some of them raised questions about relationships and the term was dropped.

Ferns are used as ornamentals, a source of "bark" for growing orchids and other plants, a source of stuffing materials for bedding, in tropical construction, as food, and in numerous folk medicinal applications. Other uses include as a source of material for basketry and weaving, as an ingredient in brewing ale, and as an ingredient in the preparation of chamois leather. One floating fern, which forms dense mats, is believed to suffocate mosquito larvae.

Fossils, which are recognizable prehistoric objects of an organic nature, are formed in different ways. Molds, casts, compressions, and imprints are formed when material buried by silt or other sediment has hardened into rock and the organic material has slowly been washed away by water. Petrifactions are uncompressed rocklike materials in which the original cell structure has been preserved. Coprolites are fossilized dungs that may contain pollen grains and other plant and animal parts. Unaltered fossils are those of plants or animals that may have fallen into bodies of oil or water or snowfields and were not subjected to decay.

REVIEW QUESTIONS

1. What basic features of the ferns and their relatives distinguish them from any organisms thus far studied?
2. How does the gametophyte of a whisk fern differ from that of a true fern?
3. Which of the fern relatives have significantly functional leaves? In those without conspicuous leaves, how are the carbohydrate needs of the plants met?
4. How does one distinguish among ground pines, spike mosses, and quillworts?
5. How did the ancient ground pines differ from those of the present day?
6. How do the spores and the female gametophyes of horsetails differ from those of any other plants studied thus far?
7. What is the location and function of carinal canals?
8. In your opinion, which have the most human relevance today: club mosses and horsetails of the present or those of the geological past? Why?
9. Define crozier, rachis, pinna, indusium, prothallus, and prefern.
10. Diagram the life cycle of a typical fern.
11. Summarize present and past human use of ferns.
12. How are fossils formed, and what different types are recognized?

DISCUSSION QUESTIONS

1. Would you assume that there is any significance to the fact that both the sporophytes and the gametophytes of whisk ferns branch in the same manner?
2. Do spike mosses, which produce two kinds of spores and gametophytes, have any advantage over ground pines, which produce only one kind of spore and gametophyte?
3. Some gametophytes of fern relatives develop underground, whereas others develop at the surface. If you were to be a gametophyte, which would you prefer? Why?
4. After looking at the internal structure of a horsetail stem, can you suggest a function for the silica in the ribs?
5. How would we be affected if all ferns were to become extinct in a few years?

ADDITIONAL READING

Bir, S. S. 1980. *Pteridophyta: some aspects of their structure and morphology.* Houston, TX: Scholarly Publications.

Cobb, B. 1977. *A field guide to the ferns and their related families.* Boston: Houghton Mifflin Co.

Dyer, A. E. 1979. *The experimental biology of ferns.* New York: Academic Press.

Gensel, P. C., and H. N. Andrews. 1984. *Plant life in the Devonian.* New York: Praeger Publishers.

Grillos, S. J. 1966. *Ferns and fern allies of California.* Berkeley: University of California Press.

Jermy, A. C., J. A. Crabbe, and B. A. Thomas, eds. 1984. *The phylogeny and classification of the ferns.* Reprint of 1973 ed. Forestburgh, N.Y.: Lubrecht and Cramer.

Lloyd, R. M. 1964. "Ethnobotanical uses of California pteridophytes by western American Indians." *American Fern Journal* 54: 76–82.

Mickel, J. 1979. *How to know the ferns and fern allies.* Pictured Key Nature Series. Dubuque, IA: Wm. C. Brown Publishers.

Puri, H. S. 1970. "Indian pteridophytes used in folk remedies." *American Fern Journal* 60: 137–43.

Sporne, K. R. 1975. *The morphology of pteridophytes. The structure of ferns and allied plants,* 4th ed. Atlantic Highlands, NJ: Humanities Press, Inc.

Stewart, W. N. 1983. *Paleobotany and the evolution of plants.* New York: Cambridge University Press.

Turner, N. J. 1975. *Food plants of British Columbia Indians. Part 1: Coastal peoples.* Victoria: British Columbia Provincial Museum.

Overview

The chapter begins by exploring the differences between ferns and seed plants, and then briefly relates the geological history of gymnosperms. The leaves, roots, and stems of pine trees are discussed. Pines are also used to portray the life cycle of a typical gymnosperm. Additional conifers such as yews, podocarps, junipers, and redwoods are given brief mention, and a short discussion of other gymnosperms such as cycads, ginkgoes, *Ephedra, Gnetum,* and *Welwitschia* follows. The chapter concludes with a digest of the human and ecological relevance of gymnosperms, with particular emphasis on the conifers.

20 *Introduction to Seed Plants: Gymnosperms*

Some Learning Goals

1. Know what features typical male and female conifer strobili have in common, and how they differ.

2. Understand what distinguishes the four divisions of living gymnosperms from one another.

3. Learn the modifications of pine leaves that adapt them to a harsh environment.

4. Indicate where the following structures occur in the life cycle of a pine tree: archegonia, eggs, sperms with flagella, male and female gametophytes, the sporophyte, integument, vessels, spore mother cells, embryo, and pollen grains.

5. Explain the function of each of the following: resin canals, albuminous cells, mycorrhizal fungi, nucellus, generative cell, and megaspore.

6. Learn a use for each of at least ten different conifers.

7. Identify three major representatives of the Division Gnetophyta.

Outline

INTRODUCTION

Some of the giant tree ferns, which have large graceful leaves held above an unbranching trunk (see figure 19.25), superficially resemble coconut palms. Anyone who has sat at the base of a coconut palm when one of its massive fruits has fallen, however, has been abruptly reminded of a fundamental difference between ferns and coconut palms. The palms are flowering plants which, like cone-bearing plants such as pines, reproduce primarily by means of **seeds,** whereas the ferns produce no seeds at all.

The oldest known seeds were produced by plants that appeared late in the Devonian period, more than 350 million years ago. Seeds provided a significant adaptation for plants that had invaded the land. Unlike spores, seeds have a protective seedcoat and a supply of food (usually endosperm) for the embryo, which may be capable of lying dormant through lengthy periods of freezing weather, drought, and in some instances may even survive fire. This survival value of seeds undoubtedly played a major role in seed plants becoming the dominant vegetation on Earth today.

The first seed plants were so fernlike in appearance that, as indicated in chapter 19, they were originally classified as ferns. When fossils with obvious seeds on the fronds were discovered, however, these *pteridosperms* ("seed ferns") were reclassified as *gymnosperms*.

The name **gymnosperm** is derived from two Greek words: *gymnos,* meaning naked, and *sperma,* a seed. The name refers to the exposed nature of the seeds, which are produced on the surface of sporophylls or similar structures instead of being enclosed within a fruit as they generally are in the flowering plants (figure 20.1). The seed-bearing sporophylls are often spirally arranged in female strobili (cones), which develop on the sporophyte along with smaller male strobili that produce *pollen grains.* The female gametophyte is produced within an **ovule** containing a fleshy, nutritive diploid **nucellus** that is itself enclosed within one or more outer layers of diploid tissue. These outer layers of tissue constitute an **integument,** which becomes a **seedcoat** after the fertilization and development of an embryo has occurred (see figure 20.6).

FIGURE 20.1 Comparison between exposed gymnosperm seeds and enclosed angiosperm seeds. *A.* Exposed seeds on a female (woody) cone. *B.* A single female cone scale with two seeds. *C.* A section through an apple, showing the enclosed seeds.

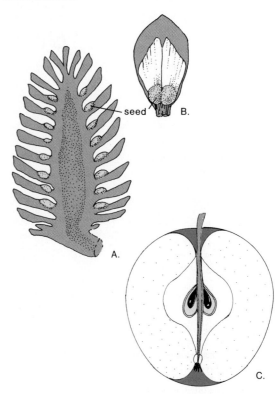

Gymnosperm sporophytes are primarily trees and shrubs, although a few are vinelike in habit. The gametophytes are even more reduced proportionally than they are in ferns and their relatives. Unlike the gametophytes of members of the Plant Kingdom discussed to this point, they are not free-living but develop within structures that, as indicated, are enclosed by parts of the sporophyte.

Four divisions of gymnosperms have living representatives at this time. The **conifers** (Division Coniferophyta) constitute the largest and most significant group by far, totaling some 575 species. Pines, firs, spruces, hemlocks, redwoods, cedars, and others belong to this division. Fossils of some conifers extend back 290 million years to the late Carboniferous era. Other gymnosperms include the *cycads* (Division Cycadophyta), which have palmlike leaves but produce their seeds in cones; *ginkgoes* (Division Ginkgophyta), which have smaller, fan-shaped leaves and seeds enclosed in a fleshy covering; and *gnetophytes* (Division Gnetophyta), whose wood contains vessels—a structural element unknown in other gymnosperms.

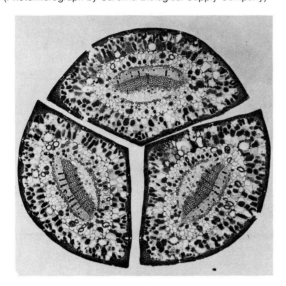

Division Coniferophyta—The Conifers

Pines

The largest genus of conifers, *Pinus* (pines) has over 100 living species. They are the predominant trees in the vast coniferous forests of the northern hemisphere. They have also been planted extensively in the southern hemisphere, but only the Merkus pine occurs there naturally, and its distribution barely extends south of the equator. They include the world's oldest known living organisms, the bristlecone pines, which occur in the White Mountains of eastern central California and in the Snake range on the central Nevada-Utah border. Some trees still standing are about 4,600 years old, and one that was, unfortunately, cut down in 1964 was found to have been about 4,900 years old.

Structure and Form

Pine leaves, which are needlelike, are arranged in clusters or bundles of two to five leaves each (a handful of species have as many as eight or as few as one to a cluster). Regardless of the number of leaves, each cluster (*fascicle*) forms a cylindrical rod if the leaves are held together (figure 20.2). Pines often occur in areas where the topsoil is frozen for a part of the year, making it difficult for the roots to obtain water. In addition, the leaves may be subjected to high winds and bitterly cold temperatures.

Accordingly, they have several modifications that enable them to withstand harsh environments. A **hypodermis,** consisting of one to several layers of thick-walled cells, occurs just below the epidermis. The epidermis itself is coated with a thick cuticle. The stomata, instead of being at the surface, are recessed or sunken in small cavities. The veins and their associated tissues are surrounded by an endodermis, and the mesophyll cells do not have the obvious air spaces typical of the spongy mesophyll of the leaves of flowering plants (see figures 7.4 and 7.8). Conspicuous **resin canals** occur in the mesophyll. These resin canals, which are found throughout other parts of the plant as well, consist of tubes lined with special cells that secrete *resin.* This substance, which is aromatic and antiseptic, prevents the development of fungi and deters certain insects from attacking the tree. In other conifers, resin canals are apparently produced primarily in response to injury. Pine fascicles usually *absciss* (i.e., fall off—see chapter 7) within two to five years of their maturing, but those of bristlecone pines persist for up to 30 years. The fascicles are lost a few at a time so that some functional leaves are always present on a healthy tree.

Unlike the xylem of woody dicots (see chapter 5), that of pines and many other gymnosperms consists primarily of tracheids and contains no vessel elements or fibers. Because of the absence of fibers, pine wood is said to be *soft* while that of broadleaf trees is described as *hard.* The degree of hardness varies considerably from species to species, however. In most conifers, the xylem is proportionately very wide, with distinct annual rings (figure 20.3) as a result of a relatively rapid rate of growth during the growing season. Resin canals are formed both vertically and horizontally throughout various tissues. The bark, which includes the secondary phloem, often becomes relatively thick. It frequently attains a width of 7.5 centimeters (3 inches) or more, and in extreme cases, such as the giant redwood (which is in a different family of conifers), it may even reach 6 decimeters (2 feet) in extent. The phloem does not have companion cells, but certain other cells called **albuminous cells** apparently perform the same function.

FIGURE 20.3 A portion of a cross section through the stem of a pine showing distinct annual rings.

cork

resin canal

phloem

vascular cambium

xylem

pith

FIGURE 20.4   A cluster of male pine cones.

Roots of pines are invariably associated with *mycorrhizal fungi* (see chapter 6). In fact, pine seedlings that germinate in sterilized soil do not grow well at all until the fungi are introduced or permitted to develop. The roots of pines adjacent to one another often interweave. New England pioneers made use of this characteristic in eastern white pines when clearing land. After trees were felled they would tip over the stumps with the roots still attached and use them for fences. The fences survived for many years.

Reproduction

As in the spike mosses, quillworts, and a few of the ferns, two kinds of spores are produced in pines. *Microspores* occur in *microsporangia* that develop in pairs toward the bases of papery or membranous scales arranged in a spiral or in whorls around an axis, forming a strobilus or male cone. Male cones (also called "pollen cones"), which are usually produced in the spring in clusters of up to 50 or more toward the tips of the lower branches (figure 20.4), are commonly not more than 1 to 4 centimeters (about 0.4 to 1.5 inches) long. They become shriveled and spent within a few weeks and then fall from the trees. *Microspore mother cells* in the microsporangia each undergo meiosis, producing four haploid *microspores*. These then develop into **pollen grains,** each consisting of four cells and a pair of external air sacs that look something like tiny water wings (figure 20.5). These sacs give the pollen grains added buoyancy and contribute to the fact that the grains frequently are carried great distances by the wind. Pines produce pollen grains in astronomical numbers. It has been estimated, for example, that each of the 50 or more pollen cones commonly found in a single cluster may produce in excess of 1 million grains, and there may be hundreds of such clusters on one tree. The grains accumulate as a fine yellow dust on cars, shrubbery, or anything else in the vicinity, and they often form an obvious scum on pools and puddles.

Megaspores are produced in *ovules* at the bases of the female cone scales. The female cones (also called "seed cones"), or strobili, are much larger than the male cones, attaining lengths of 6 decimeters (2 feet) in sugar pines, and weighing as much as 2.3 kilograms (5 pounds) in Coulter pines. They have woody scales, with inconspicuous bracts between them, arranged in a spiral around an axis. They are produced on the upper branches of the same tree on which the male cones appear. The ovules (figure 20.6), which occur in pairs toward the base of each scale, are larger and more complex than their microsporangial counterparts in the male cones. Each contains a multicellular nutritive tissue called the **nucellus,** which is surrounded and enclosed by a thick, layered **integument.** The integument has a channel or pore called a **micropyle** that opens toward the center of the cone. One of the integument layers later becomes the *seedcoat* of the seed. A single *megaspore mother cell* within the nucellus of each ovule undergoes meiosis, producing a row of relatively large *megaspores.* All except one of the megaspores soon degenerate. The remaining one slowly develops over a period of months into a *female gametophyte,* which ultimately may consist of several thousand cells. Toward the end of the gametophyte development, two to six *archegonia* become differentiated at the end facing the micropyle. Each archegonium contains a single large *egg.* Often when one examines stained thin sections of a pine ovule under the microscope, only one archegonium or egg may be seen, or the micropyle may appear to be missing. This frequently occurs because the knife that cut the section may not have sliced precisely through the middle of all the structures.

Female cones, which are green at first, commonly take two seasons to mature into the brownish woody structures that are familiar to nearly everyone. During the first spring the cone scales

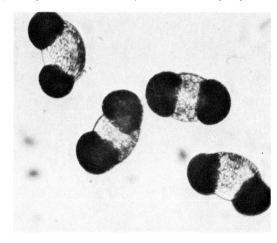

FIGURE 20.5   Pollen grains of a pine, as seen with the aid of a microscope. The pairs of dark objects attached to each pollen grain are air sacs that provide added buoyancy.

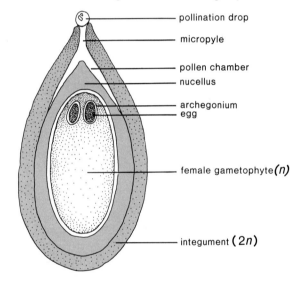

FIGURE 20.6   A longitudinal section through a pine ovule.

- pollination drop
- micropyle
- pollen chamber
- nucellus
- archegonium
- egg
- female gametophyte *(n)*
- integument *(2n)*

spread apart, and pollen grains carried by the wind sift down between the scales. There they catch in sticky drops of fluid (*pollen drops*) oozing out of the micropyles. As the fluid evaporates, the pollen is drawn down through the micropyle to the top of the nucellus. After pollination the scales grow together and close, protecting the developing ovule. It is not until about a month after pollination that meiosis occurs and the megaspores are formed. The female gametophyte is not mature with archegonia for more than a year after that. Meanwhile the pollen grain forms an outgrowth called a **pollen tube,** which slowly digests its way through the nucellus to the area where the archegonia develop. While the pollen tube is

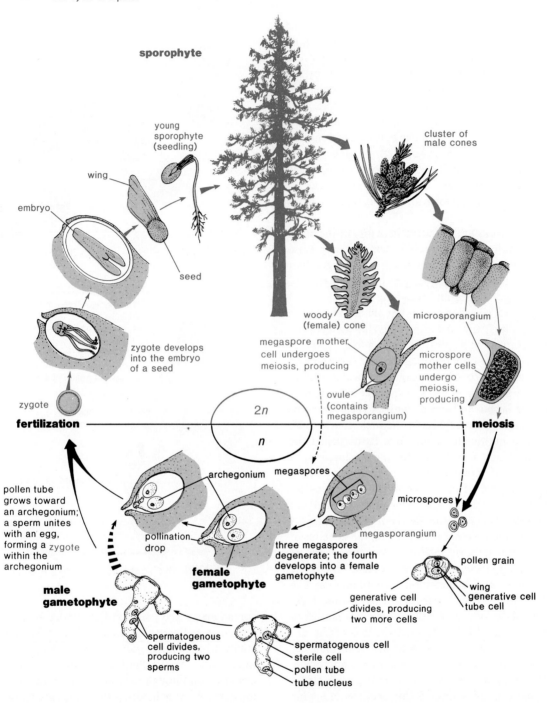

FIGURE 20.7   Life cycle of a pine.

sporophyte

young
sporophyte
(seedling)

wing

embryo

seed

zygote develops
into the embryo
of a seed

zygote

fertilization

cluster of
male cones

woody
(female) cone

megaspore mother
cell undergoes
meiosis, producing

ovule
(contains
megasporangium)

microsporangium

microspore
mother cells
undergo
meiosis,
producing

2n

n

meiosis

archegonium

megaspores

megasporangium

microspores

pollen tube
grows toward
an archegonium;
a sperm unites
with an egg,
forming a zygote
within the
archegonium

pollination
drop

**female
gametophyte**

three megaspores
degenerate; the fourth
develops into a female
gametophyte

generative cell
divides, producing
two more cells

pollen grain

wing
generative cell
tube cell

**male
gametophyte**

spermatogenous
cell divides,
producing two
sperms

spermatogenous cell
sterile cell
pollen tube
tube nucleus

growing, two of the original four cells in the pollen grain enter it. One of these, called the **generative cell,** divides and forms two more cells, called the *sterile cell* and the *spermatogenous cell.* The latter divides again, producing two male gametes, or *sperms.*

These sperms, unlike those encountered up to this point in the Plant Kingdom (and also in some other gymnosperms), have no flagella. This germinated pollen grain, with its pollen tube and two sperms, constitutes the *male gametophyte.* Notice that no antheridium has been formed (figure 20.7).

About 15 months after pollination the pollen tube reaches the archegonium. The contents of the pollen tube are discharged, and one sperm unites with the egg, forming a zygote. The other sperm and remaining cells of the pollen grain degenerate. The sperms of other pollen grains present may unite with the eggs of other archegonia, and each zygote begins to develop into an *embryo.* This may be likened to the development of fraternal twins or triplets in animals. At a later stage, an embryo may divide in such a way as to produce the equivalent of identical twins or quadruplets in animals. Normally, however, only one embryo completes development. While this development is occurring, one of the layers of the integument hardens, becoming a *seedcoat.* A thin membranous layer of the cone scale becomes a "wing" on each seed. This wing aids in the seed's dispersal, which may also be helped along by squirrels and other animals breaking open the cones (see figure 20.7 for a diagram of the life cycle of a pine).

In other species, such as the lodgepole, jack, and knobcone pines, the cones remain on the tree with the scales closed until they are seared by fire. Sometimes these cones are slowly engulfed by tissue and buried as the cambium increases the girth of the branch. Seeds of such engulfed cones have been reported to germinate after they have been dug out of the stem.

Other Conifers

Some conifers do not produce woody female cones with conspicuous scales, nor do all produce both male and female cones on the same tree. In the yews and California nutmeg, for example, ovules are produced singly at the tips of short axillary shoots. Each ovule is at least partially surrounded by a fleshy, cuplike covering called an **aril.** In yews, this is bright red and open at one end, giving the fleshy seed the appearance of a small red hors d'oeuvre olive with its stuffing removed. The fleshy seeds are produced only on female trees, while the pollen-bearing strobili, or male cones, are produced only on male trees. Podocarps, which are conifers of the southern hemisphere, are widely planted as ornamentals in regions with milder climates. Their fleshy-coated seeds, which are produced singly, are similar to those of yews, but they are not open at one end and have an additional larger appendage at the base (figure 20.8). The origin of these fleshy seeds is not clear, and has led to speculation that members of the yew and podocarp families may have diverged from other conifers very early in the evolution of gymnosperms.

The scales of the female cones of junipers tend to be fleshy at maturity, so that they look more like berries than cones. Juniper pollen, as well as that of a number of other conifers, does not have air sacs. Redwood female cone scales are flattened at the tips and narrow at the base, where they do not overlap one another as do pinecone scales.

The two species of living redwoods found in California are both renowned for their size, height, and longevity. It is not uncommon for coastal redwoods to attain a height of 90 meters (295 feet), and one tree in Humboldt County, California, with a height of 111.6 meters (366.2 feet) is believed to be the tallest tree in the world. The other species, usually referred to as "Big Trees" or Giant Redwoods and confined to the eastern slopes of California's Sierra Nevada range, does not become quite as tall as the coastal redwood but exceeds it in total size. The "General Sherman" tree in Sequoia National Park, for example, is 31 meters (101.5 feet) in circumference at the base and over 24 meters (79 feet) in circumference 1.5 meters (5 feet) above the ground; it weighs an estimated 5,594 metric tons (6,167 tons). The total timber in this single tree exceeds 600,000 board feet, sufficient to build more than 75 five-room houses (although the wood is generally not suitable for construction purposes) or enough to make 20 billion toothpicks. The tree is over 3,500 years old.

B.

Other Gymnosperms

Members of the other three divisions of gymnosperms with living representatives are not as numerous or as well known as the conifers and outwardly do not resemble them at all. In fact, some of them look more like leftover props from a science-fiction movie set. They include the *cycads,* the *ginkgoes,* and the *gnetophytes* (pronounced nee-toe-fytes).

Division Cycadophyta—The Cycads

Cycads are slow-growing plants of the tropics and subtropics that look like a cross between a tree fern and a palm. They have unbranched trunks that grow to more than 15 meters (50 feet) tall in a few species, with a crown of large pinnately divided leaves. Extinct members of this division, known as *cycadeoids,* were abundant during the Mesozoic era. Several of the approximately 100 known living species are presently facing extinction. Their life cycles are similar to those of pine trees; their sperms, however, have from 10,000 to 20,000 spirally arranged flagella. The male and female strobili, which are sometimes massive, are produced on separate plants (figure 20.9). The female strobili of some species are covered with feltlike or woolly hairs and can become as much as 1 meter (3 feet, 3 inches) long.

Division Ginkgophyta—The Ginkgoes

There is only one living species of *Ginkgo* (figure 20.10), whose name is derived from Chinese words meaning "silver apricot." The fossil record indicates *Ginkgo* and other members of its family (Ginkgoaceae) were once widely distributed, especially in the northern hemisphere. Despite isolated reports to the contrary, there are doubts that ginkgoes now exist anywhere they have not been cultivated, and the plant has often been called a living fossil.

Ginkgoes are frequently referred to as maidenhair trees because of the similarity of the notched, broad, fan-shaped leaves to the individual pinnae of maidenhair ferns. They are widely cultivated in the United States and are popular street trees in some areas. The leaves, which are produced in a spiral on short, slow-growing spurs, have no midrib or prominent veins. Instead, hairlike veins branch dichotomously (fork evenly) and are relatively uniform in their width. They are deciduous and turn a bright golden yellow before abscission in the fall. The life

FIGURE 20.10 *Ginkgo. A.* A mature tree. *B.* Seeds and leaves.

A.

B.

FIGURE 20.11 Joint fir (*Ephedra*). *A.* A single plant. *B.* A branch bearing male strobili. *C.* A branch bearing female strobili.

A.

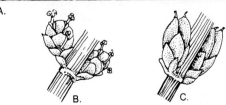

B.　　　C.

cycle of ginkgoes is also similar to that of pines. The male and female reproductive structures are produced on separate trees, and the sperms have flagella. Mature seeds are enclosed in a fleshy covering and resemble a small plum. These seeds have a rank odor and are irritating to the skin of some individuals, causing male trees (propagated from cuttings) to be preferred for ornamental purposes. Despite the odor, however, the seeds are edible and are sold in the markets of China and Hong Kong.

Division Gnetophyta—The Gnetophytes

A heterogeneous group of some 70 species of plants distributed among three genera comprise the **gnetophytes.** They are unique among the gymnosperms in having vessels in the xylem. The genus *Ephedra,* whose members are sometimes called *joint firs* (figure 20.11), accounts for more than half of the species. These shrubby plants, which inhabit drier regions, produce tiny leaves in twos or threes at a node. The leaves turn brown soon after they appear.

FIGURE 20.12 A species of *Gnetum* that climbs.

The slightly ribbed branches, which are green when they are young and carry on most of the photosynthesis, are often whorled. A tubular extension, like the neck of a tiny bottle, extends into the air from the integument of an ovule at the time of pollination. Sticky fluid oozes out of this extension, which constitutes the micropyle, and airborne pollen catches in it. Male and female strobili may be produced on the same plant or on different ones, depending on the species.

Most of the remaining species in this division are in the genus *Gnetum* (figure 20.12), which has not been given an English common name. Its members, which have broad leaves similar to those of flowering plants, occur in the tropics of South America, Africa, and Southeast Asia. Most are vinelike, but the best known species is a tree that grows up to 10 meters (33 feet) tall.

The third genus, *Welwitschia* (figure 20.13), has only one species, which is confined to the temperate Namib and Mossamedes deserts of southwestern Africa. Here the average annual rainfall is only 2.5 centimeters (1 inch), and in some years it does not rain at all. The plants apparently survive much of the time on fog and dew water absorbed through the surfaces of the leaves. *Welwitschias* are also truly extraordinary plants in appearance. The stem, which rises only a short distance above the surface, is in the form of a large shallow cup that tapers at the base into a long taproot. At maturity the plants, which may live to be 100 years old, have a crusty, barklike covering on the surface of the stem cup. The stems may exceed 1 meter (3 feet, 3 inches) in diameter. Throughout the life of the plant only two leaves are produced. These are wide and straplike, each with a meristem at the base. The meristems constantly add to the length of the leaves, but as the leaves flap about in the wind they become tattered and split, wearing off at the tips so that they seldom exceed 2 meters (6.5 feet) in actual length. Male and female strobili, which are produced on separate plants, are on axes that emerge from the axils of the leaves so that they appear to be growing around the rim of the stem cup.

HUMAN AND ECOLOGICAL RELEVANCE OF GYMNOSPERMS

What do we plant when we plant the tree?
We plant the ship, which will cross the sea.
We plant the mast to carry the sails;
We plant the planks to withstand the gales—
The keel, the keelson, the beam, the knee;
We plant the ship when we plant the tree.

What do we plant when we plant the tree?
We plant the houses for you and me.
We plant the rafters, the shingles, the floors,
We plant the studding, the lath, the doors,
The beams, the siding, all parts that be;
We plant the house when we plant the tree.

What do we plant when we plant the tree?
A thousand things that we daily see;
We plant the spire that out-towers the crag,
We plant the staff for our country's flag,
We plant the shade, from the hot sun free;
We plant all these when we plant the tree.

Henry Abbey

As a group, the gymnosperms are second only to the flowering plants in their impact on our daily lives. Space does not permit a detailed account of all they contribute, but the following sections provide an overview of some of their uses, past and present.

Conifers

In the early 1970s the late author-naturalist Euell Gibbons filmed a series of television commercials in which he mentioned uses of several wild plants for food. In one of the commercials, he rhetorically inquired if his audience had ever eaten a pine tree and added, "many parts are edible." The edibility of parts of many conifers was known to Native Americans long before Europeans set foot on the North American continent. In fact, early explorers found large numbers of pines stripped of their bark; for centuries these inner parts (phloem, cambium) had been used for emergency food. The Adirondack Mountains of New York are believed to have received their name from a Mohawk Indian word meaning "tree-eater," in reference to Native American use of the inner bark of eastern white pines. This material (specifically the phloem) contains sugars that make it sweet to the taste. Some tribes ate the material raw, some dried it and ground it to flour, and others boiled it or stored dried strips for winter food. Early settlers in New England candied strips of eastern white pine inner bark, and to prevent scurvy both they and local Native Americans drank a tea made of the needles, which are rich in vitamin C.

The seeds of nearly all pines are edible, but those of western North America include the larger and better-tasting species. The protein content of those analyzed generally ranges between 15% and 30%, with much of the remainder consisting of oils. California Indians relished digger pine seeds in particular, but even the small seeds of ponderosa pine were eaten raw or made into a meal for soups and bread. Cones of pinyons were collected by tribes of the Southwest and thrown on a fire to loosen the seeds.

These were then pounded and made into cakes or soup. The latter was often fed to infants. In Siberia, peasants obtain a nutritious oil by crushing the seeds of the Siberian white pine, but its use has declined since corn and cottonseed oils became available. In Italy and other parts of Europe, "pignolias," the seeds of the stone pine, are cooked in stews and soups. They are also used in cakes and cookies, and some are exported to the United States for this purpose. Many of the so-called nuts used by commercial American bakers in cakes and confectionery, however, are really seeds from the east Himalayan chilghoza pine. Other sources include the Mexican stone pine and a few pinyons.

Eastern white pines were favored for use as masts in sailing vessels. In colonial days, the royal surveyors marked certain trees for the use of the Crown, and severe penalties were imposed on colonists who ignored the ban on the use of any white pine not growing on private land. It was, however, legal for colonists to use white pines that had blown down, which gave rise to the term *windfall*. Eastern white pine wood contains less resin than that of other species and was extensively used for crates, boxes, matchsticks, furniture, flooring, and paneling. By the end of the nineteenth century, eastern white pines, which originally occurred over vast tracts of the northeastern United States and Canada, and bald cypress stands in the southeastern United States, had been decimated by wholesale logging done with no thought to conservation. White pine blister rust also took its toll. Although new growth is now being encouraged, most white pine lumber used today originates in large stands of western white pine in the Pacific Northwest. In California, smog has severely damaged ponderosa and other native pines. For a number of years the U.S. Forest Service has experimented with Afghanistan pine, a smog- and drought-resistant pine from Russia and adjacent areas, as a replacement for native trees. Growth rates in tests have been very rapid. Rapid growth is a desirable commercial feature, since considerably more timber can be produced in the same time needed to obtain it from slower-growing species, but the wisdom of introducing non-native plants into natural communities is in question, since there are many examples of such activities thoroughly disrupting delicate ecological balances.

FIGURE 20.14 A southern long leaf pine from which resin is being extracted. Part of the bark was cut away and a receptacle to collect the resin for commercial use was placed at the bottom of the cut area. (Courtesy FRP Company)

The resin produced in the resin canals of conifers is a combination of a liquid solvent called *turpentine* and a waxy substance called *rosin.* When a conifer tree is wounded or damaged by insects, resin usually covers the area, sometimes trapping the insects. Out in the air, the turpentine evaporates quickly, leaving a protective layer of rosin, which prevents water loss and fungal attacks. Both turpentine and rosin are very useful products, and a large industry centered in the southern United States (figure 20.14) and in the south of France is devoted to their extraction and refinement. They are often referred to as naval stores, a term that originated when the British Royal Navy used large quantities for caulking and sealing their sailing ships. Most *naval stores,* and a third or more of the lumber used in the United States, today come from a group of southern yellow pines, particularly slash pine. Pitch pine, which was also a source of naval stores before slash and other yellow pines proved more profitable, was used in the past for the construction of mill water wheels. The wood of this pine was also used as fuel for steam engines, since it produces considerable heat when it burns.

Turpentine is considered a premier paint and varnish solvent, whereas rosin is used by musicians on violin bows. Baseball pitchers use rosin to improve their grip on the ball, and batters apply pine tar to the handles of bats to minimize slippage. In the past, pine pitch was used by Native Americans for patching canoes, and it has been suggested that Noah's Ark was sealed with pitch from aleppo pine. Pine resin was used for purifying wine in the first century A.D., and today Greeks still add it to certain wines. *Colofonia* is the Spanish word for a type of resin that was produced in abundance by the numerous Monterey pines around the old Spanish capital of Monterey. The early California priest, Padre Arroyo, suggested during the first half of the nineteenth century that California received its name from this Spanish word. California was, however, the name of a mythical paradise in a Spanish novel published in the early sixteenth century, and no more than an interesting similarity between two Spanish words may be involved.

The huge kauri pines of New Zealand, which are generically different from true pines, are the source of a mixture of resins called *dammar,* which are used in high-quality colorless varnishes. This was also the resin originally used in the manufacture of linoleum. The substance, also called *amber,* is obtained primarily in fossil form from former or present kauri pine forest areas. It occurs as lumps of translucent material with a deep orange-yellow tint. Some of the lumps weigh up to 45 kilograms (100 pounds). The supply, which was at its peak at the turn of the century, is now nearing exhaustion. Remarkably lifelike preservations of prehistoric insects millions of years old have occurred in amber (figure 20.15).

Other products refined from resin are used in the manufacture of menthol for cigarettes (menthol also occurs naturally in members of the mint family), floor waxes, printer's ink, paper coatings, varnishes, and perfumes.

The chief source of pulpwood for newsprint (figure 20.16) and other paper in North America is the white spruce. Enormous quantities are used every day. A single midweek issue of a large metropolitan newspaper may use an entire year's growth of 50 hectares (123 acres) of these trees, and that amount may double for weekend editions. A large American publishing company, in an attempt to find ways of reducing paper consumption in the United States, tried trimming 2.5 centimeters (1 inch) from the

FIGURE 20.15 A prehistoric red wolf spider preserved in Dominican Republic amber. The amber is estimated to be between 20 and 30 million years old. (Photograph by John Yellen)

FIGURE 20.16 A modern newsprint factory. (Courtesy International Paper Company)

width of all rolls of toilet paper in their building facilities. They found that the employees still used the same number of rolls per month as they had previously. From this it was calculated that if all rolls of toilet paper were similarly trimmed in width, 1 million trees would be saved each year in the United States alone.

The roots of the white spruce are quite pliable and were used in split form by Native Americans for lashing canoes and for basketry. Spruce beer, brewed from young twigs and leaves with a sugar source such as honey or molasses added, was once used as a remedy for scurvy. Resin of white, red, black, and sitka spruces was used as a type of chewing gum by Native Americans, who sometimes hardened it slightly in cold water. Europeans who have tried it report that it has to be of the right consistency to be enjoyable, since it behaves like unhardened taffy if it is too soft and is bitter if it is too hard.

The tracheids of spruces have spiral thickenings on the inner walls. These apparently are responsible for giving the wood a resonance that makes it ideal for use as soundboards for musical instruments (figure 20.17). Sitka spruce of the Pacific Northwest produces a strong resilient wood that is a favorite material of manufacturers of light airplanes.

Larches, which along with the dawn redwood and bald cypress are exceptions to the rule that conifers are evergreens, have some of the toughest of all conifer woods. Fence posts of larch are known to last 20 years. In the southern and southwestern United States, posts of juniper wood and bald cypress last even longer, some remaining usable for 40 to 50 years or more. The resin of the western larch has been used in the manufacture of baking powder, and the European larch is the source of a special type of turpentine.

There are about 40 species of true firs, which are widely used in the construction, plastic, and paper industries, as ornamentals, and as Christmas trees. The balsam fir produces on its bark blisters containing a clear resin. This resin, known as Canada balsam, was used in the past for cementing optical lenses and is still used for making permanent mounts on microscope slides. It has medicinal properties, too, and was used by New England colonists in sore throat medications. Douglas fir, found in the mountains of the West, is not a true fir. In the Pacific Northwest, it grows into giant trees that are second

FIGURE 20.17   A violin with a soundboard made from red spruce.

only to the redwoods in size. It is probably the most desired timber tree in the world today. The wood, which is strong and relatively free of knots as a result of rapid growth with less branching than most other conifers, is heavily used in plywoods and is a major source of large beams. A useful wax is extracted from the bark. Exploitation has nearly eliminated old-growth stands, but large numbers of new trees are being grown in managed forests.

Coastal redwoods are also prized for their wood, which contains substances that inhibit the growth of fungi and bacteria. The wood is light in weight, strong, and soft, but it splits easily. It is used for some types of construction, furniture, posts, greenhouse benches, and for many other purposes. California wineries use it extensively for wine barrels. The Giant Redwoods are so huge that a double bowling alley was built on the log of the first specimen cut down. They are no longer logged and now serve almost exclusively as tourist attractions.

Wood of the bald cypress, found in southern swamps, is like that of the redwood in being very resistant to decay. In the past it was used for railroad ties, coffins, general construction, guttering, and shingles. The trees are well known for their "knees,"

which rise above the water as tapering growths from
the roots (figure 20.18). At one time it was widely
believed that these were a means of admitting oxygen
to the roots, but this is now in doubt. They are favored for making knickknacks such as wall ornaments and lamp bases. The leaves of the bald cypress
yield a red dye.

A dull red dye can be obtained from the younger
bark of the eastern hemlock, which is also a source
of tannins for shoe leather. The wood of these small
trees contains exceptionally hard knots, which can
chip an axe blade. It sputters and throws out sparks
freely when placed on a fire. Native Americans made
a poultice for scrapes and cuts by pounding the inner
bark. British Columbia Indians used to scrape out
the cambium of both the western and mountain
hemlocks for food.

Eastern white cedar, also known as arborvitae,
is a favorite ornamental in temperate areas. The
wood is pliable, and several Native American tribes
used it for canoes. The Atlantic or southern white
cedar was the first tree to be used for the construction of pipes for pipe organs in North America.

During World War II old logs of this species found
in a swamp in New Jersey were milled and used in
the construction of patrol torpedo boats.

The fleshy red aril surrounding the seeds of yews
is sweet and edible, but the seeds themselves and
other parts of the plants are definitely poisonous. The
wood, which is tough and resilient, is favored for
making bows. English yew, the wood of longbows,
changed history in 1415 and brought an end to the
Middle Ages at the Battle of Agincourt, when the
English longbow proved its superiority over heavily
armored cavalry. A red or purple dye can be obtained from the bark and roots of yew. Podocarps,
two species of which are valuable timber trees in
New Zealand, have edible seeds.

The wood of incense cedars is used in the manufacture of venetian blinds and pencils. Red cedar
wood is also used for pencils, as well as for cedar
chests, closet lining, fence posts, and cigar boxes. It
was used at one time in Germany for smoking hams.

An aromatic oil used in floor-sweeping compounds is extracted from red cedar wood, and the "berries" of this and related junipers are widely eaten by birds. Many Native American tribes used the berries and inner bark as survival food during bleak winters. Some roasted the dried berries and brewed a beverage from them. Western red cedar was the most important single plant of Native Americans of the northwest who used it for housing, clothing, nets, canoes, totem poles, medicines, and other purposes. Berries of the dwarf juniper are used to flavor gin. Some authorities indicate that the word *gin* may have been derived from *genievre,* the French word for juniper berry.

Other Gymnosperms

Florida arrowroot starch was once obtained from the extensive cortex and pith of a species of cycad whose northernmost distribution occurs in Florida. Before it could be used for human consumption, however, a poisonous substance had to be leached out. Since cycads grow too slowly to make continued preparation of the arrowroot starch profitable, the practice has been abandoned. Today cycads that are seen outside of their natural habitats are being grown primarily for ornamental or educational purposes. In Louisiana, however, the large compound leaves are used in Palm Sunday religious services. Some species, which are nearing extinction in the wild, may soon be known only in botanical gardens and conservatories.

Despite the foul smell of the fleshy integument of seeds of *Ginkgo,* the starchy food reserves of the seeds themselves are edible. In the Orient, the substance is widely used for food, either boiled or roasted, and is to be found imported in canned form in Chinese food stores of large metropolitan areas in the United States.

In the southwestern United States, joint firs (*Ephedra*) are grazed by livestock, and the leaves and stems are still brewed into "Mormon tea." To offset a slight bitterness, a teaspoon of sugar, honey, or jam is added to each cup of tea. Native Americans and pioneers used a concentrated version of the tea in the treatment of venereal diseases. They also ground the seeds into flour from which a bitter bread was made. The drug *ephedrine,* which is widely used in the treatment of asthma and other respiratory problems, still is extracted from a Chinese species, but most now in use is synthetically produced. The Chinese used ephedrine medicinally more than 4,700 years ago.

One species of *Gnetum* is cultivated in Java for its shoots, which are cooked in coconut milk and eaten. Fibers from the bark are made into a rope.

See appendix 1 for scientific names of species discussed in this chapter.

SUMMARY

Gymnosperms generally have exposed seeds produced on sporophylls arranged around an axis, thus forming a strobilus, or cone. The female (seed) cones and smaller male (pollen) cones are produced on the sporophytes. The female gametophyte develops within a nucellus that is enclosed in an integument. This integument later becomes the seedcoat of a seed after fertilization and development of an embryo has occurred. Four divisions of living gymnosperms are recognized: conifers (Division Coniferophyta), cycads (Division Cycadophyta), ginkgoes (Division Ginkgophyta), and gnetophytes (Division Gnetophyta).

Pines are the most numerous conifers; they have needlelike leaves arranged in clusters of two to five. The leaves have several modifications that permit the plants to withstand harsh environments. Modifications include thick-walled cells in a hypodermis, a thick cuticle, sunken stomata, and resin canals. Resin canals, which occur throughout the plants, are tubes lined with special cells that secrete resin, an aromatic substance that inhibits fungi and certain insect pests. Pine xylem contains no vessel elements or fibers, and is therefore relatively soft. The phloem lacks companion cells but has albuminous cells that apparently perform the same function. Pine roots are always associated with mycorrhizal fungi, which are essential to normal development of the plants.

Two kinds of spores are produced. Microspores are produced in papery male cones that, in turn, develop in clusters toward the tips of lower branches. The microspores, which are the products of spore mother cells undergoing meiosis in microsporangia that develop in pairs toward the base of the male cone scales, give rise to four-celled pollen grains that occur in huge numbers. Megaspores are formed in ovules at the bases of female cone scales. The integument of the ovule has a pore called the micropyle. A spore mother cell within the nucellus undergoes meiosis, producing megaspores. One megaspore develops into a female gametophyte; the remainder degenerate. The mature female gametophyte contains archegonia. During the spring before the archegonia mature, pollen grains sift down between the cone scales and catch in a sticky fluid oozing from the micropyles. Each pollen grain produces a pollen tube that digests its way down to the vicinity of the developing archegonia and two of the original four cells in the pollen grain migrate into the tube as it grows. One of the two cells (the generative cell) divides and produces a sterile cell and a spermatogenous cell that itself divides, producing two sperms. The pollen tube reaches an archegonium about 15 months after pollination, and one sperm unites with the egg, forming a zygote. The zygote develops into an embryo of a seed that has a membranous wing formed from a layer of the cone scale. Some conifers produce fleshy or berrylike "cones," whose evolutionary origin is not clear.

Cycads superficially resemble palm trees with unbranched trunks and crowns of large, pinnately divided leaves. They have strobili and life cycles similar to those of pine trees, but their sperms, unlike those of pines, have numerous flagella. Ginkgoes have small fan-shaped leaves with evenly forking veins; their life cycle also is similar to that of pines. Their seeds, which are enclosed in a fleshy covering, have a rank odor at maturity. Gnetophytes have diverse forms, but they all have vessels in their xylem. Half the species are in the genus *Ephedra,* whose members have jointed stems and leaves reduced to scales. *Gnetum* species have broad leaves and occur in the tropics, primarily in vinelike form. *Welwitschia* has only one species, which is confined to southwest African deserts. Its stem is in the form of a shallow cup with straplike leaves that extend from the rim; basal meristems on the leaves constantly add to their length.

The seeds and inner bark of pines are edible, and a tea has been made from the leaves. Eastern white pine stems were favored for use as masts for sailing vessels and were used extensively for crates, furniture, flooring, paneling, and matchsticks. Western white pine is the source of most such lumber today. Resin from pines is the source of turpentine and rosin. Turpentine is used as a solvent and rosin is used by musicians and by baseball players. Dammar from kauri pines is used in colorless varnishes; amber is fossilized kauri resin. Resin is also used in floor waxes, printer's ink, paper coatings, perfumes, and in the manufacture of menthol.

White spruce is the chief source of newsprint; it was also used for basketry and canoe lashing by Native Americans, with molasses or honey for treating scurvy, and in brewing a beer. Spruce resin was used for a type of chewing gum. The wood is used as soundboards for musical instruments and in the construction of aircraft. Larch and juniper woods are used for fence posts. Firs are used in the construction, paper, ornament, and Christmas tree industries. Douglas fir is probably the most desired timber tree in the world today. Coastal redwoods are also prized for their wood, which is resistant to fungi and insects. Bald cypress wood, used in the past for coffins and shingles, is also resistant to decay. A dye and tannins are obtained from the eastern hemlock. Native Americans used parts of hemlocks for poultices and for food. Eastern white cedar's pliable wood was used for canoes, and that of the Atlantic cedar was used for construction of pipes for pipe organs. Yew wood is used for making bows. Podocarps of New Zealand have edible seeds. Incense cedar wood is used for cedar chests, cigar boxes, pencils, and fence posts. Juniper berries are used to flavor gin and were used by Native Americans for food and a beverage.

Arrowroot starch was once obtained from a cycad; *Ginkgo* seeds are edible; Mormon tea is brewed from the leaves and stems of joint firs (*Ephedra*), which, in the past, were also a source of the drug ephedrine and a venereal disease treatment. One *Gnetum* species is cultivated in Java for food.

REVIEW QUESTIONS

1. What is a gymnosperm? How is it distinguished from any other kind of organism?
2. What is the difference between a seed and a spore?
3. How are the leaves of pines different from those of broadleaf flowering plants?
4. What is a resin canal, and what is its function? Where are resin canals found?
5. How do pines differ in their reproduction from ground pines and ferns?
6. How do pollen grains differ from spores or sperms?
7. Which conifers discussed in this chapter do not have woody female cones?
8. If you had samples of leaves of a pine, a cycad, *Ginkgo,* a joint fir, and *Welwitschia* all together, indicate how you could tell them apart by constructing a key.
9. What parts of a pine are considered edible?
10. What is resin? Discuss some of its uses, past and present.

DISCUSSION QUESTIONS

1. Ginkgoes and cycads have broad leaves, whereas those of pines are needlelike. Can you suggest any significance of this in terms of the climates and habitats involved?

2. If no distinction were made at the level of Kingdom between plants and animals, what would be the equivalent, if any, of sporophyte and gametophyte in humans?

3. Both bristlecone pines and redwoods can live to be thousands of years old. What do you suppose makes this possible?

4. Most of the old-growth stands of conifers in North America are now gone, and others will be gone soon. Much of what has been harvested is being replaced with new growth, sometimes with hybrids and non-native plants. Our forests are essential to our economy as we know it. If you had the power to change the way we manage and exploit our natural forest resources, what would you do differently, assuming you did not want to damage the economy?

5. If money were no object and you wished to landscape your yard primarily with gymnosperms, realistically what would you include, taking into account your particular geographical area? Why?

ADDITIONAL READING

Bever, D. N. 1981. *Northwest conifers: a photographic key.* Portland, OR: Binford and Mort Publishing.

Bold, H. C., C. S. Alexopoulos, and T. Delevoryas. 1987. *Morphology of plants and fungi,* 5th ed. New York: Harper & Row, Publishers.

Foster, A. S., and E. M. Gifford, Jr. 1974. *Comparative morphology of vascular plants,* 2d ed. San Francisco: W.H. Freeman and Co.

Mirov, N. T., and J. Hasbrouck. 1976. *The story of pines.* Bloomington: Indiana University Press.

Ouden, P., and B. K. Boom. 1982. *Manual of cultivated conifers.* Hingham, MA: Kluwer Academic Publishers.

Singh, H. 1978. *Embryology of gymnosperms* (*Encyclopedia of plant anatomy:* Vol X, No. 2). Forestburgh, NY: Lubrecht and Cramer.

Sporne, K. R. 1967. *Morphology of gymnosperms. The structure and evolution of primitive seed plants.* Atlantic Highlands, NJ: Humanities Press, Inc.

Weiner, M. A. 1980. *Earth medicine, earth foods,* 2d ed. New York: Macmillan.

Overview

*T*his chapter begins by calling attention to differences between angiosperms and gymnosperms, and then discusses the theoretical origin of flowering plants. The exceptional diversity of form and habit of the flowering plants is reiterated before the chapter continues with a description of the parallel development of the gametophytes in the anthers and ovules; this leads up to pollination, followed by some details of fertilization and the development of a seed.

A discussion of trends of specialization and classification in flowering plants is followed by a section on flower preservation, which deals with simple herbarium techniques and practice, and ends with a discussion of the three-dimensional drying of flowers in a box. The chapter closes with a brief survey on the uses of herbaria and a word of caution to plant collectors concerning unnecessary depletion of native floras.

21 *Flowering Plants*

Some Learning Goals

1. Understand the basic differences between angiosperms and gymnosperms.

2. Contrast two principal schools of thought concerning the origin of the flowering plants and the nature of the first flowers.

3. Diagram the life cycle of a flowering plant, indicating shifts from haploid to diploid cells and vice versa.

4. Compare two types of embryo sac development.

5. Learn how a male gametophyte develops.

6. Know major trends of specialization in the flowering plants.

7. Know the functions of a herbarium and the techniques of preparing herbarium specimens.

Outline

INTRODUCTION

Explorers in remote reaches of the Andes Mountains in Bolivia occasionally view an extraordinary sight seen by few people during the past 100 years—a rare species of *puya* plant in flower (figure 21.1). There are many rare plants, of course, but what makes this member of the bromeliad family so unusual is the exceptional manner and magnitude of its flowering. These particular puyas have a cluster of large basal leaves with spines along the margins, somewhat resembling the leaves of *agaves* (century plants). The plants grow for about 150 years, slowly producing new leaves and storing food, but they do not flower until the last year of their long existence. Then during the final months something unknown triggers a dramatic change. Within the space of a few short weeks, an enormous flowering stalk develops. This ultimately reaches a height of 10 to 12 meters (33 to 40 feet) and becomes 2.5 meters (8 feet) wide, producing a total of some 8,000 individual flowers. When flowering and fruiting are completed, the puya dies.

These rare puyas are but one of more than 240,000 known species of flowering plants that comprise by far the largest and most diverse of the divisions of the Plant Kingdom. These flowering plants are called **angiosperms.** The term *angiosperm* is derived from two Greek words, *angeion,* meaning "vessel," and *sperma,* meaning "seed." The "vessel" is the **carpel,** which may be likened to an enrolled leaf with seeds along its margins. It is the structure within which seeds develop from *ovules,* and is part of an *ovary* that ultimately gives rise to a *fruit.* Although the angiosperms generally have organs and tissues similar to those of the gymnosperms, the enclosed ovules and seeds of the angiosperms are the primary feature that distinguishes them from the exposed ovules and seeds of gymnosperms.

All angiosperms are placed in the Division Anthophyta (see chapter 14 for comments on other classifications of flowering plants), which is divided into two large classes, the Dicotyledonae (dicots) and the Monocotyledonae (monocots). Members of the two classes are distinguished from one another on the basis of features listed in table 8.1.

Since Darwin's *Origin of Species* appeared in 1859, two major schools of thought concerning the origin of angiosperms have developed. One, promulgated by the German botanist Adolph Engler and his followers, holds that flowering plants evolved

FIGURE 21.1   A rare puya plant. (Photograph by John Yellen)

from conifers, and that primitive flowers are similar in structure to those of conifers. Such flowers are inconspicuous and in clusters, like those of the grasses, oaks, willows, and cattails. Most contemporary botanists, however, tend to believe that angiosperms evolved independently from the *pteridosperms* ("seed ferns"—see chapter 20), and that a flower is really a modified stem bearing modified leaves, with the most primitive flower having a long receptacle with many spirally arranged flower parts that are separate and not differentiated into sepals and petals; in addition, the stamens and carpels are flattened. Such flowers are found amongst relatives of magnolias and buttercups, leading many modern botanists to postulate that all present day flowering plants are derived from a primitive stock with such characteristics. A further discussion of primitive and advanced characteristics is given later in the chapter.

FIGURE 21.2   Snowplant flowers. (Courtesy Robert A. Schlising)

DIVISION ANTHOPHYTA— THE FLOWERING PLANTS

As indicated in chapter 8, the plants of members of this division vary greatly in size, shape, texture, and form. The division includes, for example, the tiny duckweeds, all the grasses and palms, many aquatic and epiphytic plants, and most shrubs and trees, including the huge *Eucalyptus regnans* trees of Tasmania, which rival the redwoods in total volume. Other flowering plants are parasitic. Dodders, for example, occasionally cause serious crop losses as they twine about their hosts and intercept food and water in the host xylem and phloem by means of *haustoria* (see figure 6.12). Broomrapes also parasitize a variety of plants, as do mistletoes. The latter produce some chlorophyll and only partially depend on their hosts for food. Still others, such as the beautiful snowplant (figure 21.2) and some of the orchids, are saprophytes. The vast majority of flowering plants, however, produce their food independently via the process of photosynthesis.

Like the gymnosperms, the angiosperms are *heterosporous* (produce two kinds of spores), and the sporophytes are even more dominant than in the gymnosperms, the female gametophytes being wholly enclosed within sporophyte tissue and reduced to only a few cells, and the male gametophytes consisting at maturity of a germinated pollen grain with three nuclei.

Development of Gametophytes

While the flower is developing in the bud, a diploid **megaspore mother cell** becomes differentiated from all the other cells in the ovule (figure 21.3). This cell undergoes meiosis, producing four haploid *megaspores.* Three of these megaspores degenerate and disappear not long after they are produced, but the nucleus of the fourth undergoes mitosis and the cell enlarges. While the cell is growing larger, its two haploid nuclei divide once more; the resulting four nuclei then divide yet another time. Consequently, eight haploid nuclei in all are produced (without walls being formed between them). By the time these three successive divisions have been completed, the cell has grown to many times its original volume. At the same time, two outer layers of cells of the ovule become differentiated. These layers, called **integuments,** later become the **seedcoat** of the seed. As they develop, they leave a pore, or gap, called the **micropyle** at one end.

At this stage, the eight haploid nuclei are in two groups of four toward the ends of the large cell. Now one nucleus from each group migrates toward the middle of the cell. These two **polar nuclei** may fuse together, forming a single diploid nucleus, or they may remain separate until fertilization occurs. Cell walls begin to form around the remaining nuclei. In the group closest to the micropyle, one of the cells functions as the female gamete, or *egg;* the other two cells, called *synergids,* either degenerate later or are destroyed during events that occur later. At the other end, the remaining three cells, called *antipodals,* apparently have no function and later they also degenerate. The large sac containing the seven or eight nuclei constitutes the *female gametophyte;* at maturity, it is called an **embryo sac.**

While the female gametophyte is being produced, a parallel process eventually leading to the formation of male gametophytes takes place in the anthers. As an anther develops, four patches of tissue become distinguishable within the main mass of cells. These patches of tissue contain numerous diploid **microspore mother cells,** each of which undergoes meiosis, producing a quartet (also referred to

FIGURE 21.3   Life cycle of a typical flowering plant.

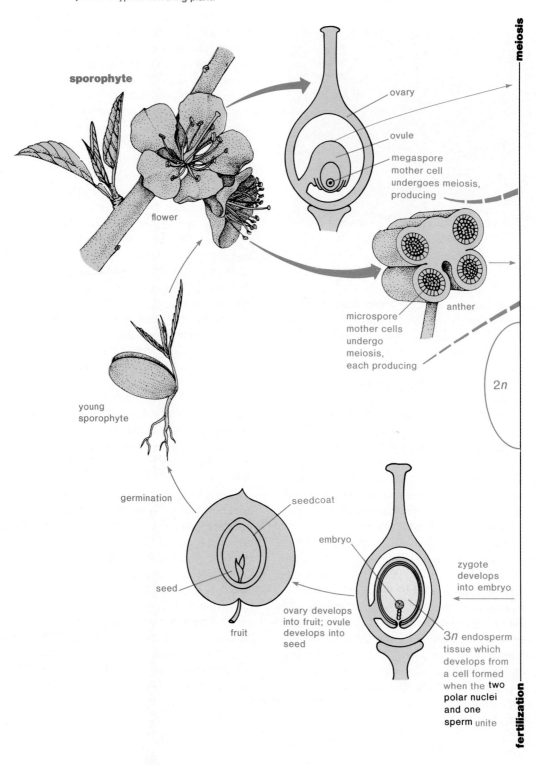

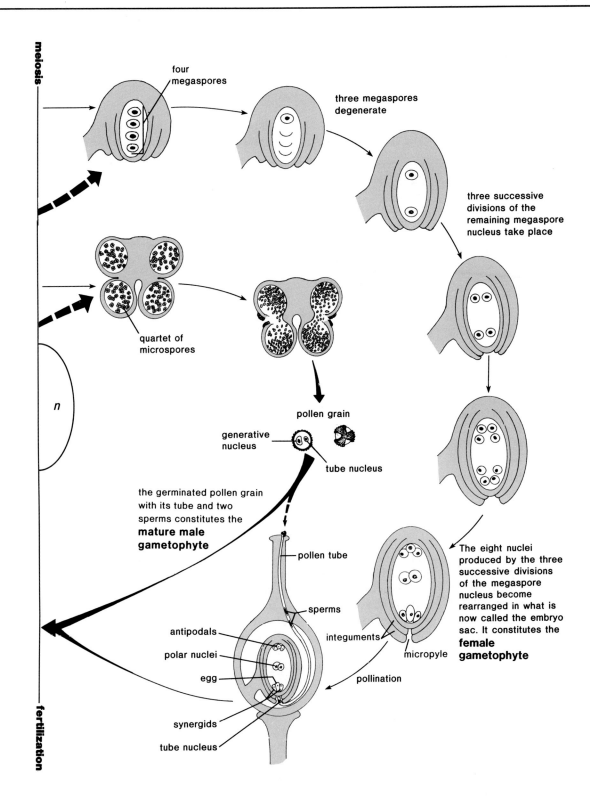

meiosis

four megaspores

three megaspores degenerate

three successive divisions of the remaining megaspore nucleus take place

quartet of microspores

n

pollen grain

generative nucleus

tube nucleus

the germinated pollen grain with its tube and two sperms constitutes the **mature male gametophyte**

pollen tube

sperms

antipodals

polar nuclei

egg

integuments

micropyle

The eight nuclei produced by the three successive divisions of the megaspore nucleus become rearranged in what is now called the embryo sac. It constitutes the **female gametophyte**

synergids

tube nucleus

pollination

fertilization

as a tetrad) of **microspores.** Four chambers or cavities lined with nutritive *tapetal* cells are visible in an anther cross section by the time the microspores have been produced. As the anther matures, the walls between adjacent pairs of chambers break down so that only two larger sacs remain. After meiosis, the microspores in these chambers undergo several changes more or less simultaneously: (1) the nucleus in each microspore divides once; (2) the members of the quartet separate from one another (in some species this separation does not occur, but this is unusual); (3) a two-layered wall, which is often finely sculptured, develops around each unit. When these three events have been completed, the microspores have become **pollen grains** (figure 21.4). The outer layer of the pollen grain wall, called the **exine,** contains substances that may later produce chemical reactions with the stigma of a flower. These reactions may result either in germination of the pollen grain or inhibition of further development, depending on whether or not it originated from the same plant, another plant of the same species, or a plant of a different species. Each pollen grain usually has three thin areas or *apertures* in the wall. Any of the apertures may function in the completion of the development of the male gametophyte discussed in the next section. One of the pollen

grain's two nuclei, the *generative nucleus,* will later divide, producing two sperms. The remaining *tube nucleus* is involved in events that occur after the pollen grain has left the anther.

Pollination

Frequently one reads in the press or popular magazines about flowers being *pollenized,* and the implication is made that the dusting of the pollen of one flower upon another is the equivalent of fertilization. Not so! **Pollination** is simply the transfer of pollen grains from an anther to a stigma, nothing more. **Fertilization** involves the union of egg and sperm; it may not occur until weeks or months after pollination has taken place or it may not follow pollination at all.

Most pollination is brought about by insects or wind, but water, birds, bats, other mammals, and gravity act as agents, or pollinators, in a number of species. The adaptations between the flower and its pollinators can be intricate and precise (see figure 23.21), and may even in certain instances involve force, drugs, or sexual enticement (see figure 23.23). A discussion of pollination ecology is given in chapter 23.

Fertilization and Development of the Seed

After pollination has occurred, further development of the male gametophyte may not take place unless the pollen grain is (1) from a different plant of the same species or (2) from a variety different from that of the flower receiving it. There are, however, some species or varieties (e.g., garden peas), that are naturally self-pollinating. Under suitable conditions, the dense cytoplasm of the pollen grain absorbs substances from the stigma and bulges out in the form of a tube through one of the apertures. This **pollen tube** then grows down between the cells of the stigma and style until it reaches the micropyle. In corn, it may have to grow more than 50 centimeters (20 inches) before it arrives at its destination, but in most plants, including peaches, the distance is considerably less. The pollen tube's journey may last less than 24 hours and usually does not take more than two days, although there are a few plants in which growth takes over a year. As the tube proceeds, the contents of the pollen grain are discharged into it. The tube nucleus stays at the tip, while the generative nucleus (cell) lags behind and eventually divides by mitosis in the tube, producing two sperms that have no fla-gella. Sometimes the generative nucleus (cell) di-vides before the pollen tube has formed. The germinated pollen grain with its tube nucleus and two sperms constitutes the *mature male gameto-phyte* (figure 21.5).

When the pollen tube reaches the embryo sac, it usually discharges its contents into a degenerating synergid, or at least destroys the synergid upon en-tering. Then an event unique to angiosperms, called **double fertilization** (or **double fusion**), occurs. One sperm migrates to and unites with the egg, forming a **zygote;** the other sperm migrates to and unites with the polar nuclei, producing a **3n** (triploid) *endo-sperm nucleus.* The endosperm nucleus divides by mitosis, forming nutritive 3n tissue called **endo-sperm.** In some plants, such as corn and other grasses, the endosperm tissue becomes an extensive part of the seed, but in the Legume Family (e.g., peas, beans) the endosperm is used up by the embryo that develops from the zygote, so that the endosperm has disappeared by the time the seed is mature. The integuments harden, becoming a *seedcoat,* and the remaining haploid nuclei, or cells (antipodals, synergids, and tube nucleus), degenerate. At the conclusion of these various events, the ovule has be-come a seed, and at the same time the ovary matures into a fruit. Seed dispersal and fruits are discussed in chapter 8.

The process of embryo sac development de-scribed for a peach occurs in about 70% of the known flowering plants. The remaining 30% exhibit vari-ations in which the embryo sac has from 4 to 16 nu-clei or cells at maturity, and the endosperm may be 5n, 9n, or even 15n. One such variation occurs in lilies, which are favorites of manufacturers of mi-croscope slides that show embryo sac development. When the megaspore mother cell undergoes meiosis in a lily, all 4 of the haploid megaspore nuclei pro-duced remain functional nuclei (not cells), with 3 of them uniting to form 1 3n nucleus while the fourth remains n. The 3n nucleus divides twice by mitosis, forming 2 3n nuclei in the first division and 4 3n nuclei in the second, while the fourth n megaspore nucleus divides in similar fashion to form 4 haploid nuclei. At the same time the embryo sac develops so that at the end of this process the sac contains 8 nu-clei—4 3n and 4 n. One polar nucleus is 3n, the other is n, so that when fertilized by an n sperm the result is a 5n endosperm nucleus (figure 21.6).

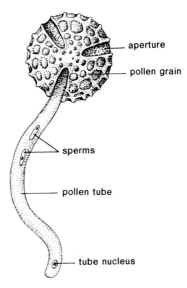

FIGURE 21.5 A mature male gametophyte of a flowering plant.

FIGURE 21.6 How the embryo sac of a lily develops. *A.* The process begins with a diploid (2n) megaspore mother cell. *B.* and *C.* The megaspore mother cell undergoes meiosis, producing four haploid (n) megaspore nuclei. *D.* and *E.* The megaspore nuclei unite, forming a triploid (3n) nucleus; the other haploid nucleus remains separate. *F.* and *G.* Both the 3n and n nuclei undergo two consecutive mitotic divisions, resulting in four 3n and four n nuclei. *H.* Three of the 3n nuclei function as antipodals; the fourth 3n nucleus and one n nucleus function as polar nuclei. The remaining haploid nuclei function as an egg and two synergids.

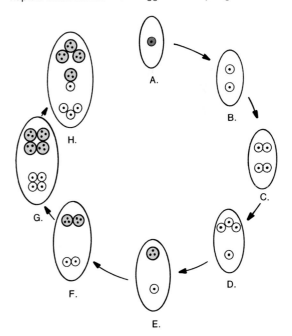

Apomixis and Parthenocarpy

Some embryos of seeds can develop *apomictically* (see discussion of *apomixis* in chapter 13), that is, without development or fusion of gametes (sex cells) but with the normal structures (e.g., ovaries) otherwise being involved. An embryo may develop, for example, from a 2*n* nutritive cell or other diploid cell of an ovule, instead of from a zygote. This makes the plant that develops after germination from such a seed the equivalent of a vegetatively propagated plant. Fruits that develop from ovaries having unfertilized *eggs* (female sex cells) are said to be **parthenocarpic.** Examples of such fruits, which are seedless, are found in navel oranges and certain varieties of figs and grapes. To complicate matters further, not all seedless fruits are parthenocarpic. In Thompson seedless grapes, for example, fertilization does occur, but the ovules fail to develop with the fruit. Parthenocarpy can be induced artificially by the application of dilute hormone sprays to flowers. Seedless tomatoes are often produced in this way. Seedless watermelons are produced by crossing varieties with different chromosome numbers. The *hybrid* (see chapter 13) that results from such a cross produces fruit, but the chromosomes cannot pair properly during meiosis (see chapter 12). Fertilization and seed formation do not occur.

Trends of Specialization and Classification in Flowering Plants

As indicated in chapter 14, various classifications of plants have been proposed ever since Theophrastus grouped them into trees, shrubs, and herbs in the fourth century B.C. Initially, classifications were merely for convenience and did not necessarily take into account natural relationships. Even today, plants may be lumped together in unnatural groupings in order to facilitate their identification. For example, some wildflower books arrange together all white-flowered species or all yellow-flowered species. There is nothing wrong with such arrangements as a means of identification, but because such schemes do not reflect natural relationships (i.e., relationships based on heredity), one does not recognize family characteristics or lineage. One should not infer that all

blondes are more closely related to each other than they are to brunettes or that all long-haired dogs bear a closer kinship to each other than they do to short-haired dogs. Modern botanists therefore try to group plants according to their natural relationships, which are based on evidence gleaned from breeding experiments, forms and structures, chemistry, fossil records, and other features. There are, however, many interpretations of trends in the specialization of flowering and other plants that are based primarily on inference from evidence currently available.

Although the information is very incomplete, the fossil record suggests that the flowering plants first appeared about 160 million years ago during the late Jurassic period (see table 19.1) and that they developed during the ensuing epochs of the Mesozoic and Cenozoic eras into the dominant elements of the flora they are today. Most botanists believe that primitive flowering plants had the following features in common. Their leaves were simple. The flowers had numerous spirally arranged parts that were not fused to each other and were indefinite in number. The flowers were also *radially symmetrical* (i.e., the flowers could be divided into two equal halves along more than one lengthwise plane) and possessed both stamens and pistils.

While primitive flowers still abound today, many specializations and modifications have occurred since their initial appearance. Parts have become fewer and definite in number. Various parts have fused, and spiral arrangements have been compressed to whorls (figure 21.7). Originally a pistil was apparently formed by the blade of a leaflike structure with ovules along its edges rolling inward and fusing along the margins. Such a fertile blade was called a *carpel*. Eventually the separate carpels of primitive flowers fused together, forming the common *compound pistil* consisting of several carpels, which is found in many of today's angiosperms (figure 21.8). Each segment of an orange, for example, represents a single carpel, and if one cuts across a cucumber, three carpels are easily distinguishable. In advanced flowers, the receptacle or other flower parts have fused to and grown up around the ovary so that the calyx and the corolla appear to be attached at the top of it. When the ovary is embedded in this way, it is said to be an **inferior ovary,** in contrast to a primitive **superior ovary,** which is produced on top of the receptacle with the other flower parts attached around its base.

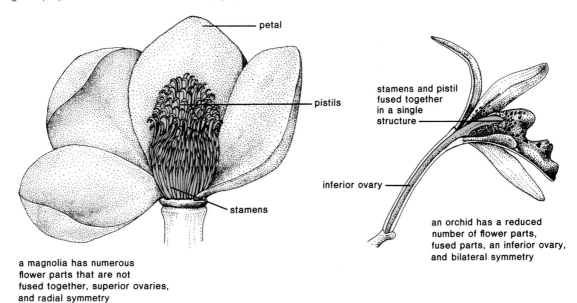

FIGURE 21.7   Comparison between a primitive flower, *Magnolia* (left), and an advanced flower, orchid (right).

petal

pistils

stamens

a magnolia has numerous flower parts that are not fused together, superior ovaries, and radial symmetry

stamens and pistil fused together in a single structure

inferior ovary

an orchid has a reduced number of flower parts, fused parts, an inferior ovary, and bilateral symmetry

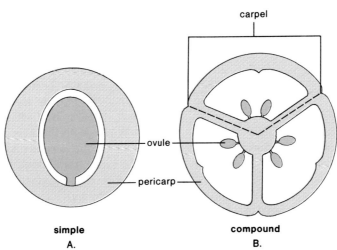

FIGURE 21.8   Ovaries in cross section. *A.* The ovary of a simple pistil as found in a peach. *B.* The ovary of a compound pistil as found in a lily.

carpel

ovule

pericarp

simple
A.

compound
B.

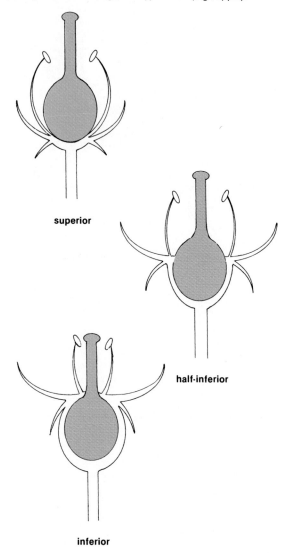

superior

half-inferior

inferior

A.

B.

Some flowers have ovaries in an intermediate or *half-inferior* position (figure 21.9). Flowers have also tended to become *bilaterally symmetrical* (i.e., capable of being divided into two symmetrical halves only by a single lengthwise plane passing through the axis), as in sweet peas and orchids. In some families, the flowers have become *unisexual;* in other words, each flower has either stamens or a pistil but not both. Such flowers are seen in the Pumpkin Family (e.g., pumpkins, squashes, watermelons, cantaloupes, cucumbers) (figure 21.10). When both male and female unisexual flowers occur on the same plant, it is said to be **monoecious.** If male or female flowers occur only on separate plants, however, the species is said to be **dioecious.**

Within both the dicots and the monocots there has been reduction and fusion of parts and a shifting of the ovary from a superior to an inferior position as hypothetical progression is made from primitive families through intermediates to those that are advanced.

Observations on the human and ecological relevance of the flowering plants occupy literally thousands of volumes around the world. A brief overview of several aspects of this subject is given in chapters 22 and 23.

HERBARIA AND PLANT PRESERVATION

The botanical resources of many universities and other institutions include **herbaria** (singular: **herbarium**), which are something like libraries of dried and pressed plants arranged in such a manner that specific plants may be readily located. Specimens that are properly prepared and maintained can remain in excellent condition for 300 or more years. Making one's own herbarium does not require formal training or experience. All that is needed are the materials, some simple equipment (which may be homemade), and the ability to follow a few relatively elementary procedures.

Flowers and other plant parts that are to be preserved need to have their moisture content reduced as rapidly as possible with a minimum of distortion. This is usually accomplished with the aid of a *plant press* (figure 21.11). This simple device consists of two pieces of plywood (or other wood materials, or thin metal plates) with dimensions of approximately 30 × 46 centimeters (12 × 18 inches) and a pair of webbing or leather straps to go around the boards. Between the boards are a number of felts or sheets of heavy blotting paper of similar dimensions. A folded page of newspaper is placed between each blotter, and a few sheets of stiff cardboard are interspersed between the blotters.

Any soil clinging to the roots of a specimen to be pressed is washed off, and the plant is laid out on one of the newspaper sheets. Leaves are carefully straightened out, as are petals and other plant parts, so that they are not folded during pressing. Notes on where, when, and by whom the specimen was collected are penciled on the newspaper, or a number corresponding to such notes in a separate field notebook is used. The newspaper is then folded over the specimen and placed between two blotters (felts). Many specimens may be placed in a press at one time or, if space permits, between the same sheets of newspaper. Only one species should be placed in a single fold of newspaper, however, as mixing species invariably leads to some being pressed better than others, due to the varying extents of woody tissues within the stems and other factors. After the newspaper and specimen are returned to the press, the straps are tightened around it as much as possible and it is placed in the sun or near a heater (but not close enough to scorch!) to dry for three or four days. Unless the leaves were succulent or wet at the time they were placed in the press, they should be

FIGURE 21.11   A plant press.

dry enough to mount on paper at this time. If a press is not available, plants may be pressed between newspapers and blotters by placing heavy weights on top of them.

If possible, 100% rag-content paper should be used for mounting the specimen, as pulp-content papers deteriorate with age. The paper should also be of heavy weight to lend some support to the now brittle specimen. The pressed plant(s) may be attached to the paper in one of several ways. Some workers spread a white library glue of good quality on a glass plate and then place the specimen on the wet glue so that one side is covered with it. They then transfer the specimen to herbarium paper, which normally measures 29 × 42 centimeters (11.5 × 16.5 inches). Others place the specimen on the paper first and add glue or liquid plastic at strategic points. The bottom right-hand corner of the paper should be kept clear for a label indicating the scientific name of the plant, collection information, the collector's name, and the collection date (figure 21.12). Professional botanists also number their individual collections, giving a collection number after the collector's name. The label should measure about 7.5 × 12.5 centimeters (3 × 5 inches), and its paper should also be of 100% rag content. Specimens are then placed in manila folders and stored in some systematic fashion so that retrieval of individual specimens is easily accomplished. Major institutional herbaria usually arrange the plants in a presumed evolutionary sequence by families, and then the genera and species are alphabetically arranged within the families. Mothballs or similar material should be added to the cabinet or storage area to prevent insect damage during storage.

FIGURE 21.12   A mounted herbarium specimen.

FIGURE 21.13   Pressed flowers framed as wall art.

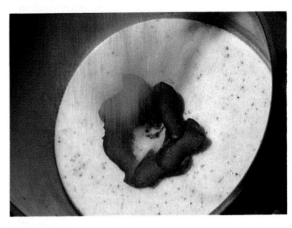

Some people are interested in pressing flowers for use in dry arrangements on cards, placemats, or other decorative items. In such cases, the pressing is done in the same manner as for the herbarium specimens, but the material is then further manipulated in one of several ways. For wall art, the specimens can be mounted on a piece of smooth cardboard, covered with clear plastic or glass, matted, and framed (figure 21.13). Decorative notepaper can be made by placing pressed flowers on the paper and covering them with rice paper or facial tissue. A mixture of one part white glue to three parts water is then brushed over the tissue, causing it to become a permanent mount, with the specimen clearly visible through the thin paper film. Clear contact paper cut to the appropriate size makes a good mount for pressed flowers placed on cards or placemats. Pressed flowers can also be embedded in clear plastic poured into molds.

Flowers can be dried without pressing and with a little patience. Relatively lifelike three-dimensional preservations can be made, although flowers with petals that are not easily detached lend themselves more readily to three-dimensional drying than others. Most such drying is done in a shoebox, but almost any type of container can be used. The bottom of the box is covered with about 2 centimeters (0.8 inch) of sand, silica gel, or borax mixture. The fresh flower, with a little of the peduncle or stem still attached, is then gently laid on the surface. After this, more sand, silica gel, or borax mixture is slowly drizzled by hand into the box until the entire flower is buried, with care being taken not to create air pockets around any parts (figure 21.14). Each of the three drying agents mentioned has certain advantages and drawbacks. Sand must be thoroughly washed several times to be certain it is perfectly clean before use. The sand also needs to be completely dry, uniformly fine, and if possible, have individual grains that are relatively rounded rather than angular. Such sand with rounded grains is found around the Great Salt Lake in Utah and is available in arts and crafts stores. It takes about two weeks to dry most flowers with sand. Silica gel is also available commercially and is the quickest drying agent of the three mentioned, usually completing the job in four to five days (if the box is placed in a microwave oven, the drying process may be completed in as little as five minutes). Its granules tend to be of different sizes, however, giving the flower surface a slightly irregular texture. It also tends to darken certain colors, and it is expensive. When borax is used, two parts of it should be mixed with one part sand or cornmeal. It dries flowers in about three weeks, but is sometimes difficult to remove completely from the dried flower surfaces.

When the drying period has been completed, the container is tilted and the drying material is slowly and gently poured out in an uninterrupted motion. If any petals have come loose, they may be glued in place later. Wire may be inserted in the stem to add rigidity, and after any clinging granules have been carefully removed, soft, powdered colored chalk may be dusted on to restore any fading of color. It is recommended that beginners use fairly large flowers whose petals are not easily detached (e.g., zinnias, roses) for their initial attempt at three-dimensional flower drying.

A WORD OF CAUTION: Dried plant collections of herbaria in particular have proved invaluable in botanical research in the past and will continue to do so in the future. They have facilitated quick identifications of plants in emergency situations where children have eaten plants or plant parts suspected of being poisonous, or have helped pinpoint specific plants that have caused allergic reactions. Herbaria also have been involved in archaeological research where the uses of plants by past cultures have been determined, and have been used for teaching purposes at various educational levels. Herbarium specimens have been useful in criminal litigation for both the prosecution and the defense, and have been the primary source of information on the distribution of plants with potential for new agricultural and horticultural crops, or those with possible medicinal values. Most of the unraveling of problems pertaining to natural relationships of plants begins in a herbarium, and without these "plant libraries" increasing our knowledge along many practical and theoretical lines would be severely restricted. Literally hundreds of plants native to North America are now on rare and endangered species lists, however, and thousands more are in similar predicaments on other continents. *The day has come when both professional and amateur persons interested in plants must discipline themselves to exercise extreme caution in collecting native plants. Collectors should first know what they are collecting or otherwise refrain, and collecting for private collections without serious purpose should be strictly limited. Except for certain types of research, a good photograph of a plant may actually be preferable to a dried specimen and aesthetically more pleasing. It is sincerely hoped that each reader will confine a collection of native plants to photographs.*

SUMMARY

The flowering plants are angiosperms, with ovules and seeds completely enclosed within carpels which, in turn, make up ovaries that become fruits. All flowering plants are placed in the Division Anthophyta, which is divided into two major classes constituting the dicots and monocots.

The Englerian school of botanists believes that flowering plants evolved from conifers, but most contemporary botanists believe that flowering plants evolved separately from pteridosperms, and that the first flowers had many separate, flattened parts spirally arranged on an elongate receptacle. Flowering plants are heterosporous.

The gametophytes of flowering plants develop in separate structures. In the ovule a megaspore mother cell undergoes meiosis, producing four megaspores. Three megaspores degenerate and the fourth develops into an embryo sac containing eight nuclei in three groups. Closest to the micropyle is an egg flanked by two synergids; in the center are two polar nuclei which may fuse, and at the other end are three antipodals. Integuments, which later become a seedcoat, surround the embryo sac, or female gametophyte. Numerous microspore mother cells undergo meiosis in an anther, producing microspores. The microspores develop into pollen grains with two nuclei—the generative nucleus and the tube nucleus.

Pollination, which is simply the transfer of pollen grains from an anther to a stigma, is brought about by insects, wind, and other agents. After pollination a pollen tube may emerge through an aperture from a pollen grain and grow down the style toward the embryo sac. As it does so, the tube nucleus remains at its tip, and the generative nucleus divides, producing two sperms (without flagella), either before or after it enters the pollen tube. The contents of the pollen tube are discharged into the embryo sac. One sperm unites with the egg and forms a zygote, and the other sperm simultaneously unites with the polar nuclei, forming a $3n$ endosperm nucleus; this nucleus divides mitotically, becoming nutritive endosperm tissue. The endosperm may become an extensive part of the seed or it may be used up by the embryo by the time the seed is mature. Nuclei that have not participated in fertilization degenerate. Some flowering plants produce endosperm that is $5n$, $9n$, or $15n$ due to variations in the manner in which the embryo sac develops.

Some books group flowers on the basis of color to facilitate identification, but such grouping does not reflect natural relationships (i.e., relationships based on heredity). Accordingly, modern botanists try to group plants naturally on the basis of evidence gleaned from a variety of sources, including fossils that suggest the flowering plants first appeared about 160 million years ago. Primitive flowering plants had simple leaves and numerous spirally arranged flower parts that were not fused to each other, and possessed both stamens and pistils. They were also radially symmetrical (capable of being divided in half along more than one lengthwise plane).

Specializations that have occurred since primitive flowers first appeared include a reduction in the number of parts; fusion of parts; appearance of compound pistils composed of several individual pistils (referred to as carpels in their fused state); fusion of the receptacle to the ovary so that the ovary appears embedded, or inferior as opposed to superior, with some flowers having half-inferior ovaries; bilateral symmetry (capable of being divided in half along only one lengthwise plane), and unisexual flowers (flowers with either a pistil or stamens). Monoecious species have both male and female flowers on the same plants, whereas dioecious species have male and female flowers on separate plants.

Herbaria may be likened to libraries of dried and pressed plants arranged so that specific plants may be readily located. Properly pressed plants may last for hundreds of years. Initially a plant to be pressed is placed in a plant press between sheets of newspaper and absorbent material, after soil has been removed and parts carefully straightened. When the specimens are dry they may be affixed to sheets of high-quality paper with a label giving collection information. Pressing may also be done for nonacademic purposes such as the preparation of artwork, placemats, and so on. Flowers may be preserved in three dimensions by placing them in a box containing drying materials such as sand, silica gel, borax, or cornmeal, but each agent has drawbacks.

Because so many plants are now on rare and endangered species lists, collectors should try to confine their future collecting to photographs.

REVIEW QUESTIONS

1. What are the basic differences between gymnosperms and angiosperms?
2. What is the function of the integuments?
3. How do the pollen grains of flowering plants differ from those of pine trees?
4. What is the rough equivalent of an archegonium in a flower?
5. What is the function of endosperm, and how does it originate?
6. What differences are there between the embryo sacs of peaches and lilies?
7. Distinguish between radial and bilateral symmetry.
8. What distinguishes primitive flowers from those that are advanced?

DISCUSSION QUESTIONS

1. The world's largest flowers occur on inconspicuous parasitic plants that resemble a fungus mycelium. If the flowers were not present, would it be possible to tell that the plant is not a fungus? How?
2. In most embryo sacs, the egg and the polar nuclei are the only ones of the eight original nuclei present that do not degenerate. What might be the consequences if the megaspore itself functioned as an egg, not producing other nuclei in an embryo sac?
3. It takes only one pollen grain to initiate the development of an ovule into a seed, yet a single flower may produce many thousands of pollen grains. Do you suppose such huge numbers of pollen grains are really necessary? Why?
4. Are such items as the name of the collector and the date significant on a herbarium specimen?
5. Is it really important that plants be classified? What would be the consequences if they were not?

ADDITIONAL READING

Bold, H. C., C. S. Alexopoulos, and T. Delevoryas. 1987. *Morphology of plants and fungi,* 5th ed. New York: Harper & Row, Publishers.

Condon, G. 1982. *The complete book of flower preservation,* rev. ed. Boulder, CO: Pruett Publishing Co.

Cronquist, A. 1979. *How to know the seed plants.* Dubuque, IA: Wm. C. Brown Publishers.

Cronquist, A. 1981. *An integrated system of classification of flowering plants.* New York: Columbia University Press.

Foster, A. S., and E. M. Gifford, Jr. 1974. *Comparative morphology of vascular plants,* 2d ed. San Francisco: W. H. Freeman & Co.

Geesink, R. 1981. *Thommer's analytical key to the families of flowering plants.* New York: State Mutual Book and Periodical Service.

Johri, B. M., ed. 1984. *Embryology of angiosperms.* New York: Springer Verlag.

Mathias, M. E., ed. 1985. *Flowering plants in the landscape.* Berkeley, CA: University of California Press.

Raven, P. H., R. F. Evert, and S. E. Eichhorn. 1985. *Biology of plants,* 4th ed. New York: Worth Publishers, Inc.

Overview

*T*his chapter begins with comments on some of the problems involved in distinguishing between fact and fancy in reported past uses of plants. It continues with a brief discussion of Vavilov's centers of origin of cultivated plants and a survey of 16 well-known flowering plant families. Miscellaneous information given for the families, which are presented in phylogenetic sequence, includes brief comments on family characteristics and some past, present, or possible future uses of the following family members:

Buttercup (Ranunculaceae)
Laurel (Lauraceae)
Poppy (Papaveraceae)
Mustard (Brassicaceae)
Rose (Rosaceae)
Legume (Fabaceae)
Spurge (Euphorbiaceae)
Cactus (Cactaceae)
Mint (Lamiaceae)
Nightshade (Solanaceae)
Carrot (Apiaceae)
Pumpkin (Cucurbitaceae)
Sunflower (Asteraceae)
Grass (Poaceae)
Lily (Liliaceae)
Orchid (Orchidaceae)

22 *Flowering Plants and Civilization*

Some Learning Goals

1. Give reasons for basing scientific evaluation on more than a single sampling.

2. Learn Vavilov's centers of distribution of cultivated plants, and identify several plants from each center.

3. Know characteristics of ten flowering plant families.

4. Understand what constitutes a primitive family.

5. Know five useful plants in the Laurel, Rose, Legume, and Spurge Families.

6. Identify medicinal plants in the Poppy and Nightshade Families.

7. Construct a simple, original key to five flowering plant families.

Outline

INTRODUCTION

One species of plants commonly known as *puccoons* grows on dry plains and slopes throughout the western United States and British Columbia. The seeds of this puccoon, which has greenish yellow flowers, are so hard that in Nevada and California the plant is known by the common name of "stoneseed."

Women of the Shoshoni tribe of Nevada reportedly drank a cold water infusion of the roots of stoneseed every day for six months to ensure permanent sterility. A biologist by the name of Clellan Ford became curious about these reports and experimented by administering extracts of stoneseed plants to mice. He found that the extracts effectively eliminated the estrous cycle of the mice and decreased the weights of the ovaries, the thymus, and the pituitary glands.

Although in this instance a Native American reported use of a plant was demonstrated experimentally to have a basis in fact, distinguishing between fact and fancy in reported past uses of plants is often difficult, particularly if the plants have become rare or extinct. But this type of research is much needed today. Indeed, it is essential if we are to save potential sources of medicinal drugs and other useful plant products before they are permanently eliminated by the bulldozer and other symbols of "progress." For despite the abundance of superstition and folklore associated with the use of plants by primitive peoples, there is no question that some of their cures and treatments were effective, even though they had no knowledge of the scientific reasons for the results. As indicated earlier, botanists who have recognized this have teamed with anthropologists, medical doctors, and interpreters to interview tribal and other medicine men in the tropics of the Americas and Africa. They are gleaning and sifting through as much of this information as possible before it is too late. It would be surprising if their work does not lead to useful new discoveries in the near future.

ORIGIN OF CULTIVATED PLANTS

In 1822 a Swiss botanist by the name of Alphonse de Candolle published a book entitled *Origin of Cultivated Plants*. After gathering data from a variety of sources, de Candolle deduced that cultivated plants probably originated in areas where their wild relatives grow. In 1916, 94 years after de Candolle's book was published, a Russian botanist named N. I. Vavilov began a followup of de Candolle's work, and in the ensuing 20 years he modified and expanded de Candolle's conclusions. Vavilov became persuaded, as a result of his research, that most cultivated plants differ appreciably from their wild relatives. He also concluded that dispersal centers of cultivated plants are characterized by the presence of dominant genes in plant populations (see chapter 13), with recessive genes becoming apparent toward the margins of a plant's distribution. Vavilov recognized eight centers of origin of cultivated plants, with some plants originating in more than one center. The centers (figure 22.1), some of which are subdivided, are as follows:

Chinese Center

The Chinese Center, the earliest independent center, consists of mountainous areas and adjacent lowlands of western and central China. Cultivated plants believed to have originated here include millets, soybean and several other legumes, bamboo, radish, eggplant, cucumber, some citrus, peach, apricot, walnut, persimmon, tea, some sugar canes, hemp, and others.

Indian Center

The Indian Center, second in importance only to the Chinese Center, consists of Burma and India (exclusive of the northwest portion). Included with the cultivated plants believed by Vavilov to have originated here are rice, sorghum, mung and several other beans, gourds, yam, mango, orange and other citrus fruits, sugar cane, bowstring hemp, black pepper, betel nut, cardamon, gum arabic, henna, senna, strychnine, rubber plant, and many others.

FIGURE 22.1 Major centers of origin of cultivated plants. (After N. I. Vavilov, *The origin, variation, immunity and breeding of cultivated plants*. Translated from the Russian by K. Starr Chester. New York: The Ronald Press Company, copyright 1951.)

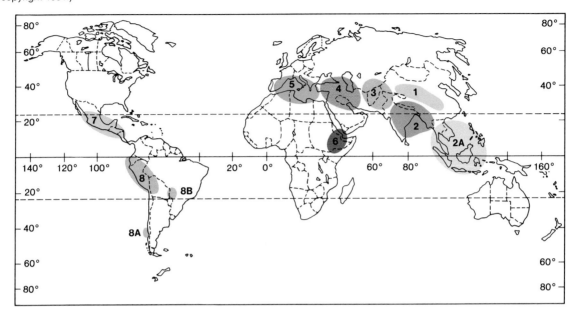

Indo-Malayan Center

The Indo-Malayan Center consists of Indochina, Malaysia, Java, Borneo, Sumatra, and the Philippines. Vavilov believed the giant bamboo, ginger, banana, breadfruit, candlenut, coconut, clove, nutmeg, Manila hemp, and many tropical fruits originated here.

Central Asiatic Center

The Central Asiatic Center consists of northwest India, Afghanistan, and adjacent Soviet provinces. Wheat, garden pea, lentil, mustard, safflower, cotton, garlic, carrot, onion, basil, pear, almond, grape, apple, and other fruits and nuts are believed to have originated here.

Near-Eastern Center

The Near-Eastern Center consists of the interior of Asia Minor, Iran, Transcaucasia, and the highlands of Turkmenistan. Cultivated plants believed to have originated here include several types of wheat, rye, barley and oats, alfalfa, vetch, anise, poppy, cantaloupe, cabbage, lettuce, fig, pomegranate, cherry, and hazelnut.

Mediterranean Center

The Mediterranean Center consists of areas bordering the Mediterranean ocean. Cultivated plants from this region include additional varieties of wheat, fava beans, clover, flax, black mustard, olive, carob, beet, parsley, leek, chive, savory, celery, parsnip, rhubarb, caraway, thyme, hyssop, lavender, peppermint, sage, rosemary, hop, and chufa.

Abyssinian Center

The Abyssinian Center consists of Ethiopia and Somaliland. Among the cultivated plants believed to have originated here are additional varieties of wheat, sesame, garden cress, coffee, okra, indigo, and fenugreek.

South Mexican and Central American Center

The South Mexican and Central American Center consists of southern Mexico, Guatemala, El Salvador, Honduras, Nicaragua, and Costa Rica. Maize (corn), common bean, lima bean, sweet potato, pepper, some cotton varieties, prickly pear, papaya, cashew, cacao (source of chocolate), cherry tomato, and annatto (a dye and spice plant) are believed to have originated here.

South American Center

The South American Center consists of the highlands of Peru, Ecuador, and Bolivia. Several species of potatoes (other than the common white or Irish potato), tomato, pumpkin, marigold, cocaine plant, Egyptian cotton, guava, tobacco, and quinine tree are believed to have originated here.

Chiloe Center

The Chiloe Center is confined to an island off the coast of southern Chile. The common white or Irish potato and the strawberry are believed to have originated here.

Brazilian-Paraguayan Center

The Brazilian-Paraguayan Center consists of Brazil and Paraguay. Despite the large diversity of species in this huge area, few cultivated plants are believed to have originated here. Manioc (source of cassava), peanut, passion fruit, pineapple, Brazil nut, and Pará rubber tree are among the cultivated plants from this area.

Each of these centers is isolated by mountain ranges, deserts, or oceans, and together they constitute less than 3% of the total land area of the Earth. Agricultural development appears to have arisen independently in each center as primitive peoples investigated various members of the diverse local floras and in time came to make use of the plants available to them.

SELECTED FAMILIES OF FLOWERING PLANTS

What follows is a survey of a few of the well-known plant families, indicating not only past uses of some of their members but also possible future uses in a few instances, along with biological notes and miscellaneous observations. The reader is cautioned in particular against assuming that past medicinal uses were always effective or without harmful side effects and is urged to refrain from experimenting with such plants.

More than 300 families of flowering plants are recognized today. Flowers have proved to be considerably more reliable and stable indicators of heredity than leaves or other vegetative parts, and so flowering plant families are distinguished from one another primarily on the basis of flower parts and structure. Some families include a mere handful of species, while others are very large, their members numbering in the thousands. Space permits only brief discussion of a few of the larger, better-known families here, and the reader is referred to the additional readings for other sources of information. The families are taken up in a more or less phylogenetic sequence, beginning with those that have more primitive flower structure and progressing to those whose flowers are considered advanced. Primitive flowers generally have numerous parts that are not fused together and the ovaries are superior. In advanced flowers, the parts are reduced in number and often fused together; the ovary is inferior (see figure 21.9 and the amplification of this subject that begins on page 434).

The following is an abbreviated key to the families that will be discussed:

1. Flowers with parts in 4s or 5s or multiples thereof; seeds with 2 cotyledons (DICOTS).
    2. Petals separate from one another, or lacking.
        3. Petals present.
            4. Stamens more than twice as many as the sepals.
                5. Stamens, petals, and sepals attached to the rim of a cup surrounding the 1 to many pistils ....*ROSE FAMILY* (Rosaceae)
            5. Stamens, petals, and sepals not attached to the rim of a cup.
                6. Pistils several to many in each flower ....
                .... *BUTTERCUP FAMILY* (Ranunculaceae)
            6. Pistil 1.
                7. Ovary superior ....*POPPY FAMILY* (Papaveraceae)
                7. Ovary inferior...*CACTUS FAMILY* (Cactaceae)
        4. Stamens not more than twice as many as the petals.
                8. Herbaceous vines; fruit a pepo ......*PUMPKIN FAMILY* (Cucurbitaceae)
                8. Primarily herbs, shrubs and trees; fruit not a pepo.
                    9. Fruit a legume.................................*LEGUME FAMILY* (Fabaceae)
                    9. Fruit not a legume.

10. Fruit a silique or silicle......*MUSTARD FAMILY* (Brassicaceae)
10. Fruit not a silique or silicle.
    11. Ovary superior; stems square in cross section; fruit of 4 nutlets.....................................*MINT FAMILY* (Lamiaceae)
    11. Ovary inferior; stems rounded in cross section; fruit not of 4 nutlets.
        12. Inflorescence composed of several to numerous florets on a common receptacle ....*SUNFLOWER FAMILY* (Asteraceae)
        12. Inflorescence an umbel; each flower with its own receptacle ....*CARROT FAMILY* (Apiaceae)
3. Petals lacking; calyx sometimes petal-like.
        13. Ovary of 3 carpels and usually elevated on a gynophore; anthers splitting lengthwise.........................*SPURGE FAMILY* (Euphorbiaceae)
        13. Ovary of 1 carpel; gynophore lacking; anthers splitting by raised flaps...*LAUREL FAMILY* (Lauraceae)
        NIGHTSHADE FAMILY (Solanaceae)
2. Petals fused together..*NIGHTSHADE FAMILY* (Solanaceae)
1. Flowers with parts in 3s or multiples thereof; seed with 1 cotyledon (MONOCOTS).
        14. Flowers inconspicuous; without petals or sepals ....*GRASS FAMILY* (Poaceae)
        14. Flowers conspicuous, the petals and sepals mostly similar in coloration.
            15. Ovary superior; petals all alike...................................*LILY FAMILY* (Liliaceae)
            15. Ovary inferior; 1 petal different in form from the other 2 ....*ORCHID FAMILY* (Orchidaceae)

The Buttercup Family (Ranunculaceae)

Flowers of the Buttercup Family have numerous stamens and pistils, and their petals are often irregular in number; the ovary is superior (figure 22.2). Most of the approximately 1,500 species, which are concentrated in north temperate and arctic regions, are herbaceous, often with dissected leaves that have no stipules and with petioles that are somewhat expanded at the base. A number of well-known representatives are ornamental plants (figure 22.3), including buttercup, columbine, larkspur, anemone, monkshood, and *Clematis*. The columbine, a species of which is the state flower of Colorado, receives its name from *columba,* the Latin word for dove. The name relates to the somewhat dovelike appearance of each of its five, spurred petals. Native Americans made a tea from boiled columbine roots for control of diarrhea, and members of at least two tribes believed the seeds to be an aphrodisiac. They would pulverize the seeds, rub them in the palms of their

FIGURE 22.2 A buttercup flower.

A.

B.

C.

hands, and would then try to shake hands with the woman of their choice. Others crushed and moistened the seeds and applied them to the scalp to repel lice.

Most members of the family are at least slightly poisonous, but the cooked leaves of cowslips have been used for food, and the well-cooked roots of the European bulbous buttercup are considered edible. In its natural state, the European buttercup does, however, cause blistering on the skin of sensitive individuals. East Indian fakirs are reported to blister their skin deliberately with buttercup "juice" in order to appear more pitiful when begging. Native Americans of the West gathered buttercup achenes, which they parched and ground into meal for bread. Others obtained a yellow dye from buttercup flower petals. Karok Indians made a blue stain for the shafts of their arrows from blue larkspurs and Oregon grape berries.

Goldenseal is a plant that was once abundant in the woods of temperate eastern North America but has become virtually extinct in the natural state because of relentless collecting by herb dealers. They sold the root for various medicinal uses, including remedies for inflamed throats, skin diseases, and sore eyes. At least one Native American tribe mixed the pounded root in animal fat and smeared it on the skin as an insect repellent. Another member of the Buttercup Family, monkshood, yields a drug complex called *aconite,* which was once used in the treatment of rheumatism and neuralgia. Although popular as garden flowers, the plants are very poisonous. Death follows within a few hours of consumption of any part of the plant. Most monkshood species have purplish to bluish or greenish flowers, but one Asian species called wolfsbane has yellow flowers. Wolf hunters in the past obtained a juice from wolfsbane root that they used to poison the animals.

The Laurel Family (Lauraceae)

The Laurel Family is also a primitive family whose flowers have no petals but whose six sepals are sometimes petal-like. The stamens, which occur in three or four whorls of three each, are a curiosity because the anthers open by flaps that lift up. The ovary is superior. Most of the approximately 1,000 species in this family are tropical evergreen shrubs and trees, many with aromatic leaves. The family received its name from the famous laurel cultivated in Europe for centuries. Its foliage was used by the ancient Greeks to crown victors in athletic events and later was employed in the conferring of academic honors.

Several important spices come from members of this family. Powdered cinnamon is the pulverized bark of a small tree that is native to India and Sri Lanka (formerly Ceylon), although it is also grown commercially elsewhere. Cassia, which is very similar, is often sold interchangeably with cinnamon today. Cinnamon oil is distilled from young leaves of the trees. Use of cinnamon and cassia dates back thousands of years. They were used in perfumes and anointing oils at the time of Moses, and other records indicate their use in Egypt at least 3,500 years ago.

Camphor is another member of the Laurel Family that has been used since ancient times. This evergreen tree, which is native to China, Japan, and Taiwan, is the main source of camphor essence still used today in cold remedies and inhalants, insecticides, and perfumes. The essence is distilled from wood chips. Because of its ability to withstand smog, camphor is becoming popular as a street tree in cities and towns located in the milder climatic areas of the United States.

Sassafras trees, which are native to the eastern United States and eastern Asia, also have spicy-aromatic wood. A flavoring widely used in toothpaste, chewing gum, mouthwashes, and soft drinks is obtained by distillation of wood chips and bark. Sassafras tea still is considered a refreshing beverage. It has in the past been used by country physicians for treating hypertension and for inducing a sweat in those with respiratory infections. Reports indicate it has a narcotic-stimulant effect in large doses. Sassafras is also an ingredient of some homemade root beers, and in the southern states an alcoholic beer has been made by adding molasses to boiled sassafras shoots and allowing the mixture to ferment. In Louisiana, powdered sassafras leaves (called *filé*) are used as a thickening and flavoring agent for gumbo.

FIGURE 22.4   Fruits and leaves of a California bay tree.

FIGURE 22.5   A prickly poppy (*Argemone*) flower.

The sweet bay, used as a flavoring agent in gravies, sauces, soups, and meat dishes, comes from the leaves of the laurel. Leaves of the related California bay (figure 22.4) are sometimes used as a substitute for sweet bay, and for making Christmas wreaths. The natural distribution of this California tree includes southwestern Oregon, where it is known as myrtle (true myrtles, however, belong to a different family). Its wood, which is hard, can be polished to a high luster and is used for making a variety of bowls, ornaments, and other smaller wooden articles. Early settlers in the West and Native Americans of the region used California bay in a number of ways. A bath for the relief of rheumatism was made by adding a quantity of leaves to hot water. The nutlike fruits (drupes) were roasted and used for winter food. A leaf was placed under a hat on the head to "cure" a headache (but even a small piece of leaf placed near the nostrils can produce an almost instant headache!). A few leaves placed on top of flour or grains in a canister will keep weevils away, and small branches have been used as chicken roosts to repel bird lice and fleas. A leaf or two rubbed on exposed skin functions as a mosquito repellent.

Avocados are also members of the Laurel Family. The fruits, which have more energy value by weight than red meats, are rich in vitamins and iron.

The Poppy Family (Papaveraceae)

Members of the Poppy Family are mostly herbs distributed throughout temperate and subtropical regions north of the equator, but several poppies occur in the southern hemisphere, and a number are widely planted as ornamentals around the world. Poppies, like buttercups, tend to have numerous stamens, but most have a single pistil (figure 22.5). Most also have milky or colored sap, and their sepals usually fall off as the flowers open. All members produce drugs.

A popular early flowering spring plant of eastern North American deciduous forests is called *bloodroot* because of the bright reddish sap produced in its rhizomes. This sap was used by some Native Americans as a facial dye, an insect repellent, and a cure for ringworm. Youngsters today still paint their nails with it. Except when ingested in minute amounts, it has a very bitter taste. This latter feature made it effective in inducing vomiting, but members of one tribe counteracted the bitterness and used it for treating a sore throat by squeezing a few drops on a lump of maple sugar so that it could be held in the mouth.

Opium poppies have had a significant impact on societies of both the past and the present. Opium itself was described by Dioscorides in the first century A.D., and ancient Assyrian medical texts refer to both opium and opium poppies. Opium smoking, which does not extend back nearly as far as the use of the drug in other ways, became a major problem in China in the 1600s. Smoking opium has given way to other forms of use in recent years. The substance is obtained primarily by making small gashes in the

FIGURE 22.7   Shepherd's purse.

green capsules of the poppies (figure 22.6). The crude opium appears as a thick, whitish fluid oozing out of the gashes and is scraped off. It contains two groups of drugs. One group contains the narcotic and addicting drugs morphine and codeine, which are best known for their widespread medicinal use as painkillers and cough suppressants. Members of the other group are neither narcotic nor addictive. They include papaverine, which is used in the treatment of circulatory diseases, and noscapine, which is used as a codeine substitute because it functions like codeine in suppressing coughs but does not have its side effects.

Heroin, a scourge of modern societies, is a derivative of morphine. It is from four to eight times more powerful than morphine as a painkiller and less than 100 years ago was advertised and marketed in the United States as a cough suppressant. Up until the early 1980s it was estimated that 75% of American drug addicts used heroin, but its use has since been surpassed by that of cocaine. The loss to society in terms of its economic, physical, and moral impact is enormous, since addicts frequently commit violent crimes to obtain the funds needed to support their habits (which often cost well over $150 per day).

The seeds of opium poppies contain virtually no opium and are widely used in the baking industry as a garnish. They also contain up to 50% edible oils, which are used in the manufacture of margarines and shortenings. Another type of oil obtained from the seeds after the edible oils have been extracted is used in soaps and paints.

The Mustard Family (Brassicaceae)

The original Latin name for the Mustard Family, still in widespread use today, was *Cruciferae*. The name describes the four petals of the flowers, which are arranged in the form of a cross. The flowers also have four sepals, usually four nectar glands, and six stamens, two of which are shorter than the other four. All members produce siliques or silicles (see figure 8.17), such fruits being unique to the family. All 2,500 species of the family produce a pungent, watery juice, and nearly all are herbs distributed primarily throughout the temperate and cooler regions of the northern hemisphere.

Many edible plants are found in the Mustard Family. Some are widely cultivated, particularly in cooler temperate climates. Such plants include cabbage, cauliflower, brussels sprouts, broccoli, radish, kohlrabi, turnip, horseradish, watercress, and rutabaga. Some edible members are also widespread weeds. The leaves of shepherd's purse (figure 22.7), for example, can be cooked and eaten like cabbage, and the seeds can be used for bread meal. Other wild edible members include several cresses, pepper-grass, sea rocket, toothwort, and wild mustard. The latter can become a weed problem in row crop planting. Their leaves are sometimes sold as vegetable greens in markets.

The seeds of wild mustard, shepherd's purse, and several other members of this family produce a sticky mucilage when wet. Biologists at the University of California at Riverside discovered a potential new use for these seeds. They fed pelleted alfalfa rabbit food to mosquito larvae in water tanks, which they were using for experiments on mosquito control. They noticed that the larvae, which had to come to the surface at frequent intervals for air, often stuck to the pellets and suffocated. Curious, the workers examined the pellets under a microscope and found that they contained mustard seeds. Evidently the field where the alfalfa had been harvested had also contained mustard plants. The scientists then tried heating the mustard seeds to kill them and found that this did not affect production of mucilage by the seeds when wet. It was calculated that 0.45 kilogram (1 pound) of such seed could kill about 25,000 mosquito larvae. Experiments are needed to determine if there is a practical way to control mosquitoes by such nonpolluting means on a large scale.

Native Americans mixed the tiny seeds of several members of this family with other seeds and grains for bread meal and gruel. To prevent or reduce sunburn, Zuni Indians applied to the skin a water mixture of ground western wallflower plants. Watercress, which is widely known as a salad plant, has had many medicinal uses ascribed to it. During the first century A.D., for example, Pliny listed more than 40 medicinal uses. Native Americans of the west coast of the United States treated liver ailments with a diet consisting exclusively of large quantities of watercress for breakfast, abstinence from any further food until noon, and then resumption of an alcohol-free but otherwise normal diet for the remainder of the day. This was repeated until the disease, if curable, disappeared.

Dyer's woad, a European plant that has become naturalized and established in parts of North America, is the source of a blue dye that was used for body markings by the ancient Anglo-Saxons. Another member of the family, camelina, has been grown in the Netherlands for the oil that is obtained from its seeds. Camelina oil has been used in soaps and was once used as an illuminant for lamps.

FIGURE 22.8   A wild rose flower.

The Rose Family (Rosaceae)

The Rose Family comprises a large number of trees, shrubs, and herbs distributed throughout much of the world. The flowers characteristically have the basal parts fused into a cup, with petals, sepals, and numerous stamens being attached to the cup's rim (figure 22.8). The more than 3,000 species of the family are divided into subfamilies on the basis of flower structure and fruits. The flowers of one group have inferior ovaries and produce pomes for fruits (see figure 8.11). Flowers of other groups have ovaries that are superior or partly inferior and produce follicles, achenes, drupes or clusters of drupelets.

The economic impact of members of the rose family is enormous, with large tonnages of stone fruits (e.g., cherries, apricots, peaches, plums), pome fruits (e.g., apples, pears), and aggregate fruits (see figure 8.12) such as strawberries, blackberries, loganberries, and raspberries being grown annually in temperate regions of the world.

Members of this family have been relevant to humans in many other ways in the past, however, and still continue to be so. Roses themselves, for example, have been favorite garden ornamentals of countless numbers of gardeners for centuries, and their elegant fragrance delights many. In Bulgaria

FIGURE 22.9   Harvesting damask roses for attar (otto) of roses perfume in Bulgaria. (Courtesy Field Museum of Natural History, Chicago)

FIGURE 22.9   Harvesting damask roses for attar (otto) of roses perfume in Bulgaria. (Courtesy Field Museum of Natural History, Chicago)

FIGURE 22.10   Rose hips.

and neighboring countries, a major perfume industry has grown up around the production of a perfume oil known as attar (or otto) of roses from damask roses (figure 22.9). In a valley near Sofia, more than 200,000 persons are involved in the industry, whose product brought more than $2,200 per kilogram ($1,000 per pound) during the 1970s. A considerable quantity of the oil is blended with less expensive substances in the perfume industry. It has been reported that perfume workers rarely develop respiratory disorders, thus suggesting that medicinal properties could be attributed to the plant extracts.

The fruits of wild roses, called *hips* (figure 22.10), are exceptionally rich in vitamin C. In fact, they may contain as much as 60 times the vitamin C of citrus fruit. Native Americans from coast to coast included rose hips in their diet (except for members of a British Columbia tribe, who believed they gave one an "itchy seat"), and it is believed that this practice contributed to the fact that scurvy was unknown among them. During World War II when food supplies became scarce in some European countries, children in particular were kept healthy on diets that included wild rose hips. The hips also contain, in addition to vitamin C, significant amounts of iron, calcium, and phosphorus. Today many Europeans eat "Nyppon Sopa," a sweet, thick purée of rose hips, whenever they have a cold or influenza.

After giving birth, the women of one western Native American tribe drank western black chokecherry juice to staunch the bleeding. Other tribes frequently made a tea from blackberry roots to control diarrhea. Once it was reported that 500 Oneida Indians cured themselves of dysentery with blackberry root tea, while many white settlers in the vicinity died from the disease. Men of certain tribes used older canes of roses for arrow shafts (presumably after removing the thorns!). Wild blackberries, raspberries, salmonberries, thimbleberries, dewberries, juneberries, and strawberries all provided food for Native Americans and early settlers, and they are still eaten today, either fresh or in pies, jams, and jellies. A spiced blackberry cordial is still a favorite for "summer complaints" in southern Louisiana. Wild strawberries are considered by many to be distinctly superior in flavor to cultivated varieties.

The Legume Family (Fabaceae)

The Legume Family is the third largest of the approximately 300 families of flowering plants, being exceeded in numbers of species only by the Sunflower and Orchid Families. Its 13,000 members, which are cosmopolitan in distribution, include many important plants. The flowers range in symmetry from radial to bilateral. The latter have a characteristic *keel*, which is a boat-shaped fusion of two

FIGURE 22.11   Parts of a legume flower.

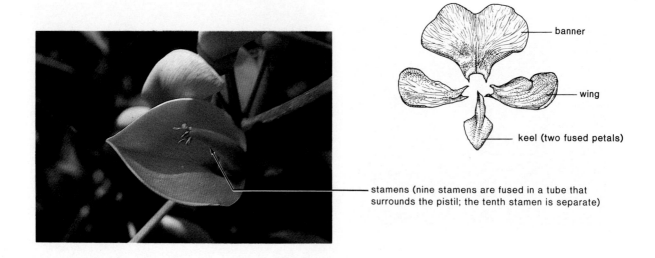

banner

wing

keel (two fused petals)

stamens (nine stamens are fused in a tube that
surrounds the pistil; the tenth stamen is separate)

petals enclosing the pistil, two *wing petals,* and a
larger *banner* petal (figure 22.11). The stamens in
such flowers are generally fused in the form of a tube
around the ovary. The common feature that keeps
the members together in one family is the fruit,
which is a legume (see figure 8.16).

Important crop plants include peas, beans of all
kinds (e.g., kidney, lima, garbanzo, broad, mung, te-
pary), lentils, peanuts, alfalfa, sweet clover, licorice,
and wattle. The latter is an Australian tree that is
grown commercially as a source of tannins for leather
tanning. Carob, which is now widely used as a chocolate substitute, is also a member of this family.
Several copals (hard resins used in varnishes and
lacquers) are obtained from certain legume plants,
as are gum arabic and gum tragacanth, which are
used in mucilages, pastes, paints, and cloth printing.

Important dyes such as indigo, logwood (used
in staining tissues for microscope slides and now
scarce), and woadwaxen (a yellow dye) come from
different legume plants. Locoweeds, which have been
responsible for the death of many horses, cattle, and
sheep, particularly in the southwestern United
States, belong to a large genus of about 1,600 species. The poisonous principle in those species affecting livestock seems to vary in concentration
according to the soil type in which the plants are
growing. Other poisonous legumes include lupines,
jequirity beans, black locusts, and mescal beans.

The sensitive plant (*Mimosa pudica*), which
grows as a weed in the tropics and the deep south of
the United States, is a popular curiosity in temperate and colder areas because of the rapid manner
in which its leaflets and leaves move in response to
touch or other disturbances (see figure 11.18). About
90% of the members of the Legume Family exhibit
leaf movements, but few are as rapid as those of the
sensitive plant (*Mimosa pudica*). Many of the
movements are correlated more with daylength than
with other factors.

Wild clovers have been widely used for food in
the past by primitive peoples. The leaves are not
readily digestible in quantity, but the rhizomes were
gathered and usually roasted or steamed in salt
water, then dipped in grease before being eaten. The
seeds of both clovers and vetches also were gathered
and either ground for meal or cooked in a little water
and eaten as a vegetable. Today seeds of several legumes, including alfalfa and mung beans, are popular for their sprouts, which are widely used in salads
and Oriental dishes. A tropical bean called *winged
bean* has unusually high levels of protein, and all
parts of the plant are edible. It is presently being
grown in several widely scattered tropical and subtropical regions, and also is being marketed on a
limited scale in some temperate zones. It is believed
to have great potential for malnourished peoples of
all tropical areas of the world.

FIGURE 22.12   A spurge. *A.* An inflorescence. *B.* An individual cyathium.

A.

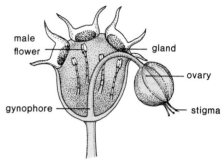

B.

The Spurge Family (Euphorbiaceae)

Although many of the members of the Spurge Family are tropical, a number are widespread in temperate regions both north and south of the equator. The stamens and pistils are produced in separate flowers, and the flowers often lack a corolla, making them inconspicuous. In true spurges (*Euphorbia*) the female flower is elevated on a stalk called a *gynophore* and is surrounded by several male flowers that each consist of little more than an anther. Both the female and male flowers are inserted on a cup composed of fused bracts; the cup has distinctive glands on the rim. This type of inflorescence is called a *cyathium* (figure 22.12). Sometimes the inconspicuous flowers are surrounded by brightly colored bracts (e.g., *poinsettia;* see figure 7.14) that give the inflorescence the appearance of a single large flower. Most members of this large family have milky sap, and a number of species are poisonous.

Sooner or later many gardeners have "urges to purge spurges," as some members of the family are exceptionally aggressive weeds that reproduce very rapidly. Other tropical spurges closely resemble cacti and become very large. Several economically important plants are cultivated, particularly in frost-free regions. For example, an estimated 90 million metric tons (100 million tons) of cassava are harvested annually from plants cultivated in the jungles of South America, Africa, and eastern Asia. The roots, which develop thickened storage areas that resemble large sweet potatoes, are the diet staple of tropical peoples, as white potatoes and cereals are of the peoples of temperate areas. Poisonous principles are removed by boiling, fermenting, or by squeezing out the juice. In dried and powdered form, the cassava is known as farinha. In western countries, tapioca is prepared by forcing heated cassava pellets through a mesh while it is being agitated. Cassava starch is also used as a base for the production of alcohol, acetone, and other industrial chemicals.

Another cultivated spurge of the tropics is the Pará rubber tree, the source of the crude rubber from which most rubber products are made today. Although wild South American trees were the original commercial source of rubber, they have been widely planted in Indonesia, Africa, and adjacent areas. The trees, which vary in height from less than 5 to over 50 meters (16 to 164 feet), produce most of the latex in the inner bark. The laticifers (see chapter 5) in which latex is secreted, spiral around inside the trunk at an angle of about 30 degrees. Accordingly, cuts are made at the same angle into the inner bark to obtain maximum yields of latex, which trickles down into collecting cups that are attached in noninjurious fashion to the tree. After collection, the latex is coagulated with the aid of either various chemicals or smoke and then shipped in sheet or crumbled form to processing plants. Sometimes an anticoagulant is mixed with it, and the liquid is transferred in tankers. Automobile and aircraft tires utilize a conspicuous share of the world's rubber, but other products manufactured from rubber are legion. The Pará rubber tree should not be confused with the broad-leaved ornamental known as a rubber plant. The rubber plant, which is popular as a houseplant, also produces latex but is a member of the Fig Family (Moraceae).

The latex of other spurges may hold a key to future sources of fuel and lubricating oils.[1] In 1976, Melvin Calvin, a University of California Nobel Prize winner, proposed the use of gopher plant latex as a source of materials for oil. He estimated that such plants, which can grow in semidesert areas, would produce 10 to 50 barrels of oil per year on 0.4 hectare (1 acre) of land at a cost of $3 to $10 per barrel.

A spurge called *candelilla* occurs in remote areas of Mexico. It produces a wax on its stems that is collected and used in the making of candles and other wax products. Still another spurge produces seeds with a special oil used in plasticizers, and castor oil (from castor beans) is used in the manufacture of nylon, plastics, and soaps. Castor beans themselves are very poisonous—as few as one to three are sufficient to kill a child. The plants, which grow very rapidly, are popular as ornamentals but are one of the leading natural causes of poisoning among American children.

A Mexican jumping bean is the seed of a certain spurge in which a small moth has laid an egg. When the egg hatches, the grub periodically changes position with a jerk, causing the seed to jump. The crown-of-thorns is an ornamental plant with somewhat flexible twisting stems bearing vicious-looking

FIGURE 22.13   A cactus flower.

spines. Some believe it to have been the plant from which the crown of thorns for the head of Christ was made. Poinsettias, or Christmas flowers (see figure 7.14), are favorite yuletide plants in various parts of the world. Tung oil, used in oil paints and varnishes, and Chinese vegetable tallow, a substance used in the manufacture of soap and candles, are two more commercially important products obtained from the seeds of cultivated members of the Spurge Family.

The Cactus Family (Cactaceae)

Cacti are native only to the Americas but include many highly regarded ornamentals that have been exported around the globe. The flowers are usually showy (figure 22.13), with numerous stamens, petals, and sepals. The sepals are often colored like the petals; the inferior ovary develops into a berry (see figure 8.10). There are possibly more than 1,500 species, most occurring naturally in drier subtropical regions. The leaves of many are reduced in size or missing, with the fleshy, flattened or cylindrical, often fluted stems carrying on the photosynthesis of the plants (figure 22.14). Many cacti can tolerate high temperatures, and some can withstand up to several years without moisture. They vary in size from pinheadlike forms to the giant saguaro (featured in figure 23.17), which can attain heights of 15 meters (50 feet) and weigh more than 4.5 metric tons (5 tons). Cacti are exceptionally slow-growing and make ideal houseplants for sunny windows because they need so little care.

1. Jojoba, a member of the Box Family (Buxaceae), is a desert shrub that has pea-sized seeds containing about 50% liquid wax. This high-grade wax and another found in the seeds of meadow foam, a member of the Meadow Foam Family (Limnanthaceae), are the only known natural substitutes for sperm whale oil, a vital ingredient in engine lubricants. The importing of sperm oil into the United States was banned in 1970 to protect the nearly extinct large ocean mammals, and an expensive synthetic substitute is presently being used. Experiments and research are now in progress to find improved strains of jojoba and to test the feasibility of its being grown on a large scale. Two California counties have been using a mixture of petroleum oil and jojoba oil since 1981 in the transmissions of public transportation buses to see if the mixture will keep them from overheating. Preliminary results look promising and may reduce the need for transmission oil changes from once every 50,000 miles to once every 100,000 miles.

FIGURE 22.14   Cacti. *A.* Prickly pear cacti. *B.* A barrel
cactus. *C.* Peyote. *D.* An organpipe cactus.

A.

B.

C.

D.

In 1944, a marine pilot was forced to bail out of his aircraft over the desert near Yuma, Arizona. He survived the intense heat and low humidity of the area by chewing the juicy pulp of barrel cacti in the vicinity until he was rescued five days later. Since then, the use of cacti for emergency fluids and food has been recommended in most survival manuals. In fact, only three cacti (peyote, living rock, and hedgehog cactus) are known to be inedible because of poisonous principles present; thus, the fruits of most cacti can be eaten. Those of prickly pears, which are occasionally sold in American supermarkets, taste somewhat like pears. Prickly pear fruits also have seeds that Native Americans of the Southwest dried and ground for flour they used in *atole,* a staple food. A good syrup is obtained from boiling the fruits of prickly pears and also those of the giant saguaros. In the past, cactus candy was made by partly drying strips of barrel cactus and boiling them in saguaro fruit syrup, but the cactus is now usually boiled in syrup made from cane sugar.

Native Americans of the Southwest used to scoop out barrel cacti, dry them, and use them for pots. They also mixed the sticky juice of prickly pear cacti in the mortar used in constructing their adobe huts. In Texas, a poultice of prickly pear stem was applied to spider bites. Hopi Indians chewed raw cholla cactus as a treatment for diarrhea, and the skeletons of these cacti were used for flower arrangements.

In the middle of the nineteenth century, Australia imported a few prickly pear cacti, which were planted in the dry interior. These cacti found no natural enemies in their new environment and multiplied rapidly, infesting more than 24 million hectares (60 million acres) within 75 years. In 1925 as an effort to control them, Australia introduced a moth from Argentina. The moth's caterpillars, which feed on prickly pear cacti, gradually brought the plants under control, and the land was made useable again.

Another cactus parasite, the cochineal insect (related to the mealybug, a common houseplant pest), feeds on prickly pear cacti in Mexico. At one time, the insects were collected for a crimson dye they produce, which was used in lipstick and rouge before aniline dyes were introduced.

Peyote cacti are small buttonlike plants that have no spines, with roots resembling those of carrots. They contain several drugs, the best known of which is *mescaline,* a powerful hallucinogen. Dried

FIGURE 22.15   Flowers of lamb's ears mint.

slices of peyote have been used in native religious ceremonies in Mexico for centuries and more recently by at least 30 tribes of Native Americans. The drug gives the user a variety of hallucinations in vivid colors.

The Mint Family (Lamiaceae)

The 3,000 members of the Mint Family are relatively easy to distinguish from those of other families in having stems that are square in cross section, opposite leaves, and bilaterally symmetrical flowers (figure 22.15). They also generally produce aromatic oils in the leaves and stems. The superior ovary is four-parted, with each of the four divisions developing into a nutlet. Included in the family are such well-known plants as rosemary, thyme, sage (not to be confused with sagebrush, which is in the Sunflower Family), oregano, marjoram, basil, lavender, catnip, peppermint, and spearmint.

FIGURE 22.16   A simple apparatus for distilling mint oil at home.

FIGURE 22.18   Chia. (Courtesy Robert A. Schlising)

Mint oils can be distilled in the home with ordinary canning equipment. Whole plants (or at least the foliage) are loosely packed to a depth of about 10 centimeters (4 inches) or more in the bottom of a large canning pot. Then a wire rack or other support is also put in the pot, and a bowl is placed in the middle on the rack. Enough water is added to cover the vegetation, the pot is placed on a range, and the lid is inverted over it. As the water is brought to a boil, ice is placed on the inverted lid. The oils vaporize, and condense when they contact the cold lid, dripping then from the low point into the bowl (figure 22.16). Of course, some moisture also condenses, but the oil, being lighter, floats on top. Peppermint oil in particular is easy to collect this way and will keep indefinitely in a refrigerator.

Mint oils have been used medicinally and as an antiseptic in different parts of the world. Mohegan Indians used catnip tea for colds, and dairy farmers in parts of the midwestern United States use local mint oils to wash their milking equipment. As a result, mastitis, a common disease of dairy cattle, is seldom encountered in their herds. Horehound, a common mint weed of Europe, has become naturalized on other continents and is cultivated in France. A leaf extract is still used in horehound candy and cough medicines. In England, it is a basic ingredient of horehound beer. Vinegar weed, also known as blue curls, is a common fall-flowering plant of western North America. Native Americans of the area used it in cold remedies, for the relief of toothaches, and in a bath for the treatment of smallpox. It was also used to stupefy fish.

Menthol, the most abundant ingredient of peppermint oil (figure 22.17), is widely used today in toothpaste, candies, chewing gum, liqueurs, and cigarettes. The primary American source is the Columbia River basin of Oregon and Washington, where it is grown commercially. Geese are sometimes used in the mint fields to control both insects and weeds, since they do not interfere with the growth of the mint plants themselves.

Ornamental mints include salvias and the popular variegated-leaf Coleus plants, neither of which has typical mint oils in the foliage. Another relatively odorless mint is *chia* (figure 22.18), whose

Flowering Plants and Civilization   459

FIGURE 22.19   A sectioned petunia flower.

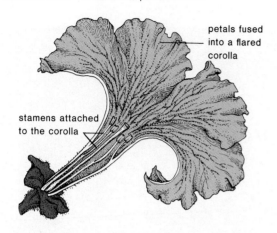

petals fused
into a flared
corolla

stamens attached
to the corolla

habitat is confined to the drier areas of western North America. Native Americans parched chia seeds and used them in gruel. The seeds, which become mucilaginous when wet, were also ground into a paste that was placed in the eye to aid in the removal of dirt particles. The paste was also used as a poultice for gunshot wounds, and Spanish Californians made a refreshing drink from ground chia seeds, lemon juice, and sugar. Chia seeds reportedly contain an unidentified substance that has effects similar to those of caffeine. Before the turn of the century, a physician by the name of Bard maintained that a tablespoon of chia seeds was sufficient to sustain a man on a 24-hour endurance hike. Since that time, backpackers have experimented with the seeds, and results tend to support the earlier claim. A thorough scientific investigation of the matter is needed.

The Nightshade Family (Solanaceae)

Flowers of the Nightshade Family, which is concentrated in the tropics of Central and South America, have their petals fused together, at least at the base, and the filaments of the stamens are fused to the corolla so that they appear to be arising from it (figure 22.19). The superior ovary develops into a berry or a capsule (see figure 8.18). The more than 3,000 species of the family are widespread in distribution, have alternate leaves, and occur as herbs, shrubs, trees, or vines. Well-known representatives include tomato, white potato, eggplant, pepper, tobacco, and petunia.

Many nightshades produce poisonous drugs, some of which have medicinal uses. One of the best-known medicinal drug producers is the deadly nightshade of Europe. A drug complex called *belladonna* is extracted from its leaves. Belladonna, which was used in the "magic potions" of the past, and also for dilating human pupils for cosmetic purposes, is now the source of several widely used drugs, including atropine, scopolamine, and hyoscyamine. Atropine is used in shock treatment, for relief of pain, to dilate eyes, and to counteract muscle spasms. Scopolamine is used as a tranquilizer, and hyoscyamine has effects similar to those of atropine. Capsicum, obtained from a pepper, is used as a gastric stimulant.

Jimson weed (see figure 8.6) is also a source of medicinal drugs that have been used in treatment of asthma and other ailments. The drugs can be fatal if ingested in sufficient quantities but have been much used in controlled amounts in Native American rituals of the past. Records indicate that users became temporarily insane but had no recollection of their activities when the effects of the drug wore off. The drug solanine is present in most, if not all members of the family. Many arthritis sufferers apparently are sensitive to solanine, and a number of arthritics have reported partial relief through total abstinence of members of this family (including potatoes, tomatoes, peppers, and eggplant).

More than 800,000 hectares (2 million acres) of American farmland is allotted to the growing of tobacco, which in its dried form contains 1% to 3% of the drug nicotine. Nicotine is used in certain insecticides, and it is also used for killing intestinal worms in farm livestock. It is, however, an addictive drug. With the mounting evidence of tobacco use by humans pointing to its being a primary cause of heart and respiratory diseases including lung cancer, and other cancers such as those of the mouth and throat, the only "benefit" it may have to humans appears to be as a killer of leeches. It is said that leeches attaching themselves to heavy smokers will drop off dead within five minutes from nicotine poisoning— a very dubious justification for continued human use!

Tomatoes are among the most popular of all "vegetables." About 18 million metric tons (20 million tons) are grown annually around the world. The plants are day-neutral (see chapter 11), and even though they require warm night temperatures (16° C or 60° F) to set fruit well, they are easily cultivated in greenhouses when natural conditions

do not favor their growth. Most tomatoes grown commercially in the United States are processed into juice, tomato paste, and catsup. In Italy, a small amount of edible oil is extracted from the seeds after the pulp has been removed. Most American tomatoes are grown in California, where they are harvested with special machinery developed during the 1960s when inexpensive labor ceased to be available (figure 22.20).

The white or Irish potato is one of the most important foods grown in temperate regions of the world, with annual production estimated at well over 270 million metric tons (300 million tons). The leading producers are the Soviet Union, which accounts for about 30% of the total, China, Poland, and the United States. It is believed that white potatoes originated on an island off the coast of Chile and were sent back to Europe by Spanish invaders of South America in the sixteenth century. In the 1840s, the late blight infestations that destroyed the potato crop in Ireland caused severe famine and subsequent immigration of Irish settlers to the United States, Australia and other parts of the world. When potato tubers are exposed to the sun, they turn green at the surface. Poisonous drugs are produced in the green areas. These have proved fatal to both animals and humans and should never be eaten.

The Carrot Family (Apiaceae)

Many members of the Carrot Family, which is abundantly represented in the northern hemisphere, have savory aromatic herbage. The flowers tend to be small and numerous and are arranged in umbels (see figure 8.7). The ovary is inferior, and the stigma is two-lobed. The petioles of the leaves, which are generally dissected, usually form sheaths around the stem at their bases. Included in the 2,000 members of the family are dill, celery, carrot, parsley, caraway, coriander, fennel, anise, and parsnip. Anise is one of the earliest aromatics mentioned in literature. It is used for flavoring cakes, curries, pastries, and candy. Pocket gophers apparently are attracted by its aroma, and some poison baits are enhanced with anise. A liqueur known as *anisette* is flavored with it.

Another liqueur, called *kümmel,* is flavored with caraway seeds, which are well known for their use in rye and pumpernickel breads.

FIGURE 22.21   Water hemlock.

FIGURE 22.22   Fruits and items associated with members of the Pumpkin Family (Cucurbitaceae). *A.* A Hawaiian ceremonial gourd with feathers attached and gourds used in South America for drinking maté. A metal straw that strains out the maté leaves is resting in one gourd. *B.* Cantaloupes. *C.* A luffa (vegetable sponge).

A.

B.

C.

A few members of the Carrot Family are poisonous. Water hemlock (figure 22.21) and poison hemlock, which are common weeds in ditches and along streams, are deadly and have thus often been fatal to unwary wild food lovers. Socrates is believed to have died as a result of ingesting poison hemlock, which should not be confused with cone-bearing hemlock trees.

Several members of the Carrot Family, such as cow parsnip, squawroot, and hog fennel, have edible roots and were used for food by Native Americans. The reader is advised, however, to be absolutely certain of the identity of such plants before experimenting with them.

The Pumpkin Family (Cucurbitaceae)

Although most species in the Pumpkin Family are tropical or subtropical, numbers occur in temperate areas of both the northern and southern hemispheres. Plants are prostrate or climbing herbaceous vines with tendrils. The flowers have fused petals and an inferior ovary with three carpels. All are unisexual. Some species have both male and female flowers on the same plant, whereas others have only male or only female flowers on one plant. In male flowers, the stamens cohere to varying degrees, depending on the species. The family has about 700 members, several of which have numerous horticultural varieties.

Many important edible plants are included in this family, and some have been cultivated for so long that they are unknown in the wild, their origins being obscure. Well-known edible members of the family include pumpkins, squashes, cucumbers, cantaloupes (figure 22.22B), and watermelons. The vegetable sponge (figure 22.22C), which resembles a

FIGURE 22.23   Parts of a sunflower. *A.* A section through a single floret. *B.* A section through an inflorescence.

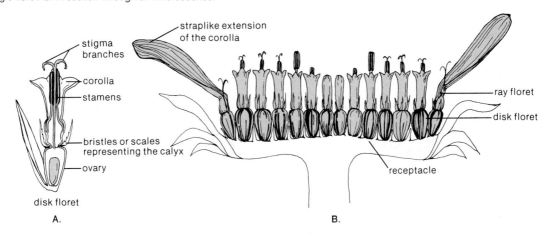

stigma branches

straplike extension of the corolla

corolla

stamens

ray floret

disk floret

bristles or scales representing the calyx

ovary

receptacle

disk floret

A.

B.

large cucumber when it is growing, has a highly netted fibrous skeleton, which serves as a bath sponge after the soft tissues have been removed.

Gourds found in Mexican caves have been dated back to 7000 B.C. Numerous types of gourds (figure 22.22A) are still grown today, and they serve many purposes. Some are scooped out and used for carrying liquids or for food storage, particularly grains. South Americans drink maté, a tea, from gourds, which are also the basis of several types of musical instruments. In parts of Africa, gourds are used to catch monkeys. A type with a narrow neck is scooped out and partly filled with corn or other grains. One end of a rope is then tied to the gourd and the other to a stake driven into the ground. When a monkey tries to grab a fistful of grain, it finds that the neck will not allow its bulging hand to be removed. Most do not realize that letting go of the grain would permit their escape, and they stubbornly hang on until they are captured.

One diminutive cucumberlike vine of the southeastern United States, melonette, has seeds that are reported to be drastically laxative (purgative). Other cucumberlike plants of the western states, manroots (see figure 6.8), produce huge water-storage roots, some weighing as much as 90 kilograms (200 pounds). These roots were crushed by Native Americans and thrown into dammed streams to stupefy fish. An oil from the seeds was applied to the scalp as a remedy for infections that caused hair loss.

The Sunflower Family (Asteraceae)

The Sunflower Family, with approximately 20,000 species, is the second largest of the flowering plant families in terms of number of species. The individual flowers are called *florets*. They are usually tiny and numerous but are arranged in a compact inflorescence so that they resemble a single flower. A sunflower or daisy, for example, consists of dozens if not hundreds of tiny flowers crowded together, with those around the margin having greatly developed corollas that extend out like straps, forming what appear to be the "petals" of the inflorescence (see figure 22.23 for details). In dandelions, all the individual florets of the inflorescence have narrow straplike extensions.

Well-known members of this family include lettuce, endive, chicory, Jerusalem artichoke, dahlia, chrysanthemum, marigold, sunflower, and thistle.

Santonin is a drug obtained from flower buds of a relative of sagebrush that is native to the Middle East. It is used as an intestinal worm remedy. Tarragon, used as a spice in meat dishes and pickles, comes from another relative of sagebrush. Pyrethrum is a natural insecticide obtained from certain chrysanthemum flowers. Fructose, a sugar, is obtained from the tubers of Jerusalem artichokes and dahlias. Dahlias are also renowned for their huge showy flowers, while Jerusalem artichokes are often eaten as a vegetable.

FIGURE 22.24 Flowers of grasses. *A.* An expanded grass spikelet. *B.* An enlargement of a single flower (floret).

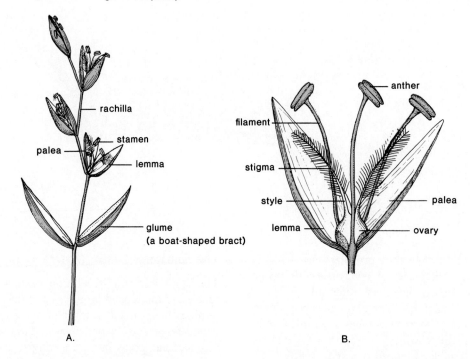

A.

B.

Marigolds are favorite plants of organic gardeners. Their roots are said to release a substance that repels nematodes, and the odor of the leaves repels white flies and other insects. Unfortunately, snails seem to be immune and consume the foliage voraciously.

Many members of this family were widely used by Native Americans for various purposes. The seeds of balsamroot and mule ears were used for food. Balsamroot plants as a whole were eaten raw or cooked, and in the West extracts of both mule ears and tarweeds were used to treat poison oak inflammations. Salsify and thistle roots were also used for food.

Young leaves and roots of dandelions have long been on lists of wild edible plants, and the flowers have been used to make wine. Roasted dandelion and chicory roots have been used as a coffee substitute. During World War II chicory was grown as a crop specifically for use as a coffee adulterant.

The dried and boiled leaves of American yarrow are said to make a nourishing broth. The European yarrow, which has become naturalized in North America, contains a drug that has been used in suppression of menstruation. This plant is believed to have been used by Achilles in treating the wounds of his soldiers.

Sunflowers themselves were widely used by Native Americans for their seeds, which were ground into a meal for bread. They are grown commercially today primarily for the edible oil extracted from the seeds, but their use in seed form by modern Americans is increasing.

The Grass Family (Poaceae)

Although there are possibly only one-fourth as many species of grasses as there are members of the Sunflower Family, it is the largest family of flowering plants in terms of numbers of individuals and is the most widely distributed. The flowers of grasses are highly specialized in structure and have a terminology all their own (figure 22.24). The calyx and corolla are represented by tiny, inconspicuous scales, and the flowers are protected by boat-shaped bracts. The stigmas, when they are exposed, are feathery, and the leaves sheathe the stem at their bases.

All of the cereals, including wheat, barley, rye, oats, rice, and corn, belong in this family which, as previously mentioned, includes nine of the ten economically most important plants in the world; indeed, civilization as we know it would be vastly different without them. More than 900 million metric tons (1 billion tons) of cereals, feeding more than half of the world's population, are harvested each year, primarily in the Orient, North America, and Europe.

Sugar cane (figure 22.25), from which about 55 million metric tons (60 million tons) of sugar is extracted annually, is grown at lower elevations throughout humid tropical areas. It is a large grass, often growing to heights of 6 meters (20 feet). After the cane is harvested, the liquid portion is squeezed out. The solid waste is sometimes made into paper or particle board. It has also been converted to gasoline in South Africa and is now used in the production of electric power in Hawaii. Through a series of steps, sugar is crystallized out of the liquid portion and separated by centrifuging. The dark remnant, known as molasses, may be used as a base for rum or alcohol production.

Grasses have been widely used by primitive peoples for making mats and baskets and for thatching huts. One variety of sorghum is grown for its fibers, which are made into brooms, although natural broom fiber has now been largely replaced by synthetic materials. Some varieties are grown for their seeds, which are processed into cereal flours. Others are a source of silage and a carnaubalike wax. Citronella oil, once a common ingredient of mosquito repellents, is obtained from a grass grown in Indonesia. It is now used in cheaper soaps, cosmetics, and perfumes. Related grasses are the source of lemon grass oil, which is used in ways similar to those of citronella oil.

Juice squeezed from fresh young grass mowings has proved to have a high protein content and is being investigated as a source of protein for future human consumption.

The Lily Family (Liliaceae)

Although the 2,000 species of lilies are particularly abundant in the tropics and subtropics, they occur in almost any area that supports vegetation. The flowers are often large, and their parts are all in multiples of three, with the sepals frequently colored the same as and resembling the petals (figure 22.26). The flowers of the related Amaryllis Family are similar. Lily flowers, however, generally have superior ovaries, whereas those of the Amaryllis Family are mostly inferior.

FIGURE 22.25   Sugar cane plants.

FIGURE 22.26   A tiger lily flower.

FIGURE 22.28   An uprooted California soaproot plant.

In addition to numerous types of lilies used widely as ornamentals, the family includes asparagus, sarsaparilla, squill, meadow saffron, bowstring hemp, and *Aloe*. Sarsaparilla, which was at one time widely used for flavoring soft drinks and medicines, is obtained from the roots of a genus of woody vines whose stems are often covered with prickles. The bulbs of squills are the source of a rodent poison and also of a drug used as a heart stimulant. Meadow saffron is the source of *colchicine,* a drug once used to treat rheumatism and gout but now much more widely used in experimental agriculture to interfere with spindle formation in cells so that the chromosome number of plants may be artificially increased. This increasing of the chromosome number can result in larger and more vigorous varieties of plants. (Meadow saffron should not be confused with true saffron, a member of the Iris Family and the source of the world's most expensive spice and a powerful yellow dye.) Bowstring hemps are related to the familiar, seemingly indestructible houseplants called sansevierias (figure 22.27) which have long, narrow, stiff leaves that stand upright from the base. The plants are cultivated in tropical Africa for their long fibers, which are used for string, rope, bowstrings, mats, and cloth. New Zealand flax, a larger plant, is grown in South America and New Zealand for similar purposes but is also widely used in ornamental plantings.

Several *Aloe* species produce juices used to treat X-ray and other burns. African *Aloe* species are prized as ornamentals in areas with milder climates. Their thick fleshy leaves have short spines along the margins; the spines were once used for phonograph needles.

Many lily bulbs are edible and were widely used for food by Native Americans (**Caution:** lily bulbs should not be confused with those of daffodils and other members of the Amaryllis Family. Daffodil and related bulbs are highly poisonous), but their use for food now should be discouraged, as such use to any great extent might render many species extinct. One member of the Lily Family confined to California and southern Oregon, the California soaproot (figure 22.28), had several uses in addition to being an important food item for Native Americans of the region. The large bulbs are covered with coarse fibers, which were removed and tied to sticks to make small brooms. The bulbs themselves produce a lather in water and were used for soap. Sometimes numbers of bulbs were crushed and thrown in a small stream

FIGURE 22.29   Bamboo orchids growing wild in Hawaii.

that had been dammed. Fish would be stupefied and float to the surface. The bulbs were generally eaten after being roasted in a stonelined pit in which a fire had been made. While they were roasting, a sticky juice would ooze out. This was used for gluing feathers to arrow shafts.

A resin used in stains and varnishes exudes from the stem of dragon's blood plants. Grass trees of Australia yield resins used in sealing waxes and varnishes.

The Orchid Family (Orchidaceae)

Members of this very large family, which, according to some authorities has more than 35,000 species, are very widely distributed although, like members of the Lily Family, they are especially abundant in the tropics. In many genera the number of individual plants at any one location may be quite small—sometimes limited to a single plant.

The flowers show exceptional variation in size and form, and the habitats of the plants are equally varied. The flowers of one Venezuelan species has a diameter of less than 1 millimeter (1 twenty-fifth of an inch), whereas those of a species native to Madagascar may exceed 45 centimeters (about 18 inches) in length. One species of *Dendrobium* orchid from Java has flowers that are so delicate they perish within five or six minutes after opening. Many orchids are epiphytic on the bark of trees, with one such species native to Malaysia and the Philippine Islands producing 10,000 flowers during a five-month flowering season on plants that weigh more than 1 metric ton. Others are aquatic or terrestrial, and a saprophytic species native to western Australia grows and flowers entirely underground (see figure 8.1).

Orchids commonly have three sepals and three petals, with one of the petals (the lip petal) differing in form from the other two (figure 22.29). The stamens and pistil are united in a single structure, the *column* (see figure 21.7), which is unique to the family. The stigma often consists of a sticky depression on the column, and the anthers, which contain sacs of pollen called *pollinia,* are usually covered with a cap until they are removed by an insect or other pollinator. The specific adaptations between orchid flowers and their pollinators is extraordinary and sometimes bizarre (see figure 23.23).

Orchids have minute seeds that are often produced in prodigious numbers (e.g., a single fruit of certain orchid species may contain up to 1 million seeds). Each seed consists of only a few cells, and, in order for a seed to germinate it must become associated with a specific mycorrhizal fungus that produces substances necessary for its development. Once a seed has germinated, it may take from 6 to 12 years or more before the first flower appears.

Contrary to popular belief, some orchids can be grown relatively easily on a windowsill that has bright light, but not direct sunlight (see appendix 4). Because orchids are among the most beautiful and prized of flowers, a large industry has grown up around their culture and propagation (see the section on "mericloning and tissue culture" in appendix 4). One species, the vanilla orchid, is grown commercially in the tropics for its fruits, which are the source of true vanilla flavoring.

See appendix 1 for the scientific names of the plants mentioned in this chapter.

SUMMARY

Distinguishing between fact and fancy in reported past uses of plants is often difficult because plants may have had medicinal or other values despite the fact that their use was based on superstition or reasoning devoid of scientific knowledge. At present, teams of specialists in several fields are interviewing tribal medicine men and women of the tropics in an attempt to save potentially useful plants from extinction.

N. I. Vavilov, building on the earlier work of A. de Candolle, determined that there were eight major centers of distribution of cultivated plants. These centers are located in China, India, and Indo-Malaysia, Central Asia, the Near East, the Mediterranean, the Abyssinian area of Africa, Central America and adjacent South Mexico, and South America (with independent centers in Peru, Ecuador, Bolivia, Chiloe Island, Brazil, and Paraguay).

The flowering plant families surveyed in somewhat phylogenetic sequence are as follows: the Buttercup Family (Ranunculaceae)—its members include buttercup, columbine, larkspur, anemone, monkshood, *Clematis,* goldenseal, and wolfsbane; the Laurel Family (Lauraceae)—its members include cinnamon, camphor, sassafras, sweet bay, California bay (or myrtle), avocado, and laurel; the Poppy Family (Papaveraceae)—its members include bloodroot and opium poppy (the source of medicinal drugs and heroin); the Mustard Family (Brassicaceae)—its members include cabbage, cauliflower, brussels sprouts, broccoli, radish, kohlrabi, turnip, horseradish, watercress, rutabaga, shepherd's purse, western wallflower, dyer's woad, and camelina; the Rose Family (Rosaceae)—its members include stone fruits (e.g., cherry, apricot, peach, plum), strawberry, raspberry, rose (whose fruits are a source of vitamin C), and related wild species; the Legume Family (Fabaceae)—its members include all legumes (e.g., pea, bean, lentil, peanut), alfalfa, clover, licorice, wattle (source of tannins), indigo, logwood, locoweed, sensitive plant (*Mimosa pudica*), and winged bean (a tropical bean with high protein levels); the Spurge Family (Euphorbiaceae)—its members include spurge, poinsettia, cassava, Pará rubber, gopher plant, candelilla, castor bean, Mexican jumping bean, crown-of-thorns, and tung oil tree; the Cactus Family (Cactaceae)—its members include all cacti (e.g., prickly pear, cholla, barrel cactus, peyote); the Mint Family (Lamiaceae)—its members include rosemary, thyme, sage, oregano, marjoram, basil, lavender, catnip, peppermint, spearmint, horehound, salvia, coleus, and chia; the Nightshade Family (Solanaceae)—its members include tomato, white potato, eggplant, pepper, tobacco, petunia, belladonna (source of medicinal drugs), and jimson weed (source of hallucinogenic drugs); the Carrot Family (Apiaceae)—its members include dill, celery, carrot, parsley, caraway, coriander, fennel, anise, parsnip, water hemlock, poison hemlock, cow parsnip, squawroot, and hog fennel; the Pumpkin Family (Cucurbitaceae)—its members include pumpkin, squash, cucumber, cantaloupe, watermelon, vegetable sponge, gourd, melonette, and manroot; the Sunflower Family (Asteraceae)—its members include sunflower, dandelion, lettuce, endive, chicory, Jerusalem artichoke, dahlia, chrysanthemum, marigold, thistle, sagebrush, pyrethrum, balsamroot, tarweed, and yarrow; the Grass Family (Poaceae)—its members include all cereals (e.g., wheat, barley, rye, oats, rice, corn), sugar cane, sorghum, citronella, and lemon grass; the Lily Family (Liliaceae)—its members include lilies, asparagus, sarsaparilla, squill, meadow saffron, bowstring hemp, *Aloe,* and New Zealand flax; the Orchid Family (Orchidaceae)—one of its members is the source of vanilla flavoring.

REVIEW QUESTIONS

1. To which flowering plant family does each of the following belong: poinsettia, lupine, columbine, peach, pear, cinnamon, sarsaparilla, belladonna, peyote, horehound, rubber, gourd, jimson weed, parsley, sorghum, asparagus, broccoli, lettuce, tomato, and opium?
2. Make a list of the poisonous plants in the families discussed.
3. Which plants mentioned are or have been used for medicines?
4. Which plants mentioned have been used for tools or utensils?
5. Native Americans made extensive use of plants for a wide variety of purposes in the past. List such uses for as many plants as possible.

DISCUSSION QUESTIONS

1. Scientific investigations often take a great deal of time and money. Is it worth the effort to check out scientifically the past uses of plants?
2. A return to exclusively herbal medicines is being advocated in some quarters. Is this a good idea? What are the pros and cons?
3. Would you expect drugs produced naturally by plants to be more effective or better than drugs produced synthetically? Why?
4. If you were asked to single out the three most important families of the 16 discussed, which would you choose? Why?
5. A number of wild edible plants were mentioned in this chapter. What would happen if a large portion of the population were to gather these wild plants as a major source of food?

ADDITIONAL READING

Advisory Committee on Technology Innovation. 1975. *Underexploited tropical plants with promising economic value.* Washington, DC: National Academy of Sciences.

Advisory Committee on Technology Innovation. 1978. *Field guide to medicinal wild plants.* Harrisburg, PA.: Stackpole Books.

Benson, L. 1979. *Plant classification,* 2d ed. Lexington, MA.: D. C. Heath and Company.

Chrispeels, M. J., and D. Sadava. 1977. *Plants, food and people.* San Francisco: W. H. Freeman and Co.

Gibbons, E. 1970. *Stalking the wild asparagus.* New York: David McKay Co.

Hardin, J. W., and J. M. Arena. 1974. *Human poisoning from native and cultivated plants.* Durham, NC: Duke University Press.

Harrington, H. D. 1974. *Edible native plants of the Rocky Mountains.* Albuquerque: University of New Mexico Press.

Kingsbury, J. M. 1964. *Poisonous plants of the United States and Canada,* 3d ed. Englewood Cliffs, NJ: Prentice-Hall.

Kirk, D. R. 1975. *Wild edible plants of the western United States,* Colored ed. Healdsburg, CA: Naturegraph Publishers.

Krochmal, A., and C. Krochmal. 1984. *A field guide to the medicinal plants.* New York: Times Books.

Kunkel, G. 1984. *Plants for human consumption: annotated checklist of edible phanerogams and ferns.* Forestburgh, NY: Lubrecht and Cramer.

Lewis, W. H., and M. P. F. Elvin-Lewis. 1982. *Plants affecting man's health.* New York: John Wiley & Sons.

Murphey, E. V. A. 1959. *Indian uses of native plants.* Fort Bragg, CA: Mendocino County Historical Society.

Radford, A. E. 1986. *Fundamentals of plant systematics.* New York: Harper & Row, Publishers.

Schery, R. W. 1972. *Plants for man,* 2d ed. Englewood Cliffs, NJ: Prentice-Hall.

Sweet, M. 1976. *Common edible and useful plants of the west,* rev. ed. Healdsburg, CA: Naturegraph Publishers.

Turner, N. J. 1975. *Food plants of British Columbia Indians, Part 1: Coastal peoples.* Victoria: British Columbia Provincial Museum.

Turner, N. J., and A. F. Szczawinski. 1980. *Wild green vegetables of Canada.* Chicago: University of Chicago Press.

Overview

*T*his chapter explores some of the ecological topics not already extensively discussed elsewhere in the text.

Populations, communities, and ecosystems are discussed first. This is followed by a brief look at producers, primary and secondary consumers, decomposers, and food chains. Then energy flow through an ecosystem is considered, and human disruption of ecosystems is briefly explored. Next, the nitrogen and carbon cycles are introduced, and discussions of successions follow. The subject of fire ecology is then broached. Major biomes of North America are described, and a telescopic view of their plants and animals is given. The chapter concludes with a discussion of pollination ecology.

23   *Ecology*

Some Learning Goals

1. Know the functions of producers, primary consumers, secondary consumers, and decomposers in an ecosystem.

2. Understand the cycling of energy in an ecosystem.

3. Know at least ten ways in which humans have disrupted ecosystems.

4. Learn how nitrogen and carbon are cycled.

5. Define succession and know a succession that begins with bare rocks.

6. Define ecotype, secondary succession, eutrophication, and climax vegetation.

7. Learn the major biomes of North America and describe the principal living members of each.

Outline

INTRODUCTION

The broad discipline of biology that deals not only with plant and animal relationships but with the relationship of organisms in general to one another and to their environment is known as **ecology.**

The word *ecology* was first proposed by the German biologist Ernst Haeckel in 1869, and ecology was recognized as a field of biological investigation at the beginning of this century. Its origins, however, date back to early civilizations when humans first learned to modify their environment through the use of tools and fire. Since the 1960s, ecology has become a household word, and it is now such a vast area of study, with so many aspects, that it is not possible in a text of this nature to do more than mention a few aspects. In the past two decades alone, topics such as the effects of pollution on lands, waters, and peoples have filled literally thousands of volumes. During this time attempts to rectify or arrest ecological damage have resulted in the requirement that various construction projects file environmental impact reports before proceeding. These reports provide information that helps various agencies evaluate the possible effects a proposed project may have on the flora, fauna, and physical environment and to determine what mitigations may be necessary before the project can or should be approved. The process has sometimes resulted in a great deal of controversy and emotional debate between persons involved in industry and those who feel that preservation of the environment should supersede material considerations.

Various aspects of ecology were treated in earlier chapters. Soils and organisms associated with them were discussed in chapter 9. Some of the adaptations and modifications of leaves associated with specific environments were discussed in chapter 7. The effects of light and temperature on plants were covered in chapters 10 and 11. Other important ecological topics comprise the bulk of this chapter, but the reader is referred to the "Additional Reading" section for a sampling of works covering a number of these themes in greater depth.

We recognize that plants, animals, and other organisms tend to be associated in various ways with one another and also with the physical environment. For example, the term *forest* is applied to *populations* (groups of individuals of the same species) of trees or other plants that form a **plant community** (unit composed of all the populations of plants occurring in a given area). The lichen and moss flora on a rock also constitutes a community, as do the various seaweeds in a tidepool (figure 23.1). These communities also invariably have animals and other living organisms associated with them; it is preferable, therefore, to refer to the unit composed of all the populations of living organisms in a given area as a **biotic community.** Considered together, the **communities** and their physical environments constitute **ecosystems,** which are interconnected by physical, chemical, and biological processes. Each of these terms may be applied to entities ranging in size from microscopic to global.

FIGURE 23.1   A tidepool community at low tide. (Courtesy I. A. Abbott)

POPULATIONS

Populations may vary in numbers, in density, and in the total mass of individuals. Depending on circumstances, a field biologist may investigate a population in various ways. If, for example, a conservation organization is concerned about the preservation of a rare or threatened species, the organization may simply count the *number* of individuals, although this may not always be feasible. If a count is not feasible, the organization may estimate population *density* (number of individuals per unit volume—e.g., five blueberry bushes per square meter). If the individuals in a population vary greatly in size or are unevenly scattered, a better estimate of the population's importance to the ecosystem may be calculated by determining the *biomass* (total mass of the individuals present). Population biology studies also include, among many things, physiological ecology and the examination and tabulation of factors such as seed dispersal, germination, survival, and pollination in the study of relationships of populations to their environment.

COMMUNITIES

Communities are composed of populations of one to many species of various types of organisms living together in the same location. Similar communities occur under similar environmental conditions, although actual species composition can vary considerably and can change between the boundaries of an area. A community is difficult to define precisely, however, because members of one community may also occur in other communities. Furthermore, members of one community may have specific genetic adaptations to that community, so that if individuals are transplanted to a second, different community where the same species occurs, the transplanted individuals may not necessarily be able to survive alongside their counterparts, which are themselves adapted to this second community. Individuals adapted to specific communities within their overall distribution are called *ecotypes*.

Analysis and classification of communities is important in the preparation of maps that form the basis of activities such as land use planning, forestry, natural resource management, and military maneuvers.

ECOSYSTEMS

Living organisms that are interacting with one another and with factors of the nonliving environment constitute an *ecosystem*. The nonliving factors of the environment include light, temperature, oxygen level, air circulation, precipitation, energy, and soil type. The distribution of a plant species in an ecosystem is controlled mostly by temperature, precipitation, soil type, and the effects of other living organisms (biotic factors). In Mediterranean climates such as those that occur in parts of California and Chile, for example, nearly all precipitation occurs during the winter months and the summers are dry. This type of climate favors spring annuals that complete their life cycles by summer. Forests may occur in areas where heavy winter snowfall soaks deeply into the soil, thus compensating to a certain extent for the lack of summer moisture.

Species of plants that occur naturally in areas of low precipitation and high temperatures (*xerophytes*) generally have modifications of leaves and other parts (see discussions under "Specialized Leaves" in chapter 7 and "Regulation of Transpiration" in chapter 9) that reduce transpiration and thus adapt them to their particular environment. Plants of arid areas may also have specialized forms of photosynthesis (such as CAM photosynthesis—see chapter 10). Similarly, plants that grow in water (*hydrophytes*) have modifications that adapt them to their aquatic environment (see chapter 7).

The mineral content of soils plays an important role in influencing the distribution of plant species. For example, serpentine soils—which contain relatively high amounts of magnesium, iron, usually nickel and chromium, and low amounts of calcium and nitrogen—often support species that are not found on nearby nonserpentine soils. Biotic factors such as competition for light, the mineral nutrients and water available, and grazing by the animal members of the biotic community also influence the distribution of plant species.

Ideally, ecosystems sustain themselves entirely through photosynthetic activity and the recycling of nutrients. Organisms that are capable of carrying on photosynthesis capture light energy and convert it, along with carbon dioxide and water, to

FIGURE 23.2　A food web.

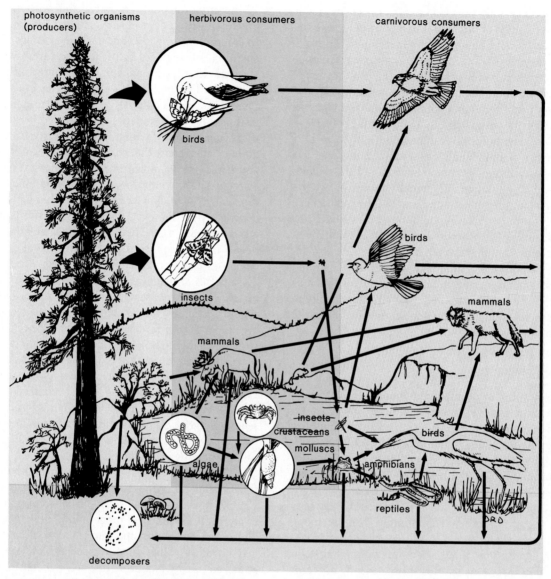

photosynthetic organisms
(producers)

herbivorous consumers

carnivorous consumers

birds

insects

mammals

birds

mammals

algae

insects
crustaceans

molluscs

birds

amphibians

reptiles

decomposers

energy-storing molecules. Such organisms are called **producers.** Animals such as cows, caribou, and caterpillars feed directly on producers. They are called **primary consumers. Secondary consumers,** such as tigers, toads, and tsetse flies, feed on primary consumers. **Decomposers** break down organic materials to forms that can be reassimilated by the producers. The foremost decomposers in most ecosystems are bacteria and fungi.

In any ecosystem, the producers and consumers interact to form food chains or interlocking *food webs,* which determine the flow of energy through the different levels (figure 23.2). Since most organisms have more than a single source of food and are

themselves often consumed by a variety of consumers, there are considerable differences in the length and intricacy of food chains or webs.

Energy itself, which enters at the producer level, cannot be recycled in an ecosystem. Only about 1% of the light energy falling on a temperate zone community is actually converted to organic material. The energy then gradually escapes in the form of heat as it passes from one level to another. It has been estimated that when cattle graze only about 10% of the energy stored by the green plants they consume is converted to animal tissue. Most of the remaining energy dissipates into the atmosphere as heat. When we eat beef, only roughly 10% of its stored energy is utilized by our bodies to manufacture new blood

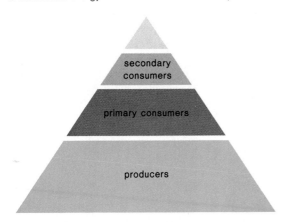

FIGURE 23.3   An energy pyramid of an ecosystem. There is much more energy at the bottom than at the top.

cells and otherwise sustain life; the remaining energy is converted to heat. Obviously, then, in a long food chain the final consumer gains only a tiny fraction of the energy originally captured by the producer at the bottom of the chain. Conversely, there is proportionately much less loss of energy between levels in short food chains. Let us assume, for example, that for every 100 calories of light energy falling on lettuce plants each day, 10 calories are converted to lettuce tissue. Then suppose that the lettuce is fed to hogs. Again only 10%, or 1 calorie, of the original energy may be converted to animal tissue. If as secondary consumers we eat such pork, our bodies, in turn, may utilize 10% of the energy available in the pork, or only 0.1% of the original energy available. If, however, we eat the lettuce directly, we end up with the hog's percentage which is ten times more energy than we would get if we ate the hog that ate the lettuce. From this it follows that a vegetarian diet makes much more efficient use of solar energy than one that relies heavily on meats, and where food is scarce or humans are very abundant (as in India or Ethiopia) humans become virtual vegetarians. It also follows that in terms of the numbers of individuals and the total mass there is a sharp reduction at each level of the food chain. In a given portion of ocean, for example, there may be billions of microscopic algal producers supporting millions of tiny crustacean consumers, which, in turn, support thousands of small fish, which meet the food needs of scores of medium-sized fish, which are finally consumed by one or two large fish (figure 23.3). In other words, one large fish may very well depend on a billion tiny algae to meet its energy needs every day.

The interrelationships and interactions among the components of an ecosystem can be quite complex, but all function together in a regulatory fashion. An increase in food made available by producers can result in an increase in consumers, but the increased number of consumers reduces the available food, which then inevitably leads to a reduction of consumers to earlier levels. The net result is sustained self-maintenance of the ecosystem. This is a basis for the concept of the *Balance of Nature*.

INTERACTIONS BETWEEN PLANTS, HERBIVORES, AND OTHER ORGANISMS

While it is easy to see that the total mass of consumers is largely determined by the total mass of food made available by the producers, the interactions among producers themselves and also between the decomposers and the other members of the ecosystem are usually more subtle. Many flowering plants produce substances that either inhibit or promote the growth of other flowering plants. Black walnut trees, for example, produce a substance that wilts tomatoes and potatoes and injures apple trees that come in contact with black walnut roots, and many other plants produce *phytoalexins* (chemicals that kill or inhibit disease fungi or bacteria), making them resistant to various diseases (see discussion under "Use of Resistant Varieties" in appendix 2). Conversely, some bacteria and fungi limit higher plant growth through the production of various inhibitory chemical compounds. Other bacteria, fungi, and flowering plants limit population size through parasitism. The degree of parasitism varies considerably with the organisms involved. Some of the species of the Figwort Family (which includes snapdragons and similar plants) have no chlorophyll and depend entirely on their flowering plant hosts for their energy and other nutritional needs. Other related species do have chlorophyll but apparently also require supplemental food from their hosts. Still other species often parasitize the roots of certain plants but are also capable of existing independently.

FIGURE 23.4   A species of *Acacia* that is host to ants that live in its hollow thorns. The ants attack any organisms, large or small, that come in contact with the plant. The plant provides food for the ants through nectaries at the bases of the thorns (*A*) and nitrogenous bodies at the tips of the leaflets (*B*).

A.

B.

Mycorrhizal fungi are intimately associated with the roots of most woody and many other plants in such a way that both organisms derive benefit (see *mutualism* in chapter 6). The fungi greatly increase the absorptive surface of the root, usually playing a major role in the absorption of phosphorus and other nutrients, while utilizing root cells as a source of energy.

More than 100 years ago a naturalist by the name of Thomas Belt first called attention to an association in the tropics between ants and thorny, rapidly-growing species of *Acacia* (figure 23.4). The *Acacia,* which has large hollow thorns at the base of each leaf, is host to ants, which feed on sugars, fats and proteins produced by petiolar nectaries and special bodies at the tip of each leaflet. The ants live within the hollow thorns and vigorously attack any other organism, from insects to large animals, that come in contact with the plant. They also kill, by *girdling,* any plant that touches the *Acacia* (see "bridge grafting" in appendix 4). It has been demonstrated experimentally that plants of the species

whose ants have been removed grow very slowly and usually soon die from insect ravages or shading by other plants.

Large herbivorous animals such as deer and moose feed on a wide variety of plants, each differing in nutritional value. Each plant species also produces different combinations, types, and amounts of chemical compounds in addition to proteins, fats, and carbohydrates. Many of the chemical compounds are poisonous to the consumers, but the animals do not display symptoms of poisoning because their digestive systems are capable of breaking down the compounds and eliminating or excreting them to a limited degree. The limitations imposed by such compounds result in the consumers varying their diet, seeking familiar foods and being wary of new ones. If a plant species did not have some natural defense, such as chemical compounds or structural modifications (for example, spines), primary consumers of all kinds from insects to elephants would soon render that species extinct. In an ecosystem, the defenses that both producers and consumers have against each other have been developed through a process of coevolution resulting from natural selection, and are maintained in delicate balance.

HUMANS IN THE ECOSYSTEM

It has been estimated that the total human population of the world was less than 20 million in 6000 B.C. During the next 7,750 years it rose to 500 million; by 1850 it had doubled to 1 billion, and 70 years later it had doubled again to 2 billion. The 4.48 billion mark was reached in 1980 and within five years had grown to 4.89 billion. Estimates for 1990 and 2000 are 5.32 billion and 6.25 billion, respectively. The Earth remains constant in size, but humans obviously have occupied a great deal more of it over the past few centuries or at least have greatly increased in density of population. In feeding, clothing, and housing themselves humans have had a major impact on their environment. They have cleared natural vegetation from vast areas of land and drained wetlands; they have dumped wastes and other pollutants into rivers, oceans, lakes, and added pollutants to the atmosphere; they have killed pests and plant disease organisms with poisonous substances. These poisonous substances have also killed natural predators and other useful organisms, and in general have thoroughly disrupted the delicately balanced ecosystems that existed before humans began their depredations.

If humans are to survive on this planet it is incumbent upon us to stop increasing in numbers, and the many unwise agricultural and industrial practices that have accompanied the burgeoning of human populations must be supplanted with practices more in tune with restoring some ecological balance. Agricultural practices of the future will have to include the return of organic material to the soil after each harvest (see the section on composting in chapter 15), instead of adding only inorganic fertilizers. Harvesting of timber and other crops will have to be done in a manner that prevents topsoil erosion, and the practice of clearing brush with chemicals will have to be abolished. Industrial pollutants will have to be rendered harmless and recycled whenever possible. Many substances that now are still largely discarded (e.g., garbage, paper products, glass, metal cans) will also have to be recycled on a much larger scale. Biological controls (see appendix 2) will have to replace the use of poisonous controls whenever possible. Water and energy conservation will have to be universally practiced, and rare plant species, with their largely unknown gene potential for future crop plants, will need to be saved from extinction by preservation of their habitats and by other means. The general public will have to be made acutely aware of the urgency for wise land management and conservation—which will be needed especially when pressures are exerted by influential forces promoting unwise measures in the name of "progress"—before additional large segments of our natural resources are irreparably damaged or lost forever. The alternative appears to be nothing less than death from starvation, respiratory diseases, poisoning of our food and drink, and other catastrophic events ensuring the premature demise of large segments of the world's population.

THE NITROGEN CYCLE

As was noted in chapter 3, much of the protoplasm of living cells consists of protein. The most abundant element in our atmosphere, nitrogen, constitutes about 18% of the protein. There are nearly 69,000 metric tons of nitrogen in the air over each hectare of land (35,000 tons per acre), but the total amount of nitrogen in the soil seldom exceeds 3.9 metric tons per hectare (2 tons per acre) and is usually considerably less. This discrepancy results from the nitrogen of the atmosphere being chemically inert, which is another way of saying that it will not combine readily with other molecules. It is, therefore, largely unavailable to plants and animals for their use in building proteins and other substances containing nitrogen.

As noted in chapter 15, most of the nitrogen supply of plants (and indirectly, therefore, of animals) is derived from the soil in the form of inorganic compounds and ions taken in by the roots. These compounds and ions include those that contain nitrogen chemically combined with oxygen or hydrogen. They are produced by bacteria and fungi as these break down the more complex molecules of dead plant and animal tissues to simpler ones. Some nitrogen from the air is also *fixed,* that is, converted to ammonia or other nitrogenous compounds by various *nitrogen-fixing bacteria* (see figure 15.7). Some of these organisms gain access to various plants, particularly legumes (e.g., peas, beans, clover, alfalfa), through the root hairs while others live free in the soil.

FIGURE 23.5   The nitrogen cycle.

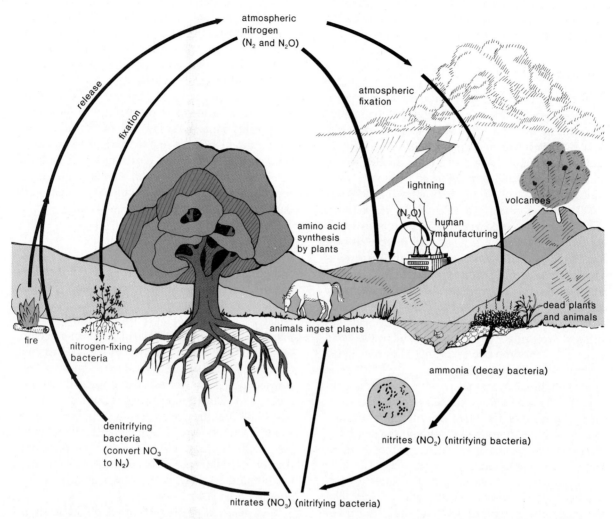

atmospheric
nitrogen
(N_2 and N_2O)

release

fixation

atmospheric
fixation

lightning

volcanoes

(N_2O)

human
manufacturing

amino acid
synthesis
by plants

dead plants
and animals

fire

nitrogen-fixing
bacteria

animals ingest plants

ammonia (decay bacteria)

denitrifying
bacteria
(convert NO_3
to N_2)

nitrites (NO_2) (nitrifying bacteria)

nitrates (NO_3) (nitrifying bacteria)

As illustrated in figure 23.5, there is a constant flow of nitrogen from dead plant and animal tissues into the soil and from the soil back to the plants. Decay bacteria and fungi can break down enormous quantities of dead leaves and other tissues to tiny fractions of their original volumes within a few days to a few months. If they were abruptly to cease their activities, the available nitrogen compounds would be completely exhausted within a few decades, and the carbon dioxide supply needed for photosynthesis seriously depleted. Forests, jungles, and prairies would die as the accumulations of shed leaves, bodies, and debris buried the living plants and shielded their leaves from the light essential to photosynthetic activity. At present, even with the various bacteria involved in the nitrogen cycle functioning normally, the total amount of nitrogen in the soil is not being increased by their activities but rather merely being recycled.

Further, significant amounts of nitrogen are continually being lost as water leaches it out or carries it away through erosion of topsoil. More is lost with each harvest, the average crop removing about 25 kilograms per hectare (25 pounds per acre) per year. This nitrogen loss from the harvesting of crops can be sharply reduced if vegetable and animal wastes are recycled and returned to the soil each year. While bacteria are decomposing tissues they use nitrogen, and little is available until they die and release their accumulations into the soil. Accordingly, crops should not be planted in soils to which only partially decomposed materials have been added until bacteria have completed their breakdown of organic matter. Likewise, when sawdust, straw, or other organic mulches are spread around plants in a garden to control weeds and conserve soil moisture, less nitrogen will be available to the growing plants until the mulches have been decomposed.

FIGURE 23.6   The carbon cycle.

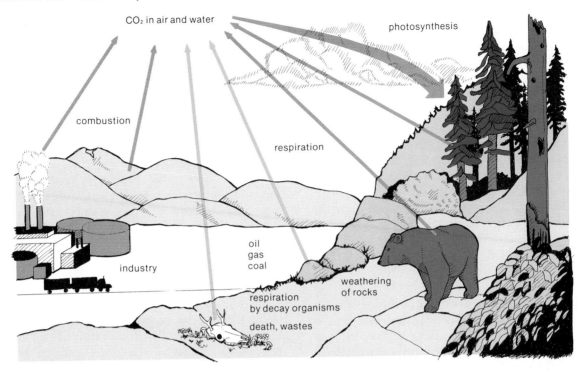

Weeds and stubble are frequently burned as a means of eradication. Fire, however, causes serious loss of nitrogen, which has to be replaced. It has been estimated that the annual combined loss of nitrogen from the soil in the United States from fire, harvesting, and other causes exceeds 21 million metric tons (23 million tons), and only 15.5 million metric tons (17 million tons) are replaced by natural means. To offset the net loss, some 32 million metric tons (35 million tons) of inorganic fertilizers are applied to the soils each year. If organic matter is not added at the same time, however, this application of inorganic fertilizers, combined with the annual burning of stubble, may result eventually in the creation of a *hardpan* soil. Hardpan develops through the gradual accumulation of salt residues, which dissolve humus and disrupt the structure of the soil, causing the clay particles to clump and also producing colloids that are impervious to moisture. In hardpan soils and others low in oxygen (e.g., flooded areas), *denitrifying bacteria* use nitrates instead of oxygen in their respiration, thus rapidly depleting the remaining soil nitrogen.

Precipitation returns a little nitrogen to the soil from the atmosphere, where it has accumulated as a result of the action of light on industrial pollutants, the conversion of nitrogen gas to nitrates by flashes of lightning, and diffusion of ammonia released through decay. The activities of nitrogen-fixing bacteria and volcanoes also contribute to the natural replenishment of nitrogen by converting it to forms that can be utilized by plants.

THE CARBON CYCLE

In the process of recycling nitrogen, bacteria also play a major role in recycling carbon and many other substances (figure 23.6). As noted in chapter 10, one of the two raw materials of photosynthesis is carbon dioxide, which constitutes 0.03% of our atmosphere. It is estimated that the combined plant life of the oceans and the land masses uses about 14.5 billion metric tons (16 billion tons) of carbon obtained from carbon dioxide every year. This is replaced through the respiration of all living organisms, with perhaps as much as 90% or more being produced by the incredible numbers of decay bacteria and fungi as they decompose tissues. Lesser amounts of carbon dioxide are released into the air as a result of the burning of fossil fuels by the internal combustion engines of industry and transportation, and a small amount originates with fires and volcanic activity. At the present rate of use by photosynthetic plants, it has been calculated that all of the carbon dioxide of our atmosphere would be used up in about 22 years if it were not constantly being replenished.

For aeons carbon, nitrogen, water, phosphorus, and other molecules have been passing through cycles. Some molecules that were a part of a primeval forest that became compressed and turned to coal may have become part of another plant after the coal burned. Then the new plant may have been eaten by an animal, which, in turn, contributed molecules to a part of yet another living organism. Just think of where the molecules in your own body may have been in the past few billion years. They may well have been a part of some prehistoric seaweed, a saber-toothed tiger, a mighty dinosaur, or even all three!

SUCCESSION

After a volcano spews lava over a landscape, or after an earthquake or a landslide exposes rocks for the first time, there is initially no sign of life on the lava or rock surfaces. Within a few years or sometimes within a few months, living organisms begin to appear and a sequence of events known as **succession** takes place. During succession the species of plants and other organisms that first appear gradually alter their environment as they carry on their normal activities such as metabolism and reproduction. In time the accumulation of wastes, dead organic material,

and inorganic debris, and other changes (such as of shade and water content in the habitat) favor different species, which may replace the original ones. These, in turn, modify the environment more so that yet other species become established.

Succession occurs whenever and wherever there has been a disturbance of natural areas on land or in water. It proceeds at varying rates, depending on the climate, the soils, and the animals or other organisms in the vicinity. Ecologists recognize a number of variations of two basic types of succession. *Primary succession* involves the actual formation of soil in the beginning stages, while *secondary succession* takes place in areas that had previously been covered with soil and vegetation.

Primary Succession

Xerosere

One of the most universal types of primary succession, called a **xerosere**, begins with bare rocks and lava (see discussion of soils in chapter 6) that have been exposed through glacial or volcanic activity or through landslides. Initially, the rocks are sometimes subjected to alternate thawing and freezing, at least in temperate to colder areas. Tiny cracks or flaking may occur on the surface as a result. Lichens often become established on such surfaces (figure 23.7). They produce acids that very slowly etch the rocks, and as they die and contribute organic matter,

FIGURE 23.7   Early stages of succession on a Hawaiian road following an earthquake. Ferns have become established in cracks less than two years after the disruption.

they are replaced by other, larger lichens. Certain rock mosses adapted to long periods of desiccation also may become established, and a small amount of soil begins to build up. This is augmented by dust and debris blown in by the wind. Eventually enough of a mat of lichen and moss material is present to permit some ferns or even seed plants to become established, and the pace of soil buildup and rock breakdown accelerates. If deep cracks appear in the rocks, the larger seeds may widen them further as they germinate and the roots expand in girth. It has been calculated that germinating seedlings can exert a force of up to 31.635 kilograms per square centimeter (450 pounds per square inch). Indeed, there are known instances of seedlings splitting rocks that weigh several tons (figure 23.8).

As soil buildup continues, larger plants take over, and eventually the vegetation reaches an equilibrium in which the associations of plants and other organisms remain the same until another disturbance takes place or climatic changes occur. Such relatively stable plant associations are referred to as **climax vegetation.** The climax vegetation of deciduous forests in eastern North America are dominated by maples and beeches, oaks and hickories, hemlocks and white pines, or other combinations of trees. In desert regions, various cacti form a conspicuous part of the climax vegetation, while in the Pacific Northwest large coniferous trees predominate. In parts of the Midwest, prairie grasses and other herbaceous plants form the climax vegetation, and in wet tropical regions a complex association of jungle plants constitutes the climax.

Occasionally when a volcano does not produce lava but instead primarily ash that buries existing landscape and associated vegetation, some of the successional stages involving lichens and mosses may be bypassed, with larger plants becoming the successional pioneers. Such has been the case following the series of ash eruptions of Mount St. Helens in the state of Washington during the early 1980s.

Hydrosere

Succession takes place in wet habitats as well as in drier ones and such succession is called a **hydrosere.** In the northern parts of midwestern states such as Michigan, Wisconsin, and Minnesota, ponds and lakes of various sizes abound. Many were left behind by retreating glaciers and often have no streams draining them. The water that evaporates from them is replaced annually by precipitation runoff. They also grow a tiny bit smaller each year as a result of succession (figure 23.9). This succession often begins with algae either carried in by the wind or

FIGURE 23.8 A live oak that grew from an acorn lodged in a small crack of the rock. The roots of the tree have split the rock apart.

FIGURE 23.9 Succession in a pond in northern Wisconsin. (Courtesy Robert A. Schlising)

transported on the muddy feet of waterfowl and wading birds. The algae tend to multiply in shallow water near the margin, and with each reproductive cycle, the dead parts sink to the bottom. Floating plants such as duckweeds may then appear, often forming a band around the body of water just offshore. Next, water lilies and other rooted aquatic plants with floating leaves become established, each group of plants contributing to the organic material on the bottom, which slowly turns to muck. Cattails and other flowering plants that produce their inflorescences above the water often take root in the muck around the edges, and the accumulation of organic material accelerates.

Meanwhile the algae, duckweeds, and other plants move farther out, and the surface area of exposed water gradually diminishes. Grasslike sedges become established along the damp margins and sometimes form floating mats as their roots interweave with one another. Dead organic material accumulates and fills in the area under the sedge mats, and herbaceous and shrubby plants then move in. As the margins become less marshy, coniferous trees whose roots can tolerate considerable moisture (e.g., tamaracks or eastern white cedars) gain a foothold, eventually growing across the entire site as the pond or lake disappears. The trees continue to aid in the formation of true soil, and in due course the climax vegetation takes over. No visible trace of the pond or lake now remains, and the only evidence of its having been there lies beneath the surface, where fossil pollen grains, bits of wood, and other material reveal the past history. Such succession may take thousands of years and has never been witnessed from start to finish. The evidence that it does occur, however, is extensive and compelling.

Under natural conditions stream-fed lakes and ponds eventually become filled with silt and debris, although this, too, may take thousands of years to occur. The streams that feed these lakes gradually bring in silt, and the nutrient content of the water slowly rises as dissolved organic and inorganic materials are brought in. This gradual enrichment, called **eutrophication,** permits the growth of algae and other organisms, which also add their debris to the bottom of the lake. When sewage and other pollutants are permitted to enter the lake, the process of eutrophication may be greatly accelerated through stimulation of the growth of aquatic organisms. Eutrophication may also be accelerated when trees are cleared from land surrounding lakes prior to the construction of summer homes and resorts. The cleared land erodes more readily, with precipitation runoff carrying soil into the water. Regardless of size, all bodies of water, including rivers, are subjected to these processes.

Secondary Succession

Secondary succession, which occurs more rapidly than primary succession, may take place if soil is already present and there are surviving species in the vicinity; in fact, survivors strongly condition subsequent succession. Many secondary successions follow human disturbances (e.g., land that was cleared when timber was harvested or land converted to farmland); other secondary successions follow fires. Grasses and other herbaceous plants become established on burned or logged land. These usually are followed by trees and shrubs that have widely dispersed seeds (e.g., aspen and sumac in the Midwest and East, and chaparral plants such as chamise and gooseberries in the West). After going through fewer stages than are typical of primary succession, the climax vegetation takes over, often within 100 years or less.

FIRE ECOLOGY

Natural fires started primarily by lightning and the activities of prehistoric humans have occurred for thousands of years in North America and other continents. Trees such as the giant redwood and ponderosa pines, although scarred by certain types of fires, often survive, and the dates of fires can be determined by the proximity of the scars to specific annual rings (see figure 5.7). In the West checks of growth rings of ponderosa pines show that forests, in the past, burned on an average of approximately every six to seven years. These fires and the climate, topography, and soil combined to have a profound effect on various biomes. But as humans became civilized, major and largely successful efforts to control fires were made and this, in turn, significantly altered vegetational patterns. As knowledge of the role of fires in the maintenance of ecosystems has accumulated, it has become apparent that trying to eliminate fires, at least in certain areas, disrupts natural habitats more in the long run than allowing them to occur, and agencies such as the U.S. National Park Service now allow fires at higher elevations to run their natural course.

Fires actually benefit grasslands, chaparral, and forests by converting accumulated dead organic material to mineral-rich ash, whose nutrients are recycled within the ecosystem. If the soil has been subjected to fire, some of its nutrients and organic matter may have been lost and the composition of microorganisms originally present is likely to have been altered. Losses are offset, however, by the fact that soil bacteria, including blue-green bacteria, which are capable of fixing nitrogen from the air, increase in numbers after a fire, and there is a decrease in fungi that cause plant diseases.

In some areas, such as the prairie states of the Midwest, grasses are better adapted to fire than woody plants. They produce seeds within a year or two after germination, and perennial grass buds, which are at the tips of rhizomes close to or beneath the surface where they escape the most intense heat of fires, usually survive, producing new growth the first season after a fire. Thus, a fire destroys only one season's growth of grass, often after growth has been completed. Shrubs, however, have much of their living tissue above ground, and a fire may destroy several years' growth. In addition, woody plants often do not produce seeds until several years after a seed germinates. Many shrubs do sprout from burned stumps, particularly in chaparral areas (figure 23.10), but repeated burning keeps them small, thus favoring grasses. There is evidence that at least some

of the North American grasslands originated and were maintained because of fire. Since grassland fires have largely been controlled, many of the areas have now been invaded by shrubs that were once confined to water courses.

Fires also play a role in the composition of forests. In the mountains of east-central California, gooseberry and deerbrush appear in abundance after a fire, but their numbers stabilize within 15 to 30 years when larger trees return to the area. Ponderosa, jack and southern longleaf pines, and Douglas firs (which do not tolerate shade) are among the species that repeatedly replace themselves after fires, and seeds of some species rarely germinate until they have been exposed to fire. The majority of chaparral species, both woody and herbaceous, are so adapted to fire that their seeds also will not germinate, as a rule, until fires remove accumulated litter and toxic wastes produced by the plants during growth.

In view of the long-range beneficial effects of fires in some ecosystems, wise land management practices of the future undoubtedly will include guiding succession in many plant communities to the *sere* (a series of ecological communities) of greatest utility and safety.

FIGURE 23.10   Secondary succession on a burned area. Chamise is resprouting the first season after a burn. (Courtesy Robert A. Schlising)

FIGURE 23.11   Major biomes of North America.

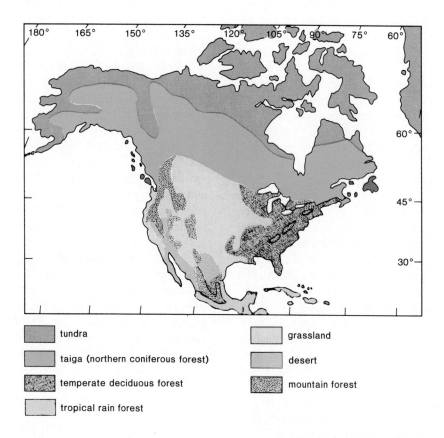

tundra

taiga (northern coniferous forest)

temperate deciduous forest

tropical rain forest

grassland

desert

mountain forest

MAJOR BIOMES OF NORTH AMERICA

Biotic communities considered on a global or at least on a continental scale are referred to as **biomes.** Several important biomes occupy the North American continent (figure 23.11), and most are also represented on other continents.

Tundra

Tundra (figure 23.12) occupies vast areas of the Earth's land surfaces (about 20% in all), primarily above the Arctic circle. Patches of alpine tundra occur above timberline in mountains below the Arctic circle. It has the appearance of being treeless, although miniature willows only 2.5 centimeters (1 inch) tall do survive in some areas. Another characteristic of arctic tundra is the presence of *permafrost* (permanently frozen soil) at a depth of from a few centimeters (an inch or two) to about 1 meter (3 feet 3 inches) below the surface. The level of the permafrost determines the depth to which plant roots can penetrate. It is generally absent from

alpine tundra areas. Precipitation averages less than 25 centimeters (10 inches) a year, but because of the permafrost, it is largely held at or near the surface of the land. There is a very short growing season of only two to three months, with frost possible on any day of the year; however, temperatures can soar to 27° C (81° F) during a long midsummer day. The vegetation of arctic tundra is dominated by lichens and grasses, with mosses and sedges abundant in some areas. Grasses and sedges tend to predominate in alpine tundra. During the brief growing season tiny perennial plants produce brightly colored flowers and form brilliant mats over the topsoil, which is largely organic and generally only 5.0 to 7.5 centimeters (2 to 3 inches) deep.

The biome is exceptionally fragile. A car driven across tundra compresses the soil sufficiently to kill plant roots, and the tracks are evident many years later. Occasionally sheep grazing tundra have been observed to pull up patches of the matted vegetation, leaving exposed edges. High winds then can catch the exposed edges and rip away larger segments of mat, leaving barren patches called *blowouts.*

FIGURE 23.12 Alaskan tundra in the summer. (Courtesy Calvin Winey, Jr.)

Animals of the tundra include lemmings, the little rodents that reproduce prodigiously every three or four years and then decline in numbers. The arctic fox and snowy owl also fluctuate in numbers, the fluctuations being correlated with the size of the populations of lemmings, which constitute their principal food. Other animals of the tundra include caribou, polar bears (near the coast), shrews, ptarmigans, loons, plovers, jaegers, and arctic terns.

Taiga

The *taiga* (figure 23.13), also referred to as northern coniferous or boreal forest, is found adjacent to and south of the arctic tundra. The vegetation is dominated by evergreen trees such as spruces, firs, hemlocks, and pines; tamaracks also occur in the taiga. Birches, aspens, and willows may be found in some of the wetter areas. Many perennials and a few shrubs occur in the taiga, but there is a dearth of annuals. Snow blankets the region during the winters, which are long and severe, with temperatures dropping to −50° C (−58° F) or lower during the coldest months. In summer the temperatures often reach 27° C (81° F). Most precipitation occurs in the summer, ranging from about 25 centimeters (10 inches) to more than 100 centimeters (39 inches) in

FIGURE 23.13 Taiga in northern Minnesota.

parts of western North America. Many lakes, ponds, and marshes dot the region, which is inhabited by a variety of birds, including jays, warblers, and nuthatches. Many rodents such as shrews and jumping mice, and larger mammals, notably moose, occur. Ermine and wolverines also make their home in the taiga.

Temperate Deciduous Forest

Deciduous trees are generally broad-leaved species that shed their leaves annually during the fall and remain dormant during the shorter days and colder temperatures of winter. Most *temperate deciduous forests* (figure 23.14) are found, like the taiga, on large continental masses in the northern hemisphere. In North America, this type of forest occurs from the Great Lakes region south to the Gulf of Mexico and extends from roughly the Mississippi River to the eastern seaboard, although some place the eastern coastal plain in a separate biome. Temperatures within the area vary a great deal but normally fall below 4° C (39° F) in midwinter and rise above 20° C (68° F) in summer. The climax trees of the forest are well adapted to subfreezing temperatures as long as the cold is accompanied by precipitation or snow cover. Most of the annual precipitation, which totals between 50 and 165 centimeters (20 to 65 inches), occurs in the summer.

Some of the most beautiful of all the broad-leaved trees are found in this biome in a variety of associations. In the upper midwest, sugar maples and American basswoods predominate. Sugar maples are also found to the northeast, where they tend to be associated with the stately American beeches. In the west-central part of the forest, oaks and hickories abound. Oaks also are abundant along the eastern slopes of the Appalachian Mountains, where American chestnuts were once a conspicuous part of the flora. The chestnuts have now virtually disappeared, having succumbed to the ravages of chestnut blight disease, which was introduced and began taking its ugly toll during the early 1900s. Oak trees extend into the southeastern United States, where they become associated with pine trees and other species such as the bald cypress.

Before the arrival of European immigrants, a mixture of large deciduous trees that included maples, ashes, basswoods, beeches, buckeyes, hickories, oaks, tulip trees, and magnolias was found on the eastern slopes and valleys of the Appalachian

FIGURE 23.15 Undisturbed western grassland in the spring.

During the summer the trees of the deciduous forests form a relatively solid canopy that keeps most direct sunlight from reaching the floor. Many of the showiest spring flowers of the region (such as bloodroot, hepatica, trilliums, and violets) flower before the trees have leafed out fully, and complete most of their growth within a few weeks. Other plants that can tolerate more shade, principally members of the Sunflower Family (e.g., asters and goldenrods), flower in succession in forest openings from midsummer through fall.

Animal life in the eastern deciduous forest includes red and gray foxes, raccoons, opossums, many rodents such as gray squirrels, snakes such as copperheads and black rat snakes, salamanders, and a wide variety of birds, including hawks, flickers, and mourning doves.

Grassland

Naturally occurring *grasslands* (figure 23.15) are found toward the interiors of continental masses. They tend to intergrade with forests, woodlands, or deserts at their margins, depending on precipitation patterns and amounts. A grassland may receive as little as 25 centimeters (10 inches) of rainfall or as much as 100 centimeters (39 inches) annually. Temperatures can range from 50° C (122° F) in midsummer to −45° C (−49° F) in midwinter.

In North America huge herds of buffalo once grazed the prairies, but these disappeared as the settlers cultivated more and more of the land and hunters slaughtered more and more of the large animals. By 1889, there were only 551 left. Large areas are now used for growing cereal crops (particularly corn and wheat) and for grazing cattle. Before it was destroyed, the American prairie was a remarkable sight. In Illinois and Iowa, the grasses grew over 2 meters (6 feet 6 inches) tall during an average season and another meter taller during a wet one. A dazzling display of wildflowers began before the young perennial grasses emerged in the spring and continued throughout the growing season. I have counted over 50 species of flowering plants in bloom at the same time on 1 hectare (2.47 acres) of protected grassland in the middle of spring.

Mountains. Some of the trees grew over 30 meters (100 feet) tall and had trunks up to 3 meters (10 feet) in diameter. Except for a few protected pockets in the Great Smoky Mountains National Park, the largest trees of this rich forest have been all but eliminated through logging. A smaller-treed extension of this forest is found in an area northeast of Baton Rouge, Louisiana, western Tennessee and Kentucky, and in southern Illinois. American elms, also once a part of the forest, are rapidly disappearing as Dutch elm disease fells both wild trees and those planted along city streets. One small midwestern town, which had some 600 elms planted along its streets, found itself left with only 6 live trees within a year or two after Dutch elm disease struck.

A mixture of deciduous trees and evergreens occurs on the northern and southeastern borders of the forest. Hemlocks and eastern white pines are found from New England west to Minnesota and south to the Appalachians. The once vast stands of eastern white pine are now largely gone, their valuable lumber having been used for construction and other purposes. Some have been lost to still another tree disease, white pine blister rust, but scattered trees remain, particularly where they have been protected. Various pines dominate the eastern coastal plain from New Jersey to Florida, and west to east Texas. Pitch pine is common in New Jersey, while some of the southern pines (e.g., long-leaf, slash, loblolly) are now cultivated in the southeastern United States for wood pulp, turpentine, lumber, and other commercially valuable products.

FIGURE 23.17   A desert community in the southwestern
United States.

Grasslands occurring in areas with a Mediterranean climate (e.g., the Great Central Valley of California), where most of the precipitation takes place during the winters, usually include *vernal pools*. These are temporary accumulations of rainwater that evaporate after the rains come to an end. Their unique floras include an orderly succession of flowering plants, some appearing initially at the pool margins, with each species forming a distinct zone or band until the water is gone (figure 23.16). Some species flower only in the damp soil and drying mud that remains. The seeds of many species germinate under water.

Grassland animals include cottontails, jackrabbits, gophers, mice, and pronghorns. As indicated earlier, buffalo were once abundant; some 15,000 now are protected in national parks, game preserves, and private ranches. Various sparrows (e.g., vesper and savannah sparrows) and other birds still find homes in uncultivated tracts of this biome.

Desert

If you ask the average person to describe a desert, the response will probably include words like *sand, heat, mirage, oasis,* and *camels.* These are indeed found in some of the world's large deserts that are located in the vicinity of 30 degrees latitude both north and south of the equator, but deserts occur wherever precipitation is consistently low or the soil is too porous to retain water (figure 23.17). Most deserts receive less than 12.5 centimeters (5 inches) of rain per year. The low humidity results in wide daily temperature ranges. On a summer day when the temperature has reached over 35° C (95° F), it will generally fall below 15° C (59° F) the same night. The light intensities reach higher peaks in the dry air than they do in areas where atmospheric water vapor filters out some of the sun's rays. Many desert plants have become adapted to these higher light intensities through the evolution of crassulacean acid metabolism (CAM) photosynthesis, a process by which certain organic acids accumulate in the chlorophyll-containing parts of the plants during the night and are converted to carbon dioxide during the day (see chapter 10). This permits much more photosynthesis to take place than would otherwise be possible, since most of the carbon dioxide of the atmosphere is excluded from the plants during the day by the stomata, which remain closed (thereby also retarding water loss).

Other adaptations of desert plants include thick cuticles, water-storage tissues in stems and leaves, leaves with a leathery texture and/or reduced in size, or even total absence of leaves. In cacti and other succulents without functional leaves, the stems take over the photosynthetic activities of the plants. Desert perennials are adapted to the biome in various ways. Cacti and similar succulents have widespread shallow root systems that can rapidly absorb water from the infrequent rains. The water can then be stored for long periods in the interior of the stems, which are modified and permit this function. Other perennials grow from bulbs that are dormant for much of the year. Annuals provide a spectacular display of color and variety, particularly during an occasional season when above-average precipitation has occurred. The seeds of the annuals often germinate after a fall or winter rain and then grow slowly or remain in rosette form for several months before producing flowers in the spring. Literally hundreds of different species of desert annuals may occur within a few square kilometers (1 or 2 square miles) of typical desert in the southwestern United States.

Desert animals are adapted, in many instances, to foraging at night when lower air temperatures prevail. These include various mice and kangaroo rats, snakes (notably rattlesnakes), chuckwallas, and lizards (e.g., gila monster). Various thrashers, doves, and flycatchers are included in the bird life.

Mountain Forest

In the geologic past, deciduous forests extended to western North America. As the climate changed and summer rainfall was reduced, conifers largely replaced the deciduous trees, although some (e.g., maples, birches, aspens, oaks) still remain, particularly

FIGURE 23.18 Coastal redwoods in northern California.

at the lower elevations. Today coniferous forests occupy vast areas of the Pacific Northwest and extend south along the Rocky Mountains and the Sierra Nevada and California coast ranges. Isolated pockets of this biome also occur in other parts of the West, particularly toward the southern limits of the mountains. The trees tend to be very large, particularly in and to the west of the Cascade Mountains of Oregon and Washington and on the western slopes of the Sierra Nevada. Part of the reason for the huge size of trees such as Douglas fir is the high annual rainfall, which exceeds 250 centimeters (100 inches) in some areas; the world's tallest trees, the coastal redwoods of California (figure 23.18), however, apparently depend more on fog, which reduces transpiration rates, than on large amounts of rain for their size and longevity.

One of the characteristics of the mountain forest is a fairly conspicuous altitudinal zonation of species. In other words, one encounters different associations of plants as one proceeds from sea level up the mountainsides. At lower elevations in both the

Rocky Mountains and the Sierra Nevada, the predominant conifer is ponderosa pine. At lower elevations in the northern part of the Cascades, Douglas firs, western red cedars, and western hemlocks are more common. At intermediate elevations in the Sierra Nevada, sugar pines, white firs, and Jeffrey pines take over, while at higher elevations other species of pines and firs predominate.

Most of the mountain forest biome has comparatively dry summers. This led to frequent forest fires, even before human carelessness became a factor. Several tree species are well adapted to fires. The Douglas fir, for example, has a very thick protective bark that can be charred without transmitting sufficient heat to the interior to damage more delicate tissues, and its seedlings thrive in open areas after a fire. When the bark of the giant redwoods of the Sierra Nevada is burned, the trees are rarely killed. This has undoubtedly contributed to the great age and size many of the trees have attained. The cones of some of the pine trees (e.g., knobcone pine) remain closed and do not release their seeds until a fire causes them to open, while seeds of several other species germinate best after they have been exposed to fire. These attributes of members of the mountain forest biome, as mentioned in the section on fire ecology, have led to the practice in some of our national parks of allowing fires at higher elevations to run their natural course. The higher-than-normal incidence of fires occurring since humans came in large numbers to the forest has made it necessary, however, for us to control them in most instances, even though in doing so we may be interfering with natural cycles that would otherwise occur in the biome.

Animal life in the mountain forest includes many different rodents (especially chipmunks and voles), bears, mountain lions, bobcats, mountain beavers (not related to true beavers), and mule deer. Large birds such as the golden eagle and many small birds, including mountain chickadees, warblers, and juncos, are an integral part of this biome.

Tropical Rain Forest

Nearly half of the forested areas of the Earth are included in the *tropical rain forest* biome which is distributed throughout those areas of the tropics where annual rainfall amounts normally range between 200 and 400 centimeters (79 and 157 inches), and where temperatures range between 25° C and 35° C (77° F and 95° F). Although monthly rainfall amounts vary, there is no dry season and some

FIGURE 23.19   A tropical rain forest scene in Puerto Rico. (Courtesy Mary Lane Powell)

precipitation occurs throughout all 12 months of the year, frequently in the form of afternoon cloudbursts; the humidity seldom drops below 80%.

Such climatic conditions support a diversity of flora and fauna so great that the number of species exceeds those of all the other biomes combined. The forests are dominated by broadleaf evergreen trees, whose trunks are often unbranched for as much as 40 or more meters (160 feet), with luxuriant crowns that form a beautiful dark green and several-layered canopy; the root systems are shallow and often buttressed (see figure 6.12). There are literally hundreds of species of such trees, each usually represented by widely scattered specimens. Most of the plants of the rain forests are woody, although not all of them are evergreen. Several of the deciduous tree species shed their leaves from some branches, retain the leaves on others, and flower on yet other branches all at the same time, while branches of adjacent trees of the same species are losing their leaves or flowering at different times. Numerous hanging woody vines and even more numerous *epiphytes*—especially orchids and bromeliads—can be seen on or attached to tree branches (figure 23.19). The epiphyte roots, which are not parasitic, have no contact with the ground and the plants are sustained entirely by their own photosynthetic activities and the rainwater with dissolved minerals that accumulates in their leaf bases. The rainwater accumulates traces of minerals, as it trickles over decaying bark and dust.

FIGURE 23.20 Flower markings on a coneflower. *A.* Under ordinary light. *B.* Under ultraviolet light.

A.

B.

The multilayered canopy is so dense that very little light penetrates to the floor, and the few herbaceous plants that survive are generally confined to openings in the forest. Despite the lush growth there is little accumulation of litter or humus, and the soil is relatively poor. Decomposers rapidly break down any leaves or other organic material on the forest floor, and the nutrients released by decomposition are quickly recycled or leached by the heavy rains.

A few larger animals, with adaptations for moving through the mesh of branches in the rain forest, are found on the forest floor. Such animals include peccaries, tapirs, and anteaters. Most of the great numbers of animals, however, live out their lives in the canopies. Tree frogs, with adhesive pads on their toes, are common in tropical rain forests, as are various monkeys, sloths, opossums, tree snakes, and lizards. Ants abound, and many of the extraordinary variety of other insects are adapted to their environment through excellent camouflages. Large flocks of parrots and other birds feed on the abundance of insects, as well as the available fruits.

POLLINATION ECOLOGY

When certain consumers forage among producers for food, they often come in contact with flowers. Many insects and other animals become dusted with pollen, and in the course of their travels they unwittingly but effectively bring about pollination of the plants they visit. Throughout the evolutionary history of the flowering plants, the pollinators have evidently coevolved with plants. In some instances, the relationship between the two has become highly specialized.

Included among pollinators of present-day flowering plants are some 20,000 different species of bees. By far the best known of these are honeybees. Their chief source of nourishment is nectar, but they also gather pollen for their larvae. The flowers that bees visit are generally brightly colored and predominatly blue or yellow in color—rarely pure red. Pure red appears black to them, and therefore they generally overlook red flowers. Flowers often have lines or other distinctive markings that may function as honey guides that lead the bees to the nectar. Bees can see ultraviolet light (a part of the spectrum not visible to humans), and some flower markings are visible only in ultraviolet light, making patterns seen by bees sometimes different from those seen by humans (figure 23.20).

Many bee-pollinated flowers are delicately sweet and fragrant. By way of contrast, flowers pollinated by beetles tend to have stronger, yeasty, spicy, or fruity odors. They are usually white or dull in color, in keeping with the diminished visual sense of their pollinators. Some beetle-pollinated flowers secrete no nectar but either furnish the insects with pollen or have food available on the petals in special storage cells, which the beetles consume.

FIGURE 23.21 A *Stapelia* (carrion flower) plant.

FIGURE 23.22 A hummingbird visiting a flower.

Certain flowers pollinated by short-tongued flies tend to be dull red or brown in color and often have foul odors resembling those of rotten meat. Such flowers include the stapelias of southern Africa (figure 23.21). These plants are related to our milkweeds, although superficially they do not resemble milkweeds at all. They are often called *carrion flowers* because of their foul odor and appearance. A number of bee-pollinated flowers are also pollinated by flies with longer tongues.

Moth- and butterfly-pollinated flowers are similar to bee-pollinated flowers in that they frequently have sweet fragrances. Many moths forage only at night, the flowers they visit tending to be white or yellow—colors that stand out against dark backgrounds in starlight or moonlight. Some of the most specialized relationships between moths and flowers occur between certain small moths and members of the genus *Yucca*. This genus includes Joshua trees and Our Lord's Candle—the striking plants with the stiff trunks and sword-shaped leaves that stand out on the landscape of some southwestern desert areas of North America. The female moth gathers balls of sticky pollen from the stamens of one plant and carries them, with the aid of specialized mouth parts, to another plant. There the moth enters the flowers and pierces the ovary walls, laying eggs in the vicinity of the tiers of ovules. Then, depending on the species, the moth either packs the pollen balls into stigma cavities or rubs the pollen across the stigma

surfaces. As the ovules develop into seeds, the eggs hatch, and the larvae consume some of the seeds for food. When the larvae are fully grown, they chew their way out of the ovaries and remain inactive in cocoons just below the surface of the ground until the next flowering season, when the adult moths emerge. The larvae seldom destroy more than a third of the seeds.

Some butterflies can detect red colors, and so red flowers are sometimes pollinated by them. The nectaries of these flowers are found at the bases of corolla tubes or spurs, where only the longer tongues of moths and butterflies can reach. Occasionally an enterprising bumblebee will bypass convention and chew through the base of a spur to get at the nectar.

Birds—particularly the hummingbirds of the Americas and the sunbirds of Africa—and the flowers that they pollinate are also adapted to one another (figure 23.22). The birds do not have a highly developed sense of smell, but they have a keen sense of vision. Their flowers are thus frequently bright red or yellow and usually have little if any odor; they are typically large or are part of a large inflorescence. Birds are highly active pollinators and tend to burn up energy rapidly; they must feed frequently to sustain themselves. Many of the flowers they visit produce copious quantities of nectar, thus assuring repeated visits by the birds. The nectar is frequently produced in long floral tubes, which prevent most insects from gaining access to it. Some native California fuchsias and their relatives have

FIGURE 23.23   A cluster of four pollen grains of California fuchsia, showing the threads that catch in the short bristle hairs at the base of a hummingbird's bill.

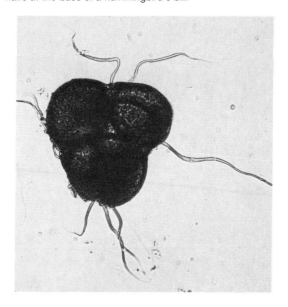

FIGURE 23.24   A bat visiting a flower of an organpipe cactus. (Photo by Donna J. Howell)

long threads extending from each pollen grain (figure 23.23). When a hummingbird inserts its long bill into such a flower, the pollen grain threads catch on some short stiff hairs located toward the base of its bill, and the bird unknowingly transfers pollen from one flower to another by this means.

Bat-pollinated flowers, found primarily in the tropics, tend to open only at night when the bats are foraging (figure 23.24). These flowers are dull in color, and like bird-pollinated flowers, they either are large enough for the animal to insert part of its head, or consist of ball-like inflorescences containing large numbers of small flowers whose stamens readily dust the visitor with pollen.

The very large Orchid Family, which has approximately 35,000 species, has pollinators among all the types mentioned. Some of the adaptations between orchid flowers and their pollinators are extraordinary (figure 23.25). Many orchids produce their pollen grains in little sacs called **pollinia** (singular: **pollinium**), which typically have sticky pads at the bases. When a bee visits such a flower, the pollinia are usually deposited on its head. The "glue" of the sticky pads drys almost instantly, causing the pollinia to adhere tightly. In some orchids, the pollinia are forcibly "slapped" on the pollinator through a trigger mechanism within the flower. Members of one genus of orchids found in Europe and in North Africa have a petal modified so that it resembles a female bumblebee or wasp. Male bees or wasps emerge from their pupal stage a week or two before

the females and apparently mistake the flowers for potential mates. They try to copulate with the flowers, and while they are doing so, pollinia are deposited on their heads. When they visit other flowers, the pollinia catch in sticky stigma cavities. The pollinia lost in the process are replaced with new ones, the switch taking place during a single visit. In orchids pollinated by moths and butterflies, the pollinia become attached to their long tongues by means of sticky clamps instead of pads. The pollinia of certain bog orchids become attached to the eyes of their pollinators, which happen to be female mosquitoes. After a few visits, the mosquitoes are blinded and unable to continue their normal activities—a striking example of biological controls within an ecosystem. Among the most bizarre of the orchid pollination mechanisms are those whose effects are to dunk the pollinator in a pool of watery fluid secreted by the orchid itself and permit the pollinator to escape underwater through a trapdoor. The route of the insect ensures contact with pollinia and stigma surfaces. In other orchids with powerful narcotic fragrances, pollinia are slowly attached to the drugged pollinator. When the transfer of pollinia has been completed, the fragrance abruptly disappears; the temporarily stupefied insect recovers and flies away.

FIGURE 23.25   Some pollinating mechanisms in orchids.

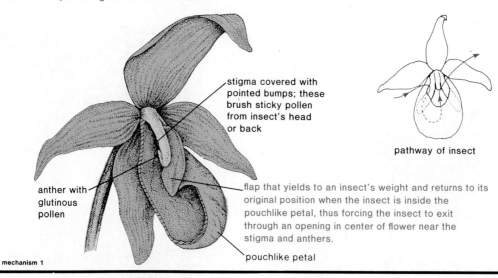

stigma covered with
pointed bumps; these
brush sticky pollen
from insect's head
or back

pathway of insect

anther with
glutinous
pollen

flap that yields to an insect's weight and returns to its
original position when the insect is inside the
pouchlike petal, thus forcing the insect to exit
through an opening in center of flower near the
stigma and anthers.

pouchlike petal

mechanism 1

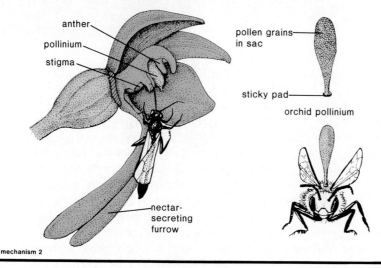

anther

pollinium

stigma

pollen grains
in sac

sticky pad

orchid pollinium

nectar-
secreting
furrow

mechanism 2

An insect follows a nectar-
secreting furrow on a petal
until it reaches the center
of the flower. Pollinia are
removed when they
adhere to the insect
following an "explosion"
initiated by the insect
bumping a sensitive point
below an anther. When the
insect flies to another
flower the pollinia on its
head catch on the sticky
surface of the stigma and
are pulled off at the same
time new pollinia are being
attached.

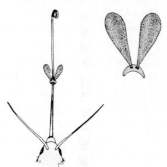

Orchids pollinated by butterflies and moths may have a clamp
at the base of the pollinia. When an insect inserts its tongue
down the throat of the flower and touches a sensitive area, the
clamp attaches the pollinia to the insect's tongue.

mechanism 3

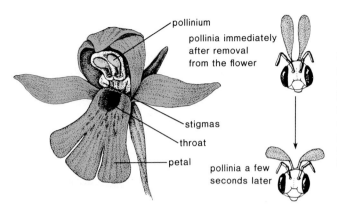

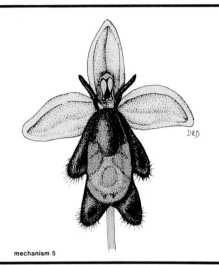

pollinium

pollinia immediately
after removal
from the flower

stigmas

throat

petal

pollinia a few
seconds later

In the showy orchids, there are two
separate stigma patches. When an insect
visits a flower the pollinia are attached to
its head; then they twist outward and
forward so that when the insect visits
another flower the pollinia are
simultaneously deposited in the separate
stigmas while new pollinia are attached to
the insect's head.

mechanism 4

One genus of orchids native to Europe and North Africa has a
petal modified to resemble a female wasp. Male wasps try to
copulate with the "female" and in doing so bump a sensitive
area containing pollinia. The pollinia are "glued" to a wasp's
head in the process, and carried to another flower. The wasps
thus effectively bring about cross-pollination.

mechanism 5

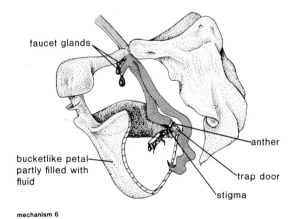

faucet glands

bucketlike petal
partly filled with
fluid

anther

trap door

stigma

The bucket orchids of South America have faucet
glands that partially fill a bucketlike petal with fluid.
In visiting the flower the insects may fall into the
fluid. They escape by a trap door. The pathway of
an insect assures contact with both the stigma and
anthers containing pollinia, since they are located
just in front of the trap door.

mechanism 6

SUMMARY

Ecology deals with the relationships of organisms to one another and to their environment. It is a vast field of study with many facets. Groups of individuals of the same species constitute populations. Communities are units composed of all the populations of organisms occurring in a given area. Ecosystems are communities and their physical environment considered together.

Populations vary in numbers, in density, and in the total mass of individuals. Numbers may be counted whereas density and biomass (total mass of living organisms) may be calculated. A community is difficult to define precisely because given members of one community may also occur in another community, and variants of a given species (ecotypes) may be specifically adapted to a single community. The environmental factors playing the greatest roles in determining the distributions of plant species in an ecosystem include precipitation, temperature, soils, and biotic factors. Xerophytes (plants adapted to areas of low precipitation) and hydrophytes (plants adapted to aquatic environments) have modifications of leaves and other organs. The mineral content of soils influences species distribution, and biotic factors such as competition for light, nutrients, and water also play roles in this regard.

Ecosystems ideally sustain themselves entirely through photosynthetic activity and the recycling of nutrients. Producers are organisms that carry on photosynthesis, whereas consumers feed on producers. Primary consumers feed directly on producers and secondary consumers feed on primary consumers. Decomposers such as fungi and bacteria break down organic materials to forms that can be reassimilated by the producers. In any ecosystem, the producers and consumers comprise food chains, which determine the flow of energy through the different levels. Food chains vary in length and intricacy, and because of their interconnections, form what are often referred to as food webs.

Energy itself is not recycled in an ecosystem. At the producer level only about 1% of the light energy available is converted to organic material, and the energy gradually escapes in the form of heat as it passes from one level to another. In a long food chain the final consumer gains only a tiny fraction of the energy originally captured by the producer at the bottom of the chain. If producers increase the amount of food available in an ecosystem, there is a corresponding increase in consumers, but the increased number of consumers reduces the available food, with the result that the ecosystem is self-maintaining. Living components of an ecosystem may also influence its composition by the secretion of growth-inhibiting substances, growth-promoting substances, and by parasitism.

Human populations have increased dramatically in the past few centuries, and the disruption of ecosystems by the activities directly or indirectly associated with the feeding, clothing, and housing of billions of people threatens the survival of not only humans but many other living organisms as well. Some restoration of ecological balance by the curbing of pollution, wise agricultural and management practices, conservation of natural resources, and education of the public to the dangers of disrupting ecosystems is essential.

Some nitrogen from the air is fixed (converted to nitrogen-containing compounds) by the nitrogen-fixing bacteria found in legumes and other plants.

In the nitrogen cycle there is a constant flow of nitrogen from dead plant and animal tissues into the soil and from the soil back to the plants. Water leaches nitrogen from the soil and carries it away when erosion occurs; other nitrogen is lost from harvesting crops, but the loss can be reduced if wastes are decomposed and annually returned to the soil; fire also causes nitrogen loss. Replacement of nitrogen loss by the application of chemical fertilizers can eventually create hardpan by altering the soil structure. Bacteria also recycle carbon and other substances such as water and phosphorus.

Succession (a directional change in the species composition of a given area over a period of time) occurs whenever there has been a disturbance of natural areas on land or in water. Primary succession involves the formation of soil in the beginning stages, whereas secondary succession takes place in areas previously covered with vegetation. Primary succession involving the conversion of bare rock or lava to soil through the activities of lichens, rock mosses, flowering and other plants, and physical forces, in an orderly progression of events over a period of time is called a xerosere. A stable vegetation (climax vegetation) becomes established at the conclusion of succession and remains until or unless a disturbance disrupts it. Succession initiated in a wet habitat (hydrosere) also proceeds in an orderly fashion and culminates in the establishment of a climax vegetation. As a lake or other body of water is filled in with silt and debris, eutrophication (the gradual enrichment of the nutrient content of the water) occurs; this permits the growth of algae and other organisms.

Secondary succession, which proceeds more rapidly than primary succession, may take place if soil is present. It may occur after fires, which alter the nutrient and organic composition of the soil and open up areas in which trees such as aspens may become established until slower growing climax species return.

Biotic communities considered on a global or at least a continental scale are called biomes. The following are the major biomes of North America. Tundra, which is found primarily above the Arctic circle, includes few trees and many lichens and grasses; it is characterized by the presence of permafrost (permanently frozen soil) below the surface. Taiga is dominated by coniferous trees, with birches, aspens, and willows in the wetter areas; many perennials and few annuals occur in taiga. Temperate deciduous forests are dominated by trees such as sugar maples, American basswoods, beeches, oaks, and hickories, with evergreens such as hemlocks and eastern white pines toward the northern and southeastern borders. Grasslands are found toward the interiors of continental masses; those located in Mediterranean climatic zones usually include vernal pools with unique annual floras. Many grasslands have been converted to agricultural use. Deserts are characterized by low annual precipitation and wide daily temperature ranges, with plants adapted both in form and metabolism to the environment.

Mountain forest occupies much of the Pacific Northwest and extends south along the Rocky Mountains and California mountain ranges. The primarily coniferous tree species tend to be in zones determined by altitude. This forest has mostly dry summers, and some of the species associated with it (e.g., Douglas fir, giant redwood) have thick bark that protects the trees from fires, which occur frequently in this type of climate. Other species (e.g., knobcone pine) depend on fires for normal distribution and germination of seeds. The tropical rain forests constitute nearly half of all forest land, and contain more species of plants and animals than all the other biomes combined. Numerous woody plants and vines form multilayered canopies, which permit very little light to reach the floor. The soils are poor,

with nutrients released during decomposition being rapidly recycled. The biome is being destroyed so rapidly that it will have disappeared within 25 years.

The relationship between pollinators and flowering plants has, in some instances, become highly specialized. Bee-pollinated flowers are delicately sweet and fragrant. Beetle-pollinated flowers tend to have stronger odors and are usually white or dull in color. Some fly-pollinated flowers emit foul odors. Moth-pollinated flowers tend to be white or yellow. Bird-pollinated flowers are usually bright red or yellow and have much nectar but little odor. Many orchids produce their pollen grains in sacs called pollinia that adhere or clamp on to parts of visiting insects. The flowers of some orchid species have developed bizarre pollination mechanisms (e.g., mechanisms that effect the dunking of the pollinator in a pool of fluid).

REVIEW QUESTIONS

1. Distinguish among a plant community, a biotic community, an ecosystem, and a biome.
2. What is the difference between a primary consumer and a secondary consumer?
3. How are short food chains more efficient in the use of solar energy than long ones?
4. Why is the diet of larger herbivorous animals varied?
5. What types of checks and balances exist in an ecosystem?
6. What adaptations have developed between flowering plants and their pollinators?
7. How are nitrogen and carbon recycled in nature?
8. What is primary succession as opposed to secondary succession?
9. What are general characteristics of climax vegetation as compared with nonclimax species?
10. Characterize desert and tundra biomes.
11. What is the function of fire in grassland and mountain forest biomes?
12. What kinds of plants predominate in tropical rain forests?

DISCUSSION QUESTIONS

1. Humans have disrupted ecosystems almost everywhere they have established themselves, at least in industrialized countries. Do you believe that humans could also improve an ecosystem?
2. If a vegetarian diet makes more efficient use of solar energy, should we all strive to become vegetarians?
3. Besides the hypothetical example given (in which it was observed that eating lettuce directly was a more efficient use of solar energy than feeding it to hogs and then eating the pork), can you think of other ways in which food chains might be shortened?
4. Peanut butter has many of the nutrients needed in human nutrition. On the basis of what you have learned about the diet of animals in an ecosystem, do you think it would be a good idea to live on peanut butter and water as a means of saving money?
5. Criticize the following statements: "Flowers give nectar to bees so they will be pollinated." "When plants need to be pollinated by night-flying moths, they produce white or yellow flowers."
6. Could succession take place in an abandoned swimming pool?
7. Fire has been a natural phenomenon in several biomes for thousands of years, and most plant species are adapted to it in various ways in the ecosystem where it occurs regularly. Should we then not extinguish forest and grassland fires when they occur?

ADDITIONAL READING

Barbour, M., J. H. Burk, and W. D. Pitts. 1980. *Terrestrial plant ecology.* Menlo Park, CA: Benjamin/Cummings Publishing Company, Inc.

Faegri, K., and L. van der Pijl. 1972. *The principles of pollination ecology,* 2d ed. Elmsford, NY: Pergamon Press.

Forsyth, A., and K. Miyata. 1984. *Tropical nature: life and death in the tropical rain forests of Central and South America.* New York: Charles Scribner's Sons.

Jensen, W. A., and F. B. Salisbury. 1984. *Botany: an ecological approach,* 2d ed. Belmont, CA: Wadsworth Publishing Co.

Kormondy, E. J. 1984. *Concepts of ecology,* 3d ed. Englewood Cliffs, NJ: Prentice-Hall.

Kozlowski, T. T., and C. E. Ahlgren, eds. 1974. *Fire and ecosystems.* New York: Academic Press.

Krebs, C. J. 1985. *Ecology: the experimental analysis of distribution and abundance,* 3d ed. New York: Harper & Row, Publishers.

Odum, E. P. 1983. *Basic ecology.* Philadelphia: Saunders College Publishing.

Proctor, M., and P. Yeo. 1972. *The pollination of flowers.* New York: Taplinger Publishing Co.

Rice, E. L. 1984. *Allelopathy,* 2d ed. New York: Academic Press, Inc.

Scientific American. 1970. *The biosphere.* San Francisco: W. H. Freeman and Co.

Whittaker, R. H. 1975. *Communities and ecosystems,* 2d ed. New York: Macmillan Co.

Appendix 1   Scientific Names of Organisms Mentioned in the Text

This is a cross-referenced alphabetical list of the scientific names of all organisms mentioned in the text. Common and scientific names of organisms mentioned in appendices 2 through 4 are provided within the respective appendices.

Common Names and Scientific Names of Organisms

Common Name	Scientific Name	Common Name	Scientific Name
Aardvark	*Orycteropus* spp.	Algae, yellow-green	members of Class Xanthophyceae, Division Chrysophyta, Kingdom Protista
Abrasives, horsetail source of	*Equisetum* spp.		
Absinthe liqueur, source of ingredients	*Pimpinella anisum, Artemisia absinthium,* and others	Almond	*Prunus amygdalus*
		Aloe juice, source of	*Aloe barbadensis, A. ferox, A. vera,* and others
Acetone-producing bacteria	*Clostridium acetobutylicum* and others		
		Amaryllis	*Amaryllis* spp.
Acidophilus bacteria	*Lactobacillus acidophilus*	Ammonifying bacteria	*Clostridium* spp., *Micrococcus* spp., *Proteus* spp., *Pseudomonas* spp., and others
Aconite, source of	*Aconitum spp.*		
Adder's tongue fern	*Ophioglossum* spp.		
Afghanistan pine	*Pinus eldarica*	Amoeba	*Amoeba proteus* and others
African sausage tree	*Kigelia pinnata*		
African violet	*Saintpaulia* spp.	Amoeba, fungal internal parasites of	*Cochlonema verrucosum* and others
Agar, source of	*Gelidium* spp. and others		
		Amoeba, fungal trappers of	*Dactylella* spp. and others
Agave	*Agave* spp.		
Air plant	*Kalanchoë* spp.	Amphibious fern	*Marsilea* spp. and others
Alder	*Alnus* spp.		
Aleppo pine	*Pinus halepensis*	Anabaena	*Anabaena* spp.
Alfalfa	*Medicago sativa*	Anemone	*Anemone* spp.
Alfalfa caterpillar	*Colias philodice*	Angelica	*Angelica archangelica*
Algae	members of Kingdom Protista—all divisions	Anise	*Pimpinella anisum*
		Annatto	*Bixa orellana*
Algae, brown	members of Division Phaeophyta, Kingdom Protista	Annual moss	*Acaulon* spp., *Ephemerum* spp., and others
Algae, golden-brown	members of Class Chrysophyceae, Division Chrysophyta, Kingdom Protista	Ant	*Formica* spp. and others
		Ant and termite nest fungi	A *Leucoagaricus* sp. has been identified, but vast majority are unknown
Algae, green	members of Division Chlorophyta, Kingdom Protista	Anteater	*Myrmecophaga jubata*
		Ants, bullhorn *Acacia*	*Pseudomyrmex ferruginea*
Algae, red	members of Division Rhodophyta, Kingdom Protista	Anthrax bacteria	*Bacillus anthracis*
		Antibiotics, source of	*Penicillium* spp., *Cephalosporium* spp., and others

Common Names and Scientific Names of Organisms

Common Name	Scientific Name	Common Name	Scientific Name
Antler and bone mosses	*Tetraplodon* spp.	Bacteria, anthrax	*Bacillus anthracis*
Aphid	*Anuraphis* spp., *Aphis* spp., and others	Bacteria, blue-green	members of Class Cyanobacteriae, Division Eubacteriophyta, Subkingdom Eubacteriophytineae, Kingdom Monera
Apple	*Pyrus malus*		
Apple rust	*Gymnosporangium juniperi-virginianum*		
Apricot	*Prunus armeniaca*	Bacteria, botulism	*Clostridium botulinum*
Aquatic floating ferns	*Azolla* spp., *Salvinia* spp., and others	Bacteria, brucellosis	*Brucella abortus, B. suis, B. melitensis*
		Bacteria, BT	*Bacillus thuringiensis*
Arbor-vitae	*Thuja occidentalis*	Bacteria, bubonic plague	*Yersinia pestis*
Archaebacteria	members of Division Archaebacteriophyta, Subkingdom Archaebacteriophytineae, Kingdom Monera	Bacteria, buttermilk	*Streptococcus lactis, S. cremoris, Leuconostoc citrovorum,* and others
Argentine moth	*Cactoblastosis cactorum*	Bacteria, butyl alcohol	*Clostridium acetobutylicum* and others
Arrowroot, Florida, source of	*Zamia floridana*		
Artichoke, globe	*Cynaria scolymus*	Bacteria, cheese—*see* Bacteria, buttermilk	
Artichoke, Jerusalem	*Helianthus tuberosus*	Bacteria, cholera	*Vibrio cholerae*
Arum lily family	Araceae	Bacteria, decay	*Clostridium* spp., *Micrococcus* spp., *Proteus* spp., *Pseudomonas* spp., and others
Ash (fig. 8.20)	*Fraxinus* spp.		
Ash (fig. 21.13)	*Fraxinus latifolia*		
Ash, blue	*Fraxinus quadrangulata*		
Ash, white	*Fraxinus americana*	Bacteria, denitrifying	*Micrococcus denitrificans, Thiobacillus denitrificans,* and others
Asian climbing fern	*Lygodium salicifolium*		
Aspen	*Populus tremuloides* and others		
Aster	*Aster* spp.	Bacteria, dextran	*Leuconostoc mesenteroides*
Astringent, fern source of	*Actiniopteris radiata, Drynaria quercifolia, Pteridium aquilinum,* and others	Bacteria, diphtheria	*Corynebacterium diphtheriae*
		Bacteria, ensilage	*Lactobacillus delbrueckii, L. plantarum,* and others
Astringent, horsetail source of	*Equisetum arvense, E. debile,* and others		
Athlete's foot fungus	*Trichophyton* spp.	Bacteria, frost-damage preventing	*Pseudomonas syringiae*
Atlantic white cedar	*Chamaecyparis thyoides*	Bacteria, gas gangrene	*Clostridium novyi, C. perfringens, C. septicum*
Australian baobab	*Adansonia gregorii*		
Avocado	*Persea americana* and others		
Azalea	*Rhododendron* spp.	Bacteria, glutamic acid-producing	*Arthrobacter* spp., *Brevibacterium* spp., *Micrococcus* spp.
Baby blue eyes	*Nemophila menziesii*		
Baby powder, ground pine source of	*Lycopodium clavatum*	Bacteria, gonorrhea	*Neisseria gonorrhoeae*
Bacteria, acetone-producing	*Clostridium acetobutylicum* and others	Bacteria, grease- and oil-dissolving	*Pseudomonas aeruginosa*
		Bacteria, green sulfur	*Chlorobium* spp., *Chloropseudomonas* spp., *Prosthecochloris* spp., and others
Bacteria, acidophilus	*Lactobacillus acidophilus*		
Bacteria, ammonifying	*Clostridium* spp., *Micrococcus* spp., *Proteus* spp., *Pseudomonas* spp., and others	Bacteria, hydrogen	*Hydrogenomonas* spp.
		Bacteria, iron	*Gallionella* spp., *Sphaerotilus* spp.

Common Name	Scientific Name	Common Name	Scientific Name
Bacteria, kefir	*Lactobacillus bulgaricus, Streptococcus lactis*	**Bacteria, sulfur**	*Desulfovibrio* spp., *Thiobacillus* spp., and others
Bacteria, lactic acid	*Lactobacillus delbrueckii* and others	**Bacteria, syphilis**	*Treponema pallidum*
		Bacteria, tetanus	*Clostridium tetani*
Bacteria, Legionnaire's disease	*Legionella pneumophilia*	**Bacteria, tularemia**	*Francisella tularensis*
		Bacteria, typhoid fever	*Salmonella typhi*
Bacteria, luminescent	*Achromobacter* spp., *Flavobacterium* spp., *Photobacterium* spp., *Pseudomonas* spp., *Vibrio* spp., and others	**Bacteria, typhus fever**	*Rickettsia prowazeki* and others
		Bacteria, vinegar	*Acetobacter* spp.
		Bacteria, whooping cough	*Bordetella pertussis*
Bacteria, meningitis	*Neisseria meningitidis* and others	**Bacteria, yogurt**	*Streptococcus thermophilus*
Bacteria, methane	*Methanobacterium* spp., *Methanococcus* spp., *Methanosarcina* spp., and others	**Bald cypress**	*Taxodium distichum*
		Balsa	*Ochroma lagopus*
		Balsam fir	*Abies balsamea*
		Balsamroot	*Balsamorhiza* spp.
Bacteria, milky spore disease	*Bacillus popilliae*	**Bamboo**	*Bambusa* spp.
		Banana	*Musa paradisiaca* and others
Bacteria, mosquito-killing	*Bacillus thuringiensis* var. *israelensis*	**Banyan tree**	*Ficus* spp.
Bacteria, nitrate (nitrifying)	*Nitrobacter* spp.	**Baobab, Australian**	*Adansonia gregorii*
		Barbasco	*Lonchocarpus nicou* var. *utilis* and others
Bacteria, nitrite (nitrosifying)	*Nitrosomonas* spp.	**Barberry**	*Berberis* spp.
Bacteria, nitrogen-fixing	*Azotobacter* spp., *Clostridium pasteurinum, Rhizobium* spp.	**Barberry, Common**	*Berberis vulgaris*
		Bark, green algae that inhabit	*Protococcus* spp.[1]
Bacteria, paratyphoid fever	*Salmonella paratyphi*	**Barley**	*Hordeum* spp.
		Barn swallow	*Hirundo rustica erythrogaster*
Bacteria, pneumonia (some forms)	*Streptococcus pneumoniae* and others	**Basil**	*Ocimum basilicum*
		Basswood	*Tilia* spp.
Bacteria, PPLO	*Mycoplasma pneumoniae*	**Bat**	*Eidolon* spp., *Epomophorus* spp., and others
Bacteria, pseudomonad	*Pseudomonas* spp.	**Bat (fig. 23.24)**	*Leptonycteris sanbornii*
Bacteria, purple nonsulfur	*Rhodomicrobium* spp., *Rhodopseudomonas* spp., *Rhodospirillum* spp.	**Bay, California**	*Umbellularia californica*
		Bay, sweet	*Laurus nobilis*
		Bean, broad	*Vicia faba*
Bacteria, purple sulfur	*Amoebobacter* spp., *Lamprocystis* spp., *Rhodothece* spp., and others	**Bean, castor**	*Ricinus communis*
		Bean, garbanzo	*Cicer arietinum*
		Bean, garden	*Phaseolus vulgaris*
Bacteria, salmonella, ("food poisoning")	*Salmonella* spp.	**Bean, jequirity**	*Abrus precatorius*
		Bean, kidney	*Phaseolus vulgaris*
Bacteria, salt	*Halococcus* spp., *Halobacterium* spp.	**Bean, lima**	*Phaseolus lunatus*
Bacteria, sauerkraut	*Leuconostoc* spp. and others	**Bean, Mexican jumping**	*Sebastiana* spp., *Sapium* spp.
Bacteria, sorbose	*Acetobacter suboxydans*		
Bacteria, spotted fever	*Rickettsia rickettsii*		
Bacteria, "strep" throat	*Streptococcus* spp.		
Bacteria, sulfolobus	*Sulfolobus* spp., *Thermoplasma* spp., *Thermoproteus* spp.		

1. These algae are known under several names (*Desmococcus, Phytoconis, Pleurococcus, Protococcus*), and uncertainty exists as to which name has priority. The green algal component of certain lichens, *Trebouxia*, also occurs independently on bark.

Common Names and Scientific Names of Organisms

Common Name	Scientific Name	Common Name	Scientific Name
Bean, mung	*Phaseolus aureus*	Blackbird	*Euphagus* spp. and others
Bean, pinto	*Phaseolus vulgaris*		
Bean, scarlet runner	*Phaseolus coccineus*	Black bread mold	*Rhizopus stolonifer* and others
Bean, tepary	*Phaseolus acutifolius* var. *latifolius*		
		Black Forest mushroom	*Lentinus edodes*
Bean, winged	*Psophocarpus tetragonolobus*	Black locust	*Robinia pseudo-acacia*
		Black mangrove	*Avicennia nitida*
Bear	*Ursus* spp. and others	Black rat snake	*Elaphe obsoleta*
Bearberry	*Arctostaphylos uva-ursi*	Black spruce	*Picea mariana*
Bear, polar	*Thalarctos maritimus*	Black stem rust of wheat	*Puccinia graminis*
Beaver, mountain	*Aplodontia rufa*		
Bedstraw	*Galium* spp.	Black walnut	*Juglans nigra*
Beech	*Fagus grandifolia*	Bladderwort	*Utricularia* spp.
Beech fern	*Thelypteris* spp.	Blazing star	*Liatris ligulistylis*
Beefsteak morel	*Helvella* spp.	Bleeding, ground pine used to arrest	*Lycopodium clavatum*
Bee, honey	*Apis mellifera*		
Beet, garden	*Beta vulgaris*	Bleeding heart	*Dicentra* spp.
Beet, sugar	*Beta vulgaris* (horticulturally selected strains)	Bleeding heart, Pacific	*Dicentra formosa*
		Bloodroot	*Sanguinaria canadensis*
		Blueberry	*Vaccinium* spp.
Beetle	member of Order Coleoptera, Class Insecta, Phylum Arthropoda, Kingdom Animalia	Blue curls	*Trichostema* spp.
		Blue-green bacteria	members of Class Cyanobacteriae, Division Eubacteriophyta, Subkingdom Eubacteriophytineae, Kingdom Monera
Beetle, elm bark	*Hylurgopinus rufipes*, *Scolytus multistriatus*		
Beetle, fungi used for food by ("ambrosia")	*Ambrosiella* spp., *Monilia* spp.	Blue-green bacteria, Lake Chad edible	*Spirulina* sp.
Beetle, scarab	member of Family Scarabaeidae—*see* Beetle	Blue jay	*Cyanocitta cristata*
		Bobcat	*Lynx rufus*
		Bolete	*Boletus* spp., and others
Begonia	*Begonia* spp.		
Belladonna, source of	*Atropa belladonna*	Bollworm	*Pectinophora gossypiella*
Bermuda grass	*Cynodon dactylon*		
Betel nut palm	*Areca catechu*	Boston fern	*Nephrolepis exaltata*
Betony, wood	*Pedicularis canadensis*	Botulism bacteria	*Clostridium botulinum*
Big tree	*Sequoiadendron giganteum*	Bowstring fibers, source of	*Sansevieria metalaea*
Birch	*Betula papyrifera* and others	Bowstring hemp, source of	*Sansevieria* spp.
		Box elder	*Acer negundo*
Bird's foot fern	*Pellaea mucronata*	Boysenberry	*Rubus* hybrids, with *R. ursinus* as one parent
Bird's nest fern	*Asplenium nidus*		
Bird's-nest fungi	*Nidularia* spp. and others	Bracken fern	*Pteridium aquilinum*
Bird's-nest fungus (fig. 17.27)	*Crucibulum levis*	Bracket fungi	*Fomes* spp., *Daedalea* spp., and others
Birth control pills, fungi used in manufacture of	*Rhizopus nigricans*, *R. arrhizus*	Bracket fungus (fig. 17.25)	*Pleurotus ostreatus*
Bittersweet	*Celastrus scandens*	Bracket fungus (fig. 17.19C)	*Polyporus sulphureus*
Blackberry	*Rubus argutus*, *R. laciniatus*, *R. procerus*, and others	Brazil nut	*Bertholettia excelsa*
		Breadfruit	*Artocarpus* spp.

Common Name	Scientific Name	Common Name	Scientific Name
Bridalwreath	*Spiraea vanhouttei* hybrids and others	Buttermilk bacteria	*Streptococcus lactis, S. cremoris, Leuconostoc citrovorum*, and others
Bristlecone pine	*Pinus longaeva*		
Bristlecone pine, Colorado	*Pinus aristata*		
		Butterwort	*Pinguicula* spp.
Broad bean	*Vicia faba*	Button snakeroot	*Eryngium* spp.
Broccoli	*Brassica oleracea* var. *botrytis*	Butyl alcohol bacteria	*Clostridium acetylbutylicum* and others
Bromeliad family	Bromeliaceae		
Broomrape	*Orobanche* spp.	Cabbage	*Brassica oleracea*
Brown algae	members of Division Phaeophyta, Kingdom Protista	Cabbage family	Brassicaceae (Cruciferae)
		Cabbage looper	*Trichoplusia ni*
Brown fruit rot fungi	*Monolinia fruiticola* and others	Cabbage worm	*Pieris rapae*
		Cacao	*Theobroma cacao*
Brucellosis bacteria	*Brucella abortus, B. suis, B. melitensis*	Cactus (figure 22.13)	*Hamatocactus setispinus*
Brussels sprouts	*Brassica oleracea* var. *gemmifera*	Cactus, barrel	*Mamillaria* spp. and others
Bryophyte (*see also* individual listings)	member of Class Anthocerotae, Class Hepaticae, or Class Musci, Division Bryophyta, Kingdom Plantae	Cactus, cholla	*Opuntia* spp. (cylindrical forms)
		Cactus, giant saguaro	*Carnegia gigantea*
		Cactus, hedgehog	*Echinocereus* spp. and others
Bryopsid	member of Class Chlorophyceae, Division Chlorophyta, Kingdom Protista	Cactus, living rock	*Ariocarpus fissuratus* and others
		Cactus, organpipe	*Lemaireocereus* spp.
		Cactus, prickly pear	*Opuntia* spp.
BT bacteria	*Bacillus thuringiensis*	Cactus family	Cactaceae
BTI bacteria	*Bacillus thuringiensis* var. *israelensis*	Cajeput, source of	*Melaleuca leucodendron*
		Calabazilla	*Cucurbita foetidissima*
Bubonic plague bacteria	*Yersinia pestis*	Calendula	*Calendula* spp.
Bucket orchid	*Coryanthes* spp.	California bay	*Umbellularia californica*
Buckeye	*Aesculus* spp.	California fuchsia	*Epilobium* spp.
Buckwheat	*Fagopyrum esculentum*	California nutmeg	*Torreya californica*
Buffalo	*Bison bison*	California poppy	*Eschscholzia californica*
Bullhorn *Acacia*	*Acacia cornigera*	California soaproot	*Chlorogalum pomeridianum*
Bur clover	*Medicago hispida*		
Burn treatment, (ashes) horsetail source of	*Equisetum hyemale* and others	Camel	*Camelus* spp.
		Camelina	*Camelina sativa*
Burn treatment, fern source of	*Polystichum munitum*	Camellia	*Camellia* spp.
		Camphor, source of	*Cinnamomum camphora*
Butcher's broom	*Ruscus* spp.	Candelilla	*Euphorbia antisyphilitca*
Buttercup	*Ranunculus* spp.	Candlenut	*Aleurites moluccana*
Buttercup, European bulbous	*Ranunculus bulbosa*	Cankerworm	*Alsophila pometaria* and others
Buttercup family	Ranunculaceae	Cantaloupe	*Cucumis melo*
Butterfly	member of Superfamily Papilionoidea, Order Lepidoptera, Phylum Arthropoda, Kingdom Animalia	Cap-thrower fungus	*Pilobolus* spp.
		Caraway	*Carum carvi*
		Cardamon	*Elettaria cardamomum*
		Caribou	*Rangifer* spp.
Butterfly iris	*Moraea* sp.	Carnation	*Dianthus caryophyllus* and others

Common Names and Scientific Names of Organisms

Common Name	Scientific Name	Common Name	Scientific Name
Carnaubalike wax, source of	*Stipa tenacissima*	Cherry	*Prunus avium* and others
Carnauba wax, source of	*Copernicia cerifera*	Chestnut	*Castanea* spp.
Carob	*Ceratonia siliqua*	Chestnut, American	*Castanea americana*
Carpetweed family	Aizoaceae	Chestnut blight	*Endothia parasitica*
Carrot	*Daucus carota*	Chia	*Salvia columbariae*
Cashew	*Anacardium occidentale*	Chickadee, mountain	*Parus gambeli*
Cassava	*Manihot esculenta*	Chickweed (Himalayan)	*Stellaria decumbens*
Cassia[2]	*Cinnamomum cassia*	Chicle, source of	*Achras sapota*
Castor bean	*Ricinus communis*	Chicory	*Cichorium intybus*
Catalpa	*Catalpa* spp.	Chilghoza pine	*Pinus gerardiana*
Caterpillar	larval stage of member of Order Lepidoptera, Phylum Arthropoda, Kingdom Animalia	Chinese vegetable tallow	*Sapium sebiferum*
		Chipmunk	*Eutamias* spp. and others
Catnip	*Nepeta cataria*	Chlamydomonas	*Chlamydomonas* spp.
Cattail	*Typha* spp.	Chlorella	*Chlorella* spp.
Cattle—*see* Cow		Chlorine-assimilating fungus	*Aspergillus terreus*
Cauliflower	*Brassica oleracea* var. *botrytis*[3]	Chocolate, source of	*Theobroma cacao*
Caussu wax, source of	*Calathea lutea*	Chokecherry	*Prunus virginiana* var. *melanocarpa*
Cedar, Atlantic white	*Chamaecyparis thyoides*	Cholera bacteria	*Vibrio cholerae*
Cedar, eastern red	*Juniperus virginiana*	Cholla	*Opuntia* spp. (cylindrical forms)
Cedar, eastern white	*Thuja occidentalis*	Christmas flower	*Euphorbia pulcherrima*
Cedar, incense	*Calocedrus decurrens*	Chrysanthemum	*Chrysanthemum* spp.
Cedar, southern white	*Chamaecyparis thyoides*	Chuckwalla	*Sauromalus obesus*
		Chufa	*Cyperus esculentus*
Cedar, western red	*Thuja plicata*	Chytrids	member of Class Chytridiomycetes, Division Mycota, Kingdom Fungi
Celery	*Apium graveolens*		
Cellular slime mold	member of Division Acrasiomycota, Kingdom Fungi	Cinnamon	*Cinnamomum zeylanicum*
Century plant	*Agave* spp.	Cinnamon fern	*Osmunda cinnamonea*
Chain fern	*Woodwardia fimbriata*	Citric acid, fungal producers of	*Aspergillus niger* and others
Chamise	*Adenostoma fasciculatum*	Citronella oil, source of	*Cymbopogon nardus*
Chard	*Beta vulgaris* var. *cicla*	Citrus	*Citrus* spp.
Cheese bacteria—*see* Buttermilk bacteria		Citrus family	Rutaceae
		Clematis	*Clematis* spp.
Cheese fungi	*Penicillium camembertii* (for Camembert cheese), *P. roquefortii* (for blue, Gorgonzola, Roquefort, and Stilton cheeses)	Climbing fern, Asian	*Lygodium salicifolium*
		Clover	*Trifolium* spp.
		Clover, bur	*Medicago hispida*
		Clover, sweet	*Melilotus* spp.
		Cloves	*Eugenia caryophyllata*
		Club fungus	member of Class Basidiomycetes, Division Mycota, Kingdom Fungi
		Club moss	member of Division Lycophyta, Kingdom Plantae
		Coastal redwood	*Sequoia sempervirens*
		Cobra plant	*Darlingtonia californica*

2. This should not be confused with the genus *Cassia,* the source of senna in the Legume Family, or cassie, a perfume oil whose source is another member of the Legume Family, *Acacia farnesiana.*

3. Broccoli and cauliflower are two different forms of the same variety.

Common Names and Scientific Names of Organisms

Common Name	Scientific Name	Common Name	Scientific Name
Cocaine, source of	*Erythroxylon coca*	Crocus	*Crocus* spp.
Cochineal insect	*Dactylopius coccus*	Crocus, autumn	*Colchicum autumnale*
Cocklebur	*Xanthium* spp.	Crown-of-thorns	*Euphorbia milii* var. *splendens* and others
Cockroach	*Blatta orientalis, Blatella germanica,* and others		
Cockroach plant	*Haplophyton cimicidum*	Crustacean	member of Class Crustacea, Phylum Arthropoda, Kingdom Animalia
Cockscomb	*Celosia* spp.		
Coconut palm	*Cocos nucifera*	Cryptophyte	member of Class Cryptophyceae, Division Chrysophyta, Kingdom Protista
Codling moth	*Carpocapsa pomonella*		
Coffee	*Coffea arabica*		
Coleus	*Coleus blumei* and others		
		Cucumber	*Cucumis sativus*
Columbine	*Aquilegia* spp.	Cucumber, squirting	*Ecballium elaterium*
Columbine (fig. 22.3A)	*Aquilegia formosa*	Cup fungus	member of Class Ascomycetes, Division Mycota, Kingdom Fungi
Compass plant	*Lactuca serriola*		
Coneflower	*Rudbeckia* sp.		
Coneflower, Asian	*Strobilanthes* spp.	Cup fungus (fig. 17.14)	*Caloscypha fulgens*
Copal, sources of	*Agathis alba, Copaifera demeussei, Hymenea coubaril, Trachylobium verrucosum,* and others	Cycad (fig. 20.9)	*Dioon edule, Encephalartos altensteinii*
		Cyclamen	*Cyclamen* spp.
		Cypress, bald	*Taxodium distichum*
Copperhead	*Ancistrodon contortrix*	Daffodil	*Narcissus* spp.
Coralline algae	*Bossiella* spp., *Corallina* spp., *Lithothamnion* spp., and others	Dahlia	*Dahlia* spp.
		Daisy	*Dimorphotheca* spp., *Layia* spp., and others
Coral tree	*Erythrina crista-galli*		
Cordage fibers, source of	*Agave sisalina, A. heterocantha, A. lophantha, Phormium tenax,* and others	Damask rose	*Rosa damascena*
		Dandelion	*Taraxacum officinale*
		Dandruff, fern(s) used in treatment of	*Adiantum capillus-veneris, Polystichum munitum*
Coriander	*Coriandrum sativum*		
Cork oak	*Quercus suber*	Date palm	*Phoenix dactylifera*
Corn	*Zea mays*	Dawn redwood	*Metasequoia glyptostroboides*
Corn rust	*Puccinia sorghi*		
Corn smut	*Ustilago maydis*	DDT-like compound, algal producers of	*Laurencia* spp. and others
Cotton	*Gossypium* spp.		
Cottonwood	*Populus deltoides* and others	Death angel	*Amanita phalloides*
		Decay bacteria	*Clostridium* spp., *Micrococcus* spp., *Proteus* spp., *Pseudomonas* spp., and others
Coughs, fern(s) used in treatment of	*Adiantum aethiopicum, A. lunulatum Polypodium glycyrrhiza,* and others		
		Deer	*Odocoileus* spp. and others
Coulter pine	*Pinus coulteri*		
Cow	*Bos* spp.	Deer brush	*Ceanothus* spp.
Cow parsnip	*Heracleum lanatum*	Deer, mule	*Odocoileus hemionus*
Cowslip	*Caltha palustris*	Denitrifying bacteria	*Micrococcus dentrificans, Thiobacillus denitrificans,* and others
Crabapple	*Crataegus* spp.		
Cranberry	*Vaccinium macrocarpus*		
Cress, garden	*Lepidium sativum, Barbarea verna,* and others		
		Desmids	*Closterium* spp., *Cosmarium* spp., and others
Crested wheat grass	*Agropyron cristatum*		

Common Names and Scientific Names of Organisms

Common Name	Scientific Name	Common Name	Scientific Name
Dewberry	*Rubus* hybrids with *R. ursinus* as one parent	Dwarf mistletoe	*Arceuthobium* spp.
		Dyer's woad	*Isatis tinctoria*
Dextran bacteria	*Leuconostoc mesenteroides*	Dyes, fern source of	*Sadleria cyatheoides* (trunk) *Sphenomeria chusana* (fronds)
Diabetes, fern used in treatment of	*Adiantum caudatum*	Dyes, sources of	see listing in appendix 3
Diarrhea, fern(s) used in treatment of	*Botrychium lunaria, B. ternatum, Pteridium aquilinum,* and others	Dysentery, fern(s) used in treatment of	*Botrychium lunaria, B. ternatum, Pteridium aquilinum,* and others
Diatom	*Biddulphia* spp., *Cymbella* spp., *Navicula* spp., and others	Eagle, golden	*Aguila chrysaëtos*
		Earth star	*Geaster* spp. and others
		Earthworm	*Lumbricus* spp. and others
Diatom (fig. 16.3)	*Cymatopleura solea*	Eastern hemlock	*Tsuga canadensis*
Dicot	member of Class Dicotyledonae, Division Anthophyta, Kingdom Plantae	Eastern larch	*Larix laricina*
		Eastern white cedar	*Thuja occidentalis*
		Eastern white pine	*Pinus strobus*
Digger pine	*Pinus sabiniana*	Ebony	*Diospyros ebenum*
Digitalis, source of	*Digitalis purpurea*	Eczema, fern used in treatment of	*Lygodium flexuosum*
Dill	*Anethum graveolens*		
Dinoflagellate	member of Division Pyrrophyta, Kingdom Protista	Eelworm (nematode)	member of Class Nematoda, Phylum Aschelminthes, Kingdom Animalia
Dinoflagellate, midnight-bioluminescent	*Gonyaulax polyedra*	Eelworms (nematodes), fungi that trap with constricting rings	*Dactylaria* spp., *Arthrobotrys dactyloides*
Diphtheria bacteria	*Corynebacterium diphtheriae*	Eelworms (nematodes), fungi that trap with passive rings	*Dactylella* spp.
Dischidia	*Dischidia rafflesiana*		
Diuretic, ferns used as	*Adiantum venustum, Lygodium japonicum*	Eggplant	*Solanum melongena*
Dodder	*Cuscuta* spp.	Elephant	*Elephas* spp., *Loxodonta* spp.
Dogbane	*Apocynum* spp.	Elephant ears	*Colocasia* spp.
Dogwood	*Cornus* spp.	Elk	*Cervus canadensis*
Douglas fir	*Pseudotsuga menziesii*	Elm	*Ulmus* spp.
Dove	member of Family Columbidae, Class Aves, Phylum Vertebrata, Kingdom Animalia	Elm bark beetle	*Hylurgopinus rufipes, scolytus multistriatus*
		Endive	*Cichorium endivia*
		English ivy	*Hedera helix*
Dove, mourning	*Zenaidura macroura*	Ensilage bacteria	*Lactobacillus delbrueckii, L. plantarum,* and others
Downy mildew of grape	*Plasmopora viticola*		
Dragon tree	*Dracaena draco*		
Dragon's blood	*Dracaena* spp., *Daemonorops* spp.	Ergot	*Claviceps purpurea*
		Ermine	*Mustela erminea*
Duckweed	*Lemna* spp., *Wolffia* spp., and others	Eucalyptus, Tasmanian giant	*Eucalyptus regnans*
Dulse	*Rhodymenia* spp.	Eucalyptus oil, source of	*Eucalyptus* spp.
Dung mosses (on dung of carnivores)	*Tayloria* spp.		
Dung mosses (on dung of herbivores)	*Splachnum* spp.	Euglenoids	members of Division Euglenophyta, Kingdom Protista
Dutch elm disease	*Ceratocystis ulmi*	European corn borer	*Pyrausta nubialis*
Dutchman's breeches	*Dicentra cucullaria*	European larch	*Larix decidua*
Dwarf juniper	*Juniperus communis* and others	European spiderwort	*Tradescantia paludosa*
		European stone pine	*Pinus pinea*

Common Names and Scientific Names of Organisms

Common Name	Scientific Name	Common Name	Scientific Name
Extinguisher mosses	*Encalypta* spp.	**Fern(s), fossil**	*Psaronius* spp., *Thamnopteris* spp., and others
Eye diseases, fern used in treating	*Asplenium adiantum-nigrum*	**Fern, goldback**	*Pityrogramma triangularis*
Fall webworm	*Hyphantria cunea*		
False morel	*Helvella* spp.	**Fern—used for hairwash**	*Dryopteris dilatata*
Fennel	*Foeniculum vulgare*		
Fenugreek	*Trigonella foenumgraecum*	**Fern, holly**	*Polystichum lonchitis*
Fern(s), adder's tongue	*Ophioglossum* spp.	**Fern(s)—used by hummingbirds**	*Cyathea arborea, Lophosoria quadripinnata, Nephelea mexicana*
Fern(s), amphibious	*Marsilea* spp. and others		
Fern(s), aquatic (floating)	*Azolla* spp., *Salvinia* spp.	**Fern—used for treating insect stings and bites**	*Adiantum capillus-veneris*
Fern(s)—source of astringent	*Actiniopteris radiata, Drynaria quercifolia, Pteridium aquilinum,* and others	**Fern—used for easing labor pains**	*Athyrium filix-femina*
		Fern, lady	*Athyrium filix-femina*
Fern(s), beech	*Thelypteris* spp.	**Fern(s)—used as laxative**	*Asplenium trichomanes, Polypodium vulgare*
Fern, bird's foot	*Pellaea mucronata*	**Fern—used in treating leprosy**	*Marsilea quadrifolia*
Fern, Boston	*Nephrolepis exaltata*		
Fern, bracken	*Pteridium aquilinum*	**Fern, licorice**	*Polypodium glycyrrhiza*
Fern—used in treating burns	*Polystichum munitum*	**Fern(s)—poisonous to livestock**	*Onoclea sensibilis, Pteridium aquilinum*
Fern, chain	*Woodwardia fimbriata*	**Fern, maidenhair (fig. 19.16B)**	*Adiantum pedatum*
Fern, cinnamon	*Osmunda cinnamomea*		
Fern, climbing (Asian)	*Lygodium salicifolium*	**Fern—edible Malaysian (relative of lady fern)**	*Athyrium esculentum*
Fern(s)—used in treating coughs	*Adiantum aethiopicum, A. lunulatum, Polypodium glycyrrhiza*		
		Fern, male	*Dryopteris filix-mas*
Fern(s)—used in treating dandruff	*Adiantum capillus-veneris, Polystichum munitum*	**Fern, mosquito**	*Azolla caroliniana*
		Fern, nest	*Asplenium nidus*
		Fern—used to arrest nosebleeds	*Pellaea mucronata*
Fern—used in treating diabetes	*Adiantum caudatum*	**Fern(s)—used for orchid bark**	*Cibotium* spp., *Osmunda* spp.
Fern(s)—used in treating diarrhea	*Botrychium lunaria, B. ternatum, Pteridium aquilinum,* and others	**Fern, Oriental water**	*Ceratopteris thalictroides*
		Fern—used as poison antidote	*Polystichum squarrosum*
Fern(s)—used as diuretic	*Adiantum venustum, Lygodium japonicum*	**Fern(s)—used in treating rickets**	*Asplenium ruta-muraria, Osmunda regalis*
Fern—sources of dyes	*Sadleria cyatheoides* (trunk), *Sphenomeris chusana* (fronds)	**Fern(s), seed (Pteridosperms)**	*Lyginopteris* spp., *Medullosa* spp., and others
Fern(s)—used in treating dysentery	*Botrychium lunaria, B. ternatum, Pteridium aquilinum,* and others	**Fern(s)—used for stuffing mattresses, pillows, upholstery**	*Cibotium* spp., *Sadleria* spp.
Fern—used in treating eczema	*Lygodium flexuosum*	**Fern, sword**	*Polystichum munitum*
Fern—used in treating eye diseases	*Asplenium adiantum-nigrum*	**Fern—used in treating toothache**	*Pityrogramma triangularis*
Fern—used to reduce fevers	*Marsilea quadrifolia*	**Fern(s), Hawaiian tree**	*Cibotium* spp., *Sadleria* spp.
Fern, five-finger	*Adiantum pedatum*	**Fern(s), tree**	*Cyathea* spp., *Ctenitis* spp., *Dicksonia* spp., *Sphaeropteris* spp., and others
Fern(s)—used as food	*Athyrium filix-femina, Dryopteris austriaca, D. filix-mas, Polystichum munitum,* and others		
		Fern, tropical tree (fig. 19.25)	*Cibotium glaucum*

Common Names and Scientific Names of Organisms

Common Name	Scientific Name	Common Name	Scientific Name
Fern—used for expelling worms	*Dryopteris filix-mas*	Fossil horsetails	*Equisetites* spp., *Hyenia* spp., *Sphenophyllum* spp., and others
Fern(s)—used for treating wounds	*Lygodium circinatum, Ophioglossum vulgatum*	Fossil horsetails, treelike	*Calamites* spp.
Fevers, fern used to reduce	*Marsilea quadrifolia*	Four-o'clock family	Nyctaginaceae
Fevers, ground pine used to reduce	*Lycopodium clavatum*	Fox, arctic	*Alopex lagopus*
		Fox, gray	*Urocyon cinereoargentus*
Field horsetail	*Equisetum arvense*	Fox, red	*Vulpes fulva*
Fig, common	*Ficus carica*	Foxglove	*Digitalis* spp.
Fig, tropical	*Ficus* spp.	Frangipanni	*Plumeria* spp.
Fig, tropical (fig. 6.12)	*Ficus macrophyllus*	Frog	*Rana* spp. and others
Figwort family	Scrophulariaceae	Fruit fly	*Drosophila melanogaster* and others
Filaree	*Erodium* spp.		
Fir, balsam	*Abies balsamea*	Fuchsia, California	*Epilobium* spp.
Fir, Douglas	*Pseudotsuga menziesii*	Fumitory, Himalayan	*Corydalis gerdae*
Fir, white	*Abies concolor*	Fungi that produce antibiotics	*Penicillium* spp., *Cephalosporium* spp., and others
Fish	member of Class Pisces, Phylum Vertebrata, Kingdom Animalia	Fungi that cause aspergilloses	*Aspergillus fumigatus, Candida albicans, Coccidiodes immitis,* and others
Fish molds	*Saprolegnia* spp. and others		
Five-finger fern	*Adiantum pedatum*	Fungi that cause athlete's foot	*Trichophyton* spp.
Flashlight fish	*Anomalops katoptron, Photoblepharon palpebratus*	Fungi—used by beetles for food	*Ambrosiella* spp., *Monilia* spp.
Flashlight powder, ground pine source of	*Lycopodium* spp.	Fungi, bird's-nest	*Nidularia* spp.
		Fungi—used in manufacturing birth control pills	*Rhizopus nigricans, R. arrhizus*
Flatworm	*Convoluta roscoffensis*	Fungi, bracket	*Fomes* spp., *Daedalea* spp., and others
Flatworm algae	*Platymonas* spp.		
Flax	*Linum* spp.	Fungi, brown fruit rot	*Monolinia fruticola* and others
Flax, New Zealand	*Phormium tenax*		
Flea	member of Order Siphonaptera, Phylum Arthropoda, Kingdom Animalia	Fungi, cap-thrower	*Pilobolus* spp.
		Fungi, cheese	*Penicillium camembertii* (for Camembert cheese), *P. roquefortii* (for blue, Gorgonzola, Roquefort, and Stilton cheeses)
Flicker	*Colaptes* spp.		
Florida arrowroot	*Zamia floridana*		
Flour, Hopi Indian horsetail source of	*Equisetum laevigatum*		
Flowerpot leaf plant	*Dischidia rafflesiana*	Fungi, citric acid-producing	*Aspergillus niger* and others
Fly	member of Order Diptera, Phylum Arthropoda, Kingdom Animalia	Fungi, flavor-producing	*Aspergillus* spp.
		Fungi, hallucinogenic	*Amanita muscaria, Conocybe* spp., *Panaeolus* spp., *Psilocybe* spp., and others
Fly agaric	*Amanita muscaria*		
Flycatcher	*Empidonax* spp., *Myiarchus* spp., and others		
		Fungi, horse dung	*Pilobolus* spp.
Fossil, compression (fig. 19.26)	*Annularia radiata*	Fungi, industrial alcohol-producing	*Aspergillus* spp.
Fossil ferns	*Psaronius* spp., *Thamnopteris* spp., and others	Fungi, insect-parasitizing	members of Order Laboulbeniales, Class Ascomycetes, Division Mycophyta, Kingdom Fungi, and others

Common Name	Scientific Name	Common Name	Scientific Name
Fungi, jelly	*Auricularia* spp., *Exidia* spp., *Tremella* spp., and others	Ginger	*Zingiber officinale* and others
Fungi, meat-tenderizing	*Thamnidium* spp.	Ginseng, source of	*Panax quinquefolium*
Fungi, shelf—*see* Fungi, bracket		Giraffe	*Giraffa camelopardalis*
		Gladiola	*Gladiolus* spp.
Fungi, *shoyu*	*Aspergillus oryzae, A. soyae*	Gloeocapsa	*Gloeocapsa* spp.
Fungi—used in silvering of mirrors	*Aspergillus* spp.	Glutamic acid-producing bacteria	*Arthrobacter* spp., *Brevibacterium* spp., *Micrococcus* spp.
Fungi—used in manufacturing soap	*Penicillium* spp.	Goat	*Capra* spp.
Fungi, soil	*Fusarium* spp. and others	Goldback fern	*Pityrogamma triangularis*
Fungi, soy sauce	*Aspergillus oryzae, A. soyae*	Golden-brown algae	members of Class Chrysophyceae, Division Chrysophyta, Kingdom Protista
Fungi, *sufu*	*Actinomucor elegans, Mucor* spp.		
Fungi, *teonanacatl* (sacred)	*Conocybe* spp., *Panaeolus* spp., *Psilocybe* spp., and others	Golden chain tree	*Laburnum anagyroides*
		Goldenrod	*Solidago* spp.
		Goldenseal	*Hydrastis canadensis*
		Goldenweed	*Haplopappus gracilis*[4]
Fungus, chlorine-assimilating	*Aspergillus terreus*	Gold violet	*Viola douglasii*
Fungus, "foolish seedling"	*Gibberella fujikuroi*	Gonorrhea bacteria	*Neisseria gonorrhoeae*
Fungus, *miso*	*Aspergillus oryzae*	Goose	*Branta* spp. and others
Fungus—used in producing plastics	*Aspergillus terreus*	Gooseberry	*Ribes* spp.
		Goosefoot Family	Chenopodiaceae
Fungus, *tempeh*	*Rhizopus oligosporus*	Gopher	*Geomys* spp., *Thomomys* spp.
Fungus—used in manufacturing toothpaste	*Aspergillus niger*	Gopher, pocket	*Geomys bursarius* and others
Fungus—used in manufacturing yellow food-coloring agent	*Blakeslea trispora*	Gopher plant	*Euphorbia lathyrus*
		Gourd	*Lagenaria siceraria* and others
		Grape	*Vitis* spp.
Funori, source of	*Gloiopeltis* spp.	Grapefruit	*Citrus paradisi*
Fur, green algae that inhabit animal	*Trentepohlia* spp.	Grass	*Bromus* spp. and others[5]
Garbanzo bean	*Cicer arietinum*	Grass, Bermuda	*Cynodon dactylon*
Garden bean	*Phaseolus vulgaris*	Grass, crested wheat	*Agropyron cristatum*
Gas gangrene bacteria	*Clostridium novyi, C. perfringens, C. septicum*	Grass family	Poaceae (Gramineae)
		Grass, Indian	*Sorghastrum nutans*
		Grass, pampas	*Cortaderia sellona*
Gentian, source of	*Gentiana* spp.	Grass tree	*Xanthorrhea* spp.
Geranium	*Geranium* spp., *Pelargonium* spp.	Gray fox	*Urocyon cinereoargentus*
Geranium family	Geraniaceae	Greenbrier	*Smilax* spp.
Giant Galápagos tortoise	*Testudo elephantopus porteri*	Green rope algae	*Codium* spp.
		Green sulfur bacteria	*Chlorobium* spp., *Chloropseudomonas* spp., *Prosthecochloris* spp., and others
Giant horsetail	*Equisetum telmateia*		
Giant kelp	*Macrocystis pyrifera*		
Giant redwood	*Sequoiadendron giganteum*		
		Ground pine	*Lycopodium* spp.
Giant saguaro cactus	*Carnegia gigantea*		
Giant waterlily	*Victoria amazonica*		
Gila monster	*Heloderma suspectum*		

4. Species with four chromosomes per cell.

5. About 4,500 species of the Grass Family, Poaceae (Gramineae), in all.

Common Names and Scientific Names of Organisms

Common Name	Scientific Name	Common Name	Scientific Name
Ground pine, fossil relative of	*Baragwanathia* spp., *Drephanophycus* spp., *Protolepidodendron* spp., and others	Hog	*Sus scrofa* and others
		Hog fennel	*Lomatium* spp.
		Holly, American	*Ilex opaca*
		Holly fern	*Polystichum lonchitis*
Ground pine—used for baby powder	*Lycopodium clavatum*	Honeybee	*Apis mellifera*
		Honey locust	*Gleditsia triacanthos*
Ground pine—used to arrest bleeding	*Lycopodium clavatum*	Hoopoe	*Upupa africana*
		Hop hornbeam	*Ostrya virginiana*
Ground pine—used as intoxicant	*Lycopodium selago*	Hops	*Humulus lupulus*
Ground pine—used for ornaments	*Lycopodium clavatum*, *L. complanatum*, and others	Horehound	*Marrubium vulgare*
		Hornwort	*Anthoceros* spp.
		Horse	*Equus caballus*
Ground pine—used to reduce fevers	*Lycopodium clavatum*	Horseradish	*Rorippa armoracia*
Guava	*Psidium guajava*	Horsetail	*Equisetum* spp.
Gum arabic, source of	*Acacia senegal* and others	Horsetail(s)—used as abrasive	*Equisetum* (all spp.)
Gum tragacanth, source of	*Astragalus echidenaeformis*, *A. gossypinus*, *A. gummifer*, and others	Horsetail(s)—used as astringent	*Equisetum arvense*, *E. debile*, and others
		Horsetail(s)—ashes used for treating burns	*Equisetum hyemale* and others
Guppy	*Lebistes reticulatus*	Horsetail—used for treating diarrhea	*Equisetum hyemale*
Gypsy moth	*Porthetria dispar*		
Hadeda ibis	*Hagedashia hagedash hagedash*	Horsetail(s)—used as diuretic	*Equisetum arvense*, *E. debile*, and others
Haircap moss	*Polytrichum* spp.	Horsetail—used for treating dysentery	*Equisetum hyemale*
Hairwash, fern used as	*Dryopteris dilatata*	Horsetail, field	*Equisetum arvense*
Hairwash, horsetail used as	*Equisetum hyemale*	Horsetail(s), fossil	*Equisetites* spp., *Hyenia* spp., *Sphenophyllum* spp., and others
Hallucinogenic fungi	*Amanita muscaria*, *Conocybe* spp., *Panaeolus* spp., *Psilocybe* spp., and others	Horsetail, Hopi Indian flour source	*Equisetum laevigatum*
		Horsetail, treelike fossil	*Calamites* spp.
		Horsetail, giant	*Equisetum telmateia*
Hawk	*Buteo* spp., *Falco* spp., and others	Horsetail—used as hairwash	*Equisetum hyemale*
Hazel nut	*Corylus* spp.	Horsetail—used as water source	*Equisetum telmateia*
Heath	*Erica* spp. and others	Hot springs, blue-green algae of	*Bacillosiphon induratus*, *Synechococcus* spp., and others
Heath family	Ericaceae		
Hedgehog cactus	*Echinocereus* spp. and others	"Human hair" slime mold	*Stemonitis* spp.
Hemlock, eastern	*Tsuga canadensis*	Hummingbird	*Archilocus* spp. and others
Hemlock, mountain	*Tsuga mertensiana*		
Hemlock, poison	*Conium maculatum*	Hummingbird, Oasis (fig. 23.22)	*Rhodopis vesper*
Hemlock, water	*Cicuta* spp.		
Hemlock, western	*Tsuga heterophylla*	Hummingbirds, ferns used by (for nest material)	*Cyathea arborea*, *Lophosoria quadripinnata*, *Nephelea mexicana*
Hemp	*Cannabis sativa*		
Hemp, Manila	*Musa textilis*		
Hemp, Mauritius	*Furcraea gigantea*		
Henbit	*Lamium amplexicaule*	Hummingbirds, tropical	*Chlorostilbon maugaeus* and others
Henna	*Lawsonia inermis*		
Hepatica	*Hepatica* spp.	Hyacinth	*Hyacinthus* spp.
Hepatica (fig. 22.3C)	*Hepatica americana*	Hyacinth, grape	*Muscari* spp.
Hickory	*Carya* spp.	Hydrogen bacteria	*Hydrogenomonas* spp.
Himalayan fumitory	*Corydalis gerdae*		

Common Names and Scientific Names of Organisms

Common Name	Scientific Name	Common Name	Scientific Name
Hyssop	*Hyssopus officinalis*	Jumping mouse	*Zapus hudsonius, Napaeozapus insignis*
Ice-minus bacteria	*Pseudomonas syringiae*		
Ice plant	*Mesembryanthemum* spp.	Junco	*Junco* spp.
		Junco, slate-colored	*Junco hyemalis*
Incense cedar	*Calocedrus decurrens*	Juneberry	*Amelanchier* spp.
India, toxic blue-green bacteria of	*Lyngbya majuscula*	Juniper	*Juniperus* spp.
Indian pipe	*Monotropa* spp.	Juniper, dwarf	*Juniperus communis* and others
Indian warrior	*Pedicularis densiflora*	Jute	*Corchorus* spp.
Indicator mosses for absence of calcium	*Andraea* spp., *Rhacomitrium lanuginosum*	Kaffir lily	*Clivia* sp.
		Kauri pine	*Agathis australis, A. robusta*
Indicator mosses for presence of calcium	*Didymodon* spp., *Desmatodon* spp., and others	Kefir bacteria	*Lactobacillus bulgaricus, Streptococcus lactis*
Indicator mosses for saline (salty) soil	*Pottia* spp.	Kelp	*Alaria* spp., *Dictyoneurum* spp., *Egregia* spp., *Laminaria* spp., *Lessoniopsis* spp., *Nereocystis* spp., and others
Indicator mosses for seasonal running water	*Fontinalis* spp.		
Indigo	*Indigofera* spp.		
Industrial alcohol, fungal producers of	*Aspergillus* spp.		
		Kelp, giant	*Macrocystis pyrifera*
Inky cap mushrooms	*Coprinus* spp.	Kidney bean	*Phaseolus vulgaris*
Insects—*see* individual entries		Knobcone pine	*Pinus attenuata*
		Knotweed	*Polygonum aviculare*
Insects, fern used for treating stings and bites of	*Adiantum capillus-veneris*	Kohlrabi	*Brassica oleracea* var. *caulorapa*
Intoxicant, ground pine used as	*Lycopodium selago*	Kudzu	*Puerania thunbergiana*
		Kumquat	*Fortunella japonica*
Ipecac, source of	*Cephaelis ipecacuanha*	Labor pains, fern used to ease	*Athyrium filix-femina*
Iris	*Iris* spp.		
Iris, butterfly	*Moraea,* sp.	Lactic acid bacteria	*Lactobacillus delbrueckii* and others
Iris family	Iridaceae		
Irish moss	*Chondrus crispus*	Lady fern	*Athyrium filix-femina*
Iron bacteria	*Gallionella* spp., *Sphaerotilus* spp.	Lake Chad region, edible blue-green bacteria	*Spirulina maxima*
Ironwood, South American	*Krugiodendron ferreum*		
		Lamb's ears	*Stachys byzantina*
Ivy, Boston	*Parthenocissus tricuspidata*	Larch, eastern	*Larix laricina*
		Larch, European	*Larix decidua*
Ivy, English	*Hedera helix*	Larch, western	*Larix occidentalis*
Jacaranda	*Jacaranda* spp.	Larkspur, blue	*Delphinium* spp.
Jack pine	*Pinus banksiana*	Larkspur, red	*Delphinium nudicaule*
Jaeger	*Stercorarius* spp.	Late blight of potato	*Phytophthora infestans*
Jeffrey pine	*Pinus jeffreyi*	Laurel	*Laurus nobilis*
Jelly fungi	*Auricularia* spp., *Exidia* spp., *Tremella* spp., and others	Laurel family	Lauraceae
		Lavender	*Lavandula officinalis*
Jequirity bean	*Abrus precatorius*	Laxative, ferns used as	*Asplenium trichomanes, Polypodium vulgare*
Jerusalem sage	*Phlomis fruticosa*		
Jimson weed	*Datura* spp.	Leaf miner	*Agromyza* spp. and others
Jimson weed (fig. 8.6)	*Datura stramonium*		
Jojoba	*Simmondsia californica*	Leaf roller	*Archips argyrospila* and others
Joshua tree	*Yucca brevifolia*		
		Leafy liverwort—*see* Liverwort, leafy	

Common Names and Scientific Names of Organisms

Common Name	Scientific Name	Common Name	Scientific Name
Legionnaire's disease bacteria	*Legionella pneumophilia*	**Lime**	*Citrus aurantifolia*
Legume family	Leguminosae (Fabaceae)	**Litmus indicator dye, source of**	*Rocella* spp.
Lemming	*Lemmus* spp., *Dicrostonyx groenlandicus*	**Liverwort**	member of Class Hepaticae, Division Bryophyta, Kingdom Plantae
Lemon	*Citrus limon*	**Liverwort, leafy (fig. 18.8)**	*Porella* sp.
Lemongrass oil, source of	*Cymbopogon citratus, C. flexuosus*	**Liverworts, leafy**	*Frullania* spp., *Jungermannia* spp., *Porella* spp., and others[7]
Lentil	*Lens esculenta*		
Leprosy, fern used for treating	*Marsilea quadrifolia*	**Liverworts, thalloid**	*Marchantia* spp. and others
Lettuce	*Lactuca sativa*	**Livestock, ferns poisonous to**	*Onoclea sensibilis, Pteridium aquilinum*
Lichen (symbiotic association of an alga and a fungus)	Class Lichens, Kingdom Fungi[6]	**Living rock cactus**	*Ariocarpus fissuratus* and others
Lichen, foliose (fig. 17.38)	*Physcia* sp.	**Lizard**	*Sceloporus* spp. and others
Lichen, foliose (fig. 17.39B)	*Parmelia* sp.	**Lobeline sulphate, source of**	*Lobelia inflata*
Lichen, fruticose (fig. 17.39C)	*Usnea* sp.	**Loblolly pine**	*Pinus taeda*
Lichen, grazed by North African sheep	*Lecanora* spp.	**Locoweed**	*Astragalus* spp.
Lichen, litmus	*Rocella* spp.	**Loganberry**	*Rubus hybrids,* with *R. ursinus* as one parent
Lichen, natural dye	*Parmelia* spp., *Usnea* spp., and others	**Longleaf pine**	*Pinus palustris*
Lichen, perfume stabilizer	*Evernia* spp.	**Loon**	*Gavia* spp.
Lichen, reindeer ("reindeer moss")	*Cladonia* spp., *Cetraria islandica*	**Louse**	Order Mallophaga, Order Anaplura, Class Insecta, Phylum Arthropoda, Kingdom Animalia
Lichens, crustose (fig. 17.39A)			
black	*Rinodina* sp.	**Luffa**	*Luffa cylindrica*
chartreuse	*Acarospora citrina*	**Luminescent bacteria**	*Achromobacter* spp., *Flavobacterium* spp., *Photobacterium* spp., *Pseudomonas* spp., *Vibrio* spp., and others
gray	*Psora* sp.		
orange-red	*Caloplaca elegans*		
yellow	*Candelariella vitellina*		
Lichens used as miniature trees and shrubs	*Cladonia* spp. and others	**Luminous moss**	*Schistostega pennata*
Licorice, source of	*Glycyrrhiza glabra*	**Lupine**	*Lupinus* spp.
Licorice fern	*Polypodium glycyrrhiza*	**Lupine, tree (with seed valves)**	*Lupinus arboreus*
Lignum vitae	*Guaiacum officinale*	**Madder family**	Rubiaceae
Lilac	*Syringa vulgaris*	**Magnolia**	*Magnolia* spp.
Lily	*Lilium* spp. and others	**Maidenhair fern**	*Adiantum* spp.
Lily, tiger	*Lilium pardalinum*	**Malaysian edible fern (relative of lady fern)**	*Athyrium esculentum*
Lily, Turk's cap	*Lilium superbum*		
Lily, wood	*Lilium superbum*	**Male fern**	*Dryopteris filix-mas*
Lily family	Liliaceae	**Mallow**	*Malva* spp.
Lima bean	*Phaseolus lunatus*	**Mango**	*Mangifera indica*
		Mangrove	*Rhizophora mangle* and others

---

6. The lichens are arbitrarily treated as a class within Kingdom Fungi on the basis of the fact that the fungal component of each species of lichen is unique to the species, whereas the algal component may be common to more than one species of lichen.

---

7. There are thousands of species of leafy liverworts assigned to about 200 genera.

Common Name	Scientific Name	Common Name	Scientific Name
Manila hemp	*Musa textilis*	Mollusc	member of Phylum Mollusca, Kingdom Animalia
Manioc—*see* Cassava			
Manroot	*Marah* spp.	Monkey	*Ateles dariensis* and others
Maple	*Acer* spp.		
Maple	*Acer saccharinum*	Monkey flower	*Mimulus* spp.
Maple, hard	*Acer saccharum*	Monkshood	*Aconitum* spp.
Maple, sugar	*Acer saccharum*	Monocot	member of Class Monocotyledonae, Division Anthophyta, Kingdom Plantae
Marigold	*Tagetes* spp.		
Marijuana	*Cannabis sativa*		
Marjoram	*Majorana hortensis*	Monterey pine	*Pinus radiata*
Maté	*Ilex paraguariensis*	Moose	*Alces americana*
Mauritius hemp	*Furcraea gigantea*	Morel	*Morchella esculenta* and others
Meadow foam	*Limnanthes* spp.		
Meadow saffron	*Colchicum autumnale*	Morning glory	*Ipomoea violacea* and others
Mealy bugs	*Pseudococcus* spp.		
Meat-tenderizing fungi	*Thamnidium* spp.	Mosquito	*Anopheles* spp., *Culex* spp., and others
Melon	*Cucumis melo*		
Melon, honeydew	*Cucumis melo* (variety)	Mosquito fern	*Azolla caroliniana*
Melonette	*Melothria pendula*	Mosquito-killing bacteria	*Bacillus thuringiensis* var. *israelensis*
Meningitis bacteria	*Neisseria meningitidis* and others	Moss	member of Class Musci, Division Bryophyta, Kingdom Plantae
Merkus pine	*Pinus merkusii*		
Mermaid's wineglass	*Acetabularia* spp.	Moss, luminous	*Schistostega pennata*
Mesquite	*Prosopis glandulosa*	Mosses, annual (bare soil)	*Acaulon* spp., *Ephemerum* spp., and others
Methane bacteria	*Methanobacterium* spp., *Methanococcus* spp., *Methanosarcina* spp., and others		
		Mosses, antler and bone	*Tetraplodon* spp.
		Mosses, dung (on dung of carnivores)	*Tayloria* spp.
Mexican jumping bean	*Sebastiana* spp., *Sapium* spp.	Mosses, dung (on dung of herbivores)	*Splachnum* spp.
Mexican jumping bean moth	*Carpocaps asaltitans*	Mosses, extinguisher	*Encalypta* spp.
Mexican poppy	*Hunnemania* spp.	Mosses, haircap	*Polytrichum* spp.
Mexican stone pine	*Pinus cembroides*	Mosses, indicator for absence of calcium	*Andreaea* spp., *Rhacomitrium lanuginosum*
Milkweed	*Asclepias* spp.		
Milkweed, swamp	*Asclepias incarnata*	Mosses, indicator for presence of calcium	*Didymodon* spp., *Desmatodon* spp., and others
Milky spore disease bacteria	*Bacillus popilliae*		
Millet	*Pennisetum glaucum*, *Setaria italica*, and others	Mosses, indicator for saline (salty) soil	*Pottia* spp.
		Mosses, indicator for seasonal running water	*Fontinalis* spp.
Millipede	member of Class Diplopoda, Phylum Arthropoda, Kingdom Animalia		
		Mosses, peat	*Sphagnum* spp.
		Mosses, pygmy—*see* Mosses, annual	
Mint—*see* Peppermint, Spearmint, etc.			
Mint family	Lamiaceae (Labiatae)	Mosses, rock	*Grimmia* spp. and others
Miso fungi	*Aspergillus oryzae*		
Mistletoe	*Phoradendron* spp.	Mosses, sphagnum	*Sphagnum* spp.
Mistletoe, dwarf	*Arceuthobium* spp.	Mosses used with splints	*Philonotis* spp., *Fontinalis* spp., and others
Mite	member of Order Acarina, Phylum Arthropoda, Kingdom Animalia		
		Moss rose	*Portulaca grandliflora*

Common Names and Scientific Names of Organisms

Common Name	Scientific Name	Common Name	Scientific Name
Moth	member of Order Lepidoptera, Class Insecta, Phylum Arthropoda, Kingdom Animalia	**Myrtle**[9]	*Umbellularia californica*[9]
		Nasturtium (garden)	*Tropaeolum majus*
		Natural dye lichens	*Parmelia* spp., *Usnea* spp., and others
Moth, Argentine	*Cactoblastis cactorum*	**Neem tree**	*Melia* sp.
Moth, Mexican jumping bean	*Carpocaps asaltitans*	**Nematode**	member of Class Nematoda, Phylum Aschelminthes, Kingdom Animalia
Moth, Yucca	*Pronuba* spp., *Tegeticula* spp.		
Moth mullein	*Verbascum blattaria*	**Nest fern**	*Asplenium nidus*
Mountain beaver	*Aplodontia rufa*	**Nettle**	*Urtica* spp.
Mountain hemlock	*Tsuga mertensiana*	**Nicotine relative (nornicotine), source of**	*Duboisia hopwoodii*, *Nicotiana tabacum*
Mouse	*Mus musculus*, *Peromyscus* spp., and others		
		Nightshade, deadly	*Atropa belladonna*
Mouse, jumping	*Zapus hudsonius*, *Napaeozapus insignis*	**Nightshade family**	Solanaceae
		Nitrate (nitrifying) bacteria	*Nitrobacter* spp.
Mulberry	*Morus* spp.	**Nitrite (nitrosifying) bacteria**	*Nitrosomonas* spp.
Mulberry, red	*Morus rubra*		
Mule ears	*Wyethia* spp.	**Nitrogen-fixing bacteria**	*Azotobacter* spp., *Clostridium pasteurinum*
Mullein	*Verbascum thapsus*		
Mung bean	*Phaseolus aureus*	**Nosebleeds, fern used to arrest**	*Pellaea mucronata*
Mushroom[8]	*Agaricus* spp. and others		
		Nostoc	*Nostoc* spp.
Mushroom, Black Forest	*Lentinus edodes*	**Nutmeg**	*Myristica fragrans*
		Nutmeg, California	*Torreya californica*
Mushroom, pore (fig. 17.24)	*Serillus pungens*	**Oak**	*Quercus* spp.
Mushroom, shaggy mane	*Coprinus comatus*	**Oak, cork**	*Quercus suber*
		Oak, Hooker	*Quercus lobata*
Mushrooms, common woods (17.19A)	*Russula* sp.	**Oak, live (fig. 23.8)**	*Quercus wislizenii*
		Oak, white	*Quercus alba*
Mushrooms, inky cap	*Coprinus* spp.	**Oats**	*Avena* spp.
Mushrooms, oyster (fig. 17.25)	*Pleurotus ostreatus*	**Oedogonium**	*Oedogonium* spp.
		Olibanum tree	*Boswellia* spp.
Mushrooms, teonanacatl (sacred)	*Conocybe* spp., *Panaeolus* spp., *Psilocybe* spp., and others	**Olive**	*Olea europaea*
		Onion	*Allium cepa*
		Orange	*Citrus sinensis*
Muskrat	*Ondatra zibesthicus*	**Orchid**	*Cattleya* spp. and others[10]
Mustard	*Brassica campestris*, *B. nigra*, and others		
		Orchid, bamboo	*Arundina graminifolia*
Mustard, cultivated	*Brassica alba*, *B. juncea*, and others	**Orchid, bletilla (fig. 8.18B)**	*Bletilla* sp.
		Orchid, underground-flowering	*Rhizanthella gardneri*
Mustard family	Brassicaceae (Cruciferae)	**Orchid "bark," fern sources of**	*Cibotium* spp., *Osmunda* spp.
Myrrh tree	*Commiphora* spp.		

8. *Mushroom* is a term generally applied to the fruiting bodies with stalked, caplike structures produced by members of the Class Basidiomycetes, Division Mycota, Kingdom Fungi; the term is also loosely applied to some of the fruiting bodies of members of other classes of true fungi. There are thousands of known species.

9. This plant, also known as the California bay, is in the Laurel Family. True myrtles are in the Myrtle Family (Myrtaceae).

10. Depending on which authorities are followed, the number of orchid species (all in the family Orchidaceae) may exceed 30,000.

Common Names and Scientific Names of Organisms

Common Name	Scientific Name	Common Name	Scientific Name
Orchid with cladophylls	*Epidendrum* spp.	Penicillin mold[11]	*Penicillium* spp.[11]
Orchid family	Orchidaceae	Pennyroyal	*Hedeoma pulegioides*
Orchids, saprophytic	*Corallorhiza* spp., *Eburophyton austinae,* and others	Peony	*Paeonia* spp.
		Peperomia	*Peperomia* spp.
		Pepper	*Capsicum anuum, C. frutescens*[12]
Orchids, vanilla	*Vanilla* spp.		
Oregano	*Origanum vulgare* and others	Peppergrass	*Lepidium* spp.
		Peppermint	*Mentha piperita*
Oregon grape	*Berberis aquifolium* and others	Perfume stabilizer lichen	*Evernia* spp.
Organpipe cactus	*Lemaireocereus* spp.	Persimmon	*Diospyros* spp.
Oriental water fern	*Ceratopteris thalictroides*	Petitgrain oil, source of	*Citrus aurantium* var. *amara*
Ornaments, ground pines used for	*Lycopodium complanatum, L. clavatum,* and others	Petunia	*Petunia* spp.
		Peyote	*Lophophora williamsii*
Osage orange	*Maclura pomifera*	Phoebe	*Sayornis phoebe*
Oscillatoria	*Oscillatoria* spp.	Pillbugs	*Cylisticus convexus* and others
Our Lord's Candle	*Yucca whipplei*		
Owl, snowy	*Nyctea scandiaca*	Pine	*Pinus* spp.
Oyster mushrooms (fig. 17.25)	*Pleurotus ostreatus*	Pine, Afghanistan	*Pinus eldarica*
		Pine, Aleppo	*Pinus halepensis*
Painted lady	*Echeveria derenbergii*	Pine, bristlecone	*Pinus longaeva*
Palm, panama hat	*Carludovica palmata*	Pine, Chilghoza	*Pinus gerardiana*
Palm, Seychelles Island	*Lodoicea maldivica*	Pine, Colorado bristlecone	*Pinus aristata*
Palm, wax	*Copernicia cerifera*		
Palm family	Palmaceae	Pine, Coulter	*Pinus coulteri*
Pampas grass (fig. 7.3)	*Cortaderia sellona*	Pine, digger	*Pinus sabiniana*
Panama hat palm	*Carludovica palmata*	Pine, eastern white	*Pinus strobus*
Pansy	*Viola tricolor*	Pine, European stone	*Pinus pinea*
Papaya	*Carica papaya*	Pine, jack	*Pinus banksiana*
Pará rubber	*Hevea brasiliensis*	Pine, jeffrey	*Pinus jeffreyi*
Paratyphoid fever bacteria	*Salmonella paratyphi*	Pine, kauri	*Agathis australis, A. robusta*
Parsley	*Petroselinum sativum*	Pine, knobcone	*Pinus attenuata*
Parsley family	Apiaceae (Umbelliferae)	Pine, loblolly	*Pinus taeda*
Parsnip	*Pastinaca sativa*	Pine, longleaf	*Pinus palustris*
Passion fruit	*Passiflora* spp.	Pine, Merkus	*Pinus merkusii*
Patchouli oil, source of	*Pogostemon cablin* and others	Pine, Mexican pinyon	*Pinus cembroides*
		Pine, Mexican stone	*Pinus cembroides*
Pea	*Pisum sativum*	Pine, Monterey	*Pinus radiata*
Pea, sweet	*Lathyrus odoratus*	Pine, pinyon	*Pinus edulis, P. monophylla, P. quadrifolia*
Peach	*Prunus persica*		
Peach leaf curl	*Taphrina deformans*	Pine, pitch	*Pinus rigida*
Peanut	*Arachis hypogaea*	Pine, ponderosa	*Pinus ponderosa*
Pear	*Pyrus communis*	Pine, shortleaf	*Pinus echinata*
Peat moss	*Sphagnum* spp.		
Pecan	*Carya illinoensis*		
Peccary	*Pecari angulatus, Tayassus pecari*		

11. The original penicillin producer discovered by Sir Alexander Fleming was *Penicillium notatum*; current commercially used penicillin producers are strains of *Penicillium chrysogenum*.

12. The drug capsicum, whose active ingredient is the oleoresin capsaicin, is derived from these species, and garden peppers include these and other species of *Capsicum*. Condiment or black pepper is derived from the unrelated *Piper nigrum*.

Common Names and Scientific Names of Organisms

Common Name	Scientific Name	Common Name	Scientific Name
Pine, Siberian white	*Pinus sibirica*	Poor man's pepper	*Lepidium virginicum*
Pine, slash	*Pinus caribaea, P. elliottii*	Popcorn	*Zea mays* (horticultural variety)
Pine, southern yellow—*see* Pine, loblolly; Pine, longleaf; Pine, shortleaf; and Pine, slash		Poplar	*Populus* spp.
		Poplar leaf spot rust	*Melampsora medusae*
		Poppy	*Papaver* spp. and others
Pine, stone—*see* Pine, European stone and Pine, Mexican stone		Poppy, bush	*Dendromecon rigida*
		Poppy, California	*Eschscholzia californica*
		Poppy, Mexican	*Hunnemannia* spp.
Pine, sugar	*Pinus lambertiana*	Poppy, opium	*Papaver somniferum*
Pine, western white	*Pinus monticola*	Poppy, Oriental	*Papaver orientale*
Pine, western yellow	*Pinus ponderosa*	Poppy, prickly (fig. 22.5)	*Argemone glauca*
Pineapple	*Ananas comosus*		
Pinedrops	*Pterospora* spp.	Poppy family	Papaveraceae
Pinto bean	*Phaseolus vulgaris*	Porcupine	*Erethizon* spp., *Hystrix* spp.[13]
Pinyon pines	*Pinus edulis, P. monophylla, P. quadrifolia*		
		Portulaca family	Portulacaceae
		Potato, Irish	*Solanum tuberosum*
Pistachio	*Pistacia vera*	Potato, sweet	*Ipomea batatas*
Pitcher plants	*Sarracenia* spp. and others	Potato, white—*see* Potato, Irish	
Pitcher plants, Asian	*Nepenthes* spp. and others	Potato vine	*Solanum jasminoides*
		Powdery mildew	*Erysiphe* spp. and others
Pitch pine	*Pinus rigida*		
Plantain	*Plantago* spp.	PPLO pneumonia bacteria	*Mycoplasma pneumoniae*
Plastic, fungus used in production of	*Aspergillus terreus*	Prayer plant	*Maranta* spp.
Plasticizers, source of oil for	*Euphorbia agascae*	Preferns	*Cladoxylon* spp., *Protopteridium* spp., and others
Platelike colonial green algae	*Pediastrum* spp., *Chaetopeltis* spp., and others		
		Prickly pear cactus	*Opuntia* spp.
Plover	*Charadrius* spp. and others	Primrose	*Primula* spp.
		Prochlorobacteria	member of Class Prochlorobacteriae, Division Eubacteriophyta, Subkingdom Eubacteriophytineae, Kingdom Monera
Plum	*Prunus domestica* and others		
Pneumonia bacteria	*Streptococcus pneumoniae* and others		
Podocarps, New Zealand timber	*Podocarpus dacrydioides, P. totara*	Pronghorn	*Antilocarpa americana*
		Pseudomonad bacteria	*Pseudomonas* spp.
		Ptarmigan	*Lagopus* spp.
Podocarps used as ornamentals	*Podocarpus macrophylla, P. nagi,* and others	Pteridosperms	*Lyginopteris* spp., *Medullosa* spp., and others
Poinsettia	*Euphorbia pulcherrima*	Puffballs	*Calvatia* spp., *Lycoperdon* spp.
Poison antidote, fern used for	*Polystichum squarrosum*		
Poison hemlock	*Conium maculatum*	Pulque, source of	*Agave* spp.
Poison ivy	*Toxicodendron radicans*	Pumpkin	*Cucurbita pepo*
Poison oak	*Toxicodendron diversilobum*	Pumpkin family	Cucurbitaceae
		Puncture vine	*Tribulus terrestris*
Polyanthus	*Primula polyanthus* and hybrids	Purple laver	*Porphyra* spp.
		Purple laver (fig. 16.24)	*Porphyra tenera*
Pomegranate	*Punica granatum*		
Ponderosa pine	*Pinus ponderosa*		

---

13. *Hystrix* is also a name for a genus of grasses.

Common Names and Scientific Names of Organisms

Common Name	Scientific Name	Common Name	Scientific Name
Purple nonsulfur bacteria	*Rhodomicrobium* spp., *Rhodopseudomonas* spp., *Rhodospirillum* spp.	**Rhododendron**	*Rhododendron* spp.
		Rhubarb	*Rheum rhaponticum*
		Rice	*Oryza sativa*
Purple sulfur bacteria	*Amoebobacter* spp., *Lamprocystis* spp., *Rhodothece* spp., and others	**Ricepaper plant**	*Tetrapanax papyrifera*
		Rickets, ferns used in treating	*Asplenium ruta-maria, Osmunda regalis*
Puya (rare)	*Puya raimondii*	**Ringworm fungi**	*Epidermophyton* spp., *Microsporium* spp., *Trichophyton* spp.
Pygmy mosses	*Acaulon* spp., *Ephemerum* spp., and others		
		Robin	*Turdus migratorius*
Quillwort	*Isoetes* spp.	**Rock cress**	*Arabis* sp.
Quillwort, fossil relatives of	*Isoetites* spp.	**Rock mosses**	*Grimmia* spp. and others
Quince	*Cydonia oblonga*	**Rock-rose, European**	*Helianthemum vulgare*
Quinine, source of	*Cinchona* spp.	**Rock tripe**	*Umbilicaria* spp.
Rabbit	*Oryctolagus cuniculus*	**Rockweeds**	*Fucus* spp., *Pelvetia* spp., and others
Rabbit, cottontail	*Sylvilagus* spp.	**Rose**	*Rosa* spp.
Rabbit, jack	*Lepus* spp.	**Rose, damask**	*Rosa damascena*
Raccoon	*Procyon lotor*	**Rose family**	Rosaceae
Radish	*Raphanus sativus*	**Rosemary**	*Rosmarinus officinalis*
Rafflesia	*Rafflesia* spp.	**Rotenone relative, source of**	*Tephrosia vogelii*
Rafflesia (fig. 8.2)	*Rafflesia arnoldii*		
Ragweed	*Ambrosia* spp.	**Rubber, Pará**	*Hevea brasiliensis*
Raspberry	*Rubus* spp.	**Rubber plant**	*Ficus elastica*
Raspberry, red	*Rubus idaeus, R. strigosus,* and their hybrids	**Ruellia**	*Ruellia portellae* and others
		Rutabaga	*Brassica campestris* var. *napobrassica*
Rat	*Rattus norvegicus, R. rattus,* and others	**Rye**	*Secale cereale*
Rat, kangaroo	*Dipodomys* spp.	**Ryegrass**	*Lolium* spp.
Rat snake, black	*Elaphe obsoleta*	**Safflower**	*Carthamus tinctorius*
Rattlesnake	*Crotalus* spp.	**Saffron, meadow**	*Colchicum autumnale*
Red algae	members of Division Rhodophyta, Kingdom Protista	**Saffron (true)**	*Crocus sativus*
		Sage	*Salvia officinalis*[14]
Red cedar	*Juniperus virginiana*	**Saguaro**	*Carnegia gigantea*
Red cedar, western	*Thuja plicata*	**Salmon**	*Oncorhynchus* spp., *salmo salar,* and others
Red fox	*Vulpes fulva*		
Red mangrove	*Rhizophora mangle*	**Salmonberry**	*Rubus spectabilis*
Red raspberry	*Rubus idaeus, R. strigosus,* and their hybrids	**Salmonella bacteria**	*Salmonella* spp.
		Salsify	*Tragopogon* spp.
		Salt bacteria	*Halococcus* spp., *Halobacterium* spp.
Red Sea, blue-green bacterium of the	*Trichodesmium erythraeum*		
Red spruce	*Picea rubens*	**Salvia**	*Salvia* spp.
Redwood, coastal	*Sequoia sempervirens*	**Sansevieria**	*Sansevieria* spp.
Redwood, giant	*Sequoiadendron giganteum*	**Santonin, source of**	*Artemisia cina*
		Saprophytic orchids	*Corallorhiza* spp., *Cephalanthera austinae,* and others
Redwood sorrel	*Oxalis oregana*		
Reindeer	*Rangifer* spp.	**Sargassum (fig. 16.19)**	*Sargassum echinocarpum*
Reindeer lichens	*Cladonia* spp., *Cetraria islandica*		
Reserpine, source of	*Rauwolfia serpentina*		
Resurrection plant	*Selaginella lepidophylla*		

14. This sage, which is in the Mint Family, should not be confused with sagebrush, which is in the Sunflower Family (Asteraceae).

Common Names and Scientific Names of Organisms

Common Name	Scientific Name	Common Name	Scientific Name
Sarsaparilla, source of	*Smilax* spp.	Snowbank algae	*Chlamydomonas nivale* and others
Sassafras	*Sassafras albidum*	Snowplant	*Sarcodes sanguinea*
Sauerkraut bacteria	*Leuconostoc* spp. and others	Snowy owl	*Nyctea scandiaca*
Sausage tree, African	*Kigelia pinnata*	Soap, fungi used in manufacturing	*Penicillium* spp.
Savory	*Satureia hortensis*	Soaproot, California	*Chlorogalum pomeridianum*
Saxifrage	*Saxifraga* spp.		
Screw pine	*Pandanus* spp.	Sorbose bacteria	*Acetobacter suboxydans*
Sea anemone	*Stephanauge* spp. and others		
Sea hare	*Aplysia californica*	Sorghum	*Sorghum* spp.
Sea lettuce	*Ulva* spp.	Sorrel	*Oxalis* spp.
Sea palm	*Postelsia palmaeformis*	Southern white cedar	*Chamaecyparis thyoides*
Sea rocket	*Cakile edentula*		
Sedge	*Carex* spp. and others	Southern yellow pine— *see* Pine, loblolly; Pine, longleaf; Pine, shortleaf; and Pine, slash	
Seed ferns (Pteridosperms)	*Lyginopteris* spp., *Medullosa* spp., and others		
Senna	*Cassia senna* and others	Soybean	*Glycine max*
		Soy sauce fungi	*Aspergillus oryzae, A. soyae*
Sensitive plant	*Mimosa pudica*		
Sesame	*Sesamum indicum*	Spanish moss	*Tillandsia* spp.
Seychelles Island palm	*Lodoicea maldivica*	Sparrow, savannah	*Passerculus sandwichensis*
Shaggy mane mushroom	*Coprinus comatus*		
		Sparrow, vesper	*Pooecetes gramineus*
Sheep	*Ovis* spp.	Spearmint	*Mentha spicata*
Shelf fungi	*Fomes* spp., *Daedalea* spp., and others	Sperm whale	*Physeter catodon*
		Spiderwort	*Tradescantia* spp.
Shepherd's purse	*Capsella bursa-pastoris*	Spiderwort, European	*Tradescantia paludosa*
Shortleaf pine	*Pinus echinata*	Spike moss	*Selaginella* spp.
Showy orchid	*Orchis* spp.	Spike moss, fossil relatives of	*Lepidodendron* spp., *Sigillaria* spp., and others
Shrimp	*Crago* spp. and others		
Siberian white pine	*Pinus sibirica*		
Silvering of mirrors, fungi used in	*Aspergillus* spp.	Spinach	*Spinacia oleracea*
		Spirogyra	*Spirogyra* spp.
Sisal	*Agave sisalina*	Sponge	*Spongilla* spp. and others
Sitka spruce	*Picea sitchensis*		
Skunk	*Mephitis* spp.	Sponge, vegetable	*Luffa cylindrica*
Slash pine	*Pinus caribaea, P. elliottii*	Sponge algae	*Chlorella* spp., *Zoochlorella* spp.
Slime mold	member of Divisions Myxomycota and Acrasiomycota, Subkingdom Myxomycotineae, Kingdom Fungi	Spotted fever bacteria	*Rickettsia rickettsii*
		Spring beauty	*Claytonia virginica*
		Spruce	*Picea* spp.
		Spruce, black	*Picea mariana*
		Spruce, red	*Picea rubens*
Slime mold (fig. 17.4A)	*Arcyria versicolor*	Spruce, sitka	*Picea sitchensis*
Slime mold (fig. 17.4B)	*Lycogala epidendrum*	Spruce, white	*Picea glauca*
Sloth	*Bradypus* spp., *Choleopus* spp.	Spurge	*Euphorbia* spp.
		Spurge (fig. 22.12)	*Euphorbia cyparissias*
Smut	*Ustilago* spp. and others	Spurge family	Euphorbiaceae
		Squash	*Cucurbita mixta, C. pepo,* and others
Snail	*Haplotrema concava* and others		
		Squawroot	*Perideridia* spp.
Snapdragon	*Antirrhinum majus*	Squill	*Scilla* spp.

Common Names and Scientific Names of Organisms

Common Name	Scientific Name	Common Name	Scientific Name
Squills	*Urginea maritima*	**Tent caterpillar**	*Malacosoma americanum* and others
Squirrel	*Citellus* spp., *Sciuris* spp., and others	**Teonanacatl (sacred) mushrooms**	*Conocybe* spp., *Panaeolous* spp., *Psilocybe* spp., and others
Squirrel, gray	*Sciurus carolinensis*		
Squirrel corn	*Dicentra canadensis*		
Squirting cucumber	*Ecballium elaterium*	**Tepary bean**	*Phaseolus acutifolius* var. *latifolius*
Stapelia (fig. 23.21)	*Stapelia similis*		
Stinkhorn	*Mutinus* spp., *Phallus* spp., and others	**Tequila, source of**	*Agave* spp.
Stinkhorn, common (fig. 17.18)	*Mutinus caninus*	**Termite**	*Odontotermes* spp., *Reticulitermes* spp., and others
Stonecrop	*Sedum* spp. and others	**Tetanus bacteria**	*Clostridium tetani*
Stone pine, European	*Pinus pinea*	**Thalloid liverworts**	*Marchantia* spp. and others
Stone pine, Mexican	*Pinus cembroides*		
Stoneseed	*Lithospermum ruderale*	**Thermal blue-green bacteria**	*Bacillosiphon induratus*, *Synechococcus* spp., and others
Stonewort	*Chara* spp., *Nitella* spp.		
Strawberry	*Fragaria* spp.		
"Strep" throat bacteria	*Streptococcus* spp.	**Thimbleberry**	*Rubus parviflorus*
String-of-pearls	*Senecio rowellianus*	**Thistle**	*Cirsium* spp. and others
Strychnine, source of	*Strychnos* spp.	**Thistle, Canada**	*Cirsium arvense*
Stuffing, ferns used for mattresses, pillows, upholstery	*Cibotium* spp., *Sadleria* spp., and others	**Thrasher**	*Toxostoma* spp.
		Thyme	*Thymus vulgaris* and others
Sugar cane	*Saccharum officinarum* and others	**Tiger**	*Panthera tigris*
		Ti plant	*Cordyline* spp.
Sugar pine	*Pinus lambertiana*	**Toad**	*Bufo americanus*
Sulfur bacteria	*Desulfovibrio* spp., *Thiobacillus* spp., and others	**Tobacco**	*Nicotiana tabaccum*
		Tomato	*Lycopersicon esculentum*
Sumac	*Rhus* spp.	**Tomato, Galápagos**	*Lycopersicon esculentum* var. *minor*, *L. pimpinellifolium*
Sunbird	*Anthodiaeta* spp., *Notiocinnyris* spp., and others		
Sundew	*Drosera* spp.	**Tomato fruitworm**	*Heliothis armigera*
Sundew relative used for flypaper	*Drosophyllum lusitanicum*	**Tomato hornworm**	*Protoparce quinquemaculata*
Sunflower	*Helianthus annuus*	**Toothache, fern chewed for**	*Pityrogramma triangularis*
Sunflower family	Asteraceae (Compositae)		
Sweet pea	*Lathyrus odoratus*	**Toothpaste, fungus used in manufacturing**	*Aspergillus niger*
Sword fern	*Polystichum munitum*		
Sycamore	*Platanus* spp.	**Toothwort**	*Dentaria* spp.
Syphilis bacteria	*Treponema pallidum*	**Tortoise, giant Galápagos**	*Testudo elephantopus porteri*
Tamarack	*Larix* spp.		
Tamarisk	*Tamarix* spp.	**Touch-me-not**	*Impatiens* spp.
Tangerine	*Citrus reticulata*	**Tree fern**	*Cyathea* spp., *Ctenitis* spp., *Dicksonia* spp., *Sphaeropteris* spp., and others
Tapir	*Tapirus* spp.		
Taro	*Colocasia* spp.		
Tarragon	*Artemisia dracunculus*	**Tree fern, Hawaiian**	*Cibotium* spp., *Sadleria* spp.
Tarweed	*Grindelia* spp.		
Tarweed, western (fig. 4.8A)	*Calycadenia* sp.	**Tree fern, small Hawaiian (19.16D)**	*Sadleria cyatheoides*
		Tree lupine	*Lupinus arboreus*
Tea	*Camellia sinensis*	**Tree-of-heaven**	*Ailanthus altissima*
***Tempeh* fungus**	*Rhizopus oligosporus*		

Common Names and Scientific Names of Organisms

Common Name	Scientific Name	Common Name	Scientific Name
Trillium	*Trillium* spp.	Virginia creeper	*Parthenocissus quinquefolia*
Truffles	*Tuber* spp.		
Tsetse fly	*Glossina morsitans, G.palpalis*	Virus[15]	
		Vole	*Microtus* spp. and others
Tularemia bacteria	*Francisella tularensis*		
Tulip	*Tulipa* spp.	Wahoo	*Euonymus alata* and others
Tulip tree	*Liriodendron tulipifera*		
Tumble mustard	*Sisymbrium altissimum*	Wake robin	*Trillium* spp.
Tumbleweeds	*Amaranthus albus, Salsola pestifera,* and others	Wallflower, western	*Erysimum capitatum*
		Walnut	*Juglans* spp.
		Walnut, black	*Juglans nigra*
Tung oil, source of	*Aleurites* spp.	Warbler	*Dendroica* spp. and others
Turk's cap lily	*Lilium superbum*		
Turmeric, source of	*Curcuma longa*	Watercress	*Nasturtium officinale*
Turnip	*Brassica rapa*	Water fern, oriental	*Ceratopteris thalictroides*
Turtle	*Chelydra* spp., *Chrysemys* spp., and others		
		Water hemlock	*Cicuta* spp.
		Water hyacinth	*Eichhornia crassipes*
Turtleback algae	*Basicladia* spp., *Dermatophyton* spp.	Water lily	*Nymphaea* spp. and others
Twinflower	*Linnaea borealis*	Water lily, giant	*Victoria amazonica*
Typhoid bacteria	*Salmonella typhi*	Watermelon	*Citrullus vulgaris*
Typhus fever bacteria	*Rickettsia prowazeki* and others	Water molds	member of Class Oomycetes, Division Mycophyta, Kingdom Fungi
Ulothrix	*Ulothrix* spp.		
Ultraviolet light, flowers seen in (fig. 23.20)	*Rudbeckia* sp.	Water net	*Hydrodictyon* spp.
		Water silk	*Spirogyra* spp.
Unicorn plant	*Proboscoidea* spp.	Water source, horsetails used as	*Equisetum telmateia*
Vanilla orchid	*Vanilla* spp.		
Vegetable sponge	*Luffa cylindrica*	Water weed	*Elodea* spp.
Venus flytrap	*Dionaea muscipula*		
Vetch	*Vicia* spp.		
Vetchling, yellow	*Lathyrus aphaca*		
Vinegar bacteria	*Acetobacter* spp.		
Vinegar weed	*Trichostema* spp.		
Violet	*Viola* spp.		
Violet, African	*Saintpaulia* spp.		
Violet, gold	*Viola douglasii*		

---

15. Depending on the classification used, viruses may not have a scientific name. Many are named after the disease they cause; e.g., tobacco mosaic virus causes tobacco mosaic disease. One classification attempts to give them at least a Latin prefix, so that the virus for warts is *Papavovirus;* for smallpox, *Poxvirus;* for polio, *Picornavirus;* for measles and mumps, *Paramyxovirus.*

Common Names and Scientific Names of Organisms

Common Name	Scientific Name	Common Name	Scientific Name
Water weed, yellow	*Ludwigia repens*	Wintergreen oil, sources of	*Gaultheria procumbens* and others
Wattle	*Acacia decurrens* and others	Wisteria	*Wisteria sinensis* and others
Weaver birds	*Anaplectes* spp., *Hyphantoris* spp., and others	Witch hazel	*Hamamelis virginiana*
		Woad, dyer's	*Isatis tinctoria*
Western hemlock	*Tsuga heterophylla*	Woadwaxen	*Genista tinctoria*
Western larch	*Larix occidentalis*	Wolfsbane	*Aconitum vulparia*
Western red cedar	*Thuja plicata*	Wolverine	*Gulo luscus*
Western wallflower	*Erysimum capitatum*	Woodpecker	*Dendrocopus* spp. and others
Western white pine	*Pinus monticola*		
Whale, sperm	*Physeter catodon*	Worms, fern used in expelling from intestinal tract	*Dryopteris filix-mas*
Wheat	*Triticum* spp. and their hybrids		
Wheel tree	*Trochodendron aralioides*	Wormwood	*Artemisia absinthium*
		Wounds, ferns used for treating	*Lygodium circinatum*, *Ophioglossum vulgatum*
Whisk fern	*Psilotum* spp.		
Whisk fern, fossil relatives of	*Asteroxylon* spp., *Psilophyton* spp., *Rhynia* spp., and others	Yam	*Dioscorea* spp.
		Yareta	*Azorella yareta*
		Yarrow, American	*Achillea lanulosum*
Whisk fern, living relatives of	*Tmesipteris* spp.	Yarrow, European	*Achillea millefolium*
White fir	*Abies concolor*	Yeast	*Saccharomyces* spp.
White fly	*Aleurocanthus woglumi* and others	Yellow-green algae	members of Class Xanthophyceae, Division Chrysophyta, Kingdom Protista
White piedra fungus	*Trichosporon beigeli*		
White pine, eastern	*Pinus strobus*		
White pine, Siberian	*Pinus sibirica*	Yellow vetchling	*Lathyrus aphaca*
White pine, western	*Pinus monticola*	Yellow water weed	*Ludwigia repens*
White pine blister rust	*Cronartium ribicola*	Yew	*Taxus* spp.
White spruce	*Picea glauca*	Yogurt bacteria	*Streptococcus thermophilus*
Willow	*Salix* spp.		
Willow family	Salicaceae	Yucca	*Yucca* spp.
Window leaves, plants with	*Fenestraria* spp. and others	Zamia	*Zamia* spp.
		Zebra	*Equus zebra* and others
Winged bean	*Psophocarpus tetragonolobus*	Zinnia	*Zinnia elegans* and others

Appendix 2   *Biological Controls*

INTRODUCTION

If you were to ask the average farmer or backyard gardener how to control a particular insect or plant pest, you might be given the name of some poisonous spray or bait that has proved "effective" in the past. Evidence that spraying with such substances yields only temporary results has been mounting for many years, however, and the spraying is frequently followed by even larger invasions of pests. In addition, the residues of poisonous sprays often accumulate in the soil and disrupt the microscopic living flora and fauna essential to the soil's health. The problem is compounded and the ecology further upset when large amounts of inorganic fertilizers are added. As increasing numbers of people develop an awareness of the devastating effects on the environment of pesticides and herbicides, they have been turning to **biological controls** as an alternative to the use of poisonous sprays. To the surprise of some, such controls are often more effective than traditional controls.

One reason why poisonous sprays actually promote pest invasions is that the sprays frequently kill numerous beneficial insects along with the undesirable ones. In addition, the pests often become resistant to the sprays through mutations. In undisturbed natural areas, one rarely sees pests destroying the community, even though such pests are present, and weeds are never a problem. Why is this so? As was observed in chapter 23, all members of a community are in ecological balance with one another. The plants produce a variety of substances that may either repel or attract insects, inhibit or promote the growth of other plants, and generally contribute to the health of the community as a whole.

Virtually all insects have their own pests and diseases, as do most living organisms, and each pest functions, at least indirectly, to ensure that the various species of a community are perpetuated. To a certain extent this principle of nature can be applied to farming and gardening. The following are some general and specific biological controls either now in widespread use or showing promise for the future in various tests.

GENERAL CONTROLS

Establishment of Beneficial Insects

Ladybugs (Family Coccinellidae)

The larval stages, in particular, of the small and often colorful beetles called *ladybugs* consume large numbers of aphids, thrips, insect eggs, weevils, and other pests. They are obtainable from various commercial sources (e.g., Bio-Control Co., Box 37C-1, Berry Creek, CA 95916; Ladybug Sales, Box 903, Gridley, CA 95948) but they probably will establish themselves without being imported if given a chance. When obtained from other sources, they should be placed in groups at the bases of plants on which pests are present, preferably after watering in the early evening.

Lacewings (Families Chrysopidae and Hemerobiidae)

Lacewings are relatively slow-flying and delicate-winged insects that consume large numbers of aphids, mealybugs, and other pests. They lay their eggs on the undersides of leaves, each egg being borne at the tip of a slender stalk. The larvae consume the immature stages of leafhoppers, bollworms, caterpillar eggs, mites, scale insects, thrips, aphids, and other destructive pests. Commercial sources include: Unique Insect Control, P.O. Box 15376, Sacramento, CA 95851; All Pest Control, 6030 Grenville Lane, Lansing, MI 48910.

Praying Mantis (Family Mantidae)

About 20 species of *praying mantis* are now established in the United States. These are voracious feeders that prey somewhat indiscriminately on flying insects, and sometimes even on other mantises. They can be established by tying their egg cases to tree branches or at other locations above the ground. The egg cases, which form compact, somewhat oblongoid masses about 2.5 to 5.0 centimeters (1 to 2 inches) long, are obtainable from various commercial sources, including Mincemoyer's Nursery, Rt. 7, Box 379, Jackson, NJ 08527; Bio-Control Co., Box 37C-1, Berry Creek, CA 95916.

Trichogramma Wasps (Family Trichogrammatidae)

Trichogramma wasps are minute insects, mostly less than 1 millimeter (1/25 inch) long; they parasitize insect eggs and are known to have significantly reduced populations of well over 100 different insect pests, including alfalfa caterpillars, armyworms, cabbage loopers, cutworms, hornworms, tent caterpillars, and the larvae of numerous species of moths. As with other insects used as biological controls, care should be exercised in the release of trichogramma wasps, as lack of sufficient pest eggs in the vicinity can result in the wasps parasitizing eggs of beneficial butterflies and other useful insects. They are available from commercial sources such as Unique Insect Control, P.O. Box 15376, Sacramento, CA 95851; Gothard, Inc., Box 370, Canutillo, TX 79835.

Ichneumon Wasps (Family Ichneumonidae)

The *ichneumon wasps* belong to a very large family of primarily stingless wasps that are mostly slender and have long ovipositors that sometimes exceed the length of the body. Most insects are parasitized by at least one species of ichneumon; many species parasitize the larval stages of insects, consuming the host internally after hatching from eggs deposited on the body; alternatively, they may complete development in a later stage. Ichneumons will usually appear naturally in a backyard or farm population of pests if toxic sprays and other unnatural conditions have not interfered with their normal activities.

Tachinid Flies (Family Tachinidae)

Many members of the large family of *tachinid flies* resemble houseflies or bumblebees. All parasitize other insects, including a large variety of caterpillars, Japanese beetles, European earwigs, grasshoppers, gypsy moths, tomato worms, sawflies, and various beetles. Contact Unique Insect Control, P.O. Box 15376, Sacramento, CA 95851, for further information.

Use of Pathogenic Bacteria

Bacillus thuringiensis (BT) is one of three pathogenic bacteria registered for use on edible plants in the United States. It reproduces only in the digestive tracts of caterpillars and is completely harmless to all other wildlife, including earthworms, birds, and mammals. It is exceptionally effective against a wide range of caterpillars such as tomato hornworms and fruitworms, cabbage worms and loopers, grape leaf rollers, corn borers, cutworms, fall webworms, and tent caterpillars. It is mass-produced and sold in a powdered spore form at nurseries and garden supply stores under the trade names of *Dipel*, *Biotrol*, and *Thuricide*. The powder is mixed with water and applied as a spray.

Establishment of Toads and Frogs

It has been estimated that a single adult toad will consume about 10,000 insects and slugs in a single growing season. Toads and frogs feed at night when snails, slugs, sowbugs, earwigs, and other common pests are active.

Use of Beneficial Nematodes

These exceedingly numerous microscopic roundworms have gained a widespread negative reputation because of the damage several species inflict on economically important crops when they invade plant roots and other underground organs. Most species, however, are either harmless or beneficial to plants. They have been used successfully in parasitizing cabbage worm caterpillars, codling moth larvae, Japanese beetle grubs, and tobacco budworms and have shown considerable potential against other pests. One species that has been particularly effective in controlling ants, beetles, bugs, flies, wasps, and numerous other insects is the caterpillar nematode (*Neoaplectana carpocapsae*). It carries a symbiotic bacterium (*Xenorhabdus nematophilus*), which multiplies rapidly in the host, killing most insects within 24 hours after initial contact. It may be obtained from Nematode Farm, Inc., 2617 San Pablo Avenue, Berkeley, CA 94702.

Use of Limonoid Sprays

Limonoids are bitter substances found in the rinds, seeds, and juice of citrus fruits (especially grapefruit). If the rinds and seeds of two or three fruits are ground up, soaked overnight in a pint of water, and the solid material is strained out, the liquid may then be sprayed on plants. The bitter principle apparently stops or reduces the feeding of larvae on the foliage. In experiments, limonoid sprays have proved effective against corn earworm, fall armyworm, tobacco budworm, and pink bollworm, but undoubtedly will deter many other pests as well.

Use of Liquefied Pest Sprays

Jeff Cox, an editor of Rodale's Organic Gardening magazine, called attention to this method of pest control in the magazine in October, 1976, and again in May, 1977. Insect pests or slugs are gathered in small quantities and liquefied with a little water in a blender. The material is then further diluted with water and sprayed throughout the infested area. It is not known why spraying with "bug juice" is effective against pests. It is known, however, that virtually all organisms are subject to viral infections, and it has been theorized that even the healthy insects and slugs may be carriers of viruses that are somehow activated in the process of liquefaction. The viruses would be spread throughout an entire yard or farm if all parts of the area were sprayed. Most viruses are highly specific, generally attacking a single species of organism.

A resident pest control entomologist in Florida, M. Sipe, who has recommended the "bug juice" technique, has also suggested the possibility that the odor of the liquefied insects attracts their predators and parasites or that the insects' distress **pheromones** (naturally produced insect chemicals that influence sexual or other behavior) are released by the blender, with the pheromones acting as an insect repellent. Possibly the observed effects of spraying "bug juice" are the result of a combination of viruses, predator attraction, and repellent pheromones. Sipe warns that if one tries this method of pest control, care should be taken to use only pest species and only those that are doing significant damage. Failure to heed this warning could disrupt the activities of natural predators and other natural controls present.

This approach still needs further testing and investigation of its safety for use by humans, but preliminary results in various areas of North America have thus far yielded impressive results with no evidence of harm to humans or beneficial organisms.

Use of Resistant Varieties

Many plants may kill or inhibit disease fungi or bacteria with chemicals known as *phytoalexins*. Phytoalexins are synthesized at the point of attack or invasion by the pathogen and are toxic to the fungus or bacterium. In selecting for improved fruit quality, vigor of growth, or other desirable characteristics, horticulturists in the past have sometimes unknowingly bred out a plant's capacity to produce certain phytoalexins, although general vigor is usually accompanied by disease resistance. Now that this aspect of a plant's defense mechanisms has become known, breeders are concentrating on developing varieties capable of producing phytoalexins against

various fungi, bacteria, and even nematodes. Several tomato varieties, for example, are listed as being *VFN*. The letters *V* and *F* indicate a resistance to *Verticillium* and *Fusarium* (common pathogenic fungi), while the letter *N* denotes a resistance to *root-knot nematodes*. Other aspects of plant disease resistance include thick cuticles, the secretion of gums, resins, and other metabolic products that may interfere with fungal and bacterial spore germination, and the presence within all the cells of the plant of chemical compounds toxic to pathogens.

Interplanting with Plants That Produce Natural Insecticides or Substances Offensive to Pests

Although no single species of plant produces substances that repel all pests, many produce substances that repel a significant number. The best known include marigolds, garlic, and members of the mint family such as pennyroyal, peppermint, basil, and lavender. See appendix 3 for an expanded discussion of the subject.

SPECIFIC CONTROLS

Weeds

In 1974 the Weed Science Society of America published a special committee report (*Weed Science* 22:490–95) on the biological control of weeds, summarizing the status of projects on the biological control of weeds with insects and plant pathogens in the United States and Canada. Table A2.1 is condensed from that report and supplemented with additional information. Many other biological controls for these and other weeds are currently under investigation.

Insects

The maintenance of ecological balance in nature includes a vast array of predator-prey relationships between animals, birds, insects, and other organisms. Specific biological controls for several types of insect pests, in addition to the general controls previously discussed, are given in table A2.2.

Specific Weeds and Agents Involved in Their Biological Control

Weed	Agent(s) of Biological Control
Alligator weed (*Alternanthera philoxeroides*)	Flea beetles (*Agasicles hygrophila*)
Bladder campion (*Silene cucubalus*)	Tortoise beetle (*Cassida hemisphaerica*)
Brazil peppertree (*Schinus terebinthifolius*)	Weevil (*Bruchus atronotatus*) and others
Brushweed (*Cassia surattensis*)	Imperfect fungus (*Cephalosporium* sp.)
Curly dock (*Rumex crispus*)	Rust (*Uromyces rumicis*)
Curse (*Clidemia hirta*)	Thrip (*Liothrips urichi*) and others
Cypress spurge (*Euphorbia cyparissias*)	Sphinx moth (*Hyles euphorbiae*)
Dalmatian toadflax (*Linaria dalmatica*)	Leaf miner (*Stagmatophora serratella*) and others
Emex (*Emex australis*)	Seed weevils (*Apion antiquum*) and others
Gorse (*Ulex europaeus*)	Seed weevils (*Apion ulicis*) and others
Halogeton (*Halogeton glomeratus*)	Casebearer (*Coleophora parthenica*) and others
Hawaiian blackberry (*Rubus penetrans*)	Sawflies (*Pamphilius sitkensis; Priophorus morio*) and others
Jamaica feverplant (*Tribulus terrestris*)	Weevils (*Microlarinus* spp.)
Joint vetch (*Aeschynomene virginica*)	Imperfect fungus (*Colletotrichum gloeosporioides*)
Klamath weed (*Hypericum perforatum*)	Leaf beetles (*Chrysolina* spp.); buprestid beetle (*Agrilus hyperici*)
Lantana (*Lantana camara*)	Seed weevil (*Apion* sp.); ghost moth (*Hepialus* sp.); plume moth (*Platyptilia pusillidactyla*); hairstreaks (*Strymon* spp.); and others
Leafy spurge (*Euphorbia esula*)	Wood-boring beetle (*Oberea* sp.) and others
Mediterranean sage (*Salvia aethiopis*)	Snout beetles (*Phrydiuchus* spp.)
Milkweed vine (*Morrenia odorata*)	Oomycete fungus (*Phytophthora citrophthora*); rust (*Aecidium asclepiadinum*)
Prickly pear (*Opuntia* spp.)	Moth (*Cactoblastis cactorum*); cochineal insects (*Dactylopius* spp.); and others
Puncture vine (*Tribulus terrestris*)	Weevils (*Microlarinus* spp.)
Scotch broom (*Cytisus scoparius*)	Seed weevil (*Apion fuscirostre*) and others
Skeleton weed (*Chondrilla juncea*)	Gall mite (*Aceria chondrillae*); root moth (*Bradyrrhoa gilveolella*); rust (*Puccinia chondrillina*); powdery mildews (*Erysiphe cichoracearum, Leveillula taurica*)
Spiny emex (*Emex spinosa*)	Seed weevil (*Apion antiquum*)
Tansy ragwort (*Senecio jacobaea*)	Seed fly (*Hylemya seneciella*); cinnabar moth (*Tyria jacobaeae*); leaf beetle (*Longitarsus jacobaeae*)
Thistles:	
Bull thistle (*Cirsium vulgare*)	Weevil (*Ceuthorrhynchidius horridus*); tortoise beetle (*Cassida rubiginosa*)
Canada thistle (*Cirsium arvense*)	Weevil (*Ceutorhynchus litura*); flea beetle (*Altica carduorum*); stem gall fly (*Urophora cardui*)
Diffuse knapweed (*Centaurea diffusa*)	Seed fly (*Urophora affinis*)
Italian thistle (*Carduus pycnocephalus*)	Flea beetles (*Rhinocyllus conicus; Psylliodes chalcomera*); weevil (*Ceutorhynchus trimaculatus*)
Milk thistle (*Silybum marianum*)	Flea beetle (*Rhinocyllus conicus*)
Musk thistle (*Carduus nutans*)	Weevils (*Ceutorhynchus trimaculatus; Ceuthorrhynchidius horridus; Rhinocyllus conicus*); flea beetle (*Psylliodes chalcomera*)
Perennial sowthistle (*Sonchus arvensis*)	Peacock fly (*Tephritis dilacerata*)
Plumeless thistle (*Carduus acanthoides*)	Tortoise beetle (*Cassida rubiginosa*); seed weevil (*Rhinocyllus conicus*); weevil (*Ceuthorrhynchidius horridus*)
Russian thistle (*Salsola kali* var. *tenuifolia*)	Casebearer (*Coleophora parthenica*) and others
Slenderflower thistle (*Carduus tenuiflorus*)	Weevil (*Ceutorhynchus trimaculatus*)
Spotted knapweed (*Centaurea maculosa*)	Seed fly (*Urophora affinis*) and others
Star thistle (*Centaurea nigrescens*)	Weevil (*Ceuthorrhynchidius horridus*)
Yellow star thistle (*Centaurea solstitialis*)	Seed fly (*Urophora siruna-seva*)
Water hyacinth (*Eichhornia crassipes*)	Weevils (*Neochetina bruchi; N. eichhorniae*); moth (*Sameodes albiguttalis*)
Water purslane (*Ludwigia palustris*)	Snout beetle (*Nanophyes* sp.)

Specific Biological Controls for Several Types of Insect Pests

Insect	Control
Ants (about 8,000 spp. within the Superfamily Formicoidea)	Ants that carry aphids into trees and consume ripening fruits can be prevented from getting farther than the trunk by applying a band of sticky material around the trunk. A commercial preparation sold under the trade name of Tanglefoot is particularly effective. A water suspension of ground hot peppers (*Capsicum* spp.) used as a spray can act as an ant deterrent. *Caution:* Many ants are beneficial to a balanced ecology; they should not be decimated indiscriminately.
Grasshoppers (there are several families of grasshoppers, but the insects that usually constitute the most serious pests are species of *Melanoplus*, Family Acrididae)	In 1980 the Environmental Protection Agency permitted private companies to begin the mass culture of a protozoan, *Nosema locustae*, for use in controlling rangeland grasshoppers. Tests have shown that properly timed applications of spores mixed with wheat bran can reduce grasshopper populations by up to 50%.
Gypsy moths (*Porthetria dispar*)	Parasitic wasps (*Apanteles flavicoxis, A. indiensis*) imported from India lay their eggs in gypsy moth caterpillars and kill large numbers.
Japanese beetles (*Popillia japonica*)	The pathogenic bacterium *Bacillus popillae*, which is sold commercially, is specific for Japanese beetle larvae. It causes what is known as "milky spore disease" in the grubs while they are still in the soil, and it is very destructive.
Mealybugs (*Pseudococcus* spp.)	The small brown beetles called *crypts (Cryptolaemus montrouzieri)* effectively control mealybugs in greenhouses and also outdoors on apple, pear, peach, and citrus trees. Order from Rincon-Vitova Insectaries, Inc., P.O. Box 95, Oakview, CA 93022.
Mosquitoes (*Culex* spp., *Anopheles* spp., and others)	The bacterium *Bacillus thuringiensis* var. *israelensis* has proved to be very effective in destroying mosquito larvae. A fungus (*Lagenidium giganteum*) has also proved highly effective against mosquito larvae if the temperature is above 20° C (68° F). The bacterium is available from several sources, including Abbott Laboratories, Dept. 95-M, 1400 Sheridan Rd., N. Chicago, IL 60064; Sandoz, Inc., 480 Camino del Rio S., San Diego, CA 92108.
Red spider mites (*Tetranychus telarius*)	Predatory mites (*Phytoleseus persimilis*, which works best when weather is not hot, and *Amblyseius californicus*, which is more effective in hot weather) effectively control populations of red spider mites.
White flies (*Trialeurodes vaporariorum*)	A minute wasp *Encarsia formosa*, parasitizes white flies exclusively. The wasps have been known to be very effective in greenhouses. They are obtainable from White Fly Control Co., Box 986, Milpitas, CA 95035; Rincon-Vitova Insectaries, Inc., P.O. Box 95, Oakview, CA 93022. White flies are attracted to the color yellow. Large numbers of white flies are trapped when a yellow board is sprayed or painted with any sticky substance and placed in the vicinity of the pests.

COMPANION PLANTING

An examination of the Additional Reading list will reveal that the literature on the chemical interactions between plants and also between plants and their consumers is already extensive. Despite the scientific evidence on the subject to date, however, a significant amount of the "backyard biological control" that is practiced today is based primarily on empirical information. Such information has been obtained from thousands of gardeners and farmers who have tried various techniques with their plantings and pest controls. As a result, they have come to conclusions that certain things work while others do not, but they have not deliberately set up controlled experiments, nor have they necessarily understood the scientific basis for what they have observed. This does not mean that their observations are not useful or that they are invalid. In fact, such empirical observations have often been the inspiration for investigations and experiments by scientists. The scientific investigations have sometimes revealed that the empirical observations were biased or not carefully made or that erroneous conclusions had been drawn, but frequently sound scientific bases for these observations have been revealed.

Further insights into the means by which plants inhibit or enhance the growth of others and into the nature of their resistance to disease or insect-repelling mechanisms continue to be discovered. Observations of such phenomena in the past have led organic gardeners and others to the practice of *companion planting*, which involves the interplanting of various crops and certain other plants in such a way that each species derives some benefit from the arrangement. The following companion planting list, based primarily on empirical information, appeared in the February, 1977 issue of *Organic Gardening and Farming* magazine. It is included here with the permission of Rodale Press, Inc.

Table A2.3 is a list of combinations of vegetables, herbs, flowers, and weeds that are mutually beneficial, according to current reports of organic gardeners and to companion planting traditions.

TABLE A2.3
Companion Plants

Plant	Companions and Effects	Plant	Companions and Effects
Asparagus	Tomatoes, parsley, basil.	**Chives**	Carrots; plant around base of fruit trees to discourage insects from climbing trunk.
Basil	Tomatoes (improves growth and flavor); said to dislike rue; repels flies and mosquitoes.		
		Corn	Potatoes, peas, beans, cucumbers, pumpkin, squash.
Beans	Potatoes, carrots, cucumbers, cauliflower, cabbage, summer savory, most other vegetables and herbs; around houseplants when set outside.	**Cucumbers**	Beans, corn, peas, radishes, sunflowers.
		Dill	Cabbage (improves growth and health), carrots.
Beans (bush)	Sunflowers (beans like partial shade, sunflowers attract birds and bees), cucumbers (combination of heavy and light feeders), potatoes, corn, celery, summer savory.	**Eggplant**	Beans.
		Fennel	Most plants are supposed to dislike it.
		Flax	Carrots, potatoes.
		Garlic	Roses and raspberries (deters Japanese beetle); with herbs to enhance their production of essential oils; plant liberally throughout garden to deter pests.
Beets	Onions, kohlrabi.		
Borage	Tomatoes (attracts bees, deters tomato worm, improves growth and flavor), squash, strawberries.		
		Horseradish	Potatoes (deters potato beetles); around plum trees to discourage curculios.
Cabbage family	Potatoes, celery, dill, chamomile, sage, thyme, mint, pennyroyal, rosemary, lavender, beets, onions. Aromatic plants deter cabbage worms.	**Lamb's quarters**	Nutritious edible weed; allow to grow in modest amounts in the corn.
		Leek	Onions, celery, carrots.
Carrots	Peas, lettuce, chives, onions, leeks, rosemary, sage, tomatoes.	**Lettuce**	Carrots and radishes (lettuce, carrots, and radishes make a strong companion team), strawberries, cucumbers.
Catnip	Plant in borders; protects against flea beetles.	**Lovage**	Plant here and there in garden.
Celery	Leeks, tomatoes, bush beans, cauliflower, cabbage.	**Marigolds**	The workhorse of pest deterrents. Keeps soil free of nematodes; discourages many insects. Plant freely throughout garden.
Chamomile	Cabbage, onions.		
Chervil	Radishes (improves growth and flavor).		
		Marjoram	Here and there in garden.

Companion Plants

Plant	Companions and Effects	Plant	Companions and Effects
Mint	Cabbage family; tomatoes; deters cabbage moth.	**Rue**	Roses and raspberries; deters Japanese beetle. Keep it away from basil.
Mole plant	Deters moles and mice if planted here and there throughout the garden.	**Sage**	Rosemary, carrots, cabbage, peas, beans; deters some insects.
Nasturtium	Tomatoes, radishes, cabbage, cucumbers; plant under fruit trees. Deters aphids and pests of cucurbits.	**Southernwood**	Cabbage; plant here and there in garden.
		Soybeans	Grows with anything helps everything.
Onion	Beets, strawberries, tomato, lettuce (protects against slugs), beans (protects against ants), summer savory.	**Spinach**	Strawberries.
		Squash	Nasturtium, corn.
		Strawberries	Bush beans, spinach, borage, lettuce (as a border).
Parsley	Tomato, asparagus.	**Summer savory**	Beans, onions. Deters bean beetles.
Peas	Squash (when squash follows peas up trellis), plus grows well with almost any vegetable; adds nitrogen to the soil.	**Sunflower**	Cucumbers.
		Tansy	Plant under fruit trees; deters pests of roses and raspberries; deters flying insects; also Japanese beetles, striped cucumber beetles, squash bugs, deters ants.
Petunia	Protects beans; beneficial throughout garden.		
Pigweed	Brings nutrients to topsoil; beneficial growing with potatoes, onions, and corn; keep well thinned.		
		Tarragon	Good throughout garden.
Potato	Horseradish, beans, corn, cabbage, marigold, limas, eggplant (as trap crop for potato beetle).	**Thyme**	Here and there in garden; deters cabbage worm.
		Tomato	Chives, onion, parsley, asparagus, marigold, nasturtium, carrot, limas.
Pot marigold	Helps tomato, but plant throughout garden as deterrent to asparagus beetle, tomato worm, and many other garden pests.	**Turnip**	Peas.
		Valerian	Good anywhere in garden.
		Wormwood	As a border, keeps animals from the garden.
Pumpkin	Corn.	**Yarrow**	Plant along borders, near paths, near aromatic herbs; enhances essential oil production of herbs.
Radish	Peas, nasturtium, lettuce, cucumbers; a general aid in repelling insects.		
Rosemary	Carrots, beans, cabbage, sage; deters cabbage moth, bean beetles, and carrot fly.		

SOME SOURCES OF HERB PLANTS AND SEEDS

Casa Yerba, Star Route 2, Box 21, Days Creek, OR 97429

Earth Star Herb Gardens, 438 West Perkinsville Road, Chino Valley, AZ 86323

Fox Hill, Box 7, Parma, MI 49269

Golden Sun Herb Gardens, 5318 North Gravenstein Highway, Sebastopol, CA 95472

Greenleaf Seeds, P.O. Box 980, Conway, MA 01341

Happy Hollow Nursery, 221 Happy Hollow Road, Villa Rica, GA 30180

Lechampion, P.O. Box 1602, Freedom, CA 95019

Liberty Seed Company, Box 806–A6, New Philadelphia, OH 44026

Lily of the Valley Herb Farm, 39690 Fox, Minerva, OH 44657

Otto Richter and Sons, Box 260, Goodwood, Ontario, LOC 1A0

PG Nursery, R18, Box 470, Bedford, IN 47421

Putney Nursery, Putney, VT 05346

Reminiscent Herb Farm, 1344 Boone Aire Road,
   Florence, KY 41042

Sanctuary Seeds, 2388 West Fourth Avenue, Vancouver,
   BC V6K 1P1

Shoestring Seeds, P.O. Box 2261, Martinsville, VA
   24113

Sunnybrook Farms Nursery, Box 6, Chesterland, OH
   44026

Sunshine Herbs and Flowers, Rt. 1, Box 234, Comer,
   GA 30629

Thompson and Morgan, Inc., P.O. Box 1308, Jackson,
   NJ 08527

Wildwood Herbal, P.O. Box 746, Albemarle, NC 28002

Willhite Seed Company, Box 23, Poolville, TX 76076

ADDITIONAL READING

Bosch, R. van den, et al. 1981. *An introduction to
   biological control.* New York: Plenum Publishing
   Company.

Burges, H. D., and N. W. Hussey eds. 1981. *Microbial
   control of pests and plant diseases, 1970-1980.*
   New York: Academic Press.

Carson, R. 1962. *Silent spring.* Boston: Houghton
   Mifflin Co.

Cook, R. J., and K. F. Baker. 1983. *Nature and practice
   of biological control of plant pathogens.* St. Paul,
   MN: American Phytopathological Society.

Debach, P. 1974. *Biological control by natural enemies.*
   New York: Cambridge University Press.

Goeden, R. D., L. A. Andres, T. E. Freeman, P. Harris,
   R. L. Pienkowski, and C. R. Walker. 1974.
   "Present status of projects on the biological control
   of weeds with insects and plant pathogens in the
   United States and Canada." *Weed Science* 22:490–
   95.

Henry, J. E. 1981. "Natural and applied control of
   insects by protozoa." *Annual Review of
   Entomology* 26:49–73.

Hoy, M., and G. L. Cunningham. 1983. *Biological
   control of pests by mites: proceedings of a
   conference.* Oakland, CA: Agricultural and Natural
   Resources, University of California.

Huffaker, C. B., and P. S. Messenger. 1977. *Theory and
   practice of biological control.* New York:
   Academic Press.

Rice, E. L. 1984. *Allelopathy,* 2d ed. New York:
   Academic Press.

Rice, E. L. 1985. *Pest control with nature's chemicals:
   allelochemics and pheromones in gardening and
   agriculture.* Norman, OK: University of Oklahoma
   Press.

Whittaker, R. H., and P. P. Feeny. 1971.
   "Allelochemics: chemical interactions between
   species." *Science* 171:757–70.

Yepsen, R. B., Jr., ed. 1976. *Organic plant protection.*
   Emmaus, PA: Rodale Press.

Appendix 3   *Useful Plants and Poisonous Plants*

WILD EDIBLE PLANTS

Words of Caution

Literally thousands of native and naturalized plants, or at least parts of them, have been used for food and other purposes by Native Americans and the immigrants who came later from other quarters of the globe. Table A3.1 has been compiled from a variety of sources; I have had opportunities to sample only a fraction of these plants myself and thus cannot confirm the edibility of all of the plants listed. *The reader is cautioned to be certain of the identity of a plant before consuming any part of it.* Cow parsnip (*Heracleum lanatum*) and water hemlocks (*Cicuta* spp.), for example, resemble each other in general appearance, but although cooked roots of cow parsnip have been used for food for perhaps many centuries, those of water hemlocks are very poisonous and have caused many human fatalities.

As was indicated in chapter 20, many species of organisms are now on rare and endangered species lists, and a number of them are doomed to extinction within the next few years. Although the wild edible plants discussed here are not included on such lists it might not take much indiscriminate gathering to endanger their existence as well. Because of this, one should exercise the following rule of thumb: *Never reduce a population of plants by more than 10% when collecting them for any purpose!*

TABLE A3.1
Wild Edible Plants

Plant	Scientific Name	Uses
Amaranth	*Amaranthus* spp.	young leaves used like spinach; seeds ground with others for flour
Arrow grass	*Triglochin maritima*	seeds parched or roasted (***Caution:*** *The plant is otherwise poisonous.*)
Arrowhead	*Sagittaria latifolia*	tubers used like potatoes
Balsamroot	*Balsamorhiza* spp.	whole plant edible, especially when young, either raw or cooked
Basswood	*Tilia* spp.	fruits and flowers ground together to make a paste that can serve as a chocolate substitute; winter buds edible raw; dried flowers used for tea
Bedstraw	*Galium aparine*	roasted and ground seeds make good coffee substitute
Beechnuts	*Fagus grandifolia*	seeds used like nuts; oil extracted from seeds for table use
Biscuit root	*Lomatium* spp.	roots eaten raw or dried and ground into flour; seeds edible raw or roasted
Bitterroot	*Lewisia rediviva*	outer coat of the bulbs should be removed to eliminate the bitter principle; bulbs are then boiled or roasted
Blackberry (wild)	*Rubus* spp.	fruits edible raw, in pies, jams, and jellies
Black walnut	*Juglans nigra*	nut meats highly edible
Bladder campion	*Silene cucubalus*	young shoots (less than 5 cm tall) cooked as a vegetable
Blueberry	*Vaccinium* spp.	fruits edible raw, frozen, and in pies, jams, and jellies
Bracken fern	*Pteridium aquilinum*	young uncoiling leaves ("fiddleheads") cooked like asparagus; rhizomes also edible but usually tough (***Caution:*** *Recent evidence indicates that frequent consumption of bracken fern can cause cancer of the intestinal tract.*)
Broomrape	*Orobanche* spp.	entire plant eaten raw or roasted
Bulrush (Tule)	*Scirpus* spp.	roots and young shoot tips edible raw or cooked; pollen and seeds also edible
Butternut	*Juglans cinerea*	nut meats edible
Caraway	*Carum carvi*	young leaves in salads; seeds for flavoring baked goods and cheeses
Cattail	*Typha* spp.	copious pollen produced by flowers in early summer is rich in vitamins and can be gathered and mixed with flour for baking; rhizomes can be cooked and eaten like potatoes

Wild Edible Plants

Plant	Scientific Name	Uses
Chicory	*Cichorium intybus*	leaves eaten raw or cooked; dried, ground roots make good coffee substitute
Chokecherry	*Prunus virginiana*	fruits make excellent jelly or can be cooked with sugar for pies and cobblers
"Coffee" (wild)	*Triosteum* spp.	berries dried and roasted make good coffee substitute
Common chickweed	*Stellaria media*	plants cooked as a vegetable
Corn lily	*Clintonia borealis*	youngest leaves can be used as a cooked vegetable
Cow parsnip	*Heracleum lanatum*	roots and young stems cooked
Cowpea	*Vigna sinensis*	"peas" and young pods cooked as a vegetable (plant naturalized in southern U.S.)
Crab apple	*Pyrus* spp.	jelly made from fruits
Crowberry	*Empetrum* spp.	fruits should first be frozen then cooked with sugar
Dandelion	*Taraxacum officinale*	leaves rich in vitamin A; dried roots make good coffee substitute; wine made from young flowers
Dock	*Rumex* spp.	leaves cooked like spinach; tartness of leaves varies from species to species and sometimes from plant to plant—tart forms should be cooked in two or three changes of water
Douglas fir	*Pseudotsuga menziesii*	cambium and young phloem edible; tea made from fresh leaves
Elderberry	*Sambucus* spp.	fresh flowers used to flavor batters; fruits used in pies, jellies, wine (**Caution:** *Other parts of the plant are poisonous.*)
Evening primrose	*Oenothera hookeri, O. biennis,* and others	young roots cooked
Fairy bells	*Disporum trachycarpum*	berries can be eaten raw
Fennel	*Foeniculum vulgare*	leaf petioles eaten raw or cooked
Fireweed	*Epilobium angustifolium*	young shoots and leaves boiled as a vegetable
Ginger (wild)	*Asarum* spp.	rhizomes can be used as substitute for true ginger
Gooseberry	*Ribes* spp.	berries eaten cooked, dried, or raw; make excellent jelly
Grape (wild)	*Vitis* spp.	berries usually tart but can be eaten raw; make good jams and jellies
Grass	Many genera and species	seeds of most can be made into flour; rhizomes of many perennial species can be dried and ground for flour
Greenbrier	*Smilax* spp.	roots dried and ground; refreshing drink made with ground roots, sugar, and water
Groundnut	*Apios americana*	tubers cooked like potatoes
Hawthorn	*Crataegus* spp.	fruits edible raw and in jams and jellies
Hazel nut	*Corylus* spp.	nuts eaten raw or roasted
Hickory	*Carya* spp.	nuts edible
Highbush cranberry	*Viburnum trilobum*	fruits make excellent jellies and jams
Huckleberry	*Vaccinium* spp.	berries eaten raw or in jams and jellies
Indian paintbrush	*Castilleja* spp.	flowers of many species edible (**Caution:** *On certain soils, plants absorb toxic quantities of selenium.*)
Indian pipe	*Monotropa* spp.	whole plant edible raw or cooked
Juniper	*Juniperus* spp.	"berries" dried, ground, and made into cakes
Labrador tea	*Ledum* spp.	tea made from young leaves

Wild Edible Plants

Plant	Scientific Name	Uses
Lamb's quarters	*Chenopodium album*	leaves and young stems used as cooked vegetable
Licorice	*Glycyrrhiza lepidota*	roots edible raw or cooked
Mallow	*Malva* spp.	leaves and young stems used as vegetable (use only small amounts at one time)
Manzanita	*Arctostaphylos* spp.	berries eaten raw, in jellies or pies, or made into ''cider''
Maple	*Acer* spp.	sugar maples (*Acer saccharum*) well known for the sugar content of the early spring sap; other species (e.g., box elder—*A. negundo*, bigleaf maple—*A. macrophyllum*) also contain usable sugars in their early spring sap
Mariposa lily	*Calochortus* spp.	bulbs edible raw or cooked
Mayapple	*Podophyllum peltatum*	fruit good raw or cooked (**Caution:** *Other parts of the plant are poisonous.*)
Maypops	*Passiflora incarnata*	fruits edible raw or cooked
Miner's lettuce	*Montia perfoliata*	leaves eaten raw as a salad green
Mint	*Mentha arvensis* and others	leaves of several mints used for teas
Mormon tea	*Ephedra* spp.	tea from fresh or dried leaves (add sugar to offset bitterness); seeds for bitter meal
Mulberry	*Morus* spp.	fruits of the red mulberry (*M. rubra*) are used raw and in pies and jellies; fruits of white mulberry (*M. alba*) edible but insipid
Mushrooms	Many genera and species	*Utmost caution* should be exercised in identifying mushrooms before consuming them. Although poisonous species are in the minority, they are common enough. Edible forms that are relatively easy to identify include morels (*Morchella esculenta*), most puffballs (*Lycoperdon* spp.), and inky cap mushrooms (*Coprinus* spp.)
Mustard	*Brassica* spp.	leaves used as vegetable; condiment made from ground seeds
Nettles	*Urtica* spp.	leaves and young stems boiled like spinach
New Jersey tea	*Ceanothus americanus*	tea from leaves
Nutgrass	*Cyperus esculentus* and others	tubers can be eaten raw
Oak	*Quercus* spp.	acorns were widely used ground for flour by native North Americans; all contain bitter tannins that must be leached out before use
Onion (wild)	*Allium* spp.	bulbs edible raw or cooked
Orach	*Atriplex patula* and others	leaves and young stems cooked as a vegetable
Pawpaw	*Asimina triloba*	fruit edible raw or cooked
Pennycress	*Thlaspi arvense*	young leaves are edible raw
Peppergrass	*Lepidium* spp.	immature fruits add zest to salads; seeds spice up meat dressings
Persimmon	*Diospyros virginiana*	fully ripened fruits can be eaten raw or cooked
Pickerel weed	*Pontederia cordata*	fruits edible raw or dried
Pigweed (*see* Amaranth)		
Pines	*Pinus* spp.	cambium and young phloem edible; tea from fresh needles rich in vitamin C
Pipsissewa	*Chimaphila umbellata*	drink made from boiled roots and leaves (cool after boiling)
Plantain	*Plantago* spp.	young leaves eaten in salads or as cooked vegetable
Poke	*Phytolacca americana*	fresh young shoots boiled like asparagus (older parts of plant poisonous)

Wild Edible Plants

Plant	Scientific Name	Uses
Prairie turnip	*Psoralea esculenta*	turniplike roots cooked like potatoes
Prickly pear	*Opuntia* spp.	fruits and young stems peeled and eaten raw or cooked
Purple avens	*Geum rivale*	liquid from boiled root has flavor similar to chocolate
Purslane	*Portulaca oleracea*	leaves and stems cooked like spinach
Raspberry (wild)	*Rubus* spp.	fruits edible raw or in pies, jams, and jellies
Redbud	*Cercis* spp.	flowers used in salads; cooked young pods edible
Rose (wild)	*Rosa* spp.	fruits ("hips") exceptionally rich in vitamin C; hips can be eaten raw, pureed, or candied
Salsify	*Tragopogon* spp.	roots edible raw or cooked
Saltbush	*Atriplex* spp.	seeds nutritious (**Caution:** *On certain soils, plants can absorb toxic amounts of selenium.*)
Sassafras	*Sassafras albidum*	tea from roots (**Caution:** *Large quantities have narcotic effect*); leaves and pith used for Louisiana filé
Serviceberry	*Amelanchier* spp.	all fruits edible (mostly bland)
Shepherd's purse	*Capsella bursa-pastoris*	leaves cooked as vegetable; seeds eaten parched or ground for flour
Showy milkweed	*Asclepias speciosa*	flowers eaten raw or cooked; young shoots cooked
Soap plant	*Chlorogalum pomeridianum*	bulbs slow-baked and eaten like potatoes after fibrous outer coats are removed
Solomon's seal	*Polygonatum* spp.	rootstocks dried and ground for bread flour
Sorrel	*Oxalis* spp.	leaves mixed in salads
Spatterdock	*Nuphar polysepalum*	seeds placed on hot stove burst like popcorn and are edible as such; peeled tubers eaten boiled or roasted
Speedwell	*Veronica americana* and others	leaves and stems used in salads
Spring beauty	*Claytonia* spp.	bulbs edible raw or roasted
Strawberry (wild)	*Fragaria* spp.	fruits superior in flavor to cultivated varieties
Sunflower	*Helianthus annuus*	seeds eaten raw or roasted; seeds yield cooking oil
Sweet cicely	*Osmorhiza* spp.	roots have aniselike flavor
Sweet flag	*Acorus calamus*	young shoots used in salads; roots candied
Thistle	*Cirsium* spp.	peeled stems edible; roots edible raw or roasted
Vetch	*Vicia* spp.	tender green pods edible baked or boiled
Watercress	*Nasturtium officinale*	leaves edible raw in salads or cooked as a vegetable
Waterleaf	*Hydrophyllum* spp.	young shoots raw in salads; shoots and roots cooked as vegetable
Water plantain	*Alisma* spp.	the bulblike base of the plant is dried and then cooked
Water shield	*Brasenia schreberi*	tuberlike roots are peeled and then dried to be ground for flour or boiled
Winter cress	*Barbarea* spp.	leaves and young stem edible as cooked vegetable
Yarrow	*Achillea lanulosa*	plant dried and made into broth (**Caution:** *The closely related and widespread European yarrow—A. millefolium—is somewhat poisonous.*)
Yellow pond lily (*see* Spatterdock)		
Yew	*Taxus* spp.	bright red pulpy part of berries edible (**Caution:** *Seeds and leaves are poisonous.*)

POISONOUS PLANTS

Literally thousands of plants contain varying amounts of poisonous substances. In many instances, the poisons are not present in sufficient quantities to cause adverse effects in humans when only moderate contact or consumption is involved and cooking may destroy or dissipate the substance.

Some plants have substances that produce toxic effects in some organisms but not in others. Ordinary onions (*Allium cepa*), for example, occasionally poison horses or cattle yet are widely used for human food, and poison ivy (*Toxicodendron radicans*) or poison oak (*Toxicodendron diversilobum*) produce dermatitis in some individuals but not in others. Table A3.2 and table A3.3 include plants that are native to, or cultivated in, the United States and Canada.

TABLE A3.2
Plants Known to Have Caused Human Fatalities

Plant	Scientific Name	Poisonous Parts
Angel's trumpet	*Datura suaveolens*	all parts, especially seeds and leaves
Azalea	*Rhododendron* spp.	leaves and flowers (however, poisoning is rare)
Baneberry	*Actaea* spp.	berries and roots
Belladonna	*Atropa belladonna*	all parts, especially fruits and roots
Black cherry	*Prunus serotina*	bark, seeds, leaves (**Caution:** *Seeds of most cherries, plums, and peaches contain a poisonous principle.*)
Black locust	*Robinia pseudo-acacia*	seeds, leaves, inner bark
Black snakeroot	*Zigadenus* spp.	bulbs
Buckeye	*Aesculus* spp.	seeds, leaves, shoots, flowers
Caladium	*Caladium* spp.	all parts
Carolina jessamine	*Gelsemium sempervirens*	all parts
Castor bean	*Ricinus communis*	seeds
Chinaberry	*Melia azedarach*	fruits and leaves
Daphne	*Daphne mezereum*	all parts
Death camas (*see* Black snakeroot)		
Dieffenbachia	*Dieffenbachia* spp.	all parts
Duranta	*Duranta repens*	berries
English ivy	*Hedera helix*	berries and leaves
False hellebore	*Veratrum* spp.	all parts
Foxglove	*Digitalis purpurea*	all parts
Golden chain	*Laburnum anagyroides*	seeds and flowers
Jequirity bean	*Abrus precatorius*	seeds
Jimson weed	*Datura stramonium* and other *Datura* spp.	all parts, especially seeds
Lantana	*Lantana camara*	unripe fruits
Lily of the valley	*Convallaria majalis*	all parts
Lobelia	*Lobelia* spp.	all parts
Mistletoe	*Phoradendron* spp.	berries
Monkshood	*Aconitum* spp.	all parts
Moonseed	*Menispermum canadense*	fruits
Mountain laurel	*Kalmia latifolia*	leaves, shoots, flowers
Mushrooms	Many genera and species, especially *Amanita* spp.	all parts
Nightshade	*Solanum* spp.	unripened fruits (**Caution:** *Common potatoes are* Solanum tuberosum; *a poisonous principle is produced in tubers exposed to light.*)
Oleander	*Nerium oleander*	all parts

Plants Known to Have Caused Human Fatalities

Plant	Scientific Name	Poisonous Parts
Poke	*Phytolacca americana*	roots and mature stems
Rhododendron (*see* Azalea)		
Rhubarb	*Rheum rhaponticum*	leaf blades (**Caution:** Although young petioles are widely eaten, dangerous accumulations of a poisonous substance can occur in leaf blades.)
Rubber vine	*Cryptostegia grandiflora*	all parts
Sandbox tree	*Hura crepitans*	milky sap and seeds
Tansy	*Tanacetum vulgare*	leaves, flowers
Tung tree	*Aleurites fordii*	all parts, especially seeds
Water hemlock	*Cicuta* spp.	roots
White snakeroot	*Eupatorium rugosum*	all parts
Yellow oleander	*Thevetia peruviana*	all parts, especially fruits
Yew	*Taxus* spp.	all parts except ''berry'' pulp

TABLE A3.3
Other Plants Producing Significant Quantities of Poisonous Substances

Plant	Scientific Name	Poisonous Parts
Amaryllis	*Amaryllis* spp.	bulbs
Autumn crocus	*Colchicum autumnale*	all parts
Bittersweet	*Celastrus scandens*	seeds
Bleeding hearts	*Dicentra* spp.	all parts
Bloodroot	*Sanguinaria canadensis*	all parts
Blue cohosh	*Caulophyllum thalictroides*	fruits, leaves
Boxwood	*Buxus sempervirens*	leaves
Buckthorn	*Rhamnus* spp.	fruits
Bushman's poison	*Acokanthera* spp.	all parts
Buttercup	*Ranunculus* spp.	all parts; toxicity varies from species to species; mostly cause blistering
Buttonbush	*Cephalanthus occidentalis*	leaves
Chincherinchee	*Ornithogalum thyrsoides*	all parts
Crown of thorns	*Euphorbia milii*	milky latex
Culver's root	*Veronicastrum virginicum*	root
Daffodil	*Narcissus* spp.	bulbs
Desert marigold	*Baileya radiata*	all parts
Dutchman's breeches	*Dicentra cucullaria*	all parts
Fly poison	*Amianthemum muscaetoxicum*	leaves, roots
Four-o'clock	*Mirabilis jalapa*	seeds, roots
Gloriosa lily	*Gloriosa* spp.	all parts
Goldenseal	*Hydrastis canadensis*	rhizomes, leaves
Holly	*Ilex aquifolium*	berries
Horse chestnut	*Aesculus hippocastanum*	seeds, flowers, leaves
Hyacinth	*Hyacinthus* spp.	bulbs
Hydrangea	*Hydrangea* spp.	buds, leaves
Jack-in-the-pulpit	*Arisaema triphyllum*	roots, leaves
Jessamine	*Cestrum* spp.	leaves, young stems
Jonquil (*see* Daffodil)		
Karaka nut	*Corynocarpus laevigata*	seeds
Kentucky coffee tree	*Gymnocladus dioica*	fruits
Larkspur	*Delphinium* spp.	young plants, seeds
Lignum vitae	*Guaiacum officinale*	fruits

Other Plants Producing Significant Quantities of Poisonous Substances

Plant	Scientific Name	Poisonous Parts
Locoweed	*Astragalus* spp.	location of poisonous principles varies from species to species; plants more of a problem for livestock than for humans
Lupine	*Lupinus* spp.	location of poisonous principles varies from species to species, primarily in pods and seeds
Marijuana	*Cannabis sativa*	resins secreted by glandular hairs among flowers
Mayapple	*Podophyllum peltatum*	all parts except ripe fruits
Mescal bean	*Sophora secundiflora*	seeds
Narcissus (*see* Daffodil)		
Ngaio	*Myoporum laetum*	leaves
Opium poppy	*Papaver somniferum*	unripe fruits
Philodendron	*Philodendron* spp.	leaves, stems
Pittosporum	*Pittosporum* spp.	fruits, leaves, stems
Poinsettia	*Euphorbia pulcherrima*	milky latex
Poison ivy	*Toxicodendron radicans*	leaves
Poison oak	*Toxicodendron diversilobum*	leaves
Poison sumac	*Toxicodendron vernix*	leaves
Prickly poppy	*Argemone* spp.	seeds, leaves
Privet	*Ligustrum vulgare*	fruits
Sneezeweed	*Helenium* spp.	all parts
Snow-on-the-mountain	*Euphorbia marginata*	milky latex
Squirrel corn	*Dicentra canadensis*	all parts
Star-of-Bethlehem	*Ornithogalum umbellatum*	all parts
Sweet pea	*Lathyrus* spp.	seeds
Tobacco	*Nicotiana tabacum*	leaves (when eaten)

MEDICINAL PLANTS

Although numerous modern medicines include synthetic substances, many still contain drugs naturally produced by plants, and as recently as 50 years ago the vast majority of medicines used in the treatment of human diseases and ailments were plant-produced. Table A3.4 includes a sampling of plants associated with medicinal uses in the past. Some of the drugs concerned are still prescribed for specific ailments by modern physicians, while others are a part of folk medicine still practiced in rural areas. *Caution: Do not use any of the plants listed here for medicinal purposes without consulting a physician.*

TABLE A3.4
Plants Associated with Medicinal Uses

Plant	Scientific Name	Uses
Aloe	*Aloe* spp. (especially *Aloe vera*)	juice from leaves used on burns, especially X-ray burns
American mountain ash	*Pyrus americana*	liquid made from steeping inner bark in water used as an astringent; tea of berries used as a wash for piles; berries eaten to prevent or cure scurvy
Anemone	*Anemone canadensis*	pounded boiled root applied to wounds as an antiseptic
Arnica	*Arnica* spp.	plants applied as a poultice to bruises and sprains
Balm of Gilead	*Populus* × *gileadensis*	buds used as an ingredient in cough syrups

Plants Associated with Medicinal Uses

Plant	Scientific Name	Uses
Balsam poplar	*Populus balsamifera*	buds made into ointment, which was placed in nostrils by Native Americans for relief of congestion
Blackberry	*Rubus* spp.	tea of roots used by northern California Native Americans to cure dysentery
Black cohosh	*Cimicifuga racemosa*	dried roots used in cough medicines and for rheumatism
Black haw	*Viburnum prunifolium*	bark used in treatment of asthma and for relieving menstrual irregularities
Bloodroot	*Sanguinaria canadensis*	Native Americans used rhizome for ringworm, as an insect repellent and for sore throat
Blue cohosh	*Caulophyllum thalictroides*	tea of root drunk by Native Americans and early settlers a week or two before giving birth to promote rapid parturition
Boneset	*Eupatorium perfoliatum*	water infusion of dried plant tops widely used to treat fevers and colds
Broom snakeweed	*Gutierrezia sarothrae*	Navajo Indians applied chewed plant to insect stings and bites of all kinds
Burdock	*Arctium lappa*	used as an insulin substitute in folklore; root extract used in seventeenth century for venereal diseases
Button snakeroot	*Eryngium* spp.	Natchez Indians inserted chewed stem in nostrils to arrest nosebleed
California bay	*Umbellularia californica*	Yuki Indians put leaves in bath of hot water and bathed for relief of rheumatism; leaves used as an insect repellent
Camphor	*Cinnamomum camphora*	oil from leaves and wood used in cold remedies and liniments
Camptotheca	*Camptotheca acuminata*	extracts from flowers and immature fruits yield camptothecin, which has given evidence of being effective against certain forms of cancer
Cascara	*Rhamnus purshiana*	bark extract widely used in laxatives
Catnip	*Nepeta cataria*	leaf tea used for treatment of colds and to relieve infantile colic
Cayenne pepper	*Capsicum frutescens*	used to reduce mucous drainage (recent evidence suggests it may be carcinogenic)
Cherry (wild)	*Prunus serotina*	tea brewed from bark used for coughs and colds
Chia	*Salvia columbariae*	mucilaginous seeds used by Spanish Californians to make a refreshing drink; seeds contain a caffeinelike principle that enabled Native Americans to perform unusual feats of endurance; seed paste used in eyes irritated by foreign matter
Cinchona	*Cinchona* spp.	bark yields quinine drugs used in treating malaria
Club moss	*Lycopodium clavatum*	spores dusted on wounds or inhaled by Native Americans to arrest nosebleeds
Coca	*Erythroxylon coca*	cocaine from leaves used as a local anesthetic; South American laborers use it as a stimulant
Cotton	*Gossypium* spp.	cotton root bark used by black slaves and Native Americans to induce abortions
Creosote bush	*Larrea divaricata*	decoction from leaves used as a cure-all by Native Americans but especially for respiratory problems
Cubebs	*Piper cubeba*	dried fruit best known as a condiment but is also used in treatment of asthma

Plants Associated with Medicinal Uses

Plant	Scientific Name	Uses
Deadly nightshade	*Atropa belladonna*	belladonna, a drug complex extracted from leaves, contains the drugs atropine, hyoscyamine, and scopolamine; these are used as an opium antidote, in shock treatments, and for dilation of pupils; scopolamine is also used as a tranquilizer and for "twilight sleep" in childbirth
Dogbane	*Apocynum androsaemifolium*	roots boiled in water and resulting liquid used as a heart medication (contains a drug similar in action to digitalis)
Dogwood	*Cornus* spp.	inner bark boiled in water and resulting liquid drunk to reduce fevers
Ephedra	*Ephedra* spp.	drug ephedrine, widely used to relieve nasal congestion and low blood pressure, obtained from stems (most ephedrine now in use is synthetic)
Ergot	(Source: *Claviceps purpurea* on cereal grains)	used to treat migraine headaches and to control bleeding after childbirth.
Flowering ash	*Chionanthus virginicus*	bark used as a laxative and in treatment of liver ailments
Foxglove	*Digitalis purpurea*	drug digitalis, widely used as a heart stimulant, obtained from leaves
Gentian	*Gentiana catesbaei*	Catawba Indians applied hot-water extract of roots to sore backs; liquid drunk as a remedy for stomachaches
Geranium	*Geranium maculatum*	dried root used for dysentery, diarrhea, and hemorrhoids
Ginger (wild)	*Asarum* spp.	extract of rhizome used as a broad-spectrum antibiotic
Ginseng	*Panax quinquefolium and other Panax* spp.	considered a general panacea, especially by the Chinese
Goldenseal	*Hydrastis canadensis*	rhizome source of alkaloidal drugs used in treatment of inflamed mucous membranes; also used as a tonic
Goldthread	*Coptis groenlandica*	Native Americans boiled plant and gargled resulting liquid for treatment of sore or ulcerated mouths
Green hellebore	*Helleborus viridis*	extract of plant used in treatment of hypertension (drug now largely synthesized); Thompson Indians used it in small amounts for treatment of syphilis
Hemlock	*Tsuga* spp.	Native Americans made tea of inner bark for treatment of colds and fevers
Horehound	*Marrubium vulgare*	extract of dried tops of plants used in lozenges for relief of sore throat and colds
Horsetail	*Equisetum* spp.	plants boiled in water and resulting liquid used as a hairwash; liquid gargled for treatment of mouth ulcers
Indigo (wild)	*Baptisia tinctoria*	Native Americans boiled plant and used resulting liquid as antiseptic for skin sores
Ipecac	*Cephaelis ipecacuana*	drug from roots and rhizome used to treat amoebic dysentery; also used as an emetic
Jimson weed	*Datura* spp.	drugs atropine, hyoscyamine, and scopolamine obtained from seeds, flowers, and leaves; drug stramonium used for knockout drops and in treatment of asthma (see Deadly nightshade)
Joe-pye weed	*Eupatorium purpureum*	dried root said to prevent formation of gallstones
Joshua tree	*Yucca brevifolia*	cortisone and estrogenic hormones made from sapogenins produced in roots

Plants Associated with Medicinal Uses

Plant	Scientific Name	Uses
Juniper	*Juniperus* spp.	tea of "berries" drunk by Zuni Indian women to relax muscles following childbirth
Kansas snakeroot	*Echinacea angustifolia*	dried roots used as antiseptic for treatment of sores and boils, periodontal disease, and sinus drainage problems
Lily of the valley	*Convallaria majalis*	all parts of plant contain a heart stimulant similar to digitalis
Lobelia	*Lobelia inflata*	drug lobeline sulphate obtained from dried leaves; drug used in antismoking preparations and in treatment of respiratory disorders
Mandrake	*Mandragora officinarum*	extracts of plant used as painkiller in folk medicine (drugs hyoscyamine, podophyllin, and mandragorin have been isolated; podophyllin experimentally used in treatment of paralysis)
Manroot	*Marah* spp.	Native Americans used oil from seeds to treat scalp problems and the crushed roots for relief from saddle sores
Marginal fern	*Dryopteris marginalis*	rhizomes contain oleoresin used in expulsion of tapeworms from intestinal tract
Marijuana	*Cannabis sativa*	tetrahydrocannabinol obtained from resinous hairs in inflorescences; ancient medicinal drug of China
Mayapple	*Podophyllum peltatum*	podophyllin obtained from roots used experimentally in treatment of paralysis; dried root powder used on warts
Maypop	*Passiflora incarnata*	dried leaves used as a sedative; Native Americans used juice for treatment of sore eyes
Mesquite	*Prosopis juliflora*	Native Americans mixed powdered leaves with water to make liquid used to treat sore eyes
Milkweed	*Asclepias syriaca*	Quebec Indians promoted temporary sterility by drinking infusion of pounded roots
Mistletoe	*Phoradendron flavescens*	Native Americans of California drank tea of leaves to induce abortions or as a contraceptive
Monkshood	*Aconitum napellus*	source of aconite once used in treatment of rheumatism and neuralgia
Mulberry	*Morus rubra*	Rappahannock Indians applied milky latex of leaf petioles to scalp for ringworm
Mullein	*Verbascum thapsus*	Native Americans smoked leaves for respiratory ailments and asthma; flowers once widely used in cough medicines
Onion (wild)	*Allium* spp.	bulbs eaten to prevent scurvy; Cheyenne Indians applied bulbs in poultice to boils
Opium poppy	*Papaver somniferum*	morphine and codeine obtained from latex of immature fruits
Oregon grape	*Berberis aquifolium*	bark tea drunk by Native Americans to settle upset stomach; used in strong doses as a remedy for venereal diseases
Pansy (wild)	*Viola tricolor*	plants ground up and applied to skin sores
Pennyroyal	*Hedeomia pulegioides*	Native Americans used leaf tea for relief of headaches and flatulence and to repel chiggers
Persimmon	*Diospyros virginiana*	liquid from boiled fruit used as an astringent; leaves said to be rich in vitamin C
Peyote	*Lophophora williamsii*	alcoholic extract of plant used as an antibiotic
Pinkroot	*Spigelia marilandica*	powdered root very effective in expulsion of roundworms from intestinal tract

Plants Associated with Medicinal Uses

Plant	Scientific Name	Uses
Pipsissewa	*Chimaphila umbellata*	Native Americans steeped plant in water and applied resulting liquid to draw out blisters
Pitcher plant	*Sarracenia purpurea*	Native Americans used root widely as smallpox cure (records indicate it was effective)
Plantain	*Plantago ovata* and others	seeds used in laxatives
Pleurisy root	*Asclepias tuberosa*	liquid resulting from roots being boiled in water taken for respiratory problems
Prickly ash	*Zanthoxylum americanum*	bark and berries widely used by Native Americans for toothache (pieces inserted in cavities) and liquid infusion drunk for venereal diseases
Quassia	*Picraea excelsa; Quassia amara*	wood extract used as pinworm remedy and as an insecticide
Rauwolfia	*Rauwolfia serpentina*	reserpine obtained from roots; drug used in treatment of mental illness and in counteracting effects of LSD
Saffron (meadow)	*Colchicum autumnale*	drug colchicine from corms once used in treatment of gout and back ailments but now largely used for experimental doubling of chromosome numbers in plants
Sarsaparilla	*Aralia nudicaulis*	cough medicines made from roots
Sassafras	*Sassafras albidum*	tea of root bark used to induce sweating; used externally as a liniment
Self-heal	*Prunella vulgaris*	Native Americans applied plants in poultices to boils
Seneca snakeroot	*Polygala senega*	bark boiled in water; liquid applied to snakebites and taken internally as an abortifacient; also used as a remedy for coughs
Senna	*Cassia senna* and others	extract of leaves used as a purgative
Skeleton weed	*Lygodesmia juncea*	widely used by Native American women to increase flow of milk
Skullcap	*Scutellaria laterifolia*	dried plant used as an anticonvulsive in treatment of epilepsy and as a sedative
Slippery elm	*Ulmus fulva*	dried inner bark contains an aspirinlike substance that soothes inflamed membranes
Spicebush	*Lindera benzoin*	berries, buds, bark, and leaves brewed for tea used to reduce fevers
Spruce	*Picea* spp.	Cree Indians ate small cones for treatment of sore throat
Squills	*Urginea maritima*	bulbs of red variety are the source of a heart stimulant; bulbs of white variety are the source of a widely used rodent killer
Stoneseed	*Lithospermum ruderale*	Shoshoni Indian women reported to have drunk water infusion of roots daily for six months to produce permanent sterility (experiments with mice suggest substance to the reports)
Strophanthus	*Strophanthus sarmentosus* and others	seeds are a major source of cortisone and also the source of a heart stimulant

Plants Associated with Medicinal Uses

Plant	Scientific Name	Uses
Strychnine plant	*Strychnos nox-vomica*	strychnine extracted from seeds widely used as an animal and insect poison and is the principal active ingredient in blowgun darts and poison arrows of primitive South American aborigines; minute amounts stimulate the central nervous system and relieve paralysis
Sumac	*Rhus glabra* and others	Native Americans applied leaf decoction as a remedy for frostbite; fruits and liquid made from leaves applied to poison ivy rash and gonorrhea sores; root chewed for treatment of mouth ulcers
Sweet flag	*Acorus calamus*	boiled root applied to burns; root chewed for relief of colds and toothache
Sweet gum	*Liquidambar styraciflua*	bud balsam used to treat chigger bites; balsam also used in insect fumigating powders
Sword fern	*Polystichum munitum*	boiled rhizome used by Native Americans to treat dandruff; sporangia applied to burns
Tamarind	*Tamarindus indica*	fruit pulp used as a laxative
Valerian	*Valeriana septentrionalis*	pulverized plants applied to wounds
Velvet bean	*Mucuna* spp.	seeds contain L-dopa used in treatment of Parkinson's disease
Virginia snakeroot	*Aristolochia serpentaria*	Native Americans used tea of plant for reducing high fevers
Wahoo	*Euonymus atropurpureus*	bark steeped in water; resulting liquid has digitalislike effect on heart
Western wallflower	*Erysimum capitatum*	Zuni Indians ground plant with water and applied it to skin to prevent sunburn
Willow	*Salix* spp.	Chickasaw Indians snuffed infusion of roots as a remedy for nosebleed; Pomo Indians boiled bark in water and applied resulting liquid for relief of skin itches; fresh inner bark contains salicin, an aspirinlike compound used to reduce fevers
Wintergreen	*Gaultheria procumbens*	oil from leaves used as a folk remedy for body aches and pains
Witch hazel	*Hamamelis virginiana*	oil distilled from twigs used as an external medicine
Wormseed	*Chenopodium ambrosioides*	oil from seeds used to expel intestinal worms
Wormwood	*Artemisia* spp.	Yokia Indians made tea from leaves to treat bronchitis; other Native American tribes used tea as a remedy for colds
Yarrow	*Achillea millefolium*	Native Americans used infusion of plant for treatment of wounds, burns, and earaches
Yellow lady's slipper	*Cypripedium calceolus*	dried root used for relief of insomnia or as a sedative
Yellow nut grass	*Cyperus esculentus*	Paiute Indians pounded tubers with tobacco leaves and applied mass in wet dressing for treatment of athlete's foot
Yerba santa	*Eriodictyon californicum*	Native Americans smoked leaves or drank tea of leaves for treatment of colds or asthma

HALLUCINOGENIC PLANTS

Although a few hallucinogenic substances produced by animals have been isolated and some have been synthesized, the majority of known hallucinogens are produced by plants. Table A3.5 is not a complete list but it includes the better known sources. The reader is referred to the Additional Reading for further information.

TABLE A3.5
Hallucinogenic Substances Produced by Plants

Plant	Scientific Name	Part Used	Principal Active Substance
Ajuca	*Mimosa hostilis*	roots	nigerine
Belladonna	*Atropa belladonna*	leaves	hyoscyamine, scopolamine
Caapi	*Banisteriopsis caapi*	wood	harmine
Canary broom	*Cytisus canariensis*	seeds	cytisine
Catnip	*Nepeta cataria*	leaves	unknown
Cohoba	*Piptadena peregrina*	seeds (snuff)	tryptamines
Coral bean	*Erythrina* spp.	seeds	unknown
Cubbra borrachera	*Methysticodendron amesianum*	leaves	scopolamine
Ergot fungus	*Claviceps purpurea*	rhizomorph	ergine (LSD)
Fly agaric	*Amanita muscaria*	mushroom cap	ibotenic acid, muscimol
Henbane	*Hyoscyamus* spp.	leaves	hyoscyamine, scopolamine
Iboga	*Tabernanthe iboga*	root bark	ibogaine
Jimson weed	*Datura* spp.	all parts	scopolamine
Kava kava	*Piper methysticum*	root (large amounts of beverage produce hallucinations)	myristicinlike compound
Mace	*Myristica fragrans*	aril of seed	myristicin
Mescal bean	*Sophora secundiflora*	seeds	cytisine
Morning glory	*Ipomoea violacea*	seeds	ergine
Nutmeg	*Myristica fragrans*	seeds	myristicin
Ololiuqui	*Rivea corymbosa*	seeds	turbicoryn
Peyote	*Lophophora williamsii*	stems	mescaline
Psilocybe mushrooms	*Psilocybe* spp., *Conocybe* spp., *Panaeolus* spp., and others	all parts	psilocybin, psilocin
Rape dos Indios	*Maquira sclerophylla*	dried plant (snuff)	unknown
San Pedro	*Trichocereus pachanoi*	stems	mescaline
Sassafras	*Sassafras albidum*	root bark (large amounts of tea)	safrole
Sweet flag	*Acorus calamus*	dried root	asarone, β-asarone
Syrian rue	*Peganum harmala*	seeds	harmine
Vygie	*Mesembryanthemum expansum*	all parts	mesebrine
Wood rose	*Argyreia nervosa*	seeds	ergoline alkaloids
Yakee (Parica)	*Virola* spp.	resin from inner surface of freshly removed bark (snuff)	tryptamine
Yohimbehe	*Corynanthe* spp.	bark	yohimbine

SPICE PLANTS

The word *spice* describes any aromatic plant or part of a plant used to flavor or season food; spices are also used to add scent or flavor to manufactured products (table A3.6). Although spices have no nutritional value, they add a pleasurable zest to meals, and before food preservation was possible they helped make palatable food that was still edible but unappealing.

The value placed on spice plants was responsible for changing the course of Western civilization as a principal motive behind the voyages of discovery.

TABLE A3.6
Plants Used to Season or Flavor

Spice	Scientific Name of Plant	Parts Used; Remarks	Principal Source
Allspice	*Pimento officinalis*	powdered dried fruit	Jamaica
Almond	*Prunus amygdalus*	oil from seed used for flavoring baked goods	Mediterranean; U.S.
Angelica	*Angelica archangelica*	stems candied; oil from seeds and roots used in liqueurs	Europe; Asia
Anise	*Pimpinella anisum*	oil distilled from fruits used for flavoring	widely cultivated
Arrowroot	*Maranta arundinacea*	powdered root used in milk puddings, baked goods	South America
Asafoetida	*Ferula asafoetida*	powdered gum from stems and roots used in minute quantities with fish	Middle East
Balm (Melissa)	*Melissa officinalis*	oil from leaves used in beverages; leaves used as food flavoring	U.S.; Mediterranean
Basil	*Ocimum basilicum*	leaves used in meat dishes, soups, sauces	Mediterranean
Bay	*Laurus nobilis*	leaves used in soups, sauces	Europe
Bell pepper	*Capsicum frutescens*	dried diced fruit used in chip dips, salad dressings	widely cultivated
Bergamot	*Monarda didyma*	leaves used with pork (*Note:* A perfume oil obtained from a variety of orange—*Citrus aurantium* var. *bergamia*—is also called bergamot.)	North America (*Monarda*); Italy (*Citrus*)
Black pepper	*Piper nigrum*	dried fruits used as a condiment	India; Indonesia
Borage	*Borago officinalis*	leaves used as a beverage flavoring	England
Burnet	*Sanguisorba minor*	used in soups and casseroles	Eurasia
Calamus	*Acorus calamus*	powdered rhizome used for flavoring	Europe; Asia; North America
Capers	*Capparis spinosa*	flower buds used for flavoring relishes, pickles, sauces	Mediterranean
Caraway	*Carum carvi*	seeds used in breads, cheeses; seed oil used in the liqueur kümmel	North America; Europe
Cardamon	*Elletaria cardamomum*	dried fruit and seeds used for flavoring baked goods (*Note:* several false cardamons—*Amomum* spp.—are sold commercially.)	India; Sri Lanka; Central America
Cassia	*Cinnamomum cassia*	powdered bark used as cinnamon substitute	Southeast Asia
Cayenne pepper	*Capsicum* spp.	powdered dried fruits used in chili powder, Tabasco sauce	American tropics
Celery	*Apium graveolens*	seeds used in celery salt, soups	Europe; U.S.
Chervil	*Anthriscus cerefolium*	used as a parsley substitute	Europe; Near East
Chives	*Allium schoenoprasum*	leaves, bulbs used with sour cream, butter	widely cultivated
Chocolate	*Theobroma cacao*	ground seeds used for flavoring	Africa; South America
Cilantro	*Coriandrum sativum*	leaves used in avocado dip and with poultry	Europe

Plants Used to Season or Flavor

Spice	Scientific Name of Plant	Parts Used; Remarks	Principal Source
Cinnamon	*Cinnamomum zeylanicum*	ground bark used for flavoring baked goods; oil from leaves used as flavoring, clearing agent	Seychelles; Sri Lanka
Citrus	*Citrus* spp.	fruits, especially rinds, source of flavoring oil	Mediterranean; South Africa; U.S.
Coffee	*Coffea arabica*	roasted seeds source of mocha-coffee flavoring	Tropics
Coriander	*Coriandrum sativum*	ground seed used in German frankfurters, curry powders	Mediterranean
Cubebs	*Piper cubeba*	dried fruits used as seasoning	East Indies
Cumin	*Cuminum cyminum*	ground seed used with meats, pickles, cheeses, curry	Mediterranean
Curry	—	a spicy condiment containing several ingredients, such as turmeric, cumin, fenugreek, and zedoary	India
Dill	*Anethum graveolens*	seeds used in pickling brines; leaves used for seasoning meat loaves, sauces	Europe; Asia
Dittany	*Origanum dictamnus*	leaves used as seasoning for poultry, meats	Crete
Eucalyptus	*Eucalyptus* spp.	oil from leaves used in toothpastes, flavoring agents	Australia
Fennel	*Foeniculum vulgare*	seeds used in baked goods	Europe
Fenugreek	*Trigonella foenumgraecum*	oil distilled from seeds used in pickle, chutney, curry powders, imitation maple flavoring	widely cultivated
Filé (*see* Sassafras)			
Garlic	*Allium sativum*	fresh or dry bulbs used for meat seasonings	widely cultivated
Ginger	*Zingiber officinale* and others	dried rhizomes used for flavoring many foods and drinks	India; Taiwan
Grains of paradise	*Afromomum melegueta*	seeds used to flavor beverages and medicines	West Africa
Hops	*Humulus lupulus*	dried inflorescences of female plants used in brewing beer	Europe; North America
Horseradish	*Rorippa armoracia*	grated fresh root used as a condiment	Europe; North America
Juniper	*Juniperus* spp.	"berries" used to season beef roasts, poultry, sauces	North America
Licorice	*Glycyrrhiza glabra*	dried rhizome and root used to flavor pontefract cakes, candies	Middle East
Lovage	*Ligusticum scoticum*	stems candied; seeds used in pickling sauces; celery substitute	Europe
Mace	*Myristica fragrans*	aril of seed used for flavoring beverages, foods	Grenada; Indonesia; Sri Lanka
Marigold	*Tagetes* spp.	petals substituted for saffron in rice dishes, stews	widely cultivated
Marjoram	*Majorana hortensis*	leaves used in stews, dressings, sauces	Mediterranean
Mustard	*Brassica* spp.	ground seeds used in meat condiment	Europe; China
Nasturtium	*Tropaeolum majus*	flowers, seeds, leaves used in salads	widely cultivated
Nutmeg	*Myristica fragrans*	seeds used for flavoring foods, beverages	Grenada; Indonesia; Sri Lanka

Plants Used to Season or Flavor

Spice	Scientific Name of Plant	Parts Used; Remarks	Principal Source
Oregano	*Origanum vulgare* and others	leaves used as seasoning with poultry, meats	Europe
Paprika (*see* Cayenne pepper)			
Parsley	*Petroselinum sativum*	leaves used as meat garnish and flavoring in sauces	widely cultivated
Peppermint	*Mentha piperita*	oil from leaves used for food, drink, dentifrice flavoring (much commercial menthol is derived from *Mentha arvensis* grown in Japan)	U.S.; Russia
Pimiento	*Capsicum* spp.	bright red fruits of a cultivated variety of pepper used in stuffing olives and in cold meats, cheeses	Central and South America
Poppy	*Papaver somniferum*	seeds used in baking	widely cultivated
Rosemary	*Rosmarinus officinalis*	oil from leaves used in perfumes, soaps	Mediterranean
Rue	*Ruta graveolens*	flavoring for fruit cups, salads	Europe
Saffron	*Crocus sativus*	dried stigmas used to flavor oriental-style dishes	Spain; India
Sage	*Salvia officinalis*	leaves used in poultry and meat dressings	Yugoslavia
Sarsaparilla	*Smilax* spp.	roots are source of flavoring for beverages, medicines	American Tropics
Sassafras	*Sassafras albidum*	bark and wood yield flavoring for beverages, toothpaste, gumbo	U.S.
Savory (summer)	*Satureia hortensis*	leaves used in green bean and bean salads, lentil soup, with fish	Mediterranean
Savory (winter)	*Satureia montana*	leaves used as seasoning in stuffings, meat loaf, stews	Europe
Scallion	*Allium fistulosum*	leaves used in wine cookery, soups	widely cultivated
Sesame	*Sesamum indicum*	seeds used in baking	Asia
Shallot	*Allium ascalonicum*	bulbs, leaves used in Colbert butter, wine cookery	widely cultivated
Southernwood	*Artemisia abrotanum*	leaves used to flavor cakes	Europe
Star anise	*Illicium verum*	fruits used in candy and cough drops	China
Stonecrop	*Sedum acre*	dried leaves (ground) used as pepper substitute	Europe
Tarragon	*Artemisia dracunculus*	leaves and flowering tops used in pickling sauces	Europe
Thyme	*Thymus vulgaris*	leaves used in meat and poultry dishes, soups, sauces	widely cultivated
Tonka bean	*Dipteryx* spp.	seeds source of flavoring for tobacco; vanilla substitute (now largely synthesized)	American tropics
Turmeric	*Curcuma longa*	rhizomes powdered and used in curry powders, meat flavoring	India; China
Vanilla	*Vanilla planifolia*	flavoring extracted from fruits; used in foods, drinks	Malagasay Republic
Wintergreen	*Gaultheria procumbens*	oil from leaves, bark used as flavoring for confections, toothpaste	U.S.
Zedoary	*Curcuma zedoaria*	dried rhizome used in liqueurs, curry powders	India

DYE PLANTS

In the recent and the ancient past, dyes from many different plants were used to color cotton, linen, and other fabrics. Since the middle of the nineteenth century, however, natural dyes have been almost completely replaced in industry by synthetic dyes, and today the use of natural dyes is largely confined to individual hobbyists.

Any reader interested in experimenting with natural dyes is encouraged to choose not only those plant materials included in table A3.7 but to try any local plants available. The experimenter will soon find that quite unexpected colors may be derived from plants, as the colors of fresh flowers, bark, or leaves often bear little relationship to the colors of the dyes. For methods of dyeing, see the footnote given under the heading of Lichens in chapter 17 and references in the Additional Reading list.

TABLE A3.7
Plant Sources of Natural Dyes

Plant or Dye	Scientific Name of Plant Source	Remarks
Acacia	Acacia spp.	brown dyes from bark and fruits
Alder	Alnus spp.	brownish dyes from bark
Alkanet	Alkanna tinctoria	red dye from roots
Annatto	Bixa orellana	yellow or red dye from pulp surrounding seeds
Bamboo	Bambusa spp.	light green dye from leaves
Barberry	Berberis vulgaris	grayish dye from leaves
Barwood	Baphia nitida	purplish dyes from wood
Bearberry	Arctostaphylos uva-ursi	yellowish dye from leaves
Bedstraw	Galium spp.	light reddish brown dyes from roots
Birch	Betula spp.	light brown to black dyes from bark
Black cherry	Prunus serotina	red dye from bark; gray to green dyes from leaves
Black walnut	Juglans nigra	rich brown dye from bark; brown dye from walnut hulls
Bloodroot	Sanguinaria canadensis	red dye from rhizomes
Blueberry	Vaccinium spp.	blue to gray dye from mature fruits (tends to fade)
Bougainvillea	Bougainvillea spp.	light brownish dyes from floral bracts
Brazilwood	Caesalpinia spp.	reddish dyes from wood
Buckthorn	Rhamnus spp.	green dyes from fruits
Buckwheat	Fagopyrum esculentum	blue dye from stems
Buckwheat (wild)	Eriogonum spp.	dark gold, pale yellow, and beige dyes from stems and flowers
Buffaloberry	Shepherdia argentea	red dye from fruit
Butternut	Juglans cinerea	yellow to grayish brown dyes from fruit hulls
Cocklebur	Xanthium strumarium	dark green dye from stems and leaves
Coffee	Coffea arabica	light brown dye from ground roasted seeds
Cudbear (Archil)	Rocella spp. (lichen)	red dye obtained by fermentation of thallus
Cutch	Acacia spp.; Uncaria gambir	brown to drab green dyes from stem gums
Dock	Rumex spp.	light brown dyes from stems and leaves
Dogwood	Cornus florida	red dye from bark; purplish dye from root
Doveweed	Eremocarpus setigerus	light to olive green dye from entire plant
Dyer's rocket	Reseda luteola	orangish dye from all parts
Elderberry	Sambucus spp.	blackish dye from bark; purple, blue, or dark brown dyes from fruits
Eucalyptus	Eucalyptus spp.	beige dyes from bark
Fennel	Foeniculum vulgare	yellow dyes from shoots
Fig	Ficus carica	green dyes from leaves and fruits
Fustic	Chlorophora tinctoria	yellow, bright orange, and greenish dyes from heartwood
Gamboge	Garcinia spp.	yellow dye from resins that ooze from cuts made on stems
Giant reed	Arundo donax	pale yellow dye from leaves
Grape	Vitis spp.	bright yellow to olive green dyes from leaves
Hawthorn	Crataegus spp.	pink dye from ripe fruits

Plant Sources of Natural Dyes

Plant or Dye	Scientific Name of Plant Source	Remarks
Hemlock	*Tsuga* spp.	reddish brown dye from bark
Henna	*Lawsonia inermis*	orange dye from shoots and leaves
Hickory	*Carya tomentosa*	yellow dye from bark
Hollyhock	*Althaea rosea*	purplish black dye from flower petals
Horsetail	*Equisetum* spp.	tan dyes from all green parts
Indigo	*Indigofera tinctoria*	bright blue dyes from leaves
Kendall green (*see* Woadwaxen)		
Larkspur	*Delphinium* spp.	blue dyes from petals
Lichens	Many genera and species	many lichens yield brilliant shades of yellows, golds, and browns with various mordants
Litmus	*Rocella tinctoria*	famous pink-to-blue pH indicator dye from thallus
Logwood	*Haematoxylon campechianum*	dark blue purple dye from heartwood
Lokao	*Rhamnus* spp.	green dye from wood
Lupine	*Lupinus* spp.	greenish dyes from flowers
Madder	*Rubia tinctorium*	bright red dye from roots
Madrone	*Arbutus menziesii*	brown dye from bark
Manzanita	*Arctostaphylos* spp.	beige to dull yellow dyes from dried fruits
Maple	*Acer* spp.	pink dye from bark
Marsh marigold	*Caltha palustris*	yellow dye from petals
Milkweed	*Asclepias speciosa*	pale yellow dyes from leaves
Morning glory	*Ipomoea violacea*	gray green dye from blue flowers
Mullein	*Verbascum thapsus*	gold dyes from leaves
Oak	*Quercus* spp.	yellow dye from bark
Onion	*Allium cepa*	reddish brown dyes from dry outer bulb scales of red onions; yellow dyes from similar parts of yellow onions
Oregon grape	*Berberis aquifolium*	yellow dyes from roots
Osage orange	*Maclura pomifera*	yellow, gray, and green dyes from fruits; yellow orange dye from wood
Peach	*Prunus persica*	green dyes from leaves
Poke	*Phytolacca americana*	red dyes from mature fruits
Pomegranate	*Punica granatum*	dark gold dye from fruit rinds
Prickly lettuce	*Lactuca serriola*	green dye from leaves
Privet	*Ligustrum vulgare*	yellow green dye from leaves; deep gray dye from berries
Quercitron	*Quercus velutina*	bright yellow dye from bark
Rhododendron	*Rhododendron* spp.	tan dyes from leaves
Safflower	*Carthamnus tinctorius*	reddish dye from flower heads
Saffron	*Crocus sativus*	powerful yellow dye from stigmas
Sage	*Salvia officinalis*	yellow dye from shoots
Sandalwood	*Pterocarpus santalinus*	red dye from wood
Sappanwood	*Caesalpinia sappan*	red dye from heartwood
Sassafras	*Sassafras albidum*	orange brown dye from bark
Scotch broom	*Cytisus scoparius*	yellow dye from all parts of plant
Smoke tree	*Cotinus coggyria*	orange yellow dye from wood (dye sometimes called ''young fustic'')
Smooth sumac	*Rhus glabra*	grayish brown dye from bark
St. John's wort	*Hypericum* spp.	light brownish dyes from leaves
Tansy	*Tanacetum* spp.	yellow, green dyes from leaves
Toyon	*Heteromeles arbutifolia*	reddish brown dyes from leaves
Turmeric	*Curcuma longa*	orangish dye from rhizome
Woad	*Isatis tinctoria*	blue dye from leaves
Woadwaxen	*Genista tinctoria*	yellow dye from all parts
Yerba santa	*Eriodictyon californicum*	rich dark brown dyes from leaves

ADDITIONAL READING

Adrosko, R. J. 1971. Natural dyes and home dyeing. New York: Dover Publications.

Bliss, A. 1980. *North American dye plants*. New York: Charles Scribner's Sons.

Craker, L. E. and J. E. Simon, eds. 1985. *Herbs, spices and medicinal plants: recent advances in botany, horticulture and pharmacology, vol 1*. Phoenix, AZ: Oryx Press.

Furst, P. E. 1985. *Mushrooms: psychedelic fungi*. Edgemont, PA: Chelsea House Publications.

Hoffer, A. and H. Osmond. 1967. *The hallucinogens*. New York: Academic Press.

Loewenfeld, C. and P. Back. 1978. *The complete book of herbs and spices*. Pomfret, VT: David and Charles, Inc.

Merory, J. 1968. *Food flavorings: composition, manufacture and use,* 2d ed. New York: Chemical Publishing Co.

Schultes, R. E. and A. Hoffman. 1980. *The botany and chemistry of hallucinogens,* 2d ed. Springfield, IL: Charles C. Thomas, Publishers.

Spoerke, D. G., Jr. 1980. *Herbal medications*. Santa Barbara, CA: Woodbridge Press Publishing Company.

Usdin, E. and D. H. Efron. 1979. *Psychocopic drugs and related compounds,* 2d ed. Elmsford, NY: Pergamon Press, Inc.

Vogel, V. J. 1977. *American Indian medicine*. Norman: University of Oklahoma Press.

See also the Additional Reading entries in chapter 22.

Appendix 4   *Houseplants and Home Gardening*

GROWING HOUSEPLANTS

If sales volume is an indication, houseplants have never been more popular in the United States than they are now. Many are easy to grow and will brighten windowsills, planters, and other indoor spots for years if a few simple steps are followed to ensure their health and vigor.

Water

Houseplants are commonly overwatered, resulting in the unnecessary development of rots and diseases (see table A4.1). As a rule, the surface of the potting medium should be dry to the touch before watering, but the medium should not be allowed to dry out completely unless the plant is dormant. Care should be taken, particularly during the winter, that the water is at room temperature. If rainwater is available, it is to be preferred over tap water, particularly if the water has a high mineral content or is chlorinated. Broad-leaved plants should periodically have house dust removed with a damp sponge (never use detergents to clean surfaces—they remove protective waxes). Many plants benefit from a daily misting with water, particularly in heated rooms.

Containers

In time plants may develop too extensive a root system for the pots in which they are growing (commonly called becoming root bound). Nutrients in the potting medium may become exhausted, salts and other residues from fertilizers and water may build up to the point of inhibiting growth, or the plants themselves may produce substances that accumulate until they interfere with the plant's growth. To resolve these problems, the plants should be periodically repotted and divided, if necessary, at the time of repotting.

TABLE A4.1
Common Ailments of Houseplants

Problem Symptoms	Possible Causes
Wilting or collapse of whole plant	lack of water; too much heat; too much water or poor drainage resulting in root rot
Yellowish or pale leaves	insufficient light; too much light; microscopic pests (especially spider mites); too much or too little fertilizer
Brown, dry leaves	humidity too low; too much heat; poor air circulation; lack of water
Tips and margins of leaves brown	mineral content of water; drafts; too much sun or heat; too much or too little water
Ringed spots on leaves	water too cold
Leaves falling off	improper watering or water too cold; excessive use of fertilizer or wrong fertilizer; too much sun or, if lower leaf drop only, too little light
Stringy growth	needs more light; too much fertilizer
Base of plant soft or rotting	overwatering
No flowers or flower buds drop	too much or too little light; night temperatures too high
Water does not drain	drain hole plugged; potting mixture has too much clay
Mildew present	fungi present--arrest with sulphur dust

Common Pests	Controls
Aphids	wash off under faucet or spray with soapy (not detergent) water; pyrethrum or rotenone sprays also effective
Mealybugs	remove with cotton swabs dipped in alcohol; spray with Volck oil
Scale insects	remove by hand; spray with Volck oil
Spider mites	use sprays containing small amounts of xylene (act as soon as possible—spider mites multiply very rapidly)
Thrips	spray with pyrethrum/rotenone or Volck oil sprays
White flies	spray with soapy water or pyrethrum/rotenone sprays every four days for two weeks (only the adults are susceptible to the sprays)

For additional controls, see the biological controls in appendix 2.

Temperatures

Most houseplants do not thrive where the temperatures are either too high or too low (see table A4.1). They generally tend to prefer minimum temperatures of about 13° C (55° F) and maximum temperatures of about 29° C (84° F). Many houseplants that prefer warmer temperatures while actively growing also benefit from a "rest" period at lower temperatures after flowering.

Light

Next to overwatering, the most common contributor to the demise of houseplants is a lack of sufficient light (see table A4.1). This does not mean that houseplants prefer direct sunlight—such light frequently damages them, but filtered sunlight (as, for example, through a muslin curtain) is usually better for the plant than the light available in the middle of the room. Plants can also thrive in artificial light of appropriate quality. Ordinary incandescent bulbs have too little light of blue wavelengths, and ordinary fluorescent tubes emit too little red light. A combination of the two, however, works very well. Generally the wattage of the incandescent bulbs should be only one-fourth that of the fluorescent tubes in such a combination. Several types of fluorescent tubes specially balanced to imitate sunlight are also available.

Humidity

Dry air is hard on most houseplants. The level of humidity around the plants can be raised by standing the pots in dishes containing gravel or crushed rock to which water has been added. The humidity level can also be raised through the use of humidifiers, which come in a variety of sizes and capacities. Daily misting, as mentioned previously, can also help.

COMMON HOUSEPLANTS

Explanation of symbols given with each plant

Water

 = needs little water (applies primarily to cacti and succulents; these plants store water in such a way that the soil can be completely dry for a week or two without their being adversely affected)

 = water regularly but not excessively; wait until the potting medium surface is dry to the touch before watering

 = immerse pot in water for a few minutes each week and water frequently, never allowing the potting medium to become dry; do not, however, leave the base of the pot standing in water

 = little to regular

 = regular to frequent

Minerals in hard water are taxing on houseplants, and commercial water softeners do not improve water for the plants. Use rainwater or filtered water if at all possible; otherwise repot more often.

Temperature

 = cool. Maximum 13–16° C (55–61° F); minimum 5–7° C (41–45° F)

 = cool to medium. Maximum 18–21° C (65–70° F); minimum 10–13° C (50–55° F)

 = medium. Maximum 30° C (86° F)

 = medium to warm

 = warm

Many houseplants are native to the tropics, where they thrive under year-round warm temperatures, while cacti and succulents prefer cool winters. The closer one is able to imitate a plant's natural environment, the better the plant will grow (see table A4.2).

Light

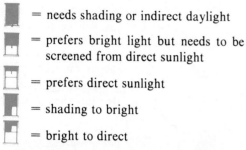

= needs shading or indirect daylight

= prefers bright light but needs to be screened from direct sunlight

= prefers direct sunlight

= shading to bright

= bright to direct

As mentioned previously, improper lighting is second only to overwatering as a cause of problems for houseplants; generally they are given too little light or occasionally too much. A southfacing windowsill may be ideal for certain plants in midwinter but excessively bright in midsummer; conversely, a north-facing windowsill may have enough light for certain plants in midsummer but not in midwinter. Adjustable screens permit manipulation of daylight to suit the plants involved.

Humidity

 = will tolerate dry air

 = dry to regular

 = will tolerate the air in most houses provided it is mist-sprayed occasionally

 = regular to humid

 = needs high humidity; use a humidifier if possible

Virtually all plants with or

symbols benefit from having a pan of gravel with water beneath the pot.

Potting Medium

 = requires a porous, slightly acid medium that drains immediately

 = requires a loam that is slightly alkaline (e.g., a mixture of sand and standard commercial potting medium)

 = requires a peaty potting mixture and acid fertilizer

TABLE A4.2
Environments Suitable for Common Houseplants

Plant	Scientific Name	Environmental Requirements	Remarks
Aechmea	*Aechmea fasciata*		*see* Bromeliad; produces side shoots that should be propagated as main plant dies after flowering
African lily	*Agapanthus* spp.		do not repot until pot is full; keep cool in winter
African violet	*Saintpaulia* spp.		let rest under cooler conditions after flowering; dislikes cold water
Agave	*Agave* spp.		keep cool and dry in winter
Algerian ivy	*Hedera canariensis*		resembles a variegated English ivy
Aloe	*Aloe* spp.		keep cool and dry in winter

Environments Suitable for Common Houseplants

Plant	Scientific Name	Environmental Requirements	Remarks
Aluminum plant	*Pilea cadierei*		plants do not usually survive long in houses
Amaryllis	*Amaryllis* spp.		let leaves die back in fall; put bulb in cool, dark place until early spring; then repot, water, and fertilize weekly
Anthurium	*Anthurium* spp.		if it grows without flowering, try putting it in a cooler location for a few weeks
Aphelandra	*Aphelandra squarrosa*		mist-spray frequently; fertilize regularly
Aralia (*see* Fatsia)			
Asparagus fern	*Asparagus plumosus*		not a true fern; repot annually; fertilize weekly
Aspidistra	*Aspidistra* spp.		sometimes called "cast iron plant" because it can stand neglect
Aucuba	*Aucuba japonica*		must be kept cool in winter
Avocado	*Persea* spp.		easily propagated from seed; provides good greenery but will not produce fruit indoors
Azalea	*Rhododendron* spp.		needs acid fertilizer; avoid warm locations
Bamboo (dwarf)	*Bambusa angulata*		needs good air circulation, bright light
Begonia	*Begonia* spp.		easily propagated from leaves; repot regularly
Bilbergia	*Bilbergia* spp.		*see* Bromeliad; tough plant that can tolerate some neglect
Birdcatcher plant	*Pisonia umbellifera*		sticky exudate on fruits attracts birds in the plant's native habitat of New Zealand; strictly a foliage plant in houses
Bird of paradise plant	*Strelitzia reginae*		can be grown outdoors in milder climates
Bird's nest fern	*Asplenium nidus*		produces a spongelike mass of roots at base; requires much water and regular fertilizing
Black-eyed Susan	*Thunbergia alata*		annual climbing vine; grow from seed
Bloodleaf	*Iresine herbstii*		easily propagated
Boston fern	*Nephrolepis exaltata*		needs regular watering and fertilizing

Environments Suitable for Common Houseplants

Plant	Scientific Name	Environmental Requirements	Remarks
Bromeliads	many species		these plants absorb virtually all their water and nutrients through their leaves; they should not be placed in regular potting soil, nor should they be watered with high calcium content water; they produce offshoots that should be propagated as main plant dies after flowering
Cacti	many species		contrary to popular belief, these slow-growing plants should not be grown in pure sand; add some humus to potting mixture and withhold water in winter; keep cool in winter
Caladium	*Caladium* spp.		must have high humidity; keep root at 18° C (65° F) in pot during winter
Calceolaria	*Calceolaria herbeohybrida*		discard after flowering
Calla lily	*Zantedeschia aethiopica*		after flowering, allow plant to dry up; repot in fall and start over
Cape jasmine	*Gardenia jasminoides*		night temperatures below 22° C (72° F) needed to initiate flowering; needs cool temperatures in winter
Carrion flower	*Stapelia* spp.		cactuslike plants with foul-smelling flowers
Century plant (see Agave)			
Chinese evergreen	*Aglaonema costatum*		needs warm temperatures and much water all year
Chrysanthemum	*Chrysanthemum* spp.		plants may be artificially dwarfed through use of chemicals; flowering initiated by short days
Cineraria	*Senecio cruentus*		needs cool temperatures; discard after flowering
Cliff brake	*Pellaea rotundifolia*		hanging basket fern; needs minimum temperature above 10° C (50° F)
Coffee	*Coffea arabica*		handsome foliage plant that will produce fruit if self-pollinating variety is obtained
Coleus	*Coleus blumei*		to control size, restart plants from cuttings annually
Copperleaf	*Acalypha wilkesiana*		seldom survives average house environment for long

Environments Suitable for Common Houseplants

Plant	Scientific Name	Environmental Requirements	Remarks
Corn plant	*Dracaena massangeana*		uses much water when large; easy to grow
Croton	*Codiaeum* spp.		needs constant high humidity and bright light
Crown of thorns	*Euphorbia milii* and *E. splendens*		deviation from watering routine may result in loss of leaves, but plant generally recovers
Cyclamen	*Cyclamen* spp.		fertilize weekly; keep cool; withhold water after flowering for few weeks, then start over
Donkey tail	*Sedum morganianum*		keep cool and dry in winter
Dracaena	*Dracaena* spp.		many kinds—all easy to grow and tolerant of some neglect
Dumbcane	*Dieffenbachia* spp.		needs regular fertilizing; keep away from small children (poisonous)
Dwarf banana	*Musa cavendishii*		will produce small edible bananas if given enough light, water, and humidity
Dwarf cocos palm	*Microcoleum weddelianum*		keep temperature above 18° C (65° F) at all times
Echeveria	*Echeveria* spp.		keep cool in winter
English ivy	*Hedera helix*		needs cool temperatures to grow at its best
False aralia	*Dizygotheca elegantissima*		benefits from frequent mist-spraying
Fatshedera	*Fatshedera lizei*		climbing plant
Fatsia	*Fatsia japonica*		also called *Aralia*
Ferns	many species		water regularly; propagate from spores or runners
Figs:			
Climbing fig	*Ficus pumila*		damp sponge leaves regularly
Fiddleleaf fig	*Ficus lyrata*		damp sponge leaves regularly
Weeping fig	*Ficus benjamina*		damp sponge leaves regularly
Fingernail plant	*Neoregelia* spp.		*see* Bromeliad; name from red tips of leaves
Fittonia	*Fittonia* spp.		strictly terrarium plants—humidity too low elsewhere
Flame violet	*Episcia cupreata*		add charcoal and peat to potting medium
Flowering maple	*Abutilon striatum*		needs bright light to flower
Fuchsia	*Fuchsia* spp.		soil must be alkaline
Gardenia (*see* Cape jasmine)			

Environments Suitable for Common Houseplants

Plant	Scientific Name	Environmental Requirements	Remarks
Geranium	*Pelargonium* spp.		make cuttings annually and discard parent plants each fall; keep cool through winter; available with scents of orange, rose, or coconut
Gloxinia	*Sinningia speciosa*		fertilize heavily and water frequently; after flowering, withhold water and keep bulb cold for few weeks
Goldfish plant	*Columnea* spp.		pot in mixture of leaf mold, fern bark, peat moss, and charcoal; use only rainwater or filtered water
Grape ivy	*Rhoicissus rhomboidea*		tolerates low light better than most plants
Haworthia	*Haworthia* spp.		aloelike plants that need minimum temperatures above 10° C (50° F) in winter
Hen and chickens	*Sempervivum tectorum*		keep cool in winter
Hibiscus	*Hibiscus rosa-sinensis*		fertilize weekly
Hippeastrum (*see* Amaryllis)			
Holly fern	*Cyrtomium falcatum*		relatively tough fern; keep cool in winter
Houseleek	*Sempervivum* spp.		keep cool in winter
Hydrangea	*Hydrangea* spp.		prune after flowering; keep cool in winter; pink-flowering plant can be converted to blue by changing the soil to acid and vice versa
Impatiens	*Impatiens* spp.		exceptionally easy to propagate from cuttings
Ivy-arum	*Rhaphidophora aurea*		can tolerate some neglect
Jade plant	*Crassula argenta* and others		keep cool in winter
Kaffir lily	*Clivia miniata*		save the plant's energy by removing flowers as they wither
Kalanchoë	*Kalanchoë* spp.		withhold water and fertilizer for a few weeks after flowering, then repot and start over
Lantana	*Lantana camara*		fertilize twice a month; can be espaliered
Madagascar jasmine (*see* Stephanotis)			
Maidenhair fern	*Adiantum* spp.		mist-spray regularly
Meyer fern	*Asparagus densiflora* var. *meyeri*		fertilize regularly; repot annually
Moneywort	*Lysimachia nummularia*		hanging pot plant; needs bright light

Environments Suitable for Common Houseplants

Plant	Scientific Name	Environmental Requirements	Remarks
Moonstones	*Pachyphytum* spp.		keep cool in winter
Moses in the cradle	*Rhoeo* spp.		can tolerate some neglect
Mother in law's tongue (*see* Sansevieria)			
Mother of thousands	*Saxifraga* spp.		also called *Saxifrage;* hanging plant; plantlets formed on runners can be removed and grown separately
Neanthe palm	*Chamaedorea elegans*		sometimes also called *Parlor palm;* stays less than 1 meter tall
Norfolk Island pine	*Araucaria* spp.		needs cold temperatures (2–3° C) (36° F–38° F) in winter
Octopus tree	*Schefflera arboricola*		does best under cool conditions
Oleander	*Nerium oleander*		keep pot cool in winter for better flowering; keep away from small children (poisonous)
Orchid	thousands of species	no single set of environmental conditions applies	contrary to popular belief, the common *Cattleya* and related orchids do not need high temperatures and humidity; most can get along with a minimum temperature of 13° C (56° F) at night and a minimum humidity of 40%. Most need bright light. They should never be placed in soil; pot them in sterilized pots with chips of fir bark or shreds of tree fern bark. *See* Additional Reading for culture references
Oxtongue	*Gasteria* spp.		can tolerate some neglect; needs a cool and relatively dry winter to flower
Palms	many species		use deep pots; fertilize regularly
Parlor palm	*Howea fosteriana*		one of the easiest palms to grow
Peperomia	*Peperomia* spp.		many kinds; keep warm, humid; fertilize regularly
Persian violet	*Exacum affine*		needs good air circulation
Philodendron	*Philodendron* spp.		relatively tough plants; repot each spring
Piggyback plant	*Tolmiea menziesii*		plantlets formed on leaves can be separated and propagated

Environments Suitable for Common Houseplants

Plant	Scientific Name	Environmental Requirements	Remarks
Pineapple	*Ananas comosus*		*see* Bromeliad; easily grown from the top of a pineapple; if plant has not flowered after one year, enclose it in a plastic bag with a ripe apple for a few days (ethylene from apple should initiate flowering); no temperature below 15° C (59° F)
Pink polka dot plant	*Hypoestes sanguinolenta*		susceptible to diseases and pests
Pittosporum	*Pittosporum* spp.		put several cuttings in one pot for bushy appearance
Poinsettia	*Euphorbia pulcherrima*		after flowering, let plant dry under cool conditions until it loses its leaves, then restart
Prayer plant	*Maranta leuconeura*		name derived from fact that leaves fold together in evening
Primrose	*Primula* spp.		needs much water; does well outside in cool weather
Purple tiger	*Calathea amabilis*		use pots that are broader than deep
Rosary plant	*Ceropegia woodii*		hanging pot plant whose potting medium must drain well or plant will not survive
Rubber plant	*Ficus elastica*		do not overwater!
Sago palm	*Cycas revoluta*		very slow growing (not a palm but a gymnosperm); never allow to dry out, but be certain water drains
Sansevieria	*Sansevieria* spp.		perhaps the toughest of all houseplants— nearly indestructible
Satin pothos	*Scindapsus pictus*		basket or pot plant
Screw pine	*Pandanus* spp.		if given space can become large; mist-spray often
Selaginella (*see* Spike moss)			
Sensitive plant	*Mimosa pudica*		leaves fold when touched; does not usually last more than a few months in most houses
Shrimp plant	*Beloperone guttata*		winter temperatures should be above 15° C (59° F)

Environments Suitable for Common Houseplants

Plant	Scientific Name	Environmental Requirements	Remarks
Spathe flower	*Spathiphyllum wallisii*		prefers warm winters and even warmer summers
Spider plant	*Chlorophytum comosum*		plantlets formed at tips of stems can be propagated separately
Spiderwort	*Tradescantia* spp.		easy to grow; do not overwater
Spike moss	*Selaginella* spp.		can become a weed in greenhouses
Splitleaf philodendron	*Monstera deliciosa*		plant adapts to various indoor locations quite well
Sprenger fern	*Asparagus densiflora* var. *sprengeri*		fertilize weekly; repot annually
Staghorn fern	*Platycerium* spp.		tough plant; immerse in water weekly
Stephanotis	*Stephanotis floribunda*		use very little fertilizer; keep cool in winter but water sparingly
Stonecrop	*Sedum* spp.; *Crassula* spp.		keep cool in winter
Stove fern	*Pteris cretica*		water and fertilize regularly
String-of-pearls	*Senecio rowellianus*		keep cool in winter
Sundew	*Drosera* spp.		sterilize pots; grow only on sphagnum moss
Syngonium	*Syngonium podophyllum*		repot annually in spring
Tillandsia	*Tillandsia* spp.		*see* Bromeliad; best known species is called *Spanish moss*
Ti plant	*Cordyline terminalis*		needs high humidity; seems to do better with other plants in pot
Treebine	*Cissus antarctica*		plant dislikes acid potting medium
Umbrella plant	*Cyperus* spp.		one of very few plants that needs to stand in water
Velvetleaf	*Gynura sarmentosa*		gets "stringy" but is easily restarted from cuttings
Venus flytrap	*Dionaea muscipula*		sterilize pots; grow only in sphagnum moss; repot annually
Vriesia	*Vriesia* spp.		*see* Bromeliad
Wandering Jew	*Zebrina pendula*		easy to grow; do not overwater
Wax plant	*Hoya* spp.		climber; leave pot in one place—does not like to be moved

GROWING VEGETABLES

General Tips on Vegetable Growing

Seed Germination

Many gardeners germinate larger seeds (e.g., squash, pumpkin) in damp newspaper. A few sheets of newspaper are soaked in water for a minute and then hung over a support for about 15 minutes or until the water stops dripping. The seeds are then lined up in a row on the newspaper, wrapped, and the damp mass is placed in a plastic bag. The bag is tied off or sealed and placed in a warm (not hot!), shaded location. Depending on the species, germination should occur within two to several days.

Tiny seeds (e.g., carrots, lettuce) may be mixed with clean sand before sowing to bring about a more even distribution of the seed in the rows.

Transplanting

Roots should be disturbed as little as possible when seedlings or larger plants are transplanted. Even a few seconds' exposure to air will kill root hairs and smaller roots. They should be shaded (e.g., with newspaper) from the sun and transplanted late in the day or on a cool, cloudy day if at all possible. To minimize the effects of transplanting, immediately water the seedlings or plants in their new location with a dilute solution of vitamin B/hormone preparation (e.g., Superthrive).

Bulb or Other Plants with Food-Storage Organs

All such plants (e.g., beets, carrots, onions) develop much better in soil that is free of lumps and rocks. If possible, the areas where these plants are to be grown should be dug to a minimum depth of 3 decimeters (12 inches) and the soil sifted through a 0.7 centimeter (approximately 0.25 inch) mesh before planting. Obviously such a procedure is not always practical, but it can yield dramatic results.

Cutworms

Cutworms forage at or just beneath the surface of the soil. Their damage to young seedlings can be minimized by a collar placed around each plant. Tuna cans with both ends removed make effective collars when pressed into the ground to a depth of about 1 to 2 centimeters (0.4 to 0.8 inch).

Protection against Cold

Some seedlings can be given an earlier start outdoors if plastic-topped coffee cans with the bottoms removed are placed over them. The plastic lids can be taken off during the day and replaced at night during cool weather. Conical paper frost caps can serve the same purpose.

Watering

Proper watering promotes healthy growth. It is much better to water an area thoroughly (e.g., for 20 to 30 minutes) every few days than to wet the surface for a minute or two daily. Shallow daily watering promotes root development near the surface, where midsummer heat can damage the root system. Conversely, too much watering can leach minerals out of the topsoil.

Fertilizers

Manures, bone and blood meals, and other organic fertilizers, which release the nutrients slowly and do not "burn" young plants, are preferred. Plants will utilize minerals from any available source, but in the long run the plants will be healthier and subject to fewer problems if they are not given sudden boosts with liquid "chemical" fertilizers.

Pests and Diseases

Biological controls are listed in appendix 2. If sprays must be used, biodegradable substances such as rotenone and pyrethrum should be used.

COMMON VEGETABLES AND THEIR NUTRITIONAL VALUES

Note: The nutritional values (NV) given are per 100 grams (3.5 ounces), edible portion, as determined by the United States Department of Agriculture.

Asparagus

NV (spears cooked in water): 20 calories; protein 2.2 gm; fats 0.2 gm; vit. A 900 I.U.; vit. B_1 0.16 mg; vit. B_2 0.18 mg; niacin 1.5 mg; vit. C 26 mg; fiber 0.7 gm; calcium 21 mg; phosphorus 50 mg; iron 0.6 mg; sodium 1.0 mg; potassium 208 mg.

Asparagus can be started from seed, but time until the first harvest can be reduced by a year or two if planting begins with one-year-old root clusters of healthy, disease-resistant varieties (e.g., "Mary Washington"). Asparagus requires little care if appropriate preparations are made before planting in the permanent location. Seeds should be sown sparsely, and the seedlings thinned to about 7.5 centimeters (3 inches) apart. Before transplantation the following spring, dig a trench 30 to 60 centimeters (12 to 24 inches) deep and about 50 centimeters (20 inches) wide in an area that receives full sun, usually along one edge of the garden. If the soil is heavy, place crushed rock or gravel on the bottom of the trench to provide good drainage. Add a layer of steer manure about 10 centimeters (4 inches) thick, followed by about 7.5 centimeters (3 inches) of rich soil that has been prepared by thorough mixing with generous quantities of steer manure and bone meal. Place root clusters about 45 centimeters (18 inches) apart in the trench and cover with about 15 centimeters (6 inches) of prepared soil (be sure not to allow root clusters to dry out). As the plants grow, gradually fill in the trench. If one-year-old roots are planted, wait for two years before harvesting tips; if two-year-old roots are planted, some asparagus may be harvested the following year. In all cases, no harvesting should be done after June, so that the plants may build up reserves for the following season. Cut shoots below the surface but well above the crown before the buds begin to expand. Cut all stems to the ground after they have turned yellow later in the season.

Beans

String or Snap Beans

NV (young pods cooked in water): 25 calories; protein 1.6 gm; fats 0.2 gm; vit. A 540 I.U.; vit. B_1 0.07 mg; vit. B_2 0.09 mg; niacin 0.5 mg; vit. C 12 mg; fiber 1.0 gm; calcium 50 mg; phosphorus 37 mg; iron 0.6 mg; sodium 4 mg; potassium 151 mg.

String or snap beans are warm-weather plants, although they can be grown almost anywhere in the United States. Wait until all danger of frost has passed and the soil is warm. Prepare the soil, preferably the previous winter, by digging to a depth of 30 centimeters (12 inches) and mixing in aged manure and bone meal. Pulverize the soil just before sowing; if soil has a low pH, add lime. Plant seeds thinly in rows about 40 to 50 centimeters (16 to 20 inches) apart; thin plants to 10 centimeter (4 inches) apart when the first true leaves have developed. Beans respond unfavorably to a very wet soil—do not overwater! In areas with hot summers, beans also prefer some light shade, particularly in midafternoon. As the bean plants grow, nitrogen-fixing bacteria invade the roots and supplement the nitrogen

supply. Early vigorous growth can be enhanced by inoculating the seeds with such bacteria, which are available commercially in a powdered form. To maintain a continuous supply of green beans, plant a new row every two to three weeks during the growing season until two months before the first predicted frost. Cultivate regularly to control weeds, taking care not to damage root systems. Do not harvest or work with beans while the plants are wet, as this may invite disease problems.

Pole Beans

NV similar to those of string beans.

Soil preparation and cultivation are the same as for string beans. Plant beans in hills around poles that are not less than 5 centimeters (2 inches) in diameter and at least 2 meters (6.5 feet) tall. As beans twine around their supports, it helps to tie them to the support with plastic tape as they grow. If harvested before the pods are mature, pole beans will produce over a longer period of time than bush varieties.

Lima Beans

NV (immature seeds cooked in water): 111 calories; protein 7.6 gm; fats 0.5 gm; vit. A 280 I.U.; vit. B_1 0.18 mg; vit. B_2 0.10 mg; niacin 1.3 mg; vit. C 17 mg; fiber 1.8 gm; calcium 47 mg; phosphorus 121 mg; iron 2.5 mg; sodium 1.0 mg; potassium 422 mg.

Lima beans take longer to mature than other beans and are more sensitive to wet or cool weather. They definitely need warm weather to do well. Prepare soil and cultivate as for string beans.

Soybeans

NV (dry, mature seeds): 403 calories; protein 34.1 gm; fats 17.7 gm; vit. A 80 I.U.; vit. B_1 1.10 mg; vit. B_2 0.31 mg; niacin 2.2 mg; vit. C—[1]; fiber 4.9 gm; calcium 226 mg; phosphorus 554 mg; iron 8.4 mg; sodium 5.0 mg; potassium 1,677 mg.

Prepare soil and cultivate as for string beans.

---

1. Values not available. See also Broad (Fava) Beans, page A–66.

Broad (Fava) Beans

NV (dry, mature seeds): 338 calories; protein 25.1 gm; fats 1.7 gm; vit. A 70 I.U.; vit. B₁ 0.5 mg; vit. B₂ 0.3 mg; niacin 2.5 mg; vit. C—¹; fiber 6.7 gm; calcium 47 mg; phosphorus 121 mg; iron 2.5 mg; sodium 1.0 mg; potassium 422 mg.

Unlike other beans, broad beans need cool weather for their development. Sow as early as possible (in mild climates they may be sown in the fall, as they can withstand light frosts). Since the plants occupy a little more space than bush beans, plant in rows about 0.9 to 1 meter (3 feet or more) apart and thin to about 20 centimeters (8 inches) apart in the rows. After the first pods mature, pinch out the tips to promote bushier development. To most palates, broad beans do not taste as good as other types of beans.

Beets

NV (cooked in water): 32 calories; protein 1.1 gm; fats 0.1 gm; vit. A 20 I.U.; vit. B₁ 0.03 mg; vit. B₂ 0.04 mg; vit. C 6 mg; fiber 0.8 gm; calcium 14 mg; phosphorus 23 mg; iron 0.5 mg; sodium 43 mg; potassium 208 mg.

Beets will grow in a variety of climates but do best in cooler weather. They can tolerate light frosts and can be grown on a variety of soil types, although they prefer a sandy loam supplemented with well-aged organic matter. As with any bulb or root crop, they develop best in soil that is free of rocks and lumps.

Beet "seeds" are really fruits containing several tiny seeds. Plant them in rows 40 to 60 centimeters (16 to 24 inches) or more apart and thin to about 10 centimeters (4 inches) apart in the rows after germination. After harvesting, the beets will keep in cold storage for up to several months. The leaves, if used when first picked, make a good substitute for spinach.

Broccoli

NV (spears cooked in water): 26 calories; protein 3.1 gm; fats 0.3 gm; vit. A 2,500 I.U.; vit. B₁ 0.09 mg; vit. B₂ 0.2 mg; niacin 0.8 mg; fiber 1.5 gm; calcium 88 mg; phosphorus 62 mg; iron 0.8 mg; sodium 10 mg; potassium 267 mg.

Broccoli is a cool-weather plant that will thrive in any good prepared soil, providing that it has not been heavily fertilized just prior to planting (fresh fertilizer promotes rank growth). The plants can stand light frosts and are planted in both the spring and fall in areas with mild climate. Although broccoli may continue to produce during the summer, most growers prefer not to keep the plants going during warm seasons because of the large numbers of pest insects they may attract.

Sow seeds indoors and transplant outdoors after danger of killing frosts has passed. Place plants about 0.9 to 1 meter (3 feet or more) apart and keep well watered. Keep area weeded and pests under control. Harvest heads (bundles of spears) while they are still compact. Smaller heads will develop very shortly after the first harvest; if these are removed regularly, the plants will continue to produce for some time, although the heads become smaller as the plants age.

Cabbage

NV (raw): 24 calories; protein 1.3 gm; fats 0.2 gm; vit. A 130 I.U.; vit. B₁ 0.05 mg; vit. B₂ 0.05 mg; niacin 0.3 mg; vit. C 47 mg; fiber 1.0 gm; calcium 49 mg; phosphorus 29 mg; iron 0.4 mg; sodium 20 mg; potassium 233 mg.

Growth requirements of cabbage are similar to those of broccoli.

Carrots

NV (raw): 42 calories; protein 1.1 gm; fats 0.2 gm; vit. A 11,000 I.U.; vit. B₁ 0.6 mg; vit. B₂ 0.5 mg; niacin 0.6 mg; vit. C 8 mg; fiber 1.0 gm; calcium 37 mg; phosphorus 36 mg; iron 0.7 mg; sodium 47 mg; potassium 341 mg.

Carrots are hardy plants that can tolerate a wide range of climate and soils, but the soil must be well prepared, free of rocks and lumps, and preferably not too acid. The seeds are slow to germinate. Plant in rows a little more than 30 centimeters (12 inches) apart and thin seedlings to about 5 centimeters (2 inches) apart in the rows. Weed the rows regularly until harvest. Carrots keep well in below-ground storage containers when freezing weather arrives.

Cauliflower

NV (cooked in water): 22 calories; protein 2.3 gm; fats 0.2 gm; vit. A 60 I.U.; vit. B₁ 0.09 mg; vit. B₂ 0.08 mg; niacin 0.6 mg; vit. C 55 mg; fiber 1.0 gm; calcium 21 mg; phosphorus 42 mg; iron 0.7 mg; sodium 9 mg; potassium 206 mg.

Growth requirements of cauliflower are similar to those of broccoli except that heavier fertilizing is required. As cauliflower heads develop, protect them from the sun by tying the larger leaves over the tender heads. Harvest while the heads are still solid.

Corn

NV (cooked sweet corn kernels): 83 calories; protein 3.2 gm; fats 1.0 gm; vit. A 400 I.U. (yellow varieties; white varieties have negligible vit. A content); vit. B_1 0.11 mg; vit. B_2 0.10 mg; niacin 1.3 mg; vit. C 7 mg; fiber 0.7 gm; calcium 3 mg; phosphorus 89 mg; iron 0.6 mg; sodium trace; potassium 165 mg.

There are several types of corn (e.g., popcorn, flint corn, dent corn), but sweet corn is the only type grown to any extent by home gardeners. It can be grown in any location where there is at least a ten-week growing season and warm summer weather.

Corn prefers a fertile soil, which should be prepared by mixing with compost and liberal amounts of chicken manure or fish meal. Since corn is wind-pollinated, it can be helpful to grow the plants in several short rows at right angles to the prevailing winds rather than in a single long row. For best results, use only fresh seeds and plant in rows 60 centimeters (24 inches) apart for dwarf varieties and 90 centimeters (36 inches) apart for standard varieties. Thin to 20 to 30 centimeters (8 to 12 inches) apart in the rows after the plants have produced three to four leaves. Cultivate frequently to control weeds. The corn is ready to harvest when the silks begin to wither.

Cucumber

NV (raw, with skin): 15 calories; protein 0.9 gm; fats 0.1 gm; vit. A 250 I.U.; vit. B_1 0.03 mg; vit. B_2 0.04 mg; niacin 0.2 mg; vit. C 11 mg; fiber 0.6 mg; calcium 25 mg; phosphorus 27 mg; iron 1.1 mg; sodium 6 mg; potassium 160 mg.

Until drought- and disease-resistant varieties were developed in recent years, cucumbers were considered rather "temperamental" plants to grow. The newer varieties are no more difficult to raise than those of most other common vegetables.

The soil should be a light loam—neither too heavy nor too sandy. It should be mixed with well-aged manure and compost and heaped into small mounds about 2 meters (6.5 feet) apart. Five to six seeds should be planted in each mound about 2.5 centimeters (1 inch) below the surface in the middle

of the spring. When the plants are about 1 decimeter (4 inches) tall, thin to three plants per mound. Cultivate regularly, and to promote continued production, pick all cucumbers as soon as they attain eating size.

Eggplant

NV (cooked in water): 19 calories; protein 1.0 gm; fats 0.2 gm; vit. A 10 I.U.; vit. B_1 0.05 mg; vit. B_2 0.04 mg; niacin 0.5 mg; vit. C 3 mg; fiber 0.9 gm; calcium 11 mg; phosphorus 21 mg; iron 0.6 mg; sodium 1 mg; potassium 150 mg.

Eggplant is strictly a hot-weather plant that is sensitive to cold weather or dry periods and needs heavy fertilizing. Since seedling development is initially slow, plant the seeds indoors about two months before the plants will be set out, which should be about five to six weeks after the last average date of frost.

Eggplants do best in enriched sandy soils that are supplemented with additional fertilizer once a month. Never permit them to dry out. Place the seedlings about 70 to 80 centimeters (28 to 32 inches) apart in rows 0.9 to 1 meter (3 feet or more) apart. Some staking of the plants may be desirable. The fruits are ready to harvest when they have a high gloss. They are still edible after greenish streaks appear and the gloss diminishes, but they are not as tender at this stage.

Lettuce

NV (crisp, cabbage-head varieties): 13 calories; protein 0.9 gm; fats 0.1 gm; vit. A 330 I.U.; vit. B_1 0.06 mg; vit. B_2 0.06 mg; niacin 0.3 mg; vit. C 6 mg; fiber 0.5 gm; calcium 20 mg; phosphorus 22 mg; iron 0.5 mg; sodium 9 mg; potassium 175 mg.

NV (leaf varieties): 18 calories; protein 1.3 gm; fats 0.3 gm; vit. A 1,900 I.U.; vit. B_1 0.05 mg; vit. B_2 0.08 mg; niacin 0.4 mg; vit. C 18 mg; calcium 68 mg; phosphorus 25 mg; iron 1.4 mg; sodium 9 mg; potassium 264 mg.

This favorite salad plant comes in a wide variety of types and forms, all of which prefer cooler weather, although a few of the leaf types (e.g., oak leaf) can tolerate some hot periods. As long as the individual plants are given room to develop and the soil is not too acid, most varieties can be grown on a wide range of soil types.

Since lettuce can stand some frost, sow the seeds outdoors as early in the spring as the ground can be cultivated. Do not cover the seeds with more than a millimeter or two of soil—they need light to germinate. Mix the soil with a well-aged manure and a general-purpose fertilizer a week or two before sowing. Plant seedlings about 30 centimeters (12 inches) apart in rows 30 to 40 centimeters (12 to 16 inches) apart. For best results, do not allow the soil to dry out, and plant only varieties suited to local conditions. The most common crisp, cabbage-head varieties found in produce markets will not form heads in hot weather, and many others will bolt (begin to flower) during hot weather and longer days. Cultivate weekly between rows to promote rapid growth and to control weeds.

Onion

NV (raw): 38 calories; protein 1.5 gm; fats 0.1 gm; vit. A 40 I.U. (yellow varieties only); vit. B_1 0.03 mg; vit. B_2 0.04 mg; niacin 0.2 mg; vit. C 10 mg; fiber 0.6 gm; calcium 27 mg; phosphorus 36 mg; iron 0.5 mg; sodium 10 mg; potassium 157 mg.

These easy-to-grow vegetables do best in fertile soils that are free of rocks and lumps, are well drained, and are not too acid or sandy.

Onions may take several months to mature from seed. The viability of the seed decreases rapidly after the first year. Bulb formation is determined by daylength rather than by the total number of hours in the ground. Because of these characteristics of onions, most gardeners prefer to purchase "sets" (young plants that already have a small bulb) from commercial growers, although green or bunching onions are still easily grown from seed.

Plant the sets upright 6 to 7.5 centimeters (2.5 to 3 inches) apart in rows and firm in place. Except for weeding, watering, and occasional shallow cultivation, they will need little care until harvest about fourteen weeks later. The onions are mature when the tops fall over. After they are pulled from the ground, allow them to dry in the shade for two days. Then remove the tops 2 to 3 centimeters (about 1 inch) above the bulbs, and spread out the bulbs to continue "curing" for two to three more weeks. After this, store them in sacks or other containers that permit air circulation until needed.

Peas

NV (green, cooked in water): 71 calories; protein 5.4 gm; fats 0.4 gm; vit. A 540 I.U.; vit. B_1 0.28 mg; vit. B_2 0.11 mg; niacin 2.3 mg; vit. C 20 mg; fiber 2.0 gm; calcium 23 mg; phosphorus 99 mg; iron 1.8 mg; sodium 1 mg; potassium 196 mg.

Peas are strictly cool-weather plants that generally produce poorly when the soil becomes too warm. Seeds should be planted in the fall or very early spring. As is the case with beans, peas receive a better start if the seeds are inoculated with nitrogen-fixing bacteria (see discussion of beans) at planting time. Prepare the soil by mixing thoroughly with liberal amounts of aged manure and bone meal. Plant the seeds about 2.5 centimeters (1 inch) deep in heavy soil or 5 centimeters (2 inches) deep in light, sandy soil, about 2.5 centimeters (1 inch) apart in single rows for dwarf bush varieties or in double files 15 centimeters (6 inches) apart for standard varieties, with intervals of 0.8 to 0.9 decimeters (32 to 36 inches) between the rows. After germination, thin the plants to 10 centimeters apart. Place support wires, strings, or chicken wire between the rows at the time of planting; peas will not do well without such support.

Green peas should be picked while still young and cooked or frozen immediately, as the sugars that make them sweet are converted to starch within two to three hours after harvest.

Peppers

NV (raw sweet or bell peppers): 22 calories; protein 1.2 gm; fats 0.2 gm; vit. A 420 I.U.; vit. B_1 0.08 mg; vit. B_2 0.08 mg; niacin 0.5 mg; vit. C 128 mg; fiber 1.4 gm; calcium 9 mg; phosphorus 22 mg; iron 0.7 mg; sodium 13 mg; potassium 213 mg.

Peppers, like eggplants, are strictly hot-weather plants for most of their growing season, but unlike eggplants they actually do better toward the end of their season if temperatures have moderated somewhat. Sweet or bell peppers are closely related and have similar cultural requirements.

Plant seeds indoors eight to ten weeks before the outdoor planting date, which is generally after the soil has become thoroughly warm. They will grow in almost any sunny location in a wide variety of soils, but to obtain the large fruits seen in produce markets the plants need to be fertilized heavily and watered regularly. Plant seedlings 5 to 6 centimeters (20 to 24 inches) apart in rows that are 60 to 90 centimeters (24 to 36 inches) apart. Sweet peppers can be harvested at almost any stage and are still perfectly edible after they have turned red.

Potatoes

NV (baked in skin): 93 calories; protein 2.6 gm; fats 0.1 gm; vit. A trace; vit. B₁ 0.10 mg; vit. B₂ 0.04 mg; niacin 1.7 mg; vit. C 20 mg; fiber 0.6 gm; calcium 9 mg; phosphorus 65 mg; iron 0.7 mg; sodium 4 mg; potassium 503 mg.

Potatoes grow best in a rich, somewhat acid, well-drained soil that has had compost or well-aged manure added to it. They are subject to several diseases, and it is advisable to use disease-free seed potatoes purchased from a reputable dealer. Two to three weeks before the average date of the last spring frost, plant the seed potatoes whole or cut into several pieces, making sure that each piece has at least one eye. Place the potato pieces about 3 decimeters (12 inches) or more apart at a depth of about 12 to 15 centimeters (5 to 6 inches) in rows 0.9 to 1 meter (about 3 feet) apart. Later plantings are feasible. Spread a thick mulch (e.g., straw) over the area after planting to keep soil temperatures down and to retain soil moisture.

Potatoes are ready for harvest when the tops start turning yellow, but they may be left in the ground for several weeks after that if the soil is not too wet. After harvest, wash the potatoes immediately and place them in a dry, cool, dark place until needed. If left exposed to light, the outer parts of the potato turn green; poisonous substances are produced in these tissues, and such potatoes should be discarded.

Spinach

NV (raw): 26 calories; protein 3.2 gm; fats 0.3 gm; vit. A 8,100 I.U.; vit. B₁ 0.10 mg; vit. B₂ 0.20 mg; niacin 0.6 mg; vit. C 51 mg; fiber 0.6 gm; calcium 93 mg; phosphorus 51 mg; iron 3.1 mg; sodium 71 mg; potassium 470 mg.

Spinach is a cool season crop that goes to seed as soon as the days become long and warm. It should be planted in the fall or early spring. If protected by straw or other mulches, it will overwinter in the ground in most areas and be ready for use early in the spring. Spinach has a high nitrogen requirement and reacts negatively to acid soils. It is otherwise easy to grow. Mix the soil thoroughly with aged manure and bone meal before planting. Plant seedlings 3 to 5 centimeters (1 to 2 inches) apart in rows 40 to 50 centimeters (16 to 20 inches) apart. Keep the plants supplied with adequate moisture, and their growing area free of weeds. Harvest the whole plant when a healthy crown of leaves develops.

Squash

NV (cooked zucchini): 12 calories; protein 1.0 gm; fats 0.1 gm; vit. A 300 I.U.; vit. B₁ 0.05 mg; vit. B₂ 0.08 mg; niacin 0.8 mg; vit. C 9 mg; fiber 0.6 gm; calcium 25 mg; phosphorus 25 mg; iron 0.4 mg; sodium 1 mg; potassium 141 mg.

All varieties of squash are warm-weather plants, and all are targets of a variety of pests. Thorough preparation of the soil before planting pays dividends in production and in the health of the plants. Mix compost and aged manure with the soil and heap the soil in small hills about 1.2 meters (4 feet) apart from one another. Plant four to five seeds in each hill and thin the seedlings to three after they are about 10 centimeters (4 inches) tall. Summer squashes (e.g., zucchini) mature in about two months, while winter squashes (e.g., acorn) can take twice as long to mature. Summer squashes should be harvested while very young—they can balloon seemingly overnight into huge fruits. Winter squashes should be harvested before the first frost; only clean, undamaged fruits will store well. Keep such squashes laid out in a cool, dry place until use and not piled on top of one another. Check them occasionally for the development of surface fungi.

Tomatoes

NV (raw, ripe): 22 calories; protein 1.1 gm; fats 0.2 gm; vit. A 900 I.U.; vit. B₁ 0.06 mg; vit. B₂ 0.04 mg; niacin 0.7 mg; vit. C 23 mg; fiber 0.5 gm; calcium 13 mg; phosphorus 27 mg; iron 0.5 mg; sodium 3 mg; potassium 244 mg.

These almost universally used fruits are easy to grow providing one understands a few basic aspects of their cultural requirements: (1) Many tomato plants normally will not initiate fruit development from their flowers when night temperatures drop below 14° C (57° F) or day temperatures climb above 40° C (104° F). For the earliest yields, seeds may be germinated indoors several weeks before the plants are to be placed outside, but little is accomplished by transplanting before the night temperatures begin to remain above 14° C (57° F); in addition, some growers insist that plants given an early start indoors do not always do as well later as those germinated outdoors. (2) Tomatoes require considerably more phosphorus than nitrogen from any fertilizers added to the soil where they are to be grown. Many inexperienced gardeners make the mistake of giving the plants lawn or general-purpose fertilizers that are proportionately high in nitrogen. As a result, the plants may grow vigorously but produce very few tomatoes. Give tomatoes bone meal, "tomato food" (Magamp is an excellent commercial slow-release preparation), or steer manure mixed with superphosphate. (3) Tomato plants seem more susceptible than most to soil fungi and to root-knot nematodes. The damage caused by these organisms may not become evident until the plants begin to bear. Then the lower leaves begin to wither, and yellowing progresses up the plant or there seems to be a general loss of vigor and productivity. Using disease- and nematode-resistant varieties (usually indicated by the letters V, F, and N on seed packets) is by far the simplest method of controlling these problems. Another effective control involves dipping the seedling roots in an emulsion of 0.25% corn oil in water when transplanting; experiments have shown that the corn oil greatly reduces root-knot nematode infestation. (4) Many garden varieties of tomatoes need to be staked to keep fruits off the ground where snails and other organisms can gain easy access to them. When using wooden stakes, be sure they have a diameter of 5 centimeters (2 inches) or more, and tie the plants securely to the stakes with plastic tape. Thinner stakes are likely to break or collapse when the plants grow to a height of 2 meters (6.5 feet) or more. Some growers prefer to use heavy wire tomato "towers" instead of stakes. (5) Hornworms and tomato worms almost invariably appear on tomato plants during the growing season. They can virtually strip a plant and ruin the fruits if not controlled. Fortunately control is simple and highly effective with the use of *Bacillus thuringiensis*, which was discussed in appendix 2. (6) The eating season for garden tomatoes can be extended for about two months past the first frost if all the green tomatoes are picked before frost occurs. Place the tomatoes on sheets of newspaper on a flat surface in a cool, dry place, where they will ripen slowly a few at a time. Generally the taste of tomatoes ripened this way is superior to that of hothouse tomatoes sold in produce markets. Be sure when picking the green tomatoes to handle them very gently, as they bruise very easily and molds quickly develop in the bruised areas.

VEGETATIVE PROPAGATION

The value of being able to take a piece of stem of a desirable plant from one locality to another and induce it to grow into a new plant identical to its "parent," or of being able to divide rhizomes, tubers, and other parts of plants and have them each develop into new plants has been recognized since ancient times. **Grafting,** which involves the permanent union of parts of two plants, also has its origins in antiquity. Plants with desirable features such as superior fruits or flowers or attractive leaves may be grafted on to related plants whose chief attraction may lie in a vigorous root system or one that is resistant to disease. The Chinese were apparently practicing grafting by the year 1000 B.C., and Theophrastus discussed grafting and other forms of vegetative propagation in his book *Causes of Plants,* written in the third century B.C. During the fourteenth and fifteenth centuries A.D. large numbers of plants were imported into European gardens and maintained by grafting. Even though the internal nature of the graft union was not fully understood, many forms of grafting were developed; no fewer than 119 different methods of grafting were described in 1821 by A. Thouin. Today the practice of various forms of vegetative propagation is almost universal and is an important part of the economies of the world. Some of the more widespread techniques of vegetative propagation are discussed in the following sections.

Stem Cuttings

Occasionally a basic change may occur in the chromosomes of a cell, and the result is a bud or a whole plant that develops differently from the rest of the tree or variety. Such a *mutation,* as it is called (see chapter 13), is likely to be undesirable or to result in the death of the cell or tissue. In a few instances, however, the mutation may produce superior characteristics such as better flowers or fruit, dwarf forms, or disease resistance. If you were to plant seeds from such a mutant form, the plant that would develop as a result could differ as much from the mutant parent as you do from either your mother or your father. Since pollen from one parent has to be transferred to a flower of another parent for fruits and seeds to form, the mutant plant carries characteristics from both parents and thus may pass on to future generations a variety of combinations of characteristics. When the mutation is desirable, the element of chance inheritance can be eliminated by cultivating tissue directly from the mutant plant, so

that each of the offspring has characteristics identical to those of the single parent; the desirable feature can thus be perpetuated indefinitely. Washington navel oranges, for example, have come from a mutation that appeared on a sweet orange tree growing near Bahia, Brazil, around 1820. The grower, whose identity is not known, apparently recognized the superior nature of the fruit on a branch of one of the trees and either by bud grafting or by making *cuttings* (pieces of a plant that are induced to produce roots and are then planted to grow on their own) multiplied the new form. Fifty years later, a missionary sent a dozen budded navel orange trees from Brazil to the United States Commissioner of Agriculture in Washington, D.C., and in 1873 two trees were shipped from Washington to Riverside, California. From these California trees, one of which is still living, almost the entire navel orange industry of the world has been developed through vegetative propagation.

A large number of plants can be started with cuttings from a few trees or shrubs in a relatively small amount of space and time, and little skill is required. Cuttings also do not have the problem of incompatibility, which is found with grafting. It is not always desirable to propagate by cuttings, however, particularly if the variety concerned does not have a sturdy, disease-resistant root system. Some plants are also very difficult to start from cuttings because they do not readily form roots. Next to planting seeds, however, cuttings have become the most widely used means of multiplying plants in the world today.

Most cuttings are made from stems that are 0.6 to 2.5 centimeters (0.25 to 1 inch) thick and contain at least two nodes. They may vary in length from 7.5 centimeters to 60 centimeters (3 inches to 2 feet). With deciduous plants, the cuttings are made—while the buds are dormant—from healthy wood of the previous season that has been growing in the sun. Pieces may be cut transversely, or a small part of wood from an older branch may be left attached at the bottom end. If a number of cuttings are being made at one time, they may be tied together with rubber bands, placed bottom ends up in cool, damp sand or sawdust from which water will drain easily, covered, and stored until spring. They should be checked from time to time to see if buds are developing. If they are, the cuttings should either be planted or refrigerated, as development of buds without corresponding root growth will kill the material. In the spring, the bottom ends of the cuttings can then be inserted 4 to 6 centimeters (3 to 4.5 inches) in damp sand (to minimize problems that may occur with soil fungi) or directly into the soil. Here adventitious roots develop, and normal growth should occur. Evergreen plants tend to produce adventitious roots much more slowly, taking from several months to a year to produce sufficient growth to become established. To keep the cuttings from drying out they need to be misted with water frequently, and they also need to be kept out of the sun and handled rapidly while they are being prepared. Leaves from the lower part of the cutting should be removed. If bottom heating is available, the additional warmth usually accelerates the development of the roots as long as high humidity can be maintained.

Leaf Cuttings

Houseplants such as African violets, peperomias, begonias, sansevierias, and others with somewhat succulent leaves can be propagated in various ways from the leaves (figure A4.1). One commonly used method is to cover a jar or pan of water with foil or plastic film, punch small holes in the covering, and insert the petioles so that they contact the water, which should be changed occasionally. After a few weeks, adventitious roots develop at the base of the petiole. The leaves may then be planted, leaving the blade exposed; new plants gradually appear from beneath the soil.

Rex begonias are sometimes propagated by punching out disks from the leaf blades with an instrument that resembles a cookie cutter; each disk should be about 1.9 centimeters (0.75 inch) in diameter. The disks are placed on damp filter paper in Petri dishes, where they develop both roots and shoots. Rex begonia leaves may also be induced to form new plants by making cuts about 3 millimeters (0.16 inch) deep across the main veins on the lower surface. Leaves are then placed lower side down on moist potting medium, and the edges of the leaf are pinned down with toothpicks. Eventually, new plants develop at the vein cuts, and the original leaf withers.

Other houseplants, most of the raspberry-blackberry group (excluding red raspberries), citrus plants, rhododendrons, and camellias may be propagated by cutting a short piece of stem containing a single node with a leaf and its axillary bud, and inserting it into a mixture of sand and peat moss, with the bud 1.3 centimeters (0.5 inch) or more below the surface. As with all evergreen or herbaceous plants, it is essential that the humidity be kept high during propagation. If a misting device is not available, the area should be covered with glass or polyethylene sheeting—allowing, however, for some air circulation—and the leaf surfaces should be sprayed occasionally with water.

Root Cuttings

A variety of plants may be propagated from root cuttings. The roots should be obtained after a growing season has been completed and the plants are dormant. The roots are cut into pieces from 2.5 to 7.5 centimeters (1 to 3 inches) long and placed in a potting medium or damp sand, either in a horizontal position or with the bottom end down. Since it is often difficult to tell the top end from the bottom once a root has been divided, some growers make transverse cuts at the top end and slanting cuts at the bottom end to avoid confusion.

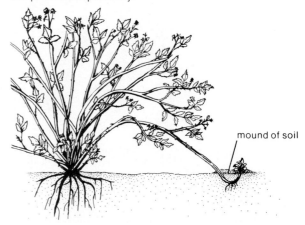

FIGURE A4.2   Tip layering. The tips of canes are bent over and covered with a small mound of soil. When a new plant has developed at the tip, it can be cut from the parent plant and planted independently.

mound of soil

The production of adventitious roots by cuttings is apparently brought about through the stimulation of auxin produced within the cells. Thus, it is common practice to add to the natural auxin of the plant by the application of either additional IAA or synthetic growth-promoting substances, particularly indole butyric acid (IBA) or naphthalene acetic acid (NAA), which often seem to be more effective than natural auxin in stimulating root growth. These substances are available at nurseries and florists in powder or paste form, and their use not only significantly increases the number of successful cuttings but also initiates more vigorous root growth. In the propagation of leaf disks, cytokinins have also been used. Since fungi are frequently a problem in vegetative propagation, fungicides have been found useful in inhibiting their growth.

Layering

Tip Layering

In blackberries, boysenberries, and other plants with somewhat flexible stems, the tips of the canes are bent over so that they touch the ground and they are covered with a small mound of soil. Roots form on the portion of buried stem, and eventually shoots appear. These new plants can then be separated from the parent stems. Variations of tip layering involve forcing a stem to lie horizontally and covering it with small mounds of soil at intervals, or heaping soil around the base of a plant so that the individual stems produce roots there. Once roots have been established, the individual plantlets or pieces of stem can be cut off from the original parent and grown independently (figure A4.2).

FIGURE A4.3   Steps in air layering. *A.* Nearly vertical cuts are made on the branch. *B.* Damp sphagnum moss is placed around the cut area. *C.* Polyethylene film is wrapped around the moss. *D.* The ends of the film are taped shut. Adventitious roots should develop in the cut area during ensuing weeks. (USDA Photos)

A.

B.

C.

D.

— stem

— plastic film wrapping

— moist sphagnum moss

Air Layering

Tropical trees and shrubs are sometimes propagated by wrapping sphagnum moss around a branch or main stem, dampening the moss with water, and then covering the area with polyethylene film. Adventitious roots eventually form on the portion of the stem covered by the moss. Sometimes the stem is girdled by removing about 2.5 centimeters (1 inch) of the bark all around the stem before the moss is applied. The moss should not be kept too wet or disease organisms may rot the stem (figure A4.3).

Propagation from Specialized Stems and Roots

In chapters 5 and 6, modified plant organs such as rhizomes, tubers, and corms were discussed. Virtually all of these may be divided, and each piece is capable of developing into a new plant (figure A4.4). If any piece of a rhizome or tuber containing at least one node is planted it should develop into an independent plant. Strawberry and saxifrage runners and the stolons of mints root at intervals and develop shoots above the roots. As soon as the new plants appear, they may be separated and grown independently. The scales of lily bulbs are quite fleshy and may be used in vegetative propagation. In late summer, while the weather is still warm, the scales are separated and planted in moist soil about 2.5 to 5 centimeters (1 to 2 inches) deep. Three to five bulblets usually form at the base of each scale in a few weeks; these can be transplanted the following spring. Bulbs of hyacinth, grape hyacinth, squills, and similar plants can be induced to form bulblets by inverting them and cutting down about a third of the way with a sterilized knife as though sectioning an apple or by scooping out the center with a sterilized melon baller. After they are cut, the bulbs are placed in dry sand in a warm place for about two weeks. During this time *callus* tissue (made up of undifferentiated cells) forms along the cuts. After the callus has formed, the bulbs are removed from the sand and stored in trays at room temperature under humid but well-aerated conditions. Small

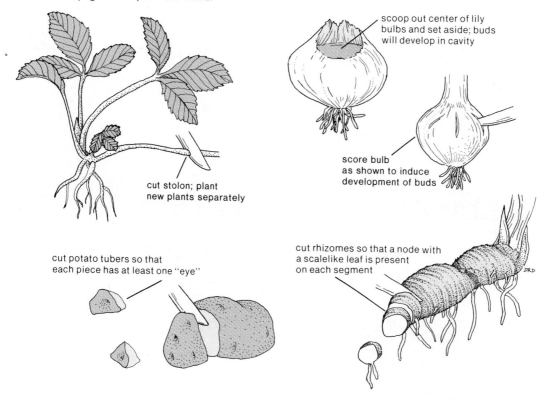

scoop out center of lily
bulbs and set aside; buds
will develop in cavity

score bulb
as shown to induce
development of buds

cut stolon; plant
new plants separately

cut potato tubers so that
each piece has at least one "eye"

cut rhizomes so that a node with
a scalelike leaf is present
on each segment

bulblets form along the cuts during the ensuing two or three months. These are left attached, and the whole original bulb is planted. After a year's growth, the bulblets are separated and replanted independently. In addition to using sterile instruments, it may be necessary to use fungicides to prevent fungal diseases from damaging the materials.

Some bulbs such as those of daffodils, tulips, and amaryllis have dry scales on the outside and fleshy scales on the inside. Unlike those of the lilies, these bulbs do not respond to being separated. These plants and plants with corms (e.g., gladioli) naturally produce bulblets or cormlets, however, which may be separated and grown independently. The number of bulblets produced appears to be correlative with depth in the soil. Most develop when the original bulb is about 7.5 centimeters (3 inches) deep. Few, if any, are produced when the bulb is more than 17 centimeters (8 inches) deep.

Plants with fleshy storage roots, such as sweet potatoes or dahlias, can be propagated by dividing the root so that each section has a shoot bud. This is done in early spring, just before planting, when shoot buds have appeared. Sweet potatoes can also be cut in pieces and induced to form new plants by

suspending each segment with toothpicks from the top of a jar partially filled with water. The segments form adventitious shoots and additional roots in a relatively short time.

Mericloning and Tissue Culture

Of all cultivated plants, orchids are among the most prized. They also produce the smallest seeds known, and it commonly takes seven years or more from seed until the first flower appears. Normally the seeds have to become associated with a specific fungus before they will germinate, and such fungi may not occur outside of the orchid's natural habitat. Orchid breeders have partially overcome the problem by substituting certain hormones and other substances for the fungus in artificial growing media. When they cross plants to produce hybrids (see chapter 13), they do not know if they have been successful in developing plants with superior characteristics until the orchids flower several years later. This is because of the variables inherited from the parents. Several techniques have, however, been developed that reduce by two years or more the time involved in producing mature orchid plants and that eliminate the often unpredictable inherited characteristics in plants grown from seeds. The techniques, which are

FIGURE A4.5 Orchid shoot meristem culture. The work is done in a chamber designed to minimize contamination. (Courtesy Rod McLellan Co., South San Francisco, California)

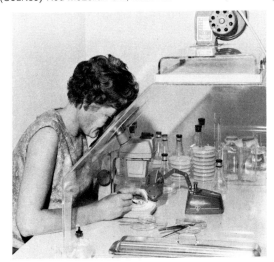

applied to a number of plants besides orchids, all involve aseptic practices (i.e., practices that seek to eliminate bacteria and other disease organisms) designed to isolate unique tissues with the potential to become new disease-free plants and to permit rapid multiplication of the materials.

The best results are obtained when the techniques are employed in a room or chamber that has been sterilized with ultraviolet light or has been steam-heated before work begins. All instruments, glassware, and media need to be sterilized in advance. An *autoclave*, which is something like a large pressure cooker, is generally used for this purpose. The ingredients for the media are measured very precisely. They usually include distilled water, various inorganic salts, organic substances such as yeast extracts, vitamins, and sugar.

Embryo Culture

In embryo culture, which is conducted under aseptic conditions, an embryo is removed from a carefully opened seed and placed on solidified agar in which nutrients have been dissolved. When shoots and roots appear, the seedlings are transferred to a sterilized growing medium.

Shoot Meristem Culture

The shoot meristem technique, also called **mericloning,** involves the careful removal of the apical meristem and the youngest one or two leaf primordia under aseptic conditions. A dissecting microscope is usually needed (figures A4.5 and A4.6). The meristem is then placed on a sterile medium and kept at room temperature under lights (figure A4.7).

FIGURE A4.6 A closeup view of cultured orchid tissue being divided. (Courtesy Rod McLellan Co., South San Francisco, California)

FIGURE A4.7 Modern laboratory for mericloning orchids. Each stoppered flask contains a piece of orchid meristem that is being cultured into a new plant. (Courtesy Rod McLellan Co., South San Francisco, California)

Adventitious roots should develop within a month. In orchid culture, the tissue that develops prior to the appearance of shoots and roots is subdivided as it grows, and each subdivision is then capable of becoming a new plant. By this process commercial growers produce thousands of plants, all genetically identical to the plant from which the meristem originated, in a relatively short time.

Tissue Culture

A *tissue culture* is essentially a mass of callus tissue growing on an artificial medium. It can be started from almost any part of the plant, although tissues taken from the vicinity of meristems usually produce the best results. With the proper media, the callus tissue eventually differentiates into shoots and roots. The technique has been used in the culture of tissues from carrot, tobacco, citrus, and other plants. See also the section on tissue culture in chapter 13.

GRAFTING

Simple grafting basically involves the insertion of a short portion of stem, called a **scion,** into another stem with a root system, the **stock** (figure A4.8). A segment of stem, called an *interstock,* is sometimes grafted between the stock and the scion, or a single bud with a little surrounding tissue may be grafted on to a stock. In all grafts, the cambia of the stock and scion must be placed tightly in contact with each other, and all parts being grafted must be related to one another (e.g., different varieties of apples may be grafted together, but it is not possible to graft elm or maple wood to an apple tree). After the graft is made, the fit may be tight enough for the graft not to need additional support, but it is customary to tie flat rubber strips, raffia, plastic tape, or waxed string around the union and cover it with grafting wax. The grafting wax is needed to prevent the tissues from drying and also to reduce the possibility of disease organisms gaining access to the interior of the stem. If the grafts are buried beneath the surface of the ground, they should be checked after growth begins to see if the wrapping is breaking down. If they are not below ground the wrapping should be removed when new growth begins to prevent the wrapping from hindering the increase in girth that occurs when the cambium adds new tissue.

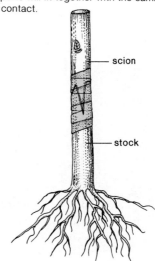

FIGURE A4.8   A simple graft. The stock (rooted portion) and scion (portion to be grafted onto the stock) are cut so that the two parts will fit together with the cambium of each part in close contact.

Whip or Tongue Grafting

Whip or *tongue grafting* is widely used for relatively small material, that is, wood between 0.7 and 1.25 centimeters (0.25 and 0.5 inch) in diameter. The stock and scion (the rooted portion and the aerial portion of the graft, respectively) should be as nearly identical in diameter as possible to bring about maximum contact between the cambia. The scion should contain two or three buds, and the cuts on both the stock and scion should be made in an internode. As shown in figure A4.9, a smooth tangential cut about 5 centimeters (2 inches) long is made with a sharp sterilized knife at the bottom of the scion and at the top of the stock. The angles of both cuts should also be as nearly identical as possible, and there should be no irregularities or undulations in the surfaces (such as those that can be produced by a dull cutting instrument). A second cut is then made in both the stock and scion about one-third of the distance from the tip of the cut surfaces. This cut is made back into the wood, nearly parallel to the first cut so that it forms a little tongue. The scion is then inserted into the stock as tightly as possible, taking care, however, not to force a split. In addition, the bottom edge of the scion should not protrude past the bottom of the cut of the stock. The process is completed by binding the materials and adding grafting wax.

If the stock and scion are not identical in diameter, it is still possible to obtain a graft if care is taken to bring the cambia in close contact along one edge of the cuts (figure A4.10).

FIGURE A4.9   Stages in whip grafting. *A.* A smooth tangential cut is made at the bottom of the scion and at the top of the stock. *B.* Cuts are made back into the wood of the stock and the scion. *C.* The cuts are widened to form a little tongue on each portion. *D.* The scion is inserted into the stock as tightly as possible without forcing a split; the graft is bound with rubber strips and sealed with grafting wax.

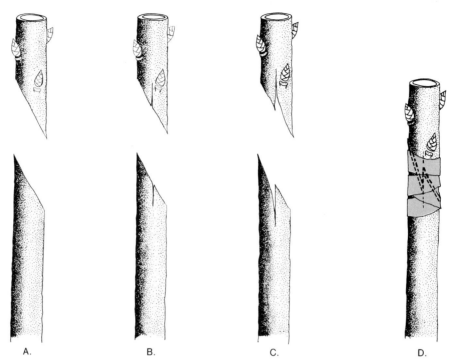

A.   B.   C.   D.

Splice Grafting

Splice grafting is sometimes used with plants where the pith is extensive. It is essentially the same as whip grafting, except that no second cut is made to form a tongue.

Cleft Grafting

Cleft grafting is a widespread method of grafting used routinely when the stock has a considerably greater diameter than that of the scion or scions. The stock branch or trunk is first cut at right angles, making sure the bark is not torn. If some bark is pulled loose by the saw, a new cut should be made. Commercial growers often minimize detachment of the bark by making a cut one-third of the way in on one side and then making a cut slightly lower on the opposite side. This usually leaves a surface with clean edges. Next, a meat cleaver or heavy knife is hammered 5.0 to 7.5 centimeters (2 to 3 inches) into the wood to make a vertical cut or shallow split. A wedge is then temporarily inserted into the cut to keep it open. Scions, usually 7.5 to 10 centimeters (3 to 4 inches) long, are inserted on each side of the cut toward the outer parts so that as much of the cambia

FIGURE A4.10   A whip graft, in which stock and scion are of different diameters.

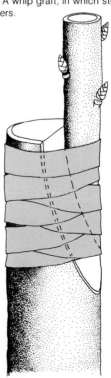

FIGURE A4.11 Stages in cleft grafting. *A.* The stock is cut transversely and a meat cleaver or a heavy knife is hammered into the wood to make a vertical cut, or split. *B.* A wedge is temporarily inserted into the vertical cut to keep it open. *C.* Scions are inserted into the vertical cut in the vicinity of the cambium and the wedge is removed. *D.* Exposed surfaces are sealed with grafting wax.

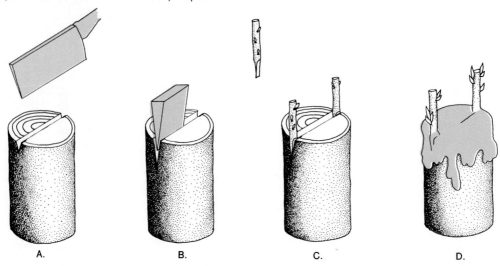

A.   B.   C.   D.

FIGURE A4.12 Stages in side grafting.

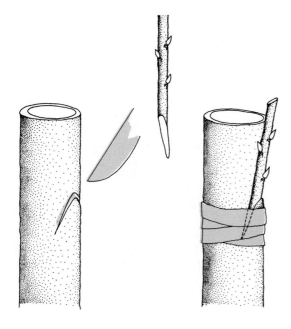

as possible are in contact. When the wedge is removed, the fit should be tight enough to prevent the scions from being easily pulled out by hand. Finally, any exposed surface is sealed with grafting wax (figure A4.11).

Side Grafting

Side grafting is often used with stocks that are about 2.5 centimeters (1 inch) in diameter. A cut about 2.5 centimeters (1 inch) deep is made with a heavy knife or chisel at an angle of 20 to 30 degrees to the surface of the stock. The bottom end of the scion, which should be 7.5 to 10.0 centimeters (3 to 4 inches) long and about 0.75 centimeter (0.25 inch) in diameter, is cut into a smooth wedge about 2.5 centimeters (1 inch) long. The stock is then bent slightly to open up the cut, and the scion is inserted, making certain that the maximum contact between each cambium is obtained; the pressure is then released to ensure a tight fit (figure A4.12). This is followed by sealing with grafting wax. Side grafting may be used to replace limbs lost through storm or other damage or for cosmetic purposes such as improving the symmetry of the plant.

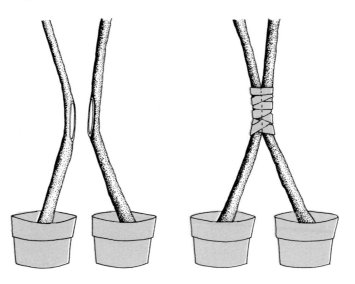

A variation of this graft, the *side tongue graft,* is often used with small broad-leaved evergreen plants. This graft involves slicing about 2.5 to 3.25 centimeters (1 to 1.5 inches) of stem out of the side toward the base to a depth of 0.75 centimeter (0.25 inch), followed by a second, smaller cut to form a tongue within the original cut. A scion, prepared in similar fashion to the side-grafted scion, is then inserted, and the graft is tied and sealed.

Approach Grafting

If two related plants tend not to form grafts very well by other means, *approach grafting* can be tried. Two independently growing plants, at least one of which is usually in a container, are prepared by making smooth cuts identical in length and depth at the same height on both stems. A tongue sometimes also is cut in both exposed areas, and then the two parts are fitted together, tied, and sealed (figure A4.13). This can be done at any time of the year but is most likely to be successful during periods of active growth. After union is achieved, the top of one plant may be cut off above the graft and the bottom of the other removed below the graft.

Inarching

Inarching is a variation of approach grafting sometimes used to save a valuable tree whose root system has been damaged. Young seedlings or trees of the same kind are planted around the base of the tree. When they have become established, the tip of each seedling, which should be 0.75 to 1.25 centimeters (0.25 to 0.5 inch) thick, is cut vertically for about 15 centimeters (6 inches) on the side nearest the main tree. Then vertical cuts of similar length should be made on the tree to the exact width of the prepared seedling tip and deep enough to expose the cambium; a small flap of bark should be left at the top to cover the tip of the seedling. Next, the prepared seedling tips are fit into the slots and nailed in with four to six flat-headed nails, and the entire area is sealed with grafting wax. If any side shoots develop from the seedlings after the grafting union has developed, they should be pruned off. Because the larger tree will be producing a considerable amount of food, the seedlings often grow very rapidly after successful inarching (figure A4.14).

FIGURE A4.14 Inarching. *A.* Established seedlings, which had been planted around the base of the tree, are cut vertically at their tips, and vertical slots of similar length are made in the tree adjacent to the seedling tips. *B.* The seedling tips are nailed into the slots. *C.* Growth of the grafted seedling bases may be very rapid if the inarching is successful.

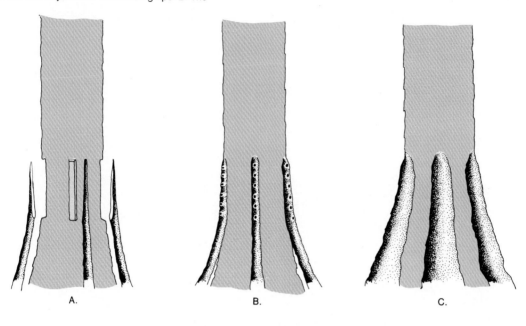

A.

B.

C.

FIGURE A4.15 Bridge grafting.

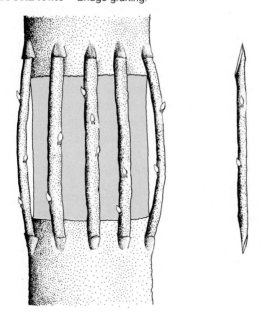

Bridge Grafting

Sometimes in temperate and colder regions, a particularly deep snowpack may prevent rodents and other animals from reaching their usual winter food. When this occurs, they may turn to the bark of trees and gnaw off a band of tissue, sometimes a decimeter or two wide. The damage usually extends through the phloem and cambium, as these tissues are the most palatable to the animals. This stripping of tissue is frequently referred to as **girdling** a tree and if left untreated will probably result in its death through starvation of the roots, since the phloem cannot conduct food past the damaged area. The tree can often be saved, however, through bridge grafting, particularly if it is done in the early spring just as new growth is beginning. The scions should be cut from dormant twigs of the same tree and kept in a refrigerator until needed. The damaged area should be cleaned and prepared by cutting out any remaining dead or tattered tissues. The scions are then inserted above and below the girdle, about 5.0 to 7.5 centimeters (2 to 3 inches) apart, with the natural bottom ends facing down and the tip ends up (figure A4.15). The graft will not succeed if a scion is put in upside down.

FIGURE A4.16   Budding (bud grafting). Leaves and side branches are removed from the stock below the point at which the graft is to be made. A "bud stick" is prepared by removing leaf material while leaving a short portion of each petiole. *A*. A bud is cut from the "bud stick" so that an oval patch of tissue surrounds it. *B*. A T-shaped incision is made in the bark of the stock. *C*. The bark is raised slightly at the corners. *D*. The bud and its surrounding tissue are inserted into the T-shaped incision on the stock. *E*. The flaps of bark are folded over the tissue, leaving the bud exposed, and the T tied shut with flat rubber strips. *F*. After growth begins, the stock is cut off just above the bud.

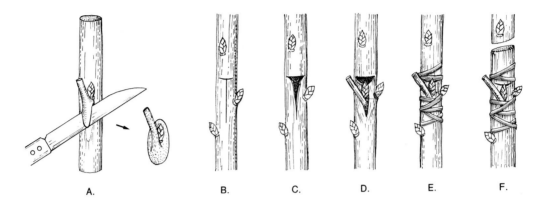

A.   B.   C.   D.   E.   F.

Bud Grafting

Bud grafting, or *budding*, is a form of grafting that utilizes a single bud. The method is widely used in commercial nurseries, partly because a single team of two to three workers can produce over a 1,000 such grafts a day and also because frequently better than 95% of the grafts are successful.

Budding is usually done in the summer when the season's axillary buds are mature and while the sap of the stocks is flowing freely. Budding is generally most successful when plump leaf buds (not flower or mixed buds) of the current season's growth are grafted to healthy stocks that are two to three years old.

The stock is prepared by removing all the leaves and side branches below and in the vicinity of the point at which the graft is to be made. Then a T-shaped incision is made through the bark with the aid of a sterilized, razor-sharp knife. The transverse cut should be roughly 1.25 centimeters (0.5 inch) wide and the vertical cut about 2.5 to 3.0 centimeters (1 to 1.2 inches) long. Both cuts should be no deeper than the cambium, and the bark should peel back easily at the junction of the two cuts.

A *bud stick* is prepared by removing all leaf material except for 1.25 centimeters (0.5 inch) of the petioles, which are left to serve as handles. Then a bud is carefully cut from the stick so that an oval piece of tissue about 1.75 centimeters (0.75 inch) in diameter surrounds it. If the bud separates from the tissue, it should be discarded and another one cut.

The bud and its oval shield is next inserted into the T-shaped cut of the stock, and the flaps of bark are folded over the shield, leaving the bud exposed. Flat rubber strips are used to tie the T shut so that only the bud remains visible. After growth begins, the stock is cut off just above the bud. The stock should not be cut off any earlier because the bud derives benefit from the transport of substances up and down the stock (figure A4.16).

Bud grafting can also be done in the spring, as soon as possible after growth of the stock begins but before the bud sticks become active. The bud material is frequently cut and stored in a refrigerator before growth begins. In areas with long growing seasons, bud grafting may also be done in early summer if the bark still peels back easily, but it should not be done any later because a young tree needs to produce sufficient growth before fall to be healthy and vigorous the following season.

In thick-barked trees, such as the Pará rubber tree and some of the nut trees, a rectangular patch of bark is cut out of the stock and replaced with a similar patch containing a bud. Normally the patch is not more than 5 centimeters (2 inches) in diameter. This method is slower than the other budding methods described but generally gives much better results in species with thick bark.

Root Grafting

Whip or tongue grafts with roots or pieces of roots used for stocks are sometimes used in the propagation of apple, pear, and other fruit trees. After the grafts have been made, they are usually stored in a cool, damp place for about two months and then refrigerated until early spring, when they are planted before growth starts. After growth begins, they are checked to make sure the scion is not producing its own roots, or the advantages of the original rootstock will be lost. These advantages may include disease resistance or dwarfing.

Double-Working

In *double-working* grafts, an *interstock* consisting of a stem segment varying in length from 2.5 centimeters to 30 centimeters (1 to 12 inches) or more is grafted between the stock and scion (i.e., three sections of stem are used for two grafts). Double-working grafts are used for special purposes. One such purpose is dwarfing, usually achieved by using special combinations of materials, but sometimes by this method of grafting. A young tree is cut off above the ground, an additional segment is cut to serve as the interstock, and the stock, interstock, and top (scion) are grafted together with the interstock inverted. This method will work only with certain varieties, as inversion normally effectively blocks the flow of materials up and down the stem. Other purposes of double-working grafts—in which the interstock is not inverted—include the influencing of growth so as to promote greater flower production than would otherwise occur, providing a disease- or cold-resistant trunk, or circumventing graft incompatibility (the failure of grafts to form permanent unions). With regard to this last purpose, if scions of certain varieties will not produce permanent unions when grafted to stocks of another variety but will form good grafts with a third variety, it may be possible to graft the third variety to the stock so that it can function as an interstock, thus circumventing the problem.

PRUNING

A good gardener or orchardist makes a habit of pruning trees, shrubs, and other plants regularly for a variety of reasons. He or she may wish to improve the quality and size of the fruits and flowers, restrict the size of the plants, keep the plants healthy, shape the shrubbery, or generally get more from the plants.

Except for spring-flowering ornamental shrubs, which should be pruned right after flowering, most maintenance pruning is done in the winter when active growth is not taking place. It usually involves removal of portions of stems, but it can also involve roots. When a terminal bud is removed, the axillary or lateral bud just below the cut will usually develop into a branch, and a bushier growth will result. Some gardeners pinch off terminal buds routinely to encourage such growth. The following sections provide a few generalities and specifics pertaining to several types of plants.

Fruit Trees

When young fruit trees are first planted, all except four or five stems and any damaged roots should be pruned off. The remaining stems should be cut back so that there is one central leader about 1 meter (3 feet) tall, with shorter side branches facing in different directions (figure A4.17). When the stems are cut, one should be careful to cut in such a way that the axillary or lateral bud just below the cut is facing outward. Each succeeding year, new growth should be pruned back to within a few centimeters of the previous year's cut. Any dead or diseased branches should be removed, and stems that have grown so that they are rubbing against each other should be pruned. The central leader should be cut out the second year in peach trees so that the interior of the tree is left relatively open. Any sucker shoots that develop from the base or along the trunk of the tree should be regularly removed.

Grapevines

There are several methods of pruning grapevines, depending on the type of vine and the circumstances under which the vines are being grown. In general, grapevines should be pruned heavily in late winter for best fruit production. After a central "trunk" has been allowed to develop, each shoot, regardless of its length, should be cut back so that no more than three axillary buds remain. Exceptions to this rule involve situations in which the vines are trained on arbors or wires, when the shoots initially may be allowed to grow longer. Even then, however, after the desired training has taken place (figure A4.18), pruning should be heavy for best results.

FIGURE A4.17   A young peach tree. *A*. Before pruning.
B. After pruning.

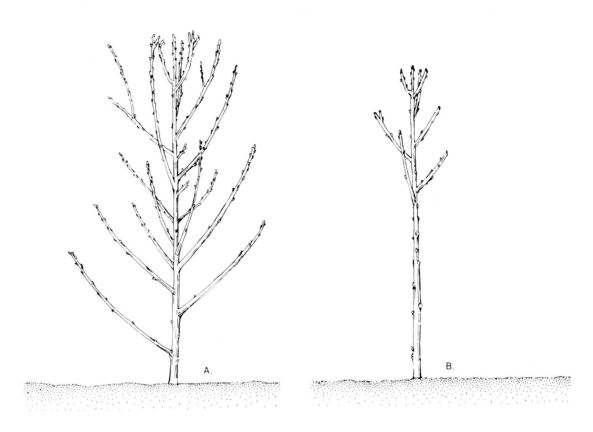

FIGURE A4.18   A grapevine, *A*. Before pruning. *B*. After
pruning.

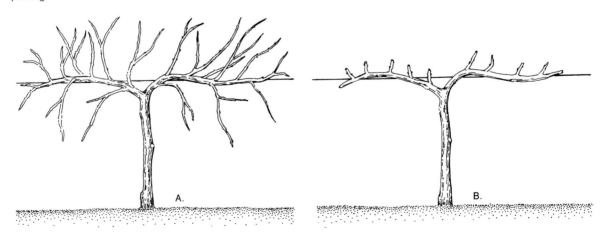

Roses

Rose bushes should be pruned heavily—they will recover! In general, new stems should be cut back to within 10 to 20 centimeters (4 to 8 inches) of their point of origin, with care taken that the top remaining axillary bud of each stem is pointing outward. This promotes growth that leaves the center open for better air circulation. Any dead or diseased canes should be removed and the number of remaining canes limited to three or four per plant.

Raspberries, Blackberries, and Their Relatives

Berry canes are biennial. They are produced from the base the first year, branch during the summer, and usually produce fruit on the branches the second year, although in milder climates they may also produce fruit on the first year's growth. The canes die after the second year.

Old, dead canes should be removed and all but three or four canes developing from each crown should be pulled out when the ground is soft. New canes should be cut back to lengths of 1 meter (3 feet) or less in the spring. Branches of one-year-old canes should be cut back in early spring to lengths of about 30 centimeters (12 inches) for larger fruits.

Bonsai

Container-grown trees that are dwarfed through the constant careful pruning of both roots and stems, the manipulation of soil mixtures, and the weighting of branches are called **bonsai** (figure A4.19). Some of these dwarfed trees attain ages of 50 to 75 years or more and may be less than 1 meter (3 feet) tall. Bonsai is an art that requires knowledge of the environmental requirements and tolerances of individual species.

In general, growers involved in bonsai pinch out new growth above a bud every few days during a growing season but never prune when a plant is dormant. The reader is referred to the Additional Reading for more information on the subject.

ADDITIONAL READING

Adams, C. F. 1981. *Nutritional value of American foods in common units.* U. S. Government Printing Office.

Arnold, J. 1983. *The illustrated encyclopedia of house plants.* New York: Dodd, Mead and Company.

Bailey Hortorium Staff. 1976. *Hortus third: a concise dictionary of plants cultivated in the United States and Canada.* New York: Macmillan Co.

Baker, J. 1985. *Jerry Baker's happy, healthy house plants.* New York: New American Library.

Bonar, A. 1982. *How to book of herbs and herb gardening.* New York: Sterling Publishing Co., Inc.

Davidson, W. 1982. *The house plant survival manual: how to keep your houseplants.* New York: W. H. Smith Publications.

Grounds, R. 1984. *Growing vegetables and herbs.* Woodstock, NY: Beekman Publishers, Inc.

Hartmann, H. T., and D. E. Kester. 1983. *Plant propagation: principles and practices*, 4th ed. Englewood Cliffs, NJ: Prentice-Hall.

Hill, L. 1985. *Secrets of plant propagation: starting your own flowers, vegetables, fruits, berries, shrubs, trees and houseplants.* Pownal, VT: Garden Way Publishing/Storey Commercial, Inc.

Kramer, J. 1984. *An illustrated guide to flowering house plants.* New York: Arco Publishing, Inc.

Lesniewicz, P. 1984. *Bonsai: the complete guide to art and technique.* Englewood Cliffs, NJ: Sterling Publishing Co., Inc.

MacCaskey, M., and R. L. Stebbins. 1982. *Pruning: how to guide for gardeners.* Tucson, AZ: HP Books.

Seddon, G., et al. 1985. *The essential guide to perfect houseplants.* New York: Summit Books.

Sunset Editors. 1983. *Pruning handbook.* Menlo Park, CA: Lane Publishing Co.

Sunset Editors. 1983. *House plants*, 4th ed. Menlo Park, CA: Lane Publishing Co.

Sunset Editors. 1977. *How to grow orchids.* Menlo Park, CA: Lane Publishing Co.

Wyman, D. 1987. *The gardening encyclopedia*, updated. New York: Macmillan Co.

Young, D. 1985. *Bonsai: the art and technique.* Englewood Cliffs, NJ: Prentice-Hall.

Appendix 5   *Metric Equivalents*

Metric System of Measurement

Application	International System of Units	English System Equivalents
Length	kilometer	0.62137 miles
	meter	39.37 inches
	centimeter	0.3937 inch
	millimeter	0.03937 inch
	micrometer	0.00003937 inch
	nanometer	0.00000003937 inch
	angstrom	0.0000000003937 inch
Mass (Weight)	metric ton	2,200 pounds
	kilogram	2.2 pounds
	gram	0.03527 ounce
	milligram	0.00003527 ounce
Volume	liter	1.06 quart
	milliliter	0.00106 quart
	cubic meter	35.314 cubic feet
	cubic centimeter	0.061 cubic inch
Area	hectare	2.471 acres
Temperature	To convert Celsius to Fahrenheit, multiply the Celsius figure by 9, divide the total by 5, and add 32.	

Temperature Conversion Scale

°F   °C

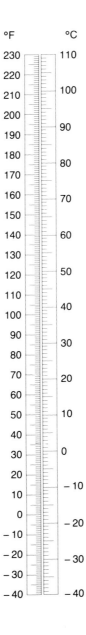

To convert Fahrenheit to Celsius use the
following formula:
$C = \frac{5}{9}(°F - 32)$

To convert Celsius to Fahrenheit use the
following formula:
$F = \frac{9}{5}(°C + 32)$

Glossary

a

abscisic acid (ab-siz'ik as'id) (ABA) a growth-inhibiting hormone of plants; it is involved with other hormones in dormancy *(p. 205)*

abscission (ab-sizh'un) the separation of leaves, flowers, and fruits from plants after the formation of an abscission zone at the base of their petioles, peduncles, and pedicels *(p. 124)*

accessory pigments (ak-ses'uh-ree pig'mentz) pigments that capture light energy and transfer it to chlorophyll *a* *(p. 181)*

achene (uh-keen') a single-seeded fruit in which the seed is attached to the pericarp only at its base *(p. 145)*

acid (as'id) a substance that dissociates in water, releasing hydrogen ions *(p. 20)*

active transport (ak'tiv trans'port) the expenditure of energy by a cell in moving a substance across a plasma membrane against a diffusion gradient *(p. 162)*

adventitious (ad-ven-tish'uss) said of buds developing in internodes or on roots, or of roots developing along stems or on leaves *(p. 83)*

aerobic respiration (air-oh'bik res-puh-ray'shun) respiration that requires free oxygen *(pp. 188, 190)*

after-ripening (af'tur-ry'pen-ing) metabolic changes that must take place in a dormant seed before germination can occur *(p. 151)*

agar (ah'gur) a gelatinous substance produced by certain red algae; it is often used as a culture medium, particularly for bacteria *(p. 201)*

aggregate fruit (ag'gruh-git froot) a fruit derived from a single flower having several to many pistils *(p. 140)*

albuminous cell (al-byu'min-uss sel) a gymnosperm phloem parenchyma cell closely associated with an adjacent sieve cell *(p. 409)*

algin (al'jin) a gelatinous substance produced by certain brown algae; it is used in a wide variety of food substances and in pharmaceutical, industrial, and household products *(p. 317)*

Alternation of Generations (ol-tur-nay'shun uv jen-ur-ay'shunz) alternation between a haploid gametophyte phase and a diploid sporophyte phase in the life cycle of sexually reproducing organisms *(p. 233)*

amino acid (ah-mee'noh as'id) one of the organic, nitrogen-containing units from which proteins are synthesized *(p. 22)*

anaerobic respiration (an-air-oh'bik res-puh-ray'shun) respiration in which the hydrogen removed from the glucose during glycolysis is combined with an organic ion (instead of oxygen) *(p. 188)*

angiosperm (an'jee-oh-spurm) a plant whose seeds develop within ovaries that mature into fruits *(p. 428)*

annual (an'you-ul) a plant that completes its entire life cycle in a single growing season *(pp. 74, 134)*

annual ring (an'you-ul ring) a single season's production of xylem (wood) by the vascular cambium *(p. 76)*

annulus (an'yu-luss) a specialized layer of cells around a fern sporangium; it aids in spore dispersal through a springlike action; also a membranous ring around the stipe of a mushroom *(p. 396)*

anther (an'thur) the pollen-bearing part of a stamen *(p. 136)*

antheridiophore (an-thur-id'ee-oh-for) a stalk that bears an antheridium *(p. 368)*

antheridium (pl. antheridia) (an-thur-id'ee-um; pl. an-thur-id'ee-ah) the male gametangium of certain algae, fungi, bryophytes, and vascular plants other than gymnosperms and angiosperms *(pp. 313, 368)*

anthocyanin (an-thoh-sy'ah-nin) a water-soluble pigment found in cell sap; anthocyanins vary in color from red to blue *(p. 45)*

antibiotic (an-tee-by-ot'ik) a substance produced by a living organism that interferes with the normal metabolism of another living organism *(p. 352)*

apical dominance (ay'pi-kul dom'i-nunts) suppression of growth of lateral buds by hormones *(p. 205)*

apical meristem (ay'pi-kul mair'i-stem) a meristem at the tip of a shoot or root *(pp. 57, 96)*

apomixis (ap-uh-mik'sis) reproduction without fusion of gametes or meiosis in otherwise normal sexual structures *(p. 252)*

archegoniophore (ahr-kuh-goh'nee-oh-for) a stalk bearing an archegonium *(p. 368)*

archegonium (pl. archegonia) (ahr-kuh-goh'nee-um; pl. ahr-kuh-goh'nee-ah) the multicellular female gametangium of bryophytes and most vascular plants other than angiosperms *(p. 368)*

aril (air'il) an often brightly colored appendage surrounding the seed of certain plants *(p. 413)*

ascospore (as'ko-spor) a spore produced in an ascus *(p. 340)*

ascus (pl. asci) (as'kus; pl. as'eye) one of often numerous, frequently fingerlike hollow structures in which the fusion of two haploid nuclei is followed by meiosis; a row of ascospores (usually eight) is ultimately produced in each ascus on or within the sexually initiated reproductive bodies of cup (sac) fungi *(p. 339)*

asexual reproduction (ay-seksh'yule ree-proh-duk'shun) any form of reproduction not involving the union of gametes *(p. 226)*

assimilation (uh-sim-i-lay'shun) cellular conversion of raw materials into protoplasm and cell walls *(p. 17)*

atom (at'um) the smallest individual unit of an element that retains the properties of the element *(p. 18)*

ATP (ay-tee-pee) adenosine triphosphate, a cell molecule with three phosphate groups; it is the principal vehicle for energy storage and exchange in cell metabolism *(p. 181)*

autotrophic (aw-toh-troh'fik) descriptive of an organism capable of sustaining itself through conversion of inorganic substances to organic material *(p. 281)*

auxin (awk'sin) a growth-regulating substance produced either naturally by plants or synthetically *(p. 201)*

axil (ak'sil) the angle formed between a twig and the petiole of a leaf; normally the site of an *axillary bud* (also called *lateral bud*) *(p. 71)*

b

backcross (bak'kross) a cross involving a hybrid and one of its parents *(p. 243)*

bacteriophage (bak-teer'ee-oh-fayj) a virus whose host is a bacterium *(p. 295)*

bark (bahrk) tissues of a woody stem between the vascular cambium and the exterior *(p. 80)*

base (bayss) a substance that dissociates in water, releasing hydroxyl (OH⁻) ions *(p. 20)*

basidiospore (bah-sidd'ee-oh-spor) a spore produced on a basidium *(p. 345)*

basidium (pl. basidia) (buh-sid'ee-um; pl. buh-sid'ee-ah) one of usually numerous, frequently club-shaped hollow structures in which the fusion of two haploid nuclei is followed by meiosis, the four resulting nuclei becoming externally borne basidiospores; basidia are produced on or within sexually initiated reproductive bodies of the club fungi (e.g., mushrooms, puffballs) *(p. 344)*

berry (bair'ee) a thin-skinned fruit that usually develops from a compound ovary and commonly contains more than one seed *(p. 139)*

biennial (by-en'ee-ul) a plant that normally requires two seasons to complete its life cycle, the first season's growth being strictly vegetative *(pp. 105, 134)*

biological controls (by-oh-loj'i-kull kun-trohlz') the use of natural enemies and inhibitors in combating insect pests and other destructive organisms *(pp. 101, A–25)*

biome (by'ohm) similar biotic communities considered on a worldwide or at least continental scale (e.g., desert biome, grassland biome) *(p. 484)*

biotic community (by-ot'ik kuh-myu'nit-ee) an association of plants, animals, and other organisms (e.g., woodland) *(p. 472)*

bivalent (biv'uh-lint or by-vay'lent) an associated pair of homologous chromosomes as seen in prophase I of meiosis *(p. 227)*

blade (blayd) the conspicuous, flattened part of a leaf (also called *lamina*) *(pp. 111, 316)*

bond (bond) a force that holds atoms together *(p. 19)*

bonsai (bon-sy') container-grown plants (usually trees) that have been dwarfed artificially through skillful pruning and manipulation of the growing medium *(p. A–84)*

botany (bot'an-ee) science involving the study of plants *(p. 6)*

botulism (bot'yu-lizm) poisoning from consumption of food infected by botulism bacteria *(p. 285)*

bract (brakt) a structure that is usually leaflike and modified in size, shape, or color *(p. 120)*

bryophyte (bry'oh-fyt) a photosynthetic, terrestrial, aquatic, or epiphytic plant without xylem and phloem (e.g., mosses, liverworts, hornworts) *(p. 364)*

budding (bud'ing) a form of grafting in which the scion is a bud surrounded by a small shield of tissue; also a form of asexual reproduction in which a new cell develops to full size from a protuberance arising from a mature cell, as in yeasts *(pp. 339, A–81)*

bulb (buhlb) an underground food-storage organ that is essentially a modified bud consisting of fleshy leaves that surround and are attached to a small stem *(p. 83)*

bundle scar (bun'dul skahr) a small scar left by a vascular bundle within a leaf scar when the leaf separates from its stem through abscission *(p. 71)*

bundle sheath (bun'dul sheeth) the parenchyma and/or sclerenchyma cells surrounding a vascular bundle *(p. 115)*

c

callose (kal'ohs) a complex carbohydrate that develops in sieve tubes following an injury; it is commonly associated with the sieve areas of sieve-tube elements *(p. 16)*

callus (kal'uss) undifferentiated tissue that develops around injured areas of stems and roots; also the undifferentiated tissue that develops during tissue culture *(p. 16)*

Calvin cycle (kal'vin sy'kuhl) see *dark reactions*

calyptra (kuh-lip'truh) tissue from the enlarged archegonial wall of many mosses that forms a partial or complete cap over the capsule *(p. 370)*

calyx (kay'liks) collective term for the sepals of a flower *(p. 134)*

cambium (kam'bee-um) a meristem producing secondary tissues; see *vascular cambium, cork cambium* *(p. 48)*

capillary water (kap'i-lair-ee waw'tur) water held in the soil against the force of gravity *(p. 170)*

capsule (kapp'sool) a dry fruit that splits in various ways at maturity, often along or between carpel margins; also the main part of a sporophyte, in which different types of tissues develop *(pp. 145, 370)*

carbohydrate (kahr-boh-hy'drayt) an organic compound containing carbon, hydrogen, and oxygen, with twice as many hydrogen as oxygen atoms per molecule *(p. 21)*

carpel (kahr'pul) an ovule-bearing unit that is a part of a pistil *(pp. 137, 428)*

caryopsis (kare-ee-op'siss) a dry fruit in which the pericarp is tightly fused to the seed; it does not split at maturity *(p. 145)*

Casparian strip (kass-pair'ee-un strip) the band of suberin around the radial and transverse walls of an endodermal cell *(p. 98)*

cell (sel) the basic structural and functional unit of living organisms; in plants it consists of protoplasm usually surrounded by a cell wall *(p. 34)*

cell biology (sel by-ol'uh-jee) the biological discipline involving the study of cells and their functions *(p. 10)*

cell cycle (sel sy'-kul) sequence of events involved in the division of a cell *(p. 46)*

cell membrane (sel mem'brayn) see *plasma membrane*

cell plate (sel playt) the precursor of the middle lamella; it forms at the equator during telophase *(p. 49)*

cell sap (sel sap) the liquid contents of a vacuole *(p. 45)*

cellulose (sel'yu-lohss) a complex, insoluble carbohydrate that constitutes the principal component of plant cell walls *(p. 21)*

cell wall (sel wawl) the relatively rigid boundary of cells of plants and certain other organisms *(p. 38)*

centromere (sen'truh-meer) the dense constricted portion of a chromosome to which a spindle fiber is attached (also called *kinetochore*) *(p. 48)*

chiasma (pl. chiasmata) (ky-az'mah; pl. ky-az'mah-tah) the X-shaped configuration formed by two chromatids of homologous chromosomes as they remain attached to each other during prophase I of meiosis *(p. 227)*

chemiosmosis (kem-ee-oz-moh'siss) a theory that energy is provided for phosphorylation by protons being "pumped" across inner mitochondrial and thylakoid membranes *(p. 193)*

chlorenchyma (klor-en'kuh-mah) tissue composed of parenchyma cells that contain chloroplasts *(p. 58)*

chlorophyll (klor'uh-fil) green pigments essential to photosynthesis *(p. 42)*

chloroplast (klor'uh-plast) an organelle containing chlorophyll, found in cells of most photosynthetic organisms *(p. 42)*

chromatid (kroh'muh-tid) one of the two strands of a chromosome; they are united by a centromere *(p. 227)*

chromatin (kroh'muh-tin) a readily staining complex of DNA and proteins found in chromosomes *(p. 45)*

chromoplast (kroh'muh-plast) a plastid containing pigments other than chlorophyll; the pigments are usually yellow to orange *(p. 42)*

chromosome (kroh'muh-sohm) a body consisting of a linear sequence of genes and composed of DNA and proteins; chromosomes are found in cell nuclei and appear in contracted form during mitosis and meiosis *(p. 45)*

cilium (pl. cilia) (sil'ee-um; pl. sil'ee-uh) a short hairlike structure usually found on the cells of unicellular aquatic organisms, normally in large numbers and arranged in rows; the most common function of cilia is propulsion of the cell *(p. 44)*

circadian rhythm (sur-kay'dee-an rith'um) a daily rhythm of growth and activity found in living organisms *(p. 212)*

cladophyll (klad'uh-fil) a flattened stem that resembles a leaf *(p. 83)*

clamp connection (klamp kuh-nek'shun) a looplike connection between adjacent cells of a dikaryotic hypha of a club fungus *(p. 344)*

class (klas) a category of classification between a division and an order *(pp. 264, 266)*

climax vegetation (kly'maks vej-uh-tay'shun) vegetational association that perpetuates itself indefinitely at the culmination of ecological succession *(p. 481)*

coenocytic (see-nuh-sit'ik) having many nuclei not separated from one another by crosswalls, as in the hyphae of water molds *(p. 334)*

cohesion-tension theory (koh-hee'zhun-ten'shun thee'uh-ree) theory that explains the rise of water in plants through a combination of cohesion of water molecules in capillaries and tension on the water columns brought about by transpiration *(p. 163)*

coleoptile (koh-lee-op'tul) a protective sheath surrounding the emerging shoot of seedlings of the Grass Family (Poaceae) (e.g., corn, wheat) *(p. 150)*

coleorhiza (koh-lee-uh-ry'zuh) a protective sheath surrounding the emerging radicle (immature root) of members of the Grass Family (Poaceae) (e.g., corn, wheat) *(p. 150)*

collenchyma (kuh-len'kuh-muh) tissue composed of cells with unevenly thickened walls *(p. 59)*

colloid (kol'oyd) substance consisting of a medium in which fine particles are permanently dispersed *(p. 169)*

community (kuh-myu'nit-ee) collective term for all the living organisms sharing a common environment and interacting with one another *(p. 472)*

companion cell (kum-pan'yun sel) a specialized cell derived from the same parent cell as the closely associated sieve-tube element immediately adjacent to it (in angiosperm phloem) *(p. 65)*

compound (kom'pownd) a substance whose molecules are composed of two or more elements *(p. 19)*

compound leaf (kom'pownd leef) a leaf whose blade is divided into distinct leaflets *(p. 111)*

conidium (pl. conidia) (kuh-nid'ee-um; pl. kuh-nid'ee-uh) an asexually produced fungal spore formed outside of a sporangium *(p. 339)*

conifer (kon'i-fur) a cone-bearing tree or shrub *(p. 408)*

conjugation (kon-juh-gay'shun) a process leading to the fusion of isogametes in algae, fungi, and protozoa; also, the means by which certain bacteria exchange DNA *(p. 312)*

conjugation tube (kon-juh-gay'shun t(y)oob) a tube permitting transfer of a gamete or gametes between adjacent cells, as in *Spirogyra* or desmids *(p. 311)*

cork (kork) tissue composed of cells whose walls are impregnated with suberin at maturity; the outer layer of tissue of an older woody stem; cork is produced by the cork cambium *(p. 66)*

cork cambium (kork kam'bee-um) a narrow cylindrical sheath of cells between the exterior of a woody root or stem and the central vascular tissue; it produces *cork* to its exterior and *phelloderm* to its interior; it is also called *phellogen* *(pp. 48, 57, 73)*

corm (korm) a vertically oriented, thickened food-storage stem that is usually enveloped by a few papery nonfunctional leaves *(p. 83)*

corolla (kuh-rahl'uh) collective term for the petals of a flower *(p. 134)*

cortex (kor'teks) a primary tissue composed mainly of parenchyma; the tissue usually extends between the epidermis and the vascular tissue *(pp. 71, 97)*

cotyledon (kot-uh-lee'dun) an embryo leaf ("seed leaf") that usually either stores or absorbs food *(pp. 74, 150)*

covalent bond (koh-vay'luhnt bond) a force provided by pairs of electrons that travel between two or more atomic nuclei, holding atoms together and keeping them at a stable distance from each other *(p. 19)*

crossing over (kross'ing oh'vur) the exchange of corresponding segments of chromatids between homologous chromosomes during prophase I of meiosis *(p. 227)*

crozier (kroh'zhur) the spirally coiled "fiddlehead" of an immature fern frond; also, a hook-shaped hyphal tip that precedes the development of an ascus in sac fungi *(p. 396)*

cuticle (kyut'i-kul) a waxy or fatty layer of varying thickness on the outer walls of epidermal cells *(pp. 61, 114)*

cutin (kyu'tin) the waxy or fatty substance of which a cuticle is composed *(pp. 61, 114)*

cyclosis (sy-kloh'sis) the flowing or streaming of cytoplasm within a cell *(p. 46)*

cytochrome (sy'toh-krohm)   iron-containing pigment involved in molecule transfer in an electron transport system   *(p. 185)*

cytokinesis (sy-toh-kuh-nee'sis)   cell division   *(p. 46)*

cytokinin (syt-uh-ky'nin)   a growth hormone involved in cell division and several other metabolic activities of cells   *(p. 204)*

cytology (sy-tol'uh-jee)   see *cell biology*

cytoplasm (sy'tuh-plazm)   the protoplasm of a cell exclusive of the nucleus   *(p. 37)*

cytoplasmic streaming (sy-tuh-plaz'mik streem'ing)   see *cyclosis*

d

dark reactions (dahrk ree-ak'shunz)   a cyclical series of chemical reactions that utilizes carbon dioxide and energy generated during the light reactions of photosynthesis, producing sugars, some of which are stored as insoluble carbohydrates while others are recycled; the reactions are independent of light and occur in the stroma of chloroplasts   *(p. 181)*

day-neutral plant (day new'trul plant)   a plant that is not dependent on specific daylengths for the initiation of flowering   *(p. 217)*

deciduous (di-sij'yu-wuss)   shedding leaves annually   *(pp. 71, 124)*

decomposer (dee-kuhm-poh'zur)   organism (e.g., bacterium, fungus) that breaks down organic material to forms capable of being recycled   *(p. 474)*

development (di-vel'up-ment)   changes in the form of a plant resulting from growth and differentiation of its cells into tissues and organs   *(p. 200)*

dicotyledon (dy-kot-uh-lee'dun)   a class of angiosperms whose seeds commonly have two cotyledons; the term is frequently abbreviated to *dicot*   *(p. 74)*

dictyosome (dik'tee-uh-sohm)   see *Golgi body*

differentially permeable membrane (dif-uh-rensh'uh-lee pur'mee-uh-bul mem'brayn)   a membrane through which different substances diffuse at different rates   *(p. 159)*

differentiation (dif-uh-ren-shee-ay'shun)   the change of a relatively unspecialized cell to a more specialized one (e.g., the change of a cell just produced by a meristem to a vessel element or fiber)   *(p. 200)*

diffusion (dif-fyu'zhin)   the random movement of molecules or particles from a region of higher concentration to a region of lower concentration, ultimately resulting in uniform distribution   *(p. 158)*

digestion (dy-jes'jin)   enzyme-controlled conversion of complex, usually insoluble substances to simpler, usually soluble substances   *(pp. 17, 21, 194)*

dihybrid cross (dy-hy'brid kross)   a cross involving two different pairs of chromosomes in the parents   *(p. 243)*

dikaryotic (dy-kair-ee-ot'ik)   having a pair of nuclei in each cell or a type of the mycelium in club fungi   *(p. 344)*

dioecious (dy-ee'shuss)   having unisexual flowers or cones, with the male flowers or cones confined to certain plants and the female flowers or cones of the same species confined to other different plants   *(p. 436)*

diploid (dip'loyd)   having two sets of chromosomes in each cell; the 2n chromosome number characteristic of the sporophyte generation   *(p. 232)*

diuretic (dy-yu-ret'ik)   a substance tending to increase the flow of urine   *(p. 392)*

division (duh-vizh'un)   the largest category of classification of organisms within a kingdom; equivalent of phylum in animals   *(p. 266)*

DNA (dee-en-ay)   standard abbreviation of deoxyribonucleic acid, the carrier of genetic information in cells and viruses   *(p. 24)*

dominance (dom'uh-nintz)   the Mendelian principle that one member of a pair of genes may mask or inhibit the expression of the other   *(p. 240)*

dominant (dom'uh-nint)   the member of a pair of genes that masks or suppresses the phenotypic expression of the recessive gene   *(p. 240)*

dormancy (dor'man-see)   a period of growth inactivity in seeds, buds, bulbs and other plant organs even when environmental conditions normally required for growth are met   *(pp. 151, 219)*

double fusion (dub'ul fu'shun)   the more or less simultaneous union of one sperm and egg (forming a zygote) and union of another sperm and polar nuclei (forming a primary endosperm nucleus) that occurs in the embryo sac of flowering plants   *(p. 433)*

drupe (droop)   a simple fleshy fruit whose single seed is enclosed within a hard endocarp   *(p. 138)*

e

ecology (ee-kol'uh-jee)   the biological discipline involving the study of the relationships of organisms to each other and to their environment   *(p. 472)*

ecosystem (ee'koh-sis-tim)   a system involving interactions of living organisms with one another and with their nonliving environment   *(p. 472)*

egg (eg)   a nonmotile female gamete   *(pp. 226, 313)*

elater (el'uh-tur)   straplike appendage (usually occurring in pairs) attached to a horsetail (*Equisetum*) spore; also, a somewhat spindle-shaped sterile cell occurring in large numbers in liverwort sporangia; both types of elaters facilitate spore dispersal   *(pp. 370, 392)*

electron (ee-lek'tron)   a negatively charged particle of an atom   *(p. 18)*

element (el'uh-mint)   one of 107 types of matter, most existing naturally but some man-made, each of which is composed of one kind of atom   *(p. 17)*

embryo (em'bree-oh)   immature sporophyte that develops from a zygote within an ovule or archegonium after fertilization   *(pp. 94, 370)*

embryo sac (em'bree-oh sak)   the female gametophyte of angiosperms, which, in approximately 70% of the species investigated, contains eight nuclei   *(p. 429)*

enation (ee-nay'shun)   one of the tiny green leaflike outgrowths on the stems of whisk ferns (*Psilotum*)   *(p. 380)*

endocarp (en'doh-kahrp)   the innermost layer of a fruit wall   *(p. 137)*

endodermis (en-doh-dur'mis)   a single layer of cells surrounding the vascular tissue (stele) in roots and some stems; many of the cells have Casparian strips   *(p. 98)*

endoplasmic reticulum (en-doh-plaz'mik ruh-tik'yu-lum)   a complex system of interlinked double membrane channels, subdividing the cytoplasm of a cell into compartments; parts of it are lined with ribosomes   *(p. 40)*

endosperm (en'doh-spurm)   a food-storage tissue that develops through divisions of the primary endosperm nucleus; it is digested by the sporophyte after germination in some species (e.g., corn) or before maturation of the seed in other species (e.g., beans)   *(p. 433)*

energy (en'ur-jee) the capacity to do work; various forms of energy include heat, light, and kinetic *(p. 20)*

enzyme (en'zym) one of numerous complex proteins that speeds up a chemical reaction in living cells without being used up in the reaction (i.e., it catalyzes the reaction) *(pp. 24, 35)*

epicotyl (ep'uh-kaht-ul) the part of an embryo or seedling above the attachment point of the cotyledon(s) *(p. 150)*

epidermis (ep-uh-dur'mis) the exterior tissue, usually one cell thick, of leaves and young stems and roots *(pp. 60, 113)*

epiphyte (ep'uh-fyt) an organism that is attached to and grows on another organism without parasitizing it *(p. 312)*

ergotism (ur'got-izm) a disease resulting from consumption of goods made with flour containing ergot fungus *(p. 341)*

essential element (uh-sen'shul el'uh-mint) one of 18 elements generally considered essential to the normal growth, development, and reproduction of most plants *(p. 171)*

ethylene (eth'uh-leen) a simple gaseous substance that inhibits plant growth and promotes the ripening of fruit *(p. 206)*

etiolation (ee-tee-oh-lay'shun) a condition characterized by long internodes, poor leaf development, and pale, weak appearance, due to a plant's having been deprived of light *(p. 216)*

eukaryotic (yu-kair-ee-ot'ik) pertaining to cells having distinct membrane-bound organelles, including a nucleus with chromosomes *(p. 46)*

eutrophication (yu-troh-fuh-kay'shun) the gradual enrichment of a body of water through the accumulation of nutrients, resulting in a corresponding increase in algae and other organisms *(p. 482)*

exine (ek'seen or ek'syne) the outer layer of the wall of a pollen grain or spore *(p. 432)*

exocarp (ek'soh-kahrp) the outermost layer of a fruit wall *(p. 137)*

eyespot (eye'spot) a small, often reddish structure within a motile unicellular organism; it appears to be sensitive to light (also called *stigma*) *(p. 306)*

f

F₁ first filial generation; the offspring of a cross between two parent plants *(p. 240)*

F₂ second filial generation; the offspring of F plants family; a category of classification between an order and a genus *(p. 240)*

FAD (eff-ay-dee) flavin adenine dinucleotide, a hydrogen acceptor molecule involved in the Krebs cycle of respiration *(p. 190)*

family (famm'uh-lee) a classification category between genus and order *(p. 266)*

fat (fat) an organic compound containing carbon, hydrogen, and oxygen but with proportionately much less oxygen than is present in a carbohydrate molecule *(p. 22)*

fermentation (fur-men-tay'shun) respiration in which the hydrogen removed from the glucose during glycolysis is transferred back to pyruvic acid, creating substances such as ethyl alcohol or lactic acid *(p. 188)*

fertilization (fur-til-i-zay'shun) formation of a zygote through the fusion of two gametes *(pp. 233, 432)*

fiber (fy'bur) a long thick-walled cell whose protoplasm often is dead at maturity *(p. 59)*

filament (fil'uh-mint) threadlike body of certain algae and fungi; also the stalk portion of a stamen *(pp. 136, 276, 310)*

fission (fish'un) the division of cells of bacteria, blue-green algae, and related organisms into two new cells *(p. 276)*

flagellum (pl. flagella) (fluh-jel'um; pl. fluh-jel'uh) a fine threadlike structure protruding from a motile unicellular organism or the motile cells produced by multicellular organisms; it functions primarily in locomotion *(p. 276)*

floret (flor'et) a small flower that is a part of the inflorescence of members of the Sunflower Family (Asteraceae) and the Grass Family (Poaceae) *(p. 463)*

florigen (flor'uh-jen) one or more hormones believed from circumstantial evidence to initiate flowering but not yet isolated or actually proved to exist *(p. 219)*

follicle (foll'uh-kuhl) a dry fruit that splits along one side only *(p. 143)*

food chain (food chayn) a natural chain of organisms of a community wherein each member of the chain feeds on members below it and is consumed by members above it, with autotrophic organisms (producers) being at the bottom; interconnected food chains are referred to as *food webs* *(p. 474)*

foot (foot) the basal part of the embryo of bryophytes and other plants; it is attached to and absorbs food from the gametophyte *(p. 370)*

fossil (fos'ul) the remains or impressions of any natural object that has been preserved in the Earth's crust *(p. 401)*

frond (frond) a fern leaf *(p. 395)*

fruit (froot) a mature ovary usually containing seeds; term also somewhat loosely applied to the reproductive structures of groups of plants other than angiosperms *(p. 137)*

fucoxanthin (fyu-koh-zan'thin) a brownish pigment occurring in brown and other algae *(p. 305)*

fundamental tissue (fun-duh-ment'ul tish'yu) parenchyma tissue of monocots (equivalent of cortex and pith in dicots) *(p. 80)*

g

GA3P (jee-ay-three-pee) glyceraldehyde 3-phosphate, a molecule produced during an intermediate step of glycolysis in the process of respiration *(p. 192)*

gametangium (pl. gametangia) (gam-uh-tan'jee-um; pl. gam-uh-tan'jee-ah) any cell or structure in which gametes are produced *(p. 317)*

gamete (gam'eet) a sex cell; one of two cells that unite, forming a zygote *(p. 226)*

gametophore (guh-me'toh-for) a stalk on which a gametangium is borne *(p. 368)*

gametophyte (guh-me'toh-fyte) the haploid (*n*) gamete-producing phase of the life cycle of an organism that exhibits Alternation of Generations *(p. 233)*

gemma (pl. gemmae) (jem'uh; pl. jem'ee) a small outgrowth of tissue that becomes detached from the parent body and is capable of developing into a complete new plant or other organism; gemmae are produced in cuplike structures on liverwort thalli, and are also produced by certain fungi *(p. 368)*

gene (jeen) a unit of heredity; part of a linear sequence of such units occurring in the DNA of chromosomes *(p. 240)*

generative cell (jen'uh-ray-tiv sel) the cell of the male gametophyte of angiosperms that divides, producing two *sperms;* also, the cell of the male gametophyte of gymnosperms that divides, producing a *sterile cell* and a *spermatogenous cell* *(p. 412)*

genetic engineering (juh'net'ik en-juh-neer'ing)   the introduction, by artificial means, of genes from one form of DNA into another form of DNA   *(p. 296)*

genetics (juh-net'iks)   the biological discipline involving the study of heredity   *(pp. 10, 238)*

genotype (jeen'oh-typ)   the genetic constitution of an organism; it may or may not be visibly expressed, as contrasted with *phenotype*   *(p. 240)*

genus (pl. genera) (jee'nus; pl. jen'er-ah)   a category of classification between a family and a species   *(p. 263)*

gibberellin (jib-uh-rel'in)   one of a group of plant hormones that have a variety of effects on growth; they are particularly known for promoting elongation of stems   *(p. 203)*

gill (gil)   one of the flattened plates of compact mycelium that radiate out from the stalk on the underside of the caps of most mushrooms   *(p. 344)*

girdling (gurd'ling)   the removal of a band of tissues extending inward to the vascular cambium on the stem of a woody plant   *(p. A–80)*

gland (gland)   a small body of variable shape and size that may secrete certain substances but that also may be functionless   *(pp. 62, 114)*

glycolysis (gly-kol'uh-sis)   the initial phase of all types of respiration in which glucose is converted in the absence of free oxygen to pyruvic acid   *(p. 188)*

gnetophytes (nee'toh-fytes)   members of the Division Gnetophyta   *(p. 415)*

Golgi body (gohl'jee bod'ee)   an organelle consisting of disc-shaped, often branching hollow tubules that apparently function in accumulating and packaging substances used in the synthesis of materials by the cell; also called *dictyosome;* collectively, the Golgi bodies of a cell may be referred to as the *Golgi apparatus*   *(p. 42)*

graft (graft)   the union of a segment of a plant, the *scion,* with a rooted portion, the *stock*   *(p. A–70)*

grain see *caryopsis*

granum (pl. grana) (gra'num; pl. gra'nuh)   a series of stacked thylakoids within a chloroplast   *(p. 42)*

gravitational water (grav-uh-tay'shun-ul waw'tur)   water that drains out of the pore spaces of a soil after a rain   *(p. 170)*

gravitropism (grav-uh-troh'pism)   growth response to gravity   *(p. 210)*

ground meristem (grownd mair'i-stem)   meristem that produces all the primary tissues other than the epidermis and stele (e.g., cortex, pith)   *(pp. 57, 71, 96)*

growth (grohth)   progressive increase in size and volume through natural development   *(pp. 16, 200)*

guard cell (gahrd sel)   one of a pair of specialized cells forming a stoma   *(pp. 62, 114)*

guttation (guh-tay'shun)   the exudation of water in liquid form from leaves due to root pressure   *(p. 166)*

gymnosperm (jim'noh-spurm)   a plant whose seeds are not enclosed within an ovary during their development (e.g., pine tree)   *(p. 408)*

h

haploid (hap'loyd)   having one set of chromosomes per cell, as in gametophytes; also referred to as having *n* chromosomes (as contrasted with 2*n* chromosomes in the *diploid* cells of sporophytes)   *(p. 232)*

haustorium (pl. haustoria) (haw-stor'ee-um; pl. haw-stor'ee-uh)   a protuberance of a fungal hypha or plant organ such as a root that functions as a penetrating and absorbing structure   *(p. 102)*

heartwood (hahrt'wood)   nonliving, usually darker-colored wood whose cells have ceased to function in water conduction   *(p. 79)*

herbaceous (hur-bay'shuss *or* ur-bay'shuss)   referring to nonwoody plants   *(p. 74)*

herbal (hur'bul *or* ur'bul)   a sixteenth- and seventeenth-century botany book that emphasized medicinal uses, edibility, and other utilitarian functions of plants   *(p. 7)*

herbarium (pl. herbaria) (hur-bair'ee-um *or* ur-bair'ee-um; pl. hur-bair'ee-uh)   a collection of dried pressed specimens, usually mounted on paper and provided with a label that gives collection information and an identification   *(p. 437)*

heterocyst (het'uh-roh-sist)   a transparent, thick-walled, slightly enlarged cell occurring in the filaments of certain blue-green bacteria   *(p. 290)*

heterospory (het-uh-ross'por-ee)   the production of both microspores and megaspores   *(p. 384)*

heterotrophic (het-ur-oh-troh'fick)   incapable of synthesizing food and therefore dependent on other organisms for it   *(p. 281)*

heterozygous (het-uh-roh-zy'guss)   having a pair of genes with contrasting characters at the same location on homologous chromosomes   *(p. 241)*

holdfast (hold'fast)   attachment organ or cell at the base of the thallus or filament of certain algae   *(pp. 310, 316)*

homologous chromosomes (hoh-mol'uh-guss kroh'muh-sohmz)   pairs of chromosomes that associate together in prophase I of meiosis; each member of a pair is derived from a different parent   *(p. 227)*

homozygous (hoh-moh-zy'guss)   having a pair of genes with identical characters at the same location on a pair of homologous chromosomes   *(p. 241)*

hormone (hor'mohn)   an organic substance generally produced in minute amounts in one part of an organism and transported to another part of the organism where it controls or affects specific metabolic processes   *(pp. 60, 200)*

hybrid (hy'brid)   offspring of two parents that differ in one or more genes   *(p. 242)*

hydathode (hy'duh-thohde)   structure at the tip of a leaf vein through which water is forced by root pressures   *(p. 166)*

hydrolysis (hy-drol'uh-sis)   the breakdown of complex molecules to simpler ones as a result of the union of water with the compound; the process is usually controlled by enzymes   *(pp. 21, 194)*

hydrosere (hy'droh-sear)   a primary succession that is initiated in a wet habitat   *(p. 481)*

hygroscopic water (hy-gruh-skop'ik waw'tur)   water that is chemically bound to soil particles and therefore unavailable to plants   *(p. 170)*

hypha (pl. hyphae) (hy'fuh; pl. hy'fee)   a single, usually tubular, threadlike filament of a fungus; *mycelium* is a collective term for many hyphae   *(p. 330)*

hypocotyl (hy-poh-kot'ul)   the portion of an embryo or seedling between the radicle and the cotyledon(s)   *(p. 150)*

hypodermis (hy-poh-dur'mis)   a layer of cells immediately beneath the epidermis and distinct from the parenchyma cells of the cortex in certain plants   *(pp. 117, 409)*

hypothesis (hy-poth'uh-sis)   a postulated explanation for some observed facts that must be tested experimentally before it can be accepted as valid or discarded if it proves to be incorrect   *(p. 6)*

i

imbibition (im-buh-bish′un) adsorption of water by nonliving materials and subsequent swelling because of the adhesion of the water molecules to the internal surfaces *(p. 161)*

indusium (pl. indusia) (in-dew′zee-um; pl. in-dew′zee-uh) small, membranous, sometimes umbrellalike covering of a developing fern *sorus* *(p. 396)*

inferior ovary (in-feer′ee-or oh′vuh-ree) an ovary to which parts of the calyx, corolla, and stamens have become more or less united so that they appear to be attached at the top of it *(pp. 136, 434)*

inflorescence (in-fluh-res′ints) collective term for a group of flowers attached to a common axis in a specific arrangement *(p. 136)*

inorganic (in-or-gan′ik) descriptive of compounds having no carbon atoms *(p. 21)*

integument (in-teg′yu-mint) the outermost layer of an ovule; it usually develops into a seedcoat; a gymnosperm ovule usually has a single integument, and an angiosperm ovule usually has two integuments *(pp. 408, 411, 429)*

intermediate-day plant (in-tur-me′dee-ut - day plant) a plant that has two critical photoperiods so that it will not flower if the days are either too short or too long *(p. 217)*

internode (in′tur-nohd) a stem region between nodes *(p. 71)*

interstock (in′tur-stok) a segment of stem that is grafted between a stock and a scion *(p. A–76)*

introgression (in-troh-gresh′uhn) backcrossing between hybrids and parents *(p. 252)*

ion (eye′on) a molecule or atom that has become electrically charged through the loss or gain of one or more electrons *(p. 19)*

isogamy (eye-sog′uh-me) sexual reproduction in certain algae and fungi having gametes that are alike in size *(p. 310)*

isotope (eye′suh-tohp) one of two or more forms of an element that have the same chemical properties but differ in the number of neutrons in the nuclei of their atoms *(p. 19)*

k

kinetochore (kih-net′uh-kor) see *centromere*

kingdom (king′dom) the highest category of classification (e.g., Plant Kingdom, Animal Kingdom) *(p. 266)*

knot (not) a portion of the base of a branch enclosed within wood *(p. 87)*

Krebs cycle (krebz′ sy′kul) a complex series of reactions following glycolysis in aerobic respiration that ultimately results in the conversion of pyruvic acid to hydrogen, carbon dioxide, and electrons *(p. 190)*

l

lamina (lam′uh-nuh) see *blade*

lateral bud (lat′uh-rul bud) see *axil*

laticifer (luh-tis′uh-fur) specialized cells or ducts resembling vessels; they form branched networks of *latex*-secreting cells in the phloem and other parts of plants *(p. 80)*

leaf (leef) a flattened, usually photosynthetic structure arranged in various ways on a stem *(p. 56)*

leaf gap (leef gap) a parenchyma-filled interruption in a stem's cylinder of vascular tissue immediately above the point at which a branch of vascular tissue (*leaf trace*) leading to a leaf occurs *(p. 72)*

leaflet (leef′lit) one of the subdivisions of a compound leaf *(p. 111)*

leaf scar (leef skahr) the suberin-covered scar left on a twig when a leaf separates from it through abscission *(p. 71)*

leaf trace (leef trays) see *leaf gap*

lenticel (lent′uh-sel) one of usually numerous, slightly raised, somewhat spongy groups of cells in the bark of woody plants; lenticels permit gas exchange between the interior of a plant and the external atmosphere *(pp. 66, 73)*

leucoplast (loo′kuh-plast) a colorless plastid commonly associated with starch accumulation *(p. 43)*

light microscopy (lite my-kross′kuh-pee) the use of a microscope where light (as opposed to electrons) is passed through or impinges on the object being viewed *(p. 45)*

light reactions (lyt ree-ak′shunz) a series of chemical and physical reactions through which light energy is converted to chemical energy with the aid of chlorophyll molecules; in the process water molecules are split, with hydrogen ions and electrons being produced and oxygen gas being released; ATP and NADPH$_2$ also are created *(p. 181)*

lignin (lig′nin) an organic hardening substance with which certain cell walls (e.g., those of wood) become impregnated *(pp. 39, 59)*

ligule (lig′yool) the tiny tonguelike appendage at the base of a spike moss (*Selaginella*) or quillwort (*Isoetes*) leaf; also, the outgrowth from the upper and inner side of a grass leaf at the point where it joins the sheath; also, the conspicuous straplike portion of the corolla of an outer floret in the flower head of a member of the Sunflower Family (Asteraceae) *(p. 384)*

linkage (link′ij) the tendency of two or more genes located on the same chromosome to be inherited together *(p. 243)*

lipid (lip′id) a general term for fats, fatty substances, and oils *(p. 22)*

locule (lok′yool) a cavity within an ovary or a sporangium *(p. 145)*

long-day plant (long-day plant) a plant in which flowering is not initiated unless exposure to more than a critical daylength occurs *(p. 217)*

m

megaphyll (meg′uh-fill) a leaf having branching veins; it is associated with a leaf gap *(p. 380)*

megaspore (meg′uh-spor) a spore that develops into a female gametophyte *(p. 384)*

megaspore mother cell (meg′uh-spor muth′ur sel) a diploid cell that produces megaspores upon undergoing meiosis *(p. 429)*

megasporophyll (meg-uh-spor′uh-fil) a leaf, usually reduced in size, on or within which megaspores are produced *(p. 384)*

meiosis (my-oh′sis) the process of two successive nuclear divisions through which segregation of genes occurs and a single diploid ($2n$) cell becomes four haploid (n) cells *(p. 226)*

mericloning (mair′i-kloh-ning) multiplication of plants through culturing and artificial dividing of shoot meristems *(p. A–75)*

meristem (mair′i-stem) a region in which undifferentiated cells divide *(p. 48)*

mesocarp (mez′uh-karp) the middle region of the fruit wall that lies between the exocarp and the endocarp *(p. 137)*

mesophyll (mez′uh-fil) parenchyma (chlorenchyma) tissue between the upper and lower epidermis of a leaf *(p. 114)*

metabolism (muh-tab′uh-lizm) the sum of all the interrelated chemical processes occurring in a living organism *(p. 17)*

microphyll (my′kroh-fil)   a leaf having a single unbranched vein not associated with a leaf gap *(pp. 380, 382)*

micropyle (my′kroh-pyl)   a pore or opening in the integuments of an ovule through which a pollen tube gains access to an embryo sac or archegonium of a seed plant *(pp. 411, 429)*

microspore (my′kroh-spor)   a spore that develops into a male gametophyte *(pp. 384, 432)*

microspore mother cell (my′kroh-spor muth′ur sel)   a diploid cell that produces microspores upon undergoing meiosis *(p. 429)*

microsporophyll (my-kroh-spor′uh-fil)   a leaf, usually reduced in size, on or within which microspores are produced *(p. 384)*

middle lamella (mid′ul luh-mel′uh)   a layer of pectic material that cements two adjacent cell walls together *(p. 38)*

midrib (mid′rib)   the central (main) vein of a pinnately veined leaf or leaflet *(p. 113)*

mitochondrion (pl. mitochondria) (my-toh-kon′dree-un; pl. my-toh-kon′dree-uh)   an organelle containing enzymes that function in the Krebs cycle and the electron transport chain of aerobic respiration *(p. 41)*

mitosis (my-toh′sis)   nuclear division, usually accompanied by cytokinesis, during which the chromatids of the chromosomes separate and two genetically identical daughter nuclei are produced *(p. 46)*

mixture (miks′chur)   a substance containing two or more ingredients, the atoms and molecules of which retain a separate identity and are not in a fixed proportion to one another *(p. 20)*

molecule (mol′uh-kyul)   the smallest unit of an element or compound retaining its own identity; it consists of two or more atoms *(pp. 17, 19)*

monocotyledon (mon-oh-kot-uh-lee′dun)   a class of angiosperms whose seeds have a single cotyledon; the term is commonly abbreviated to *monocot* *(p. 74)*

monoecious (moh-nee′shuss)   having unisexual male flowers or cones and unisexual female flowers or cones both on the same plant *(p. 436)*

monohybrid cross (mon-oh-hy′brid kross)   a cross involving a single pair of genes with contrasting characters in the parents *(p. 242)*

monokaryotic (mon-oh-kair-ee-ot′ik)   having a single nucleus in each cell or unit of the mycelium in club fungi *(p. 344)*

motile (moh′tul)   capable of independent movement *(p. 276)*

mulch (mulch)   any substance (e.g., straw, polyethylene sheeting) placed on the ground to protect plants from the environment, to conserve soil moisture, or to keep fruit clean

multiple fruit (mul′tuh-pul froot)   a fruit derived from several to many individual flowers in a single inflorescence *(p. 142)*

mutation (myu-tay′shun)   an inheritable change in a gene or chromosome *(pp. 248, 251)*

mycelium (my-see′lee-um)   a mass of fungal hyphae *(p. 330)*

mycorrhiza (pl. mycorrhizae) (my-kuh-ry′zuh; pl. my-kuh-ry′zee)   a symbiotic association between fungal hyphae and a plant root *(p. 104)*

n

NAD (en-ay-dee)   nicotinamide adenine dinucleotide, a molecule that temporarily accepts hydrogen, protons, and electrons during respiration *(p. 189)*

NADP (en-ay-dee-pee)   nicotinamide adenine dinucleotide phosphate, a high-energy storage molecule that temporarily accepts electrons for Photosystem I in the light reactions of photosynthesis *(p. 181)*

nastic movement (nass′tik moov′mint)   a nondirected movement of a flat organ (e.g., petal, leaf) in which the organ alternately bends up and down *(p. 209)*

net venation (net ve-nay′shun)   network of veins characteristic of a dicot leaf *(p. 115)*

neutron (new′tron)   an uncharged particle in the nucleus of an atom *(p. 18)*

node (nohd)   region of a stem where one or more leaves are attached *(pp. 57, 71)*

nucellus (new-sel′us)   ovule tissue within which an embryo sac develops *(pp. 408, 411)*

nuclear envelope (new′klee-ur en′vuh-lohp)   a porous double membrane enclosing a nucleus *(p. 44)*

nucleic acid (new-klee′ik as′id)   see *DNA, RNA*

nucleolus (pl. nucleoli) (new-klee′oh-luss; pl. new-klee′oh-ly)   a somewhat spherical body within a nucleus; it contains primarily RNA and protein; there may be more than one nucleolus per nucleus *(p. 45)*

nucleotide (new′klee-oh-tyd)   the structural unit of DNA and RNA *(p. 24)*

nucleus (new′klee-uss)   the organelle of a living cell that contains chromosomes and is essential to the regulation and control of all the cell's functions; also, the core of an atom *(pp. 18, 44)*

o

oil (oyl)   a fat in a liquid state *(p. 22)*

oogamy (oh-og′uh-mee)   sexual reproduction in which the female gamete or egg is nonmotile and larger than the male gamete or sperm, which is motile *(p. 314)*

oogonium (pl. oogonia) (oh-oh-goh′nee-um; pl. oh-oh-goh′nee-ah)   a female sex organ of certain algae and fungi; it consists of a single cell that contains one to several eggs *(p. 313)*

operculum (oh-per′kyu-lum)   the lid or cap that protects the peristome of a moss sporangium *(p. 374)*

orbital (or′buh-till)   a volume of space in which a given electron occurs 90% of the time *(p. 18)*

order (or′dur)   a category of classification between a class and a family *(p. 266)*

organelle (or-guh-nel′)   a membrane-bound body in the cytoplasm of a cell; there are several kinds, each with a specific function (e.g., mitochondrion, chloroplast) *(p. 37)*

organic (or-gan′ik)   pertaining to or derived from living organisms, and to the chemistry of carbon-containing compounds *(p. 21)*

osmosis (oz-moh′sis)   the diffusion of water or other solvents through a differentially permeable membrane from a region of higher concentration to a region of lower concentration *(p. 159)*

osmotic potential (oz-mot'ik puh-ten'shul)   potential pressure that can be developed by a solution separated from pure water by a differentially permeable membrane (the pressure required to prevent osmosis from taking place) *(p. 159)*

osmotic pressure (oz-mot'ik presh'ur)   see *osmotic potential*   *(p. 159)*

ovary (oh'vuh-ree)   the enlarged basal portion of a pistil that contains an ovule or ovules and develops into a fruit   *(p. 136)*

ovule (oh'vyool)   a structure of seed plants that contains a female gametophyte and has the potential to develop into a seed   *(pp. 136, 408)*

p

palisade mesophyll (pal-uh-sayd' mez'uh-fil)   mesophyll having relatively uniform rows of tightly packed parenchyma (chlorenchyma) cells located beneath the upper epidermis of a leaf   *(p. 114)*

palmately compound; palmately veined (pahl'mayt-lee kom'pownd; pahl'mayt-lee vaynd)   having leaflets or principal veins radiating out from a common point   *(p. 113)*

papilla (pl. papillae) (puh-pil'uh; pl. puh-pill-ay)   a small, usually rounded or conical protuberance   *(p. 311)*

parallel venation (pair'uh-lel ve-nay'shun)   venation in which the principal veins are more or less parallel to one another; characteristic of monocot leaves   *(p. 113)*

parenchyma (puh-ren'kuh-muh)   thin-walled cells varying in size, shape, and function; the most common type of plant cell   *(p. 58)*

parthenocarpic (par-thuh-noh-kar'pik)   developing fruits from unfertilized ovaries; the resulting fruit is, therefore, usually seedless   *(p. 434)*

passage cell (pas'ij sel)   a thin-walled cell of an endodermis   *(p. 98)*

pectin (pek'tin)   a water-soluble organic compound occurring primarily in the middle lamella; when combined with organic acids and sugar it becomes a jelly   *(p. 38)*

pedicel (ped'i-sel)   the individual stalk of a flower that is part of an inflorescence   *(p. 136)*

peduncle (pee'dun-kul)   the stalk of a solitary flower or the main stalk of an inflorescence   *(p. 134)*

peptide bond (pep'tyd bond)   the type of chemical bond formed when two amino acids link together   *(p. 23)*

perennial (puh-ren'ee-ul)   a plant that continues to live indefinitely after flowering   *(p. 134)*

pericarp (per'uh-karp)   collective term for all the layers of a fruit wall   *(p. 137)*

pericycle (per'uh-sy-kul)   tissue sandwiched between the endodermis and phloem of a root; it is often only one or two cells wide in transverse section and is the site of origin of branch roots   *(p. 98)*

periderm (pair'uh-durm)   outer bark; it is composed primarily of cork cells   *(p. 66)*

peristome (per'uh-stohm)   one or two series of flattened, often ornamented structures (teeth) arranged around the margin of the open end of a moss sporangium; the teeth are sensitive to changes in humidity and facilitate the release of spores   *(p. 374)*

permanent tissue (purr'man-ent tish'yu)   a tissue composed of cells that have assumed various shapes and sizes related to their functions as they matured following their production by a meristem   *(p. 57)*

petal (pet'ul)   a frequently flattened, often colored unit of a corolla   *(p. 134)*

petiole (pet'ee-ohl)   the stalk of a leaf   *(pp. 71, 111)*

P$_{far-red}$ or P$_{fr}$ (pee-far-red or pee-ef-ahr)   a form of phytochrome (which see)   *(p. 215)*

pH (pee-aitch)   a symbol of hydrogen ion concentration indicating the degree of acidity or alkalinity   *(p. 20)*

phage (fayj)   see *bacteriophage*

phellogen (fel'uh-jun)   see *cork cambium*

phenotype (fee'noh-typ)   the physical appearance of an organism   *(p. 240)*

pheromone (fer'uh-mohn)   a substance produced by an organism that facilitates chemical communication with another organism   *(p. A–27)*

phloem (flohm)   the food-conducting tissue of a vascular plant   *(p. 65)*

photon (foh'ton)   a unit of light energy   *(p. 183)*

photoperiodism (foh-toh-pir'ee-ud-izm)   the initiation of flowering and certain vegetative activities of plants in response to relative lengths of day and night   *(p. 216)*

photosynthesis (foh-toh-sin'thuh-sis)   the conversion of light energy, water, and carbon dioxide in the presence of chlorophyll to carbohydrate, with oxygen being released as a by-product   *(pp. 17, 179)*

photosynthetic unit (foh-toh-sin-thet'ik yew'nit)   one of two groups of about 250 to 400 pigment molecules each that function together in chloroplasts in the light reactions of photosynthesis; the units are exceedingly numerous in each chloroplast   *(p. 181)*

photosystem (foh'toh-sis-tum)   collective term for a specific functional aggregation of photosynthetic units   *(p. 184)*

phragmoplast (frag'moh-plast)   set of microfibrils forming a ring at the margin of the developing cell plate at the close of mitosis   *(p. 49)*

phytochrome (fy'tuh-krohm)   pigment associated with the absorption of light; it is found in the cytoplasm of cells of green plants   *(p. 215)*

pilus (pl. pili) (py'lis; pl. py'lee)   the equivalent of a conjugation tube in bacteria   *(p. 276)*

pinna (pl. pinnae) (pin'uh; pl. pin'ee)   primary subdivision of a fern frond; the term is also applied to a leaflet of a compound leaf   *(p. 396)*

pinnately compound; pinnately veined (pin'ayt-lee kom'pownd; pin'ayt-lee vaynd)   having leaflets or veins on both sides of a common axis (e.g., rachis, midrib) to which they are attached   *(p. 113)*

pistil (pis'tul)   the central unit of a flower; it is composed of one or more carpels and consists of an ovary, style, and stigma   *(p. 136)*

pit (pit)   a depression or cavity in a cell wall   *(p. 50)*

pith (pith)   central tissue of a dicot stem and certain roots; it usually consists of parenchyma cells that become crushed in woody plants as cambial activity increases the organ's girth   *(p. 71)*

plankton (plank'ton)   free-floating aquatic organisms that are mostly microscopic   *(p. 309)*

plant anatomy (plant uh-nat'uh-mee)   the botanical discipline that pertains to the internal structure of plants   *(p. 9)*

plant community (plant kuh-myu'nuh-tee)   an association of plants inhabiting a common environment and interacting with one another   *(p. 472)*

plant ecology (plant ee-koll'uh-jee) the science that deals with the relationships and interactions between plants and their environment *(p. 9)*

plant geography (plant jee-og'ruh-fee) the botanical discipline that pertains to the broader aspects of the space relations of plants and their distribution over the surface of the Earth *(p. 9)*

plant morphology (plant mor-fol'uh-jee) the botanical discipline that pertains to plant form and development *(p. 10)*

plant physiology (plant fiz-ee-ol'uh-jee) the botanical discipline that pertains to the metabolic activities and processes of plants *(p. 9)*

plants (plantz) members of Kingdom Plantae *(p. 364)*

plant taxonomy (plant tak-son'uh-mee) the botanical discipline that pertains to the classification, naming, and identification of plants *(p. 9)*

plasma membrane (plaz'muh mem'brayn) the outer boundary of the protoplasm of a cell; also called *cell membrane,* particularly in animal cells *(p. 39)*

plasmid (plaz'mid) one of up to 30 or 40 small, circular DNA molecules usually present in a bacterial cell *(p. 277)*

plasmodesma (pl. plasmodesmata) (plaz-muh-dez'muh; pl. plaz-muh-dez'mah-tah) minute strands of cytoplasm that extend between adjacent cells through pores in the walls *(p. 50)*

plasmodium (pl. plasmodia) (plaz-moh'dee-um; pl. plaz-moh'dee-ah) the multinucleate, semi-viscous liquid, active form of slime mold; it moves in a "crawling-flowing" motion *(pp. 330, 332)*

plasmolysis (plaz-mol'uh-sis) the shrinking in volume of the protoplasm of a cell and the separation of the cytoplasm from the cell wall due to loss of water via osmosis *(p. 160)*

plastid (plas'tid) an organelle associated primarily with the storage or manufacture of carbohydrates (e.g., *leucoplast, chloroplast) (p. 42)*

plumule (ploo'myool) the terminal bud of the embryo of a seed plant *(p. 150)*

polar nuclei (poh'lur new'klee-eye) nuclei, frequently two in number, that unite with a sperm in an embryo sac, forming a primary endosperm nucleus *(p. 429)*

pollen grain (pahl'un grayn) a structure derived from the microspore of seed plants that develops into a male gametophyte *(pp. 136, 411, 432)*

pollen tube (pahl'un t(y)oob) a tube that develops from a pollen grain and conveys the sperms to the female gametophyte *(pp. 411, 432)*

pollination (pahl-uh-nay'shun) the transfer of pollen from an anther to a stigma *(p. 432)*

pollinium (pl. pollinia) (pah-lin'ee-um; pl. pah-lin'ee-ah) a cohesive mass of pollen grains commonly found in members of the Orchid Family (Orchidaceae) and the Milkweed Family (Asclepiadaceae) *(p. 493)*

polyploidy (pahl'i-ploy-dee) having more than two complete sets of chromosomes per cell *(p. 247)*

P$_{red}$ or P$_r$ (pee-red or pee-ahr) a form of phytochrome (which see) *(p. 215)*

pressure-flow hypothesis (presh'ur - floh hy-poth'uh-sis) the theory that food substances in solution in plants flow along concentration gradients between the sources of the food and sinks (places where the food is utilized) *(p. 165)*

pressure potential (presh'ur poh-ten-shul) the turgor pressure that develops as a result of water entering the vacuole of a cell *(p. 159)*

primary consumer (pry'-mer-ree kon-soo'-mur) animals that feed directly on producers *(p. 474)*

primary endosperm nucleus (pry'mer-ee en'doh-spurm new'klee-uss) the nucleus produced by the fusion of a sperm and polar nuclei in the embryo sac of angiosperms; it gives rise to *endosperm* tissue

primordium (pry-mord'ee-um) an organ or structure (e.g., leaf, bud) at its earliest stage of development *(p. 71)*

procambium (proh-kam'bee-um) the primary meristem that gives rise to primary xylem and phloem *(pp. 57, 71, 96)*

producer (pruh-dew'sur) an organism that manufactures food through the process of photosynthesis *(p. 474)*

progametangium (proh-gam-uh-tan'jee-um) a cell or structure that becomes a gametangium *(p. 336)*

prokaryotic (proh-kair-ee-ot'ik) having a cell or cells that lack a distinct nucleus and other membrane-bound organelles (e.g., bacteria) *(pp. 46, 276)*

proplastid (proh-plas'tid) a tiny cytoplasmic body from which a plastid develops *(p. 43)*

protein (proht'ee-in *or* proh'teen) an organic compound containing carbon, hydrogen, oxygen, nitrogen, and frequently sulphur in complex molecules composed of numerous amino acids linked together by peptide bonds *(p. 22)*

prothallus (pl. prothalli) (proh-thal'us; pl. proh-thal'eye) the gametophyte of ferns and their relatives; also called *prothallium (p. 398)*

protoderm (proh'tuh-durm) the primary meristem that gives rise to the epidermis *(pp. 57, 71, 96)*

proton (proh'ton) a positively charged particle in the nucleus of an atom *(p. 18)*

protonema (proh-tuh-nee'muh) a green, usually branched, threadlike or sometimes platelike growth from a moss spore; it gives rise to "leafy" gametophytes *(p. 371)*

protoplasm (proh'tuh-plazm) the living substance of a cell (includes the cytoplasm and nucleus) *(p. 21)*

pruning (proon'ing) removal of portions of plants for aesthetic purposes, for improving quality and size of fruits or flowers, or for elimination of diseased tissues *(p. A-82)*

pyrenoid (py'ruh-noyd) a small body found on the chloroplasts of certain green algae and hornworts; pyrenoids are associated with starch accumulation; they may occur singly on a chloroplast or they may be numerous *(p. 309)*

pyruvic acid (py-roo'vik as'id) the organic compound that is the end product of the glycolysis phase of respiration *(p. 188)*

q

quartersawed (kwor-tur-sawd') wood that has been cut radially, i.e., along the rays *(p. 86)*

quiescence (kwy'ess-ens) a state in which a seed or other plant part will not germinate or grow unless environmental conditions normally required for growth are present *(p. 219)*

r

rachis (ray'kiss) the axis of a pinnately compound leaf or frond extending between the lowermost leaflets or pinnae and the terminal leaflet or pinna (corresponds with the midrib of a simple leaf) *(p. 113)*

radicle (rad'i-kuhl) the part of an embryo in a seed that develops into a root *(pp. 94, 150)*

ray (ray)   radially oriented tiers of cells that conduct food, water, and other materials laterally in the stems and roots of woody plants; they are generally continuous across the vascular cambium between the xylem and the phloem; the portion within the wood is called a *xylem ray*, whereas the extension of the same ray in the phloem is called a *phloem ray*   *(pp. 65, 78)*

receptacle (ree-sep'tuh-kul)   the commonly expanded tip of a peduncle or pedicel to which the various parts of a flower (e.g., calyx, corolla) are attached   *(p. 134)*

recessive (ree-ses'iv)   descriptive of a member of a pair of genes whose phenotypic expression is masked or suppressed by the dominant gene   *(p. 240)*

red tide (red tyd)   the marine phenomenon that results in the water becoming temporarily tinged with red due to the sudden proliferation of certain dinoflagellates that produce substances poisonous to animal life and humans; the occurrence of the phenomenon is unpredictable   *(p. 306)*

reproduction (ree-proh-duk'shun)   the development of new individual organisms through either sexual or asexual means   *(p. 16)*

resin canal (rez'in kuh-nal')   a tubular duct of many conifers and some angiosperms that is lined with resin-secreting cells   *(pp. 79, 409)*

respiration (res-puh-ray'shun)   cellular breakdown of sugar and other foods, accompanied by release of energy; in aerobic respiration oxygen is utilized   *(pp. 17, 188)*

rhizoid (ry'zoyd)   delicate root- or root-hairlike structures of algae, fungi, the gametophytes of bryophytes, and certain structures of a few vascular plants; they function in anchorage and absorption but have no xylem or phloem   *(p. 334)*

rhizome (ry'zohm)   an underground stem, usually horizontally oriented, which may be superficially rootlike in appearance but which has definite nodes and internodes   *(p. 83)*

ribosome (ry'boh-sohm)   granular particles each composed of two subunits consisting of RNA and proteins; they lack membranes and are very numerous in living cells   *(p. 41)*

RNA (ar-en-ay)   the standard abbreviation for ribonucleic acid, an important cellular substance that occurs in three forms, all involved in the synthesis of proteins   *(p. 27)*

root (root)   a plant organ that functions in anchorage and absorption; most roots are produced below ground   *(p. 56)*

root cap (root kap)   a thimble-shaped mass of cells at the tip of a growing root; it functions primarily in protection   *(p. 95)*

root hair (root hair)   a delicate protuberance that is part of an epidermal cell of a root; root hairs occur in a zone behind the growing tip   *(p. 97)*

runner (run'ur)   see *stolon*

s

saprobe (sap'rohb)   an organism that obtains its food directly from nonliving organic matter   *(p. 281)*

sapwood (sap'wood)   outer, usually functional layers of wood in a tree trunk; sapwood is usually lighter in color than heartwood   *(p. 79)*

science (sy'ints)   a branch of study involved with the observation, recording, organization, and classification of facts, and the establishment of verifiable laws, chiefly through induction and hypotheses   *(p. 6)*

scion (sy'un)   a segment of plant that is grafted onto a stock   *(p. A–76)*

sclereid (sklair'id)   a sclerenchyma cell with pits; it may vary in shape but usually one axis is not conspicuously longer than the other   *(p. 59)*

sclerenchyma (skluh-ren'kuh-muh)   tissue composed of cells with thick walls; the tissue functions primarily in strengthening and support   *(p. 59)*

secondary consumer (sek'-on-dair-ee kon-soo'-mer)   an animal that feeds on other consumers   *(p. 474)*

secondary tissue (sek'un-der-ee tish'yu)   a tissue produced by the vascular cambium (or the cork cambium (e.g., virtually all the xylem and phloem in a tree trunk)   *(p. 71)*

secretory cell, tissue (see'kruh-tor-ee sel, tish'yu)   cell or tissue producing a substance or substances that are moved outside the cells   *(p. 60)*

seed (seed)   a mature ovule containing an embryo and bound by a protective seedcoat   *(pp. 136, 226)*

seedcoat (seed'koht)   the outer boundary layer of a seed; it is developed from the integument(s)   *(pp. 408, 429)*

senescence (suh-ness'ints)   the breakdown of cell components and membranes that leads to the death of the cell   *(p. 207)*

sepal (see'puhl)   a unit of the calyx that frequently resembles a reduced leaf; sepals often function in protecting the unopened flower bud   *(p. 134)*

seta (see'tuh)   the stalk of a moss sporophyte   *(p. 370)*

sexual reproduction (seksh'yule ree-proh-duk'shun)   reproduction involving the union of gametes   *(p. 226)*

short-day plant (short-day plant)   a plant in which flowering is initiated when the days are shorter than its critical photoperiod   *(p. 217)*

sieve cell (siv sel)   a food-conducting cell found in the phloem of gymnosperms, ferns, and other lower vascular plants   *(p. 65)*

sieve plate (siv playt)   area of the wall of a sieve-tube element that contains several to many perforations that permit cytoplasmic connections between similar adjacent cells, the cytoplasmic strands being larger than plasmodesmata   *(p. 65)*

sieve tube (siv t(y)oob)   column of sieve-tube elements arranged end to end; food is conducted from cell to cell through sieve plates   *(p. 65)*

sieve-tube element (siv t(y)oob el'uh-mint)   a single cell of a sieve tube   *(p. 65)*

simple leaf (sim'pul leef)   a leaf with the blade undivided into leaflets   *(p. 111)*

solvent (sol'vent)   a substance (usually liquid) capable of dissolving another substance   *(p. 159)*

soredium (pl. soredia) (sor-ee'dee-um; pl. sor-ee'dee-uh)   an asexual reproductive structure of lichens; it consists of several algal cells surrounded by fungal hyphae   *(p. 357)*

sorus (pl. sori) (sor'uss; pl. sor'eye)   a cluster of sporangia; the term is most frequently applied to clusters of fern sporangia   *(p. 396)*

species (spee'seez)   the basic unit of classification; a population of individuals capable of interbreeding freely with one another but not with members of another species   *(p. 263)*

sperm (spurm)   a male gamete; it is usually motile and smaller than the corresponding female gamete *(pp. 226, 313)*

spice (spyss)   an aromatic organic substance used to season or flavor food or drink   *(p. A–47)*

spindle (spin'dul)   an aggregation of fiberlike threads (microtubules) that appears in cells during mitosis and meiosis; some threads are attached to the centromeres of chromosomes, whereas other threads extend directly or in arcs between two invisible points designated as *poles* (see also *phragmoplast*)   *(p. 49)*

spine (spyn)   a relatively strong, sharp-pointed woody structure usually located on a stem; it is frequently a modified leaf or stipule   *(p. 118)*

spongy mesophyll (spun'jee mez'uh-fil)   mesophyll having loosely arranged cells and numerous air spaces; it is generally confined to the lower part of the interior of a leaf just above the lower epidermis   *(p. 114)*

sporangiophore (spuh-ran'jee-uh-for)   the stalk on which a sporangium is produced   *(p. 336)*

sporangium (pl. sporangia) (spuh-ran'jee-um; pl. spuh-ran'jee-uh)   a structure in which spores are produced; it may be either unicellular or multicellular   *(p. 330)*

spore (spor)   a reproductive cell or aggregation of cells capable of developing directly into a gametophyte or other body without uniting with another cell (**Note:** a bacterial spore is not a reproductive cell but is an inactive phase that enables the cell to survive under adverse conditions); *sexual spores* formed as a result of meiosis occurring are often called *meiospores;* spores produced by mitosis may be referred to as *vegetative spores*   *(pp. 233, 330)*

spore mother cell (spor muth'ur sel)   a diploid cell that becomes four haploid spores or nuclei as a result of undergoing meiosis   *(pp. 233, 370)*

sporophyll (spor'uh-fil)   a modified leaf that bears a sporangium or sporangia   *(p. 383)*

sporophyte (spor'uh-fyt)   the diploid (2*n*) spore-producing phase of the life cycle of an organism exhibiting Alternation of Generations   *(p. 233)*

stamen (stay'min)   a pollen-producing structure of a flower; it consists of an anther and usually also a filament   *(p. 136)*

starch (starch)   an insoluble carbohydrate composed of numerous glucose units; it is the principal food-storage substance of plants   *(p. 37)*

stele (steel)   the central cylinder of tissues in a stem or root; it usually consists primarily of xylem and phloem   *(p. 73)*

stem (stem)   a plant axis with leaves or enations   *(p. 56)*

stigma (stig'muh)   the pollen receptive area of a pistil; also, the eyespot of certain motile algae   *(p. 136)*

stipe (styp)   the supporting stalk of mushrooms and other stationary organisms   *(p. 316)*

stipule (stip'yool)   one of a pair of appendages of varying size, shape, and texture present at the base of the leaves of some plants   *(pp. 71, 111)*

stock (stok)   the rooted portion of a plant to which a scion is grafted   *(p. A–76)*

stolon (stoh'lun)   a stem that grows horizontally along the surface of the ground; it typically has relatively long internodes; also called a *runner*   *(p. 83)*

stoma (pl. stomata) (stoh'muh; pl. stoh'mah-tuh)   a minute pore or opening in the epidermis of leaves, herbaceous stems, and the sporophytes of hornworts (*Anthoceros*); it is flanked by two guard cells that regulate its opening and closing and thus regulate gas exchange and transpiration; the guard cells and pore are also collectively referred to as a stoma   *(pp. 62, 114)*

strobilus (pl. strobili) (stroh'buh-luss; pl. stroh'buh-leye)   an aggregation of sporophylls on a common axis; it usually resembles a cone or is somewhat conelike in appearance   *(pp. 383, 391)*

stroma (stroh'muh)   a colorless fluid substance constituting the bulk of the volume of a chloroplast or other plastid; it contains enzymes that in chloroplasts play a key role in photosynthetic reactions   *(p. 42)*

style (styl)   the tissue that connects a stigma and an ovary   *(p. 136)*

suberin (soo'buh-rin)   fatty substance found primarily in the cell walls of cork and the Casparian strips of endodermal cells   *(pp. 66, 73)*

succession (suk-sesh'un)   an orderly progression of changes in the composition of a community from the initial development of vegetation to the establishment of a climax community   *(p. 480)*

sucrose (soo'krohs)   the primary form in which sugar produced by photosynthesis is transported throughout a plant; it is a disaccharide composed of glucose and fructose   *(p. 21)*

superior ovary (soo-peer'ee-or oh'vuh-ree)   an ovary that is free from the calyx, corolla, and other floral parts so that the sepals and petals appear to be attached at its base   *(pp. 136, 434)*

symbiosis (sim-by-oh'siss)   an intimate association between two dissimilar organisms that benefits both of them   *(p. 276)*

syngamy (sin'gam-mee)   union of gametes; fertilization   *(p. 233)*

t

tendril (ten'dril)   a slender structure that coils on contact with a support of suitable diameter; it usually is a modified leaf or leaflet, and aids the plant in climbing   *(pp. 84, 117)*

thallus (pl. thalli) (thal'uss; pl. thal'eye)   a multicellular plant body that is usually flattened and not organized into roots, stems, or leaves   *(pp. 317, 354, 367)*

3n (three-en)   having three sets of chromosomes; triploid   *(p. 433)*

thylakoid (thy'luh-koyd)   coin-shaped membranes containing chlorophyll and arranged in stacks that form the *grana* of chloroplasts   *(p. 42)*

tissue (tish'yu)   an aggregation of cells having a common function   *(p. 56)*

tissue culture (tish'yu kult'yur)   the culture of isolated living tissue on an artificial medium   *(p. 248)*

tracheid (tray'kee-id)   a xylem cell that is tapered at the ends and has thick walls containing pits   *(p. 64)*

transpiration (trans-puh-ray'shun)   loss of water in vapor form; most transpiration takes place through the stomata   *(pp. 110, 162)*

tropism (troh'pizm)   response of a plant organ or part to an external stimulus, usually in the direction of the stimulus   *(p. 209)*

tuber (t(y)oo'bur)   a swollen, fleshy underground stem (e.g., white potato)   *(p. 83)*

turgid (tur'jid)   firm or swollen because of internal water pressures resulting from osmosis   *(p. 159)*

turgor movements (turr'gor moov'mint)   movement that results from changes in internal water pressures in a plant part   *(p. 211)*

turgor pressure (tur'gur presh'ur)   pressure within a cell resulting from the uptake of water   *(p. 159)*

tylosis (pl. tyloses) (ty-loh'sis; pl. ty-loh'seez)   a ballooning of material through a pit into a vessel, thus preventing the vessel from functioning; tyloses are common in heartwood   *(p. 78)*

u

unisexual (yu-nih-scksh'yu-ul)   a term usually applied to a flower lacking either stamens or a pistil   *(p. 436)*

v

vacuolar membrane (vak-yu-oh'lur mem'brayn)   a membrane between the cytoplasm and a vacuole of a cell   *(p. 45)*

vacuole (vak'yu-ohl)   a pocket of cell sap that is separated from the cytoplasm of a cell by a membrane; it may occupy more than 90% of a cell's volume in plants; also, food-storage or contractile pockets within the cytoplasm of unicellular organisms   *(p. 45)*

vascular bundle (vas'kyu-lur bun'dul)   a strand of tissue composed mostly of xylem and phloem and usually enveloped by a bundle sheath   *(p. 74)*

vascular cambium (vas'kyu-lur kam'bee-um)   a narrow cylindrical sheath of cells that produces secondary xylem and phloem in stems and roots   *(pp. 48, 57)*

vascular plant (vas'kyu-lur plant)   a plant having xylem and phloem   *(p. 380)*

vein (vayn)   a term applied to any of the vascular bundles that form a branching network within leaves   *(p. 115)*

venter (ven'tur)   the site of the egg in the enlarged basal portion of an archegonium   *(p. 373)*

vessel (ves'uhl)   one of usually very numerous cylindrical tubes occurring in the xylem of most angiosperms and a few other vascular plants; each vessel is composed of *vessel elements* laid end to end; the perforated or open-ended walls of the vessel elements permit water to pass through freely   *(p. 64)*

vessel element (ves'uhl el'uh-mint)   a single cell of a vessel   *(p. 64)*

virus (vy'riss)   a minute particle consisting of a core of nucleic acid, usually surrounded by a protein coat; it is incapable of growth and can reproduce only within, and at the expense of, a living cell   *(p. 293)*

vitamin (vyt'uh-min)   a complex organic substance produced primarily by photosynthetic organisms; various vitamins are essential in minute amounts for normal metabolism and growth   *(p. 200)*

volva (vol'vuh)   a cuplike or egglike structure at the base of the stalk of certain mushrooms   *(p. 344)*

w

water potential (waw'tur puh-ten'shul)   the chemical potential of water expressed in pressure units—a measure of the tendency of water molecules to diffuse (*or* osmotic potential and pressure potential combined—the term is usually expressed mathematically)   *(p. 159)*

whorled (wirld)   having three or more leaves at a node   *(p. 71)*

wilting percentage (wilt'ing pur-sent'ij)   percentage of soil water not available to the roots; plants wilt when soil moisture is reduced to the wilting percentage   *(p. 171)*

x

xerosere (zer'roh-sear)   a primary succession that initiates with bare rock   *(p. 480)*

xylem (zy'lim)   the tissue through which most of the water and dissolved minerals utilized by a plant are conducted; it consists of several types of cells   *(p. 64)*

z

zoospore (zoh'uh-spor)   a motile spore occurring in algae and fungi   *(p. 310)*

zygospore (zy'goh-spor)   a zygote that develops a thick protective wall   *(p. 336)*

zygote (zy'goht)   the product of the union of two gametes   *(pp. 226, 433)*

Index

Boldfaced type indicates major treatment of a topic. Italic type indicates illustrations.

D

Daffodil, 466
Dahlia, 217, 463
Daisy, 252, 463
Damask rose, *453*
Dammar, 418
Dandelion, 80, *94,* 102, *146,* 217, 252, 463, 464
Dandruff, 400
Dark reactions, **181, 182, 186, 187**
Darwin, Charles, 10, 201, 207, 208, **248–50,** 255, 428
Darwin, Francis, 201
Darwin, Robert, 249
Date, 6, 139
Date palm, 6
Davallia, 397
Dawn redwood, 420
Day-neutral plant, 217
DDT, 179, 324, 342
DDT-like compound, 324
Deadly nightshade, 125
Death Angel, 344, *345,* 349
Decay bacteria, 283, 393, 478, 479
Deciduous plants, 71, 124, 489
Decomposer, 283, 474, 491
Deer, 147, 476
Deer, mule, 490
Deerbrush, 483
Deerfly, 286
Deficiency symptoms, 172, 173
Dehydration movements, 215
Deletion, chromosomal, *251*
Dendrobium, 467
Denitrifying bacteria, 478
Deoxyribonucleic acid, **24–28,** 204, 231, 294, 295–97, 306
  bacterial, 276–78, 295–97
  recombinant, 297
  replication of, *26,* 46, *277*
  structure, **24–25**
  viral, 294
Desert, 116, 118, 151, 166, 168, 182, 219, **488–89,** 492
Desmids, 314
Desmococcus, A3
Deuteromycetes, **352–54.** *See also* Imperfect fungi
Development, 200
Devonian period, 382, 398, 408
Dewberry, 453
Dextran, 288
Diabetes, 297, 400
Diarrhea, 324, 393, 400, 447, 458
Diatom (Bacillariophyceae), 290, **304–306,** 321
Diatomaceous earth, 321
Dicentra, 254, 262
Dichasium, *136*
Dichotomous key, **268–70**
Dichotomous venation, *112,* 113, 414
Dicot, 74, 81, 95, 134, 202, 263, 428, 446
Dicots and monocots compared, **134**
Dicotyledon, 74

Dicotyledonous stems, **74–80**
Dictyosome. *See* Golgi body
Differentially permeable membrane, 159, 160
Differentiation, 200, 204, 429
Diffusion, **158–60,** 164, 200, 202
Diffusion gradient, 158, 159, 210
Digester, organic, *281*
Digestion, 17, 21, 122, 123, **194**
Digger pine, 417
Digitalis, 126
Dihybrid cross, 242–43
Dikaryotic, 344, 345
Dill, 125, 146, 288, 461
Dimethyl sulfoxide, 39
Dinobryon, 304
Dinoflagellates, 212, **306–307,** 321
Dinosaurs, 16, 480
Dioecious, 436
Dioscorides, 7, 450
Dioxin, 203
Diphtheria, 284, 296
Diploid (definition), **232**
Disaccharide, 21, 194
Dischidia, 119
Disease,
  bacteria and, **283–87**
  fungal, 204, 335, 336, 341, 342, 347–49
  resistance to, 248, 349, A27
  viruses and, 295, 296
Disk floret, *463*
Dispersal, seed, **146–49**
Diuretic, 392
Divisions (plant or other organismal), **266**
DMSO. *See* Dimethyl sulfoxide
DNA. *See* Deoxyribonucleic acid
Dock, curly, 153
Doctrine of Signatures, 8, 367
Dodder, 102, *104,* 429
Dog, 292, 339, 349, 434
Dogbane, 80
Dogwood, *112,* 120
Dollar plant, *143*
Dominance
  in heredity, 240–45
  incomplete, *241*
  law of, 240
Dominant, 240–45, 444
Dormancy, **150–51,** 201, 205, **219–21,** 489
Double fertilization or fusion, 433
Douglas fir, 80, 420, 483, 489, 490
Dove, 447, 489
Dove, mourning, 487
Downy mildew, 336
Dracaena, 81
Dragon's blood, 467
Drugs, 105, 195, 353, 400, 432, 450, 451, 460, 464, 466
Drupe, drupelet, **138,** 140, 145, 450, 452
Dry fruits, **143–46**
Duckweed, 94, 110, 133, 401, 429, 482
Dulse, 323
Dung mosses, 364

Dutch elm disease, 342, 487
Dutchman's breeches, 253, 254, *262. See also Dicentra.*
Dutrochet, R. J. H., 34
Dwarfing, A82
Dwarf juniper, 422
Dwarf mistletoe, 149, 215
Dye, dye plants, 84, 105, 125, 357, 421, 449, 450, 452, 454, 458, 466, **A50–A52**
Dyeing, natural, 354, 357, 449, A50–A52
Dyer's woad, 452
Dysentery, 284, 400, 453

E

Eagle, golden, 490
Earth star, *343,* 346
Earthworm, 167, 287, 395
Eastern deciduous forest, 342, *A86*
Eastern hemlock, 421, 481
Eastern white cedar, 421, 482
Eastern white pine, *84,* 410, 417, 481, 487
Ebony, 75
Ecology, 255, 366, **470–98**
Ecosystem, **472–77,** 482, 493
Ecotype, 473
Eczema, 400
Edible plants, wild, 417, 422, 447–68, **A34–A37**
Eelworm, 377. *See also* Nematodes
Egg. *See* Gametes
Egg (bird's-nest fungus), 343, *347*
Eggplant, 139, 444, 460, A67
Egyptian cotton, 446
Einstein, Albert, 183
Elaioplast, 43
Elaiosome, *148*
Elaters
  hornwort, *371*
  horsetail, *392*
  liverwort, *369, 370*
Elderberry, 34, *66*
Electromagnetic spectrum, 183
Electron, 18
  acceptor molecule, 182, 184, 185, 189
  carrier, 185
  flow, 185
  microscope, 10, 35, 169
  transport chain, 182, 185, *189,* 190, 193
Electrotropism, 211
Element, 17
  deficiency symptoms, **172–73**
  essential, 171–73
Elephant, 38, 476
Elm, 145, 205, 487
Elm bark beetle, 342
Elodea, 160
Embryo, 152, 364, 380, 383, 408
  angiosperm, 433
  bryophyte, 370, 373
  culture, A75
  gymnosperm, 394, 395, 413
  of seed, 94, 148, 151, 408, 430

Embryophyta, 364
Embryo sac, **429,** *431,* 433
Empirical information, A29
Enation, 270, **380,** 381
Endive, 463
Endocarp, 137, 138
Endodermis, *95, 98,* 99, 116, 160, 164, 211, *390,* 409
Endoplasmic reticulum, 37, 38, **40–41**
  rough, 40
  smooth, 41
Endosperm, 138, 150, *151,* 204, 408, 430, 431, 433
Endosperm nucleus, 431, 433
Endospores, 279
Energy, **20–21,** 202
  ecosystems and, 473–75
  forms of, **20–21**
  living organisms and, 473–75
  pyramid, 475
Engelmann, T. W., 183
Engler, Adolph, 428
English ivy, 84, 95
English yew, 421
Ensilage, 288, 465
Envelope, nuclear, 44
Enzyme, **24,** 35, 122, 123, 125, 152, 173, 184, 186, 188, 192, 194, 220, 248, 330, 342, 353
Enzyme, repair, 297
Enzymes, restriction, 296
Ephedra, 415, 422
Ephedrine, 342, 422
Epicotyl, *150*
Epidermis, 60, 61, 71, 73, 80
  of leaf, *61,* 110, 113, 114
  of root, *96,* 101, 160
  of stem, **71,** 73
Epiphyte, 312, 429, 467, 490
Equator (of cell), 49, 227–32
Equational division, 232
Equilibrium, 159
Equisetum, **389–94**
Ergosterol, 350
Ergot, 339, 341
Ergotism, 341, 342
Ermine, 485
Escherichia coli, 27
Essential elements, **171–73**
Estrous cycle, 444
Ethyl alcohol, 188–91, 342
Ethylene, 201, **205–208,** 221
Etiolation, 216
Eucalyptus, 114, 125, *283,* 429
Eucalyptus regnans, (Tasmanian), 429
Eucalyptus oil, 125
Eucapsis, 289
Euglena, 307
Euglenoids, 265, 269, 304, **307,** 321
Euglenophyta, 269, **307,** 321
Eukaryotic cells, 45, 46, 277, 290, 303
  eukaryotic vs. prokaryotic, 45, 46
Eumycota, 269, **339–60**
Euphorbia, 455
Euphorbiaceae. *See* Spurge Family